Printed from e-media with permission by:

Curran Associates, Inc.
57 Morehouse Lane
Red Hook, NY 12571

Some format issues inherent in the e-media version may also appear in this print version.

Printed by Curran Associates, Inc. (2016)

For permission requests, please contact the Institute of Physics
at the address below.

Institute of Physics
Dirac House, Temple Back
Bristol BS1 6BE UK

Phone: 44 1 17 929 7481
Fax: 44 1 17 920 0979

techtracking@iop.org

Additional copies of this publication are available from:

Curran Associates, Inc.
57 Morehouse Lane
Red Hook, NY 12571 USA
Phone: 845-758-0400
Fax: 845-758-2634
Email: curran@proceedings.com
Web: www.proceedings.com

VI International Scientific Practical Conference on Innovative Technologies and Economics in Engineering

IOP Conference Series: Materials Science and Engineering
Volume 91

Yurga, Russia
21 - 23 May 2015

ISBN: 978-1-5108-1312-0
ISSN: 1757-8981

Table of contents

Volume 91

VI International Scientific Practical Conference on Innovative Technologies and Economics in Engineering
21–23 May 2015, Yurga, Russia

Accepted papers received: 28 July 2015
Published online: 14 September 2015

Preface

011001

VI International Scientific Practical Conference on Innovative Technologies and Economics in Engineering OPEN ACCESS D A Chinakhov

011002

Peer review statement OPEN ACCESS

Papers

Topic 1: Innovative technologies for producing and testing permanent joints in mechanical engineering

012001

Nickel and titanium nanoboride composite coating OPEN ACCESS K A Efimova, G V Galevsky, V V Rudneva, N A Kozyrev and E G Orshanskaya pg. 1

012002

Synthesis and properties of nanoscale titanium boride OPEN ACCESS K A Efimova, G V Galevskiy and V V Rudneva pg. 9

012003

Silicon nano-carbide in strengthening and ceramic technologies OPEN ACCESS V V Rudneva, G V Galevsky and N A Kozyrev pg. 19

012004

Peculiarities of Production of Chromium Carbonitride Nanopowder and Its Physical-Chemical Certification OPEN ACCESS L S Shiryaeva, I V Nozdrin and G V Galevsky pg. 30

012005

Conceptual model of resource assessment of fuel and energy complex long working equipment on the basis of structural criteria OPEN ACCESS A N Smirnov and N V Ababkov pg. 39

012006

Physical Nature of the Processes in Forming Structures, Phase and Chemical Compositions of Medium-Carbon Steel Welds OPEN ACCESS D P Il'yaschenko, D A Chinakhov, V I Danilov, G V Schlyakhova and Yu M Gotovshchik pg. 44

012007

Increasing Strength and Operational Reliability of Fixed Joints of Tubes by MMA Welding OPEN ACCESS D P Il'yaschenko, D A Chinakhov, V I Danilov, G V Schlyakhova and Y M Gotovschik pg. 50

012008

New Tendencies in Development of Carbonaceous Additives for Welding Fluxes OPEN ACCESS N A Kozyrev, R E Kryukov and O A Kozyreva pg. 57

012009

Surfacing With Tungsten-containing Ores OPEN ACCESS N A Kozyrev, G V Galevsky, D V Valuev, V M Shurupov and O E Kozyreva pg. 65

012010

Corrosion and Mechanical Properties of Austenic Steel Weld Joints OPEN ACCESS M A Kuznetsov, E A Zernin, V I Danilov, D E Kolmogorov and L N Zoubenko pg. 71

012011

Influence of Frequency and Induction of Longitudinal Magnetic Field on The Electrode Metal Loss and its Spattering During MAG-Welding OPEN ACCESS D G Nosov and V V Peremitko pg. 79

012012

Influence of Protective Gas Content on Quality of Welded Joint While Welding With Impulse Supply of Electrode Wire OPEN ACCESS N V Pavlov, A V Kryukov, E A Zernin and V V Gritsenko pg. 87

012013

Development of New Materials and Technologies for Welding and Surfacing at Research and Production Center "Welding Processes and Technologies" OPEN ACCESS N A Kozyrev, R E Kryukov, G V Galevsky, D A Titov and V M Shurupov pg. 95

012014

Effect of Heat Input Pulse on the Structure and Properties of Welded Joints of Steels Ferritic-Pearlitic Class, Operating Under Low-Frequency Temperature-Force Loading OPEN ACCESS Y N Saraev, V P Bezborodov and E A Putilova pg. 104

012015

X-Ray Structural Study of 09Nn2Si Steel Welded Joints OPEN ACCESS N I Golikov, A A Platonov and Y N Saraev pg. 113

012016

Some Aspects of Oxidation-Reduction Reactions under Carbon-Bearing Flux Welding OPEN ACCESS R E Kryukov, N A Kozyrev, G V Galevsky, Y V Bendre, V F Goryushkin and D V Valuev pg. 119

012017

Titanium Carbide: Nanotechnology, Properties, Application OPEN ACCESS G V Galevsky, V V Rudneva, A K Garbuzova and D V Valuev pg. 125

012018

Possibilities of Application of Carbon-Fluorine Containing Additions in Submerged-Arc Welding OPEN ACCESS N A Kozyrev, N E Kryukov, R E Kryukov, V F Igushev and I I Kovalskii pg. 130

012019

A New the Chnology for Producing Carbide Alloys With Gradient Structure OPEN ACCESS T N Oskolkova pg. 137

012020

Wear Resistant Coating on Tungsten Carbide Hard Alloy OPEN ACCESS T N Oskolkova pg. 144

012021

Systems to Control Molten Metal Transfer in Arc Welding OPEN ACCESS A V Filonov, A V Kryukov and D E Gusarov pg. 151

012022

Study of the Effect of Shrinkage Porosity on Strength Low Carbon Cast Steel OPEN ACCESS E Ol'khovik pg. 157

012023

Prospects for Establishment of Technological Complexes in Machine Building Industry on The Basis of Electromechatronic Propulsion Systems OPEN ACCESS Yu M Osipov, O Yu Osipov and V A Trifonov pg. 166

Topic 2: Innovative technologies for producing and machining materials in mechanical engineering

012024

Patterns of folded structure formation in the maximum bending zone of [111] FCC single crystals OPEN ACCESS E A Alfyorova, D V Lychagin and A A Popov pg. 172

012025

Preferred Orientation Evolution of Olivine Grains as an Indicator of Change in the Deformation Mechanism OPEN ACCESS D V Lychagin, P A Tishin, A S Kulkov, A I Chernyshov and E A Alfyorova pg. 178

012026

Methods of noninvasive electrophysiological heart examination basing on solution of inverse problem of electrocardiography OPEN ACCESS M Grigoriev and L Babich pg. 184

012027

Results of research on causes of the defects of laser marking articles OPEN ACCESS V V Markov, L I Lebedeva and N V Kanatnikov pg. 191

012028

Thermocapillary model of formation of surface nanostructure in metals at electron beam treatment OPEN ACCESS V D Sarychev, S A Nevskii, S V Konovalov, I A Komissarova and E V Chermushkina pg. 198

012029

Fatigue life of silumin irradiated by high intensity pulsed electron beam OPEN ACCESS S V Konovalov, K V Alsaraeva, V E Gromov and Yu F Ivanov pg. 207

012030

Investigation of defect copper substructure disrupted in creep condition under the action of magnetic field OPEN ACCESS S V Konovalov, N G Yaropolova, D V Zaguyliaev, V E Gromov, Yu F Ivanov and I A Komissarova pg. 213

012031

Influence of layer-by-layer laser sintering conditions on the quality of sintered surface layer of products OPEN ACCESS N A Saprykina, A A Saprykin, I F Borovikov and Y P Sharkeev pg. 220

012032

Reduction of longitudinal ultrasonic oscillations in $NaClO_3$ crystals OPEN ACCESS E G Soboleva, A L Igisheva and T B Krit pg. 227

012033

Estimation of Temperature Conductivity Coefficient Impact upon Fatigue Damage of Material OPEN ACCESS V Bibik and A Galeeva pg. 234

012034

Simulation of thermal fields using different types of wide burnishing OPEN ACCESS N M Bobrovskij, P A Melnikov, S N Grigoriev and I N Bobrovskij pg. 240

012035

Aspects of thermal field by wide burnishing OPEN ACCESS N M Bobrovskij, P A Melnikov, S N Grigoriev and I N Bobrovskij pg. 247

012036

Influence of Cooling Rate in High-Temperature Area on Hardening of Deposited High-Cutting Chrome-Tungsten Metal OPEN ACCESS N N Malushin, D V Valuev, A V Valueva, A Serikbol and I F Borovikov pg. 254

012037

Oxidation Processes in Blowing Steel With Inert Gas into the Ladle OPEN ACCESS R A Gizatulin, D V Valuev, V A Trifonov, A V Valueva and A Serikbol pg. 260

012038

Extraction of Chromium from Carbon Ferrochromium Residual Wastes OPEN ACCESS P P Lazarevskiy, R A Gizatulin, Yu E Romanenko, D V Valuev, A V Valueva and A Serikbol pg. 266

012039

Optimizing Technological Parameters of the Reduction Processes in Treating Steels in a Ladle Furnace OPEN ACCESS R A Gizatulin, D V Valuev, A V Valueva, A Serikbol and I F Borovikov pg. 273

012040

Ultrasonic Surface Treatment of Titanium Alloys. The Submicrocrystalline State OPEN ACCESS V A Klimenov, V A Vlasov, V Y Borozna and A A Klopotov pg. 280

012041

Forecasting of operational indicators of grinding tools with the controlled form and orientation of abrasive grains OPEN ACCESS V A Korotkov and E M Minkin pg. 286

012042

Entering the operative correction machining processes CNC OPEN ACCESS R Yu Nekrasov, A I Starikov and A A Lasukov pg. 294

012043

Reliability Modelling while Designing and Improving Working Members of Agricultural Implements OPEN ACCESS V I Myalenko pg. 300

012044

Manganese concentrate usage in steelmaking OPEN ACCESS O I Nokhrina and I D Rozhihina pg. 307

012045

The use of coal in a solid phase reduction of iron oxide OPEN ACCESS O I Nokhrina, I D Rozhihina and I E Hodosov pg. 314

012046

Processing of energy materials in electromagnetic field OPEN ACCESS A P Rodzevich, L V Kuzmina, E G Gazenaur and V I Krasheninin pg. 321

012047

Cutting capacity and strength of single grinding grains OPEN ACCESS V S Lyukshin, A V Barsuk and R R Fazleev pg. 327

012048

On the Problem of Wear Resistant Coatings Separation From Tools and Machine Elements OPEN ACCESS S I Petrushin, R H Gubaidulina, S V Gruby and A V Likholat pg. 334

012049

Influence of boric anhydride upon the physical and chemical properties of ferrosilicon slag OPEN ACCESS A M Yefimets, E P Tesleva and A V Solovyan pg. 340

012050

Increasing the Operational Stability of Railway Core by Modifying OPEN ACCESS S N Fedoseev and Z M Mukhtar pg. 346

012051

Research of influence of electric conditions of the combined electro-diamond machining on quality of grinding of hard alloys OPEN ACCESS A S Yanyushkin, D V Lobanov and P V Arkhipov pg. 353

012052

Finite Element Model of Trenchless Pipe Laying OPEN ACCESS P V Burkov, S P Burkova and A N Kravchenko pg. 359

012053

Analysis of the influence of warming on the quality of soldered instruments OPEN ACCESS V V Konovodov, A V Valentov and I S Kukhar pg. 366

012054

Influence of thick-walled cylinders length on the residual stresses generated during the single-cycle mandrelling OPEN ACCESS V F Skvortsov, A O Boznak, A B Kim and A Yu Arlyapov pg. 371

012055

Stress and Strain State Analysis of Defective Pipeline Portion OPEN ACCESS P V Burkov, S P Burkova and S A Knaub pg. 376

012056

Ultrasonic plastic deformation of steels OPEN ACCESS K Zaitsev, V Klimenov, M Loshchilova and C Polovnikov pg. 382

012057

Research of surface activating influence on formation of adhesion between gas-thermal coating and steel substrate OPEN ACCESS Z Kovalevskaya, V Klimenov and K Zaitsev pg. 389

012058

New design of cutters for coal mining machines OPEN ACCESS S Prokopenko, A Sushko and I Kurzina pg. 397

012059

Estimation of cutting ability of cubic nitride boron wheels by surface microhardness criteria for pendulum grinding of high-speed steel plates OPEN ACCESS Ya I Soler and A I Shustov pg. 405

012060

Estimating the cutting force when skiving with a radius cutter OPEN ACCESS A V Filippov and E O Filippova pg. 410

012061

Experimental estimation of chip shrinkage under cup-tip cutting with straight and radius cutters OPEN ACCESS E O Filippova and A V Filippov pg. 418

012062

Processing of brittle materials in the nanometer range of thickness of layers cut OPEN ACCESS M A Shavva, V V Lapshin and S V Grubyy pg. 426

Topic 3: Automation and informatization, economy and management of a company

012063

Review of modern biometric user authentication and their development prospects OPEN ACCESS Z V Boriev, S S Sokolov and A P Nyrkov pg. 433

012064

Using information technologies in dredging OPEN ACCESS D G Mamunts, S S Sokolov and A P Nyrkov pg. 439

012065

Microsoft Business Solutions-Axapta as a basis for automated monitoring of high technology products competitiveness OPEN ACCESS G O Tashchiyan, A V Sushko and S V Grichin pg. 444

012066

Forecasting risk of bankruptcy for machine-building plants OPEN ACCESS E Telipenko, A Zakharova and Svetlana Sopova pg. 450

012067

Software selection based on analysis and forecasting methods, practised in 1C OPEN ACCESS A N Vazhdaev, T Y Chernysheva and E I Lisacheva pg. 458

012068

Strategic Concept of Competition Model in Knowledge-Based Logistics in Machinebuilding OPEN ACCESS O V Medvedeva pg. 465

012069

Methods for control over learning individual trajectory OPEN ACCESS A A Mitsel and N V Cherniaeva pg. 472

012070

Formalization model of expert knowledge about a technical index level of engineering products OPEN ACCESS A A Zakharova and V V Ostanin pg. 479

012071

Increase of stability of functioning of a production system of machine-building enterprise OPEN ACCESS A A Rudenko, D V Antipov and M O Iskoskov pg. 487

Topic 4: Environmental protection, industrial labor safety

012072

Analysis of air pollution from industrial plants by lichen indication on example of small town OPEN ACCESS K N Orlova, I R Pietkova and I F Borovikov pg. 495

012073

Environmental management on the basis of Complex Regional Indicators Concept: case of the Murmansk region OPEN ACCESS A Kozlov, S Gutman, I Zaychenko, E Rytova and P Nijinskaya pg. 502

012074

Engineering of centrifugal dust-collectors based on parallel comparing tests applying computer modelling OPEN ACCESS Y I Bulygin, D A Koronchik and A A Abuzyarov pg. 511

012075

Methodology of research for qualitative composition of municipal solid waste to select an optimal method of recycling OPEN ACCESS M V Kravtsova and D A Volkov pg. 515

012076

The Influence of Various Factors on the Methane Fermentation Process OPEN ACCESS M G Kurbanova, E A Egushova and OG Pozdnjakova pg. 522

012077

Updating of sewage - purification facilities of electroplating enterprises with counterflow ion-exchange filters OPEN ACCESS V F Torosyan, E S Torosyan, P D Sorokin and A A Telitsyn pg. 530

012078

The Study of Soil Protection in the System of the Cultivated Lands of Kemerovo Region OPEN ACCESS M A Yakovchenko, O B Konstantinova and A A Kosolapova pg. 538

012079

Statistic analysis of acoustic noise in the industrial and living areas OPEN ACCESS I N Kirillov and V V Bulkin pg. 545

012080

The problems of reducing the noise level on open sites of machine building plants and living areas OPEN ACCESS V V Bulkin and M V Kalinichenko pg. 551

012081

Investigations of physicochemical properties of bottom-ash materials for use them as secondary raw materials OPEN ACCESS A G Mal'chik, S V Litovkin and P V Rodionov pg. 556

012082

Carbon Nanotubes Influence the Enzyme Activity of Biogeochemical Cycles of Carbon, Nitrogen, Phosphorus and the Pathogenesis of Plants in Annual Agroecosystems OPEN ACCESS O B Vaishlya, N N Osipov and N V Guseva pg. 563

Topic 5: Advanced technologies and equipment for agro-industrial complex (AIC) and mining

012083

Comparison of mathematical methods of geochemical data processing OPEN ACCESS A Marfin, D V Lychagin, A Shapovalov and E A Alfyorova pg. 572

012084

The distribution of stresses and strains in the mating elements disk tools working bodies of roadheaders OPEN ACCESS A A Khoreshok, L E Mametyev, A Yu Borisov and A V Vorobyev pg. 578

012085

Development of math model of geokhod bladed working body interaction with geo-environment OPEN ACCESS V Yu Sadovets, V Yu Beglyakov and V V Aksenov pg. 586

012086

The application of the analytic hierarchy process when choosing layout schemes for a geokhod pumping station OPEN ACCESS R V Chernukhin, A A Dronov and M Y Blashchuk pg. 594

012087

Radial deformations of working cylinder of hydraulic Legs depending on their extension OPEN ACCESS G D Buyalich, K G Buyalich and V V Voyevodin pg. 602

012088

Classification of geokhod units and systems based on product cost analysis and estimation for a prototype model production OPEN ACCESS V V Aksenov, A V Walter, A A Gordeyev and A V Kosovets pg. 609

012089

Spectral analysis of granular material reaction to long-term weak dynamic effect OPEN ACCESS A A Kazantsev, V P Kosykh and A F Revuzhenko pg. 621

012090

On the theory of plasticity, associated with a new integral characteristic of shearing stresses OPEN ACCESS A A Kazantsev, O A Mikenina and A F Revuzhenko pg. 629

012091

Improving of the operation efficiency of the vehicle due to using of the neodymium magnets inside the vibration isolation devices OPEN ACCESS E G Gurova pg. 636

012092

Research of the pre-launch powered lubrication device of major parts of the engine D-240 OPEN ACCESS M Korchuganova, A Syrbakov, A Tkachev and T Zorina pg. 643

012093

Formation of the composition and properties of dumps on the open-pit mines of Kuzbass OPEN ACCESS Y V Lesin, S Y Luk'yanova and M A Tyuleneva pg. 650

IOP Conf. Series: Materials Science and Engineering **91** (2015) 011001 doi:10.1088/1757-899X/91/1/011001

In these conference proceedings we present papers from the 6th International Scientific Practical Conference on Innovation Technology and Economics in Engineering held at the Yurga Institute of Technology branch of the National Research Tomsk Polytechnic University from 21st – 23rd May 2015. The proceedings contain the selected scientific reports submitted to the conference. Having started in 1996, the scientific conference at the Yurga Institute of Technology branch of the National Research Tomsk Polytechnic University acquired international status in 2010. This year, scholars from Russia, Kazakhstan, Belarus, Ukraine, Azerbaijan, China, Germany and Poland have submitted their papers to the conference.

The scientific reports published in these proceedings have been revised and approved by the editorial team of the conference. All of the reports exhibit clear, concise, and precise expositions that appeal to a broad international readership interested in mechanical engineering, welding, metallurgy, materials science as well as in computer-aided manufacturing and economics. The reports present original ideas or results of general significance supported by clear reasoning and compelling evidence, and employ methods, theories and practices relevant to the research. The authors state clearly the questions and the significance of their research to theory and practice, describe how the research contributes to new knowledge, and provide tables and figures that meaningfully add to the narrative.

The organizing committee of the conference thanks all the participants for their fruitful work and personal contribution to the development of these conference proceedings.

Dmitry A. Chinakhov, PhD,
Conference Convenor,
Vice-Chancellor, Yurga Institute of Technology,
National Research Tomsk Polytechnic University

Peer review statement

All papers published in this volume of *IOP Conference Series: Materials Science and Engineering* have been peer reviewed through processes administered by the proceedings Editors. Reviews were conducted by expert referees to the professional and scientific standards expected of a proceedings journal published by IOP Publishing.

Nickel and titanium nanoboride composite coating

K A Efimova[1], G V Galevsky[2], V V Rudneva[2], N A Kozyrev[2], E G Orshanskaya[3]

[1]Postgraduate student, Siberian State Industrial University, 654007, Novokuznetsk, 42, Kirov str.
[2]Doctor of Engineering, Professor, Siberian State Industrial University, 654007, Novokuznetsk, 42, Kirov str.
[3]Doctor of Pedagogy, Professor, Siberian State Industrial University, 654007, Novokuznetsk, 42, Kirov str.

e-mail: kafcmet@sibsiu.ru

Abstract. Electrodeposition conditions, structural-physical and mechanical properties (microhardness, cohesion with a base, wear resistance, corrosion currents) of electroplated composite coatings on the base of nickel with nano and micro-powders of titanium boride are investigated. It has been found out that electro-crystallization of nickel with boride nano-particles is the cause of coating formation with structural fragments of small sizes, low porosity and improved physical and mechanical properties. Titanium nano-boride is a component of composite coating, as well as an effective modifier of nickel matrix. Nano-boride of the electrolyte improves efficiency of the latter due to increased permissible upper limit of the cathodic current density.

Introduction

Electrodeposited or galvanic composite coatings (GCC) are formed as a thin layer in the course of metals deposition on the products with electro-conductive surface made up of electrolytes-suspensions with a dispersed phase. Service characteristics of coating are improved due to particles. Improvement of GCC service characteristics necessitates enhancing the dispersion level of strengthening phase. For this reason experts dealing with GCC technology always use materials with high dispersion including nano-level ones as a strengthening phase [1 – 8].

Purpose and objectives of research

This research is aimed at revealing peculiarities of nickel GCC formation and properties, its strengthening phase is made up of nano- and micro-powders of titanium boride, which is a synthetic superhard, high heat, and high temperature material needed in protective coatings technology. Galvanic nickel coatings are the most utilized ones in various industries; their usage amounts to 75% of overall total electroplated deposited metals.

Materials and methods of experiments

For hardening nickel matrix we used nano-powder of titanium boride (NP TiB_2), which was obtained through boriding metal powder of titanium in nitrogen plasma flow, and its micro-powder (MP TiB_2), produced in the course of self-spreading high-temperature synthesis. TiB_2 concentration in nanopowder is 92.81%, and 94.12% in micro-powder, specific surface area of powders is 46000 and

1000 m^2/kg, dimensional range of particles is 0.01–0.07 and 3–7 μm. Micro-photographs of nano- and micro-powders, made by transmission and scanning electron microscopy (TEM, SEM), are depicted in Figure 1.

The research into parameters of GCC deposition was carried out in standard nickel-plating electrolyte, containing $NiSO_4 \cdot 7H_2O$ – 245, H_3BO_3 – 30, NaCl – 20, NaF – 6 kg/m^3 in the following conditions: pH 5.0–5.5, temperature 323 K, cathodic density of current 0.1–1.2 kA/m^2, powder concentration 1–100 kg/m^3, constant mixing of electrolyte. Steel (St. 3) samples with surface area $2.0 \cdot 10^{-3}$ m^2 were used as cathodes. 0.08 x 0.1 m nickel plates were used as anodes.

Fine structure of GCC with nano-powder was tested by electron microscopy and energy-dispersive spectrometry. For investigation of basic morphological types of micro- and nano-powders, identification of their form and linear features we used field emission scanning electron microscope JSM – 6700F with an appliance for energy-dispersion spectrometry JED – 2300F, equipped by high-power conic objective lens. This objective lens provides guaranteed resolution 1.0 nm at accelerating voltage 20 kW and 2.2 nm at 1 kW. The first stage of investigation focused on formation of surface microrelief in secondary electrons. Then, in conditions of relatively small but sufficient (x 2000 – x 3000) magnification for identifying all objects we checked the surface of prepared samples totally in order to find their most typical zones. These areas were tested more precisely to define form, length and diameter of micro- and nano-objects. Identification of element composition of GCC samples was performed by energy-dispersion spectrometry with scanning electron microscope JSM – 6480 LV with an appliance for energy-dispersion spectrometry INCA. This method taken with scanning electron microscopy makes it possible to carry out a quantitative elementary analysis, ranging from boron to uranium and in quantity 1–3 cubic μm. Sensitivity of the method is 0.1 atomic %. Relative error of measurement is 5 %.

Cathodic current efficiency of nickel is determined gravimetrically by copper coulometer, connected in-series with electrolyzer.

Titanium boride concentration in composite coatings was determined gravimetrically after their dissolution in 10 % HNO_3 solution (GOST 5744 – 94). Percentage of boride concentration in matrix was calculated as relation of insoluble residual mass to the mass of coating.

Microhardness of coating was measured by microhardness testing instrument PMT – 3 and involved static hardness indentation perpendicularly to coating layer subject to 0.49 N load. Microhardness of coatings was determined on samples 40 μm thick according to 5 – 6 measurements of imprint diagonal.

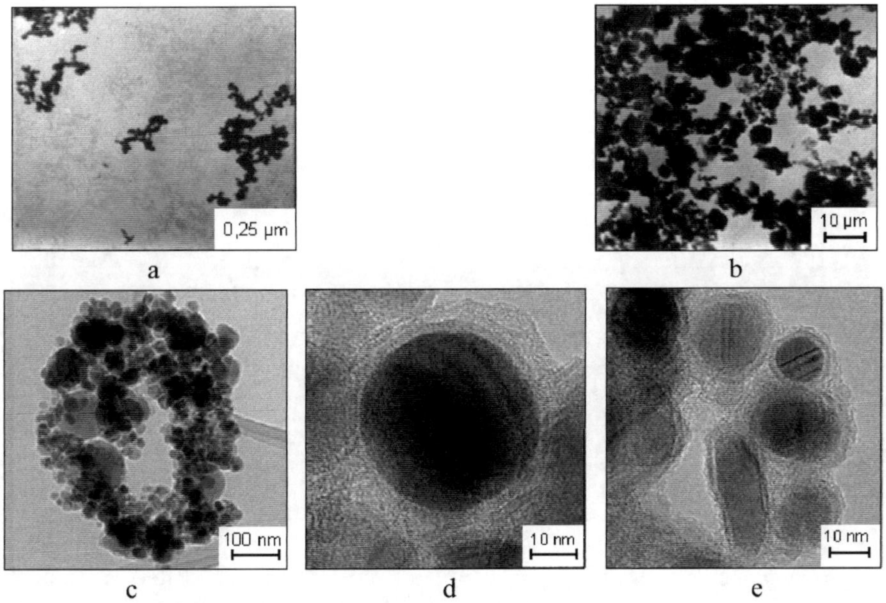

Figure 1. PEM microphotographs of titanium boride nano-powder (a) and micro-powder (b), SEM of titanium boride nano-powder (c – ensemble of particles and aggregates; d – morphological structure of aggregate; e – separate particles)

Internal stresses of coatings were measured by deformation method of elastic cathode. Coating cohesion with steel base was assessed by the method of shear loading; samples were glued together with BK – 9 glue, bonding area was $0.35 \cdot 10^{-4}$ m^2, layer thickness – $0.5 \cdot 10^{-3}$ m, after drying they were tested by P – 0.5 tension testing machine for 24 hours. Protective capability of coatings was assessed according to coating-base corrosion currents, determined by Rosenfeld method in a neutral electrolyte.

For investigation of changing GCC properties when isothermal annealing GCC samples were placed into quartz vessels, vacuumized and sealed. Then, the vessels were placed into electric-tube furnace, which provided heating rate up to needed temperature about 0.1 degree/s, there samples were kept for 120 minutes at each temperature. The research focused on the effect of temperature and temporal annealing conditions on microhardness and strength of GCC-steel base cohesion.

Methods to determine titanium boride concentration in coatings, their microhardness, fine structure, internal stresses, and cohesion with base, corrosion currents between them and the base are described in papers [9–11] in detail.

Results of research and discussion

The investigation focused on effect of cathode density of current and powders concentration in electrolyte on titanium boride concentration in coatings and their microhardness.

Table 1. The relation of strengthening phase concentration to cathode density of current in GCC

Cathode density of current, kA/m^2	Strengthening phase concentration in GCC, %	
	NP TiB_2	MP TiB_2
0.1	0.31	0.59
0.3	0.36	0.73
0.5	0.41	0.91
0.7	0.52	1.32
1.0	0.57	defective coatings
1.2	defective coatings	

IOP Conf. Series: Materials Science and Engineering **91** (2015) 012001 doi:10.1088/1757-899X/91/1/012001

Table 2. The relation of strengthening phase concentration in GCC and their microhardness to powder concentration in electrolyte

NP, MP concentration in electrolyte, kg/m^3	Strengthening phase concentration in GCC, % / Microhardness ± 0.3 GPa	
	NP TiB_2	MP TiB_2
2.0	0.22/3.0	0.33/2.9
5.0	0.30/3.4	0.60/2.9
10.0	0.51/4.8	1.18/3.1
15.0	0.60/5.3	1.32/3.3
30.0	0.61/5.2	1.73/3.3
45.0	0.59/5.1	2.01/3.6
60.0	0.52/4.7	2.32/3.8
80.0	0.52/4.8	3.01/4.1
100.0	not tested	3.12/4.0

The research into the effect of cathode density on strengthening phase concentration in a coating was conducted in the following conditions: pH of electrolyte 5.0, temperature 323 K, and strengthening phase concentration 15 kg/m^3. The results are provided in Table 1. As one can see, the increase in cathode density of current within the limit 0.1 – 1.0 kA/m^2 for nano-powder and 0.7 kA/m^2 for micro-powder causes growth of strengthening phase concentration in a coating: for NP TiB_2 from 0.31 to 0.60; MP TiB_2 from 0.59 to 1.32 %. With strengthening nano-disperse phase the upper limit of operating current density of electrolyte is 1.0 kA/m^2, it's higher than that for making nickel coatings in this electrolyte (0.5 kA/m^2). Therefore, the electrolyte with nano-powder is more efficient; and more promising one for worn surfaces restoration, respectively.

The influence of powders concentration in electrolyte on strengthening phase concentration in a coating was tested in the following conditions: pH of electrolyte – 5.0, temperature 323 K, and cathodic density of current 1.0 kA/m^2 for nano-powder; and 0.7 kA/m^2 – for micro-powder. The results are provided in Table. 2. As one can see, growing concentration of nanopowder in electrolyte-suspension up to 15 kg/m^3 results in its increase in GCC, further, there is practically no change when concentration amounts to 15–45 kg/m^3, and a slight decrease is possible in the range 45–80 kg/m^3. When micro-powder is added to electrolyte in concentration 2–100 kg/m^3 its concentration goes up in a coating from 0.53 to 3.12 %. As the consequence, the bigger coarseness of titanium boride powder is the higher concentration of strengthening phase is needed for nickel matrix saturation.

Thus, we suggest that nano-powder is more reasonable to use than micro-powder for composite coatings manufacture. Optimal conditions of GCC nickel – NP titanium boride deposition are as follows: TiB_2 concentration 15 kg/m^3, cathodic density of current 0.9–1.0 kA/m^2 temperature 323 K, pH – 5.0–5.5, and constant mixing electrolyte. Nickel current efficiency in these conditions is 92–94%.

The structure and main characteristics (hardness, cohesion with steel base, internal stresses, corrosion currents) of nickel composite coating containing nano-and micro-powder of titanium boride were tested, as well as those of pure nickel coating, i.d. nickel matrix. Deposition of GCC Ni – NP TiB_2 was performed with nano-powder concentration in electrolyte 15 kg/m^3 and cathodic density 1.0 kA/m^2. When GCC Ni – MP TiB_2 depositing concentration of micro-powder in electrolyte was 80 kg/m^3, cathodic density of current – 0.7 kA/m^2. For deposition of pure nickel coatings electrolyte of the same composition was used, without disperse phase.

Analysis of energy-disperse spectra of electrodeposited nickel and GCC confirms that they contain Ni, O and Ni, Ti, B, O. Quantitave relations are quite acceptable ones (Table 3). Here, for elements to be identified rather even distribution is typical in 30 μm coating (Fig. 2), i.d. its composite character is confirmed.

Table 3. Element chemical composition of electrodeposited nickel (specter 1) and GCC Ni – NP TiB_2 (spectra 2-4)

Element	Element concentration in particular points of coatings, %			
	Specter 1	Specter 2	Specter 3	Specter 4
Ni	98.20	97.67	97.65	97.69
O	1.80	1.71	1.79	1.72
Ti	–	0.42	0.40	0.42
B	–	0.20	0.18	0.18

Ni Ti B O

Figure 2. Results of micro-roentgen spectral analysis of basic elements distribution in GCC Ni – NP TiB_2

Microhardness of GCC is determined by strengthening phase concentration in it and dimensions of its particles (Table 2). Microhardness of GCC Ni – TiB_2 is 5.1–5.3 GPa, it's 2–2.3 higher than that of nickel matrix, and 1.4 times higher than that of GCC with MP TiB_2. Cohesion of GCC Ni – NP TiB_2 with steel base is 29.5–33.1 MPa. Isothermal annealing of GCC Ni – NP TiB_2 in vacuum in temperature range 473 – 1273 K facilitates microhardness increase in GCC (Fig. 3). However, the character of temperature dependence of microhardness for GCC under consideration is different. For instance, for GCC Ni – MP TiB_2 (curve 2, Figure 3) microhardness rises monotonously from 3.8 to 5.2 GPa when annealing temperature rising from 473 to 1273. Microhardness of GCC Ni – NP TiB_2 (curve 1, Fig. 3) in temperature range 473 – 873 K grows from 5.1 to 6.2 GPa, but in temperature range 873 – 1273 K hardly changes practically. Here, to achieve maximum microhardness of GCC Ni – NP TiB_2 60 – 75 minute annealing is sufficient (curve 1, Fig. 4), whereas for GCC – MP TiB_2 annealing for 120 minutes is required (curve 2, Fig. 4).

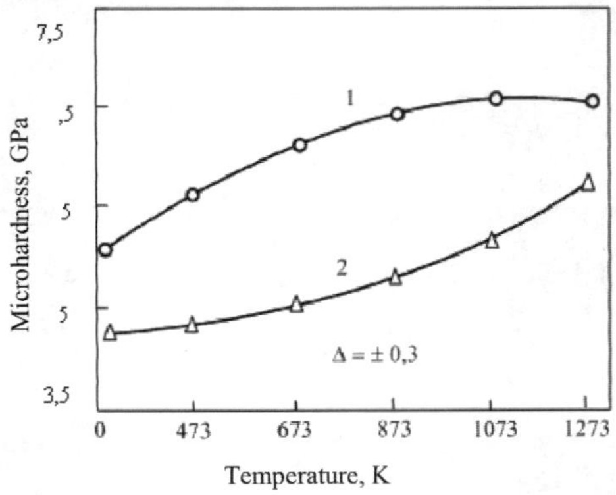

Figure 3.The dependence of microhardness of GCC Ni – NP TiB_2 (1) and Ni – MP TiB_2 (2) on temperature of 2 hour annealing in vacuum

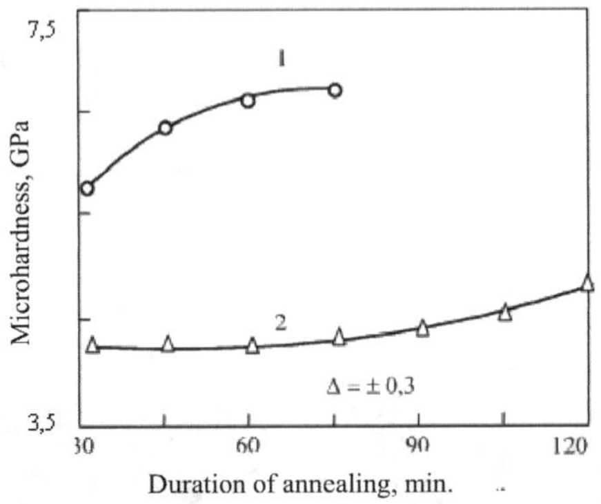

Figure 4. The dependence of microhardness of GCC Ni – NP TiB_2 (1) and Ni – MP TiB_2 (2) on annealing duration in vacuum at temperature 873 K

The strength of GCC Ni – NP TiB_2 cohesion with the base, which was annealed in vacuum at temperature 873 K for 75 minutes, increased 1.4 times in comparison to non-annealed ones and was 42.3 – 45.2 MPa.

Disperse particles in a coating reduce internal stresses, especially when titanium boride nano-powder is used. For instance, internal stresses of 40 μm thick composite coating with micro-powder TiB_2 are 1.5 times and with nano-particles 3.4 times smaller than those of pure nickel (Table 4). Obviously, the reason for it is a more fine-disperse structure of growing sediment due to evenly distributed particles of disperse phase in it. Coarse particles of micro-powder as if compared with NP cause uneven deformation of a matrix including big groups of grains, therefore, increasing internal stresses and reducing corrosion characteristics.

Corrosion currents arising in coating – steel base spacing are far lower than in nickel matrix (Table 4). In 40 μm coating, made of micro-powder, corrosion currents are 2.3 times lower than in pure nickel, and in a coating made of nano-powder – 11.9 times lower, it means, that practically pore-free GCC Ni – NP TiB_2, form and provide high protective characteristics (Figure 5).

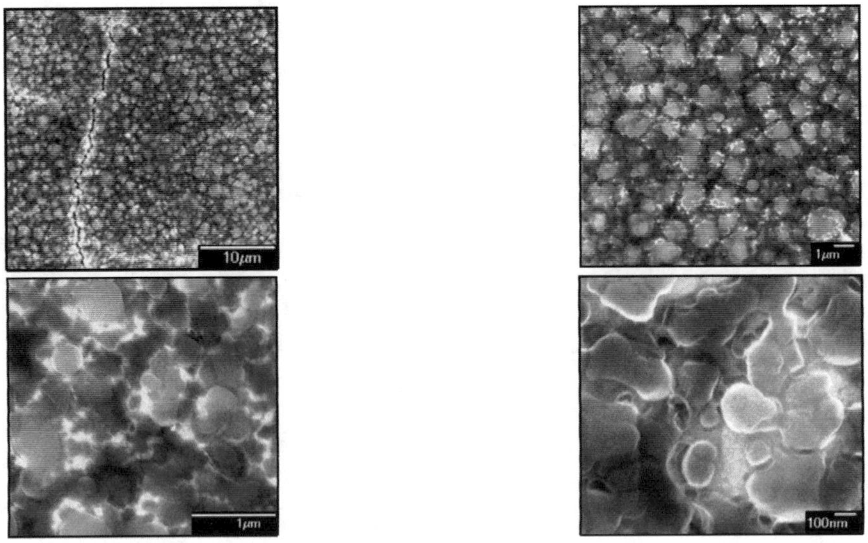

a

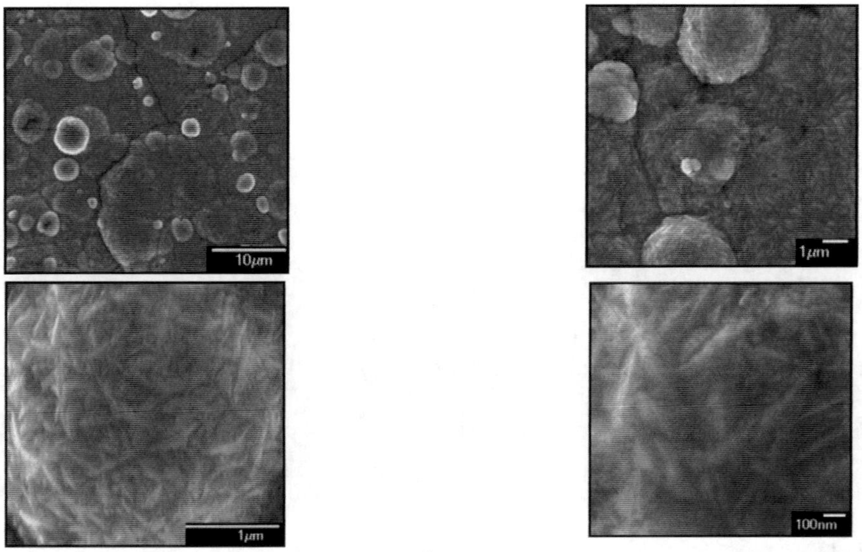

b

Figure 5. Microphotographs of electrodeposited nickel surface (a) and ГКП Ni – NP TiB_2 (b)

Table 4. The dependence of internal stresses and corrosion currents on coating thickness

Coating thickness, μm	Internal stresses, MPa			Corrosion currents, mcA/cm^2		
	Ni	Ni – NP TiB_2	Ni – MP TiB_2	Ni	Ni – NP TiB_2	Ni – MP TiB_2
5	936	351	578	0.304	0.088	0.167
10	692	159	323	0.265	0.061	0.154
20	361	62	146	0.205	0.047	0.139
30	127	54	79	0.173	0.022	0.102
40	101	30	69	0.167	0.014	0.072

Thus, nano-particles of titanium boride have significant chemical and adsorptive energy, form sedimentation and coagulation stable electrolytes-suspensions and can be easily transferred to surfaced area due to low mass. In course of electrodeposition of nickel nano-particles of boride are centers of crystallization. As a great number of boride nano-particles take part in this process, the character of crystallization is mass multinuclear one. Coatings have small dimensions of structure elements, typical matt colour and low porosity. Small particles of boride and nickel crystallite provide exact imitation of surface micro-relief. Relatively low concentration of strengthening phase in coating Ni – NP TiB_2 is the reason of its considerably low consumption per 1 m^2 of processed surface and helps to keep valuable properties of a matrix. Low concentration of nano-powder in electrolyte simplifies maintenance of electroplating bathes and reduces titanium boride losses caused by electrolyte carry-over on workpieces. Titanium boride raises electrolyte efficiency due to increased upper limit of cathodic density of current. These properties of GCC Ni – NP TiB_2 are the reason to recommend them for anticorrosion protection of workpieces, and after thermal treatment – for strengthening workpieces, running at full capacity in conditions of sliding friction at mean circular velocities (1.5–2.5 m/s) and low specific loads (5–6 MPa) (shafts, bushes, sliding bearings, spindles etc.).

Conclusions

When electrodepositing nickel from electrolyte-suspension, nano-powder of titanium boride is a component of a composite coating nickel – titanium boride and an efficient modifier at the same time. As the result, coatings are formed with fine-grain pore-free structure and high operational services.

Technological advantages of nickel galvanic composite coatings including nano-titanium boride have been revealed, as well as engineering spheres of their most efficient application.

The work was performed in Siberian State Industrial University within the project of State Order of Ministry of Education and Science of Russian Federation № 11.1531.2014/k.

References

[1] Sayfulin R S The research into composite electrochemical coatings including disperse phase of micro-and nano-particles *Vestnik of Kazan Technological University* 2009 № 6 p 80 – 90
[2] Desyatkova G I Yagodkina L M Savochkina I E Khaldeev G V Nickel composite electrochemical coatings *Protection of metals* 2002 V 38 #5 p 525 – 529
[3] Tsupak T E High-efficient electrodeposition processes of nickel and nickel-phosphorus alloy from electrolytes, containing carboxylic acids: author's abstract of Doctor Thesis … in Chemistry Moscow 2008 p 40
[4] Dolmatov V Yu Contemporary industrial technology of explosive nano-diamonds and main spheres of their application *Nano-technologies to industry* 2006 Proceedings of international academic and practical conference M: Yanus-K. 2006 p 113-151
[5] Nozdrin I V Rudneva V V Galevsky G V Electroplated composite coatings zinc – chromium boride: electrodeposition and properties *Vestnik of mining and metallurgical department of Russian Academy of Natural Sciences* Department of Metallurgy: collective papers Siberian State Industrial University Moscow-Novokuznetsk 2014 Issue 32 p 106-112
[6] Shiryaeva L S Nozdrin I V Galevsky G V Rudneva V V Investigation of composite electrochemical coatings nickel – chromium nano-carbonitride *Electroplating and surface machining* 2014 # 2 V XXII p 51-57
[7] Shiryaeva L S Nozdrin I V Galevsky G V Rudneva V V Investigation of physical and mechanical properties of electrodeposited composite material nickel – chromium nano-carbonitride *Perspective materials* 2014 # 7 p 62 – 67
[8] Shiryaeva L S Nozdrin I V Galevsky G V Rudneva V V Application of chromium nano-carbonitride in composite electroplated nickel coatings *Vestnik of Siberian State Industrial University* 2014 # 3(9) p 48 – 52
[9] Nozdrin I V Galevsky G V Rudneva V V Special characteristics of composite material nickel – nano-disperse chromium boride *Blank production in mechanical engineering* 2011 #9 p 46-48
[10] Polyakh O A Rudneva V V Galevsky G V Plasma-metallurgical production of silicon carbide for composite nickel-plating and chromium-plating M.: *Flinta: Science* 2006 p 188
[11] Gamburg Yu D Electroplated coatings *Application reference book* M: Technosphere 2006 p 216

Synthesis and properties of nanoscale titanium boride

K A Efimova[1], G V Galevskiy[2], V V Rudneva[2]

[1]Postgraduate, Siberian State Industrial University, 654007, Novokuznetsk, ul. Kirova 42

[2]D.Eng.Sc., Prof, Siberian State Industrial University, 654007, Novokuznetsk, ul. Kirova 42, tel. +7 3843 74 8913

e-mail: kafcmet@sibsiu.ru

Abstract. This work reports the scientific and technological grounds for plasma synthesis of titanium diboride, including thermodynamic and kinetic conditions of boride formation when titanium and titanium dioxide are interacting with products resulting from boron gasification in the nitrogen - hydrogen plasma flow, and two variations of its behavior using the powder mixtures: titanium - boron and titanium dioxide - boron. To study these technology variations, the mathematical models were derived, describing the relation between element contents in the synthesized products of titanium and free boron and basic parameters. The probable mechanism proposed for forming titanium diboride according to a "vapour - melt – crystal" pattern was examined, covering condensation of titanium vapour in the form of aerosol, boriding of nanoscale melt droplets by boron hydrides and crystallization of titanium - boron melt. The comprehensive physical - chemical certification of titanium diboride was carried out, including the study of its crystal structure, phase and chemical composition, dispersion, morphology and particle oxidation. Technological application prospects for use of titanium diboride nanoscale powder as constituent element in the wettable coating for carbon cathodes having excellent physical and mechanical performance and protective properties.

Introduction

Titanium diboride TiB_2 is a synthetic super hard, heat resistant, high melting and wear resistant material, which is popular on the market for the production of cermets, refractory and other coatings for various purposes [1, 2]. The relatively new and important field as to application prospects is its use in aluminum electrolytic melting as a component for the wettable coating for electrolysis' cathodes, protecting them against destructive effects of cryolite-alumina melting [3,4]. For obtaining such a coating, an aqueous slurry containing 68 - 70% of solids (90% - TiB_2, 10% -Al_2O_3) is prepared. Physical and chemical bindings between the Al_2O_3 colloidal particles and TiB_2 particles in the slurry lead to the formation of jelly-like viscoelastic state. This material does not bleed and behaves as a solid after drying. The slurry is applied by spraying or staining with intermediate drying in air after applying every layer. The total drying time is 24 hours. The coating 1.0-2.0 mm in thickness provides the wetting of a cathode with aluminum; has a high resistance to sodium dissolution; at the same time combines a sufficient hardness, bending strength, wear resistance, and adhesion to the base material; reduces the voltage dropping in cathodes and increases the cathode efficiency relative to aluminum yielding.

For producing aluminum in foreign countries, basic materials for cathode coatings to protect aluminum electrolytic cells are supplied by JSC "Moltek" under trademarks TINOR A, TINOR M and thick TINOR. When the coating is 1 mm thick, the TiB_2 consumption is 2.3 kg/m^2 of a cathode surface [5, 6] that indicates the need for further development of technological capacities to produce it. Currently, the basic process to obtain TiB_2 intended for wettable cathode coatings is a self-propagating high-temperature synthesis [7]. Despite the relative simplicity as to technological solutions, this technology is low efficient, and allows obtaining TiB_2 in the form of a coarse powder with a particle size in the 3-10 microns range. There are good grounds to suggest that adding TiB_2 to the slurry as a fine powder having particles with a size equitable or smaller than the size of Al_2O_3 particles (0.1-1 microns) will improve mechanical and physical properties and protective capabilities of the coating.

The aim of the research

In this connection, the aim of this study is to develop scientific and technological foundations for titanium diboride plasma synthesis with its physical and chemical certification and identify basic technological parameters for the technology.

The reactor for the synthesis of titanium diboride and its characteristics.

For the synthesis process of titanium diboride, an industrial plasma metallurgical complex was used based on a triple torch direct current plasma reactor having a power of 150 kW [8]. This complex together with the reactor includes plasma torches (plasmatrons), with the structural design improved by the authors, a mixing chamber, a dosing device for fine-dispersed non-free-flow basic material, a continuous sleeve filter; and systems for electrical, gas, water supply and ventilation. Three electric arc plasma torches EDP - 104A of 50 kW each used to generate a plasma flux are fitted in the mixing chamber at an angle of 30° to the reactor's axis. The powder and gaseous basic material is introduced through a water cooled lance into the zone of plasma jets' collision. In this case, the reaction mixture of a specified composition is formed; the nano dispersed products are synthesized and formed. The main parameters of the reactor are given in Table 1. In terms of its parameters, the reactor can be classified as modern electrothermal equipment.

Table 1. Plasma reactor basic parameters

Parameters	Value
Power, [kW]	150
Reactor type	Triple-jet, direct-current, vertical,
Type of plasma torch (plasmatron), Power, [kW]	EPD-104A, 50
Plasma forming gas	Nitrogen
Mass of heated gas, [kg/h]	32.5
Internal diameter, [m]	0.054
Reactor capacity, [m^3]	0.001
Refractory lining of reactor channel	Zirconium dioxide
Temperature of plasma flow, [K]	5400 (L* = 0) – 2200 (L = 12)
Temperature of refractory lining, [K]	1549 (L = 0) – 770 (L = 12)
Specified electrical output power, [MBW/m^3]	2142
Service life, [h]: -anode -cathode	3125 112
Impurity of titanium diboride with eroded material, % -anode -cathode	Cu – 0.0000954 W – 0.0000002

*L – relation between the reactor length and its internal diameter

IOP Conf. Series: Materials Science and Engineering **91** (2015) 012002 doi:10.1088/1757-899X/91/1/012002

Powered basic materials for the synthesis of titanium diboride.
We simulated the interaction of plasma flow and feedstock flow in order to choose appropriate titanium-containing basic materials. The relevance of this approach is confirmed in [9-11].

Due to the availability of titanium, titanium dioxide and boron powders, hydrocarbons - methane and plasma forming gas – nitrogen, the following mixtures: B - H- N, Ti - O - C - B - H - N, Ti - B - H – N were taken as objects in the thermodynamic simulation. The compositions of gaseous and condensed products required for the analysis were calculated using the "const" programming, which is based on the simultaneous solution to the following equations: the law of mass action, mass balance, the total number of moles of gas mixture, existence of the condensed phase, Dalton's law. The calculations were performed using a computer simulation program for high temperature complex chemical equilibria «PLASMA», having an integral database of interacting products for oxide -, boride -, carbide- and nitride forming systems. When calculating, the 1000 - 6000 K temperature range at a total pressure of 0.1 MPa in the system was taken into consideration.

Results of calculating the equilibrium compositions in the B - H - N system are shown in Figure 1 a, b. It can be seen that in the investigated system gasification of boron occurs that increases with the dilution of the system with hydrogen. Gasification of boron is caused by the high thermodynamic stability of boron hydride in BH_2 composition in the temperature range 2650 - 3250 K. When the ratio of B: H = 1: 6, the conversion degree of boron into BH_2 reaches 1. Condensation of boron from the gaseous state is thermodynamically possible at temperatures below 2650 K. Thus, it is necessary to accept the dilution of the system with hydrogen efficient and predict the effect of hydrogen concentration within the gaseous state on completeness of boride formation in the plasma flow. Quasi-equilibrium compositions of the B - H - N system (Figure 1 c, d) are described by significant extension of temperature limits required for the stability of boron hydride BH_2, corresponding to 2150 - 3250 K.

In the boride forming systems, titanium diboride can be obtained in the range of temperatures 3500-2300 K and when all relations between components are considered. The following gaseous chemical reactions in the boride formation process are thermodynamically possible:

$$Ti_г + BH_{2г} = TiB_{2к} + 2H_{2г} \quad (1), \qquad Ti_г + 2B_г = TiB_{2к} \quad (2), \qquad Ti_г + 2BH_г = TiB_{2к} + H_{2г} \quad (3),$$

$$Ti_г + 2BO_г + 2HCN_г = TiB_{2к} + 2CO_г + H_{2г} + N_{2г} \quad (4),$$

$$2TiO_г + 2HCN_г + 2BH_{2г} = 2TiB_{2к} + 2CO_г + 5H_{2г} + N_{2г} \quad (5).$$

The fractional conversion of titanium into diboride may be 0.85 according to reaction (1). The 100% conversion of titanium into diboride is achieved at the stoichiometric ratio of components in the temperature range 3200 - 2450 K (Figure 2). With the excessive boron and carbon in the systems in equilibrium when in the condensed phase, boron nitride and free carbon with a concentration for both dependent on the ratio between B/Ti and C/O can be present together with diboride at a temperature below 2800 K. The nitride formation process is thermodynamically possible at a temperature of 2300 K and below

$$2TiB_{2к} + 3N_{2г} = 2TiN_к + 4BN_к \quad (6)$$

The formation of titanium nitride together with diboride is possible, but at higher temperatures and deficient boron in the system.

Conditions for the effective processing of boron- and titanium-containing basic materials were defined using an improved mathematical model relative to the plasma and material flow interaction, with this model allowing for the influence of heat transfer gas on the heat exchange with walls, basic materials and artificial channel insulator [11]. The model is based on a simultaneous solution to equations of motion for basic material particles, the inter-component heat exchange and heat exchange between the plasma flow and reactor walls.

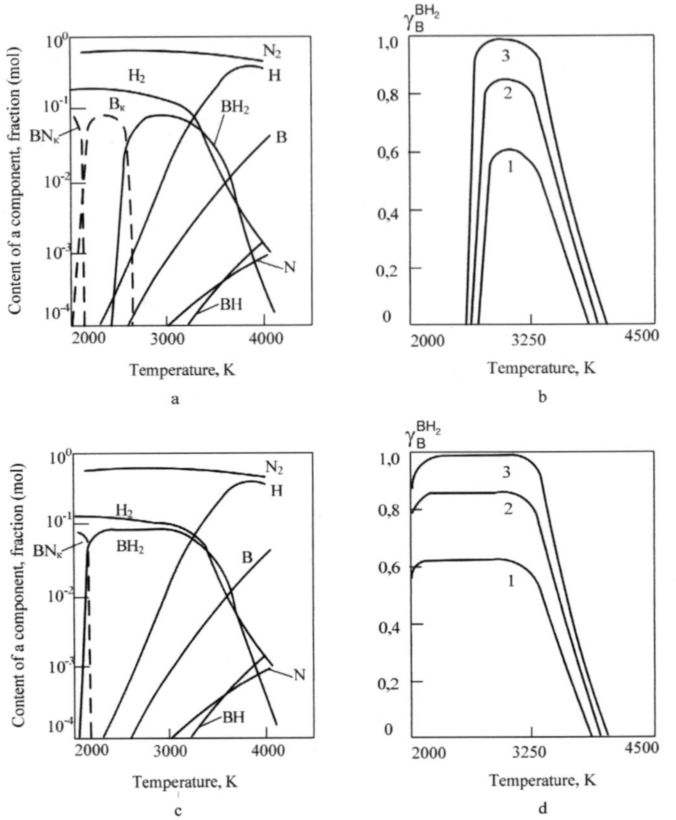

a) the equilibrium composition of gaseous and condensed phases depending on a temperature at the ratio of B:H:N = 1: 6: 20; b) the equilibrium dependence of fractional conversion B into BH_2 on the ratio of B:H = 1: 2 (1); 1: 4 (2); 1: 6 (3) and temperature; c) the quasi-equilibrium composition of gaseous and condensed phases depending on a temperature at the ratio of B:H:N = 1: 6: 20; d) the quasi-equilibrium dependence of fractional conversion B into BH_2 on the ratio of B: H = 1: 2 (1); 1: 4 (2); 1: 6 (3) and temperature

Figure 1. Results of thermodynamic calculations in the B - H – N system

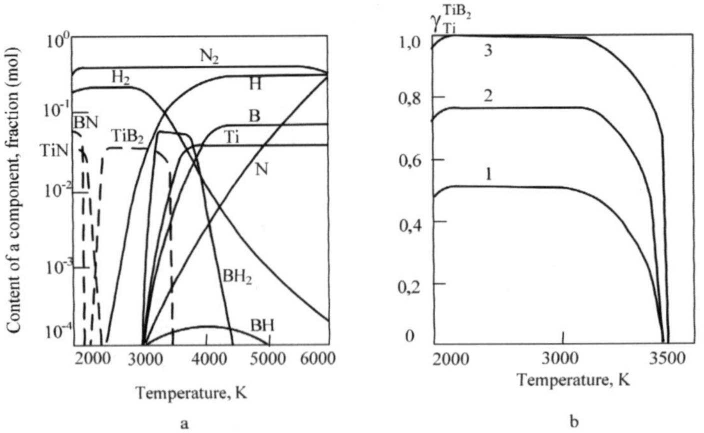

a) the equilibrium composition of gaseous and condensed phases depending on a temperature at the ratio of Ti:B:H:N = 1:2:12:20; b) the dependence of fractional conversion Ti into TiB_2 on the ratio of Ti:B = 1:1,5 (1); 1:1,75 (2); 1:2 (3) and temperature

Figure 2. Results of thermodynamic calculations in the Ti – B – H – N system

A distinctive feature of the implemented model variant is the use of calculation dependency for heat exchange between the plasma flow and the reactor's channel walls, with this calculation dependency obtained by the authors for real life conditions used in the formation of titanium boride in a plasma reactor and not included into the known equations (7): an industrial reactor power of 150 kW; a triple - jet mixing chamber with an angle of plasma jets at 30°, providing forced turbulence of the plasma flow; a 0.005 m thick lining of the reactor channel in zirconium dioxide, reducing the heat transfer by 20% from the plasma flow; introduction of fine dispersed basic powders (titanium, titanium oxide, boron) different in terms of phase composition, thermal and physical properties and dispersion of the plasma flow; the consumed mass concentration of fine dispersed basic powders 0.12 kg / kg, near to the maximum value and reducing the heat transfer by 15% from the plasma flow. The equation has the following form:

$$St = (0.672 \pm 0.155) R_{f,x}^{-0.425 \pm 0.081} . \quad (7)$$

Where St, Re, Pr are the Stanton, Reynolds, Prandtl numbers;

x and f are the indices corresponding to a bulk temperature of flow and an axial coordinate.

A rate of plasma evaporation of boron- and titanium-containing dispersed materials is determined by such factors as energy and hydrodynamic parameters of the plasma flow, particle sizes and their heat properties, mass flow concentration. The studies were carried out on the three-torch reactor under the following operation conditions: the mass flow rate of plasma forming gas (nitrogen) is 9 · 10-3 kg /s, the gas is fed to the mixing chamber of 75 kW. The results of the study are given in Figure 3.

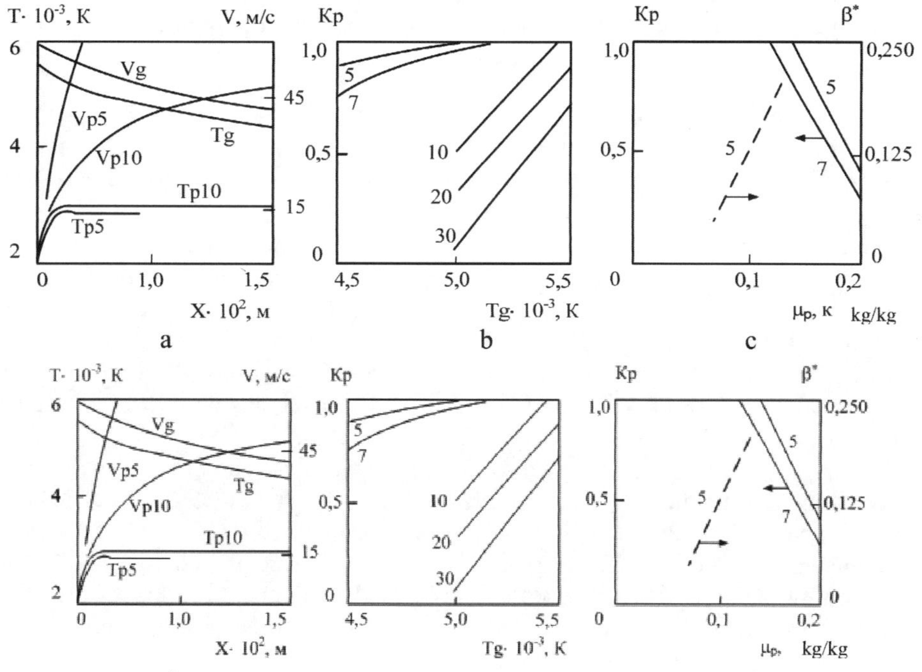

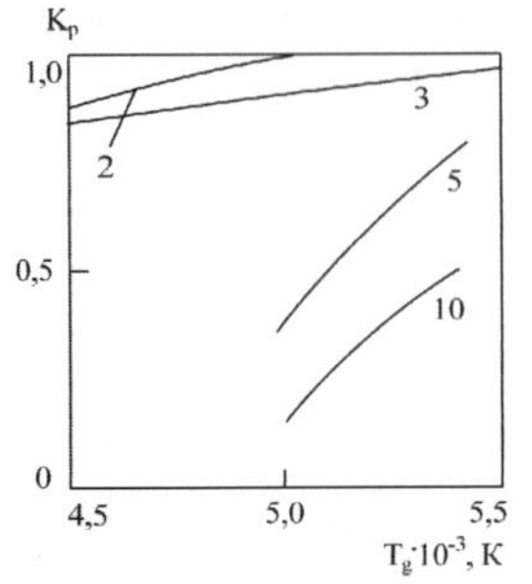

g

a) variations in bulk temperature, plasma flow rate (T_g,V_g) and titanium particles (T_p,V_p) lengthwise of the reactor; b) the influence of initial temperature of the plasma flow (T_g) on the evaporation rate of titanium particles (K_p); c) the influence of mass flow concentration (μ_p) on the evaporation rate of titanium particles (K_p) and the energy fraction transferred to particles (β^*); d) variations in bulk temperature, plasma flow rate (T_g,V_g) and titanium dioxide particles (T_p,V_p) lengthwise of the reactor; e) the influence of initial temperature of the plasma flow (T_g) on the evaporation rate of titanium dioxide particles (K_p); f) the influence of mass flow concentration (μ_p) on the evaporation rate of titanium dioxide particles (K_p) and the energy fraction transferred to particles; g) the influence of initial temperature of the plasma flow (T_g) on the evaporation rate of boron particles (K_p). 2, 3, 5, 7, 10, 15, 20, 30 are particle sizes in microns.

Figure 3. Hydrodynamic and energy conditions for processing titanium (a-c), titanium dioxide (d-f), and boron (g)

When titanium metal powder is used as a basic material, an optimal combination of thermal and physical characteristics, primarily coefficients of conductivity, heat of melting and evaporation, makes possible the process with the acceptable for processing flow concentration (0.10-0.14 kg/kg) of particles sized 5-10 microns. And, a coefficient of thermal efficiency for the flow with particles of up to 5 micron reaches 0.20. The established behaviors do not depend on the initial velocity of particles when it varies in the range 1-3 m/s. The operating reactor parameters ensure complete evaporation of titanium dioxide particles with a size of from 3 to 5 microns, with the particles interacting with the flow at similar velocities, making 32-60 m/s for 3 micron particles and 28-55 m/s for 5 micron particles, in a time respectively $15 \cdot 10^{-5}$ - $18 \cdot 10^{-5}$ and $54 \cdot 10^{-5}$-$85 \cdot 10^{-5}$s. It should be noted that an increase to a certain value of mass flow concentration for dioxide has no influence on its evaporation rate. The maximum mass flow concentration when the complete processing of TiO_2 is possible is 0.12 kg/kg for 5 micron particles and 0.14 kg/kg for 3 micron particles. The optimal loading of the reactor enables increasing a fraction of energy transferred to particles, i.e., the coefficient of thermal efficiency for the flow, for example, with 5 micron particles, increases from 0.075 to 0.20. Varying in feed rate of powder TiO_2 from 1 to 3 m/s practically does not cause any changes in the hydrodynamic behavior of the flow, and, therefore, does not affect its inter-component heat exchange. When the mass flow concentration is 0.077 kg/kg, boron particles of a 2 - 3 micron size are completely evaporated with the extreme energy characteristics for the plasma flow, corresponding to an initial temperature of 5400.

Plasma synthesis and titanium diboride characteristics.

The basic powders and gases – a reducing agent and an exchange gas – were identified according to the results of simulating the interaction of the plasma and feedstock flows. Their characteristics are given in Table. 2.

Table 2. Main characteristics of powdered basic materials and processing gases.

Powdered basic material and processing gases	Content of base material , [%] min	Dispersivity, [μm]
Finely dispersed powder of titanium PTMk	99.9	0,5-5
Titanium dioxide pigment R-1 GOST 9808-84, rev.	99.0	-1
Amorphous Boron B99	99.0	-1
Methane (natural gas)	93.6 (ethane -3.0; propane -2.18; butane -1.18)	-
Hydrogen for industrial use GOST 3022 – 80, rev.	99.8	-
Nitrogen for industrial use GOST 9293-74, rev.	99.5 (a content of oxygen is 0.5 max)	-

With photomicrographs of powdered basic material samples taken for the study, an ensemble of particles having a spherical shape is clearly observed, and the linear dimensions can be measured (Figure 4). Titanium powder (Figure 4 a) is largely represented by spherical or elliptical shape particles with the linear dimensions from 0.4 to 4.5 microns. The powder of titanium dioxide is characterized by spherical particles having a size ranging from 0.1 to 1 micron (Figure 4 b). Boron powder significantly differs from other basic material components in terms of high dispersivity and some tendency to aggregate (Figure 4 c). Hence, the boron powder particles of a predominantly spherical shape with a size widely ranging 50 to 250 nm may form aggregates ranging in size from 250 to 400 nm, containing up to 10 particles.

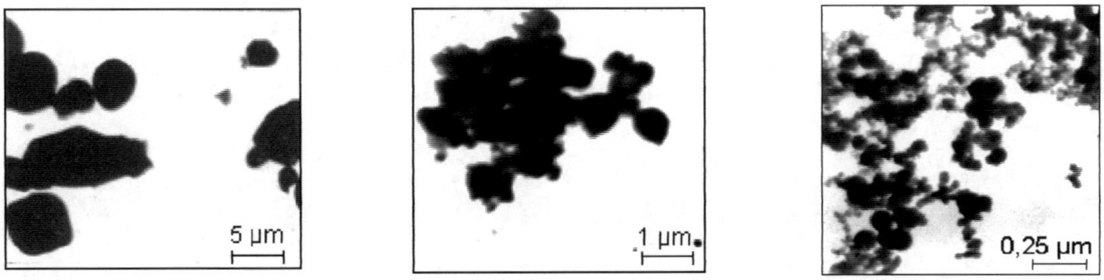

a b c

Figure 4. Photomicrographs of powdered basic materials used for the titanium diboride synthesis: (a) titanium powder, (b) titanium dioxide powder, (c) boron powder

The dispersed composition of powdered basic material is shown in Figure 5. Examining an array with 985 particles shows that the titanium powder is characterized by a particle size of 0.5 - 4 microns, the average particle size is 2.0 microns (Figure 5 a) and the particles are distributed as to fractions as follows: +0.5 - 2 microns - 48.2%, + -4 microns - 51.8%. The titanium dioxide powder has an average particle size of 0.5 micron (Figure 5 b), and the studied particles in quantity of 770 are within the size range 0.2 - 1.0 micron, with particles of a + - 0.5 micron fraction making 48.7% and + 0.5 - 1.0 micron - 51.3 %. The boron powder is the most highly dispersed basic material used (Figure. 5): 760 particles in the size range 50 - 250 nm, an average size of 112 nm and they are distributed in terms of fractions: + 50 - 125 nm - 53.2%, + 125 - 250 nm - 46.8 %.

IOP Conf. Series: Materials Science and Engineering **91** (2015) 012002 doi:10.1088/1757-899X/91/1/012002

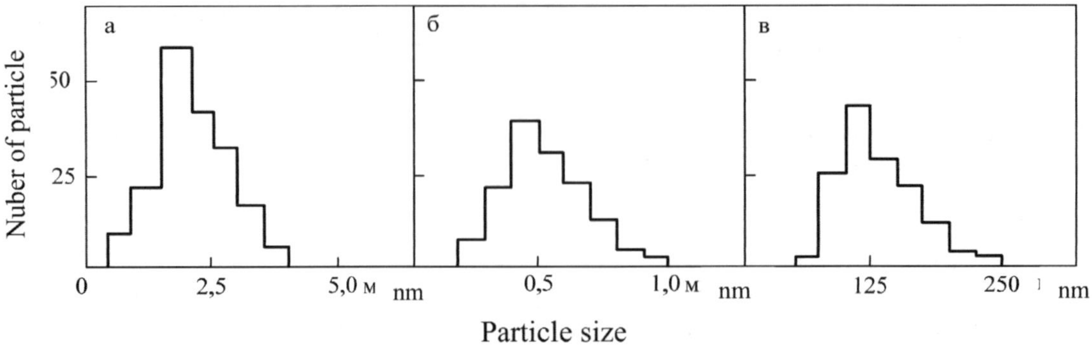

Figure 5. The dispersed composition of powdered materials:
(a) titanium powder, (b) titanium dioxide powder, (c) boron powder

The synthesis process of titanium diboride was examined using an arc power of 80 - 150 kW at a flow rate of the plasma forming gas (4.6 - 9.0) • 10^{-3} kg/s, that corresponds to the initial temperature of the plasma flow 5400 - 5500 K. The mass flow concentration was 0.10 kg of powder per 1 kg of plasma forming gas. The synthesized products were hardened by nitrogen injected through a hardening ring mounted on the reactor outlet. The hardening temperature was controlled by varying the reactor's length. The hardening gas flow was (1.0 - 2.0) · 10^{-3} kg / s.

Solid synthesis products were investigated by the following methods: X-ray, chemical, mass spectrometry, electron microscopy, thermal gravitational analysis and BET method. For the various problems of the research to solve, we examined the samples taken from the plasma flow using a water-cooled metal probe to avoid any contact with air as well as the nano powders' samples taken from the recovery system after filter depressurizing. The gaseous and thermal desorption products were analyzed using the chromatographic method.

Two process variants were investigated using the following mixtures: 1 - ($Ti + B + H_2$), 2 - ($TiO_2 + CH_4 + B$). With this, their contents are optimized in the products of titanium diboride synthesis. The following equations describing the dependence of the titanium diboride content on main technological factors:

$$[TiB_2\,(1)] = -\,412.41 + 0.09489T_0 + 2.196[B] + 0.1597\{H_2\} - 0.00061T_0[B]; \quad (8)$$

$$[TiB_2\,(2)] = 4.59 + 0.0156\,T_0 + 0.00213T_з - 0.0688\{CH_4\} - 0.214[B]\cdot\{H_2\} \quad (9)$$

Where T_0 is the initial temperature of the plasma flow (5000 – 5400 K);

$T_з$ is the hardening temperature (2600 – 2800 K);

[B] is the content of boron in the charging mixture (100 – 120% of stoichiometrically required);

$\{CH_4\}$ is the amount of the reducing agent (methane) (100 – 120% of stoichiometrically required);

$\{H_2\}$ is the concentration of hydrogen in the plasma forming gas (0-0.25 %).

The basic parameters of the synthesis process and characteristics of titanium diboride are described in Table. 3. Two variations of synthesis are compared with the purpose of identifying the most suitable titanium powder for boriding. The titanium diboride nano scale powder is evidenced by spherical aggregates in 120 - 200 nm size range formed by combining round particulates in the appreciably wide size range 10 to 60 nm (Figure 6). The round shape of titanium diboride nanoparticles proves the possibility of forming according to the "vapour - melt - crystal" pattern, supposedly in interacting titanium aerosol with boron hydrides.

It is assumed that the proposed method for producing titanium diboride is competitive as to the phase and chemical composition, dispersion, main technological parameters, and this method can be accepted as a principle technology used for diboride intended for protective aluminum-wettable cathode and galvanic composite coatings [3-6].

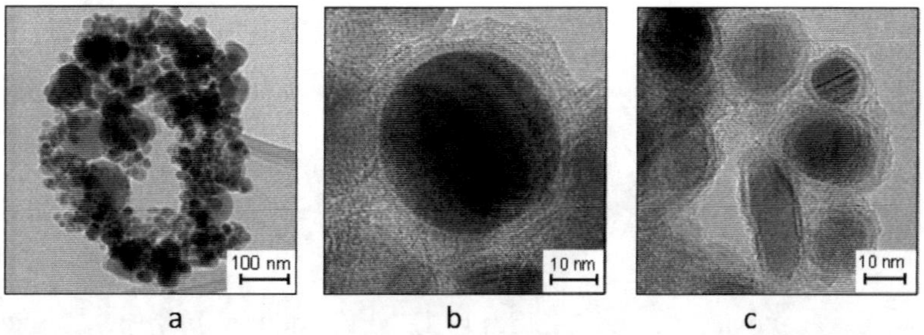

Figure 6. Photomicrographs of titanium diboride nanopowder (a – ensemble of particles and aggregates; b – morphological pattern of aggregates; c – single particles)

Table 3. Basic parameters of the synthesis and characteristics of titanium diboride

Synthesis parameters and characteristics of titanium diboride	**Synthesis technology variants**	
	1(Ti + B +H_2)	**2(TiO_2 + B + CH_4)**
Exchange gas composition, [% об.] - Nitrogen - Hydrogen - Natural gas (methane)	 74.0 25.0 1.0	
Fineness of titanium-containing basic material, [μm]	0.5…4.0	0.2…1.0
Fineness of boron-containing basic material, [μm]	0.25…0.40	
Boron content in charging powder mixture, [% of stoichiometric]	100 – 120	
Amount of reducing agent (methane), [% of stoichiometric]	-	100-120
Initial temperature of plasma flow, [K]	5400	
Hardening temperature, [K]	2600 - 2800	
Chemical analysis, [%] TiB_2 Free boron Free titanium Free carbon Oxygen Nitrogen Volatiles	 92.0 – 93.0 1.30 – 1.15 1.91 – 1.45 - 2.29 – 1.83 2.05 – 1.92 0.45 – 0.65	 90.05 – 91.30 1.91 – 1.04 - 1.42 – 0.92 3.72 – 3.52 2.26 – 2.11 0.64 – 1.11
Specific surface area, [m^2/kg]	46000 –48000	35000 - 37000
Particle shape	round	
Oxidation degree of powders x10^7, [kg O_2/m^2 of specific surface area]	3.82 – 4.98	9.51 – 10.63
Basic material flow capacity, [kg/h]	3.6	
Capacity TiB_2, [kg/h]	3.42	2.35
Rate, [kg/h·m^3]	1556	1070

Conclusions

It has been established that titanium diboride can be obtained in the form of nano particles using titanium - boron and titanium dioxide - boron powder mixtures under the nitrogen - hydrogen and nitrogen - hydrocarbon plasma flows. The behavior of the boride formation process has been identified and discussed. The two variants of the synthesis process are compared with the purpose of identifying the most suitable titanium powder for boriding. The method proposed for titanium diboride production is competitive in terms of its phase and chemical composition, dispersion, main technological parameters, and this method can be accepted as a principle technology to produce this material used in protective aluminum-wettable cathode and galvanic composite coatings.

This work has been done at Siberian State Industrial University in the framework of government contract No. 11.1531/2014/K with the Ministry of Education and Science of the Russian Federation.

References

[1] Brodov A A Kazakova T I and Molotilov B V Economic analysis of the industrial use of nanomaterials *Still in Translation* 2006 T 36 #5 p 67-71

[2] Skachkov O A Firstov M N Zmii V I Pozharov S V Ezhov A Yu Astakhov S I and Chuikova V B High-temperature protective coatings on refractory metals *Still in Translation* 2006 T 36 #4 p 75-80

[3] Serlire M Oye H A Cathodes in Aluminum Electrolysis – Dusseldorf: Aluminum – Verlag 2010 p 698

[4] Zang H, V de Nora Sekhar J A Materials used in the in the Hall–Heroult cell for Aluminum Production *Warrendale*: 1994 TMS p 153

[5] Li Q Liu Y Li J Yang J Fang J LiuY Adhesion between Ramming Pastes and Cathode Blocks in Aluminum Electrolysis Cells *Light Metals* 2004 p 327 - 332

[6] Ban Y Shi Z Wang Z Kan H Yang S Cao X Qiu Z Assessment of the impact of the titanium coating associated colloidal alumina to improve the behavior of carbon cathode baths Era – Hall *Light Metals* 2007 p 1051 - 1056

[7] Ren B Xu J Shi Z Ban Y Dai S Wang Z Gao B A method for producing titanium diboride powder wettable cathode material for aluminum electrolysis *Light Metals* 2007 p 1047 - 1052

[8] Levashov E A Senatulin B R Leyland A and Matthews A Production of composite targets in the system Ti-B-Si-Al by self-propagating high-temperature synthesis *Russian Journal Of Non-Ferrous Metals* 2006 T 47 #1 p 39 - 44

[9] Nozdrin I V Shiryaeva L S Galevski G V Rudneva V V Reactor for Plasmometallurgical Production of refractory Borides and Carbides. Still in Translation 2011 T 41 #8 p 644-648

[10] Nozdrin I V Galevski G V Shiryaeva L S Terenteva M A Particle Size of Vanadiun and Chromium Borides and Carbides in a Plasma Flux *Steel in Translation* 2011 T 41 #10 p 799-804

[11] Rudneva V V Galevski G V Galevski S G Yurkova E K Effective Processing of Disperse RAW Materials in a Plasma Reactor *Still in Translation* 2007 T 37 #5 p 425-428

[12] Nozdrin I V Terenteva M A Galevski G V Rudneva V V Features of Electrodiposition of «Nickel – Chromium Diboride Nanopowder» Composite Coatings *Russian Journal of Non – Ferrous Metals* 2013 V 54 #5 p 383 - 387

IOP Conf. Series: Materials Science and Engineering **91** (2015) 012003 doi:10.1088/1757-899X/91/1/012003

Silicon nano-carbide in strengthening and ceramic technologies

V V Rudneva, G V Galevsky and N A Kozyrev

Siberian State Industrial University
654007, Novokuznetsk, 42, Kirov str., tel. (3843) 74-89-13

e-mail: kafcmet@sibsiu.ru

Abstract. Technological advantages and conditions of new quality assurance of coatings and products, provided by silicon nano-carbide, have been ascertained in the course of composite electrodeposition of coatings, structural ceramics patterning, and surface hardening of steels via electro-explosive alloying. Silicon nano-carbide has been recommended to be used as a component of wear and corrosion resistant chromium carbide electrodeposited coatings, which can be operated at high temperatures and used for strengthening tools and equipment including those with a complex microrelief of functional surfaces. Silicon nano-carbide as a component of composite "silicon carbide – boron – carbon" can be applied to produce ceramic half products via solid phase sintering in argon under pressure of 0.1 MPa and temperature 2273 K. Application of silicon nano-carbide in technology of tool steel surface hardening via electro-explosive alloying ensures obtaining of a high micro-hard, wear and heat resistant shielding layer which is about 20 μm deep.

Introduction

Silicon carbide production is one of the most important in the structure of contemporary metallurgy. It is a combination of extraordinary thermo physical, mechanical, physical and technical properties of silicon carbide that makes it the unique material for numerous branches of engineering and predetermines such basic applications as couplant materials, ceramics, composite materials, and coatings. World production of silicon carbide is approximately 800 thousand tons annually. The main spheres of silicon carbide application are metallurgy (about 45 % of world demand), abrasive materials production (up to 30 %), and refractory materials (up to 25 %). Product markets of silicon-carbide powder materials: micronized powder with maximum 1 μm particles (so called micronized carbide) used for ceramics and nano-powder with maximum 100 nm particles (so called nano-carbide) used for high quality structural and impact resistant ceramics and galvanics are small-scale in terms of real volume (maximum 1%) but intensively expanding, they also have high money measure. Introduction of silicon carbide in form of nano-carbide opens new horizons of its application, including surface and three-dimensional modifying metal alloys and polymers, developing special components of paints etc. [1, 2]. Manufacturers of micronized carbide are enterprises of world leader companies "Saint – Gobian" and "Exolon – ESK". Among nano-carbide manufacturers there are research and production companies "Tokyo Tekko Co" (Japan), "Hefei Kaier Nanotechnology & Development Ltd. Co" (China), "NEOMAT Co" (Latvia) etc.

Purpose of research

The work focuses on assessment of application efficiency of silicon nano-carbide in technology of composite electrochemical coatings (CEC), structural ceramics, and electro-explosive alloying (EEA) of steels.

Silicon nano-carbide in technology of composite electrochemical coatings

In technology of composite electrochemical coatings metal (nickel, chromium, ferrum, and copper) is crystallized from electrolytes – suspensions containing an additive which is called a strengthening agent (filler or modifying agent) and is made as a powdery substance. Its particles integrate with a metallic matrix forming on a surface of a product. Operational characteristics of the coating get improved due to particles integrated with it, so composite electrochemical coatings can facilitate solution of numerous practical problems focused on surface hardening of structural components and tool equipment, as well as on renovation of their quick-wear components. The degree of dispersibility of a hardening agent is to be increased in order to improve characteristics of composite electrochemical coatings. For the same purpose a hardening agent is to be applied with essentially smaller particles than the matrix grains. In this case, first, electrochemical coatings quality can be improved because porosity and micro-roughness of the surface are reduced, second, a matrix with a balanced sub-grain structure and proportional concentration of particles is formed, third, physical and mechanical properties of coatings are improved due to the effect of disperse hardening and reducing internal stresses, forth, technological possibilities to produce composite electrochemical coatings are widened because of insignificant sedimentation of nano-dimensional particles in electrolytes-suspensions and increased adsorption of ions and other additives. These facts pre-determine constant attempts of experts dealing with composite electrochemical coatings technology to use materials of high dispersion, as well as nano-materials as hardening agents [3, 4].

While selecting a hardening agent diamond and cubic boron nitride, silicon carbide, high melting point, hard, wear and scale resistant oxides, metal like carbides, borides, nitrides, silicides and their composites are the most preferable substances. However, a 25-year monitoring of academic and applied investigations in the field of composite electrochemical coatings has revealed that the requirement for such materials is met basically by micro-powders of abrasive purpose, which usually have M1…10 grain fineness, or finely-dispersed materials of chemical technology. The latter are distinguished by low concentration of main agents, typical for mechanically grinded materials fission form of particles, and a great number of admixtures. A hardening agent of this kind, average particles of which are comparable with the 5-20 μm grain of coating metallic matrix, significantly improves hardness and wear-resistance of composite electrochemical coatings, which contain ferrous metals, copper and some low-melting point alloys. However, it doesn't have a positive effect on strength characteristics of a coating, reduces its resistance to corrosive attacks, and worsens heat-resistance and high-temperature strength. The reason for it is in dissimilarity of characteristics of composite electrochemical coatings, caused by unevenly distributing particles of more than 1 μm, as well as by an increased porosity of coatings and hydrogen embrittlement. For a long period of time these factors have necessitated the development and implementation of special technologies to produce materials – efficient hardening agents in order to solve urgent problems of galvanics via composite electrodeposition.

Silicon nano-carbide was used as a hardening agent in the course of electrochemical deposition of chromium-plating electrolytes. To solve the problems of composite hardening one used nano-carbide, obtained via carbonization of silicon micro-powder by methane– SiC (1) and plasma modification of micro-powder – SiC (2) [5]. The characteristics of nano-powders (NP) of silicon carbide SiC (1) and SiC (2), their physical and chemical properties are listed in Table 1, and electron micro-photographs are shown in Figures 1 and 2.

Table 1. Basic characteristics of silicon carbide powders

Characteristics of nano-powders	SiC (1)	SiC (2)
Phase composition	β-SiC	β-SiC
Chemical composition of the main substance	$SiC_{0.95}N_{0.05}$	$SiC_{0.95}N_{0.05}$
Concentration of the main substance, %	99.21*	99.30*
Concentration of free carbon admixture, %	0.06	0.05
Specific surface, m^2/kg	38000	37000
Dimension** of particles, nm	62	61
Form of particles	Faceted	
Corrosion resistance in solutions of electrolytes	Stable in electrolytes of different acidity	
Thermal-oxidative stability in air, K	to 913	
Notes: * After refining. ** It is calculated according to the size of specific surface.		

For electrodeposition of composite electrochemical coatings chromium – nano-carbide chromium-plating electrolyte was used, containing NP of silicon carbide SiC (1) or SiC (2) with the following chemical composition, kg/m^3: chromium anhydride 250, sulfuric acid 3, trivalent chromium 8. For electrolyte – suspension some pure electrolytes were added to a sample powder, the components were mixed up to complete wetting of the powder, put into electrolyzer and dissolved till a required concentration was obtained. pH was adjusted with solutions of NaOH or H_2SO_4.

On the basis of analyzed results of experiments and industrial application of silicon nano-carbide SiC (1) and SiC (2) a possibility of parametric, concentration, structural and technological effects has been revealed. So, the following change in cathodic density of current was characteristic for the used chromium-plating electrolyte in dependence on dispersion of the filler agent (MP – micro-powder of silicon carbide with particles of maximum 5 µm):

Without a filler agent (5 kA/m^2) → MP SiC (5.5-6.0) → NP SiC (1), SiC (2) (8-10).

Obviously, the increase in acceptable cathodic density of current is caused by stabilization of pH in cathode layer when the second agent is added to the electrolyte. A coating is saturated with carbide nano-particles when their mass concentration is low in a coating and in electrolyte (Table 2); that is because the quantity of nano-particles rises per a unit of volume and facilitates dispersive hardening of metallic matrix. Micro-hardness of chromium containing composite coatings with nano-components is 1.80-1.90 times higher than that of chromium, and 1.10-1.20 higher than micro-hardness of composite coatings containing micro-powders (Table 2), and in all cases microhardness grows as the quantity of particles increases in a composite material. Increase in hardness is caused by high efficiency of nano-particles acting as a barrier to distribution of dislocations.

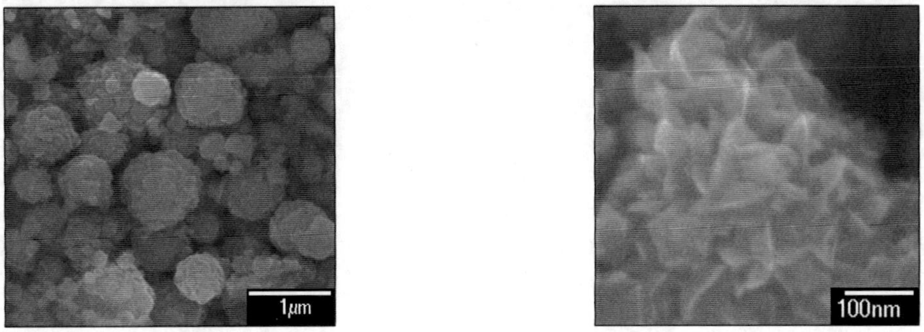

a b

IOP Conf. Series: Materials Science and Engineering **91** (2015) 012003 doi:10.1088/1757-899X/91/1/012003

c

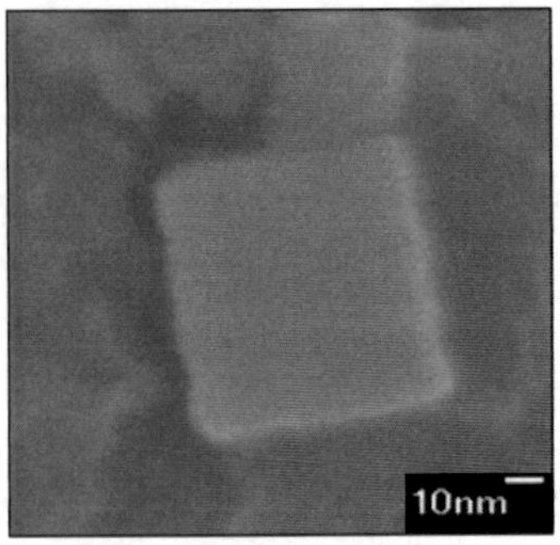

d

Figure 1 Micro-photographs of silicon carbide nano-powder SiC (1): a – in as received condition after refining; b – morphological picture of an aggregate; c – ensemble of nano-particles; d – a cubic nano-particle

a

b

c

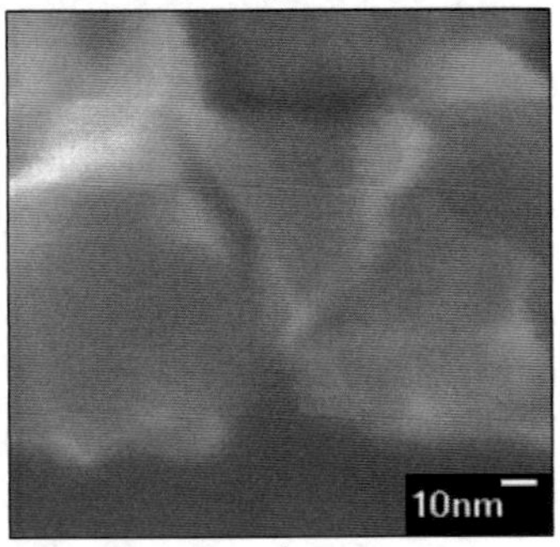

d

Figure 2. Micro-photographs of silicon carbide nano-powder SiC (2): a – in as received condition after refining; b – morphological picture of an aggregate; c – ensemble of nano-particles; d – octahedral nano-particle

Nano-particles in chromium matrix reduce the intensity of its wear 1.25-2.5 times when concentration of nano-powder changes from 0.15 to 0.45 %. Considerable improvement of wear resistance of Cr – NP SiC material is possible because there are no exciters of local stress concentration in the structure of the layer – micro-cracks; their number is rather significant in chromium layers. Corrosion stability of composite electrochemical coatings Cr – NP SiC, determined in a liquid medium, shows that composite electrochemical coatings excel 1.3-1.5 times conventional chromium coatings (frequency index of corrosion 0.3 %) and can be considered as stable ones.

IOP Conf. Series: Materials Science and Engineering **91** (2015) 012003 doi:10.1088/1757-899X/91/1/012003

Table 2. Dependence of hardening agent concentration (a) and micro-hardness (N) of composite electrochemical coatings* on concentration of NP and MP in electrolyte

Concentration of NP, MP in electrolyte, kg/m^3	a, % / N, GPa		
	Cr – NP SiC (1)	Cr – NP SiC (2)	Cr – MP SiC
1.0	0.20/9.0	0.24/9.1	0.47/8.4
3.0	0.41/9.3	0.43/9.2	0.72/8.6
10.0	0.59/9.7	0.58/9.4	0.89/8.9
20.0	0.82/10.1	0.84/10.1	1.1/9.0

Note: * Conditions of electrodeposition: CEC Cr – NP, MP SiC – cathodic density of current 8.0 kA/m^2, pH 2.0, T = 323 K.

To assess the possibility of replacing explosively synthesized nano-diamonds by silicon nano-carbide, we compared characteristics of chromium – carbide and chromium – diamond coatings (Table 3). As you can see, wear-resistance, microhardness, corrosion-resistance of chromium – carbide coatings are comparable with those of chromium-diamond ones, and the life time is even longer in operating conditions at temperatures above 473-573 K, that is provided by lower concentrations of carbide nano-powder in electrolyte and significant cost reduction of 1 m^3 of electrolyte – suspension.

Experimental data demonstrate technological reasonability of nano-powders application with a specific surface, which doesn't exceed 40,000 m^2/kg (Table 4); and has a restricted concentration (maximum 1.0 %) of free carbon (Table 5).

Table 3. Comparative characteristics of coatings containing chromium with nano-powders of silicon carbide SiC (1) and nano-diamond [6]

Conditions of electrodeposition and achieved results		Technological variants					
		1	2	3	4	5	6
Chemical composition of electrolyte, kg/m^3	Chromium anhydride	250	250	250	250	250	260
	Sulfuric acid	3	3	3	3	3	0.34
	Trivalent chromium	8	8	8	8	8	–
	Potassium fluorsilicate	–	–	–	–	–	19
	Barium sulfuric	–	–	–	–	–	7
	NP of silicon carbide – SiC (1)	4	6	7	8	10	–
	NP of diamond	–	–	–	–	–	20
Deposition mode	Temperature, K	328	328	328	328	328	331
	Density of current, A/dm^2	55	55	55	55	55	55
Results	Wear-resistance*	0.96	1.0	1.05	1.07	1.08	1.0
	Micro-hardness, GPa	8.9	9.0	9.1	9.15	9.15	9.0
	Corrosion resistance*	0.96	1.05	1.07	1.10	1.10	1.0
	Life time at temperature above 473-573 K*	1.1	1.50	1.70	2.00	2.06	1.0
	Cost of 1 m^3 of electrolyte suspension*	0.13	0.16	0.18	0.20	0.23	1.0

Note: * Wear and corrosion resistance, life time in operating conditions at temperatures above 473-573 K, cost of 1 m^3 of electrolyte suspension are given according to variant 6, the values of these characteristics for it are thought of to be equal to 1.0.

Table 4. Comparative characteristics of coatings containing chromium with NP of silicon carbide SiC (1) having different specific surfaces (for conditions of technological variant 4, Table 3)

Specific surface of silicon carbide NP, m^2/kg	11000	22000	32000	38000	51000
Wear resistance	0.89	0.99	1.0	1.03	0.93
Micro-hardness, GPa	7.92	9.27	9.32	9.39	8.51
Corrosion resistance	0.82	1.0	1.0	1.0	0.91
Life time in conditions of operating at temperature above 473-573 K	0.72	1.0	1.0	1.0	0.88

Table 5. Comparative characteristics of coating containing chromium and NP of silicon carbide SiC (1) with different concentration of free carbon in it (for conditions of technological variant 4, Table 3)

Concentration of free carbon in nano-powder of silicon carbide, %	0.8	1.5	2.4
Wear resistance	1.0	0.86	0.79
Micro-hardness, GPa	932	760	710
Corrosion resistance	1.0	0.89	0.79
Life time in conditions of operating at temperatures above 473-573 K	1.0	not recommended to use in these conditions	

Utilization of more dispersed nano-materials makes the technology of production and application of electrolyte-suspension more complicated. This fact has been also mentioned in other researchers' papers. For instance, authors in paper [6], who used diamond nano-powders with a specific surface 400.000-500.000 m^2/kg (particles 4-5 nm) for composite chromium-plating, noted, that it is rather difficult to add nano-powders to the electrolyte, if electrolyte-suspension is made in a conventional way (it takes about 50-80 hours to stir the mixture), therefore they had to design a special multi-operational technology. Authors [7], who used nano-powder of aluminum oxide for composite chromium-plating with a specific surface 100.000-150.000 m^2/kg (particles 15-18 nm), also mentioned difficulties of making electrolyte-suspension, and recommend to add nano-powder to the electrolyte in form of a concentrated suspension, processed ultrasonically with 22 kHz frequency. All the mentioned peculiarities demonstrate that there are processes of nano-particles coagulation in electrolyte, which cause coarsening of particles, and get stronger if nano-powders with specific surface 50.000-500.000 m^2/kg are applied. Reasons for coagulation processes are as follows: instability of nano-systems with an excessive surface energy and an insufficient wetting of nano-powders, caused by the state of their particles surface and a surface charge available or unavailable on them. As a result, it is necessary to elecrodeposite coatings from electrolytes – suspensions with higher concentration of nano-powders.

On the basis of research authors developed a method that helps to obtain composite electrochemical coatings containing chromium [8] and technological processes of composite hardening by silicon nano-carbide. CEC containing Cr – NP SiC are recommended to be used to protect components and technological tools subjected to wear (for example, threadlike sensors of twisting frames, shock strut pistons etc), simultaneous wear and corrosion (for example, die molds to press products of thermal plastic materials, components of gas and oil-field equipment etc.). However, economically and technologically they are more reasonable to be used instead of chromium-diamond CEC for surface hardening of tools, equipments, components of machines, which are operated in high temperatures (above 473-573 K): extrusion toolage for powder metallurgy, components of tool production – puncheons, stamps, matrixes, holders of puncheons etc., elements of pistons in internal combustion engines etc.

Silicon nano-carbide in technology of structural ceramics

Polycrystalline materials produced via sintering non-metallic powders of natural or artificial origin are thought of as ceramic materials. The advantages of ceramics over alternative metallic materials are as follows: availability of raw materials, low power-intensity of production processes, environmental safety, complex of good physical and chemical, mechanical properties (hardness, refractoriness, corrosion, thermal and radiation resistance etc.). Up to date types of ceramic are divided into two groups: structural and functional ones. Structural ceramics includes ceramic materials used to produce mechanically stable structural elements of devices, and functional one is ceramics with special electrical, thermal, optical and magnet functions. Structural ceramics is further subdivided into oxide and oxygen free ones (containing carbides, nitrides, borides).

Silicon carbide in form of micro-powder is used to manufacture products of wear resistant mechanical engineering ceramics – components of stop and control valves. Ceramic control and stop valves can be resistant to hydro and abrasive wear, corrosion and temperature effects, as well as to corrosive liquids influence. Therefore, it can be used in oil and chemical equipment to control technological processes. The valves consist of gates, butterfly dampers, valves, control flap balls (balls and seats), bushes of stuffing units, stuffing rings etc. Ceramic bearings are among the most promising types of ceramic products. Being hard and capable to stand temperatures up to 1000 °C, arising in friction, ceramic bearings 100 times outperform steel ones in terms of lifetime and run without being lubricated. As density of ceramics is low, application of ceramic bearings have mass advantage up to 60%, at high speeds it reduces centrifugal forces. Chemical stability, radiation resistance, good dielectric characteristics, lack of magnetism, and low density make it possible to use ceramic bearings to kit equipment in oil and chemical industries, in nuclear and power facilities. Ceramic bearings are also efficient in precision and navigation instruments, in defense industry, for example, in gyroscopes and in high-speed machines – turbines, compressors, machining centers.

Enterprises in Russia produce wear resistant silicon carbide ceramics oriented towards ceramic raw materials – micronized carbide, and imported technological equipment characterized by high thermal power parameters of compaction. Here, it is an important scientific and technical issue, relevant to various branches as well, to develop an industrial technology to produce silicon carbide with properties providing its efficient application in manufacturing structural ceramics for mechanical engineering (so called structural carbide). Structural carbide, containing additives, which activate sintering process, can be used to develop more promising manufacture of ceramics – by solid-phase sintering, implemented by means of relatively affordable Russian equipment.

One made use of silicon nano-carbide in technology of structural ceramics, which was produced via carbonation of silicon micro-powder with methane – SiC (1) and modification of carbide micro-powder in plasma flow of nitrogen – SiC (2) with further refining.

Silicon nano-carbide is similar to structural silicon-carbide powders produced by overseas companies (Table 6) in basic characteristics (dispersion degree, phase and chemical composition). Technological advantage of the introduced material is its possibility to deliver it as a composition with pre-set components, containing additives, which activate the sintering process (reactive boron and carbon).

Table 6. Chemical and phase compositions, modes of solid-phase sintering and physical and mechanical properties of silicon-carbide ceramic materials

Characteristics of silicon carbide powders, conditions of preparing and modes of solid-phase sintering	Trade mark, manufacturer			
	SiC (1)	SiC (2)	A-10 “Hermann Starck Co.”	B-10 “Hermann Starck Co.”
Phase composition	β-SiC	β-SiC	α-SiC	β-SiC
Chemical composition, %				
SiC	90.16	90.00	97.0	95.0
C_{free}	2.70	2.69	1.0	1.9

Characteristics of silicon carbide powders, conditions of preparing and modes of solid-phase sintering	Trade mark, manufacturer			
	SiC (1)	SiC (2)	A-10 "Hermann Starck Co."	B-10 "Hermann Starck Co."
B_{free}	2.52	2.63	–	–
Si_{free}	0.03	0.02	0.01	0.02
SiO_2	4.51	4.60	1.88	3.02
Al	0.04	0.03	0.1	0.04
Fe	0.04	0.04	0.01	0.02
Correlation SiO_2 : C in mixture, mole/mole (stoichiometric "for carbide" 1 : 3)	1 : 3.01	1 : 2.98	1 : 2.69	1 : 3.17
Specific surface, m^2/kg	38000	37000	11000	15000
Dimension* of particles, nm	67	66	700	600
Heat treatment of mixture T = 1073 K, P = 100 Pa, τ = 1 h	+	+	+	+
Plasticizing of mixture with 5 % solution of plasticizing substance	+	+	+	+
Cold press of half products in steel collapsible press form P = 50 MPa	+	+	+	+
Relative hardness before sintering	0.63	0.62	0.64	0.61
Solid-phase sintering in argon T = 2273 K, P = 0.1 MPa, τ = 2 h	+	+	+	+
Relative density of sintered samples	0.95	0.96	0.86	0.87
$\sigma_{изг.}^{20^\circ C}$, MPa	340	380	310	320
Note: * Calculated according to specific surface.				

Silicon nano-carbide was proofed and applied in solid-phase sintering of half products needed to manufacture ceramic sealing rings of grades 156.017 and 156.073 from 16 to 30 mm in outer diameter and 18 to 30 mm high intended for pumps and submersible electrical engines, dispensing oil and oil products, and condensed carbonic gases in order to substitute imported non-grinded A-10 and B-10 grades silicon carbide synthesized in furnace by companies "Hermann Starck Co.". Here, technological advantages of silicon nano-carbide were confirmed and following results were obtained (Table 6). Silicon nano-carbide makes it possible to avoid adding organic substances to the mixture and their carbonation to provide evenly distributed reactive carbon and restoration of silica film on carbide nano-particles, and mixing on rolls. It simplifies and reduces the cost of the technology and facilitates its implementation according to the following acceptable variant: heat treatment of powder in vacuum at 1073 K, adding a plasticizer, mixing and pressing under the pressure of 50 MPa, annealing for 2 hours in argon under the pressure of 0.1 MPa and at temperature 2273 K. If one follows the technology, relative density after forming 0.62-0.63, after sintering 0.95-0.96 and stable shrinkage 26-29% are provided.

Silicon nano-carbide in technology of steel surface hardening via electro-explosive alloying

Changing operational standards of materials constantly necessitate research into the field of surface modification. That is why, an urgent issue these days is development of environmentally safe new technologies to modify materials in order to change their structure and phase state, physical and mechanical properties with number of advantages over both conventional technological processes of thermomechanical, chemical and thermal treatment; and methods of impact based on such types of concentrated flows of energy that include laser irradiance, high-current electron and ion beams etc.

IOP Conf. Series: Materials Science and Engineering **91** (2015) 012003 doi:10.1088/1757-899X/91/1/012003

Here, it is worth mentioning that silicon carbide has always attracted attention of the authors of innovative technologies in the field of modification, including the following ones: alloying surface layer of steel cast products to improve their wear resistance via treatment of cast forms from inside by mixtures, containing silicon carbide; electrical discharge alloying by electrodes made of various materials of treated surfaces, preliminary coated by mixtures, containing silicon carbide; electro-explosive alloying along with adding silicon carbide to explosion zone.

The essence of electro-explosive alloying is that the surface is influenced by heterogeneous plasma beams forming in the course of electric explosion of conductors [9 – 11]. Here, the following processes take place: melting, saturation of thin surface layers of metals with products of electric explosion and self-annealing of alloying zone. The main advantage of electro-explosion alloying over other similar pulse plasma methods of surface processing is that any electricity-conductive material can be used to produce plasma. Technological possibilities of the method can be considerably widened, if powder additives of various substances are added to the zone of conductor explosion. In recent years electro-explosion alloying has been developed as an independent research domain mainly; its technological implementation is a multivariate one. Layer by layer tests of the structure, phase composition and characteristics of alloying zone demonstrate that, at least, a near surface layer has a nano-composite structure. There are intermediate and border layers below the near surface layer; constantly reducing degree of alloying and growing dimensions of structural components are their distinctive features. Nano-dimensional phases are typical to all layers. Therefore, one can insist that nano-composite bounded with the base layers are obtained via electro-explosion. As the consequence of structural changes we can consider the increased several fold microhardness, wear resistance, stability to high temperature oxidation, and corrosion resistance of processed surfaces. The main structural component of equipment, which is used for electro-explosive alloying, is a plasma generator consisting of coaxial electrodes and compression chamber with the nozzle, directing the products of explosion; the latter is located in a technological chamber connected to a vacuum pump. When a conductor explodes an ultrasonic beam of explosion products is formed, and while reacting with the surface it provides necessary conditions for pulse liquid-alloying (for approximately 100 microseconds). The mode of processing usually depends on charge voltage of storage condenser, and the surface to be processed is oriented perpendicularly to the axis of plasma beam. Technological information available nowadays, relies on results of research and implementation of electro-explosive carburizing, boriding, aluminizing, carbon-boriding, but no data are practically at disposal of researchers concerning high-melting carbides, borides, nitrides, silicides, applied traditionally in strengthening technologies due to good physical and chemical properties and complex nature of modifying effect [12 – 13].

Silicon nano-carbide SiC (1) (Table 1), containing 99.21 % β-SiC with specific surface 38000 m^2/kg was first applied as a powdery additive in the technology of surface hardening of steels via electro-explosive alloying. Forming composite surface layers were tested on the samples of tool steel X12 in conditions of electro-explosion of aluminum foil and adding silicon nano-carbide to the zone of explosion. This steel was selected because of its wide application in production of tools aimed to form materials (stamps, puncheons etc.), so high resistance to tear is relevant for this steel. It is worth paying attention to silicon nano-carbide. Phase composition of surface layers of steel after processing was tested, as well as their properties – micro-hardness, wear and heat resistance.

30 mg of aluminum foil were used as an exploded conductor; a sample of 7.5 mg silicon nano-carbide was placed on the foil. 25 x 20 x 5 mm samples of steel were tested. The surface was impacted within 100 microseconds, efficient accepted power density was 6.0 $GWatt/m^2$, and dynamic pressure of jet on the surface was 14.2 MPa. In these conditions a zone of alloying was approximately 20 μm deep.

Electro-explosion alloying in high-intensive conditions of plasma impact on a surface caused melting and saturation of steel surface layer both with products of aluminum foil explosion and nano-particles of silicon carbide, then it was self-annealed and austenite structure was formed. Microhardness of the surface of non-processed samples was 2513±205 MPa, and after processing it increased 2.8 times in average. Mass loss of steel samples in as received condition was 27.4 ± 7.3 mg while wear resistance was tested, and after electro-explosive processing – 3.3 ± 2.5 mg, i.d. wear resistant increased 8 times. Tests on heat resistance demonstrated that mass corrosion of samples in initial state was 2.7; 5.2 and 10.6 g/(m^2·h) at temperatures 800, 850 and 900 °C, respectively. After shielding processing the following values were obtained 0.3; 1.5 and 5.3 g/(m^2·h). Heat resistance increased 9; 3.5 and 2 times at given temperatures, that is approximately two times higher than corresponding values of heat resistance of this steel after two-component electro-explosion alloying, for example, by boron or gadolinium.

Therefore, the zone of alloying is characterized by high micro-hardness and is stable to abrasive wear and high temperature oxidation in the air.

Conclusions

High efficiency of silicon nano-carbide has been confirmed in the technology of composite chromium-plating, structural ceramics, surface modifying of steels via electro-explosion alloying. Chromium-carbide coatings are distinguished by wear resistance, micro-hardness, corrosion resistance comparable with chromium-diamond coatings, and a prolonged lifetime. Sintered silicon-carbide ceramic half products have high density and stable shrinkage. Zone of alloying containing silicon nano-carbide is characterized by high micro-hardness, abrasive wear and thermal-oxidative resistance.

The work is performed in Siberian State Industrial University as a project of the State Order of Ministry of Education and Science of Russia № 11.1531/2014/K.

References

[1] Galevsky G V Rudneva V V Yurkova E K Nano-materials and nanotechnologies: assessment, tendencies, and forecasts *Russian Journal of Non-Ferrous Metals* 2007 V. 48 No 2 p 157 - 160

[2] Brodov A A Kazakova T I Molotilov B V Economic analysis of the industrial use of nano-materials *Steel in Translation* 2006 V. 36 No 5 p 67 – 71

[3] Nozdrin I V Terenteva M A Galevsky G V Rudneva V V Features of Electrodepositing of «Nickel – Chromium Diboride Nanopowder» Composite Coatings *Russian Journal of Non-Ferrous Metals* 2013 V. 54 No 5 p 383 – 387

[4] Skachkov O A Firstov M N Zmii V I Pozharov S V Ezhov A Yu Astakhov S I Chuikova V B High-temperature protective coatings on refractory metals *Steel in Translation* 2006 V. 36. No 4 p 75 – 80

[5] Polyakh O A Rudneva V V Yakushevich N F Galevsky G V Anikin A E Silicon carbide production from steel plant wastes *Steel in Translation* 2014 Vol. 44 No 8 p 565 – 572

[6] Pat. 2156838 RF, MPK C25D15/00. Way of receiving composite metal-diamond coverings, E.V. Nikitin, L.A. Polyakov, N.A. Kalugin; combine «Elektrokhimpribor». No. 99108896/02; declar. 21.04.99; publ. 27.09.2000.

[7] Pat. 2148109 RF, MPK C25D3/04. Way of receiving heat-resistant chromic coverings, I.G. Idrisov, V.V. Kovalyov. BTI AGTU. No. 99102009/02; declar. 01.02.99; publ. 27.04.2000.

[8] Pat. 2318083 RF, MPK C25Д 15/00. Way of receiving composite electrochemical coverings on the basis of chrome, G. V. Galevsky, V. V. Rudneva, O. A. Polyakh. No. 2006129821/02; declar. 17.08.2006; publ. 27.02.2008. Bull. 6. 5 p.

[9] Ivanov Y F Denisova Y A Teresov A D Petrikova E A Gromov V E Soskova N A Budovskikh E A Electron-beam surface treatment of alloys based on titanium, modified by plasma from an electrical explosion of conducting material *Bulletin Of The Russian Academy Of Sciences* Physics 2012 Vol. 76 No 11 p 1246 – 1252

[10] Ivanov Y F Budovskikh E A Gromov V E Bashchenko L P Soskova N A Raikov S V Formation of nano-composite layers at the surface of vt1-0 titanium in electro-explosive carburization and electron-beam treatment *Steel In Translation* 2012 Vol. 42 # 6 p 499- 501

[11] Karpii S V Morozov M M Ivanov Y F Budovskikh E A Gromov V E Formation of nano-phases

in electro-explosive alloying with aluminum and boron and electron- beam treatment of titanium surfaces *Steel in Translation* 2010 Vol. 40 # 8 p 723-728

[12] Oskolkova T N Budovskikh E A Pulse plasma treatment of surface of alloy VK10KS *Metal Science and Heat Treatment* 2012 Vol. 53 #11-12 p 608 - 610

[13] Oskolkova T N Budovskikh E A Electric explosion alloying of the surface of hard alloy VK10KS with titanium and silicon carbide *Metal Science and Heat Treatment* 2013 Vol. 55 # 1 p 96 - 99

IOP Conf. Series: Materials Science and Engineering **91** (2015) 012004 doi:10.1088/1757-899X/91/1/012004

Peculiarities of Production of Chromium Carbonitride Nanopowder and Its Physical-Chemical Certification

L S Shiryaeva, I V Nozdrin, G V Galevsky

Siberian State Industrial University
654007, Novokuznetsk, ulitsa Kirova, 42, tel. 8-384-3-74-89-13

E-mail: kafcmet@sibsiu.ru

Abstract. Scientific and technological basics of plasma synthesis of chromium carbonitride have been developed, including analysis of the current production state and application of chromium carbon compounds, defining characteristics of three-jet plasma reactor, modeling-mathematical study of interaction of raw materials and plasma streams, prediction of technological parameters of plasma stream based on the modeling results, selection of optimal technological option, implementation of plasma-metallurgical technology of chromium nitride production, its physical-chemical certification and defining technical-economical production factors.

Introduction

Carbide Cr_3C_2 that is close in composition and structure to chromium-carbonitride $Cr_3(C_{0,80}N_{0,20})_2$ is wear- and corrosion-resistant, hard, chemically-inert material which is in demand in modern engineering and technologies for manufacturing protective coatings of metals and cermets, and as components and alloy additives of tungsten-free hard alloys. Analyzing current state of national and world production and application of chromium carbide three groups of its production methods can be distinguished (regarding aggregative state of raw materials and main application areas): (i) carbothermal reduction of chromium oxide in powder or compact form in non-oxidizing atmosphere (N_2, H_2, vacuum, inert gas); (ii) carbothermal reduction of chromium oxide in molten metal; (iii) carbothermal reduction of chromium oxide in gas phase. Chromium carbide obtained by the first group of methods is applied in powder metallurgy in cermet technology, the second is used for sputtering and welding deposition of protective coatings, the third one is used in nanomaterials. Further perspectives of chromium carbide application are associated with its production in nanostate. Technology of plasma metallurgical synthesis refers to the third group of chromium carbide obtaining methods. It was developed and utilized in 80-s within integrated scientific and technical program of national scale "Siberia" headed by M.F. Zhukov, academician of the Russian Academy of Sciences and Professor G.V. Galevsky. Technology implies using three-jet direct-flow plasma reactor (with power up to 50 kW) and nitrogen as plasma-forming gas. Implementation of the suggested plasma-metallurgical technology is environmentally-friendly, providing obtaining of chromium nanocarbide close in composition to stoichiometric, presented as nanopowders with particles size 30–70 μm. It also makes possible to separate nanodispersed pyrolitic carbon and product of hydrogen cyanide collecting – technical sodium cyanide. Along with benefits this technology also has disadvantage, i.e. technical and economical inexpedience of using propane-butane mix as carbidizer as it requires for processing nitrogene-ammonium-hydrogen plasma which is complex in its composition and difficult to generate. It is also hard to generate this process in laboratory conditions.

The present work is aimed at developing scientific and technological basics of plasma synthesis of chromium carbonitride and its physic-chemical certification.

Experimental methodology

Siberian State Industrial University has significant experience in research and running of three-jet direct-flow plasma reactor with nitrogen-based arc plasma jets and its application for synthesis of hard-melting carbides, borides and their composites. The optimum plasma reactor structure is to have uniform placement of plasma jets along the periphery, with angle of slope of plasma streams to the reactor axis 30–45° and heat protection of intensively cooled walls, providing optimal operation mode of mixing chamber, high uniformity of radial temperature distribution and velocity while providing minimum heat losses. However these results are obtained for laboratory plasma reactor (power up to 30-50 kW). The results lack design and technological proposals and recommendations on its industrialization. There also lacks complete information on its thermal and technological characteristics.

Thermo-technical study has been carried out to define industrial power level of three-jet reactor (power range 80-250 kW). The study included determining heat efficiency of plasma jets, specific enthalpy of plasma flow at the reactor input, mass flow rate of plasma-forming gas which is generated in these conditions. It was established that reactor power of 150 kW corresponded to a combination of heat efficiency, required enthalpy of plasma-forming gas and its mass flow rate close to optimal. Further increase in power level is inappropriate since the required enthalpy of plasma flow is not provided within the available plasma jets options. For a three-jet reactor with internal channel diameter of 0,054 m and power 150 kW, the mean mass temperature of plasma flow at length 12 calibres varies within (5500÷2650) K for an unlined channel and (5500÷3200) K while its thermal insulation with zirconium dioxide cylinder (0,005 m thick). At the same time the internal surface temperature varies in the range of (800÷350) K and (1900÷850) K correspondingly. Specific electric power reaches 1214 MW/m^3, significantly exceeding this value for traditional electro-thermal equipment (generally about 0.2 MW/m^3). The calculated working life of electrodes for copper anode is 4700 h and 111 h for tungsten cathode. Predicted contamination of chromium carbide with erosion products does not exceed 0.0001 % of copper and 0.00002 % of tungsten. Considering all the characteristics three-jet plasma reactor (power 150 kW) can be referred to high-efficient and reliably operating electro-thermal equipment [1].

Modeling-mathematical study of raw materials and plasma flows interaction included carrying out thermo-dynamic and kinetic analysis. Due to use of chromium, oxide Cr_2O_3 and trichloride $CrCl_3$, hydrocarbon (methane) and plasma-forming gas (nitrogen) in synthesis as chromium-containing raw materials, systems under study were considered as C–H–N, Cr–O–C–H–N, Cr–Cl–C–H–N, Cr–C–H–N.

Thermo-dynamic analysis of synthesis processes was carried out in order to predict optimum parameters of chromium carbide production (correlation of components and temperature); to determine characteristics of the process (degree of conversion of raw materials into carbide, compositions of gaseous and condensed products) for quasi-equilibrium conditions in the absence of condensed carbon; and to assess the input of gas-phase reactions into carbide-formation processes, providing effective processing of dispersed raw materials in plasma processes. Compositions of gaseous and condensed products necessary for the analysis were calculated by "constant" method which is based on joint solution of the following equations: law of mass action, material-balance equation, and equation for the total number of gas mixture moles, equation for existence of condensed phase and Dalton law. Equations were solved by means of software program for modeling of high-temperature complex chemical equilibrium "PLASMA" (Institute of Solid State and Mechanical Chemistry, Siberian Branch, the Russian Academy of Sciences), with embedded database of interaction products for oxide-, boride-, carbide- and nitride- forming systems [2]. Temperature range considered in calculations: 1000-6000 K, total pressure in the system: 0.1 MPa.

Multivariant modeling-mathematical study of macro-kinetic parameters of dispersed chromium-containing raw materials evaporation was carried out using modified mathematical modeling of interaction between plasma and raw materials flows of A.L. Mosse – I.S. Burov – G.V. Galevsky, based on a joint solution of equation of raw particles motion, equation of inter-component heat transfer and heat transfer of plasma flow with reactor walls. It also takes into consideration influence on the intensiveness of heat transfer of dispersed raw materials and artificial thermal insulation of channel [2].

Technological studys was carried out in NPF Polimet plant on the base of industrial plasma-technological complex (power 150 kW), including three-jet direct-flow plasma reactor, systems of electricity-, gas- and water-supply, batching of burden materials and catching of synthesis products.

Three electric arc heaters (plasma jest) EDP-104A (power up to 50 kW) are used to generate plasma flow in reactor. They are placed in mixing chamber under angle 30 ° to the reactor axis.

Mixing chamber is connected with sectional water-cooled channel with internal diameter of 0.064 m. In order to reduce radial temperature gradient in near-wall zone, reactor is lined with cylindrical insertions of zirconium dioxide (wall thickness 0.005 m), outer diameter of 0.064 m, that decreases the diameter of its channel up to 0.054 m. Alteration of reactor length and supply of cold gas (nitrogen) into plasma flow through hardening ring (placed at the output from reactor) enable to perform hardening of synthesis products in different temperature zones. Supply of high-dispersed raw materials into mixing chamber is performed by means of water-cooled lance.

Electric power supply of industrial plasma jet is made by thyristor of series ATP-750/600. Batch meter of combined electro-mechanical and gas-vortex type with demountable cylinder – receiver of powder raw-materials is used for batching. Catching system of synthesis products includes collection chamber, collecting around 10 % of nanopowder where process gas temperature reduces up to 873 K, and two filter bags working by turn, collecting up to 85 % of nanopowders.

Nitrogen of technical purity was used as plasma-forming gas (oxygen content up to 0.5 % vol.), metal chromium PK-1M (tech. spec. 14-1-1474-75, updat.) was used as chromium-containing raw material, natural gas with methane content up to 94 % vol. (the rest is ethane, propane, butane).

Technological studies allowed finding dependence of chromium carbonitride content (close in composition to carbide) and corresponding admixtures in synthesis products on the defining factors, presented as following equations:

$$[Cr_3(C_{0,8}N_{0,2})_2] = 66,12 + 0,03T_0 - 0,42\,C_{H_2} - 0,14\,C_{NH_3} - 0,00002T_з\cdot C_{NH_3} \quad (1)$$

$$[Crfree] = 147,95 - 0,027T0 - 0,34^{C_{H_2}} - 1,37^{C_{NH_3}} + 0,0003T0\cdot^{C_{NH_3}} \quad (2)$$

$$[Cfree] = -150,30 - 0,002T0 + 1,3^{C_{H_2}} + 13,5^{C_{NH_3}} - 0,00008T0\cdot^{C_{NH_3}} + 0,0002T3\cdot^{C_{H_2}} \quad (3)$$

$$[Cr2O3] = 41,80 - 0,0069T0 - 0,208^{C_{CH_4}} + 0,00004T0\cdot^{C_{CH_4}} \quad (4)$$

$$[Ncomb.] = 1,74 + 0,001T0 - 0,012^{C_{H_2}} - 0,03^{C_{NH_3}} - 0,00001T3\cdot^{C_{NH_3}}, \quad (5)$$

where $[Cr_3(C_{0,8}N_{0,2})_2]$, $[Cr_{free}]$, $[C_{free}]$, $[Cr_2O_3]$, $[N_{comb.}]$ – content, in % wt, of carbonitride, free chromium and carbon, chromium oxide (III) and combined nitrogen in synthesis products; T_0 and $T_з$ – initial temperature of plasma flow and hardening temperature, K; C_{CH_4} – amount of hydrocarbon from stoichiometrically necessary for chromium carbidization, %; C_{H_2} – hydrogen concentration in plasma-forming gas, % vol.; C_{NH_3} – amount of atomic nitrogen in plasma-forming gas, N (NH_3) / C (CH_4), % from stoichiometric.

The average size of chromium nanocarbide particles formed in plasma flow, calculated by the value of specific surface is proportional to mean mass temperature of flow in degree -2,96 (equation 6):

$$\bar{d} = (726 \pm 35{,}1)\ T^{(-2{,}96 \pm 0{,}436)}. \quad (6)$$

Complex physico-chemical certification of synthesis products by means of X-ray and chemical analysis requires study of crystal lattice and phase and chemical composition. Dispersiveness and particles morphology was studied by transmission electron microscopy and scanning electron microscopy (SEM). Similar approach for synthesis products identification is described in works [3-5].

Results and discussion

Results of thermo-dynamic modeling of high-temperature interaction in chromium-carbon-containing systems are given in Fig. 1-3. Thermal-dynamic analysis of the selected technological options showed that in systems Cr:O:C:H:N and Cr:C:H:N 100 % output of chromium carbide was possible by stoichiometric ratio Cr:C and C:H and temperature 2000–2200 K. In the system Cr–C–Cl–H–N 100% output of chromium carbide is reached by stoichiometric ratio Cr:C, 3-fold excess amount of hydrogen and temperature 2000-2200 K. Formation of chromium carbide occurs with condensed chromium and hydrogen cyanide according to mechanism “vapour – molten metal – crystal” thus enabling to suppose that it can reach high output in real synthesis conditions.

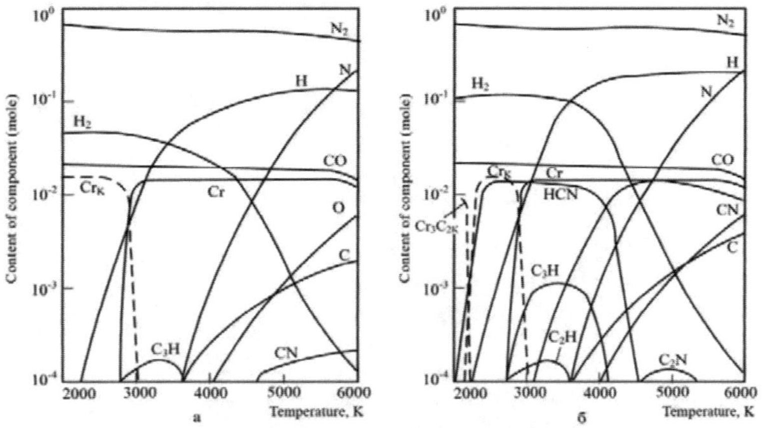

a) equilibrium compositions of gas and condensed phases depending on the temperature at ratio Cr:O:C:H:N = 0,25:0,375: 0,375:1,5:20

b) quasi-equilibrium compositions of gas and condensed phases depending on the temperature at ratio Cr:O:C:H:N = 0,25:0,375: 0,54:2,16:20

Figure 1. Results of thermo-dynamic calculations of system Cr:O:C:H:N

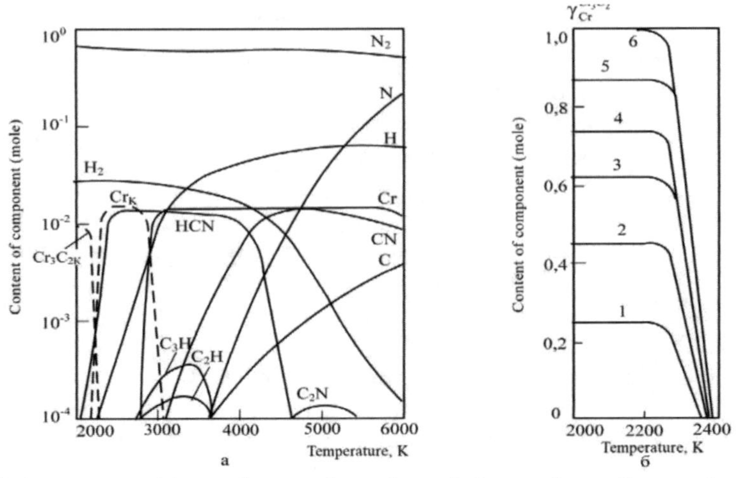

a) quasi-equilibrium compositions of gas and condensed phases depending on the temperature at ratio Cr:C:H:N = 0,25:0,16: 0,64:20

b) dependence of degree of conversion Cr in Cr_3C_2 of the ratio

Cr:C = 0,25:0,042 (1); 0,25:0,082 (2); 0,25:0,116 (3); 0,25:0,125 (4); 0,25: 0,145 (5); 0,25:0,16 (6)

Figure 2. Results of thermo-dynamic calculations of system Cr:C:H:N

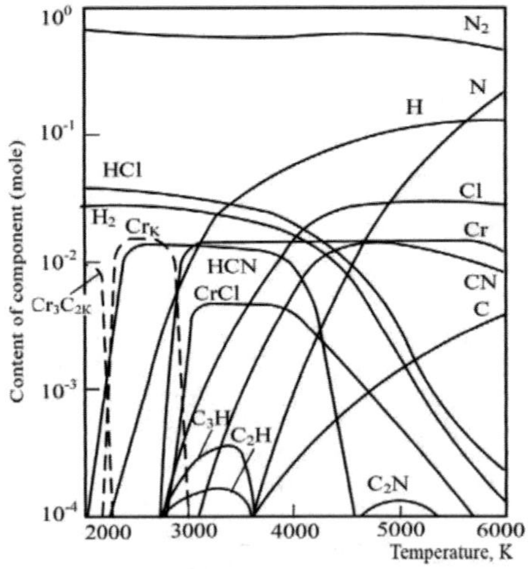

Figure 3. Quasi-equilibrium compositions of gas and condensed phases of system Cr–C–Cl–H–N at components ratio Cr:C:Cl:H:N = 0,25:0,16:0,75:2,0:20

The degree of raw materials evaporation is calculated for interaction of dispersed chromium-containing raw materials with highly heated gas flow in plasma jet depending on the energy parameters of reactor, fineness, speed of input into plasma flow and mass-expendable concentration. With power 75 kW, 100 % evaporation degree for chromium particles with fineness 10…30 μm while mass-expendable concentration 0,14…0,12 kg of chromium powder / kg of exchange gas. For chromium oxide (III) and chromium chloride (III) these values comprise 5…20 μm and 0,14…0,10 kg/kg, 30…70 μm and 0,16…0,14 kg/kg correspondingly. Based on the obtained results raw materials for chromium carbide synthesis were selected.

Results of modeling-mathematical study enabled to predict technological parameters of plasma-metallurgical production of chromium carbide for different types of chromium-containing raw materials and to select optimal technological option (Table 1).

Table 1. Predicted technological parameters of plasma-metallurgical production of chromium carbide

Process variables	Technological options		
	Carbidization of chromium with methane	Recovery of Cr_2O_3 with methane	Recovery of $CrCl_3$ with methane
Degree of conversion of raw materials	0.95	0.95	0.95
Output of chromium carbide, %	92.0	90.0	91.8
Production capacity of chromium carbide, kg/h	3.42	2.79	1.69
Production intensity, kg/h·m^3	1368	1116	676
Chromium-containing raw materials consumption, kg/kg	0.9	1.3	2.76
Consumption of exchange gas, kg/kg	9.47	11.6	19.7
Methane consumption, kg/kg	0.175	0.58	0.18
Electrical energy consumption, kW·h /kg	65.79	80.65	133.14

While calculating the variables, losses of raw materials in reactor were considered as 4 % and losses of synthesis products in catching system were considered in the amount of 5 %. It can be seen that optimal technological option is chromium carbidization with natural gas.

Optimal values of technological factors and acceptable limits of their alteration, as well as basic characteristics of synthesized by these conditions chromium carbonitride are given in Table 2 (five-fold repetition of experiments).

Table 2. Acceptable limits in alteration of parameters of chromium carbonitride synthesis in industrial reactor (power 150 kW) and its main characteristics

Synthesis parameters and characteristics of chromium carbonitride	Value
Composition of exchange gas, % vol. – nitrogen	100
Composition of carbidizer, % vol. – methane – ethane – propane – butane	 93.6 3.0 2.2 1.2
Composition of chromium-containing raw materials, % wt. – metal chromium	 99.3
Fineness of chromium-containing raw materials, μm	–10
Production capacity of raw materials, kg/h	3.11
Amount of carbidizer, % from stoichiometric	120–140
Initial temperature of plasma flow, K	н.м. 5400
Hardening temperature, K	2000÷2200
Phase composition	$Cr_3(C_{0,80}N_{0,20})_2$
Chemical composition, % wt. – $Cr_3(C_{0,80}N_{0,20})_2$ – free chromium – free carbon – chromium oxide (III)* – volatiles	 91.8–93.45 2.10–1.80 1.30–1.10 4.00–3.00 0.80–0.60

IOP Conf. Series: Materials Science and Engineering **91** (2015) 012004 doi:10.1088/1757-899X/91/1/012004

Synthesis parameters and characteristics of chromium carbonitride	Value
Output of chromium carbonitride, % wt	92.0
Specific surface, m^2/kg	32000–37000
Size of** particles, µm	30–35
Particles shape	spherical
Degree of oxidation chromium carbonitride nanopowder*** $x10^7$, kg O_2/m^2	2.56
Production capacity, kg/h	3.42
Production intensity, $kg/h{\cdot}m^3$	1368
Note: *– defined by oxygen content; **– calculated by value of specific surface; ***– defined after exposure on air during 24 hours	

Resulting from integrated physical-chemical certification of synthesis products it was established that product of plasma synthesis is ternary compound – chromium carbonitride. X-ray diagram $Cr_3(C_{0,80}N_{0,20})_2$ is given in Figure 4.

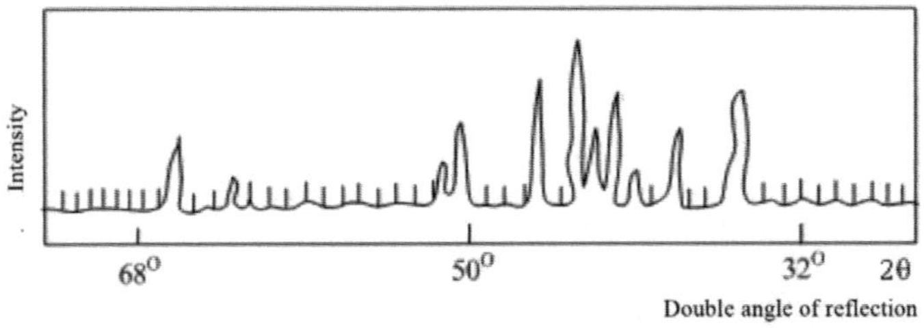

Figure 4. Fragment of X-ray diffractogram of chromium carbonitride

$Cr_3(C_{0,80}N_{0,20})_2$, is identified as rhomb-shaped carbonitride with crystal structure which differs from carbide structure by presence of octahedral structural elements, nitrogene atoms being placed inside them. Within the studied range of synthesis parameters alteration, chromium carbonitride composition is almost not changed and within limit of accuracy of analysis and calculation it corresponds to $Cr_3(C_{0,80}N_{0,20})_2$. Content of chromium nanocarbonitride in synthesis products makes 91.8 – 93.5 % wt. Microphotographs of chromium nanocarbonitride are given in Fig. 5.

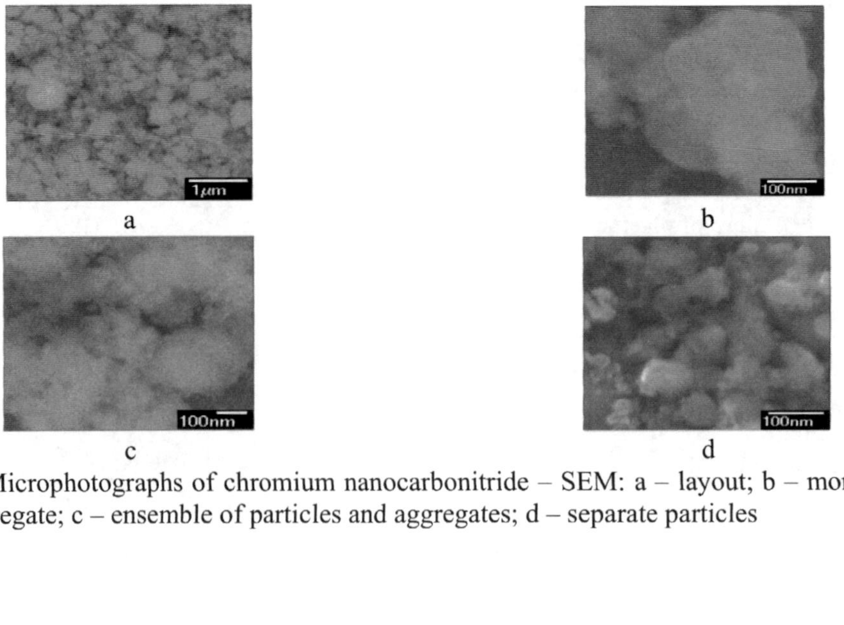

a b

c d

Figure 5. Microphotographs of chromium nanocarbonitride – SEM: a – layout; b – morphological pattern of aggregate; c – ensemble of particles and aggregates; d – separate particles

Nanopowder of chromium carbonitride is presented as aggregates of spherical or close to that form, size from 600 to 150 μm, formed by association of globular particles of quite broad size range – from 20 to 80 μm, their number in aggregate depends on its fineness. Nanolevel and morphology of particles allow considering them as products of carbonization of chromium microdroplets, formed while volume condensation of its vapors, liquid drop coalescence and crystallization. Presence of aggregates of different volume in the studied objects demonstrates high probability of further coarsening of nanoparticles at temperature fall by means of their coagulation.

Evaluation of the economical efficiency of plasma-metallurgical production of chromium carbonitride shows that while production output 3.7 t of chromium carbonitride per year (calculating per one reactor) and production cost 47 $/t retail price is 220 $/kg, that testifies its competitiveness on a world market of nanopowder materials. Currently leading foreign producers of carbide nanopowders, including scientific-industrial "Nanostructured & Amorphous Materials. Inc." (USA); "Tokyo Tekko Co" (Japan); "Hefei Nanotechnology & Development htd. Co" (China); "Neomat Co" (Latvia); "Plasma Chem Gmbh" (Germany) set the price range for carbides naniopowders within 400–2000 $/kg. Payback period on investments is 1.5 year, thus proving economical and technological expedience of chromium carbonitride production. One of the most preferable application areas of chromium carbonitride is technology of composite electro-deposition of protective coatings, as well as some materials, as described in [6-10].

Conclusions

Based on the conducted modeling-mathematical and technological studies optimal values for synthesis parameters of chromium nanocarbide and its physical-chemical parameters were defined. It is established that while carbidization of chromium powder with methane in conditions of nitrogen plasma flow in temperature range 5400 – 2000 K chromium carbonitride $Cr_3(C_{0,8}N_{0,2})_2$ is formed. Phase and chemical compositions were studied, as well as dispersiveness, morphology and degree of oxidation of synthesis products. The main technical-economical parameters of the suggested technology have been defined, such as production capacity, production intensity, production cost and retail cost.

The work is conducted in Siberian State Industrial University within project part of state task of Ministry of Education and Science of Russia, No. 11.1531.2014/K.

References

[1] Nozdrin I V Shiryaeva L S Galevskii G V Rudneva V V Reactor for plasmometallurgical production of refractory borides and carbides *Steel in Translation* 2011 T 41 № 8 p 644-648
[2] Nozdrin I V Galevskii G V Shiryaeva L S Terenteva M A Particle size of vanadium and chromium borides and carbides in a plasma flux *Steel in Translation* 2011 T 41 № 10 p 799-804
[3] Saviak M P Mel'nik A B Solonin Yu M Kotko A V Mechanosynthesis of Nanodispersed Titanium Diboride *Powder Metallurgy and Metal Ceramics* January 2015 Volume 53 Issue 9-10 pp 497-504
[4] Lavrenko V A Chekhovskii A A Particularities of Electrochemical Production of Nanocoatings and Nanopowders of Metals Oxides and Nitrides by Electrolysis of Acetonitrile at High Voltage Gradient *Powder Metallurgy and Metal Ceramics* November 2014 Volume 53 Issue 7-8 pp 399-403
[5] Osvaldo Mittsuyuki Cintho Eliane Aparecida Peixoto Favilla Jose Deodoro Trani Capocchi Mechanical-thermal synthesis of chromium carbides *Journal of Alloys and Compounds* V 439 (2007) pp189-195
[6] Oskolkova T N Budovskikh E A Electric explosion alloying of the surface of hard alloy VK10KS with titanium and silicon carbide *Metal Science and Heat Treatment* 2013 V 55 № 1-2 p 96-99

[7] Oskolkova T N Budovskikh E A Puls plasma treatment of the surface of alloy VK10KS *Metal Science and Heat Treatment* 2012 T 53 № 11-12 p 608-610
[8] Popov V A Chernov B B Prosviruyakov A S Cheverikin V V New mechanichal-alloying-based technological scheme for producing electrochemical composite coatings reinforced with non-agglomerated nanodiamond particles
[9] Barbat N Zangenen-madar K Preparation of Ti-Ni Binary powder via electroless nickel plating of titanium powder *Powder metallurgy* V 57 (2014) p 97-102
[10] Dong Z Zhang J Yuan W Xie D Du G ZNO Nanocrystals anchored Graphene: in situ solvothermal synthesis and enhanced photocatalytic performance *Journal of nanoscience and nanotechnology* V 14 (2014) p 4264-4268

IOP Conf. Series: Materials Science and Engineering **91** (2015) 012005 doi:10.1088/1757-899X/91/1/012005

Conceptual model of resource assessment of fuel and energy complex long working equipment on the basis of structural criteria

A N Smirnov[1] and N V Ababkov [2]
[1] Professor, Department of Engineering Technology KuzSTU, Kemerovo, Russian Federation
[2] Associate Professor, Department of Engineering Technology KuzSTU, Kemerovo, Russian Federation

E-mail: n.ababkov@rambler.ru

Abstract: The main approaches to the technical diagnosis and assessment of residual life of potentially dangerous equipment FEC, currently in use. A new conceptual model resource assessment of parts and units of fuel and energy complex. On the issue of technical diagnostics are considered in complex positions.

Introduction

Currently, over 80% of Russian industrial equipment worked his expected life for the fuel and energy complex (FEC) – more than 90%. And the destruction of the technical devices or their components may result and lead to major technological catastrophe with human victims. It is therefore particularly relevant is the task of providing (control) the safe operation of potentially dangerous equipment [1–3]

There are many complex physical and chemical processes related primarily to the collapse of part of the pearlite microstructure, coagulation and spherodization carbide formation of micropores, or wedge-shaped microcracks in the metal during long-term operation of equipment FEC at high temperatures and pressures.

The decay rate of the microstructure formation and growth of micro depends on the temperature and the operating pressure, in addition, an important role has the number of starts and stops of the equipment.

Highly topical issue of developing new effective methods to assess the resource elements of technical devices based on the evaluation of structural-phase state and the strength characteristics of the metal power equipment after prolonged use. More widespread receives concept based on "prediction and prevention" instead used the concept of "find and fix" [4].

There are a large number of theoretical and experimental studies to evaluate resource efficiency, reliability, durability and other characteristics of long working metal potentially dangerous equipment currently. The fundamental work of schools academics Makhutov N.A. [5] and Mitenkov F.M. [6] are of particular note among of them.

Makhutov N.A. and Gadenin M.M. consider questions of a technical diagnosis of residual life and security in terms of assessing the risks of dangerous conditions, which are defined as the product of the probabilities of their occurrence in the corresponding damage arising from reserves of strength, stability, resource, survivability in statistical or deterministic setting [5]. In other words, the analysis methods of mathematical statistics and probability theory is subjected to large amounts of data on a particular hardware. On the calculation of added analysis of possible equipment damage from an accident. According to these calculations carried ranking objects by hazard class.

The concept of residual life assessment of equipment and systems of nuclear power plants are developed by Mitenkov F.M. et al [6]. The definition of the actual model of operation (FME) and the physical law of damage accumulation ω is at the heart of this concept:

$$\omega = \sum \frac{\Delta V}{V_f}; \tag{1}$$

where V – volume fraction of defects, V_f – critical volume fraction of defects.

From Eq. 1, the use of a physical law of damage accumulation accounts for only a fraction of defects that can detect modern methods and means of nondestructive testing. The concept of residual life assessment of equipment and systems of nuclear power plants are not talking about changing the structural and phase state of the metal in the process of long-term operation.

As we know, no technique, no algorithm for calculating the residual resource does not take into account the changes that occur in the metal structure in the long-term operation of technical devices of hazardous production facilities. As an example, the calculation of resource drums high-pressure boilers. According to regulatory documents [7] resource drums high-pressure boilers is determined by the value of the accumulated damage to the metal (2):

$$A = 2\frac{n_{ss}}{N_{ss}} + 0{,}08\frac{n_{ss}}{N_{hs}} + \frac{n_{ss}}{N_h} + \frac{40n_{ss} + \omega\tau}{N_t}, \tag{2}$$

where n_{ss} – the number of start-stop drum; N_{ss}, N_{hs}, N_h, N_t – number of cycles for the start-stop mode, heat stroke, hydro and temperature fluctuations, respectively; ω – frequency thermal cycling drum during the operation of the boiler at steady state; τ – time between the drum.

The Eq. 2 shows that the extension of the resource drums high pressure boiler is taken into account only the factors relevant to the modes of operation of the equipment. Structural changes in the metal for long term use are not included.

Theoretically part

The authors propose a slightly different approach to equipment life assessment based on structural factors.

It is obvious that the total time of the safe operation of the equipment (3) consists of equipment operating time before the survey (current state) τ_{cur} and further operating time before reaching the limit state τ_{res} (residual life):

$$\tau_{l.s.} = \tau_{cur} + \tau_{res}, \tag{3}$$

There are complex physic-chemical and structural changes in metal equipment, spent a long time in a complex stress conditions, often in hostile environments (before the survey). The structure is an important indicator of the metal is characterized by its performance. All kinds of thermal effects and loads are recorded In memory of the metal at the time of diagnosis, which was exposed to the metal, since its fabrication and finishing a certain period of operation (in our case – τ_{cur}.).

It is very important to understand and appreciate these changes. However, structural-phase state of the metal after hours (τ_{cur}). Is one side of the issue. On the other hand, to assess the real resources must also take into account the macro defects (sometimes they are valid, often they are not eliminated, but just missed), resulting in the manufacture, repair, equipment operation during τ_{cur}. These defects have a significant impact on the accuracy of the estimates of the resource.

The current state of the equipment at the time of diagnosis characterizes the ratio of its current condition of K_C in this case, which can be represented as

$$K_C \to f(K_{str.c}, K_{op}, K_{def}) \to \tau_{cur}, \quad (4)$$

where $K_{str.c}$ – structural metal ratio in the current state, K_{op} – coefficient taking into account operating modes (temperature, pressure (load), cyclicity, environment) K_{def} – coefficient taking into account the presence of defects in workmanship, installation and repair of the base metal and welded joints. $K_{str.c}$ determined by the nature of changes in the structural-phase state and the internal stress fields, the redistribution of sources of stress fields and their density. $K_{str.c}$ can be expressed as a function of parameters of the structure in the form of

$$K_{str} \to f(\sigma_{in}, \tau, \rho_{\pm}, \rho, \mu, d_{gr}, \gamma_c), \quad (5)$$

where σ_{in} – the amplitude of the internal stress, MPa; τ – shear stresses; $\rho\pm$ – excess dislocation density; ρ – scalar dislocation density; μ – source density fields of internal stresses; d_{gr} – grain size; γ_c – coefficient taking into account the size and nature of the distribution of carbide phases in the metal.

For each class of materials we are developing engineering methods for calculating the structural factor.

For earlier [8] studied the authors proposed structural steels factor which is fulfilled in the long running heat-resistant steels of fuel and energy complex at all stages of the life cycle (from the initial state to the ultimate state and destruction, which has the following form

$$K_{str} = (\frac{\sigma_{in}}{\mu} + \frac{\tau}{\mu})Ч\frac{\rho}{\rho\pm}\gamma_c \cdot d_{gr}^{-1/2}, \quad (6)$$

The structural factor must be determined on equipment operated in difficult stressful conditions at high voltages and temperatures, cyclic and shock loads in corrosive environments, in periods of routine maintenance (specified period of diagnosis, developed a certain period).

To determine the actual condition of the equipment, except for the structural factor must be considered and modes and installation, repair and metallurgical defects, and accumulated operating defects, and already the basis upon which the coefficients. Therefore, to determine the need to define the K_c, K_{def} and K_{op}.

The operating conditions of the equipment, K_{op} can be represented as a coefficient that depends on the number of cycles (N), pressure or load (P), temperature (T), corrosive environments (L)

Hence $K_{op} = K_N{\cdot}K_P{\cdot}K_T{\cdot}K_L$, the numerical values of the coefficients are presented in the table.

$K_{def} = K_{met}{\cdot}K_{rep}{\cdot}K_{inst}$, the numerical values of the coefficients are presented in table. The values obtained experimentally.

Table. Symbols and values of the coefficient

№	Symbol of the coefficient	Value of the coefficient	Condition
1.	K_T – ambient temperature	1.0	Operation in stationary mode at design parameters
		1.2…1.4	Significant violations temperature regimes, overheating
		1.1	Slight temperature changes
2.	K_N – cyclic recurrence	1.0	Calculated modes, start-stop
		1.2…1.3	Violations cycling modes
		1.05…1.1	Minor violations
3.	K_L – aggressive environment	1.0	Calculated environment
		1.2…1.4	Aggressive (major changes), unplanned environment

№	Symbol of the coefficient	Value of the coefficient	Condition
		1.1	Minor violations of environment parameters
4.	K_P – pressure or load	1.0	Estimated operating modes
		1.2…1.4	Significant variations in pressure or load
		1.05…1.1	Low pressure drop or load
5.	K_{inst} – installation defects	1.0	No visual defects welds normalized hardness
		1.3	Unrepaired defects mounting form welds, hardness, kinks, bends (on the verge of admission)
		1.05	Minor defects (valid), small dents, scratches, allowable weld defects
6.	K_{met} – defects of metallurgy	1.0	The suggested structure, normal mode of heat treatment, the hardness in the normal
		1.1	Large metallurgical defects, bundles, nonmetallic inclusions, stripes, polishing defects, etc.
		1.05	Slight defects
7.	K_{rep} – repair defects	1.0	No defects
		1.1	Removable repair defects, weld shape, hardness, bending
		1.05	Removable, allowable defects

As previously mentioned – K_c, the coefficient that characterizes the technical condition of the object of control at the time of inspection. While maintaining at least a relative constancy K_{op} and K_{def} and throughout the life of the equipment and in the first approximation can be written

$$\frac{\tau_{res}}{\tau_{cur}} \approx \frac{K_{l.s.} - K_c}{K_c}, \tag{7}$$

where $K_{l.s.}$ – coefficient of the technical status of the equipment at the time of the limit state. It requires a highly individual approach, but given the level of exploitation and accumulation of defects in the same equipment at close lifetime. The determination of this factor analysis was performed on the basis of operating conditions (calculated coefficients K_{op} and K_{def}).

The structural factor limiting state was evaluated by the results of the research base metal and welded joints of equipment overage. From the Eq. 7, having a small mathematical transformations and substituting in the formula calculation results and research, we get the desired result – the remaining time of the equipment.

$$\tau_{res} = \frac{\tau_{cur}(K_{l.s.} - K_c)}{K_c}, \tag{8}$$

The authors created a database on equipment FEC of Kuzbass, have a lot of research on the study of mining equipment at present. The proposed methodology has a significant disadvantage in the complexity of determining the structural criteria. Necessary to satisfy the costly electron microscopy studies.

There are several ways free from this drawback. The most promising is the development of non-destructive evaluation of physical methods of structural changes. For a number of steel (such as Cr-Mo-Va steel), we have developed acoustic and magnetic structural damage assessment criteria [8–10]. There is the need to perform standard electron microscopy studies for each grade of steel, but in any case (to avoid large errors).

For practicing engineers developed an atlas of histograms, by which, with the knowledge of the structural factor, the operating conditions of equipment (after an analysis of the repair and maintenance documentation) expert finds the histogram lifetime (resource) of the technical device.

This methodology does not exclude holding standard NDT methods, they complement the results of our tests, at their expense introduced clarifying amendments to the corresponding coefficients.

Conclusion

1. The main approaches to the technical diagnosis and assessment of residual life of potentially dangerous equipment FEC, currently in use. It is shown that the existing methods and techniques of assessment of the resource does not take into account the behavior of the structural-phase state, methane (decay of the structure, changes in local internal stress fields) of both primary and weld metal during prolonged operation in difficult and stressful conditions in hostile environments.

2. A conceptual model for evaluating the long-term resource working metal objects FEC, based on the study of the evolution of structural and phase state of metal. The structural factor taking into account the nature of changes in the basic parameters of the structure of metal in the process of long-term use, which together with the regime coefficients and imperfection makes it possible to reliably estimate the share of thermal power equipment.

3. For various FEC equipments held a range of additional studies involving non-destructive research methods are developed and the coefficients of the histogram using which increase the reliability of resource estimates.

The work is financially supported by project RSF №14-19-00724

References

[1] Kljuev V V 2012 *Degradacija diagnostiki bezopasnost* (Moscow: Spektr) p. 128.

[2] Smirnov A N, Vasil'ev A G and SHevelev E V 2000 Ocenka stepeni povrezhdennosti dlitel'no rabotajushhego metalla jenergooborudovanija akusticheskim metodom *Vestnik Kuzbasskogo gosudarstvennogo tehnicheskogo universiteta* **5** p 46.

[3] Smirnov A N and Ababkov N. V 2010 Kompleksnyj podhod k ocenke rabotosposobnosti jelementov jenergeticheskogo oborudovanija *Izvestija Samarskogo nauchnogo centra RAN*, T. **12 №1 (2)**. – pp 520–524.

[4] Smirnov A N 2004 Strukturnaja povrezhdennost' stalej i ee ocenka spektral'no-akusticheskim i jelektronno-mikroskopicheskim metodami *Kontrol'. Diagnostika* **4** pp. 13–18.

[5] Mahutov N A and Gadenin M M 2011 *Tehnicheskaja diagnostika ostatochnogo resursa i bezopasnosti* (Moscow: Spektr) p. 187.

[6] Mitenkov F M, Kajdalov V B, Korotkih JU. G and etc. 2007 *Metody obosnovanija resursa jadernyh jenergeticheskih ustanovok* (Moscow: Mashinostroenie) p. 448.

[7] SO 153-34.17.442-2003. *Instrukcija po porjadku prodlenija sroka sluzhby barabanov kotlov vysokogo davlenija*.

[8] Uglov A L, Erofeev V I, and Smirnov A N 2009 *Akusticheskii kontrol' oborudovaniya pri izgotovlenii i ekspluatatsii* (Moscow: Nauka).

[9] Ababkov N V, Kashubskij N I, Knjaz'kov V L, Knjaz'kov A F, Kozlov JE V, Koneva N A, Makarov N M, Murav'ev V V, Popova N A, Smirnov A N, Fol'mer S V 2011 *Diagnostika, povrezhdaemost' i remont barabanov kotlov vysokogo davlenija* (Moscow: Mashinostroenie) p. 256.

[10] Smirnov A N and Ababkov N. V 2013 Kriterii ocenki sostojanija i resursa dlitel'no rabotajushhih barabanov kotlov vysokogo davlenija *Svarka i diagnostika* **4**. pp. 55–58.

IOP Conf. Series: Materials Science and Engineering **91** (2015) 012006 doi:10.1088/1757-899X/91/1/012006

Physical Nature of the Processes in Forming Structures, Phase and Chemical Compositions of Medium-Carbon Steel Welds

D P Il'yaschenko[1,a], D A Chinakhov[1, b], V I Danilov[1,2], G V Schlyakhova[2,3], Yu M Gotovshchik[1]

[1]Yurga Institute of Technology National Research Tomsk Polytechnic University Leningradskaya 26, Yurga, Russian Federation, 652055
[2] Institute of Strength Physics and Materials Science, Siberian Branch of Russian Academy of Sciences
Akademichesky, 2/4, Tomsk, Russian Federation, 634055
[3] Seversk Institute of Technology, National Research Nuclear University MEPhI Kommunistichesky, 65, Seversk, Russian Federation, 636036

e-mail: [a] mita8@rambler.ru, [b] chinakhov@tpu.ru

Abstract. This work presents peculiarities of forming a structure, phase and chemical composition while welding medium-carbon steels (Steel 45) depending on a heat content of molten electrode metal droplets when using welding power sources having different power parameters. It was experimentally established that the power inverter provides the decreased heat input into droplets of electrode metal during the welding process. This stimulates obtaining a fine-grained structure of the deposited metal and heat affected zone, reduces the extent of the HAZ that enhances working properties of welded joints.

1. Introduction

Soundness of welds in steel fabrications used in all industries has a significant influence on the safety and economic efficiency of production, therefore it is of supreme concern to both manufacturers and scientific communities.

Welding is a continuous process that is required to result in uniform strength throught base metal, weld metal and a heat-affected zone (HAZ), and to provide the complete transfer of alloying elements from an electrode to a molten weld pool. These requirements are met by using modern materials and applying advanced welding techniques, namely:

- Materials used for welding include nanoscale components (coated electrodes, flux cored wire);
- New inverter power sources fitted with controls providing a variety of energy supply schemes and parameters for welding processes.

Welding consumables having nanocomponents stimulate the refinement of weld metal structures [1, 2], improve mechanical properties of weld metals and increase a value of alloy transfer efficiency [3]. However, their use may have a negative impact on a welder's health [4, 5], and the health effects as a constraining factor keep these welding materials from the widespread use.

At present time, a number of Russia's and foreign manufacturers of welding equipment offer a wide range of power sources for manual arc welding requiring coated wire electrodes, for example, such companies as Fronius (Austria), Lincoln Electric (USA), ESAB (Sweden), Tekhnotron, (Russia),

ZAO NPF ITS (Russia), etc. This means that almost all the world leaders in the field of welding equipment are mainly focused on the development and production of inverter welding power sources [6, 7]. However, welding diode rectifiers are still widely employed by industrial enterprises. And whereas the new welding equipment is available, but at the moment there is no complex methodology enabling the objective assessment of changes in heat input in terms of weldments from various types of equipment for MMA welding.

The authors of the work in reference [8] proved the positive effect of energy characteristics of an inverter power source on the transfer of alloying elements from an electrode to the weld metal in the case of welding steel 12Cr18Ni10Ti. At the same time, such medium carbon steels as Steel 45 (Russia), S45C, LS (Japan), 1.0503, DIN (Germany) and 1.0503, EN (European Union) find wider use in manufacturing welded fabrications because of their higher sensibility to a thermal cycle.

This work gives the results obtained in the comparative study to analyze the welded joints made of steel 45 using both an inverter and diode rectifier relative to chemical composition, microstructure and mechanical properties, including microhardness.

2. Materials, research procedures and results

Plates having a 10 mm thickness were welded taking 4 layers (see Fig. 1) and using the following electrodes: the UONI 13/55 electrode (d = 3 mm) with a welding current of I = 80-90 A for root welding and UONI 13/55 electrode (d = 4 mm) with a welding current of I = 120-130 A for filling. The parts were preheating to 300oC and then slowly cooled (by covering with insulator and asbestos fiber till full cooling down).

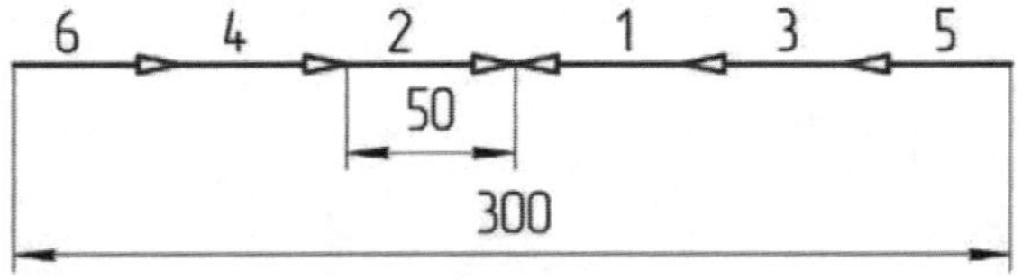

Figure 1. Schematic Welding Sequence

Analyzing the chemical composition of the welded joint in Fig. 1 showed that energetic parameters of the power supplies made an effect on chemical composition of the welded joint. When using the inverter as a power source, the workpiece is less overheated due to the decreased heat content in molten droplets of the electrode metal [9], this results in minor burning out of alloying elements, in particular manganese. This fact is indirectly confirmed by the work of the author in reference [10].

Table 1. The Statistically Average Chemical Composition of Welds in Steel 45 Using Electrodes UONI 13/55

Source of Power	Chemical Analysis, %						
	C	Si	Mn	P	Cr	Ni	Cu
Diode rectifier	0.12	0.30	0.92	0.019	0.06	0.05	0.09
Inverter	0.12	0.31	1.00	0.02	0.06	0.06	0.10

Macro- and microstructural studies based on optical metallography were performed to analyze cross-sections taken from the weld samples (Fig. 2). The Neophot-21 microscope and Genius VileaCam digital camera were used. The metallographic speciments were prepared using the following processes: mechanical grinding, mechanical polishing with diamond paste ASM 10/7 NVL and chemical etching in nitrohydrochloric acid (40% HCl + 40% HNO_3 +

10% C_2H_5OH). The DURAMIN 5 tester was used to measure microhardness in the weld metal, heat affected zone and base metal with the same metallographic specimens. The microhardness value was determined according to the Vickers test method, with a load of 50 g at a holding time of 10 s.

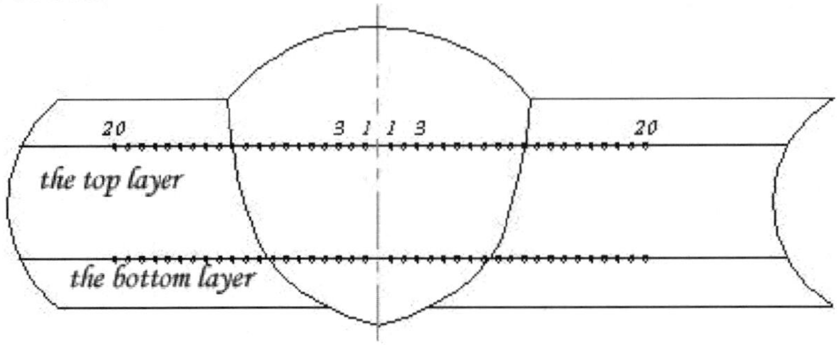

Figure 2. Diagram for micro structure studies and microhardness measuring (0,5 mm spot - to-spot interval)

Figure 3 shows how the microhardness values detected on welds are distributed. The test procedure is according to the diagram in Fig. 2.

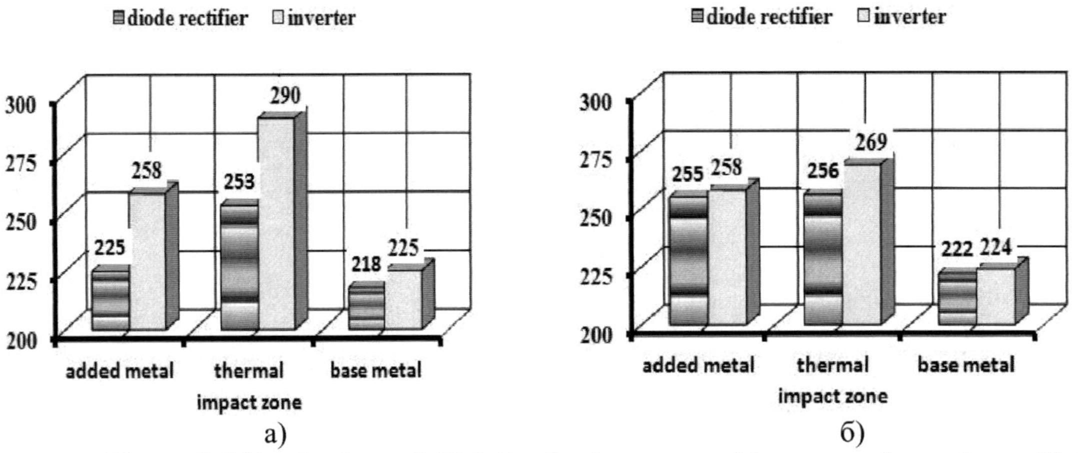

Figure 3. Microhardness (MPa) distribution over weld cross-sections when utilizing different power supplies:
a - the top layer in Fig. 2; b – the bottom layer in Fig. 2

It is obvious that in the case of the welded joints with a diode rectifier being utilized as a power source, the microhardness is maximal within the heat-affected zone. The microhardness increases in going from the weld metal to the HAZ and then decreases again in going to the base metal. Considering the bottom layer (Fig. 3, b), the difference in microhardness between the weld metal and HAZ is little, but with the top layer (Fig. 3, a) it is statistically significant. A pattern of microhardness distribution over the zones in the welded joint made using the inverter power source is qualitatively the same. The heat affected zone also has the maximum microhardness; the same seen in the bottom layer where the difference between the HAZ and weld metal is insignificant, but it is essential for the top layer (Figure 3). However, the values of microhardness are higher in the same areas of welds made using the inverter. This is particularly visible referring to the top layer (Fig. 3, a). We point out that the statistically insignificant difference in terms of the base metal microhardness detected in the top and bottom layers (Fig. 3) of both weld types is reasonably predictable.

IOP Conf. Series: Materials Science and Engineering **91** (2015) 012006 doi:10.1088/1757-899X/91/1/012006

The observed patterns are consistent with structural conditions of the welded joints under study. The weld metal in the welding process using a diode rectifier (Fig. 4, b) does not reflect a dendritic structure typical for the molten condition. Ferrite grains are polyhedral and almost equiaxial. The average size of ferrite grains is ≈ 12 microns. At a high magnification it is seen there are conglomerates of very small (less than one micrometer) ferrite grains in the intergranular spaces. Other phases and structural constituents, except for ferrites, can not be detected, corresponding to the composition of rod electrodes UONI 13/55 (0.07%).

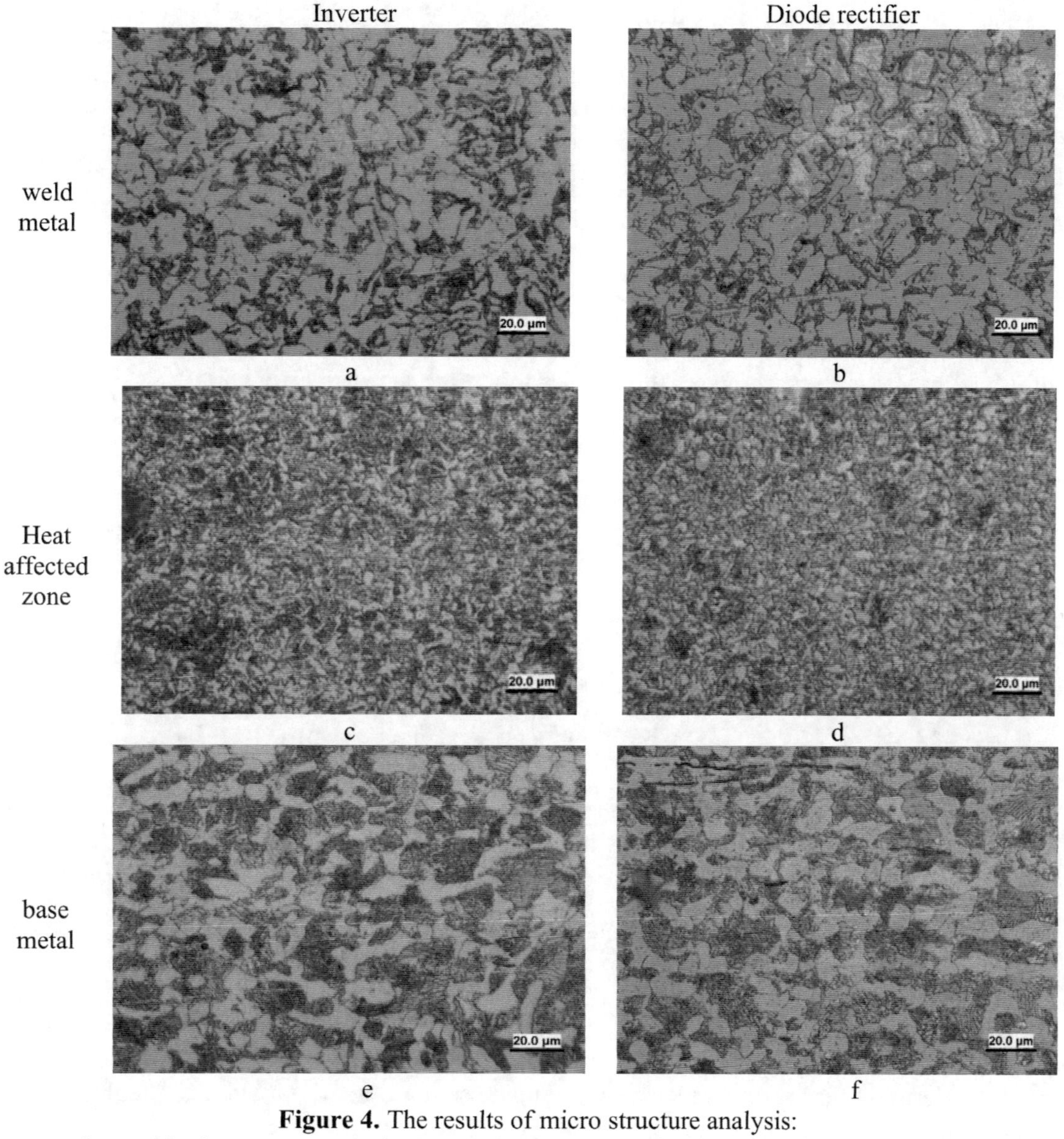

Figure 4. The results of micro structure analysis:
a, b – weld microstructure, c, d – HAZ microstructure of the welded joint; e, f - base metal microstructure

The structure of the deposited metal in the welded joint made using the inverter power supply (Fig. 4, a) is also purely ferritic and more homogeneous. The average size of ferrite grains is slightly finer and is ≈ 9,2 µm. There is an inconsiderable amount of fine powder fractions as well.

The transition from the deposited metal to the heat-affected zone and then to the base metal is smooth without sharp changes when using either a diode rectifier (Fig. 4, d) or an inverter (Figure 4, c). In both cases the heat affected zone is a polycrystalline aggregate of ferrite grains and pearlite colonies. A pearlite content increases gradually with respect to the distance from the deposited metal to the base metal and reaches ≈ 50 % in terms of volume.

When using the diode source supply (Fig. 4, d), the average ferrite grain size is 4 microns. The average grain size in pearlite colonies is the same. The heat affected zone width reaches 8 mm.

In the heat affected zone of the welded joint made using the inverter power supply (Fig. 4, c), the ferrite grain size is almost the same and makes 4.4 micron. The total width of the heat affected zone is visibly smaller (≤ 6 mm). This indicates a lower heat input and the weldable product is overheated.

The base metal is of a ferrite and pearlite structure (Fig. 4, e, f). The volume fraction of pearlite is 55% that corresponds to the chemical composition of Steel 45. Ferrite grains are polyhedral with well-defined and clean boundaries. The average size in both cases as expected is the same ≈ 12,5 μm. It is important to emphasize that forming welded joints causes significant (almost three times) refinement in the HAZ structure.

Thus, the application of an inverter insures the weldment with the uniform and fine grain structure in the deposited metal and narrower gradient heat affected zone.

Referring to the differences identified during the microstructure analysis, we assume mechanical properties vary as to the welded joints made using various power supplies.

Toughness tests on the notched weld metal test specimens were performed for impact properties according to the standard testing procedure. The test results are shown in Figure 5.

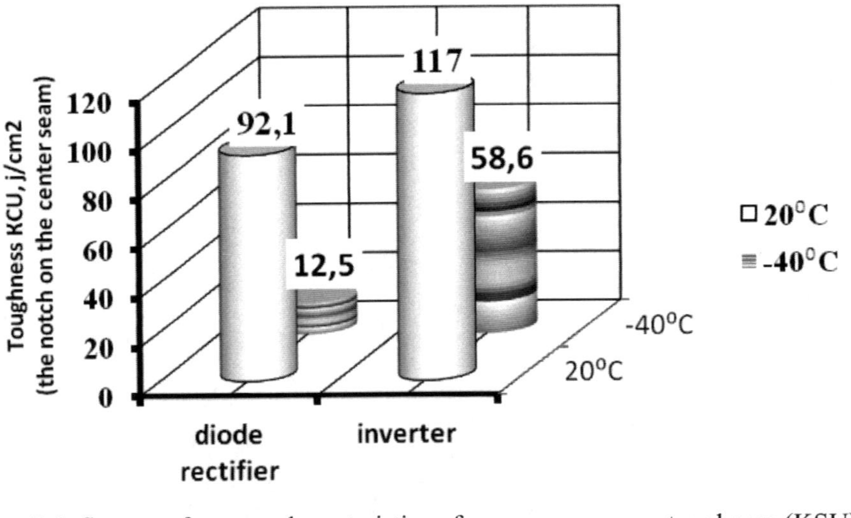

Figure 5. Influence of energy characteristics of power sources on toughness (KSU) of welded joints made of steel 45 using the UONI 13/55 electrodes

Analyzing the results of mechanical tests performed on weld specimens (Fig. 5) indicates that the decrease in heat content and overheating of electrode metal droplets while welding from the inverter power source enhance the toughness, particularly at sub-zero temperatures. This is essential for weldments designed for the use at low temperatures.

3 Conclusion

The results of the comprehensive research revealed that, when using inverters as a welding power supply, a drop of molten electrode metal has a less heat content, resulting in the decreased heat input and overheating of weldments. This insures the reduction in manganese burning out, improvement of microstructure in all weld zones and enhanced impact properties of welded joints at negative temperatures.

This work has been partially financed by the Russian Foundation for Basic Researches, SFBR grant No. 15-38-50600.

References

[1] Kuznetsov M A, Zernin E A, Danilov V I and Kartsev D S 2013 Applied Mechanics and Materials Application of nanostructured powders to control characteristic of electrode metal transfer and the process of weld structurization 379 pp 199-203.
[2] Kuznetsov M A, Zhuravkov S P, Zernin E A, Kolmogorov D E and Yavorovsky N A 2014 Advanced Materials Research Influence of Nanostructured Powder Modifiers on the Structure of a Welding Bead 872 pp 118-122.
[3] Makarov S V and Sapozhkov S B 2014 World Applied Sciences Journal Production of Electrodes for Manual Arc Welding Using Nanodisperse Materials 29 pp 720-723.
[4] Zhang M, Jian L and Bin P 2013 J Nanopart Res Workplace exposure to nanoparticles from gas metal arc welding process 11 pp 37/1-14.
[5] Guerreiro C J, Gomes F P, Santos T J and Miranda R M 2014Inhal Toxicol Characterization of airborne particles generated from metal active gas welding process 26 pp345–352.
[6] Yushin A A 2012 Development of Criteria for Evaluating Welding Properties of Arc Welding Machines with Controlled Metal Transfer: Author's Abstract PhD thesis in Eng.Sc.: 05.02.10 (Moscow) p 16.
[7] Zemskov A V 2013 High Frequency Invertors for AC Welding and their Operation in Parallel: PhD thesis in Eng.Sc. 05.09.12 Yuri Gagarin State Technical University of Saratov (Saratov) p 16.
[8] Kuskov V N, Mamadaliev R A and Obukhov A G Fundamental Researches Penetration of Alloying Elements to Weld Metal in 12Cr18Ni10Ti Steel Welding 11-9 pp 1794-1797. (in Russian).
[9] Ilyashchenko D P and Chinakhov D A 2011 Materials Science Forum Intestigating the Influence of the Power Supply the Weld Joints Properties and Health Characteristics of the Manual Arc Welding 12 pp 704-705.
[10] Ignatova A M 2013 Nauchno Tekhnicheskij Vestnik Povolzhja Mineral formation in particles of solid component in welding fumes at high temperatures and short pyrogenic welding processes 5 pp 166-173. (in Russian).

IOP Conf. Series: Materials Science and Engineering **91** (2015) 012007 doi:10.1088/1757-899X/91/1/012007

Increasing Strength and Operational Reliability of Fixed Joints of Tubes by MMA Welding

D P Il'yaschenko[1,a], D A Chinakhov[1,b], V I Danilov[1,2], G V Schlyakhova[2,3], Y M Gotovschik

[1]Yurga Institute of Technology National Research Tomsk Polytechnic University
Leningradskaya 26, Yurga, Russian Federation, 652055
[2] Institute of Strength Physics and Materials Science,
Siberian Branch of Russian Academy of Sciences
Akademichesky, 2/4, Tomsk, Russian Federation, 634055
[3] Seversk Institute of Technology, National Research Nuclear University MEPhI
Kommunistichesky, 65, Seversk, Russian Federation, 636036

e-mail: [a] mita8@rambler.ru, [b] chinakhov@tpu.ru

Abstract. This paper presents peculiar properties of structure formation, phase and chemical composition while welding of low-alloy steel 09MnSi2-1 depending on the dynamic characteristics of power sources of different types. Proper selection of power sources enables to decrease burning of alloy elements in metal of weld (Mn by 14% and Si by 17% of the weight ratio), to obtain more homogenous structure of deposited metal, to reduce length of heat-affected zone by 50% and to improve impact strength by 4-9%.

1. Introduction

Providing the required physical-mechanical properties and uniform strength of jointing areas, preventing from cold cracks formation and formation of structures reducing joints resistance to delayed and brittle failure raise the question of increasing strength and operational reliability of weld joints of tubes made from low-alloy steel [1].

Results in issues related to efficiency enhancement of the process of manual arc welding by coated electrodes (MAW) are marked by appearing of modulated current welding (MCW). MCW contributes to improvement of mechanical properties of weld joint [2,3,4]. The main technical facilities of MCW methods implementation are modulators-adapters [2] and inverter-type welding converters. However, irrespective of technical facility structure, these methods do not find wide application [5].

For quality welding of fixed joints of pipelines in conditions of constantly changing space position of welding pool current source should have dynamic properties, provide rapid growth of stresses while transition from short circuit to arc burning, and provide optimal velocity of current change (to reduce sputtering connected with transition of melted electrode droplets into welding pool) as well as to meet other requirements [6].

Welding process stability while using power sources with different energy parameters can be evaluated by the ratio $\frac{I_{s.c.}}{I_2} \rightarrow 1$ [7] , where $I_{s.c.}$- short circuit current, A: I_2 – arc current, A, according to which the value of spatter of electrode metal can be characterized. Spatter of electrode metal depends on energy parameters of welding current supply having influence on rate of rise of short circuit current, and thereby influencing the values of: power of electric explosion at the time of droplet detachment after short circuit; amplitude value of current [8].

Works [9] study the impact of welding current magnitude upon the amount of electrode metal spatter when applying coated electrodes of various types but no such dependences were revealed when using different power supplies. Rectifier further diode rectifier and inverter power source were used as power supplies. The results of the completed studies are presented in Fig. 1.

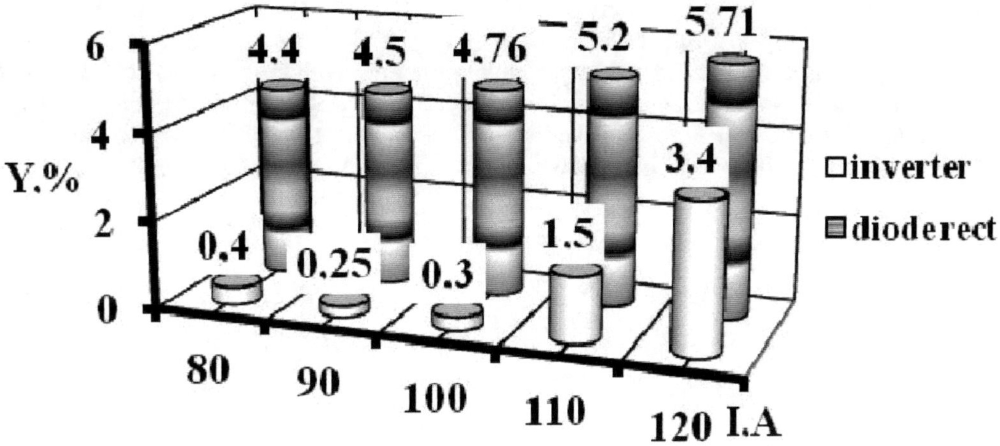

Figure 1. Impact of current magnitude upon the amount of electrode spatter under MAW with coated electrodes 3.2 mm in diameter – LB 52U (basic coating)

Performed studies [10] showed that while using inverter power source thermal impact of arc on the droplet of electrode metal is changing. Therefore use of inverter power source leads to alteration of kinetics of metallurgical process in droplet, and as a consequence to change of chemical composition of added metal [11].

The present work is devoted to comparative integrated study of chemical composition, microstructure and mechanical properties, including microhardness of welded joints from steel 09G2S made using inverter power source and diode rectifier.

2. Materials, research and results

The following samples were manufactured to conduct integrated study: weld joint of tubes Ø159 ×6 made from steel 09G2S by electrodes: root - LB-52U (d=2.6 mm), welding current I = 50-60 A, filling - LB-52U (d=3.2 mm), welding current I = 80 – 90 A from different power supplies (diode rectifier VD-306 and inverter Nebula-315).

Analysis of weld chemical composition (Table 1) showed that while using inverter power supply alloy burning of Si decreases by 14% and Mn by 17%. It is conditioned by the fact that energy parameters of inverter power supply provide less rate of rise of arc welding current and that in its turn confines the value of short-circuit current [9]. This statement is indirectly proved by the data in [12].

IOP Conf. Series: Materials Science and Engineering **91** (2015) 012007 doi:10.1088/1757-899X/91/1/012007

Table 1. Average chemical composition of metal of weld, made from tube Ø159×6 (steel 09MnSi2-1) welded with LB 52U electrodes

Power supply	Chemical composition, %							
	C	Si	Mn	S	P	Cr	Ni	Cu
diode rectifier	0.10± 0.012	0.52± 0.03	1.03±0.05	0.010	0.014	0.03±0.01	0.05±0.01	0.03
inverter	0.09± 0.005	0.60± 0.03	1.23±0.05	0.010	0.014	0.03±0.01	0.06±0.01	0.03

Macro- and microstructural studies were carried out by means of optical metallography on sections, made from weld joints (Fig. 2). Optical microscope Olympus GX-71 was used. For manufacturing of sections abrasing machining was applied, mechanical polishing with diamond paste ASM 10/7 NVL and chemical milling in 4% solution of nitric acid. Microhardness testing in added metal was carried on these sections, in thermal impact zone and in the basic metal using microhardness tester DURAMIN 5. Vickers hardness was defined while loading 50 g and soaking time 10 s.

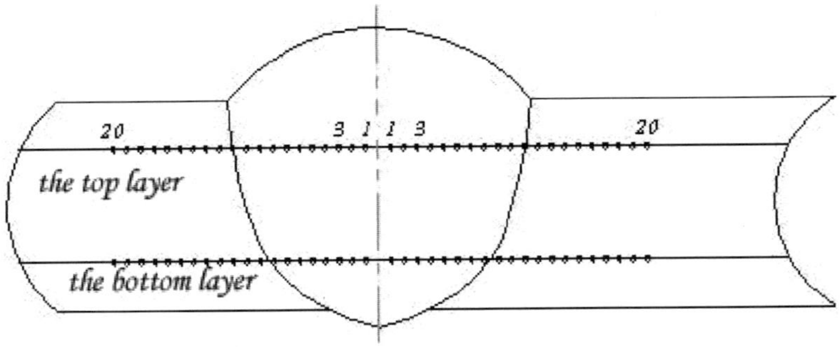

Figure 2. Schematic plan of microstructural studies and measuring microhardness (step between points – 0.5 mm)

Figure 3 demonstrate microhardness of weld joints from steel 09G2S (according to scheme given in Fig. 2).

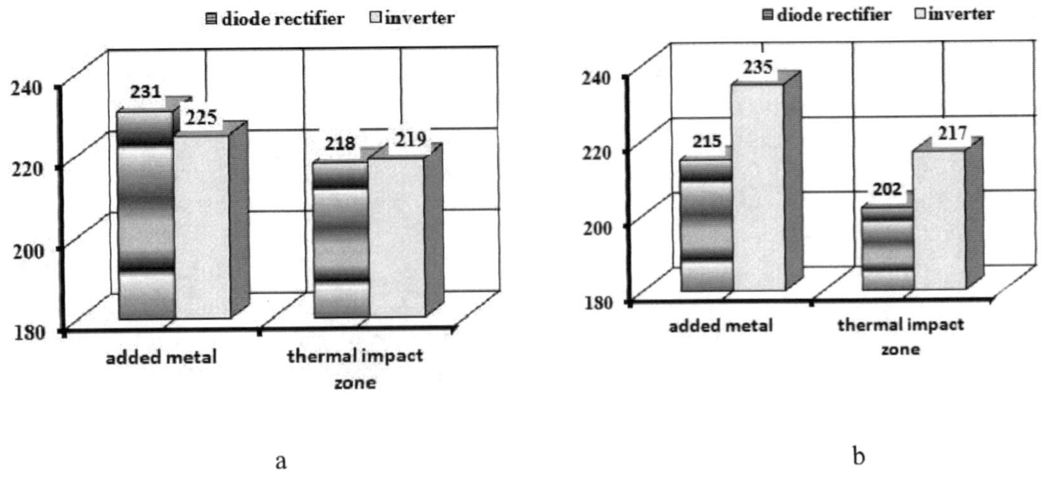

Figure 3. Average distribution bar charts of microhardness (MPa) in cross-section of weld joints made by different power supply sources: a – the top layer on Fig. 2; b – the bottom layer on Fig. 2

IOP Conf. Series: Materials Science and Engineering **91** (2015) 012007 doi:10.1088/1757-899X/91/1/012007

It can be seen that in weld joint when welding with diode rectifier, added metal has maximum hardness (Fig. 3a). During transition from added metal to the thermal impact zone microhardness decreases. In the bottom layer (Fig. 3b) difference of microhardness of added metal and thermal impact zone is insignificant, but in the top layer (Fig. 3b) it is statistically significant. Distribution pattern of microhardness along zones in weld joint obtained when welding with inverter source is qualitatively the same. Added metal also has maximum microhardness, again the difference in bottom layer between thermal impact zone and added metal is insignificant while it is significant in the top layer (Fig. 3). However the values of microhardness in the respective zones of weld joint made when using inverter source is higher. It is most noticeable for top layer (Fig. 3a). It should be noted that statically insignificant difference of microhardness of basic metal both in top and bottom layers in both joints is fairly expected (Fig. 3).

The noted patterns conform to structural conditions of the studied weld joints. Microphotographs (Fig. 4) demonstrate structure of added metal, thermal impact zones and basic metal.

Obviously added metal (Fig. 4a) while welding using diode rectifier has inhomogeneous structure of rodlike dendrite, distinctive of cast-condition. Ferrite plate reaches 1 mm in length and 20 μm in width. Inter-plate space is occupied with dendrites of smaller size. By larger scaling it was established that ferrite plates consist of polyhedral grains, about 12μm in size. Other phases and structural components except for ferrite are not observed, this corresponds with the composition of electrode core LB 52U.

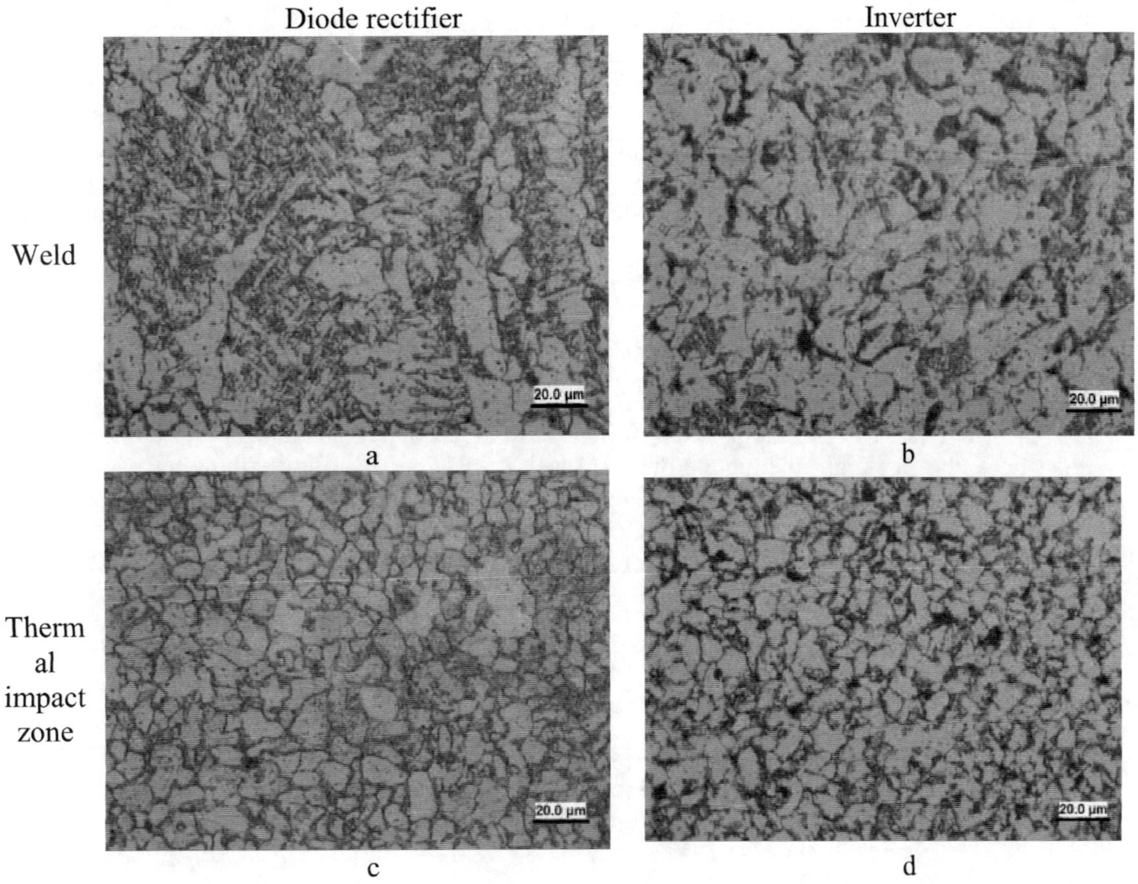

Basic metal

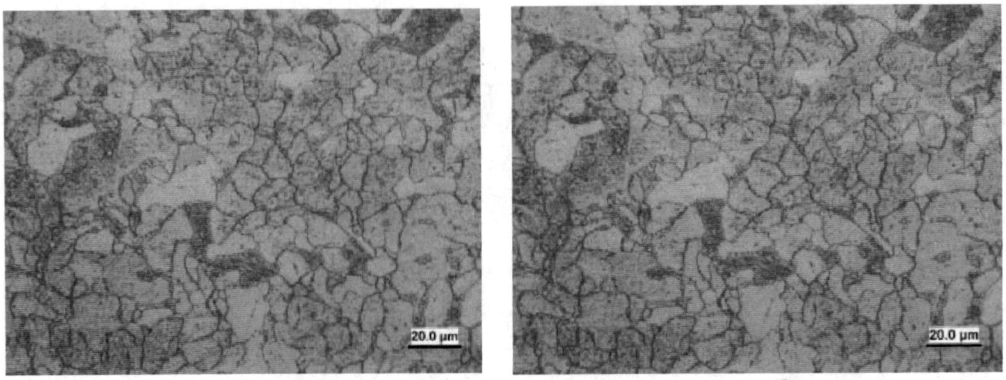

e f

Figure 4. Results of microstructural analysis: a, b – microstructure of weld; c, d – microstructure of thermal affected zone of weld joint; e, f – microstructure of basic metal

Structure of added metal in weld made using inverter source (Fig. 4b) is significantly more homogeneous. Columnar behavior is faintly observed. Length of ferrite plates does not exceed 50 µm, while thickness reaches 20 µm as in the previous case. The average size of ferrite grains that dendrites are made of is 14.5 µm.

Transition from added metal to the thermal impact zone and afterwards to the structure of basic metal occurs gradually, without sharp changes both in case of using source VD-306 (Fig. 4c) and source Nebula-315 (Fig. 4d). In both cases thermal impact zone is presented by polyhedral ferrite grains.

In thermal impact zone of weld joint obtained when welding using diode source (Fig. 4c) size of ferrite grain is insignificantly larger and reaches 10.3 µm. Pearlite is found in the structure, in the amount corresponding to basic metal (steel 09G2S). Total width of thermal zone impact reaches 2 mm.

In case of inverter using (Fig. 4d) the average grain value is 8.5 µm. There are almost no signs of pearlite inclusions. The reason might be carbon diffusion in decarbonized added metal. Width of thermal zone impact is insignificant and does not exceed 1 mm.

The structure of basic metal is ferrite-pearlite (Fig. 4, e, f). Volume ratio of pearlite is 10-12%, corresponding to the chemical composition of steel 09G2S. Ferrite grains are polyhedral with well defined clear boundaries. Thus application of inverter source enables obtaining weld joint with more homogeneous and fine grained structure of added metal. Differences of microstructures of weld joint obtained in the conducted study can be explained by less heat content of droplet of melted electrode metal while welding using inverter and therefore less burning of Si and Mn (Table 1). Microstructural differences define different mechanical properties of weld joints as well, given in Table 2.

a) b)

Figure 5. Photographs of samples after testing KCU (+20 °C): a – inverter; b – diode rectifier

Weld joints (Fig. 5) in both cases have high viscosity and less probability of crack formation, which is the necessary condition for production of pipelines which will be used in conditions of low temperatures.

Table 2. Mechanical properties of weld joints made from tube Ø159×6 (steel 09MnS2-1) by LB 52U electrodes

Power supply	Limit of ultimate tensile resistance σB, MPa	Bend angle outward, inward and edgeway, deg.	Impact strength KCU, J/sm2 (cut along center of the weld)			
			+20 °C	0 °C	-20 °C	-40 °C
diode rectifier	541-543 542	120-120 120	201-220 210	212-223 216	200-233 219	143-230 182
inverter	550-560 556	120-120 120	208-226 217	215-254 235	224-250 237	150-258 193

Data analysis of Table 2 shows increase of impact strength of weld joints made when welding using inverter source in comparison to samples obtained by diode rectifier by 4-9% while different temperature of operation. This can be explained by difference of microstructure in weld joints, decrease of length of ferrite plates by 50% and thickness by 24% and reduction of length of thermal impact zone by 50%.

3. Conclusion

Proper selection of power supply enables to:
1. Reduce burning of alloy elements in metal of weld: Mn by 14% and Si by 17% of the weight ratio;
2. Obtain more homogenous structure of deposited metal;
3. Reduce length of thermal impact zone by 50%;
4. Increase impact strength by 4–9% under different operation temperature.

This work has been partially financed by the Russian Foundation for Basic Researches, SFBR grant No. 15-38-50600.

References

[1] Saraev Y N Sleptsov O I, Bezborodov V P, Nikonova I V and Tyutev AV 2005 Ohysical Mesomechanics Influence of impulse welding on structure and properties of weld joints of tubes from high-strength steel 8 pp 141-144.
[2] Knyazkov V L and Knyazkov A F 2008 Enhancing effectiveness of manual arc welding of pipelines (Kemerovo: Publishing Office of Kuzbass State Technical University) 2008 pp 104.
[3] Krampit A G 2012 Welding International Welding with double modulation of the main welding 26 pp 867-869.
[4] Krampit N Y and Krampit A G 2011 Welding International Mechanical properties of welded joints in welding with continuous and pulsed arcs 25 pp 626-628.
[5] Dementsev K I 2010 Enhancing effectiveness of inverter-type welding converters by means of welding current modulation, PhD thesis: 05.02.10 Tomsk pp 16.
[6] Mustafin F M, Blekherova G M, Kvyatkovskii O P et al 2002 Nedra-Business Center LLC Welding of pipelines (Study book Moscow).
[7] Fedko V T 1998 Theory, technology and means of reducing spatter and labor input when welding in carbon dioxide (Tomsk, Tomsk State University) pp 156.
[8] Fedko V T, Brunov O G, Solodsky S A, Krukov A V and Sokolov P.D Engineering technology 2005 The methods of spatter reduction when welding in CO2 5 pp 24 – 30.

IOP Conf. Series: Materials Science and Engineering **91** (2015) 012007 doi:10.1088/1757-899X/91/1/012007

[9] Il'yashchenko D P, Chinakhov D A and Gotovshchik Yu M 2015 Applied Mechanics and Materials Influence of Power Supply Energy Characteristics upon the Stability of MMA Process 756 pp 97-100.

[10] Il'yashchenko D P, Chinakhov D A and Gotovshchik Yu M Applied Mechanics and Materials Calculation of the Heat Content of the Electrode Metal Droplet When Applying Power Supplies for Manual Arc Welding With Different Volt-Ampere Characteristic. 756 pp 101-104.

[11] Livshits L S 1979 Metallurgy for welders (welding of steel) (Moscow^ Machine Engineering Publishing Office pp 235.

[12] Ignatova A M 2013 Nauchno Tekhnicheskij Vestnik Povolzhja Mineral formation in particles of solid component in welding fumes at high temperatures and short pyrogenic welding processes 5 pp 166-173. (in Russian).

IOP Conf. Series: Materials Science and Engineering **91** (2015) 012008 doi:10.1088/1757-899X/91/1/012008

New Tendencies in Development of Carbonaceous Additives for Welding Fluxes

N A Kozyrev, R E Kryukov, O A Kozyreva

Federal State Budgetary Educational Institution of Higher Professional Education «Siberian State Industrial University»,
Research and Development Center «Welding Processes and Technologies»
654007, Russia, Novokuznetsk, 42, Kirov str.

E-mail: kozyrev_na@mtsp.sibsiu.ru

Abstract. The paper provides results of comparative analysis of the effect of carbonaceous components introduced into welding fluxes on molten metal – slag interaction. Thermodynamical calculations of dehydrogenization are presented for submerged arc welding.

A positive influence of carbonaceous additives on gas content and mechanical properties of welds is demonstrated. Carbon and fluorine containing additives are emphasized to be promising for automatic submerged arc welding.

Introduction

The issue of new fluxes and their additives development has been attracting much attention currently, as well as research into their influence on welding and technological characteristics of a weld and on the concentration of oxygen and non-metallic impurities in a weld [1-5].

Submerged arc welding is attended by intensive mass transfer of liquid molten metal and slag, forming from welding flux. Reactions of oxidation and deoxidation of manganese, ferrum, and silicon, i.d. exchange processes involving oxygen are typical for this process. The most grades of domestically produced fluxes, which are applied for welding low-alloyed steels are oxidizing ones and ground on silicon-manganese oxidation-reduction processes. Here, the products of these reactions are oxide compounds of silicon, manganese, ferrum, aluminum etc., which often can't surface and assimilate to slag, forming from welding flux, the level of impurity of weld metal by non-metallic admixtures increases consequently; as the result, the complex of physical and mechanical characteristics deteriorates. Apparently, restoratives, which form gaseous products of reactions, are advisable to apply in order to avoid impurity of weld metal. It is carbon that can be a restorative of this kind, and forms gaseous compounds CO_2 and CO when reacting with oxidizers.

Materials and methods of research

Shielding is usually provided through pushing atmospheric gases aside from weld zone by forming gases CO_2 (CO); that helps to reduce or even exclude the probability of molten metal saturation with oxygen, nitrogen or hydrogen from atmosphere. Gas-forming compounds of carbonates like $CaCO_3$, $MgCO_3$, $FeCO_3$, $MnCO_3$ and their derivatives are usually used for this purpose. Gas shielding is possible due to CO_2 as high-temperature decomposition of carbonates takes place according to the following reactions and temperatures [6]:

$$CaCO_3 \rightarrow CaO + CO_2 \text{ (900-1200 °C)}, \quad (1)$$

$$MgCO_3 \rightarrow MgO + CO_2 \text{ (>650 °C)}, \quad (2)$$

$$FeCO_3 \rightarrow FeO + CO_2 \text{ (280-490 °C)}, \quad (3)$$

$$MnCO_3 \rightarrow MnO + CO_2 \text{ (330-500 °C)} \quad (4)$$

According to stoichiometric calculations the results of decomposition are as follows: 1 kg $CaCO_3$ – 0.224 m^3 CO_2 , 1 kg $MgCO_3$ – 0.267 m^3, 1 kg $FeCO_3$ – 0.192 m^3, 1 kg $MnCO_3$ – 0.194 m^3.

Here CO_2 emits at high pressure of off-gases.

In terms of Mendeleev – Clapeyron law at 1600 °C (1873K)

the pressure of CO_2 is

$$p = \frac{m}{M} \cdot \frac{RT}{V} \quad (5)$$

According to equation (5) CO_2 pressure is 694.84 kPa for $CaCO_3$, 688.93 kPa for $MgCO_3$, 698.26 kPa for $FeCO_3$, and 696.54 kPa for $MnCO_3$.

Without taking into account the costs of carbonates decomposition, $MgCO_3$ and $CaCO_3$ are the most optimal components, which help to get most CO_2 when decomposing 1 kg of material, succeeded by $MnCO_3$ и $FeCO_3$.

Furthermore, when decomposing $CaCO_3$ and $MgCO_3$ basic oxides CaO and MgO are formed and improve basicity of welding flux, and that of a forming slag, respectively, whereas, when $MnCO_3$ and $FeCO_3$ decomposing oxides FeO and MnO are formed, which raise the degree of oxidation in slag systems and oxygen concentration in a weld. The latter causes all negative consequences – increasing level of impurity by non-metallic oxide components in a weld and deterioration of mechanical properties.

Having followed all mentioned pre-conditions we have developed a flux – ANK additive, protected it by a patent of the Russian Federation and applied in production process at Open Joint Stock Company "Novokuznetsk Plant of Reservoir Metalware named after N.E. Kryukov" [7]. For its manufacturing ferrosilicon FS75 (GOST 1415-78), marble M92- M97 (GOST 4416-73 (92-97% $CaCO_3$)), and liquid glass (GOST 13078-81) were used. Production technology was as follows. Marble and ferrosilicon were grinded to less than 1 mm fraction. Grinded marble and silicon were mixed in 50 to 50% mass proportion. It was dried at temperature 100-200 ^{0}C for 10 - 20 minutes, succeeded by grinding and size grading to 2.5 mm. 3-5% of additive was introduced into fluxes. Before a flux with an additive is used its 40 – 60 minutes annealing in the furnace is recommended at temperature 250-350 ^{0}C.

This additive is used for roll welding of tanks. The technology involves assembling, welding, controlling and rolling plates of tanks walls, all the processes are performed on special roll facilities with upper and down rolling. Two-side submerged arc welding of butt joints of wall plates is applied in the process, first on the upper tier, then on the lower one, after the plate is rolled. An additive helped to avoid pore formation and improve quality of welds.

However, shielding gases CO and CO_2 can form due to carbon, added to the flux, according to the reactions:

$$(C) + [O_2] = \{CO_2\} , \quad (6)$$

$$(C) + ½[O_2] = \{CO\} \quad (7)$$

Here 1.863 m^3 CO_2 and 1.864 m^3 CO release per each kg of carbon (in normal conditions), and design pressure of gases is 311.66 kPa for CO_2 and 694.84 kPa for CO at 1600 °C.

The second important issue is that of weld metal dehydrogenization. As a rule, it is carried out by introducing fluorine-containing additives (fluorite or cryolite), hydrogen combines with fluorine and is further removed as a compound HF.

The following chemical transformations can be considered as probable reactions of removal:

$$1/2\,(CaF_2) + [H] + 1/2\,[O] = (CaO) + HF_g , \quad (8)$$

$$1/6(Na_3AlF_6)+ [H]+ 1/2\,[O] = 1/6NaAlO_{2\,s}+ HF_г + (Na_2O)\,, \quad (9)$$

As well as reactions:

$$2(CaF_2) + 3(SiO_2) = 2CaSiO_{3\,s} + SiF_{4g}\,, \quad (10)$$

$$2/3(Na_3AlF_6) + 5/3(SiO_2) = SiF_{4g} + 2/3NaAlO_{2\,s} + 2/3\,(Na_2SiO_3), \quad (11)$$

succeeded by reactions of dehydrogenization with SiF_4:

$$1/2\,SiF_{4g} + [H] = 1/2SiF_{2\,g}+ HF_g \quad (12)$$

$$1/4\,SiF_{4g} + [H]+ 1/2\,[O] = 1/4\,(SiO_2) + HF_g \quad (13)$$

$$1/3\,SiF_{4g} + [H] = 1/3SiF_g + HF_g \quad (14)$$

$$1/2\,SiF_{4g} + [H] = 1/2\,SiF_2g + HF_g \quad (15)$$

Thermodynamical characteristics in standard conditions [$\Delta_rH°(T)$, $\Delta_rS°(T)$, $\Delta_rG°(T)$] needed to assess reaction probability were calculated by well-known methods [8] in the temperature range of welding processes 1700 – 2200 K [9] in terms of thermodynamic properties of reagents [[H°(T)-H°(298,15 K)], S°(T), $\Delta_fH°$(298,15 K)] [10,11]. Here, chemical states Na_3AlF_{6l}, SiO_{2l}, SiF_{4g}, $NaAlO_{2\,s}$,Na_2SiO_{3l}, $CaF_{2ж,}$ $CaSiO_{3\,s}$, $H_{2г}$, SiF_{2g}, HF_g, O_{2g}, SiF_g, H_g were selected as standard ones for substances – reagents in the range 1700 – 2200 K according to fact aggregate states of phases in the system under consideration.

The results of calculations are provided in the Table. The graphs of standard Gibbs energy reactions (8) - (11) according to temperature are depicted in Figure 1, those of reactions (12) - (15) in Figure 2.

Table. Standard Gibbs energy of reactions (8) – (15) according to temperature

Reaction	$\Delta_rG°(T)$, kJ					
	1700K	1800K	1900K	2000K	2100K	2200K
8	-16.22	-18.61	-20.93	-23.17	-25.36	-27.47
9	-32.32	-33.82	-35.20	-36.46	-37.62	-38.68
10	41.80	35.98	30.62	25.71	21.22	17.18
11	82.41	76.11	70.40	65.22	60.56	56.38
12	86.62	78.13	69.68	61.27	52.90	44.57
13	-90.16	-89.83	-89.51	-89.21	-88.91	-88.63
14	113.04	104.93	96.86	88.82	80.80	72.82
15	-38.07	-40.60	-43.08	-45.49	-47.84	-50.14

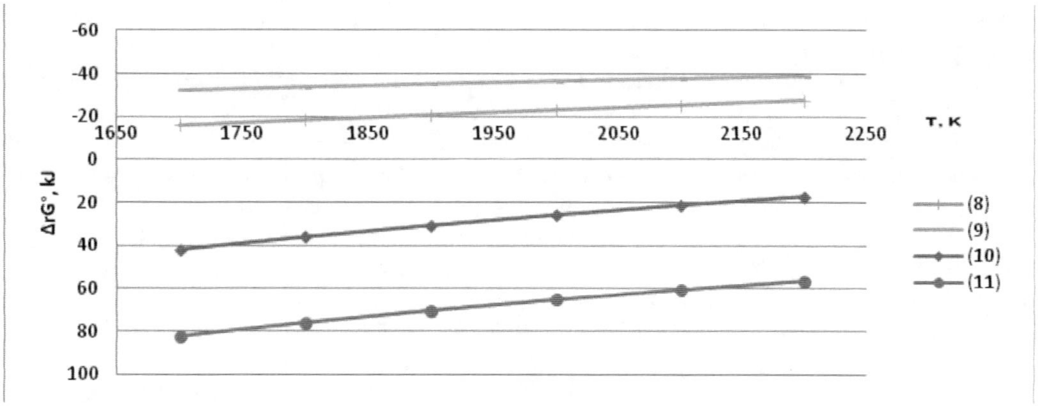

Figure 1. Standard Gibbs energy of reactions (8) - (11)

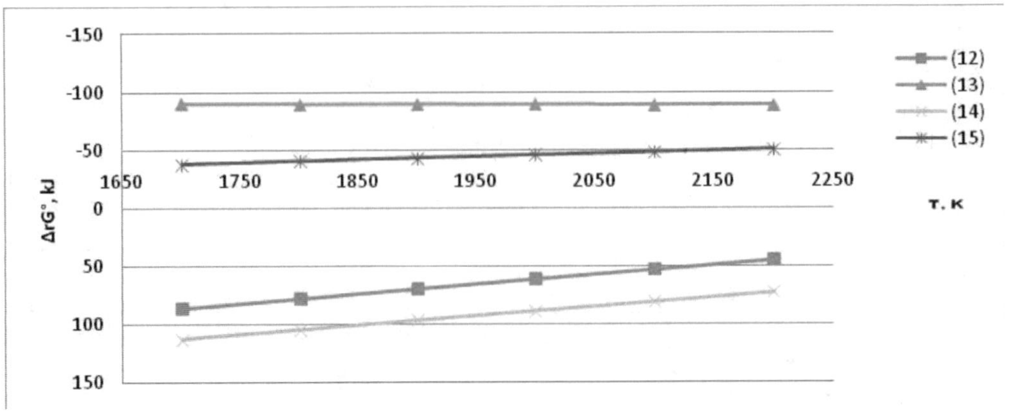

Figure 2. Standard Gibbs energy of reactions (12) - (15)

Figure 1 demonstrates that reaction (9) is thermodynamically the most probable (cryolite dehydrogenization), the second one is reaction (8) (fluorite dehydrogenization), followed by reactions (10, 11), where silicon tetrafluoride is formed as an intermediate product of further reactions (12) - (15); the latter result in formation of gaseous compound HF. Here, reaction (13) is thermodynamically the most probable (SiF_4 combines with hydrogen and oxygen). The stoichiometric reactions (15), (12), (14) are the least probable ones.

Therefore, Na_3AlF_6 is the most reasonable to use for dehydrogenization when submerged arc welding as if compared with fluorite.

Having taken into account the aforementioned preconditions, we have developed a technology of submerged arc welding with carbonaceous additives. As the basis of carbon and fluorine containing additive we took metallurgical production wastes. It was dust with the following chemical composition (mass %): Al_2O_3 = 21 – 46.23; F = 18 – 27; Na_2O = 8 – 15; K_2O = 0.4 – 6; CaO = 0.7 – 2.3; SiO_2 = 0.5 – 2.48; Fe_2O_3 = 2.1 – 3.27; C = 12.5 – 30.2; MnO = 0.07 – 0.9; MgO = 0.06 – 0.9; S = 0.09 – 0.19; P = 0.1 – 0.18.

Mineralogical makeup of dust was determined according to the data of X-ray structural analysis made by difractometer DRON-2 in the mode: Fe – K α radiation, voltage 26 kV, electrical current 30 mA.

The research into the dust of electrostatic precipitators revealed that the material consisted of bi-dimensionally ordered carbon (d_{002}=3.47Å, L_c=45.8Å), X-ray amorphous substance, cryolite, corundum, hyolithe, and various admixtures. Diffraction patterns of roasted at 700°C material demonstrate no indication of graphite, that is caused by nearly complete burning out of carbon-containing mass in this temperature range, as well as significant curve flattering on the diffraction pattern, and decrease in X-ray amorphous substance. The reason of the latter is probably chemical composition of X-ray amorphous substance, which carbon compounds are main components of. At 700°C the change in indication intensity of mineralizing components (cryolite, corundum, X-ray amorphous substance, fluorite, hematite and various admixtures) was recorded.

From the theoretical point of view the additive makes possible: 1) dehydrogenization by fluorine-containing compounds (like Na_3AlF_6,), decomposing at the temperatures of welding processes and isolating fluorine, which combines with dissolved in steel hydrogen and forms gaseous HF; 2) intensive carbon "boiling" due to forming CO and CO_2, when fluoric carbon CF_x ($1 \geq x > 0$) combines with dissolved in steel oxygen, here, as carbon is in a bound state steel carbonization is hardly possible; 3) improvement of arc stability due to potassium and sodium, facilitating ionization in arc column.

To make an additive to flux carbon and fluorine containing substance was mixed with liquid glass, then this mixture was dried, cooled down and grinded. Afterwards this additive was mixed with flux in a special mixer according to a definite, strictly determined proportion. AN-348A, AN-60, AN-67 fluxes were taken as basic ones and their mixtures with flux-additives.

The experiments were carried out on 200× 500 mm 09G2S steel samples 16 mm in thickness. Fay welding of butt joints was made on two sides, as when welding wall plates of tanks on roll facility. Sv-08GA wire 5 mm in diameter was used as a filler metal.

Submerged arc welding of samples was made in similar modes. The samples were cut of welded plates and subject to the following tests: X-ray spectral analysis of weld metal chemical composition, metallographic tests of welds; total concentration of oxygen in welds, mechanical properties, strength of joint welds and impact strength of welds were determined at temperatures -20^{0}C and -40^{0}C. Concentration of carbon, sulphur, phosphorus was determined in chemical composition of weld metal by chemical methods in terms of GOST 12344-2003, GOST 12345-2001, and GOST 12347-77, respectively. Concentration of alloying elements in weld metal; that of calcium oxide, silicon, manganese, aluminum, magnesium, ferrum, potassium, sodium and fluorine-compounds in fluxes with additives and slag, obtained after welding was determined by SHIMADZU roentgen-fluorescent spectrometer XRF-1800.

The experiments demonstrated that maximum 6% carbon and fluorine containing additive provided carbon concentration in weld similar to its concentration in original metal (Figure 3), whereas concentration of oxygen, hydrogen and nitrogen dropped (Figures 4, 5, 6).

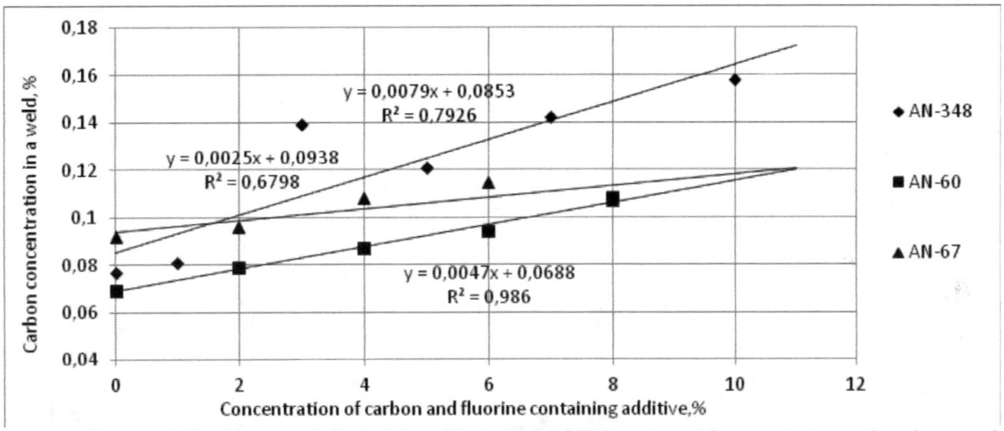

Figure 3. Influence of carbon and fluorine containing additive on carbon concentration in a weld

Metallographic research into polished sections of joint welds was carried out by optical microscope OLYMPUS GX-51 in bright field and zooming ×100, ×500. The microstructure of metal was found out by etching in 4 % HNO_3 solution in ethanol. The structure of base metal in all samples consists of ferrite grains and lamellar pearlite (4-5 µm). In base – to – added metal zone a fine-grain structure occurs (1-2 µm), which was formed as the result of re-crystallizing when heating in course of welding. In the microstructure of a weld there are ferrite grains stretched towards heat rejection because of heating and speeded up cooling down. Structures of welds didn't differ much irrespectively of used fluxes. The level of impurity by non-metallic substances decreased in samples, which were welded with fluxing agents, containing carbon and fluorine additives; it was caused by reduction of total oxygen concentration.

The research into mechanical properties (yield point, strength, modulus of elongation, impact strength at temperatures below zero) carried out on cut according to GOST 6996-66 samples, demonstrated that the level of properties went beyond the values required in GOST 31385-2008 and increased as the concentration of carbon and fluorine containing additive rose. Increasing impact strength KCV and KCU at temperatures -20°C and -40°C, respectively (Figures 7, 8) is worth mentioning. Flux-additives, which were developed, have been protected by the Russian Federation patents [12, 13].

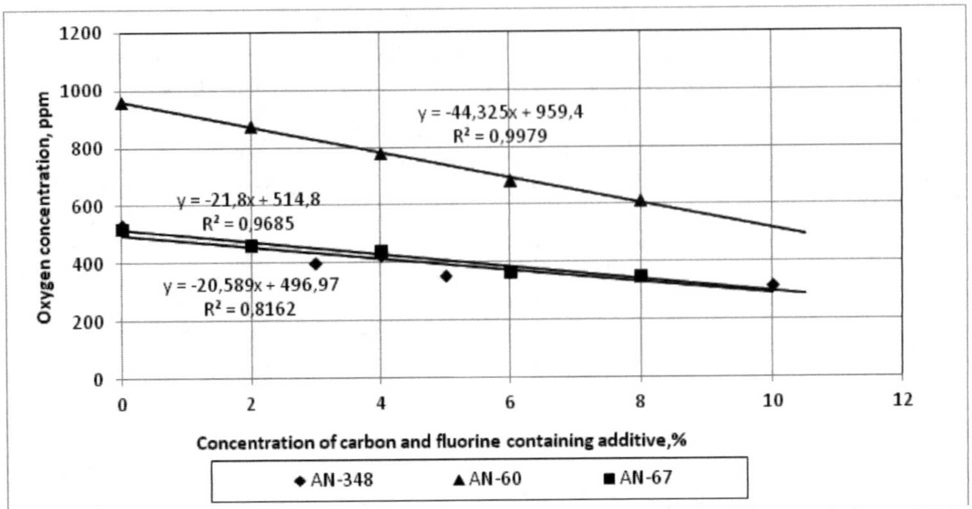

Figure 4. The change in oxygen in dependence on carbon and fluoride containing additive concentration

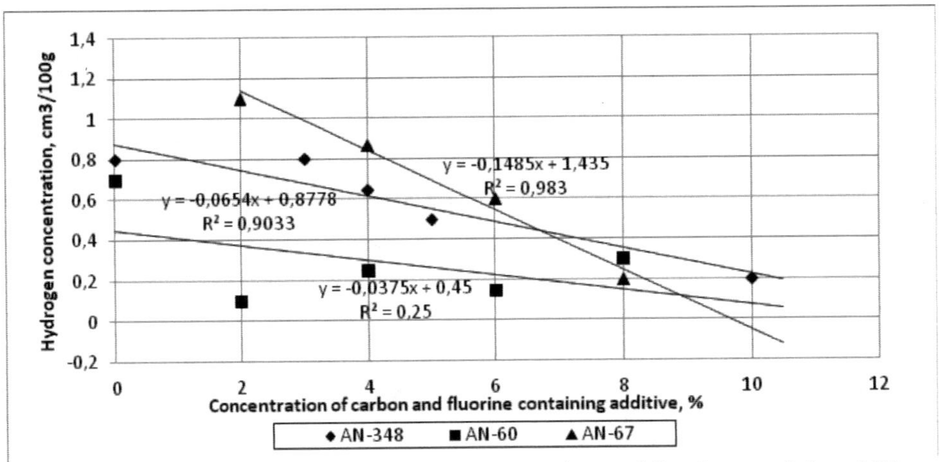

Figure 5. The change in hydrogen in dependence on carbon and fluorine containing additive

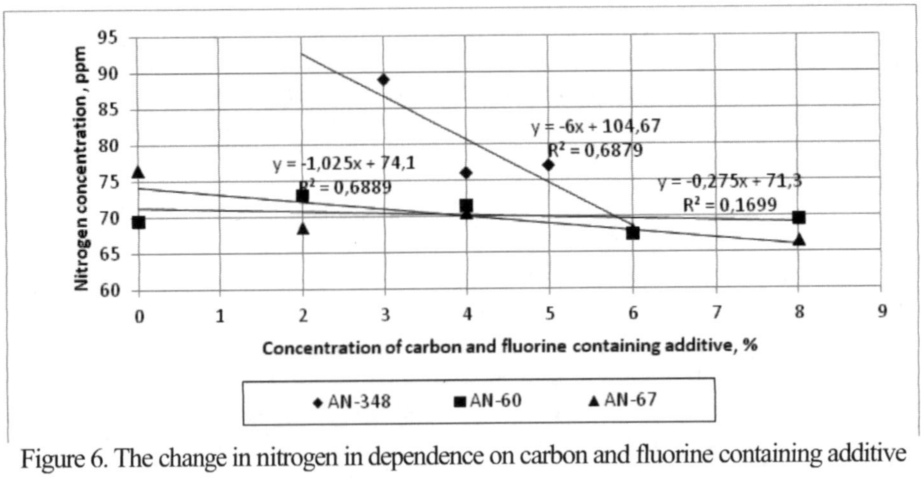

Figure 6. The change in nitrogen in dependence on carbon and fluorine containing additive

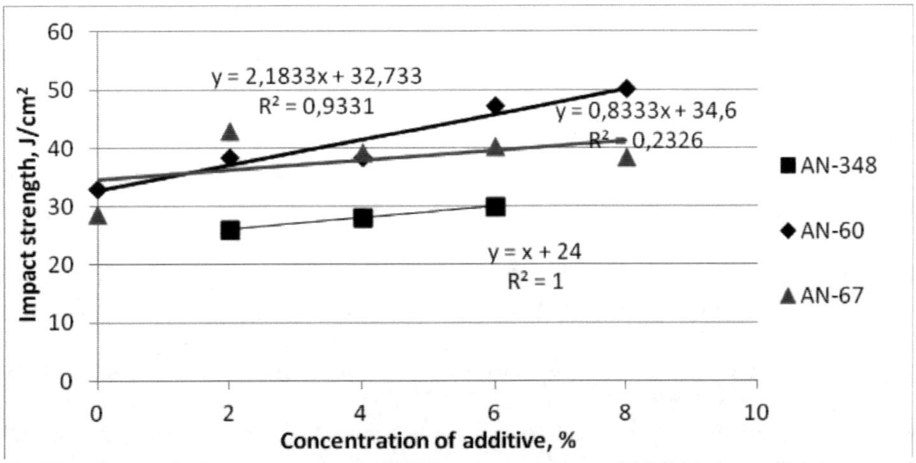

Figure 7. The change in impact strength KCV at temperature -20°C in dependence on carbon and fluorine containing additive.

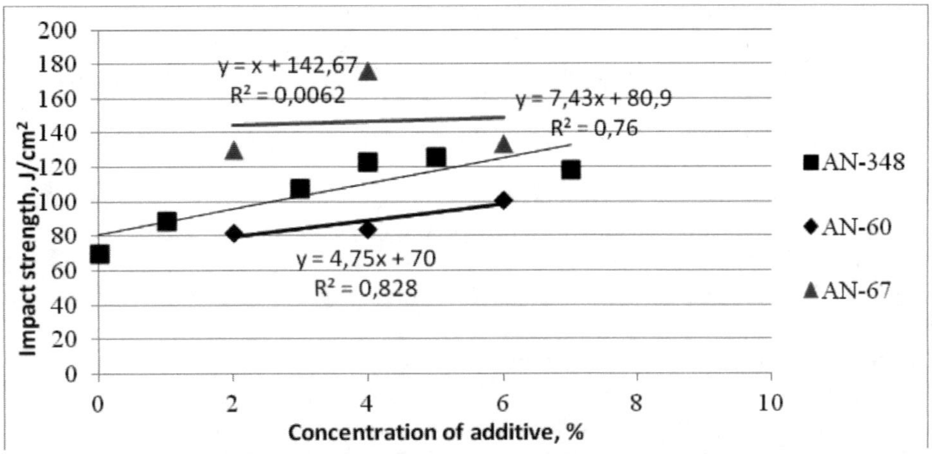

Figure 8. The change in impact strength KCV at temperature -40°C in dependence on carbon and fluorine containing additive

Conclusions

1. On the ground of made calculations and carried out experiments we can conclude that carbon containing additives to welding fluxes are possible and promising ones in order to improve welding and technological characteristics of welded metalware.

2. The probability of dehydrogenization of a weld in fluorine containing submerged arc welding has been assessed thermodynamically in the temperature range 1700 – 2200 K. Here, Na_3AlF_{6l}, SiO_{2l}, SiF_{4g}, $NaAlO_{2s}$, Na_2SiO_{3l}, CaF_{2l}, $CaSiO_{3\ s}$, H_{2r}, SiF_{2g}, HF_g, O_{2g}, SiF_g, H_g. were selected as standard states for substances - reagents. In terms of calculation of standard Gibbs energy reactions it has been found out that the reaction of gaseous hydrogen fluorine direct formation by cryolite is thermodynamically the most probable one, the second probable is the group of reactions resulting in formation of silicon tetrafluoride as an intermediate product for further HF formation. In this group the most thermodynamically probable reaction is that of SiF_4 with hydrogen and oxygen. In terms of calculations Na_3AlF_6 is more reasonable to use for dehydrogenization when submerged arc welding in comparison to fluorite.

3. Introduction of developed carbon and fluorine containing additive into fluxes AN-348A, AN-60

and AN-67 reduces gas content of a weld, the level of impurity by oxide non-metallic substances, and improves required mechanical properties and impact strength (at temperatures below zero, especially).

References

[1] Amado Cruz Crespoa, Rafael Quintana Puchola, Lorenzo Perdomo Gonzáleza, Carlos R. Gómez Péreza, Gilma Castellanosa, Eduardo Díaz Cedréa & Tamara Ortíza Study of the relationship between the composition of a fused flux and its structure and properties *Welding International* 2009 Volume 23 №2. p 120-131

[2] Volobueva Yu S Volobueva O S Parkhomenko A G Dobrozhelac E I & Klimenchuk O S Using a new general-purpose ceramic flux SFM-101 in welding of beams *Welding International* 2012 Volume 26 №8 p 649-653

[3] Golovko V V & Potapov N N Special features of agglomerated (ceramic) fluxes in welding *Welding International* 2011 Volume 25 №11 p 889 - 893

[4] Rafael Quintana Puchola, Jeily Rodríguez Blancoa, Lorenzo Perdomo Gonzaleza, Gilma Castellanos Hernándeza & Carlos Rene Gómez Péreza The influence of the air occluded in the deposition layer of flux during automatic welding: a technological aspect to consider in the quality of the bead *Welding International* 2009 Volume 23 №2 p 132-140

[5] Obtaining a submerged arc welding flux of the MnO–SiO2–CaO–Al2O3 – CaF2 system by fusion / A.C. Crespoa, R.Q. Puchola, L.P. Goncaleza, L.G. Sanchezb, C.R. Gomez Pereza, E.D. Cedrea, T.O. Mendeza & J.A. Pozola//Welding International.– 2007.- Volume 21. - №7. - p. 502-511.

[6] Lidin R A Molochko V A Andreeva L L Reaction of non-organic substances M: Drofa, 2007 p. 637

[7] Kryukov N E Koval'skii I N Kozyrev N A Igushev V F Kryukov R E Manufacture of vertical bulk –oil storage tanks for northern climates using special welding materials *Steel in Translation* 2012 V 42 № 2 p 118-120

[8] Thermodynamical properties of substances: Reference book. V.1. Issue 1 / Edited by V.P. Glushko, L.V. Gurvich et al. M.: Nauka 1978 p 22

[9] Konishchev B P Kurlanov S A Potapov N N et al. Welding materials for arc welding: CReference book in 2 volumes. V. 1. *Shielding gases and welding fluxes* Edited by Potapov N.N. - M.: Machinebuilding 1989 p 104

[10] John L Haas Jr Gilpin R Robinson Jr and Bruse S Hemingway J. Phys. Chem. Ref. Data. – 1981. – Vol. 10. – № 3. – p. 575 – 669

[11] NIST-JANAF Thermochemical Tables 1985. Version 1.0 [Electronic resource] : data compiled and evaluated by M.W. Chase, Jr., C.A. Davies, J.R. Dawney, Jr., D.J. Frurip, R.A. Mc Donald, and A.N. Syvernd. – Available at: http://kinetics.nist.gov/janaf.

[12] Patent 2467853 RF, MPK 8 V23 K35/362 Ceramic flux-additive / Kryukov N.E., Kovalsky I.N., Kozyrev N.A., Igushev V.F., Krykov R.E.; Open Joint Stock Company OAO «Novokuznetsk Plant of Reservoir Metalware» named after N.E. Kryukov.- № 201112341602/02(034654), Application 08.06.2011

[13] Patent 2484936 PF, MPK 8 V23 K35/362 Ceramic flux-additive / Kozyrev N.A., Igushev V.F., Kryukov R.E., Goldun S.V.; FSBEI HPE "Siberian State Industrial University".- №2012104939/02(007484), Application 13.02.2012

IOP Conf. Series: Materials Science and Engineering **91** (2015) 012009 doi:10.1088/1757-899X/91/1/012009

Surfacing With Tungsten-containing Ores

N A Kozyrev, G V Galevsky, D V Valuev, V M Shurupov, O E Kozyreva

Federal State Budgetary Educational Institution «Siberian State Industrial University» 654007, Novokuznetsk, 42, Kirov str.

E-mail: kozyrev_na@mtsp.sibsiu.ru

Abstract. Experimental research has been carried out into AN 26C flux surfacing with flux cored wires made of dust-like carbon and fluorine containing components and tungsten-containing ores. The investigations have revealed the fact that tungsten can be reduced immediately from tungsten-containing oxide materials. The chemical composition of added metal has been determined, as well as slag compositions after surfacing; hardness, wear resistance have been investigated and metallographic tests have been carried out. Thermodynamic calculations of reactions of WO3, as well as W2C and WC carbides reduction to W by carbon and carbon oxide have been made in standard conditions at temperature T = 1500 – 6000 K. The obtained thermodynamic characteristics of reactions have demonstrated that in the course of WO3 reduction tungsten is most likely to be obtained in terms of thermodynamics, subsequently W2C and WC carbides. After these reactions those of W2C and WC carbides obtaining are thermodynamically possible via reduced tungsten and carbon composition.

Introduction

Efficient use of tungsten is of particular importance due to its high value, shortage and high production cost. Flux cored wires containing tungsten are widely applied for surfacing high wear resistant steels; reduced tungsten is used as a filling material in these wires in the form of ferroalloys, alloys and metal powder of various grades. Moreover, silicides, borides, carbides and other compounds of tungsten are applied to achieve definite service characteristics. W_2C and WC are tungsten carbides, which are used in wires of this kind. The technology of flux cored surfacing is relevant for practical application, which presupposes tungsten oxide and carbon-containing reducer as a filler material. Theoretically, tungsten and its carbides are possible to be obtained in direct reduction.

This paper submits thermodynamic calculations and provides consideration of the possibility to reduce tungsten oxides from flux cored wires by a carbon-containing reducer in the course of flux surfacing. Tungsten concentrate of KSH-4 grade according to State Standard 213-83 produced at the open joint stock company "Mining company "AIR" with 54% of WO_3 concentration was used in experiments, as well as ores containing 0.372% of WO_3 and enriched rejects with 0.07% concentration of WO_3. Dust-like carbon and fluorine containing wastes of metallurgical production were used as a carbon-containing reducer; chemical composition in mass concentration (%) is as follows: Al_2O_3 =21-43.27; F=18-27; Na_2O = 8-13; K_2O =0.4 – 6%; CaO= 0.7- 2.1; SiO_2 = 0.5-2.48; Fe_2O_3 = 2.1-2.3; C_{gen} = 12. 5-28.2; MnO = 0.03-0.9; MgO = 0.04-0.9; S= 0.09-0.46; P=0.1-0.18.

The purpose of research is to determine the potential of tungsten reduction by carbon as the result of which W_2C and WC carbides form when surfacing.

IOP Conf. Series: Materials Science and Engineering **91** (2015) 012009 doi:10.1088/1757-899X/91/1/012009

Results and discussion

Previously conducted investigations [1-6] have demonstrated that carbon and fluorine containing dust-like wastes of metallurgical production make it possible: 1) to extract hydrogen by fluorine-containing compounds (like Na_3AlF_6, CF_x $(1 \geq x > 0)$ etc.), as the result, a gaseous compound HF forms; 2) to deoxidize a weldpool by carbon in form of CO and CO_2, which are generated when carbon fluoride CF_x $(1 \geq x > 0)$ reacts with dissolved in steel oxygen; 3) to improve the stability of arc combustion by means of potassium and sodium making ionization in arc column easier. Therefore, tungsten reduction can be made by carbon according to the reaction and reacts with tungsten, as the consequence, W_2C and WC carbides are generated.

Table 1 provides compositions of fusion mixtures of flux cored wires under consideration, as well as block coefficients of manufactured wires. The selected correlation is accepted on the ground of stoichiometric calculations provided that complete reduction from tungsten-containing ores both of tungsten, silicon and manganese is possible to be carried out by carbon, which is a part of dust-like metallurgical wastes. The following correlation in the wire is accepted: 111 – according to stoichiometric calculations, 112 – double surplus of carbon, 331 – double shortage of carbon, 441 – triple surplus of carbon, 551, 661 – according to stoichiometric calculations for ore and rejects.

AN 26C flux surfacing was carried out by tractor ASAW 1250. Surfacing conditions are provided in Table 2.

Chemical composition of added metal was determined by roentgen-fluorescent method with spectrometer XRF-1800 and atomic emission method with spectrometer DFS-71. Hardness was measured with ultrasonic hardness testing instrument UZIT - 3.

Chemical composition of added metal and hardness (after surfacing and thermal treatment) are provided in Table 3. As one can see from the aforementioned data of Tables, tungsten has been reduced completely in all samples of wires, except sample 331 (extraction was approximately 96.31%) because of the shortage of a reducer.

Metallographic investigations into micro-sections were carried out with optical microscope OLYMPUS GX-51 in a bright field at various zooming in alcoholic solution of nitric acid, and in solution of hydrofluoric acid.

Metallographic analysis of added layers has revealed the following facts: sample 111 (Figure 1) has a definitely pronounced Widmanstatten pattern with thin needles deviating from ferritic net, which is located on the edges of initial austenite grains. The dimensions of grains correspond to 3 - 4 points.

Table 1. Component composition and block coefficient of manufactured wires

Wire	Component, mass fraction, %.				Block coefficient of a wire, %
	1	2	3	4	
111	48.54	51.46	-	-	12.37
112	31.85	68.15	-	-	9.36
331	66.67	33.33	-	-	12.86
441	25	75	-	-	10.76
551	-	50	50	-	8.72
661	-	50	-	50	10.95

1 - WO_3 concentrate, 2 – carbon and fluorine containing dust-like wastes of metallurgical production, 3 – tungsten ore, 4 – rejects after reduction of tungsten ores.

Table 2. Surfacing conditions

Stamp	$I_н$ (A)	$U_д$(W)	$V_н$(m/h)	The number of added layers
111	490	30	28	5
112	490	32	27	5
331	410	27	24	5

Stamp	$I_н$ (A)	$U_д$(W)	$V_н$(m/h)	The number of added layers
441	430	28	24	5
551	420	28	24	5
661	420	28	25	5

Table 3. Chemical composition and hardness of added metal

Sample	Mass fraction of elements, %										
	C	Si	Mn	Cr	Ni	Cu	Ti	W	V	Mo	Co	Hardness HRC*
111	0.35	0.31	0.49	0.07	0.12	0.13	0.001	0.67	0.008	0.005	0.007	27.2
112	0.29	0.25	0.50	0.07	0.12	0.12	0.001	1.29	0	0.002	0.006	19.2
331	0.13	0.15	0.43	0.06	0.12	0.11	0.001	1.90	0.01	0.03	0.007	30.4/ 52
441	0.26	0.31	0.43	0.06	0.12	0.10	0.002	0.58	0.01	0.03	0.006	26.2/ 47
551	0.11	0.34	0.38	0.05	0.12	0.09	0.001	0.02	0.01	0.03	0.005	21.6
661	0.23	0.46	0.33	0.05	0.12	0.08	0.001	0.01	0.01	0.02	0.005	22.1

*- numerator: hardness after surfacing, denominator – after thermal treatment

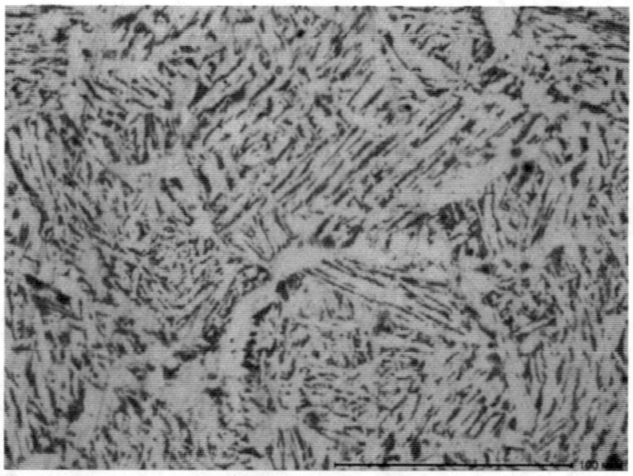

Figure 1. Sample 111

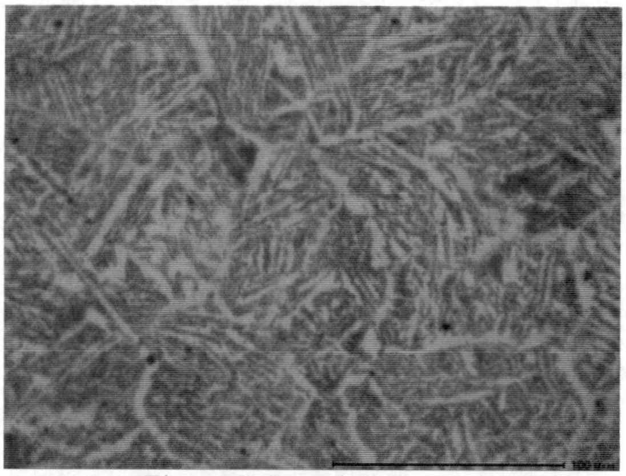

Figure 2. Sample 112

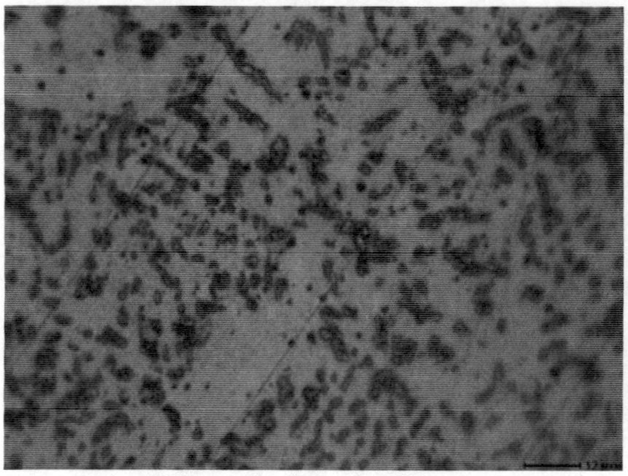

Figure 3. Sample 331

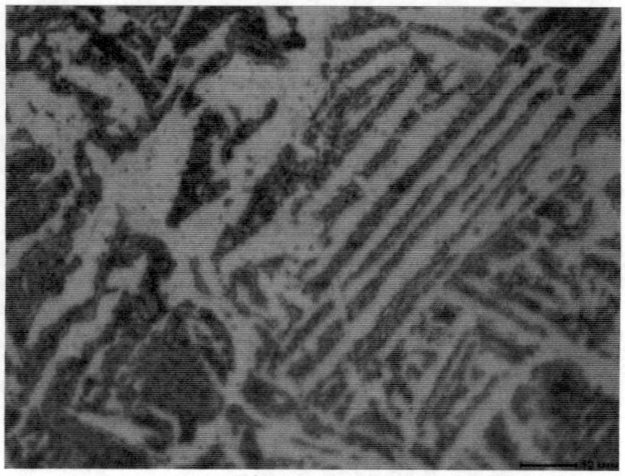

Figure 4. Sample 441

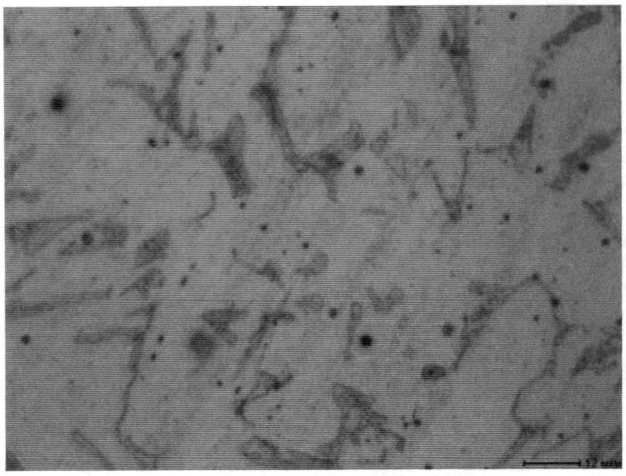

Figure 5. Sample 551

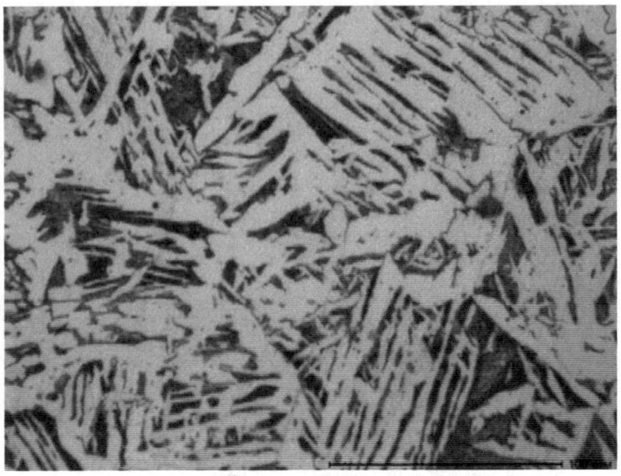

Figure 6. Sample 661

Sample 112 (Figure 2) has Widmanstatten pattern with thin needles deviating from ferritic net, which is located on the edges of initial austenite grains. The dimensions of grains correspond to 4 points.

Sample 331 (Figure 3) has a granular ferrite – carbide structure. The dimensions of carbides are within the range of 1 - 5 micrometer.

Sample 441 (Figure 4) has Widmanstatten pattern with a significant number of thin needles inside the grains. The dimensions of grains correspond to 3 points.

Sample 551 (Figure 5) has a structure, which consists mainly of ferrite with small localizations of perlite.

Sample 661 (Figure 6) has Widmanstatten structure with bulky needles and thick ferrite net. The dimensions of grains correspond to 3 points.

Therefore, the investigations have revealed a principal potential of tungsten oxides reduction from ores by a carbon-containing reducer and generation of tungsten carbides in an added layer within the range of 1-5 micrometers.

The following reactions have been considered to obtain a thermodynamical assessment of WO_3 oxide reduction by carbon and carbon oxide CO:

$$WO_3 + 3C = W + 3CO \quad (1)$$
$$WO_3 + 3/2C = W + 3/2CO_2 \quad (2)$$
$$WO_3 + 3CO = W + 3CO_2 \quad (3)$$
$$W + C = WC \quad (4)$$
$$W + 1/2C = 1/2W_2C \quad (5)$$
$$WO_3 + 5/2C = WC + 3/2CO_2 \quad (6)$$
$$WO_3 + 2C = 1/2W_2C + 3/2CO_2 \quad (7)$$
$$WO_3 + 5CO = WC + 4CO_2 \quad (8)$$
$$WO_3 + 4CO = 1/2W_2C + 7/2CO_2 \quad (9)$$

Thermodynamic characteristics of reactions necessary to evaluate the properties of reduction (1) – (9) in standard conditions [$\Delta_rH°(T)$, $\Delta_rS°(T)$, $\Delta_rG°(T)$] were calculated by the known methods [7] in the range of arc temperatures (1500 – 6000) K according to thermodynamic characteristics [[H°(T) - H°(298.15 K)], S°(T), $\Delta_fH°$(298.15 K)] of reagents WO_3, W, C, CO, CO_2, as they are in reference book [8]. Thermodynamic properties of WC and W_2C carbides depending on the temperature as they are in Tables, accepted in [8], were calculated separately with regard to the data from reference books [9, 10]. Here, the following states were selected as standard ones for reacting substances in the range (1500 – 6000) K: $W_{(s,l)}$, $WO_{3(s,l)}$, $WC_{(s)}$, $W_2C_{(s)}$, $C_{(s)}$, $CO_{(g)}$, $CO_{2(g)}$.

Calculated standard Gibbs energies of reactions (1)–(9) are shown in Table 4 and in Figure 7.

Table 4. Standard Gibbs energy of reactions (1) – (9) in dependence on temperature

Reaction	$\Delta_r G°(T)$, kJ									
	1500K	2000K	2500K	3000K	3500K	4000K	4500K	5000K	5500K	6000K
(1)	-258.35	-490.24	-703.90	-911.09	-1113.40	-1315.15	-1514.59	-1710.28	-1902.54	-2091.62
(2)	-121.56	-226.64	-315.92	-400.80	-482.60	-565.47	-647.50	-727.11	-804.56	-880.01
(3)	15.23	36.97	72.06	109.48	148.20	184.21	219.60	256.05	293.41	331.59
(4)	-47.76	-54.97	-63.57	-73.34	-83.87	-91.57	-98.83	-107.52	-117.72	-129.45
(5)	-35.11	-47.77	-61.80	-76.87	-92.47	-104.97	-116.69	-129.50	-143.40	-158.42
(6)	-169.32	-281.61	-379.48	-474.15	-566.48	-657.04	-746.32	-834.64	-922.29	-1009.46
(7)	-156.67	-274.41	-377.71	-477.67	-575.08	-670.44	-764.19	-856.61	-947.96	-1038.43
(8)	58.66	157.73	267.15	376.33	484.86	592.43	698.84	803.97	907.67	1009.88
(9)	25.72	77.06	139.60	202.71	265.99	329.13	391.94	454.28	516.00	577.05

The data show that reaction (1) is the most thermodynamically possible reaction of WO_3 reduction – reduction by carbon to tungsten and CO, followed by reactions (7) and (6) – these ones of obtaining W_2C and WC carbides, respectively, and CO_2. Reaction (2) of WO_3 reduction by carbon to tungsten and CO_2 is the forth likely one. Then, carbide formation follows from tungsten and carbon to W_2C and WC – reactions (5) and (4). The next likely possible reactions are reaction (3) of WO_3 reduction by carbon oxide (II) to tungsten, as well as reaction (9) to W_2C. The least possible reaction is reaction (8) – WO_3 reduction by CO to WC carbide. Thermodynamic probability of reactions (1), (2), (4) – (7) increases as temperature grows, and that of reactions (3), (8), (9), falls, on the contrary.

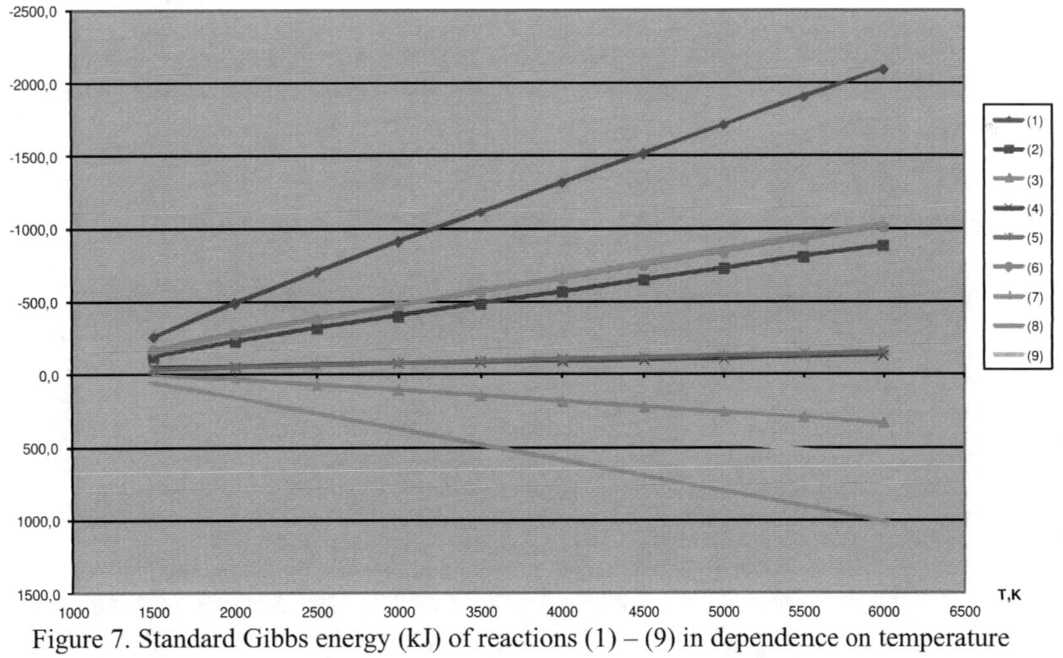

Figure 7. Standard Gibbs energy (kJ) of reactions (1) – (9) in dependence on temperature

Conclusions

1. In terms of carried out calculations obtaining of tungsten is thermodynamically the most likely one as the result of WO_3 reduction by carbon, followed by W_2C and WC carbides. Then, W_2C and WC carbides are thermodynamically possible to be obtained via reaction of tungsten with carbon.

2. Provided that a reducer is in a plenty tungsten carbide can be obtained when surfacing with flux cored wire containing tungsten oxides. Here, practical experience has revealed the potential of

tungsten oxides reduction from ores by a carbon containing reducer, as the result, tungsten carbides of grade $(Fe,W)_6$ are obtained, the dimensions are in the range 1-5 micrometers.

The paper is written at Siberian State Industrial University as a project part of State Order of the Ministry of education and Science of the Russian Federation № 11.1531.2014/k.

References

[1] Kozyrev N A Igushev V F Starovatskaya S N Kryukov R E Goldun Z V Influence of carbon and fluorine containing additives for welding fluxes on properties of welds *Proceedings of universities Ferrous metallurgy* 2012 № 6 p 26 – 29

[2] Kozyrev N A Igushev V F Goldun Z V Kryukov R E Shurupov V M Application of carbon containing additives for welding fluxes *Proceedings of universities Ferrous metallurgy* 2012 № 10 p 35 – 38

[3] Kozyrev N A Igushev V F Kryukov R E Goldun Z V Kovalsky I N Influence of carbon and fluorine containing additives in fluxes on concentration of non-metallic impurities and characteristics of welds *Welding production* 2012 № 12 p 3-6

[4] Kozyrev N A Igushev V F Kryukov R E Starovatskaya S N Roor A V Influence of AN-60 flux and carbon and fluorine containing additive on quality of welds of 09G2S steel *Proceedings of universities Ferrous metallurgy* 2013 № 4 p 30-33

[5] Kozyrev N A Igushev V F Kryukov R E Roor A V Kovalsky I N Design of additives for welding fluxes in conditions of low-alloyed steels *Welding production* 2013 № 5 p 9 - 12

[6] Kozyrev N A Igushev V F Kryukov R E Roor A V Investigation in influence of carbon and fluorine containing additive in flux AH-67 on properties of weld metal of steel 09G2S *Proceedings of universities Ferrous metallurgy* 2013 № 8 p 33-36

[7] Thermodynamic properties of separate substances: Reference book V 1 Issue 1 / Edited by V P Glushko L V Gurvich et al M : Science 1978 p 22

[8] NIST-JANAF Thermochemical Tables 1985 Version 1 0 [Electronic resource]: data compiled and evaluated by M W Chase Jr C A Davies J R Dawney Jr D J Frurip R A Mc Donald and A N Syvernd – Mode of access: http://kinetics nist gov/janaf

[9] Thermal constants of substances: Reference book Issue 7 / Edited by V P Glushko V A Medvedeva et al M: Science 1978 343p

[10] Ruzinov L P Gulyanitsky B S Balanced transformations of metallurgical reactions – M: Metallurgy 1975 416p

IOP Conf. Series: Materials Science and Engineering **91** (2015) 012010 doi:10.1088/1757-899X/91/1/012010

Corrosion and Mechanical Properties of Austenic Steel Weld Joints

M A Kuznetsov [1,a], E A Zernin [1,b], V I Danilov [1,2,b], D E Kolmogorov [1,d], L N Zoubenko [1,e]

[1]Yurga Institute of Technology National Research Tomsk Polytechnic University
26, Leningradskaya st., Yurga, Kemerovo region, 652055, Russia,
Ph. (38451)-5-09-06
[2]Institute of Strength Physics and Materials Science 634055, SB RAS Tomsk, Russia, Academic Prospect 2/4

e-mail: [a]kyznechik_85@mail.ru, [b]yuti_sp@bk.ru, [c]dvi@ispms.tsc.ru, [d]dek-79@mail.ru
[e]dmitkarcev@mail.ru

Abstract. This paper presents results of experiments on how tungsten, molybdenum and aluminum oxyhydroxide nanopowders, imbedded into the weld pool, affect corrosion resistance and mechanical properties of welded joints. It is shown that nanopowders have a significant effect on the intergranular corrosion of the weld.

1. Introduction

Most intensive corrosive damages occur in welded joints of high-alloyed, corrosion-resistant steels when compared with the base metal; this is due to the high electrochemical heterogeneity of weld metal chemical composition as well as its structure, properties and stress state.

The main difficulties of weldability of these steels are based on their multicomponent alloying and variety of operating conditions of welded products. The main disadvantage is the tendency to hot cracking, intergranular with character in the weld metal and heat affected zone (HAZ) [1]. This is the main cause of damage of pipelines, chemical, metallurgical, power, nuclear industry equipment, heat transfer fluids [2, 3].

The austenite phase is exposed to corrosion destruction. A corrosion attack develops in the weld zone more intensively if ferrite grains are larger and the austenite phase is smaller. Austenitic steels (chrome-nickel) are destroyed by corrosion along the fusion line [4]. This steel is unstable to local destructions in the HAZ and corrosive wear in comparison with other steels of this class [5, 6, 7]. In [8] it is confirmed that mechanism of HAZ metal embrittlement is associated with development of softening of the austenite grain boundaries, because this initial structure damages occur in the area of large grains during arc welding.

During operation, there is an active corrosion wear flowing particularly intense in the weld areas [9].There are several ways to improve the corrosion wear resistance of welded joints of austenitic steels as well as to restore the equipment operating in corrosive environments.

Intergranular corrosion (IGC) can be reduced if welding energy parameters are controlled by applying a high-frequency pulses [10], transition of droplets into the molten pool is regulated using a pulsed wire feed [11], inverter power sources are applied [12], the weld pool is protected with gas [13].

Welds will have a high resistance to hot cracking and intergranular corrosion if the weld metal has a two-phase austenitic-ferrite structure. Weld metal alloying with elements-ferritizers was one of the first ways of control over intergranular corrosion; but experiments have shown that this alloying does not eliminate development of corrosion, if the metal is subjected to a prolonged heating [14].

Resistance to intergranular corrosion after prolonged heating can be saved by applying a stabilizing annealing [15] or, for example, weld metal alloying with rare earth and alkaline earth elements, nickel [16]. One more method of controlling over corrosion is a laser processing of welds. [17].

Also [1] presents researches on how electrode coating affects the corrosion resistance of the weld metal, which showed that the deposited weld metal, made with basic electrodes, has a higher corrosion resistance. This paper proposes a method to improve corrosion resistance of welded joints by dispensing nanostructured particles through the shielding gas to the weld pool by means of the device described in [18]. It is also necessary to evaluate the mechanical properties of welded joints, as they also depend on the efficiency of welded steel structures.

2. Research methods

Experimental study used nanopowders (NP) of oxide Al (Al2O3), tungsten (W) and molybdenum (Mo) produced in Institute of High Technology Physics of National Research Tomsk Polytechnic University. Powders are produced by electro-explosive technology which has been developed in the above Institute [19].

Conductor electric explosive (CEE) technology is a pulsed fast processes and has several advantages such as: ability to transfer a high-density energy of desired amount to the substance; energy supplied in a pulsed mode is used with great efficiency; high rate of system thermodynamic parameters change; ability of subtle influence on material structure and formation of the structure of individual particles.

Of particular interest are the NP, obtained by the CEE. Technique and fundamentals of technology for production of NP with controlled properties by CEE method are developed at the Institute of High Technology Physics of TPU (formerly High Voltage Research Institute) [19].

This method provides pure metal powders, and powders of various metal-based compounds (carbides, oxides, nitrides, sulfides, et al.). Particles have spherical shapes, particle sizes are distributed according to the normal-logarithmic law and an average particle size is in the range 100 - 500 nm. Particles are poly-crystals; the magnitude of structural fragments is in the range of 20-30 nm. Much of the material is in the X-ray amorphous state.

NP of aluminum (Al), tungsten (W), molybdenum (Mo), obtained by the of CEE technology, are among the most interesting representatives of this class of NP. At present a number of practical applications of Al NP has been successfully tested as: a starting material for nanoscale fibers of Al oxide-hydroxide phases as the main active components of filtration and sorption materials for purification of water [20]; active supports for catalysts in oil refining and petrochemical industries [21, 21]. NP of W and Mo are used as feedstocks for disulfides of the metals having high heat resistance and good tribological properties [21] as modifying additives for catalysts in oil refining and petrochemical industries [22-24]

To produce the metal NP, an aluminum wire of "AM" grade W=0.35 mm, a tungsten and a molybdenum wire of "VA" grade W=0.31 mm were used [20, 21]

The NP specific surface area W was 2.6 m2/g. That corresponds to the surface particle average diameter of 122 n. Al NP had primary 10 nm-sized particles, which combined into aggregates with dimensions up to 500 nm, and agglomerates up to 5 μm with weak coupling between the units. Average surface size of Al NP particles was 100 nm. Al NP was previously passivated in air wherein the content of metallic aluminum was not less than 91% by weight. Then Al NP was subjected to thermal hydrolysis to produce Al oxyhydroxide (AlOOH) nanofibers. A procedure for preparation of AlOOH nanofibers, used in this work, is described in [25].

To obtain NP by electro-explosive method, pulse current of high density (1010A/m2) is passed through a metal wire, whereby the conductor explosively collapses; combustion products condense in an inert gas atmosphere to form nanoscale particles.

Powders obtained by this method have a particle size of 10 ÷ 150 nm (agglomerates of sizes up to 500 nm), the specific surface area of 2 ÷ 50 m2/g and have a high chemical activity. Depending on the kind of gas surrounding the conductor, it is possible to produce powders of metals and alloys, powders of chemical compounds and powders of composite structures [25]. Procedure for preparation of Al oxide-hydroxide phases, used in this work, is described in [25].

Steel plates of 5 mm in thickness (chemical composition of steel: 0.12% carbon, 18% chromium, 10% nickel, 1-15% titanium) were used to study corrosion resistance of welded joints;

Steel plates of 10 mm in thickness welded with consumable electrode, in argon, with welding wire (chemical composition of the wire: 0.12% carbon, 18% chromium, 9% nickel; 1-15% titanium) were used to study mechanical properties of the joints.

Equipment used: welding rectifier with rated current of 300A, a pendant welding head, an automatic control unit.

The samples were processed in four ways: №1 - welding in argon with solid section wire; №2 - welding in argon with solid section wire, with addition of tungsten nanopowder (W) to the protective gas; №3 - welding in argon with solid section wire, with addition of aluminum oxide nanopowder (Al2O3) to the protective gas; №4 - welding in argon with solid section wire, with addition of molybdenum nanopowder (Mo) to the protective gas.

Control samples and test samples were selected from each group. Control samples were previously subjected to chemical etching at an ambient temperature in aqueous solution of nitric acid and ammonium fluoride to remove welding slag.

The corrosion resistance of samples was tested in a solution of sulfuric acid and copper sulfate in the presence of metallic copper and sodium fluoride or potassium fluoride. Reagent and mode of action was chosen according to the recommendations for this grade of steel. Test duration was 2 hours. Whereupon a condition of intergranular corrosion was tested by metallographic method; for this metallographic thin sections were prepared from control samples not subjected to the test and after the test samples. A plane of the section was perpendicular to the weld and included the weld metal, HAZ and the base metal. Microstructure was revealed by etching in an electrolyte until the grain boundaries. Optical microscope NEOPHOT-21 and confocal laser scanning microscope LEXTOLS4000 were used to record the results of metallographic examination.

Mechanical properties of welded joints were determined in accordance with generally accepted techniques. Such mechanical properties of welded joints were tested as tensile ultimate strength, yield strength and relative elongation. Equipment used: universal testing machine, at t +20°C and +500°C.

3. Results and Discussion

Microstructures analysis shows that all the samples have austenitic grain structure with a grain size approximately identical. However, at an equal heat-on period, grain boundaries of sample №1 were etched maximally, grain boundaries of sample №2 were etched minimally. Tendency to IGC was evaluated by determining the number of grains with clearly identified boundaries, the width of which could be up to 30 microns. Grain boundaries were not etched in the heat affected zone. This means that the standard test revealed no intergranular corrosion in any sample. The etching process occurred over the entire surface of grains, not only at grain boundaries. Depending on grains orientation their surfaces were etched stronger or weaker. Thus, some kinds of steps were formed between the grains (Fig. 1).

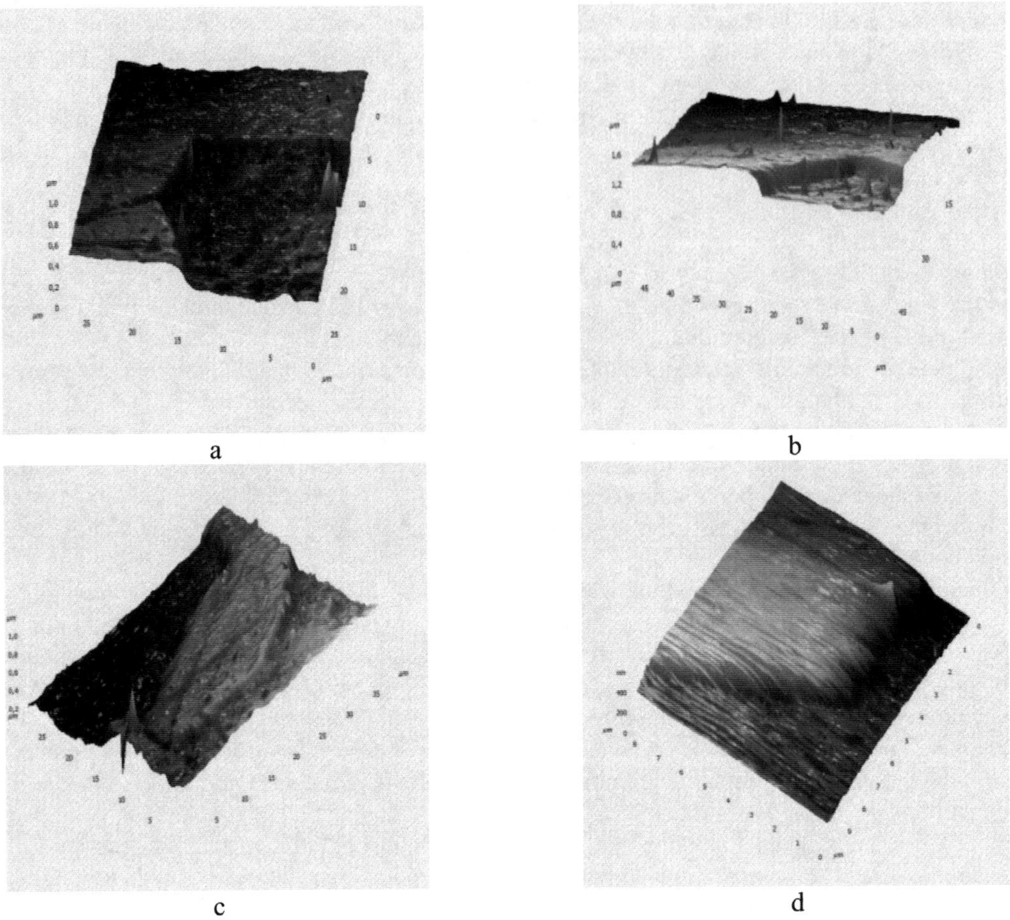

Figure 1. Microstructure of samples after test: a- sample №1; b - sample №2; c - sample №3; d - sample №4.

The average height of these steps on the samples: №1 - 320 nm, №2 - 200 nm, №3 - 275 nm, №4 - 250 nm. The finest grain boundaries were in sample №2, which also shows the greatest similarity between the reference sample and the microstructure of the sample after the test in a corrosive environment. When analyzing the weld metal, it was possible to evaluate the interdendritic spaces. They were strongly etched and grooves were formed as a result.

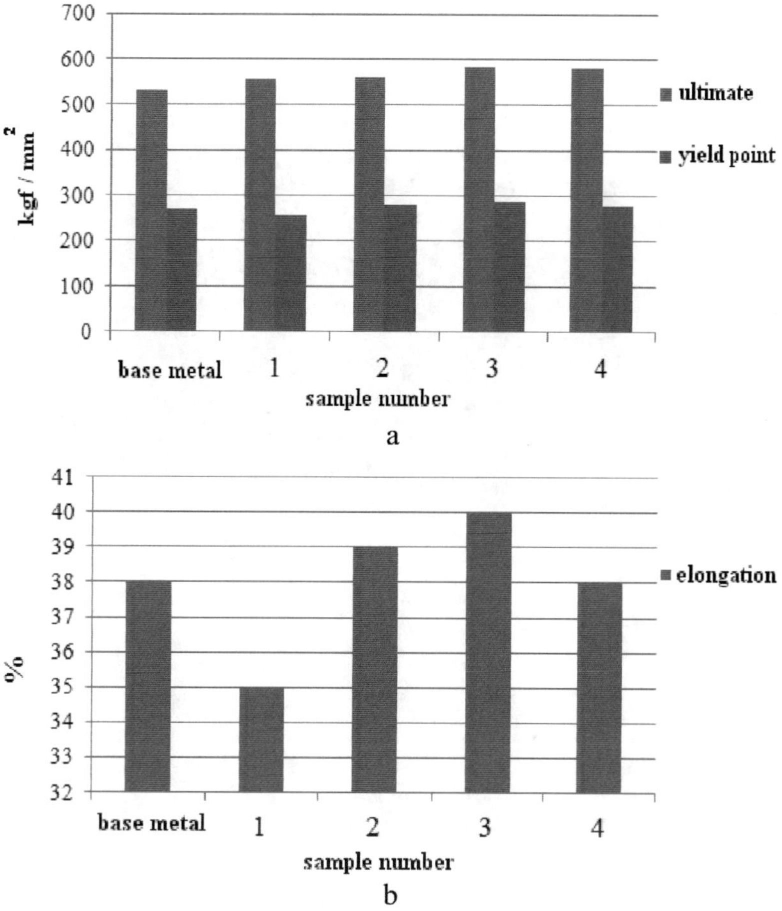

Figure 2. Mechanical properties of welded joints at +20°C:
a – tensile ultimate strength, yield strength; b – relative elongation.

Evaluation of mechanical properties (tensile ultimate strength, yield strength and relative elongation) of weld joints was made at a temperature of +20°C and +500°C. Mechanical properties of welded joints at +20°C are presented in Figure 2.

One of the characteristics of these steels is that they operate at higher temperatures, so studies were conducted at a temperature of +500°C. Mechanical properties of welded joints at +500°C are shown in Figure 3.

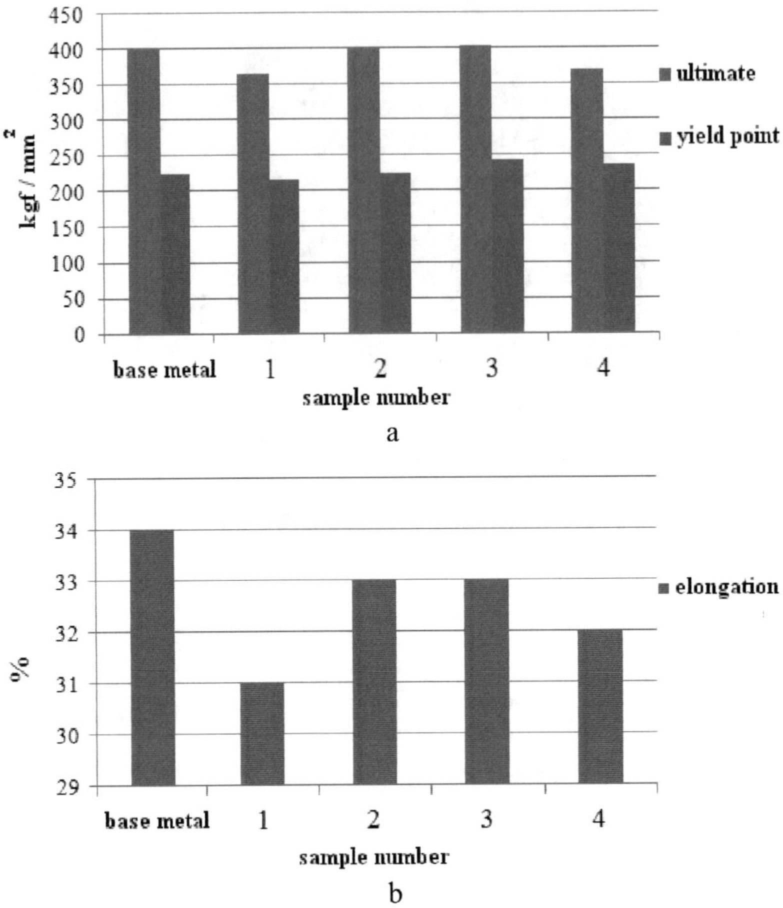

Figure 3. Mechanical properties of welded joints at +500°C:
a – tensile ultimate strength, yield strength; b – relative elongation.

Figures 2 and 3 show that introduction of nanostructured powder modifiers to the liquid weld pool increases the mechanical properties of welded joints; at t +20°C an average increase is 4 - 11%, at a temperature of +500°C an average increase is 3 - 10%. The sample with Al2O3 content showed the best mechanical properties. Study [26] proves that mechanical properties are better for weld metal samples with less dendrite structure and less coarse dendrite structure. The most equilibrium dendrite structure is achieved with the use of nanostructured powder Al2O3.

4. Conclusion

It is found that, if any of the abovementioned nanostructured powders is added into the liquid weld pool, intergranular corrosion does not occur in a single sample. Corrosion resistance of welded joints increases by 15-20%, mechanical properties of welded joints increases at +20°C by 7.5%, at +500°C by 6.5%, if compared with consumable electrode welding in argon without nano-agents.

References

[1] Krivonosova E A, Sinkina E A and Gorchakov A I 2012 Welding production Influence of electrode coating type on corrosion resistance of weld steel 08X18H10T 4 pp 38-41.

[2] Kolomiitsev E V 2012 Automatic welding Corrosion fatigue strength of 12X18H10T steel T-joints and methods of its improvement 2 pp 41-43.

[3] Makhnenko V I, Makhnenko O V, Kozlitina S S and Dzyubak LI 2012 Automatic welding Weldments of austenitic steel 10X1H10T under radiation swelling 2 pp 7-12.
[4] Kakhovsky N I and Yushchenko K A 1963 Automatic welding Influence of welding heat on microstructure and properties of steels 21-3 and 21-5 10 pp 16-25.
[5] Poletaev Y V and Zubchenko A S 1989 Welding production Tendency of from chromium-manganese and chromium-nickel austenitic steel welded joints to local destruction 10 pp 11-13.
[6] Poletaev Y V, Penkov V B, and Feklistov S I 1989 Automatic welding Tendency of austenitic steels to hot cracking during welding and local damage at low cycle creep 2 pp 8-14.
[7] Poletaev Y V and Zubchenko A S 1987 Welding production Structure and properties of chromium-manganese steel welded joints 7 pp 20-22.
[8] Poletaev Y V 2012 Welding production Effect of chemical microinhomogeneity on tendency of HAZ to local destruction 3 pp 10-13.
[9] Saraev Y N and Selivanov Y V 2011 Welding production Evaluation of performance and quality of repair of equipment made of 12X18H10T steel 1 pp 17-22.
[10] Krampit A G, Krampit N Y and Krampit M A 2013 Welding International Effect of parameters of pulsed arc welding on formation of root layers in narrow gaps 66(8) pp 3-5.
[11] Pavlov N V, Kryukov A V, Zernin E A and Polishchuk V A 2014 Applied Mechanics and Materials Modelling of welded joint geometry in welding of metal sheets 682 pp 443-447.
[12] Ilyashchenko D P, Chinakhov D A and Gotovshchik Yu M 2015 Applied Mechanics and Materials Calculation of Heat Content of Electrode Metal Droplet for Manual Arc Welding if Applying Power Supplies With Different Volt-Ampere Characteristic 56 pp 101-104.
[13] Chinakhov D A 2015 Applied Mechanics and Materials Dependence of silicon and manganese content in weld metal on welding current and method of gas shielding 756 pp 92-96.
[14] Kopersak N I 1963 Automatic welding Effect of alloying elements on 475° fragility of austenitic-ferritic weld metal 7 pp 16-20.
[15] Nechayev V A and Juncker S V 1963 Welding production Effect of 1Kh18N9T steel and its weld content on intergranular corrosion resistance after 500°C heating. // Automatic welding 7 pp 34-40.
[16] Alexandrov A G 1990 Welding production Corrosion resistance of chromium-nickel steel welded joints in alkaline environment (review) 5 pp 12-13.
[17] Safonov A N 1998 Welding production Structure and corrosion resistance of austenitic steel welded joints after surface laser treatment 7 pp 5-9.
[18] Kuznetsov M A, Zhuravkov S P, Zernin E A, Kolmogorov D E and Yavorovsky N A 2014 Advanced Materials Research Effect of Nanostructured Powder Modifiers on Structure of Welding Bead 872 pp 118-122.
[19] Yavorovsky N A 1996 Proceedings of universities. Physics Production of ultrafine powders using electric explosion 4 pp 114 - 136.
[20] Yavorovsky N A, Shiyan L N, Savelyev G G and Galanov A I 2008 Nanotechnics Modification of polymeric membranes with nano aluminum oxyhydroxide 3 (15) pp 40-45.
[21] Irtegov Y A, Ahn V V and Yavorovsky N A 2012 Proceedings of universities. Physics Producing of nanostructured sulfides of tungsten, molybdenum, iron, copper 5/2 pp 140 - 145.
[22] Vosmerikov A V, Echevsky G V, Korobitsina L L, .Barbashin J E, Zhuravkov S P, Arbuzov N V and Kodenev E G 2005 Kinetics and Catalysis Decontamination of Mo-containing zeolites during non-oxidative conversion of methane 5 (46) pp 769-772.
[23] Vosmerikov A V, Echevsky G V, Korobitsina L L, Arbuzov N V, Zhuravkov S P, Kodenev E G, and L M 2007 Kinetics and Catalysis Velichkina Non-oxidative conversion of methane to aromatics hydrocarbons on W-containing pentasils 3 (48) pp 432-437.
[24] Vosmerikov A V, Zaikovskii V I, Korobitsina L L, Kozlov V V, Arbuzov N V and Zhuravkov S P 2011 Kinetics and Catalysis Study on conversion of methane to aromatic hydrocarbons

over catalysts Ag-Mo / ZSM-5 3 (52) pp 438-444.

[25] Yavorovsky N A, Pustovalov A V, Lobanov G L, and Zhuravkov S P 2012 Proceedings of universities. Physics Study on properties of aluminum powders produced in argon with addition of oxygen 6/2 (55) pp 236 - 244.

[26] Kuznetsov M A, Zernin E A, Danilov V I and Kartsev D S 2013 Applied Mechanics and Materials Application of nanostructured powders to control characteristic of electrode metal transfer and process of weld structuring 379 pp 199-203.

Influence of Frequency and Induction of Longitudinal Magnetic Field on The Electrode Metal Loss and its Spattering During MAG-Welding

D G Nosov, V V Peremitko

Dneprodzerzhinsk State Technical University, Dneprodzerzhinsk
51918 Dneprodzerzhinsk, Ukraine, Dneprostroevskaya 2 street,
tel. (+380569) 55-21-07

e-mail: ndweld@rambler.ru

Abstract. It is possible to increase the MAG-welding efficiency by controlling the electrode metal mass transfer at the reduction of discharge coefficient on spattering by influence of longitudinal magnetic field on the arc. The paper identifies a range of longitudinal magnetic field frequencies and induction which provide the discharge coefficient reduction of the electrode metal; it has also been found the characteristics of their mutual influence on electrode metal mass transfer process; mathematical models correlating the frequency and induction of longitudinal magnetic field length with loss coefficient of electrode metal on spattering are presented; technological recommendations, the implementation of which will allow to improve the efficiency of MAG-welding in industrial environments, are given.

Introduction

Mechanized welding method in active shielding gas by solid cross-section wire - MAG-welding – certainly has a number of advantages over the manual arc welding process. However, along with the advantages this method has drawbacks that reduce the effectiveness of its use: the low productivity compared to the automatic welding methods; quality dependence of the welded joint on the skill of the welder; significant metal loss on spattering, which constitutes up to 10 ... 20% of welding wire weight.

Spattering is accompanied by the ejection from arc zone the sprays of molten metal of various size, which come into physical and chemical interaction with the surface layers of the weld metal [1]. The protection of metal surface from spray sticking and / or its cleaning leads to the need of additional works in a volume of 20 ... 40% of the overall complexity of the welding operations [2].

Among the main reasons of metal droplets ejection from the welding zone the following are pointed out: unstable metal transfer, when the power which separetes a drop from the electrode is directed away from the bath and a drop is ejected beyond its bounds; local explosive gases expelling in the volume of metal of welded bath caused by metallurgical reactions; destruction of molten metal bridge formed during metal transfer with the short circuits as a result of a sharp increase in current density under narrowing of the tie plate (pinch effect) [3].

The metal spattering intensity depends on the metal wire composition, the protective mixture components and the surface state of the base metal edges; the size and the ratio of welding variables;

characteristics of the power supply (the relationship between the dynamic characteristics of the power supply and loss on spattering of the electrode metal), and others. [4,5].

Therefore, we have made an assumption about the possibility of a control over the process of electrode metal mass transfer by the influence of the external magnetic field on the arc. However, the accurate data about the spattering indicators under MAG-welding with the use of additional influence of an external electromagnetic field on the arc are practically absent or are controversial [4,6].

Problem setting

The research object is the process of arc welding by smelting electrode in the medium of active protective gases with an additional influence of an external electromagnetic field on the arc.

The subject of study is the influence of the frequency and the induction of a longitudinal magnetic field (*further in the paper - LgMF*) on the processes of drop-formation and mass transfer of electrode metal during MAG-welding.

The aim of this work is to study the impact of LgMF induction and frequency on the loss coefficient of the electrode metal on spattering ψ and droplet distribution of electrode metal into fractions during MAG-welding, as well as the development of technological recommendations that will improve the technological and economic efficiency of this welding process.

The main objectives of the study are:

- determination of the frequency range and inductions of LgMF, providing the reduction of ψ coefficient;
- characterization of their mutual influence on the process of mass transfer of electrode metal;
- development of mathematical models relating the LgMF frequency and induction with a discharge coefficient on electrode metal spattering.

The introduction of new technologies in a production environment will improve the efficiency of MAG-welding and will lead to material, labor and energy reduction.

Materials and methods

In the experiments we use the retrofit of UD-209-UHL4, universal rectifier VDU-506 (i.e. Multi-operated Arc Rectifier-506); analytical laboratory scales. Rollers were fused on the flat carbon steel St3sp (i.e. dead-melted steel, carbon content – 0,14-0,22 %) GOST 380 (all-Union State Standard) by welding wire of brand Sv-08G2S (i.e. welding wire, 0,8 – carbon content, manganese – 2%, silicon – less than 1 %) in the carbon dioxide medium of the highest grade (GOST 8050 (all-Union State Standard). To create an external electromagnetic field we used a specialized nozzle of welding head with integrated solenoid, powered by a laboratory device LATR-1 and specialized mode control unit BKRN-2 [7].

The electrode metal droplet ranging into fractions and the evaluation of their diameters have been done with the use of sizing screen (grain size: 1,5; 2,0; 2,5; 3,0; 3,5; 5 mm).

The induction components of LgMF B_Z, were measured near the plate surface at a distance from the flat end of the electrode to the $5\cdot10^{-3}$m plate. The induction of constant magnetic field was determined by teslameter of EM-4305 type with the Hall sensor and the base of 1 × 1 mm, and the alterative - by teslameter of F-4356 type with a Hall sensor and the base 4 × 4 mm.

The level of electrode metal spattering was estimated by a coefficient, which is numerically equal to the percentage of the wire mass, spent respectively on spattering ($m_p = m_п - mweld$) and on the formation of a roll of 10^{-1} m ($m_n = \gamma_п \cdot l_п \cdot \pi \cdot d_п / 4$) where: *mweld* - weld mass, equal to the difference of sample masses before and after welding, kg; γ_n - wire material density, in kilograms; l_n - the length of wire needed for the formation of roll of 10^{-1} m, m.

$$\psi = (m_p / m_п)\cdot 100\ \% \quad (1)$$

The parameter change of the metal electrode mass transfer during welding with an additional influence of external LgMF of different frequency was investigated by method based on welding on a copper disc rotating at a predetermined angular velocity. In experimental conditions the angular velocity was 0,185c^{-1}. Nonstick coating was periodically applied on its upper surface with a diameter of 0.5 m. To ensure the same arcing process time (15 seconds), the welding power source, the pulse generator of LgMF and the specialized drive burner were put into operation with the aid of a control device with an electronic timer. We began the experimental data processing with measuring the distances between the adjacent droplets (burns) as well as for determining the transfer frequency.

Work results

In the study the following welding conditions were taken:
I= 220…250A; U=20…22V; V=0,085 m/s; direct-current of reversed polarity (DCRP). Metal thickness 10^{-2} m, diameter of the electrode wire $2 \cdot 10^{-3}$ m.

In drawing up the experiment matrix two factors, affecting the result, were taken - LgMF induction and its frequency. The loss coefficient of the electrode metal, the number of drops of electrode metal were the consequence of this matrix. To determine the mathematical models the orthogonal central composite second-order plan was chosen. Induction B_Z and pulse rate *fimp* were changed within 0 ... 80 mT (T-Tesla) and 0 ... 50 Hz, respectively (Table. 1).

Table 1 - Matrix of the experimental plan and research data

Induction of magnetic fiel B_Z, mT	*Frequency of magnetic field fimp, Hz*	*Loss coefficien t, ψ, %*	*Drops number, $n_{кп}$*	*Induction of magnetic field B_Z, mT*	*Frequency of magnetic field fimp, Hz*	*Loss coefficient , ψ, %*	*Drops number, $n_{кп}$*
+	+	15.3	212	+	0	14.9	200
-	+	12.6	172	-	0	10.9	162
+	-	14.2	189	0	+	11.2	168
-	-	14.0	78	0	-	10.6	168
0	0	10.1	181				

Experimental data processing was performed using the software StatSoft Statistica. The received dependence between the parameters was shown in the form of three-dimensional graphs (Figs 1 and 2).

A greater impact on ψ has the induction value of LgMF than its frequency according to the analysis of the dependence of electrode metal loss coefficient from LgMF induction and frequency (Fig. 1). With increasing LgMF induction from 0 to 18 mT the loss coefficient decreases in average by 25%; in absolute values (equation (1)) from 14 to 10%.

With increasing frequency of LgMF from 0 to 36 Hz with the values of its induction up to 35 mT we also observed a reduction in the loss coiefficient of electrode metal. A further increase in LgMF frequency has the opposite effect - the ψ coefficient increased from a minimum of 10% to 12.5 %. That is, the additional influence on the arc by outer LgMF is positive, as in welding without LgMF implementation under the same operation conditions the loss coefficient is equal to 14.2% as compared to 12.5%.

The influence on welding arc by a longitudinal magnetic field by 40 mT induction increases the loss coefficient of the electrode metal. For values of B_Z 40 ... 70 mT the ψ coefficient verges towards the average data for the case of MAG-welding without LgMF implementation. Further increase of B_Z causes a sharp increase in the loss coeffiient from 14 to 17.5%.

Thus, the minimum ψ values are observed in conditions of influence of alternate LgMF with 16 ... 34 Hz frequency and 18 ... 38 mT induction on the arc.

Mathematically, the above-mentioned dependence can be expressed by the following equation:

$$\Psi=13,414-0,1377 \cdot B_Z-0,1043 \cdot f_i+0,0019 \cdot B_Z^2+0,0006 \cdot B_Z f_i+0,0016 \cdot f_i^2 \quad (4)$$

IOP Conf. Series: Materials Science and Engineering **91** (2015) 012011 doi:10.1088/1757-899X/91/1/012011

It was experimentally proved that a significant impact on the number of electrode metal droplets has LgMF induction value and as well as its frequency (see Figure 2).

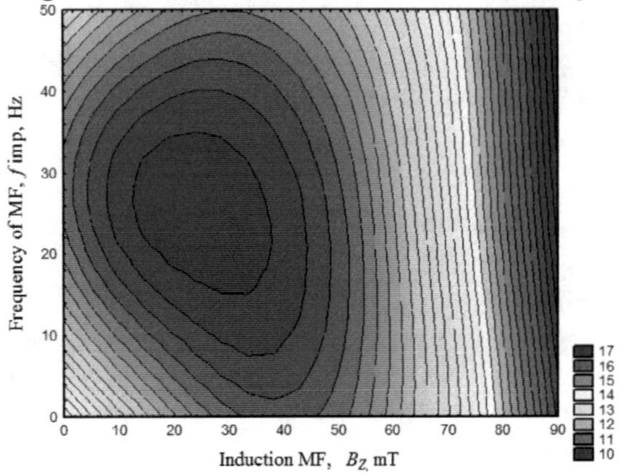

Figure 1 – Dependency graph of the loss coefficient of the electrode metal ψ from the induction B_Z and frequency f_i of LgMF

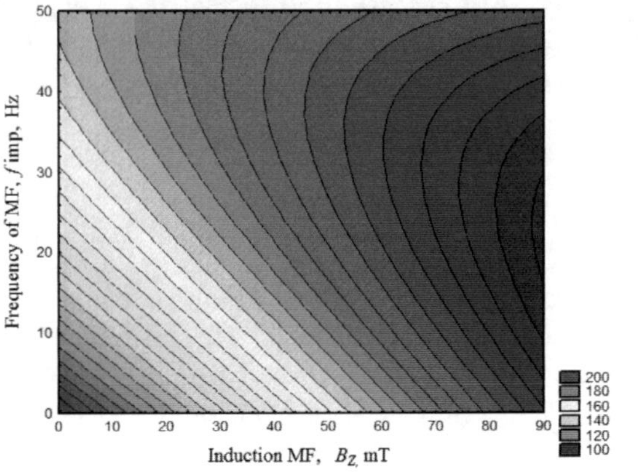

Figure 2 – Dependency graph of drops number *n* of electrode metal from the B_Z and frequency f_i of LgMF

With induction increasement of LgMF drops number of electrode metal almost doubles (from 100 to 200). With frequency increasement of LgMF the increase in the drops number of electrode metal is also observed. However, the influence of LgMF frequency on the value *n* is less significant than the effect of induction of this field. For example, at LgMF induction index up to 20 mT the *n* increase was 60%, at $B_Z = 30 \ldots 50$ mT *n* increase was 80%.

The maximum droplets number is formed under the following conditions: LgMF induction $B_Z = 75 \ldots 90$ mT, LgMF frequency $f_i = 12 \ldots 36$ Hz. At this mass transfer of electrode metal acquires jet nature.

Mathematically, the dependence data can be expressed by the equation:

$$n = 94{,}5833 + 1{,}4063 \cdot B_Z + 2{,}81 \cdot f_i + 0{,}0022 \cdot B_Z^2 - 0{,}0178 \cdot B_Z \cdot f_i - 0{,}0264 \cdot f_i^{\,2} \tag{5}$$

From the viewpoint of mass transfer of the metal electrode the number and the size of formed droplets have the importance in the process of MAG-welding. It is known that with decrease in droplet size the stability of arc welding process increases and the metal electrode loss on spattering reduces. In order to establish an optimum mode range of LgMF (frequency and induction) under which a minimum number of large (> 2 $d_э$) drops with the increase of the total number, a series of experiments were conducted, their results are shown in the Table 2 and 3.

Table 2 – The dependence of the electrode metal drops in number and fractions from the frequency of LgMF

Frequency,	The fraction size, mm						
Hz	1,5	2,0	2,5	3,0	3,5	5	Σ
0	7	4	5	24	31	7	78
10	37	26	19	18	21	5	126
20	56	37	26	17	19	3	158
30	61	42	29	17	17	0	166
40	63	45	31	14	16	0	169
50	67	54	28	13	10	0	172

Table 3 – The dependence of electrode metal drops distribution in number and fractions from induction of LgMF

Induction,	The fraction size, mm						
mT	1,5	2,0	2,5	3,0	3,5	5	Σ
0	7	4	5	24	31	7	78
20	30	26	24	18	28	6	132
40	44	39	37	17	26	5	168
60	64	51	46	12	4	2	179
80	71	55	52	8	3	0	189

Histograms of the distribution of the electrode metal droplets at MAG-welding with additional LgMF influence (Fig. 3).

Fig. 3 shows the dependence of the electrode metal amount from LgMF frequency where the columns 1-6 correspond to LgMF frequencies from 0 to 50 Hz with a step 10. As we can see with increasing LgMF frequency from 0 to 10 Hz the number of droplets almost doubled. What is more the number of small droplets fraction significantly increases: fractions from 1.5 and 2.0 mm increase in 5 times; 2.5 mm - in 4 times. A number of droplets of a larger fraction is reduced: 3.0 mm and 3.5 to 30%.

With increasing LgMF frequency from 10 to 20 Hz the number of droplets also increases and reaches 158 units, i.e., in 2.15 times. Moreover, as in the previous case, the number of small fraction droplets increases significantly: 1.5 mm fractions increase in 8 times; 2.0 mm - in 9.2 times; 2.5 mm - in 5 times respectively. A number of drops of large fraction is also reduced, on average by 30%.

With further increase in the LgMF frequency (from 30 to 50 Hz) we observed the stable high rates of droplets formation of small fraction, reducing the number of 3.5 mm fraction drops in three times and the lack of droplets with a diameter of 5 mm.

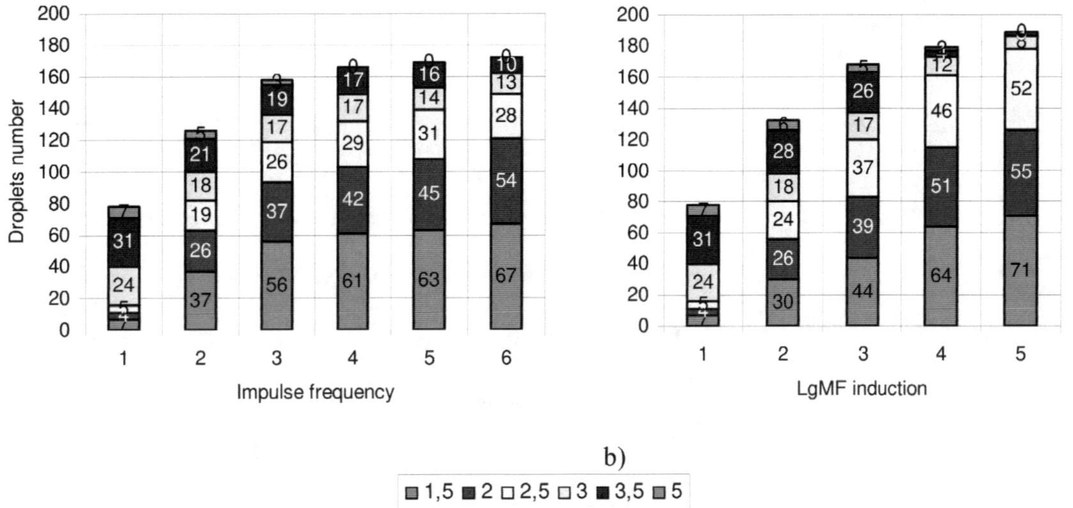

a) b)

Figure 3 - Electrode metal droplets distribution duting MAG-welding with the additional influence of LgMF (further explanation is in the text)

Fig. 3b shows the number of electrode metal by LgMF induction where 1-5 columns correspond to the LgMF induction values from 0 to 80 mT with the step 20. As we can see with increasing the LgMF induction from 0 to 20 mT the number of droplets is increased in 1.7 times. The number of small droplets fraction significantly increases - 1.5 in 4.3 times; 2.0 mm in 6.5 times; 2.5 mm in 5 times. And the number of fraction drops of 3.0; 3.5 and 5.0 mm reduces on average by 12 ... 16%.

For 40 mT values of LgMF induction the number of drops also increases in 2.15 times and reaches 168 units. What is more, as in the previous case, the number of small fraction droplets significantly increases - 1.5 mm in 6.3 times; 2.0 mm in 10 times; 2.5 mm in 7.4 times. A number of large fraction drops is also reduced on average by 15%.

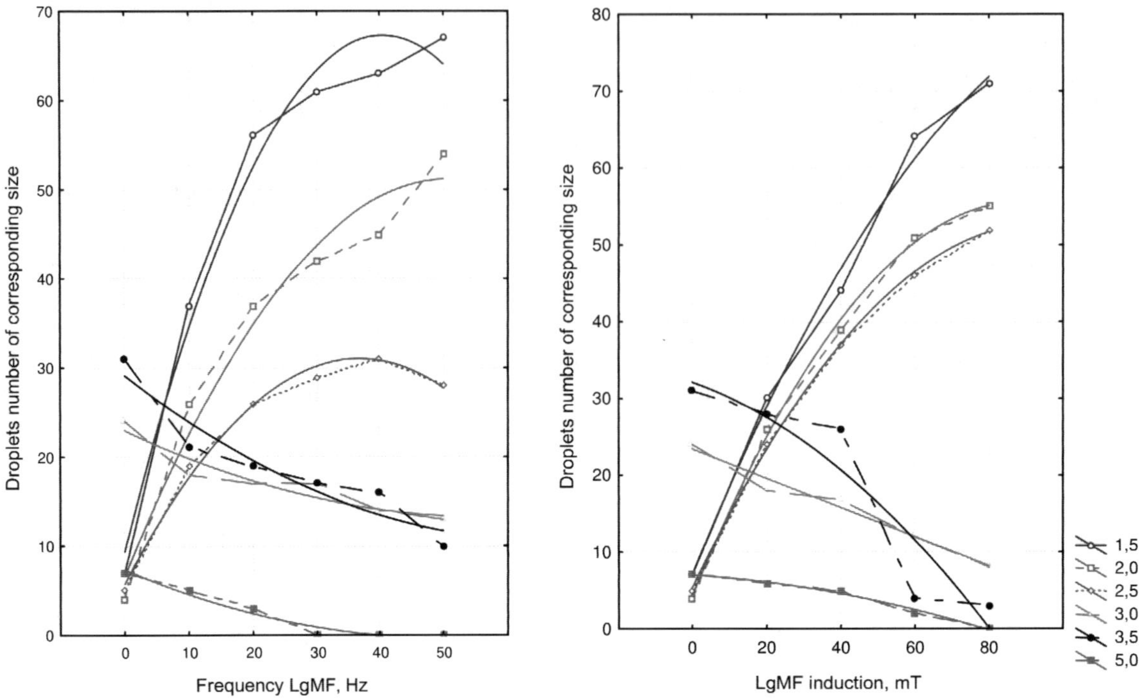

Figure 4 - Number of electrode metal drops during MAG-welding with LgMF influence

With further increase of LgMF induction (from 40 to 60 Hz) we observed consistently high indices of small droplets formation and the reduction in number of 3.0 mm drops fraction, which halved in number. Droplets with the diameter of no more than 5 mm are absent when an arc is subjected to a magnetic field induction from 60 mT.

During the analysis of LgMF frequency and induction influence it was found that a greater influence on droplet size reduction of electrode metal had a frequency of the electromagnetic impulse than its (magnetic field) amplitude values.

The dependence graphs of the drops number of the electrode metal during MAG-welding with the additional LgMF influence (Fig. 4) are plotted; we obtained the mathematical relationships (Table. 4) which afford the determination of formed droplets number (*nfr*) at the individual fractions (index *fr.* indicates the size of the droplets), where f_i - frequency electromagnetic pulse Hz; B_Z - LgMF induction, mT.

Table 4

Dependence of droplets number of electrode metal by fractions on LgMF frequency	Dependence of droplets number of electrode metal by fractions on LgMF induction
$n_{1,5} = -22,8+35,7 \cdot f_i - 3,54 \cdot f_i^2$	$n_{1,5} = -18,4+1,37 \cdot B_Z - 0,0045 \cdot B_Z^2$
$n_{2,0} = -12,7+2,1 \cdot f_i - 0,017 \cdot f_i^2$	$n_{2,0} = -21,6+1,43 \cdot B_Z - 0,006 \cdot B_Z^2$
$n_{2,5} = -9,9+1,75 \cdot f_i - 0,02 \cdot f_i^2$	$n_{2,5} = -17+1,23 \cdot B_Z - 0,005 \cdot B_Z^2$
$n_{3,0} = 26,7 - 0,4 \cdot f_i + 0,003 \cdot f_i^2$	$n_{3,0} = 27,2 - 0,19 \cdot B_Z$
$n_{3,5} = 35,2 - 0,65 \cdot f_i + 0,004 \cdot f_i^2$	$n_{3,5} = 34,4 - 0,06 \cdot B_Z - 0,003 \cdot B_Z^2$
$n_{5,0} = 10,8 - 0,37 \cdot f_i + 0,003 \cdot f_i^2$	$n_{5,0} = 7,4 - 0,004 \cdot B_Z - 0,0007 \cdot B_Z^2$

Thus, it was found that:

- the loss coefficient of electrode metal is minimal on condition that the influence on the arc by alternate LgMF is 16 ... 34 Hz frequency and induction is 18 ... 38 mT;
- the maximum number of droplets formed while influence on the arc of alternate LgMF with 12 ... 36 Hz frequency and 70 ... 90 mT induction, in this case a transition to the jet electrode metal transfer is observed, and the total number of drops is doubled compared to the welding process without the LgMF use;
- a greater influence on the reduction of droplet size of electrode metal has a frequency of the electromagnetic impulse than its (magnetic field) amplitude values.

However, it should be noted that the additional impact on the arc of external/outer LgMF with values of B_Z = 70 ... 90 mT is negative, as the loss coefficient of the electrode metal is greatly increased. The LgMF modes optimization in terms of the lowest values of the loss coefficient and the maximum number of drops of electrode metal were performed graphically. Optimization results are shown in Fig. 5.

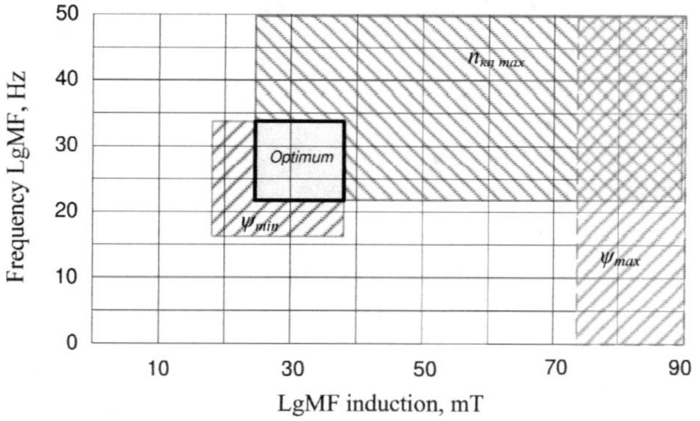

Figure 5 – Optimizing graph of LgMF frequency and induction as for the loss coefficient - ψ and the number of electrode metal drops - *n*

As we can see in Figure 3 the achievement of the jet electrode metal transfer during MAG-welding with minimum losses is possible by influence of longitudinal magnetic field induction of 26 ... 38 ... 22 mT and frequency of 34 Hz on the arc. Appliance of LgMF induction of more than 75 mT is not recommended due to the significant loss of electrode metal.

Thus, the proposed model can be used to determine the LgMF conditions for the arc MAG-welding by steel solid wire with a diameter up to $2\cdot10^{-3}$ m to minimize losses of welding materials.

Conclusions

It was determined that the efficiency of MAG-welding process can be achieved by jet electrode metal transfer while minimizing its losses spray with an additional influence of longitudinal magnetic field on the arc. The mathematical dependences for determining the number of droplets formed in separate factions, and a map of LgMF modes optimization are obtained. The recommendations are presented which allow to achieve the best results saving of welding materials with MAG-welding by low-carbon and low-alloy steels using external LgMF, namely:

- the core wire diameter of 1.2 ... 2.0 mm;
- current intensity I = 240 ... 280 A;
- arc voltage U = 20 ... 24 V;
- wire feed speed V = 0.065 ... 0.085 m/s;
- direct-current reversed polarity;
- LgMF induction of B_Z, 26 ... 38 mT;
- impulse frequency *fi* 22 ... 34 Hz.

IOP Conf. Series: Materials Science and Engineering **91** (2015) 012011 doi:10.1088/1757-899X/91/1/012011

The affect of longitudinal magnetic field induction of more than 75 mT on the arc is not recommended due to the significant loss of electrode metal during spttering.

References

[1] Potap'yevsky A G Spattering at welding CO2 with a wire Sv-08G2S A G Potapevsky V Ya Lavrischev *Automatic welding* 1972 # 8 p 39-42.

[2] Chinakhov D A Calculation of Gas-dynamic Impact of the Active Shielding Gas on the Electrode Metal Drop in Gas Jet Shielded Welding *Applied Mechanics and Materials* Vol. 379 (2013) p 188-194

[3] Chinakhov D A Zuev A V Filimonenko A G Gas-dynamic Impact of a Shielding Gas Jet on the Drop Transfer When Welding with a Consumable Electrode *Advanced Materials Research* Vol. 1040 (2014) p 850-853

[4] Razmyshlyaev A D Magnetic control of rollers and weld formation in arc surfcing and welding A D Razmyshlyaev, M V Mironov Mariupol: Publishing Priazovsky State Technical University 2009 p 242

[5] Nosov D G Maltsev V V The influence of magnetic fields by a melting rate of wire for arc surfacing under flux *Applied Mechanics and Materials* Vol. 379 (2013) p 178-182

[6] Patskevitch I R Longitudinal magnetic field influence on the surfacing and transfer of electrode metal I R Paskevich A V Zernov V S Serafimov *Welding production* 1973 #7 p 8-10

[7] Nosov D G Power source of electromagnetic system for arc welding [1] using external magnetic fields control D G Nosov *Welder* 2010 # 4 p 18-19

IOP Conf. Series: Materials Science and Engineering **91** (2015) 012012 doi:10.1088/1757-899X/91/1/012012

Influence of Protective Gas Content on Quality of Welded Joint While Welding With Impulse Supply of Electrode Wire

N V Pavlov [a], A V Kryukov, E A Zernin, V V Gritsenko

Yurga Institute of Technology (branch) of National Research Tomsk Polytechnic University
Leningradskaya 26, Yurga, Russian Federation, 652055

E-mail: pavlin123@rambler.ru

Abstract. Currently one of the advanced ways of obtaining quality welded joint while welding of medium alloy martensitic-bainitic steel is the one with impulse supply of electrode wire in gas mixture Ar(70%±3%)+CO2(30%±3%). Results of experimental studies proved that application of protective gas Ar(70%±3%)+CO2(30%±3%) in comparison with CO2(100%) enables to increase strength properties of the welded joint by 10-15% and enlarge the transition coefficient of chemical elements.

1. Introduction

Medium alloy martensitic-bainitic steel is characterized by a large number of operational properties and is used for manufacture of critical structures [1,2]. By critical structures particularly meant are high-pressure vessels, heavy loaded mechanical products and other structures as they provide high strength of structure after the appropriate heat treatment, at the same time providing reduced metal consumption. Steels of that group are marked by carbon content up to 0,5% while integrated alloy addition in total 5...9 %. Steels with carbon content being that high due to high sensitivity to welding thermal cycle are prone to formation of hardening structures and cold cracks.

Obtaining reliable welded joints is also problematic due to high sensitivity to stress raisers under static loads and dynamic loads in particular. Larger alloy addition with carbon makes welded joints even more sensitive [2].

The required strength of steel together with keeping high ductility is reached by integrated alloy addition with different elements, the main of which being chrome, nickel, molybdenum, etc. The given elements strengthen ferrite and increase steel hardening capacity. Increase of alloy addition level while high carbon content increases austenite stability and almost at any cooling rate of weld-affected zone and welding modes providing satisfactory weld formation, austenite decomposition occurs in martensitic area [1].

Different process solutions also find broad application, such as use of heatsink, different kinds of forming base surfaces, two-stream gas protection, use of nanostructured powders-modifiers while welding [3,4].

Research direction connected with application of impulse-arc welding methods has been actively developing recently, realized by:

1) tools controlling electrode metal transition by means of pulsed supply of welding arc [5,6];

Published under licence by IOP Publishing Ltd

2) mechanical programming of droplet transition into welding pool by means of impulse supply of welding wire [7,8,9].

Variety of ways to realize the first direction gives an opportunity to obtain almost any algorithms of changing the energy characteristics of welding arc.

However there is a number of disadvantages:

- complexity of design solutions;
- higher cost in comparison with standard equipment;
- inability of use together with mass-produced power supply sources.

The second direction provides controlled transition of electrode metal into welding pool by means of using the mechanisms of impulse supply of electrode wire.

While realizing this research direction it is reasonable to use pulling-type mechanisms, as they provide not only system portability, but more importantly transfer the form of impulse more accurately [7,8,9].

Therefore welding with impulse supply of electrode wire with pulling-type mechanisms is a relevant objective. This process provides advantages of impulse-arc welding methods and also does not have any significant disadvantages.

The other factor influencing the quality of weld joint is protective gas medium.

Having analyzed the existing and applied protective gases and mixtures on their basis, it was proved to be rational to use the mixture Ar+CO2 [10]. This mixture is actively applied in the process of manufacturing, has a positive effect on processing properties of welding arc (increased stability of its combustion), splashes size decreases and spatter losses reduce, weld reinforcement decreases with sharp transition to the base metal (Figure 1) [11].

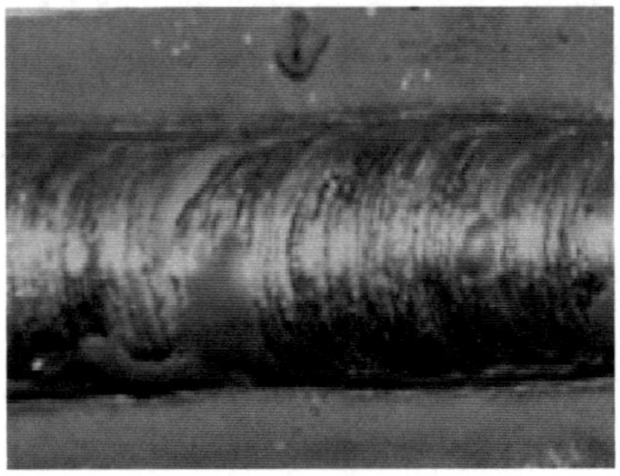

a)

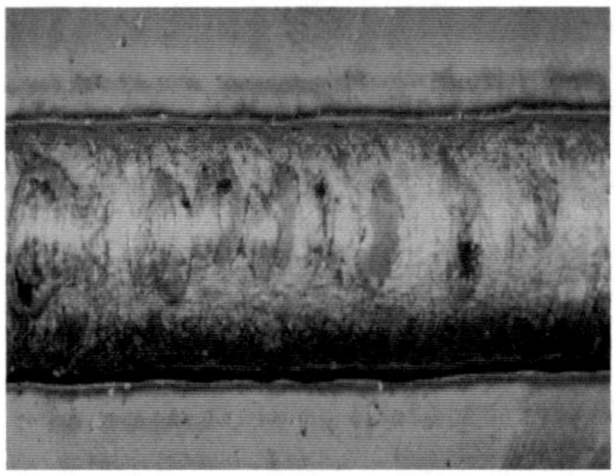

b)

Figure 1. Layout of welds obtained by welding with impulse supply of electrode wire in CO2 (a) and by welding with impulse supply of electrode wire in gas mixture Ar+CO2 (b).

Following the conducted experiments it was established that application of mixture Ar(70%±3%)+CO2(30%±3%) while steel welding of standard quality (structural carbon steel of regular quality) reduces the value of electrode metal losses by burning and spattering up to 2% [12] (Figure 2).

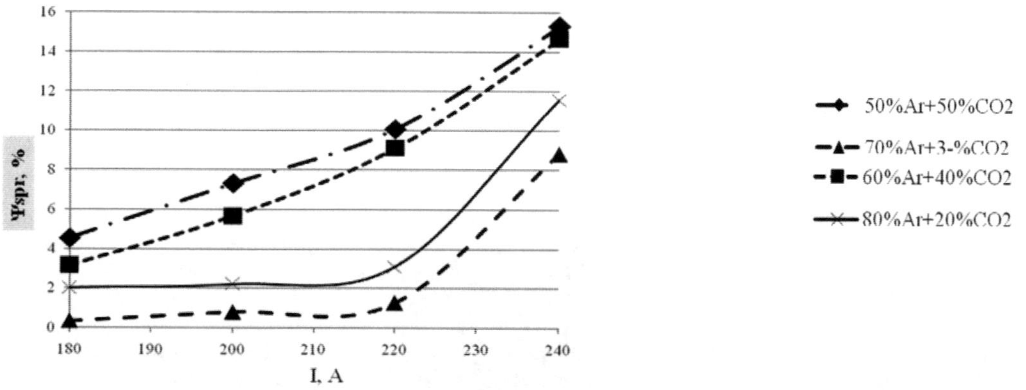

Figure 2. Dependency diagram of value of electrode metal losses by burning and spattering (ψspat.) of the value of arc welding current (I).

Based on the data described in work of Novozhilov N.M. [13] the obtained ratio is determined by the fact that addition of carbon to argon up to 70 % Ar + 30 % CO2 is accompanied by decrease in size of electrode droplets and corresponding increase of their number. Increasing of carbon content in argon beyond this limit at some conditions is accompanied by sharp decrease of the number of droplets, and by other conditions is accompanied by gradual decrease of the number of droplets formed per unit time.

However the issues of applicability of welding mode with impulse supply of electrode wire in gas mixture Ar(70%±3%)+CO2(30%±3%) while welding of medium alloy martensitic-bainitic steel are still not well studied.

Resulting from that the objective was set to increase strength properties of welded joint made of medium alloy martensitic-bainitic steel by means of using a mixture of protective gases in the ratio 70 % Ar + 30 % CO2 while welding with impulse supply of electrode wire.

2. Methods investigations

Experimental testing was held to study influence of protective gas media composition on structure and operational properties of welded joints made from medium alloy martensitic-bainitic steel:

1) traditional method – welding with impulse supply of electrode wire in CO2(100%);

2) suggested method – welding with impulse supply of electrode wire in gas mixture Ar(70%±3%)+CO2(30%±3%).

In both cases the experimental facility included: automatic welding head, equipped with impulse supply of electrode wire [14], Power supply welding rectifier (rated current – 300A), mixing equipment including three flow meters and mixing chamber. Welding was held from sheets made of steel (0.3% - C; up to 1 percent – Cr, Mn, Si) thickness 10mm, in X-shaped splicing by welding wire ER70S-6 (diameter 1,2 mm).

Welding methods of samples are given in Table 1.

Table 1. Welding methods of samples.

Welding method	I, A	U, B	Vw, mm/s	f, Hz
traditional method	200-210	23-24	3.6	64
suggested method	220-230	24-25	3.7	64

Difference in value of energy parameters is connected with the fact that in order to reach the same geometrical parameters it is necessary to increase welding modes with impulse supply of electrode wire in gas mixture Ar+CO2 to 5-10% in comparison with impulse supply of electrode wire in CO2, as addition of argon leads to its reduction [12].

Evaluation of chemical compound of weld was carried out using consequent x-ray fluorescent spectrometer LabCenter XRF-1800.

Experimental samples were taken in two points in the basic metal and in weld metal (Figure 3). Point diameter is 3mm.

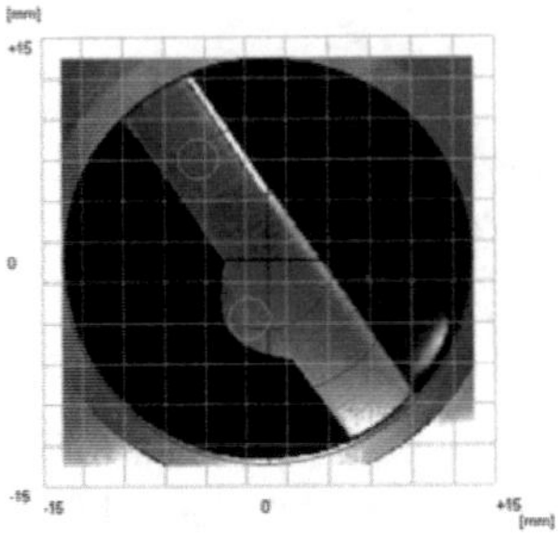

Figure 3. Points of measurement to study chemical compound.

Mechanical properties of welded joints were defined in accordance with common methods, given in Russian State Standard (GOST) No.6996-66 "Welded joints. Ways to define mechanical properties".

Microstructure of weld metal was studied and observed by means of microscope ES METAM RV (Russian State Standard (GOST) 15150-69) complete with digital camera Fuji Film Fine Pix S6500fd, providing recording of data from microscope.

3. Results and discussion

Experimental studies implied exposing of the obtained welded samples to quantitative estimation of chemical compound, mechanical testing and structural analysis of welded joint (Tables 2, 3 and Figure 4).

Table 2. Chemical compound of weld metal.

Welding method	Chemical elements				
	C, %	Si, %	Mn, %	S, %	P, %
traditional method	0.24-0.26	0.72-0.74	0.61-0.63	0.013	0.017
suggested method	0.22-0.23	0.78-0.8	0.66-0.68	0.011	0.015

Table 3. Results of mechanical testing of welded samples.

Welding method	Impact strength, KCU, J/sm2		
	T=0°C	T=20°C	T=-20°C
traditional method	62…66	72…76	52..56
suggested method	71…75	82…86	60…64

As seen in Table 2 while welding in gas mixture transition coefficient of chemical elements-deoxidizing agents (silicon and manganese) into welding pool increases by 4-6%. Such change occurs due to change in the value of specific speed of metal oxidation ωspec.

Based on the methodology described in N.M. Novozhilov's work specific speed of metal oxidation is calculated according to the formula below [13]:

IOP Conf. Series: Materials Science and Engineering **91** (2015) 012012 doi:10.1088/1757-899X/91/1/012012

$$\omega_{spec} = \frac{\sum O_w}{100t_{av}} g/gmet \cdot s, \tag{1}$$

where tav – average time of metal being in liquid state in seconds;
ΣOw – specific oxidation of metal, g/100 g met.

Due to short term of electrode droplets this value is accepted as time of metal being in liquid state equal to metal being in welding pool [13]:

$$t_{av} = \frac{G_w}{g} sec, \tag{2}$$

where Gw – amount of molten metal in welding pool, g;
g – amount of metal, melted by arc in 1 s. in g/s.

Amount of molten metal in welding pool is defined by approximate formula based on allowance that volume of welded pool equals half of the ellipsoid volume [13]:

$$G_w = 0{,}523K_f LBh\gamma_t, gmet, \tag{3}$$

where Kf – filling coefficient of welding pool with molten metal, accepted as 0,8-0,9;
L – length of welding pool, accepted as equal to crater length, mm;
B, h – width of weld and depth of weld penetration of the base metal, mm;
γt – density of molten metal in welding pool;

Amount of metal melted by arc in 1 second, is defined by formula [13]:

$$g = Fv\gamma, gmet/sec, \tag{4}$$

where F – cross section area of weld, mm2;
v – welding velocity, mm/s;
γ – density of molten metal.

Calculations performed by this method (Table 4) showed increase in specific velocity of metal oxidation by 5-6% and reduction of specific amount of oxidized metal up to 10%.

Table 4. Evaluation of intensity of metal oxidation.

Welding method	g, gmet/sec	Gw gmet	tav, s	Wspec
traditional method	0.262	0.327	1.248	0.03
Suggested method	0.279	0.327	1.172	0.028

Reduction of heat input value into droplet of electrode metal contributes to keeping elements-deoxidizing agents in welded joint, as heating capacity of welding arc is different due to value of effective heating.

Combination of these factors enables silicon dissolving in ferrite to increase yield limit and decrease tendency to cold brittleness, manganese forms solid solution with iron, increases hardness and strength insignificantly that contributes to increase in the value of impact strength by 10-15% (Table 3) compared to welding in CO2.

IOP Conf. Series: Materials Science and Engineering **91** (2015) 012012 doi:10.1088/1757-899X/91/1/012012

Conducted metallographic studies of two methods demonstrated similarity of the obtained results (Figure 4).

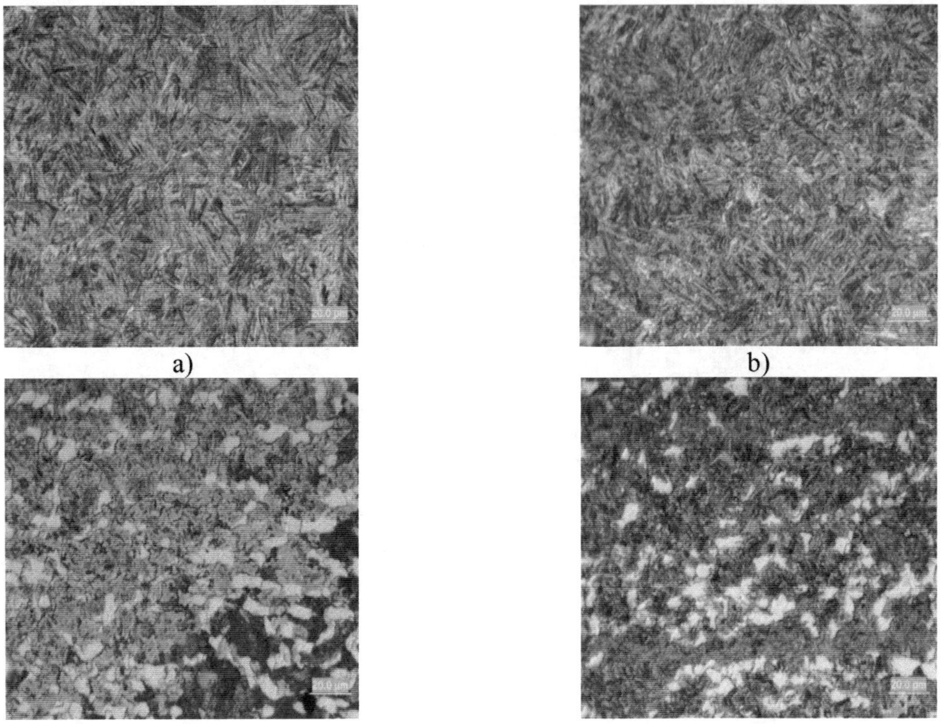

Figure 4. Microstructure (increase 20.0 μm) a) area in the center of weld metal while welding by traditional method; b) area in the center of weld metal while welding by the suggested method; c) area of thermal cycle boundary zone while welding by traditional method; d) area of thermal cycle boundary zone while welding by the suggested method.

In the center of weld metal dark-brown sheets of bainit in lighter martensite matrix are seen (Figure 4 a,b), the structure is homogeneous and has echinulate martensite composition. In the structure of heat-affected zone of steel (0.3% - C; up to 1 percent – Cr, Mn, Si) areas of hardening structures are also observed, however deformation texture is not well observed in Figure 4,c. While removing from weld area the amount of martensite decreases, intensive grain refining occurs and recrystallized zone is formed (Figure 4,d). Recrystalized zone has fine grain and structure close to equilibrium state (Figure 4,d). While welding by traditional method recrystallized zone is characterized by better marked transition to the base metal (Figure 4,c).

It should be noted that welded joints did not undergo heat treatment. Elimination of heating and after-welding heat treatment from welding procedure of (0.3% - C; up to 1 percent – Cr, Mn, Si) steel and reduction of time for stripping of product will enable to increase labor efficiency in comparison with traditional method by 5-10%.

The obtained changes are connected with different temperature increment on the surface of product (Figure 5).

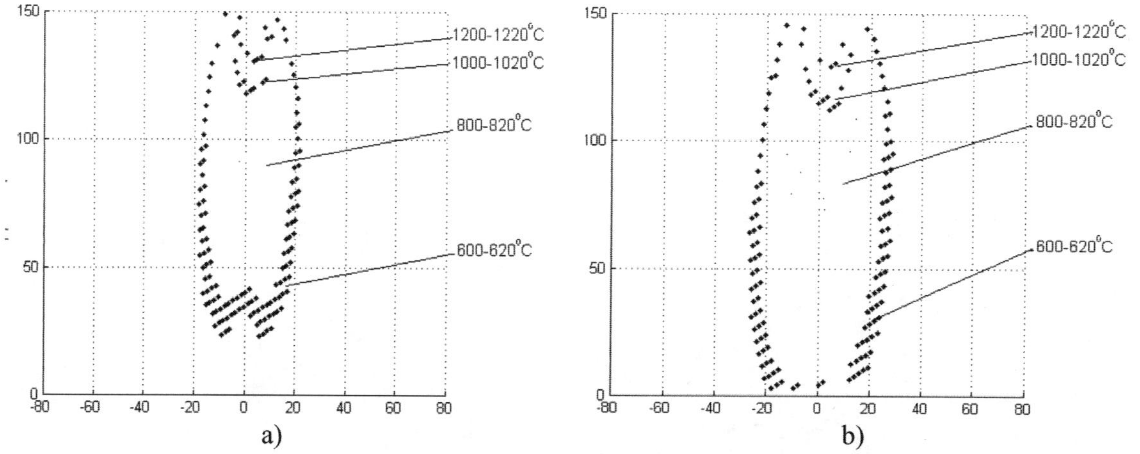

Figure 5. Experimental thermograms obtained after treatment: a) isotherm of temperature increment on the surface of weld while welding by traditional method; b) isotherm of temperature increment on the surface of weld while welded by the suggested method

During welding by the suggested method isotherm (Fig. 5,b) has more oblong form along weld (by 40%) while keeping weld width that proves more homogeneous distribution of temperature fields on the surface of product.

4. Conclusion

Resulting from experimental studies it was established that application of gas media Ar(70%±3%)+CO2(30%±3%) compared to CO2(100%) enables to increase strength properties of welded joint by 10-15% and increase transmission coefficient of chemical elements.

References

[1] Lakhtin Y M and Leontieva V P 1990 Materials Science (Moscow: Machine Engineering Publishing Office).

[2] Chinakhov D A and Brunov O G 2006 Welding with impulse supply of electrode wire of ring connections from steel 30HGSA *Bulletin of Tomsk Polytechnic University* 309 pp 136-138.

[3] Chinakhov D A, Vorobyov A V and Tomchik A A 2013 *Simulation of active shielding gas impact on heat distribution in the weld zone* Materials Science Forum 762 pp 717-721.

[4] Kuznetsov M A, Zhuravkov S P, Zernin E A, Kolmogorov D E and Yavorovsky N A 2014 Advanced Materials Research *Influence of nanostructured powder modifiers on the structure of a welding bead* 872 pp 118-122.

[5] Krampit A G and Krampit N Y 2013 Method for the determination of the geometrical dimensions and area of the welded joint *Welding International* 27 pp 834-836.

[6] Paton B E, Lebedev V A and Poloskov S I 2013 Application of mechanical impulses to control processes of automatic and mechanized welding by consumable electrode *Welding and diagnostics* 6 pp 16-20.

[7] Durgerov N G, Sagirov K N and Lenivkin V A 1985 Equipment for impulse-arc welding by consumable electrode (Moscow: Energoizdat Publishing Office).

[8] Lebedev V A 1998 Mechanized arc welding in CO2 with regulated impulse supply of welding wire *Welding production* 5 pp 30-33.

[9] Brunov O G, Fedko V T and Slistin A P 1999 Mechanisms of impulse supply of welding wire *Metals technology* 11 pp 7-10.

[10] Asnis A E, Gutman L M and Pokladii V P 1982 Welding in active gases mixture (Kiev: Naukova Dumka Publishing Office).

[11] Yazykov Y F and Aleksina I V 2008 Advantages of welding in protective gas mixtures *Welding production* 9 pp 29-30.
[12] Pavlov N V, Kriukov A V and Zernin E A 2010 Welding with impulse supply of wire in gas mixture *Welding production* 4 pp 27-28.
[13] Brunov O G, Fedko V T and Kriukov A V 2005 Mechanism of impulse supply of welding wire *Russian Federation patent for invention No.2254969.*
[14] Novozhilov N M 1972 The basics of metallurgy of arc welding in active protective gases (Moscow: Machine Engineering Publishing Office).

IOP Conf. Series: Materials Science and Engineering **91** (2015) 012013 doi:10.1088/1757-899X/91/1/012013

Development of New Materials and Technologies for Welding and Surfacing at Research and Production Center "Welding Processes and Technologies"

N A Kozyrev, R E Kryukov, G V Galevsky, D A Titov, V M Shurupov

Federal State Budgetary Educational Institution
of Higher Professional Education «Siberian State Industrial University»,
Research and Production Center «Welding Processes and Technologies»
654007, Russia, Novokuznetsk, 42 Kirova str.

E-mail: kozyrev_na@mtsp.sibsiu.ru

Abstract. The paper provides description of research into the influence of new materials and technologies on quality parameters of welds and added metal carried out at research and production center «Welding processes and technologies».

New welding technologies of tanks for northern conditions are considered, as well as technologies of submerged arc welding involving fluxing agents AN - 348, AN - 60, AN - 67, OK.10.71 and carbon-fluorine containing additives, new flux cored wires and surfacing technologies, teaching programs and a trainer for welders are designed.

Introduction

The basic requirement of industrial safety consists in assuring needed level of mechanical characteristics of a weld. Here, mechanical characteristics are pre-determined by non-metallic impurity degree and total content of gas in a weld. Most non-metallic impurities in welds are oxide ones irrespectively to their exogenous or endogenous nature, the quantitative and qualitative composition of the latter depends on the total concentration of oxygen. The issue of new fluxes and their additives development has been attracting much attention currently, as well as research into their influence on welding and technological characteristics of a weld and on the concentration of oxygen and non-metallic impurities in a weld [1-5].

Materials and methods of research

The most grades of domestically produced fluxes, applied for welding low-alloyed steels are oxidizing ones and ground on silicon-manganese oxidation-reduction processes. Here, the products of these reactions are oxide compounds of silicon, manganese, ferrum, aluminum etc., which often can't surface and assimilate to formed of welding flux slag when welding; as the result, the level of impurity of weld metal by non-metallic admixtures increases, causing deterioration of physical and mechanical characteristics. Obviously, restoratives, which form gaseous products of reactions, are advisable to apply in order to avoid impurity of weld metal. It is carbon that can be a restorative of this kind, forming gaseous compounds CO_2 and CO when reacting with oxidizers.

Taking into account these pre-conditions a series of research activities has been carried out into the influence of additives to various fused and ceramic fluxes on quality parameters of welds.

In recent years exploration of oil fields in the North has necessitated manufacturing of wall plates for bulk-oil tanks suitable for northern conditions. At Open Joint Stock Company «Novokuznetsk Plant of Reservoir Metalware» manufacture of tanks suitable for northern conditions has been developed according to «Standards of vertical cylindrical steel tanks for oil and oil products» PB-03-605-03 and SNIP II – 23-81. The technological process of assembling, welding, controlling and rolling wall plates is performed on special roll facilities. Silicon-manganese steel 09G2S (GOST 6713-91) is used for tanks manufacturing. As the result, an optimal technology has been designed and developed for two-side welding tank wall plates maintained in low temperature conditions: inside the tank it is welded by Sv-08GA wire (on the upper tier) with mixed fluxing agents AN-67B and AN-348A taken in proportion 1:1, outside the tank it is welded by Sv-10NMA wire (on the lower tier) with welding fluxes AN-60 and AN-348A taken in proportion 1:1. Flay welding was carried out up to 18 mm thick plate. On upper tiers flat butt welds are made in reduced power conditions, which provide penetration into metal up to 0.55 plate thickness. On the lower tier welding was carried out in conditions of high current providing penetration up to 0.7 plate thicknesses.

Weld metal consists principally of re-melted base 09G2S and added Sv-10NMA metals. Mechanical properties of joint welds and welds were determined at temperature 20^0C, impact strength of weld metal and thermal impact zone KCU was valued at temperature -50^0C. Mechanical characteristics and impact strength of a weld and thermal impact zone exceeded required rated values.

The developed technology made it possible to obtain the whole range of required mechanical properties and impact strength of tank metalware at temperatures below zero, and avoid defects – cracks when manufacturing [6]. This technology is patented in the Russian Federation [7, 8].

The influence of carbon and fluorine containing additive to AN-348, AN-60, AN-67 fluxing agents and that of imported fluxing agent OK.10.71 on refinement of weld metal is experimentally tested [9,10]. As the result, it has been found out, that the growing content of carbon and fluorine containing additive in various fluxing agents at permitted slight increase of carbon concentration (Figure 1) causes the drop of total concentration of oxygen in a weld (Figure 2), and considerable improvement of mechanical properties, especially that of impact strength at temperatures below zero (Figure 3).

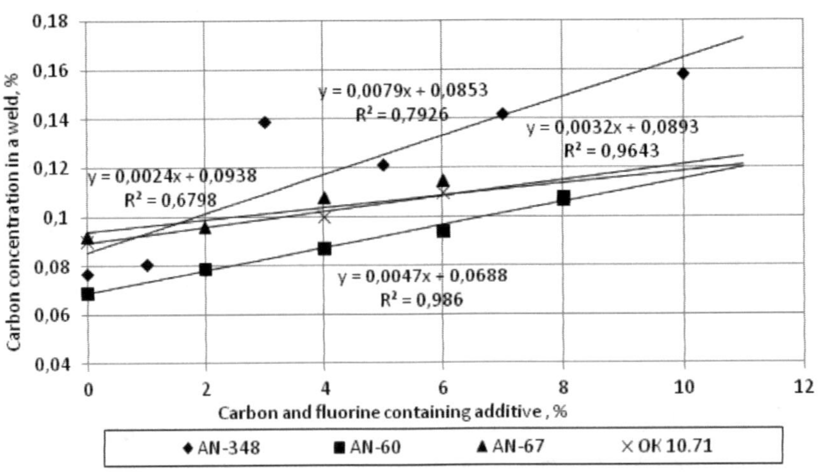

Figure 1. Influence of carbon and fluorine containing additive to carbon concentration in a weld

Fraction gas analysis conducted by reduction melting with gas analyzer «LECO» TS-600 revealed, that oxygen redistributed in impurities depending on oxidation and basicity of slag system.

The designed submerged arc welding technology of metalware maintained in conditions of extremely low temperatures rely on theoretical pre-conditions and practical try-out of carbon and fluorine containing additive technology. The flux of FD – UFS additive is produced at Open Joint Stock Company «Novokuznetsk Plant of Reservoir Metalware named after N.E. Kryukov». The designed welding technologies of metalware involving flux-additive are patented in Russian Federation [11], the composition of new basic welding flux is also patented in the Russian Federation [12].

Surfacing forming rolls is widely applied at manufacturing enterprises in Russia for restoration and improvement of resistance. Despite the significant number of systems used for surfacing, two systems are applied more often: **C—Si—Mn—Cr—W—V** and **C—Si—Mn—Cr—V—Mo**.

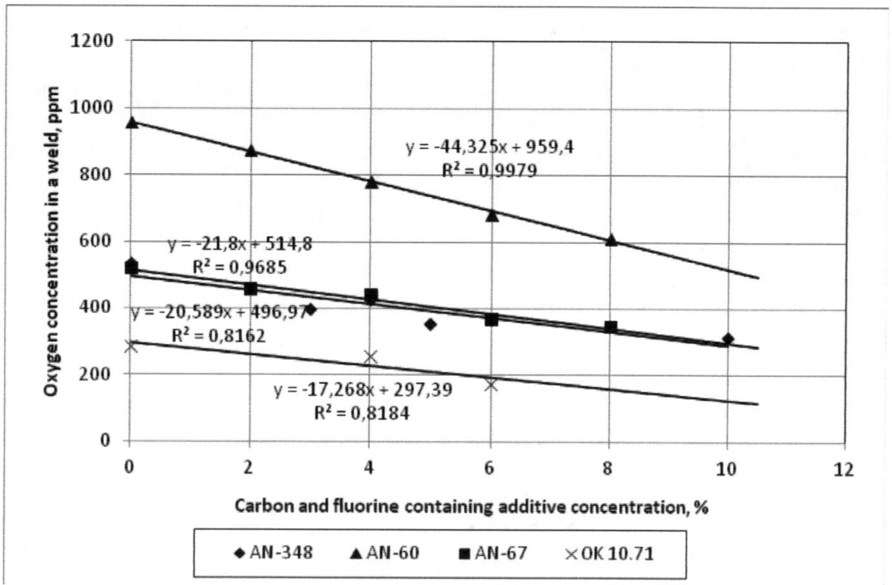

Figure 2. The change in total oxygen concentration in fluxing agents in dependence on carbon and fluorine additive.

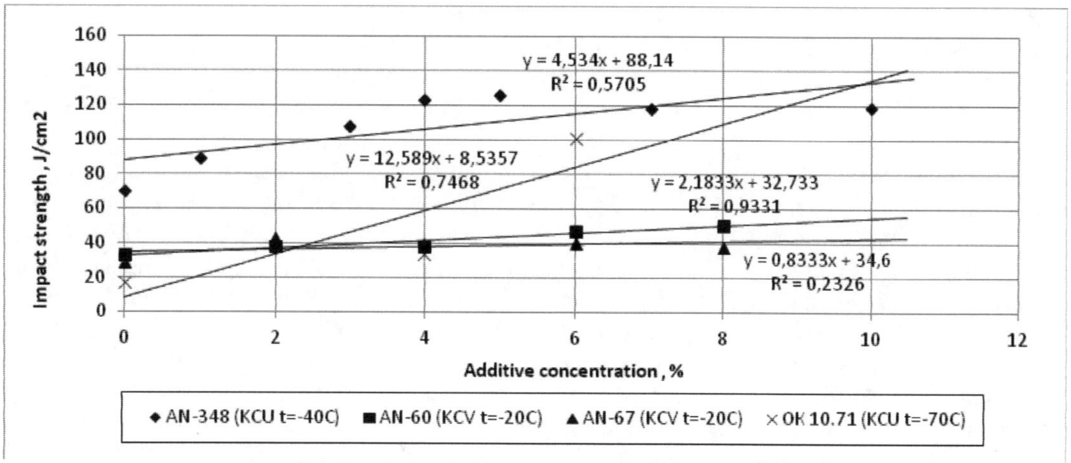

Figure 3.The change in impact strength in dependence on carbon and fluorine additive concentration

One of the elements in the first system is PP-Np-35W9Cr3SiV flux cored wire containing up to 10% valuable and expensive tungsten and it is applied for wear-resistant surfacing steel rolls of different hot-rolled mills. Metal, which is surfaced with this wire, is distinguished by significant abrasion resistance at high temperatures, but is quite low thermally resistant; therefore, rolls surfaced with the wire often fail because of erosion and chipping cracks. Sometimes stripe wear is possible caused by areas of added metal with inhomogeneous structure and hardness. Such areas can be found when multilayer surfacing alloyed steels with overlapping of a bead made before.

A surfaced metal similar to economically alloyed chromium-molybdenum and chromium-tungsten-molybdenum steels demonstrated good results when restoring steel rolls of hot-rolled mills. These steels are quite similar to chromium-tungsten ones in terms of heat resistance, and outperform them regarding thermal fatigue resistance. Vanadium alloying is also applied to improve the thermal resistance and abrasion strength of added metal.

Machining of added layer is rather difficult when surfacing rolls with various passes because of its relatively high hardness. Surfacing materials like martensite-ageing or precipitation hardening steels are promising ones for above mentioned rolls. After being surfaced hardness of these steels can amount to 28...35 HRC, and they are easy to be machined. After tempering the hardness increases up to 48...55 HRC, and added metal gains sufficient service qualities. However, special equipment is needed for heat treatment.

The second system provides the highest thermal fatigue resistance of metal. This system represented by PP-Np-25Cr5VMoSi flux cored wire makes it possible to gain significant impact strength of added metal at high temperatures.

From the standpoint of laboratory and experimental-industrial research it is recommended to use PP-Np-35W9Cr3SiV flux-cored wire for surfacing rolls of continuous bar mills, pipe-rolling, and wire mills; and PP-Np-25Cr5VMoSi flux cored wire on blooming mills (blooming, slabbing), on heavy section and rail and structural steel mills, as well as on medium and small section sheet mills.

The samples of standard PP-Np-25Cr5VMoSi flux cored wire were fabricated in laboratory conditions. Multilayer surfacing of samples was carried out as follows: sheets were pre-heated up to 350^0C and cooled down slowly after surfacing. 5-layer surfacing of 09G2S steel sheets was performed by welding tractor ASAW-1250 with fabricated flux cored wire, surfacing conditions: welding current I_a=400A, arc voltage U_a = 32W, rate of welding υ_n=0.8 cm/s.

When wire fabricating concentration of silicon, manganese, chromium, vanadium was varied more enormously than values of these elements required for PP-Np-25Cr5VMoSi wire by GOST 26101-84. Nickel was added to chemical composition of some samples, and amorphous carbon was changed into carbon and fluorine containing dust (wastes of metallurgical production), chemical composition was as follows, mass %: Al_2O_3 = 21 – 46; F = 18 – 27; Na_2O = 8 – 15; K_2O = 0.4 – 6%; CaO = 0.7 – 2.3; SiO_2 = 0.5 – 2.5; Fe_2O_3 = 2.1 – 3.3; C = 12.5 – 30.2; MnO = 0.07 – 0.9; MgO = 0.06 – 0.9; S = 0.09 – 0.19; P = 0.10 – 0.18. The dispersion of this material enables good mixing with metal constituent of flux cored wire burden. Previously conducted research into application of this material as an additive to welding flux agents revealed that carbon as a component of carbon and fluorine containing dust is super active, it deoxidizes metal and reduces oxides, which are there in metal and slag, as the result, carbon oxides form. To simulate oxidation and reduction processes, taking place when surfacing, weakly oxidizing flux agent AN–67 was used according to GOST R 52222-2004. Manganese was not added to chemical composition of the wire because of its reduction from forming slag. Silicon was reduced partially from slag. The experiments demonstrated that forming slag hardly ever oxidized alloying elements of added metal.

Chemical composition of added metal was determined by roentgen-fluorescent method with spectrometer XRF-1800 and by atom-emission method with spectrometer DFS-71. Metallographic research into micro-sections was carried out by optical microscope OLYMPUS GX-51 in bright field in different zooming modes after etching in alcohol solution of nitric acid, as well as in solution of hydrofluoric acid. Hardness was measured by ultrasonic hardness testing instrument UZIT -3. The information on chemical composition of flux cored wires, hardness of added metal and volume concentration of retained austenite is provided in Table 1.

If amorphous graphite was used as an additive in the structure of added metal (samples 1 and 2) there is martensite and retained austenite in axle spacing. Some oxygen inclusion lines are found.

Point vanadium and chromium carbides are distributed over the grain. It is worth noting, that adding more than 3% vanadium caused a considerable drop in added metal hardness.

Metallographic analysis has demonstrated that metal structure, which was surfaced with flux cored wire, includes an acicular troostite with martensite and formed along the borders of grains separate thin austenite fringes containing some carbide impurities. Carbon and fluorine containing additive to wire burden facilitated removing of oxide non-metallic impurities, smoother distribution of carbon in weld metal and formation of fine-dispersed carbides, all this may improve thermal strength in practice.

Austenite forming nickel added to flux cored wire facilitated additional grinding of grains. If the system C—Si—Mn—Cr—V—Mo of the wire was added carbon and fluorine containing additive and nickel as well the degree of added metal impurity by oxide non-metallic inclusions dropped, grain grinding was improved, and fine-disperse carbides formed. In case of application it can improve thermal resistance of forming rolls.

Table 1. Chemical composition of tested flux cored wires and added metal hardness of the system C—Si—Mn—Cr— V—Mo

№	Mass concentration of elements, %											HRC (HB)	CRA
	C	Si	Mn	Cu	Cr	Mo	Ni	Al	W	V	Ti		
1	0.16	1.77	2.25	0.26	3.51	1.51	0.16	0.044	0.107	3.25	0.018	(186.2)	-
2	0.15	1.59	1.67	0.18	0.35	1.11	0.16	0.032	0.06	0.35	0.006	43.2	4.3
3	0.20	1.33	2.16	0.17	3.52	1.51	0.13	0.04	0.09	0.36	0.020	48.1	5.6
4	0.24	0.93	2.11	0.15	3.18	1.31	0.21	0.054	0.071	0.46	0.021	51.8	15.6
5	0.31	1.19	2.13	0.20	3.50	1.34	0.32	0.064	0.074	0.47	0.022	53.4	14.7
6	0.24	0.92	1.93	0.22	3.00	1.85	0.38	0.034	0.081	0.43	0.014	54.14	8.4

Notes: 1) CRA – volume concentration of retained austenite,
2) samples 1 and 2 included amorphous graphite;
3) samples 3 and 6 included carbon and fluorine containing dust with different concentration.

We have mentioned above that the most stressed heat forming rolls are strengthened by surfacing with PP-Np-35W9Cr3SiV flux cored wire according to GOST 26101-84, which contains more valuable tungsten. From the experience of surfaced forming rolls maintenance we know that their surface doesn't wear evenly. In the process forming rolls are subject to cyclic thermomechanical stresses, corrosion and abrasive wear. Stripe wear caused by added metal with different structure is possible and caused by added metal areas of different structure and hardness. Such areas can be found when multilayer surfacing of alloyed steels with overlapping previously surfaced bead. 35W9Cr3SiV added metal is well-wear resistant at high temperatures, but its thermal strength isn't sufficient enough, therefore, beads, surfaced with this wire can often fail because of heat cracks and chipping.

The samples of standard PP-Np-35W9Cr3SiV flux cored wire, which is used for surfacing hot forming rolls and rolls of roller conveyers were fabricated in laboratory conditions. Multilayer surfacing was carried out as follows: sheets were pre-heated up to 350 ^{0}C and then slowly cooled down. 5-layer surfacing of 09G2S steel sheets was performed by welding tractor ASAW-1250 with fabricated flux cored wire, surfacing conditions: welding current I_a=400A, arc voltage U_a = 32W, rate of welding υ_n=0.8 cm/s.

When wire fabricating concentration of valuable tungsten and chromium was calculated lower than required for PP-Np-35W9Cr3SiV wire according to GOST 26101-84. A number of wire samples were added nickel and amorphous carbon was changed into carbon and fluorine containing dust. Chemical composition of added metal was tested by roentgen-fluorescent spectrometer XRF-1800 and atom-emission spectrometer DFS-71. Metallographic tests of micro-sections were conducted by optical microscope OLYMPUS GX-51 in bright field in different zooming conditions after etching in alcohol solution of nitric acid, as well as in solution of hydrofluoric acid. The information on chemical composition of flux cored wires, and hardness of added metal is provided in Table 2.

Provided that amorphous graphite is used as an additive (sample 1) added metal has a dendrite structure (martensite with retained austenite in axial spacing). There are a lot of oxygen inclusion lines, which are stress concentrators and areas of brittle cracks formation. Point tungsten and chromium carbides are distributed over grains.

Metallographic analysis has demonstrated that metal structure, which was surfaced with flux cored wire, includes an acicular troostite with martensite and formed along the borders of grains separate thin austenite fringes containing some carbide impurities. Carbon and fluorine containing additive to wire burden facilitated removing of oxide non-metallic impurities, smoother distribution of carbon in weld metal and formation of fine-dispersed carbides, all this may improve thermal strength in practice. Nickel was added to flux cored wire to improve thermal resistance of metal through stabilization of austenite and grain grinding. The change in volume concentration of retained austenite in a surfaced layer with various concentration of nickel is provided in Table 3.

Therefore, experiments demonstrated that carbon and fluorine containing additive and nickel added to wire of the system C—Si—Mn—Cr—W—V could reduce considerably the degree of added metal impurity by oxide non-metallic substances, and facilitated formation of fine-dispersed carbides, thus, thermal resistance of forming rolls can improve in practice. Chemical composition of these flux cored wires and their modifications is protected by patents of Russian Federation [13, 14].

Table 2. Chemical composition of tested flux cored wires and added metal hardness of the system C—Si—Mn—Cr— W—V *

№	Mass concentration of elements, %											
	C	Si	Mn	Cu	Cr	Mo	Ni	Al	W	V	Ti	HRC
1	0.27	1.62	1.42	0.19	2.93	0.14	0.14	0.04	11.39	0.47	0.02	49
2	0.23	0.69	1.04	0.21	1.92	0.09	0.16	0.039	5.58	0.15	0.019	50.8
3	0.24	0.83	1.83	0.24	2.16	0.09	0.29	0.054	7.49	0.27	0.019	53.3
4	0.26	0.77	1.72	0.26	2.17	0.09	0.33	0.042	6.12	0.29	0.016	53.0
5	0.33	1.37	1.06	0.22	2.80	0.10	0.61	0.07	10.71	0.42	0.04	54.8

* sample 1 included amorphous graphite,
samples 2-5 were fabricated of carbon and fluorine containing dust and nickel powder with different concentrations.

Table 3. Volume concentration of retained austenite in an added layer (the mean in three fields)

Test	2	3	4
Volume concentration of austenite, %	15.4	5.2	6.8

The most defective goods when welding are caused by low qualification of welders. Taking in account, that the most significant fact is the length of the arc and constant tracking results in ocular fatigue and decreasing training efficiency, various training devices are used at the first stage of welders training.

To develop correct psychomotor skills of the trained person focusing on keeping standard parameters (the length of arc space, angle of welding electrode incline, displacement from the grooving center of workpieces to be welded, welding rate) in different attitude positions, to increase training efficiency through controlling required parameters of welding and reducing costs a training device is used in the course of training at the department (Figure 4); it consists of the following elements: 1 – screen; 2 – lamp; 3 – simulator of electrode holder; 4 – headphones; 5 – unit of signal receiving.

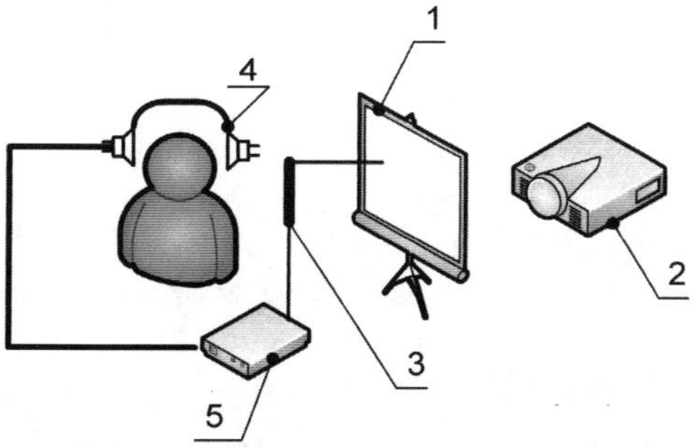

Figure 4. Principal plan of welder training device

Welder training with the training device is based on the following methods. On the screen there are transparent lines imitating the trajectory of arm movement in conditions of various welding techniques. If the simulator of electrode holder is over the transparent line the light from the source gets into the photoreceiver, which is placed on the top of electrode model, at the same time there is no acoustic signal meaning that the technique of welding is correct. Provided that the top of electrode model is displaced for a period of time within the limits 0.5 – 1 s, an acoustic signal is given meaning the displacement from the set trajectory, an error, respectively. On the light-proof screen there are various light-proof lines, imitating different depths of base metal fusion and weld formation, depending on transverse vibrations of the electrode. After the skills of movement are mastered by a welder he improves them through complicating the trajectory of a light-proof line. Training device makes it possible to consolidate and unify the technical skills of electrode movement and create a correct electrode inclination angle towards the product and welding direction; in this way quality of weld formation is predominated.

From the experience of training device application in groups of welders while pre-training for welder certification we know that after training in groups without training device the number of junctions in reference samples of welds meeting the requirements of quality amounted to 45-50 %, whereas after training with a training device it was 87-95%. The training device is protected by patent of Russian Federation [15].

Conclusions

1. The technologies of new carbon containing additives to welding fluxes have been developed, which can reduce the level of steel oxide non-metallic impurity, as well as gas content in a weld, and improve required mechanical properties. These technologies are introduced into production process and used for welding bulk-oil tanks maintained in conditions of low temperatures. Flux-additive FD-UFS is produced and protected by patent of the Russian Federation.
2. Flux cored wires of the systems C—Si—Mn—Cr—W—V and C—Si—Mn—Cr—V—Mo have been developed and protected by patent of the Russian Federation with the added carbon and nickel containing elements used for surfacing forming rolls. Application of flux cored wires when surfacing provides efficient removing oxide non-metallic impurities, even distribution of carbon in weld metal, and improvement of thermal resistance of surfaced product.
3. For training and formation of correct psychomotor skills when manual arc welding a training device of welder has been developed and introduced into the training process, as well as patented in the Russian Federation.

The work was performed in Siberian State Industrial University within the project of State Order of Ministry of Science Russian Federation № 11.1531.2014/k. For tests, investigations, measuring we used equipment of Collective Center "Material Science" of Siberian State Industrial University.

References

[1] Rafael Quintana Puchola, Jeily Rodríguez Blancoa, Lorenzo Perdomo Gonzaleza, Gilma Castellanos Hernándeza & Carlos Rene Gómez Péreza The influence of the air occluded in the deposition layer of flux during automatic welding: a technological aspect to consider in the quality of the bead *Welding International* 2009 Volume 23 №2. p 132-140.

[2] Crespoa A C Puchola R Q Goncaleza L P Sanchezb L G Gomez Pereza C R Cedrea E D Mendeza T O & Pozola J A Obtaining a submerged arc welding flux of the MnO–SiO2–CaO–Al2O3 – CaF2 system by fusion *Welding International* 2007 Volume 21 №7 p 502-511.

[3] Amado Cruz Crespoa, Rafael Quintana Puchola, Lorenzo Perdomo Gonzáleza, Carlos R. Gómez Péreza, Gilma Castellanosa, Eduardo Díaz Cedréa & Tamara Ortíza Study of the relationship between the composition of a fused flux and its structure and propertie Welding *International* 2009 Volume 23 №2 p 120-131

[4] Using a new general-purpose ceramic flux SFM-101 in welding of beams / Yu. S. Volobueva, O. S. Volobueva, A. G. Parkhomenko, E. I. Dobrozhelac & O. S. Klimenchukd // Welding International.– 2012. - Volume 26. - №8. - pp. 649-653

[5] Special features of agglomerated (ceramic) fluxes in welding / V. V. Golovko & N. N. Potapov //Welding International. – 2011. - Volume 25. - №11. - pp. 889-893.

[6] Manufacture of vertical bulk-oil storage tanks for northern climates using special welding materials/ N.E. Kryukov, I.N. Kovalsky, N.A. Kozyrev, V.F. Igushev, R.E. Kryukov // Steel in Translation. -2012. - V. 42. - № 2.- pp. 118-120.

[7] Pat. 2465108 RF, MPK 8 B23 K9/18, B23 K35/362 Submerged arc welding / N.E. Kryukov, I.N. Kovalsky, N.A. Kozyrev, V.F. Igushev, R.E. Kryukov; Open Joint Stock Company «Novokuznetsk Plant of Reservoir Metalware» named after N.E. Kryukov.- № 2011123342/02(034573), Application. 08.06.2011.

[8] Pat. 2467853 RF, MPK 8 B23 K35/362 Ceramic flux-additive/ N.E. Kryukov, I.N. Kovalsky, N.A. Kozyrev, V.F. Igushev, R.E. Kryukov; Open Joint Stock Company «Novokuznetsk Plant of Reservoir Metalware» named after N.E. Kryukov.- № 201112341602/02(034654), Application. 08.06.2011.

[9] Effect of carbon- and fluorine-containing additions in the composition of fluxes on the content of nonmetallic inclusions and properties of welded joints / N.A. Kozyrev, V.F. Igushev, R. Ev. Krukov, Z.V. Goldun, I.N. Kovalsky // Welding International. – 2013. – Vol. 27. - № 12, pp. 963 – 965.

[10] Kozyrev N A Krukov R E Kolmogorov D E New carbon-fluorine containing additive for the welding fluxes *Mechanics and materials* vol.682(2014) p 495-498.

[11] Pat. 2484936 RF, MPK 8 B23 K35/362 Ceramic flux-additive / N.A. Kozyrev, V.F. Igushev, R.E. Kryukov, S.V. Goldun; Federal State Budgetary Educational Institution of Higher Professional Education «Siberian State Industrial University. - №2012104939/02(007484), Application. 13.02.2012.

[12] Pat. 2492983 RF, MPK 8 B23 K35/36 Welding flux / N.A. Kozyrev, V.F. Igushev, R.E. Kryukov, S.V. Goldun, O.E. Kozyreva; Federal State Budgetary Educational Institution of Higher Professional Education «Siberian State Industrial University.- №2012108414/02 (012721), Application. 05.03.2012.

[13] Pat. 2518211 RF, MPK 8 B23 K35/368 Flux cored wire / N.A. Kozyrev, V.F. Igushev, D.A. Titov, O.E. Kozyreva, S.N. Starovatskaya; Federal State Budgetary Educational Institution of Higher Professional Education «Siberian State Industrial University.- №2013121468/02 (031580), Application. 07.05.2013.

IOP Conf. Series: Materials Science and Engineering **91** (2015) 012013 doi:10.1088/1757-899X/91/1/012013

[14] Pat.. 2518035 РФ, МПК 8 B23 K35/368 Flux-cored wire / N.A. Kozyrev, V.F. Igushev, D.A. Titov, O.E. Kozyreva; Federal State Budgetary Educational Institution of Higher Professional Education «Siberian State Industrial University.- №2013128792/02 (042890), Application. 24.06.2013.

[15] Pat. 2447515 RF, MPK 7 G09B19/24 Training device of arc welding / V.M. Shurupov N.A. Kozyrev. -№ 2011104516/12 (006345), Application. 08.02.2011.

Effect of Heat Input Pulse on the Structure and Properties of Welded Joints of Steels Ferritic-Pearlitic Class, Operating Under Low-Frequency Temperature-Force Loading*

Y N Saraev [1, 3, a], **V P Bezborodov** [1,3,b], **E A Putilova** [2,c]
[1]Institute of the Physics of Strength and Materials Science SB RAS, Tomsk, Russia
[2]Institute of Engineering Science, Ural branch of RAS, Ekaterinburg, Russia
[3]National Research Tomsk Polytechnic University, Tomsk, Russia

[a]e-mail: litsin@ispms.tsc.ru; [b]e-mail: val@ispms.tsc.ru; [c]e-mail: tuevaevgenya@mail.ru.

Abstract. We have investigated the influence of the modes of adaptive pulse-arc welding and surfacing on the structure and physical-mechanical properties of welded joints of steel 09Mn2Si and the surfaced composition of this steel coated with modified powder material of chromium carbide with the submicrocrystalline structure. It is shown that the pulsed mode of welding and surfacing can improve the homogeneity of the structure of the welded joint of steel and surfaced coating and reduce the grain size of metals in both of them. Structural changes lead to the increase in ductility and toughness of the weld metal.

Introduction

The problem of increasing the serviceability of the products containing welded joints and surfaced coatings is quite urgent. One of the most effective trends of enhancing the performance characteristics of high-duty constructions is improvement of the welding procedure and surfacing due to application of new welding materials, modified by multiphase structural compositions [1-10], and methods of adaptive pulsed-arc welding and surfacing that provide the opportunity to control the nature of micrometallurgical processes behavior due to the pulsed modification of power mode parameters [11-15]. At the same time owing to a programmable input of heat into the area of a welded joint, control of the melting processes and transfer of every droplet of the electrode metal, it is possible to obtain a disperse structure in weld metals, coatings and their heat-affected zones in the areas of permanent joints of engineering systems. This allows a significant decrease in the degree of residual strains and enhancement of product operational reliability.

When surfacing carbon steel the hardening occurs owing to formation of a new surface layer. The surface properties after surfacing the steel depend on the type of alloying elements determining the phase constitution, boundaries of phase transitions and mechanical properties. Hard alloys possess the properties determined by the chemical composition (carbides, borides, nitrides). One of the methods of increasing the strength of steel is grain refining. During this process, the steel yield strength increases and simultaneously the cold-shortness threshold decreases. The process-dependent parameters of surfacing as well as the quantity and the size of alloying elements influence the structure and properties of the weld metal. When the mode of surfacing is changed, the process of material melting and chemical homogeneity of the pad weld. In particular, the wear resistance and impact resistance of

IOP Conf. Series: Materials Science and Engineering **91** (2015) 012014 doi:10.1088/1757-899X/91/1/012014

the hardened parts are influenced by the fact of retaining of the strengthening phases (carbides, borides, nitrides) and their form, sizes, and disposition in a matrix material in the process of surfacing. The matrix material is characterized by the degree of fixedness of solid inclusions and the capability to take alternating loads. The resistance of the matrix material to different loads depends on the operational conditions of the product, nature of load actions, and operating environment.

The object of the paper:
Enhancement of the structure and properties of weld metals and surfaced coatings by means of application of the methods of pulsed welding procedure and surfacing during modification of the melted metal by composite powder materials possessing submicrocrystalline structure.

Materials and research methodology
In the paper samples of weld joints and surfaced coatings of steel 09Mn2Si have been studied. Welding and surfacing were conducted by means of the electrodes of UONI 13/55 model, FEB-315 "MAGMA" power supply with control panel "Pulse" for realization of the pulsed-arc welding process. The welding procedure and surfacing were conducted with the use of direct current (DCW) and welding with current modulation (CMW) which is known as adaptive pulsed-arc method. The registration of the parameters of welding and surfacing processes was conducted by means of apparatus AWR-224 MD. Welded joints were subjected to radiographic inspection by X-ray unit "PION-2M" according to GOST 7512-82 methodology. The macrostructure was studied by means of the optical microscope. The microstructure was investigated by means of the optical microscope NEOPHOT-21. The measurements of microhardness of welded joints and surfaced coatings were conducted with the use of the microhardness tester Leika. The presence of chromium carbides was determined by the X-ray structural analysis on plant DRON 3M. Chromium distribution in the composition "surfaced coating– steel base (09Mn2Si)" was determined by means of X-ray microanalysis (XRMA). Bending impact tests were conducted according to GOST 1497-84 using testing machine UMM-5. The studies of fatigue crack life were carried out according to GOST 5264-80. The cycling tests of samples were conducted with the use of machine INSTRON 8802 at frequency of cycling equal to 5 kHz. Magnetic characteristics of different areas of welded joints of steel 09Mn2Si were registered by means of the complex Remagraph C-500.

The results of the experiment
In the initial state the base metal – steel 09Mn2Si possesses hardness equal to ~250 HV0.05, welding material hardness which is ~ 270 -300 HV0.05, weld-adjacent zone (WAZ) hardness equal to - 210-230 HV0.05. The distribution of microhardness in the studied samples corresponds to the twenty-percent criterion. Steel 09Mn2Si in the initial state has a lineage ferritic-pearlitic structure with grain sizes equal to ~10 μm, which corresponds to the 10th grain size index. The grain size of WAZ exceeds the size of initial grain, which is connected to metal overheating during the welding process. During welding with the use of direct current the area with the enlarged grain size is wider than it is during welding using current modulation which allows pointing out high heat inputs and long-term thermal influence of the glowing arc [2]. Metallographic studies of the welded joints revealed refining of the structural constituents of WAZ material ~1,5 times less after CMW. It is explained by the peculiarities of the pulsed mode of welding when an intensive mixture of the melt with formation of new crystallization centers occurs. After CMW the width of heat-affected zone is less, and there is not any grain growth in it, which positively influences the quality of the weld joint in general. The ferrite-carbide structure of the weld after DCW resembles Widmanstätten structure (Fig. 1, a). However, there is not any observable reduction in strength properties, which is characteristic for steel having the same structure. In the central part of the weld coarse bainite grains with ferrite grains on the boundaries are typical for the structure.

The grain size after DCW is up to ~1,5 times greater than after CMW (Fig. 2), which is due to the fact that the structure of the root and filling pad of the weld after crystallization is subjected to reheating during subsequent layup.

As a result of this heating recrystallization (normalization) takes place. During DCW the heating temperature of the weld pads is higher than the temperature of the phase recrystallization, and, as a consequence, the grain growth occurs.

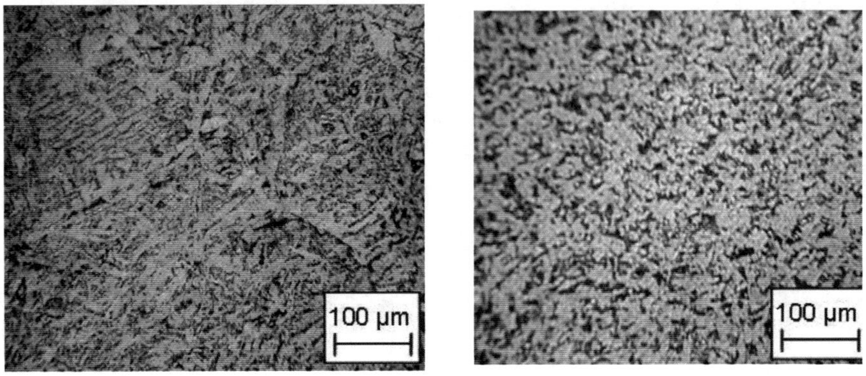

a b

Figure 1. Microstructure of the weld metal after welding: a) DCW; b) CMW

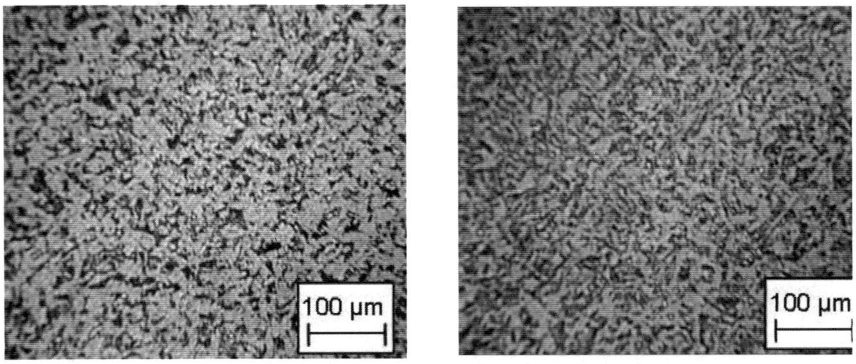

a b

Figure 2. WAZ microstructure of the weld at different modes of welding: a) DCW; b) CMW

During CMW the secondary heating takes place at the temperature close to that of polymorphic transformation of steel which contributes to structure refining. The increase in hardening in the weld zone correlates with the results of the mechanical tests for uniaxial tension. The properties of materials of the weld joint after CMW are given in table 1.

Table 1. Mechanical and magnetic properties of materials

Material	$\sigma_{0.2}$, MPa	σ_B, MPa	δ, %	ψ, %	$\sigma_{0.2}/\sigma_B$	H_c, A/sm	B_r, T	μ_{max}	M_{max}, T
WAZ	340	450	15	44	0,76	4,37	1,28	1430	2,13
Weld seam	375	480	11	51	0,78	5,95	1,00	824	2,23

It is important to note that the index of ratio $\sigma_{0.2}/\sigma_B$, which growth characterizes the reduction in the ductility of metal, for the weld material is lower than for the base metal and WAZ. WAZ material possesses maximum values of the ratio $\sigma_{0.2}/\sigma_B$, which indicates an increased probability of brittle structure taking place in this zone. A weld joint obtained by means of CMW is the most equal in strength: the change in the ratio of strength properties from zone to zone does not exceed 2,5 %. Samples of weld joints after tests at room temperature have plastic fracture in a zone of the rupture area, and at negative temperature - there is brittle fracture. The tests have shown that the impact strength of CMW metal of all the samples mentioned above is higher than it is in the base metal. Maximum impact strength (KCV=1,05 MJ/m^2) is reached on samples belonging to WAZ subjected to CMW, and the lowest values are obtained when KCV=0,86 MJ/m^2 with the use of DCW. Fractures of all tested samples have viscous (fibrous) structure. Fractograms of sample fractures after tests at room temperature justify that the fracture is tough with the pit size of 10-40 μm (Fig. 3).

This pattern is observed both in the case of the base metal and in the heat-affected zone. The parameter values of dynamic fracture strength correlate with the magnitude of toughness very well. They allow best revealing the advantages of WAZ metal after CMW in resistance to brittle fracture of metal in comparison to DCW mode. The values of the work on fracture propagation, obtained using the results of processing of impact diagrams, are 2-5 times higher than the values of work of fracture initiation for all tested samples that justifies the presence of significant structural strength margins in metal of different areas of weld.

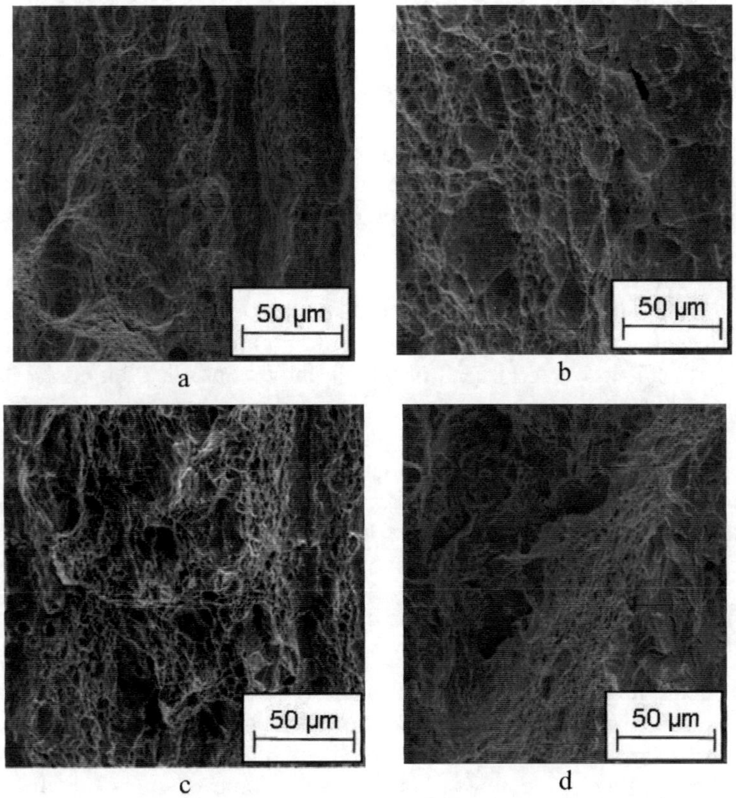

Figure 3. Fractographs of sample fractures after testing at the following temperatures: 1) 20 °C, base metal (a) and WAZ (b), 2) - 60 °C, base metal (c) and WAZ (d).

It is important to note that with the decrease in testing temperature up to –60°C the impact characteristics of the base metal change insignificantly. WAZ metal after CMW obtains a higher level of impact strength characteristics at -60°C. Sample fracture mechanism obtained by means of DCW after tests at a temperature of –60 °C for base metal is viscous and dimple, and for the weld adjacent zone it is quasi-spalling. However, for samples after CMW the rupture mechanism of the heat-affected zone is mixed: there are areas containing viscous dimple fracture. This fact indicates that this kind of welding mode decreases the risk of fracture occurrence in the weld adjacent zone.

The conducted comparative tests have revealed the advantages of fatigue crack growth in the field of stress intensity factor exceeding 20 MPa $\cdot m^{1/2}$ of the WAZ metal of the welded joint obtained by CMW.

The studies of magnetic properties of the base metal, weld materials and WAZ have shown that in the initial state the coercive force of the base metal is less than that of the metals of other zones. Increased magnetic hardness of WAZ metal and weld is conditioned by a greater number of alloying elements in the electrode material as well as by the presence of the dispersed bainite in their structure and an increased level of internal stresses in weld metals and WAZ metals. The magnetic texture of stresses (induced magnetic anisotropy) is formed [3]. If material magnetostriction and external stresses have like signs we can observe the preferred orientation of magnetic domains in the direction of the applied load, and therefore there is a decrease in values of coercive force, increase in magnetic permeability and residual induction along the axis of elongation.

When ready-made carbide compounds are introduced into the coating formed during surfacing, it is possible to obtain the structure of weld metal with strengthening carbide phase located in the immediate region of the grain. Powder materials on the basis of submicrocrystalline particles of chromium carbides were applied as modifying mixtures. For implementation of surface hardening different schemes of introduction of powder compositions containing submicrodimensional-hardening particles have been developed to apply coatings.

Alloying with ready-made compounds significantly simplifies the adjustment of alloy matrix structure, as in this case redistribution of alloying elements between the high hardness phase and matrix slows down. In the alloy structure there are fine carbide impurities spread all over the coating volume. In the process of coating the embedment of such phases in the metal matrix leads to formation of dispersion strengthened pads and enhancement of physical-mechanical and service properties of the obtained materials. The study and optimization of crystallization processes, temperature control of the melt on the account of the introduction of the dispersed refractory compounds, modifying coating material including refinement of their structure and enhancement of their properties, into the composition of the surfacing materials, all of which allows optimization of the technology of surface coating.

Hardening is achieved through a directed high energetic and modifying influence of dispersed high melting compounds on the structure, physical-mechanical and auxiliary properties of coatings made of metals and alloys applied by the surfacing methods. During pulsed-arc surfacing as a result of influence of repetitive live pressure of the arc, while surfaced coating is being formed, the metal of the welding bath carries out reciprocal motions owing to periodic power influence of the arc with modulation frequency of current. This kind of surfacing process behavior provides repetitive cycling recurrence of physical processes at the stages of formation of weld bath and crystallization of the coating metal out of the melt and enables its dynamic stirring. Such stirring of the melt contributes to levelling its heat content and provides setting of the required quantity of the melted metal under the arc by time of starting of current pulse operation thus contributing to the reduction of the weld penetration. Periodic motion of the metal in melt enables levelling of its heat content and more uniform distribution of alloying elements throughout the whole volume of the melted metal.

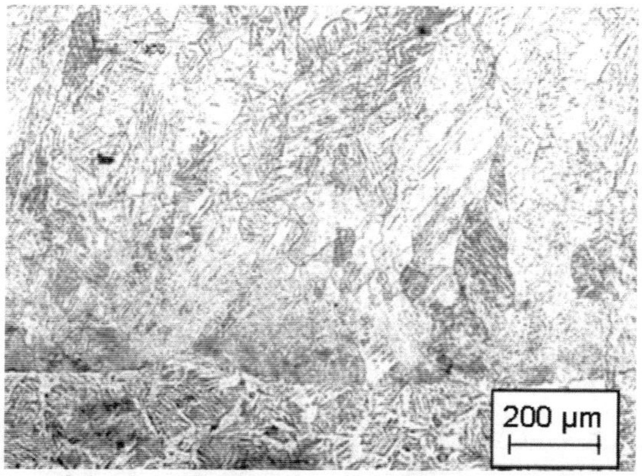

Figure 4. The structure of the interface region “surfaced coating – steel base (09Mn2Si)”. (Coating - top.).

The interface region “surfaced coating – steel base (09Mn2Si)” and the traces of microhardness imprints after pulsed-arc weld surfacing are shown in figures 4 and 5. Chromium distribution on the boundary of interface “surfaced coating – steel base (09Mn2Si)” after pulsed-arc weld surfacing obtained by means of XRMA and diffractogram of surfacing are shown in figure 6. The fact of chromium distribution is evidence of a more significant content of this element in the surfaced pads.

Thus, a new complex approach to improvement of properties of welded joints and surfaced coatings of the products made of low-carbon steel 09Mn2Si by means of application of the method of adding materials modification using superdispersed high melting particles and pulsed technology has been proposed.

The use of the consumables in the batch – in electrodes and wires of the components containing microalloying additives allows influence on the structure and properties of the molded metal – the welded joint and surfaced coating. During modification the structure of weld metal is refined and, as a consequence, physical-mechanical properties and performance characteristics are improved.

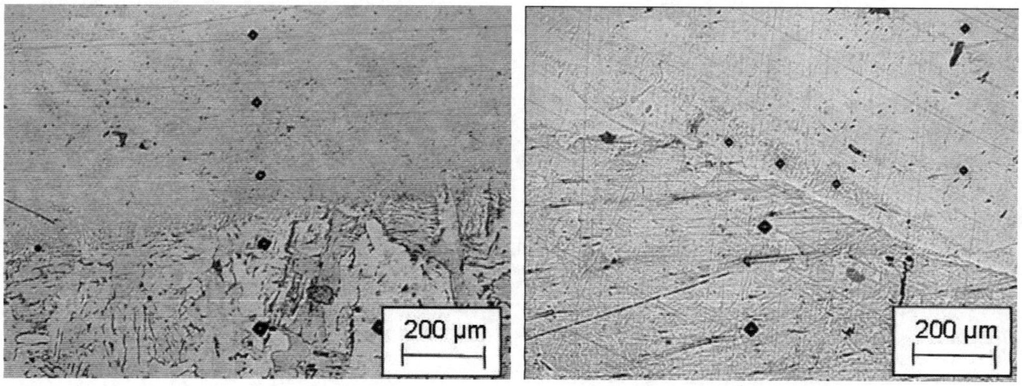

Figure 5. Traces of microhardness imprints in the interface region coating – base. The base has been etched (a) and slightly etched (b). The coating (top) has not been etched

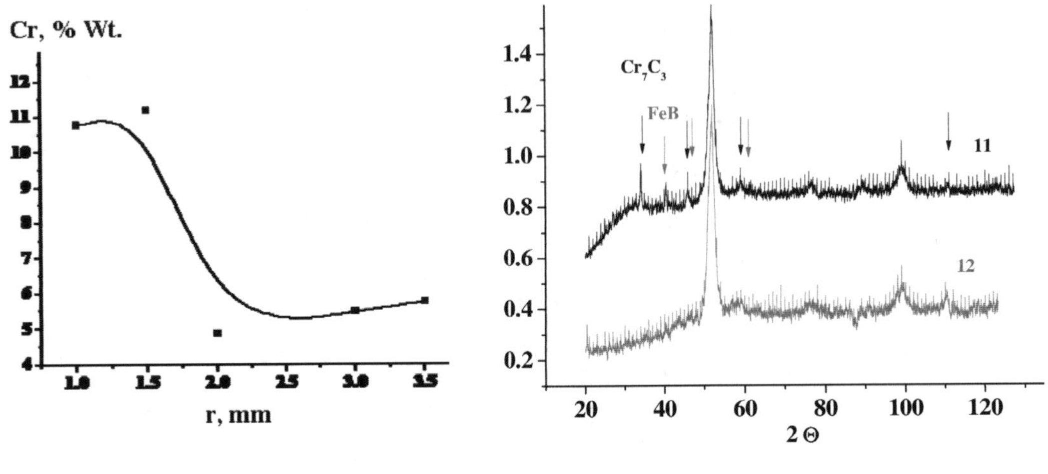

a b

Figure 6. Chromium distribution at the interface "surfaced coating - steel base (09Mn2Si)" (a) and the diffractogram of coating (b)

The novelty of the approach under development is in providing of the continuous monitoring over metallurgical processes at the stage of formation of permanent joints by means of complex application of new additive materials, modified by multiphase compositions, application of the equipment and technologies of adaptive pulsed-arc welding and surfacing, which allows control over the processes of melting and transfer of each drop of electrode metal by means of programmable heat input into the area of welded joint, and, as a consequence, providing formation of the dispersed structures in the areas of welded joints of technical systems, which leads to an increased resistance of these joints to delayed and brittle fracture. This allows a substantial decrease in the degree of the residual deformations in the areas of structural heterogeneity and, as a consequence, enhancement in operational reliability of the products.

Application of the pulsed mode of welding provides significant refinement of the structure of weld metal and HAZ, which is most subjected to structural changes, owing to adjustable heat input, as a result of which there is plummeting in microhardness and strength in this zone and thereby the degree of softening in the fusion zone of the weld is reduced. If combined these changes provide the reduction in postwelding deformations and an increase in toughness of welded joints.

Conclusions

1. Application of the method of pulsed-arc welding of low-carbon steel 09Mn2Si allows formation of welded joints with more fine-grained structure in weld metals and HAZ, as well as with uniform distribution of hardness in them in comparison with compounds obtained by means of arc welding using direct current.
2. The result of mechanical tests for uniaxial tension correlates with hardness increase in the weld zone. The maximum values of the ratio of yield and tensile strengths belong to WAZ material. The most full-strength weld is that obtained by CMW technology, for which the change in the ratio of strength properties of separate zones of the welded joint does not exceed 2,5%.
3. The maximum value of toughness at room and negative temperatures of tests (-60°C) is reached after application of CMW. The values of the parameter of dynamic fracture toughness properly correlate with the value of toughness, which allows revealing neatly the advantages in brittle fracture resistance of WAZ metal after CMW.
4. Comparative tests have revealed the advantages of the fatigue crack growth of WAZ metal of the welded joint obtained by CMW in comparison with the base metal. It has been established

that at room temperature in a low-cycle region the durability of weld metal of welded joints of 09Mn2Si steel obtained by means of CMW is higher than those obtained by means of DCW.

5. It has been found that the analysis of field dependencies of differential magnetic permeability allows receiving data on the condition of separate components of the welded joint. The obtained characteristics can be applied for estimation of the degree of deformation at the stage of elastic stresses in the products during operation.

6. Application of the method of pulsed-arc surfacing along with modification of the melted metal with composite powder material of chromium carbide, which possesses a submicrocrystalline structure, allows obtaining such dispersed strengthening phase in the coating of low-carbon 09Mn2Si steel.

***Acknowledgements**
This work was done in the framework of the basic research of the state academies of sciences of the Russian Federation for 2013-2020 years. and Basic Research Program of the Presidium of the Russian Academy of Sciences' Search basic research for the development of the Arctic zone of the Russian Federation ", code:" Arctic-2014 ".

References

[1] Paton B Ye Lebedev V A Pichak V G Analysis of technical and technological capabilities of the pulsed feed of electrode wire in the processes of arc welding and surfacing, Welding engineering 2002 # 2 p 24-31
[2] Paton B Ye, Sheiko P P Zhernosekov A M Stabilization of the process of pulsed-arc welding by means of the consumable electrode, *Automatic welding*, 2003 # 8 p 3-6
[3] Ryzhov R N Influence of pulsed electromagnetic control on the formation and crystallization of welds, *Automatic welding* 2007 # 2 p 56-58
[4] Lebedev V A Some peculiarities of the arc mechanized welding of aluminum using controlled pulsed feed of electrode wire, *Welding engineering* 2007 # 11 p 26-30
[5] Lobanov L M Lebedev V A Maksimov S Yu New capabilities of mechanized arc spot welding using impulse actions, *Automatic welding* 2012 # 5 p 17-22.
[6] Mashin V S Pashulya M P Shonin V A Pulsed-arc welding using a consumable electrode in argon of light-gage aluminum alloys *Automatic welding* 2010 # 5 p 49-53
[7] Mozok V M Lebedev V A Additional peculiarities of the technologies of arc mechanized and automatic welding with pulsed feed of electrode wire *Welding engineering* 2010 # 2 p 34-38
[8] Chang Yunlong Che Xiaoping He Youyou Hi Guangxu Influence of ongitudinal magnetic field on metal transfer in MIG arc welding *China Weld* 2008 Vol. 17 # 2 p 27-31
[9] Huang Jiankang Han Rihong Shi Yu Fan Ding Zhang Yuming J Mech. Eng 2012 Vol. 48 # 8 p 44-48
[10] Yao Ping Xue Jiaxiang Huang Wenchao Zhang Rui Analysis of droplet transfer of pulsed MIG welding based on electrical signal and high-speed photography *China Weld* 2009 Vol. 18 # 1 p 67-71
[11] Saraev Y N Adaptive pulsed-arc methods of mechanized welding in the construction of transmission pipelines *Welding International* 2002 # 16 (6) p 475 – 480
[12] Saraev Y N Sleptsov O I Bezborodov V P et al Formation of the structure and properties of welded joints in the circumstances of controlled heat input during pulsed-arc welding *Physical mechanics* 2005 # 8 (Spec. issue) 3 p 137-140
[13] Saraev Y N Bezborodov V P Grigoryeva A A Golikov N I Dmitriev V V and. Sannikov I I Distribution of residual stresses in welded joints in 09G2S steel produced by adaptive pulsed-arc welding *Welding International* 2015 Vol. 29 # 2 p 131–134
[14] Grigorieva A A Saraev Y N Bezborodov V P Control of structure and properties of welded connections of alloy steels by application highly concentrated pulse energy actions *Applied Mechanics and Materials* Vol. 682 (2014) p 261-265, Submitted: 07.07.2014. Online

available since 2014/Oct/27 at www.scientific.net Accepted: 31.07.2014 © (2014)
[15] Zaikova V A, Shur Ya S Concerning the influence of strain on magnetic properties and magnetostriction curves of silicon iron *Metal Science* 1966 Vol. 51 # 5 p 664-673

X-Ray Structural Study of 09Nn2Si Steel Welded Joints

N I Golikov [1,a], A A Platonov[1], Y N Saraev [2,3,b]

[1]Institute of North Physical and Technical Problems named after V.P. Larionov, SB RAS, October st.1, Yakutsk, 667980, Russia
[2]Institute of Strength Physics and Material Science RAS, av. Academic, 2/4, Tomsk, 634055. Russia
[3] National Research Tomsk Polytechnic University, av. Lenin, 30, Tomsk, 634050, Russia

[a]e-mail: administration@iptpn.ysn.ru; [b]e-mail: litsin@ispms.tsc.ru.

Abstract. The article is devoted to handling a vital scientific and technical problem of improving operational reliability and safety of critical constructions, exploited in Siberia and Far North, by developing of new technological approaches to welding. In the article results of X-ray diffraction examinations of 09Mn2Si steel welded joints are given, produced by different welding operations. Resulting from researches, the authors have concluded that pulse-arc welding is the most preferred welding process as compared with direct current welding.

Introduction

It is known that production of cold-resistant welds of high strength steels is greatly hindered due to a number of specific factors. This is due, first of all, to arc burning when welding at low ambient temperatures, as well as to crystallization kinetics of the weld metal and to stresses and deformations in welded structures. Practice of exploitation of machines and constructions, employed in Siberia and Russia's Far North , has shown that performance of different types of machinery decreases in winter to 1.5 times if compared with a summer period. MTBF drops 2-3 times; an actual service life is reduced by 2-3 times as compared with a rated life [1].

Analysis of the data on failures of the equipment and structures operated at low temperatures shows that damages occur more frequently in welds as they have lower cold resistance if compared with the base metal [2]. Welding enhances grain growth, and further hydriding, which increases the cold cracking of welded joints. Furthermore, heat welding enhances phase transformations and secretion of impurities at grain boundaries, which also increases brittleness of steel [3]. It is known that cementite (iron carbide), as a phase component, occurs in the iron-carbon alloys even there is small carbon content [4]. In [5, 6] processes of carbide formation during thermal tempering of structural steels are examined.

It is shown that compositions of carbides, formed at various tempering temperatures of structural steels, are unstable. Carbides, generated at low and medium tempering temperature, contain less alloying elements than such carbides generated at higher tempering temperatures [5].

[6] shows a change in cementite volume fraction in 30CrNi3MoV cast steel, depending on tempering duration at 660° C; it is noted that an increase in the cementite volume fraction occurs in the first minute of tempering. Thus, it can be assumed that secretion of cementite phases are possible

when welding, even after brief exposure to temperatures; these could reduce the resistance to cracking in weld joints.

To solve the problem development of fundamentally new provisions of the welding theory is required, which in practical application of welding and surfacing will increase the operational reliability of welded structures and products for the North. The basis of these provisions, in our opinion, may be methods of adaptive control over complex electro-dynamic system: power supply - arc - weld pool – product, which provides conditions for formation of equal strength in zones of cold-resistant material joints through optimal heat input at welding [7,8].

In recent years, fundamental studies of welding processes and the nature of various defects when welding in the cold made it possible to establish physical nature of arc acting when welding at low temperatures [9]. This allowed developing a conceptual approach to reducing unsoundness and increasing strength and operational performance of welded joints stated in adaptive pulse arc welding method. It lies in the control over all stages of the formation of the welded joint: arc burning, melting, electrode metal transfer to the weld pool and weld metal crystallization through feedback channels that control the basic energy characteristics of the process, taking into account perturbing actions [7, 10]. This method became the basis of innovative technological solutions in welding and surfacing that increase the safety performance, survivability and operational reliability of technical systems operating in Siberia and the High North.

The purpose of the work is to develop concepts of properties of 09Mn2Si steel welded joints produced by DC arc welding or by pulse-arc welding at above-zero and sub-zero temperatures, performed by X-ray diffraction examination.

Research technique

Butt welds of 09Mn2Si steel were tested; welded samples according to State Standard 5264-80 were prepared for tests, connection of S17 type. Pre-welded plates with dimensions of 100x200x6 mm were lengthwise processed with a bevel angle of 30° and 2 mm dulling. Before welding the plates, for to avoid post-weld deformation, were assembled in a special jig with a clearance of 2.5 - 3.0 mm; tacks were arranged on the ends of the plates. One-side weld of sheet samples was performed in two runs, a root run and a facing run, with LB-52U electrodes. When welding the root run an electrode with diameter of 3.2 mm was used, and for the facing run – that of 4 mm.

The electrodes are certified by NAWC (National Association of Welding Control) and are widely used to weld various low-carbon and low-alloy steel constructions. Carbon content in LB-52U electrode wire is equal to carbon content in 09Mn2Si steel (Table 1).

Table 1. Chemical content (%)

Ø, mm	C	Si	Mn	P	S	Ni*	Cr*	Mo*	V*
3.2	0.06	0.51	1.02	0.011	0.006	0.01	0.02	0.01	trace
4.0	0.06	0.49	1.01	0.013	0.004	0.01	0.03	0.01	trace

* These elements are not added on purpose.

Welding of the samples was made with the inverter power supply FEB-315 «MAGMA» with modes available for DC welding and adaptive pulse-arc welding. Temperature conditions of ambient air were changed from +18°C, up to -48°C.Registration of welding parameters was performed using the device AWR-224 MD.

The average parameters of welding conditions are shown in Tables 2 and 3.

X-ray diffraction examinations were performed with a high-precision X-ray diffractometer ULTIMA IV Rigaku, using Co-κ_α radiation. The diffractograms were obtained from etched side surfaces of the samples. 4 samples of 09Mn2S steel welds were examined:

- №1 - modulated current welding at + 18°C;
- №2 - DC welding at + 18°C;

- №3 - modulated current welding at -48°C;
- №4 - DC welding at -48°C

X-ray exposure of the samples was performed in three areas: №1 - metal joint (MJ); №2 - at a distance of 1.5 mm from the fusion line; №3 - at a distance of 15 mm from the fusion line (see Figure 1). The results were compared with the reference diffractogram obtained by X-ray exposure of the annealed sample of the 09Mn2Si steel.

The first series of surveys was made in the angle range $42.00° \leq 2\theta \leq 162.00°$ at a scanning speed of 1 deg/min and scan step 0.02° with the aim of finding all peaks obtained from the sample. The second series of surveys was made in the angle range $50.00° \leq 2\theta \leq 58.00°$ at a speed of 0.1 deg/min and step scan 0.01° for more exact definition of the line of cementite (121) and the peak heights.

Table 2. Modes of DC arc welding

Run	I, A	U, B	$V_{св}$, m/h	Heat input, kJ / m
Root	90	25	4,9	1405
Facing	150	25	6,99	1642

Table 3. Modes of adaptive pulse-arc welding

Run	$I_и$, A	$I_п$, A	$t_и$, c	$t_п$, c	U, B	$V_{св}$, m/h	Heat input, kJ / m
Root	90	40	0,3	0,3	20	4,29	1017
Facing	180	50	0,3	0,3	24	6	1606

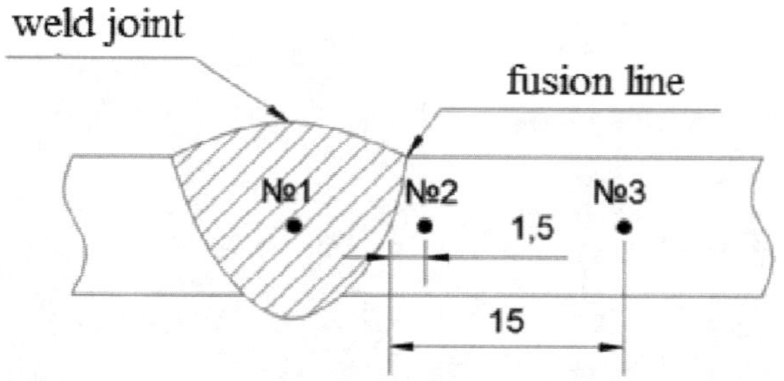

Figure 1. Diagram of X-ray exposure of samples and numbers of areas

Results and discussion

It is known that sensitivity of the X-ray examination is often insufficient for qualitative and quantitative determination of cementite in low-alloy steels. This statement is confirmed by the diffraction pattern shown in Figure 2. Steel 09Mn2Si refers to ferrite-pearlite class, but X-ray examination of the samples in the initial state does not show the presence of cementite phases. Obviously, sensitivity of the diffractometer is insufficient to identify a low content of cementite phase in the steel.

During welding, re-melting of welded material and its subsequent crystallization occurs with formation of zones of structural heterogeneity. Behavior of the material in this zone during operation determines the properties of permanent joints, especially if thermal-strength loading.

Study of the weld metal and heat affected zone of welded joints of steel, obtained at different welding modes, revealed significant differences in the diffraction patterns of the metal. Evidently, change in the welding thermal cycle facilitates quantitative growth of cementite phases; the percentage of which can be estimated by X-ray examination.

X-ray phase analysis is based on the fact that each phase has a specific crystal lattice with certain parameters and the corresponding system of lines on the radiograph. Therefore, in general, when

IOP Conf. Series: Materials Science and Engineering **91** (2015) 012015 doi:10.1088/1757-899X/91/1/012015

radiographing of the metal, which is a mixture of several phases, a radiograph is obtained, which contains all lines of the phases forming the sample [6].

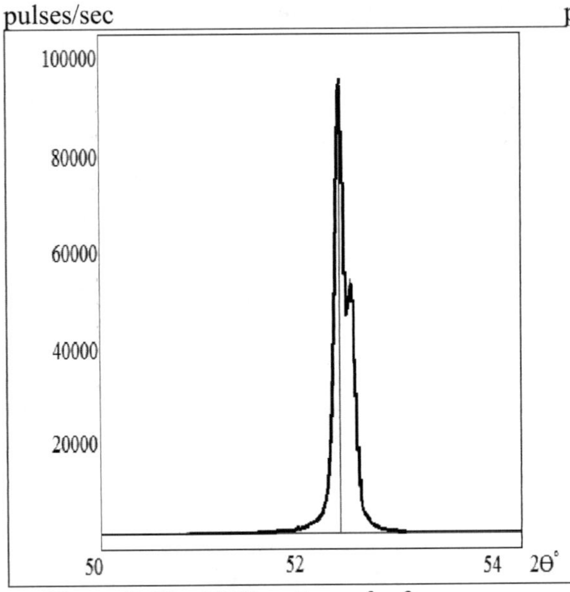

Figure 2. The XRD pattern of reference sample

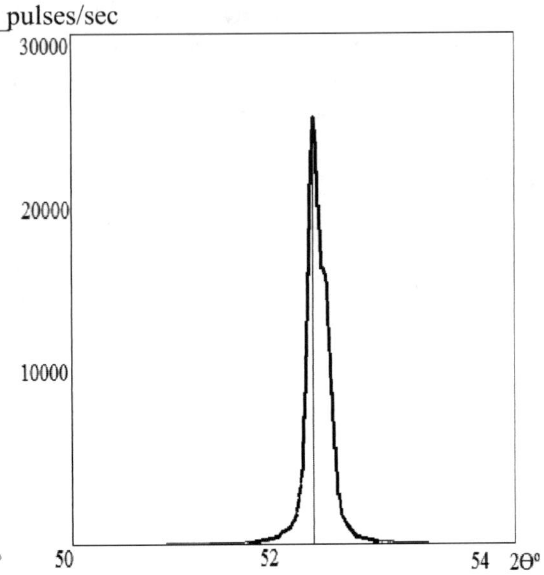

Figure 3. XRD pattern of area №3, sample №1

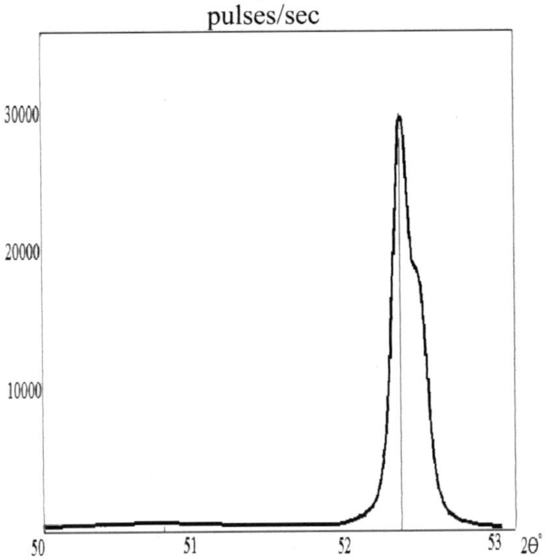

Figure 4. XRD pattern of the sample №2 weld metal

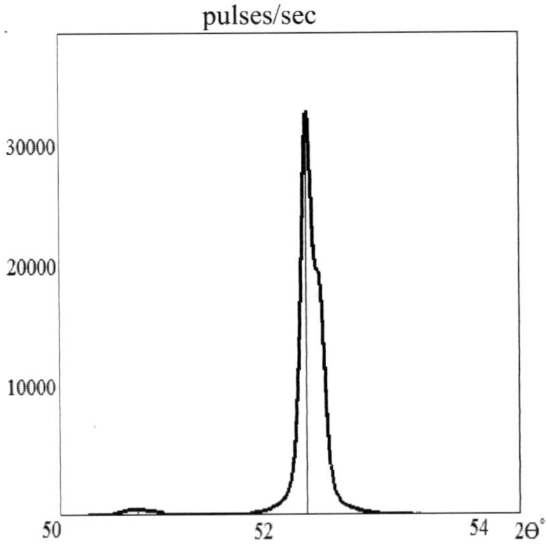

Figure 5. XRD pattern of the sample №4 weld metal

Intensity of different phase lines on radiographs depends on many factors, including quantity of a particular phase. With increase of phase content in the mixture, intensity of the lines, corresponding to it, increases. However, a reliable quantifying of a particular phase in a mixture is possible only if there are certain minimum quantities of it. Further reducing the amount of this phase will lead to almost complete disappearance of its lines on radiographs [11].

X-ray diffraction results are shown in Table 4, where θ - angle of reflection; I - intensity of diffraction lines from the reflection plane (121) and (110). In the reference sample there are no cementite lines (Fig. 2), and in the welds there are lines of cementite (iron carbide - Fe_3C), except section 3 of the sample1 (Figure 3).

For analysis, X-ray diffraction examination of the samples welded with different welding methods at above-zero and sub-zero temperatures was carried out (see Table 3). Analysis of the cementite diffraction lines showed that the highest intensity (and thus content) is found in the weld metal, area 1 (Fig. 1), and the smallest in area 3 at a distance of 15 mm from the fusion line (Fig. 1). Figures 4 and 5 show the diffraction patterns of the weld metal samples № 2 and № 4. Depending on the welding conditions the intensity of cementite lines changes. Thus, the DC welds (samples № 2 and № 4), show larger intensity than the pulse-arc welds (samples № 1 and № 3).

It was found that increasing of the cementite line intensity during welding at sub-zero temperatures as compared with welding at above-zero temperatures. As it is known, cementite is a brittle substance, and its high content in the steel reduces cold resistance. Table 4, column 7, shows the percentage of iron and cementite line intensities where the greatest value (1.4%) was found in the weld metal produced by DC welding at above-zero temperature.

Table 4. The results of X-ray diffraction study

Sample	№ area	2θ Fe_3C (121) [deg]	I Fe_3C (121) [cps]	2θ α-Fe (110) [deg]	I α-Fe (110) [cps]	% Fe_3C (121) от α-Fe (110)
1	2	3	4	5	6	7
№1	1	50.78	164	52.32	25380	0.65
	2	50.89	131	52.33	25151	0.52
	3	-	-	52.33	24424	-
№2	1	50.79	398	53.34	28574	1.4
	2	50.70	230	52.34	27908	0.82
	3	51.12	66	52.33	26945	0.24
№3	1	50.67	222	52.33	25171	0.88
	2	50.86	103	52.34	25309	0.41
	3	50.94	44	52.34	25737	0.17
№4	1	50.73	289	52.32	31427	0.92
	2	50.89	283	52.33	25120	1.13
	3	51.15	106	52.34	25469	0.42
Reference sample		-	-	52.36	91391	-

Conclusion

The study has established that in 09Mn2Si steel welded joints, in areas of weld metal and heat affected zone, iron carbides increase noticeably; their content is entirely dependent on the methods and welding conditions as well as temperature conditions of formation of welded joints.

In general, the use of adaptive pulse-arc welding enables to reduce the content of cementite in the joints welded at above-zero and sub-zero temperatures; that demonstrates advantages of these methods for welding ferritic-pearlitic steel structures if compared with methods of DC welding.

Acknowledgements

This work was done in the framework of the basic research of the state academies of sciences of the Russian Federation for 2013-2020 years. and Basic Research Program of the Presidium of the Russian Academy of Sciences' Search basic research for the development of the Arctic zone of the Russian Federation ", code:" Arctic-2014 ".

IOP Conf. Series: Materials Science and Engineering **91** (2015) 012015 doi:10.1088/1757-899X/91/1/012015

References

[1] Larionov V P Arc welding structures for northern *Nauka* Novosibirsk 1986

[2] Golikov N I Ammosov A P The strength of welded joints of tanks and pipelines operated in the North *NEFU* Yakutsk 2012 p 232

[3] Makarova E L (Ed) Welding and welding materials, walkability of materials *Metallurgy* 1991 Vol. 1 p 528

[4] Kalinin B D Rudnev A V The effects of diffraction X-ray fluorescence analysis and their use for analytical purposes *Analysis and control* 2011 15(3) p 332-338

[5] Tihonkova O V Popova N A Tsellermaer V V Gromov V E Kozlov E V Influence of tempering on the phase composition of cast steel structural intermediate-alloy steel *Proceedings of the higher educational institutions Iron and steel* 2007 p 52-56

[6] Mirkin L I X-ray inspection of engineering materials *Mechanical Engineering* Moscow 1979

[7] Saraev Y N Adaptive pulse-arc mechanized welding in the construction of main pipelines *Welding production* 2002 No 1 p 4-11

[8] Saraev Y N Sleptsov O I Bezborodov V P Nikonova I V Tyutev A V Fatigue failure and properties of welded joints of pipelines in Siberia and the Far North *Mounting and special works in construction* 2005 No 8 p 18-22

[9] Saraev Yu N Bezborodov V P Grigorieva A A Golikov N I Dmitriev V V and Sannikov I I Distribution of residual stresses in welded joints in 09G2S steel produced by adaptive pulsed-arc welding *Welding International* 2015 Vol. 29 No. 2 p 131–134 http://dx.doi.org/10.1080/09507116.2014.897804

[10] Lvov N S Gladkov E A Automation and automation of welding processes *Mechanical Engineering* Moscow 1982 p 302

[11] Umansky Y S Radiography of metals and semiconductors *Metallurgy* Moscow 1969 p 496

Some Aspects of Oxidation-Reduction Reactions under Carbon-Bearing Flux Welding

R E Kryukov [1,a], N A Kozyrev [1,b], G V Galevsky [1,c], Y V Bendre [1,d], V F Goryushkin [1,e], D V Valuev [2,f]

[1]Siberian State Industrial University, Novokuznetsk, Russia
[2]National Research Tomsk Polytechnic University, Yurga Technological Institute branch of Tomsk Polytechnic University 652050, Kemerovo region

[a]e-mail: Kozyrev_na@mtsp.sibsiu.ru, [f]e-mail: valuevden@rambler.ru

Abstract. The authors have completed thermodynamic calculations of oxidation-reduction processes under submerged arc welding with application of carbon containing additive. The calculations have shown that carbon, due to its high reducing properties under T = 1950 – 2200 K, introduced into the system can significantly decrease the amount of non-metallic inclusions in the weld metal and so improve its mechanical properties.

Introduction

Flux welding is associated with intensive mass transfer processes between the molten metal and slag formed out of the welding flux. The process is accompanied by the oxidation-reduction reactions of manganese, iron and silicon, i.e. oxygen involving transfer processes. Currently most grades of domestic fluxes applied for low-alloy steel welding are oxidizing and are based upon the principles of silicon-manganese oxidation-reduction processes. The resultants of the given reactions are the oxides of manganese, silicon, iron, aluminum, etc. which normally lag to come to the surface to be assimilated by the slag formed from the welding fluxes and this increases weld metal contamination with non-metallic inclusions leading to significant reduction of physical and mechanical properties. To avoid weld metal contamination it seems reasonable to use reducing agents forming reaction gases. Such reducing agent can be carbon which forms gaseous compounds CO_2 and CO when reacting with oxidizers.

We estimated the thermodynamic probability of reactions (1) – (24) which result in reduced non-metallic inclusions containment in the weld metal. The reducing properties of carbon were compared to those of other reducing agents in the studied system of liquid weld metal – oxide melt – gas:

$$(FeO) + [Mn] = [Fe] + (MnO), \quad (1)$$
$$(FeO) + 1/2[Si] = [Fe] + 1/2(SiO_2), \quad (2)$$
$$(FeO) + C_{(solid)} = [Fe] + CO_{(gas)}, \quad (3)$$
$$(FeO) + 1/2C_{(solid)} = [Fe] + 1/2CO_{2(gas)}, \quad (4)$$
$$(FeO) + CO_{(gas)} = [Fe] + CO2g, \quad (5)$$
$$(FeO) + 2/3[Al] = [Fe] + 1/3(Al_2O_3), \quad (6)$$
$$(MnO) + 1/2[Si] = [Mn] + 1/2(SiO_2), \quad (7)$$
$$(MnO) + [Fe] = [Mn] + (FeO), \quad (8)$$
$$(MnO) + C_{(solid)} = [Mn] + CO_{(gas)}, \quad (9)$$

$$(MnO) + 1/2C_{(solid)} = [Mn] + 1/2CO_{2(gas)}, \quad (10)$$
$$(MnO) + CO_{(gas)} = [Mn] + CO2g, \quad (11)$$
$$(MnO) + 2/3[Al] = [Mn] + 1/3(Al_2O_3), \quad (12)$$
$$(SiO_2) + 2[Mn] = [Si] + 2(MnO), \quad (13)$$
$$(SiO_2) + 2[Fe] = [Si] + 2(FeO), \quad (14)$$
$$(SiO_2) + 2C_{(solid)} = [Si] + 2CO_{(gas)}, \quad (15)$$
$$(SiO_2) + C_{(solid)} = [Si] + CO_{2(gas)}, \quad (16)$$
$$(SiO_2) + 2CO_{(gas)} = [Si] + 2CO_{2(gas)}, \quad (17)$$
$$(SiO_2) + 4/3[Al] = [Si] + 2/3(Al_2O_3), \quad (18)$$
$$(Al_2O_3) + 3/2[Si] = 2[Al] + 3/2(SiO_2), \quad (19)$$
$$(Al_2O_3) + 3[Fe] = 2[Al] + 3(FeO), \quad (20)$$
$$(Al_2O_3) + 3[Mn] = 2[Al] + 3(MnO), \quad (21)$$
$$(Al_2O_3) + 3C_{(solid)} = 2[Al] + 3CO_{(gas)}, \quad (22)$$
$$(Al_2O_3) + 3/2C_{(solid)} = 2[Al] + 3/2CO_{2(gas)}, \quad (23)$$
$$(Al_2O_3) + 3CO_{(gas)} = 2[Al] + 3CO_{2(gas)}, \quad (24)$$

The thermodynamic characteristics of reactions (1) – (24) necessary for reducing properties assessment under the standard conditions [ΔrH°(T), ΔrS°(T), ΔrG°(T)] were calculated by the known methods [1] in the interval of welding processes temperature of 1700 – 2200 K according to the thermodynamic properties of the reagents [[H°(T)- H°(298,15 K)], S°(T), ΔfH°(298,15 K)] taken from the reference books [2, 3].

1- The work was completed in SibSIU within the project part of the State task of Ministry of Education and Science of the Russian Federation № 11.1531.2014/k.

The following states $FeO_{(liquid)}$, $MnO_{(liquid)}$, $SiO_{2(liquid)}$, $Al_2O_{3(liquid)}$, $Al_{(liquid)}$, $Si_{(liquid)}$, $C_{(solid\ graphite)}$, $Mn_{(liquid)}$, $Fe_{(liquid)}$, $CO_{(gas)}$, $CO_{2(gas)}$ were chosen as standard for the reacting substances according to the real aggregate state of the phases in the studied system.

The calculated standard Gibbs energies of the reactions are provided in Table 1 and in Figures 1 – 4.

The analysis of the table data and diagrams shows that the highest reduction activity in the system is demonstrated by Al, C, Si. The reducing ability of silicon decreases as the temperature grows and the reducing ability of silicon increases with the growth of temperature. As a result carbon demonstrates stronger reducing ability than silicon at temperatures over T ≈ 1940 K corresponding to the lower limit of the temperature interval in the weld pool under automatic submerge-arc welding (1973 – 2133 K) [4]. Manganese as a reducing agent can be efficient only for iron oxide in the temperature interval up to 1800 K. Over this temperature the reducing ability of carbon exceeds that of manganese.

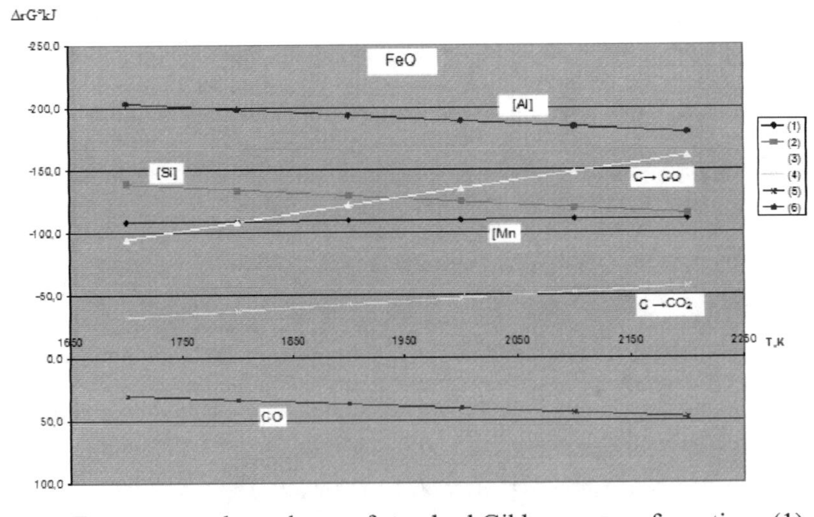

Figure 1. Temperature dependence of standard Gibbs energy of reactions (1) – (6)

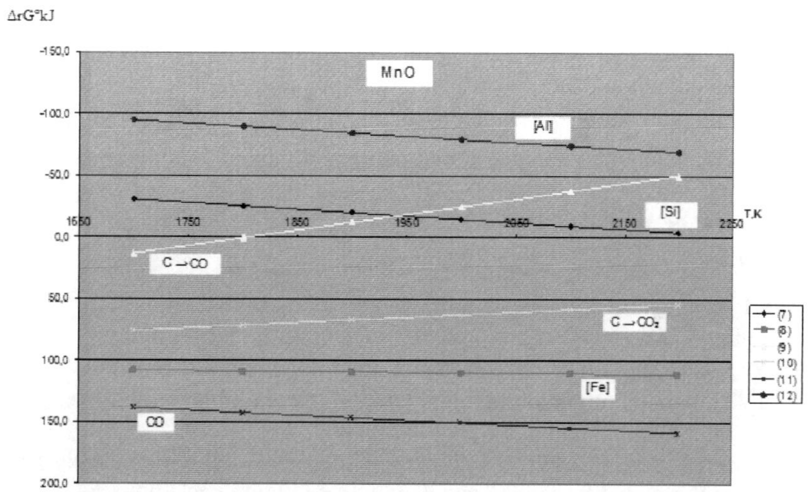

Figure 2. Temperature dependence of standard Gibbs energy of reactions (7) – (12)

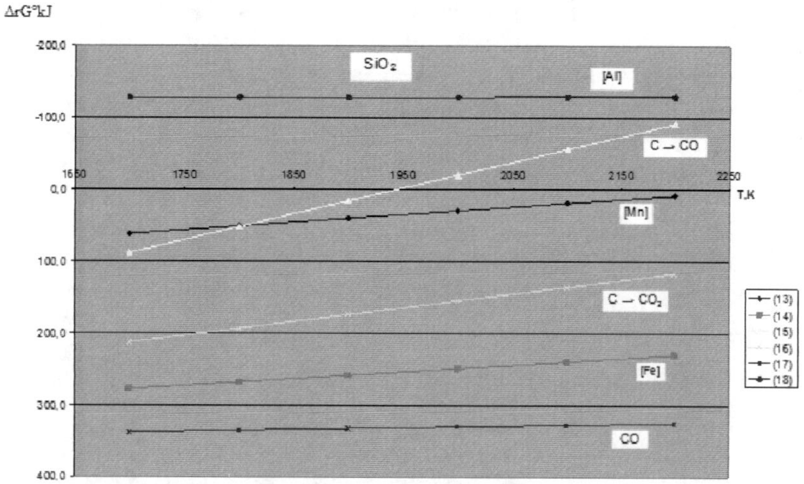

Figure 3. Temperature dependence of standard Gibbs energy of reactions (13) – (18)

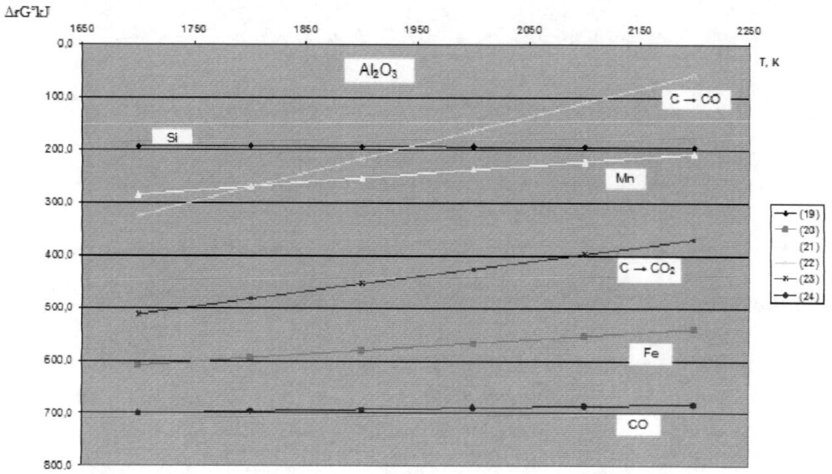

Figure 4. Temperature dependence of standard Gibbs energy of reactions (19) – (24)

Under the temperature over 2200 K aluminum can be formed in the weld metal as a result of reaction between Al_2O_3 oxide with carbon (reaction 22). Under these conditions $\Delta rG°(22) \approx 0$ we can expect that equilibrium constant of the reaction:

$$K = \frac{P_{CO}^3 \cdot a_{[Al]}^2}{a_{(Al_2O_3)}}, \quad (25)$$

(where P_{CO} – equilibrium partial pressure CO in the gas phase; $a_{[Al]}$ – activity of aluminum in the liquid weld metal; $a_{(Al_2O_3)}$ – activity of aluminum oxide in the oxide melt) becomes close to one. Under such equilibrium constant activity of aluminum in the weld metal and, thus, its concentration will be largest for the considered system and the reducing role of aluminum may be important. Aluminum is the best reducing agent for all oxides in the system but it can be formed in the weld metal only in small quantities, thus, concentration of aluminum oxide formed in this case in reactions (6), (12), (18) will be insignificant.

Works [5-14] provide the studies of the influence of carbon-fluorine-containing flux additive upon the process of weld metal refining and weld metal properties. The experiments were completed for the samples from steel 09G2S 16 mm thick 200×500 mm in size. Production of double square butt welds was completed in the same way as that of sheets for tank shells at the rolling stand. Sv-08GA wire 5 mm in diameter was used as additive material. Welding of samples under fluxes was completed under similar welding conditions. Samples were cut out of the welded sheets and the following tests were completed: X-ray spectral analysis of the weld metal composition, metallographic study of the welds, determining general oxygen content in the welds, determining mechanical properties and impact strength of the welds under the temperatures -200 C and – 400 C. The chemical analysis of the welds for carbon was completed according to National State Standard 12344-2003.

The experiments showed that application of up to 6% of carbon-fluorine-containing additive provides concentration of carbon in the weld similar to that in the base metal (Figure 5).

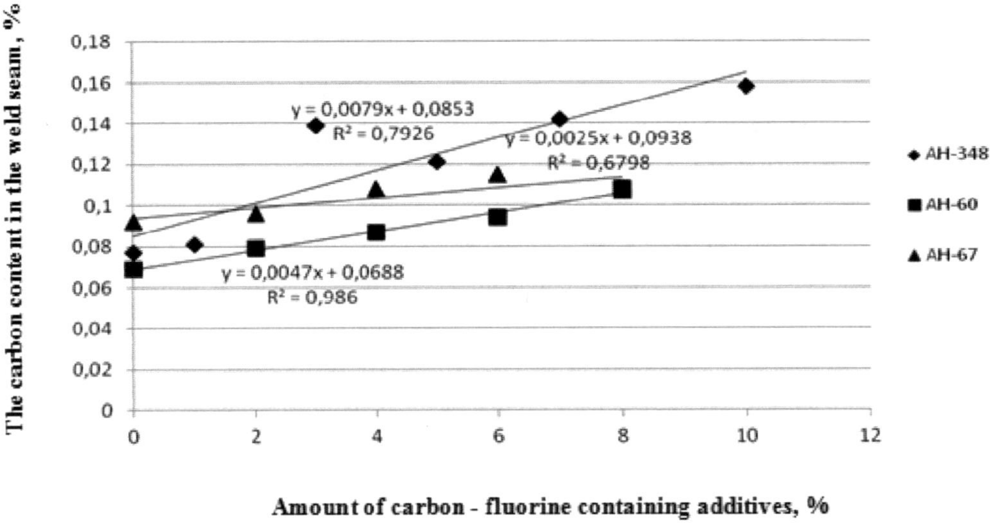

Figure 5. Influence of carbon-fluorine-containing additive upon carbon content in the weld

Determining the oxygen by the method of reduction melting on the gas analyzer "LEGO" TC-600 showed that the weight ratio of the given gas decreased with the growth of additive in the flux (Figure 6). The completed fraction analysis showed that oxygen redistribution in the inclusions depends upon degree of oxidation and basicity of the slag system. The largest percentage of aluminates and aluminum silicates which are detrimental to the physical and chemical properties of the weld joint was observed under flux AH-60 welding but introduction of the additive reduced their amount. The changes under AH-348 and AH-67 fluxes welding were insignificant.

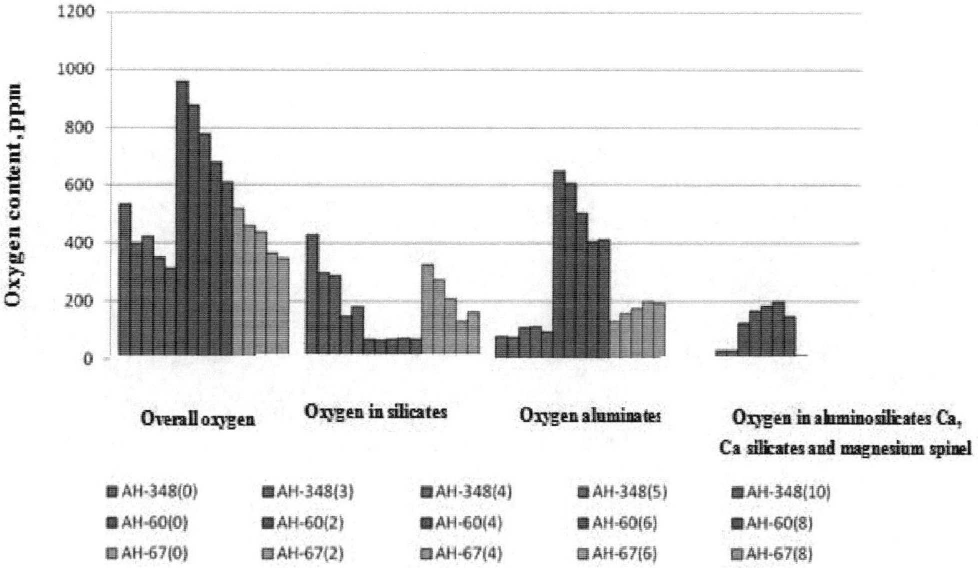

Note: *- the information in brackets specifies the content of carbon-fluorine-containing additive
Figure 6. Changes of concentration of general and fractional oxygen depending upon the amount of carbon-fluorine-containing additive

Metallographic bright-field examination of polished microslices of the weld joints were completed with optical microscope OLYMPUS GX-51 under the magnifications by ×100, ×500. The microstructure of metal was revealed by etching in the solution of 4% HNO_3 in ethanol. The basic metal structure of all samples consists of ferrite grains and lamellar pearlite (4-5 μ). In the interface region from the basic to the weld metal fine grained structure (1-2 μ) formed after recrystallyzing due to the heating in the process of welding. The structure of welds completed under various fluxes did not show noticeable differences.

In the samples welded under fluxes with carbon-fluorine-containing additives we observed lower amount of non-metallic impurities due to the general decrease of general amount of oxygen. Study of mechanical properties (flow limit, strength, percentage of elongation and impact strength under subzero temperatures) of the samples cut according to National State Standard 6996-66 showed that the properties improved significantly. As a result we observed lower amount of non-metallic inclusions and that of general oxygen in the weld joint as the amount of carbon-fluorine-containing additive in the flux increases.

Conclusions

Thus, the thermodynamic analysis of reactions in the system of molten weld metal – molten oxide – gas proved that, due to its high reducing properties under T = 1950 – 2200 K, carbon introduced into the system can significantly decrease the amount of non-metallic inclusions in the weld metal and so improve the complex of mechanical properties and impact strength (especially under subzero temperatures).

References

[1] Glushko V P Gurvich L V Thermodynamic properties of individual substances Moscow Nauka 1978
[2] Information on http://www chem msu ru/rus/tsiv/

IOP Conf. Series: Materials Science and Engineering **91** (2015) 012016 doi:10.1088/1757-899X/91/1/012016

[3] Information on http://kinetics nist gov/janaf
[4] Konishchev B P Kurlanov S A Potapov N N Welding materials for arc welding: Reference book in 2 vol V 1 *Welding gases and welding fluxes* Moscow Mashinostroyeniye 1989
[5] Kozyrev N A Igushev V F Starovatskaya S N Kryukov R E Goldun Z V News of Higher Educational Institutions J *Ferrous metallurgy* 6 (2012) p 26 – 29
[6] Kozyrev N A Igushev V F Goldun Z V Kryukov R E Shurupov V M News of Higher Educational Institutions J *Ferrous metallurgy* 10 (2012) 35 p 38
[7] Kozyrev N A Igushev V F Kryukov R E Goldun Z V Kovalsky I N J Welding production 12 (2012) p 3-6
[8] Chinakhov D A Zuev A V Filimonenko A G Gas-dynamic impact of a shielding gas get on the drop transfer when welding with a consumable electrode J *Advanced Materials Research* 1040 (2014) p 850-853
[9] Chinakhov D A Calculation of gas-dynamic impact of the active shielding gas on the electrode metal drop in gas jet shielded welding J *Applied Mechanics and Materials* 379 (2013) p 188-194
[10] Iljyashchenko D P Chinakhov D A Gotovshchik Y M Investigating the influence of the power supply type upon the properties of the weld joints under manual arc welding J *Advanced Materials Research* 1040 (2014) p 837-844
[11] Kozyrev N A Igushev V F Kryukov R E Starovatskaya S N Poop A B News of Higher Educational Institutions J *Ferrous metallurgy* 4 (2013) p 30-33
[12] Kozyrev N A Igushev V F Kryukov R E Poop A B Kovalsky I N J Welding production 5 (2013) p 9 - 12
[13] Kozyrev N A Igushev V F Kryukov R E Poop A B News of Higher Educational Institutions J *Ferrous metallurgy* 8 (2013) p 33-36
[14] Kozyrev N A Igushev V F Kryukov R E Goldun Z V R F Patent 2012104939/02 (007484) (2012)

Titanium Carbide: Nanotechnology, Properties, Application

G V Galevsky [1,a], V V Rudneva [1,b], A K Garbuzova [1,c], D V Valuev [2,d]

[1]Siberian State Industrial University, Novokuznetsk, Russia
[2]National Research Tomsk Polytechnic University, Yurga Technological Institute branch of Tomsk Polytechnic University 652050, Kemerovo region

[a]e-mail: kafcmet@sibsiu.ru, [d]e-mail: valuevden@rambler.ru

Abstract. The paper develops scientific and technological bases for fabrication of titanium carbide which is a nanocomponent of composite materials. The authors determine optimum technology specifications and the main titanium carbide properties: fineness of titaniferous raw materials, carbide-forming agent quantity, set temperature of plasma flow, tempering temperature, titanium carbide yield, productivity, specific surface, size and shape of particles. The paper includes equations to describe how the major specifications of the fabrication technique influence the content of titanium carbide and free carbon in the end product.

Introduction

Titanium carbide is a synthetic, super-hard, high-melting, heat-resistant material which is widely used for manufacturing metal-working tools, protective coatings and carbide steel. Producing nanocondition titanium carbide allows new applications including different types of composite materials [1 – 3].

The tests were carried out by means of an industrial plasma assembly based on a three-spool direct-flow vertical plasma reactor with the capacity of 150kW [4 – 6]. The plasma assembly consists of a mixing chamber with 3 plasmatrons EDP-104A which are fixed at an angle of 300, a reactor and a sedimentation chamber. Alongside with the reactor the plasma assembly includes a charge-conveying system, electricity-, gas- and water-supply systems, a synthesis product trapping system and a ventilation system. Powder and gas raw materials are fed into the plasma jet impact area with the help of a water-cooled lance. As this takes place a reaction mixture with predetermined composition is produced and nanodispersed particles are synthesized. Table 1 shows the main engineering specifications of the reactor.

Table 1. Reactor specifications

Feature	Values
Capacity, kW	150
Reactor type	three-spool direct-flow vertical
Plasmatron type, capacity, kW	EDP-104A, 50
plasma-supporting gas	nitrogen
Heated gas weight, kg/h	32.5

Feature	Values
Inside diameter, m	0.054
Reactor volume, m^3	0.001
Reactor channel lining	zirconium dioxide ZrO_2
Plasma flow temperature, K	5400 (L=0) – 2200 (L=12)
Lining temperature, K	1549 (L=0) – 770 (L=12)
Specific electric power, MW/m^3	2142
Operational life of anode cathode	3125 112
Titanium carbide erosion contamination, % anode weight cathode weight	 Cu-0.0000954% W-0.0000002%

The specifications of the reactor prove that it is an up-to-date electro-thermal device.

The experiments were carried out in the reactor heat-insulated by zirconium dioxide lining with wall thickness of 0.005m and outside diameter of 0.066m which reduces its diameter to 0.056m. Titanium powder (alternative 1) and titanium dioxide (alternative 2) powder were chosen as raw materials, natural gas with methane content of 96% was used as carbide-forming agent. Table 2 shows properties of raw materials and plasma-supporting gas for titanium carbide synthesis.

Table 2. Raw materials and plasma-supporting gas

Raw materials	Base material content, not less than %	Dispersity, μm
Fine titanium powder PTMk	99.9	0.5-5
Titanium dioxide pigment R-1 (National State Standard 9808-84)	99.0	-1
Methane (natural gas)	93.6 (ethane-3.0; propane -2.18; butane - 1,18)	-
Nitrogen (National State Standard 9293-74)	99.5 (oxygen content not more than 0.5)	-

Having taken into consideration properties of raw materials and plasma-supporting gas as well as reactor operating conditions we have developed the necessary specifications for thermodynamic and kinetic process simulation which are described in papers [7, 8]. The results of the simulation and mathematical calculations let us project specifications for production of titanium carbide by plasma method.

The experimental research demonstrated that the content of titanium carbide and free carbon in the end products has increased. The following equations for titanium carbide production are obtained. They allow estimating the influence of different factors on synthesis parameters, optimizing and controling the processes.

$$[TiC(1)]=17.3211+0.0105\cdot T_0-0.0156\cdot T_3+0.1859\cdot[CH_4]-3.432\cdot\{H_2\}-0.4078\cdot[N]+0.000004562\cdot T_0\cdot T_3+0.000782\cdot T_0\cdot\{H_2\}-0.0000567\cdot T_3\cdot[CH_4]-0.000435\cdot T_3\cdot\{H_2\}+0.0001353\cdot T_3\cdot[N]; \quad (1)$$

$$[TiC(1)-C_{free}(1)]=-53.95+0.01152\cdot T_0+0.0476\cdot[CH_4]+0.1325\{H_2\}+0.09257\cdot[N]-0.00000576\cdot T_0\cdot$$

$\cdot T_з - 0.002938 \cdot T_з - 0.00000876 \cdot T_0 \cdot [CH_4] -$
$- 0.00000588 \cdot T_0 \cdot \{H_2\} + 0.00000728 \cdot T_з \cdot [CH_4] - 0.000053 \cdot T_з \cdot [N] +$
$+ 0.000417 \cdot \{H_2\} [N];$ (2)
$[TiC(2)] = -182.277 + 0.05187 \cdot T_0 + 0.000927 \cdot T_з + 0.9428 \cdot [CH_4] -$
$- 0.4464 \cdot \{H_2\} - 0.1208 \cdot [N] - 0.0001878 \cdot T_0 \cdot [CH_4];$ (3)
$[TiC(2) - C_{free}(2)] = -13.162 + 0.01157 \cdot T_з + 0.01588 \cdot [CH_4] -$
$- 0.1244 \cdot \{H_2\} + 0.00013 \cdot [N] - 0.000001162 \cdot T_0 \cdot T_з + 0.00279 \cdot T_0 +$
$+ 0.000057 \cdot T_з \cdot \{H_2\} + 0.005707 \cdot \{H_2\} \cdot [N]$ (4)

where T_0 is set temperature of plasma flow, К; $T_з$ is tempering temperature, K; $[CH_4]$ is hydrocarbon percentage of the amount stoichiometricaly needed for producing titanium carbide, %; $\{H_2\}$ is concentration of hydrogen in plasma-supporting gas, % of volume; [N] –percentage of atomic nitrogen in plasma-supporting gas of the amount stoichiometricaly needed for producing hydrogen cyanide, %. Titanium carbide content in both cases depends on carbide-forming agent concentration, set temperature of plasma flow and tempering temperature.

Table 3 shows the main synthesis specifications and properties of titanium carbide. Having compared the two alternatives, we determined that carbidising titanium powder with methane is preferable. General properties of the synthesized titanium carbide meet requirements to nanodisperse components of composite materials.

Table 3. Synthesis indices and properties of titanium carbide

Synthesis indices and properties of titanium carbide	Alternative I (TiO_2 + natural gas)	Alternative II (Ti + natural gas)
Exchange gas composition, % of volume – nitrogen	99.5	99.5
Carbide-forming agent composition, % of volume -methane -ethane -propane -butan	93.6 3.0 2.2 1.2 -	93.6 3.0 2.2 1.2 -
Titaniferous raw material fineness, μm	1-3	5
Percentege of carbide-forming agent, % of stoichiometric one	120-140	120-140
Set temperature of plasma flow, K	5400	5400
Tempering temperature, K	2600-2800	2600-2800
Chemical composition, % of weight -TiC - titanium dioxide -free titanium (Ti+O) -free hydrogene - volatile matter	91.62-92.13 6.56-6.82 - 1.31-1.56 0.82-0.97	92.88-93.42 - 5.37-5.68 1.21-1.44 0.97-1.11
Titanium carbide yield, % of weight	92	92
Specific surface, m^2/kg	29000-32000	33000–35000
Particle size*, nm	38-42	34-36
Particle shape	facetted, cubic	facetted, cubic
Titanium carbide nonopowder oxidation $^{**}\times10^7$, kg O_2/m^2	12.5	8.6
Productivity, kg/h	2.8	3.7
Note: *this index is calculated according to the size of specific surface **this index was measured after exposure of the air during 24h.		

The end-product contains titanium monocarbide with cubic face-centered space lattice in which a = 0.4323 nm, which is by 0.0004nm less than that with massive crystals. It is stipulated by nonequilibrium conditions of near-surface layers particles less than 100 nm in size, which leads to deformation (compression) of space lattices, displacement of atoms from ideal positions, generating microstresses. With the first alternative titanium carbide is accompanied by dioxide with rutile and anatase space lattice. Free pyrolytic carbon which accompanies carbide is generated due to amorphous methane decomposition and it cannot be seen on the diffractograms. With alternatives I and II, the end products contain the following percentage of the weight (%): TiC – 91.88 and 93.15; $C_{free.}$ – 1.44 and 1,33; N – 0.90 and 1.00, and have specific surface 30000 and 34000 m^2/kg.

Titanium carbide nanopowder consists of round-shape aggregates from 150 to 600 nm in size formed by cubic particles with quite wide range of dimensions: from 10 to 60 nm. Facet shape of titanium carbide particles proves that they are formed according to "vapor – crystal" pattern when titanium and hydrogen cyanide vapors undergo a reaction. Aggregates of different sizes which can be found in the end product indicate that further aggregation of nanoparticles by coagulation at lowered temperatures is highly prabable.

A characteristic feature of pyrolytic carbon nanopowder is its ability to form micro-aggregates up to 100 – 150 nm in size with 6 – 10 particles 30 – 40 nm in size. Evidently the obtained titanium carbide greatly differs by its nanolevel and morphology from its micropowder obtained by mechanical dispersion of carbide of carbomechanical furnace synthesis.

Conclusion

Scientific bases of titanium carbide fabrication in plasma reactor including thermodynamic conditions of carbide forming and kinetic laws of evaporating titanium raw material are worked out.

It is proved that carbidization of titanium is a preferable method of titanium carbide plasma fabrication. This technology applies PTMk metallic titanium, nitrogen (National State Standard 9293-74) as plasma-supporting gas, methane as carbide-forming agent. Optimal technological indices and titanium carbide properties are determined.

References

[1] Rudneva V V Galevsky G V Yurkova E K Nanomaterials and nanotechnologies: estimation tendencies forcast J *News of Higher Educational Institutions Non-ferrous metallurgy* 2 (2007) p 73 – 76

[2] Garbuzova A K Rudneva V V Galevsky G V Nozdrin I V Reinforcing galvanic composite coatings on the basis of nickel with nano- and micropowders of titanium carbide J *Metallurgy: technologies management innovations quality*: edited volume of XVIII scientific conference SibSIU Novokuznetsk (2014) p 191 – 194

[3] Chinakhov D A Zuev A V Filimonenko A G Gas-dynamic impact of a shielding gas get on the drop transfer when welding with a consumable electrode J *Advanced Materials Research* 1040 (2014) p 850-853

[4] Garbuzova A K Rudneva V V Galevsky G V Shiryaeva L S Analysis of current condition of fabrication and application of titanium carbide J *Reporter of Siberian State Industrial University* 1 (2014) p 58 – 66

[5] Chinakhov D A Calculation of gas-dynamic impact of the active shielding gas on the electrode metal drop in gas jet shielded welding J *Applied Mechanics and Materials* 379 (2013) p 188-194

[6] Galevsky G V Shiryaeva L S Nozdrin I V Research in reactors for plasma-metallurgical fabrication of infusible borides and carbides J *News of Higher Educational Institutions Ferrous Metallurgy* 8 (2011)p 27–32

[7] Rudneva V V Plazma reactor for nanotechnologies: research expluatation improvement J Reporter of RANS: *Problems of metallurgy development in Russia* Vol 6 3 (2006) p 18 – 30

[8] Chinakhov D A Study of thermal cycle and cooling rate of steel 30ХГСА single-pass weld

IOP Conf. Series: Materials Science and Engineering **91** (2015) 012017 doi:10.1088/1757-899X/91/1/012017

joints J *Applied Mechanics and Materials* 379 (2013) p 188-194

[9] Rudneva V V Research in heat thermotechnical characteristics of three-spool plasma reactor J *News of Higher Educational Institutions Ferrous metallurgy* 2 (2007) p 57 – 60

[10] Garbuzova A K Rudneva V V Galevsky G V Shiryaeva L S Thermodynamic simulation of titanium carbide plasma synthesis J *Reporter of RANS SibSIU* 32 (2014) p 136-145

[11] Garbuzova A K Rudneva V V Galevsky G V Shiryaeva L S Simulating and mathematical research in effective processing of titaniferous raw material in plasma reactor J *Special metallurgy: yesterday today tomorrow*: edited volume of International scientific conference (2014) p 221 – 229

Possibilities of Application of Carbon-Fluorine Containing Additions in Submerged-Arc Welding

N A Kozyrev [1], N E Kryukov[2], R E Kryukov[1], V F Igushev[3], I I Kovalskii[2]

[1] Siberian State Industrial University, Novokuznetsk, Russia.
[2] Novokuznetsk reservoir metal structure plant, Novokuznetsk, Russia.
[3] Baltic ship-building college, Kaliningrad, Russia.

E-mail: kozyrev_na@mtsp.sibsiu.ru

Abstract. The paper provides results of comparative analysis of the effect of carbonaceous components introduced into welding fluxes on molten metal – slag interaction. A positive influence of carbonaceous additives on gas content and mechanical properties of welds is demonstrated. Carbon and fluorine containing additives are emphasized to be promising for automatic submerged arc welding.

Introduction

The development of oil deposits in the North regions (in conditions of low temperatures) is connected with the necessity of oil tank manufacture for negative temperature exploitation. According to the "Rules of installation of vertical cylindrical steel tanks for oil and oil products", Safety rules 03-605-03 and Construction codes 23-81, the increased demands are made on welded joints, with particular attention to the impact strength at negative temperatures. At Russian enterprises submerged-arc welding (SAW) has been widely used, and technological process of assembling, welding, controlling and rolling of tank wall sheets into coils is done on special rolling unit with upper and lower rolling.

Reservoir metal structure plant in Novokuznetsk has developed and mastered the reservoir manufacture technology for oil products for negative temperature exploitation [1 – 4]. In the given technology a two-side welding of butt joints of wall sheets is done automatically in SAW first on the top layer, then, after drawing the sheet by the drum, on the lower layer. To ensure a high productivity in manufacturing, crack formation elimination and to improve the mechanical properties we have developed and patented in Russia [5, 6] a technology of welding for tanks wall sheets. According to the technology a welding of tank is done with the welding wire Sv-08GA under the mixture of fluxes AN-348A and AN-67B with the ratio of 1:1. Welding is done without bewel to the sheet thickness of 16 mm. On the top layer the butt joints are welded at reduced power conditions, ensuring a weld metal penetration up to 0.55 of sheet thickness. On the lower layer welding is done at high power conditions, permitting to get a weld metal penetration up to 0.7 of sheet thickness.

However, the required level of mechanical properties is not always ensured when using some flux grades, namely, the impact strength at lower temperature due to the formation of a large number of nonmetallic inclusions in the welded joints. Most of the nonmetallic inclusions in the joints are oxide, irrespective of whether they are exogenic or endogenic. A content of nonmetallic inclusions in a joint, in turn, depends on the quantity of total oxygen. As the oxygen solubility in a solid metal is small, the total oxygen content in a joint characterizes the level of its contamination by nonmetallic inclusions.

The existing fluxes for low alloy steel welding show the increased total oxygen content, and, consequently, the increased number of nonmetallic inclusions in the joints. Today the reduction of nonmetallic inclusion content in welded joints in automatic SAW is obtained with low oxidizing flux application. But these fluxes have unsatisfactory welding and technological properties and are rarely used in low alloy steel welding. In the case of oxidizing flux application the metal of a joint1) to do the removal of hydrogen at the expense of fluorine containing compounds {like Na_3AlF_6; CFx = 25 – 35 (1≥x>0) etc.}, disintegrating at temperatures of welding processes with the liberation of fluorine, which, in turn, reacts to hydrogen dissolved in steel with the formation of gaseous compound HF;

2) to carry out the intensive carbon boiling at the expense of CO and CO_2 being formed in integration of carbon fluoride CFx (1≥x>0) with the dissolved oxygen in steel, and as carbon is in a bound state a carbonization of steel should not practically occur;

3) to increase the arc stability at the expense of the elements facilitating the potassium and sodium ionization in the arc column.

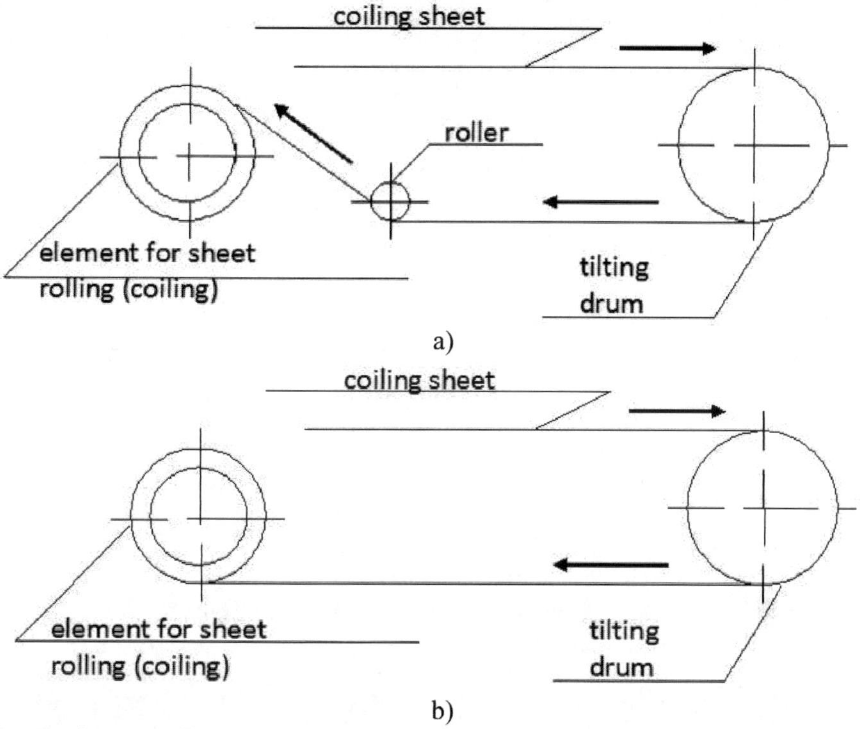

Figure 1. Diagram of installation for rolling of reservoirs with upper (a) lower rolling (b)

Methodology

The addition to flux was prepared in the following way: a carbon and fluorine containing components was mixed with glass, and after it a drying, cooling and crushing was done. Then the addition was mixed with a flux in a special mixer in a strictly prescribed ratio. For investigation the flux grades AN -348, AN -60, AN -67 and OK 10.71 were taken as base variant and their mixtures with admixture of flux-additions.

Experiments were carried out on the specimens made from 09Mn2Si steel, 16 mm thick and of size 200x500 mm. Welding of butt joints without edge preparation was done from two sides, as in welding of sheets of walls on the stand for rolling. A wire Sv-08Mn of 5 mm diameter was used as a filler metal.

Submerged-arc welding of specimens was done in identical conditions. Specimens were cut from welded plates and the following investigations were carried out: X-ray spectroscopic analysis of composition of joint metal, metallographic analysis of welded joints, determination of total oxygen content in joints, mechanical properties, hardness of welded joints and impact toughness of joints at temperatures of 200C, minus 400C and minus 700C.

The metal chemical composition determination of welded joints by carbon, sulphur and phosphorus content was done by chemical methods according to State Standard 12344-2003, State Standard 12345-2001, State Standard 12347-77 respectively. Content of alloying elements in a metal of joint, calcium oxide, sodium and fluorine compounds in fluxes with additions and slags obtained after welding was determined on X-ray fluorescence spectrometer XRF-1800 by SHIMADZU firm.

Results

The test showed that the carbon concentration in a joint corresponding to the carbon concentration in the initial metal (Fig. 2) is insured upon use of carbon and fluorine containing addition in quantity up to 6%.

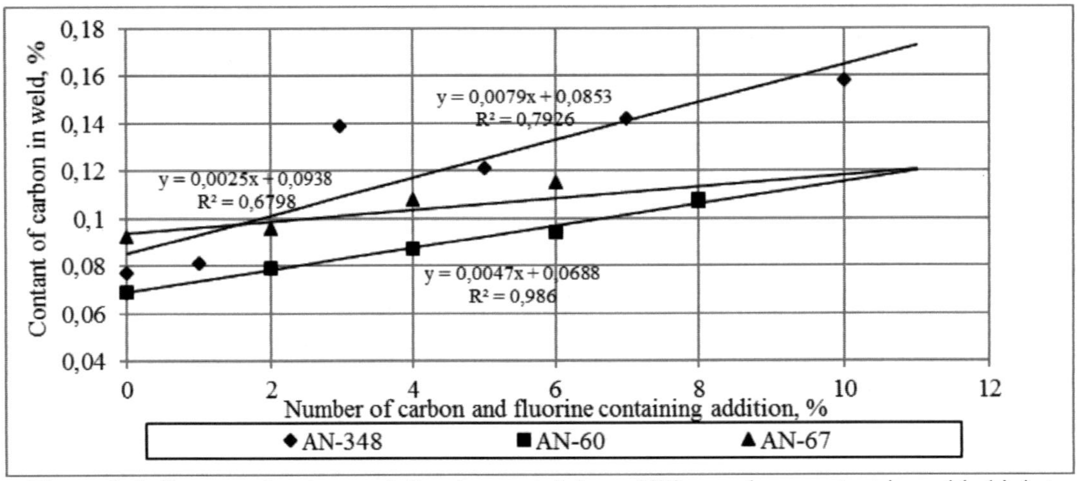

Figure 2. Influence of carbon and fluorine containing addition carbon content in welded joint

Sulphur and phosphorus content in a metal of joint obtained in submerge-arc welding with additions and without additions did not change and was in the range of S = 0.015 – 0.019%, P – 0.013 – 0.015%, i.e. the application of fluxes with additions had no effect on the transition of sulphur and phosphorus from the formed slag into metal.

Determination of oxygen by method of reduction smelting on gas analyzer TC-600 by LECO firm showed that mass fraction of the given gas with the increase of addition content in flux has decreased (Fig. 3), and a fractional gas analysis has showed that depending on a state of oxidation and basicity of a slag system a redistribution of oxygen in inclusions occurs. Distributions of oxygen in silicates, aluminates, alumosilicates is likely to be connected with basicity of the obtained slag and assimilation of nonmetallic inclusions by slag depending on the obtained slag viscosity.

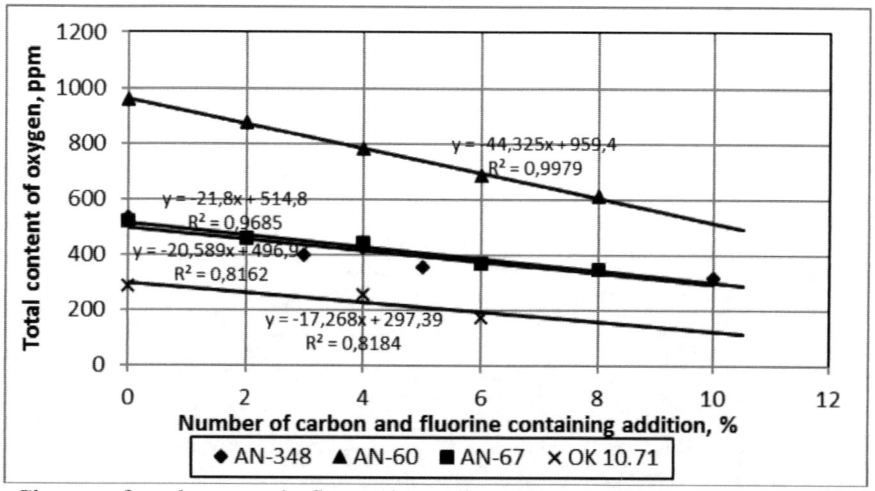

Figure 3. Change of total oxygen in fluxes depending on the introduction of carbon and fluorine containing addition

The largest number of aluminates and alumosilicates, effecting adversely the physical and chemical properties of the welded joint, was contained in submerged-arc welding with flux AN -60, when the addition was introduced the decrease in number of these compounds was observed. In fluxes AN -348 and AN -67 the changes were negligible (Fig. 4).

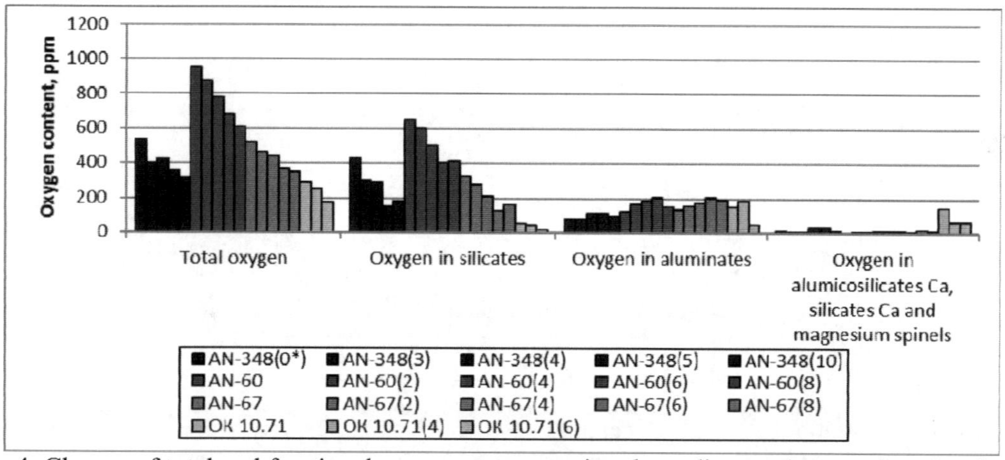

Figure 4. Change of total and fractional oxygen concentration depending on the quantity of carbon and fluorine containing addition

The carbon and fluorine containing addition influenced the reduction of the hydrogen in the welded joint according to the mechanism described above at the expense of fluorine (Fig. 5), and sodium concentration was slightly decreased (Fig. 6).

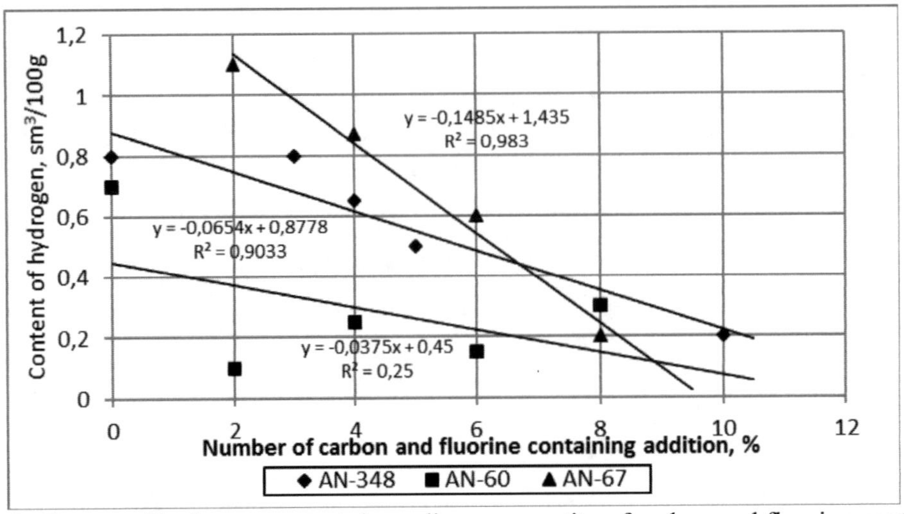

Figure 5. Change of hydrogen content depending on quantity of carbon and fluorine containing addition

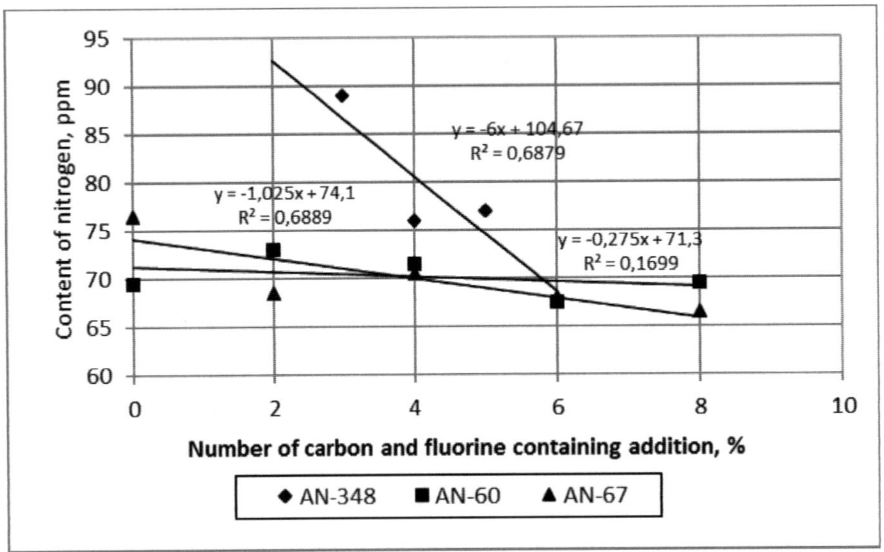

Figure 6. Change of nitrogen content depending on quantity of carbon and fluorine containing addition

A metallographic examination of polished microsections of welded joints were done with the help of optical microscope OLYMPUS GX-51 in a light field at magnifications x 100, x 500. Metal microstructure was revealed by etching in a solution of 4% HNO_3 in ethyl alcohol. The structure of base metal of all specimens consists of ferrite grains and lamellar pearlite (4 – 5 mcm). In the transition zone from the base metal to the deposited one a fine-grained structure is observed (1 – 2 mcm), being formed as a result of recrystallization on heating in welding process. Ferrite grains elongated in the direction of heat removal due to heating and accelerated cooling are present in the microstructure of welded joint. A detectable difference in structure of joints welded under different fluxes was not revealed. In specimens welded by submerged-arc welding with carbon and fluorine containing additions a decrease of contamination level by nonmetallic inclusions, connected with the decrease of total oxygen content was observed.

Study of mechanical properties (yield point, strength, unit elongation and impact toughness at negative temperatures) of specimens, cut according to State Standard 6996-66, has shown that a level of properties exceeds significantly the required values of State Standard 32385-2008 and normative values of Safety Rules 03-605 (rules for manufacture of vertical cylindrical tanks for oil and oil products) and rises with the increase of carbon and fluorine containing addition (Fig. 7). The increase of impact toughness values at negative temperatures (Fig. 8) should be especially noted.

Conclusion

Thus, upon introduction of the developed carbon and fluorine containing addition to the fluxes AN-60, AN-67, OK 10.71 the gas saturation of welded joint is increased, contamination by oxide nonmetallic inclusions is decreased, a complex of the required mechanical properties and impact strength (especially at negative temperatures) is enhanced.

The production of carbon and fluorine containing addition protected by the patent [13], is mastered in the conditions of the joint-stock company N.E. Kryukov Novokuznetsk reservoir metal structure plant.

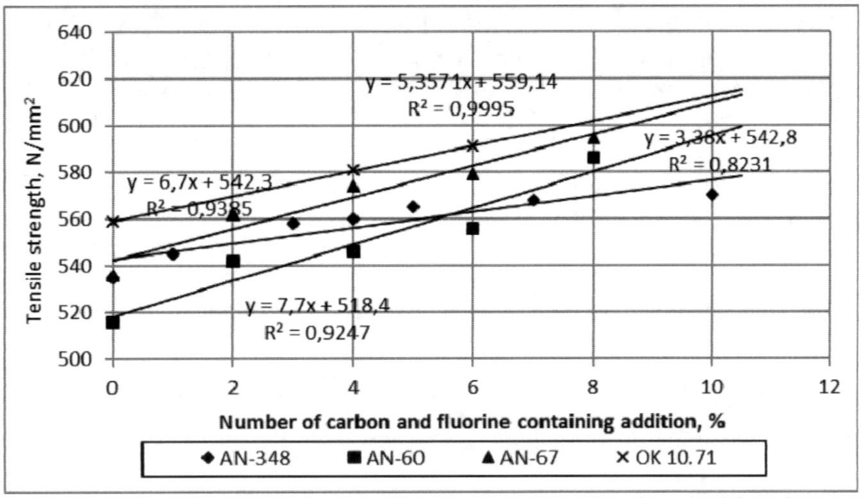

Figure 7. Change of tensile strength depending on quantity of carbon and fluorine containing addition.

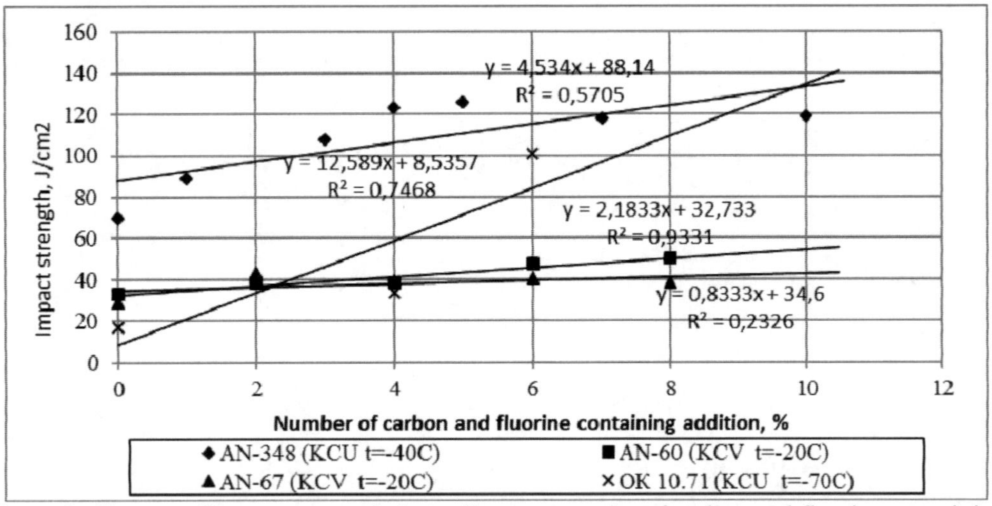

Figure 8. Change of impact strength depending on quantity of carbon and fluorine containing addition.

IOP Conf. Series: Materials Science and Engineering **91** (2015) 012018 doi:10.1088/1757-899X/91/1/012018

References

[1] Kryukov N E Kovalskii I N Kozyrev N A Igushev V F Kryukov R E Electrometallurgiya 2011 № 10 P 28 – 31
[2] Kryukov N E Kovalskii I N Kozyrev N A Igushev V F Kryukov R E Izvestiya vuzov Chernaya metallurgiya 2012 № 2 P 49 – 52
[3] Kryukov N E Kovalskii I N Kozyrev N A Igushev V F Kryukov R E Svarochnoye proizvodstvo 2012 № 5 p 35 – 38
[4] Kryukov N E Kovalskii I N Kozyrev N A Igushev V F Kryukov R E Steel in translation 2012 V 42 № 2 P 118 – 120
[5] Patent 2465108 RF MPK 8 V 23 K9/18 B23 K35/362 Method of submerged-arc welding / Kryukov N E Kovalskii I N Kozyrev N A Igushev V F Kryukov R E ; Joint-stock company N E Kryukov Novokuznetsk reservoir metal construction plant - № 2011123342/02(034573) appied 08 06 2011
[6] Patent 2467853 RF MPK 8 V 23 K35/362 Ceramic flux-addition/ Kryukov N E Kovalskii I N Kozyrev N A Igushev V F Kryukov R E ; Joint-stock company N E Kryukov Novokuznetsk reservoir metal consruction plant - № 201112341602(034573) appied 08 06 2011
[7] Kozyrev N A Igushev V F Starovatskaya S N Kryukov R E Goldun Z V Izvestiya vuzov Chernaya metallurgiya 2012 № 6 P 26– 29
[8] Kozyrev N A Igushev V F Goldun Z V Kryukov R E Shurupov V M Izvestiya vuzov Chernaya metallurgiya 2012 № 10 P 35– 38
[9] Kozyrev N A Igushev V F Kryukov R E Goldun Z V Kovalskii I N Svarochnoye proizvodstvo 2012 № 12 – P 3 – 6
[10] Kozyrev N A Igushev V F Kryukov R E Starovatskaya S N Roor A V Izvestiya vuzov Chernaya metallurgiya 2013 № 4 P 30– 33
[11] Kozyrev N A Igushev V F Kryukov R E Roor A V Kovalskii I N Svarochnoye proizvodstvo 2013 № 4 P 30 – 33
[12] Kozyrev N A Igushev V F Kryukov R E Roor A V Izvestiya vuzov Chernaya metallurgiya 2013 № 8 P 33– 36
[13] Patent 2484936 RF MPK 8 V 23 K35/362 Ceramic flux-addition/ Kozyrev N A Igushev V F Kryukov R E ; Goldun Z V ; FGBOU VPO Siberian State Industrial University - № 2012104939/02(007484) applied 13 02 2012

A New the Chnology for Producing Carbide Alloys With Gradient Structure

T N Oskolkova

Russia, Novokuznetsk, ul. Kirova,42, tel. (3843)74-87-63

E-mail: oskolkova@kuz.ru

Abstract. A carbide alloy with gradient structure consisting of viscous core of WC10KS alloy and more wear-resistant surface of alloy WC6-OM has been obtained by means of electrodischarge hardening by alloying. The core of this alloy can bear considerable shock loads, and its surface serves as wear-resistant cutting tools.

Introduction

WC–Co alloys are the strongest of available sintered carbide alloys but they do not always meet service durability standards. In the bulk of depreciated tools wearing and failures of carbide parts make up 80%. In this regard one of the promising trends in the development of carbide alloys is inventing technologies which provide higher wear resistance without reducing viscosity. Possible ways of development in this sphere are modifying carbide alloy surface with carbides, borides and the like, as well as treatment of the surface with concentrated energy flows [1-6]. In Russia WC–Co carbide alloy plates with TiC and TiN ion-plasma coating are produced. There are some other ways to increase durability of carbide alloy tools alongside with coatings. A method of producing carbide alloy plates by means of sintering with gradient strength is among them.

Results and discussion

The Institute of Ultra-Hard Materials of the National Academy of Sciences of Ukraine worked out a method of producing an alloy with variable content of cobalt heightwise a plate by means of impregnation of a sintered carbide alloy. This alloy is to be employed in mining machines. The possibility of obtaining gradient strength through a carbide alloy from viscous core to wear resistant surface is theoretically grounded. Thereby it is possible to alter the content of tungsten and cobalt in the alloy 8cm heghtwise from 80% and 20% (WC20) to 98% and 2% (WC2) correspondently. As a result, wear resistance of the working part of the plates is as high as that of WC2 and the core can bear considerable bending stress.

Similar research took place in Russia. There is a method of obtaining carbide alloy with variable content of bonding, in which the content of tungsten and cobalt is changed from the surface to the core in the process of pressing plates: WC3 – WC6 – WC10 – WC15. As a technical solution, it was necessary to find the way of producing laminated gradient plates with viscous core and harder surface in order to provide their wear resistance to different materials. The solution is the following: compression of refractory powders and cobalt, preliminary sintering, surface alloying, final sintering. The preliminary sintering is carried out at the temperature of 830 - 900°C. For surface alloying, carbide alloy powder with low cobalt content is impregnated in the mixture of spirit and glycerin in the ratio 2:1 in

IOP Conf. Series: Materials Science and Engineering **91** (2015) 012019 doi:10.1088/1757-899X/91/1/012019

ultrasonic field during 4 – 8 minutes. Ultrasonic field allows balancing particles in the suspension, making them movable, introducing disperse component into carbide alloy pores.

Thus, the gradient plate consists of a viscous core, which serves as a damper and buffers shock loads, but it can lack wear resistance. It is the transition section (plate surface – coating) that provides the properties which correspond to a certain application of a plate. The transition section must possess the most efficient ratio of wear resistance, hardness and viscosity, its thickness being not less than 300 mkm, because the effective wearing of the transition section takes place in this range. Manufacturing a gradient plate of carbide alloy can serve as an example. A plate of WC20 alloy was previously sintered at 850°C and then it was impregnated in the mixture of spirit and glycerin in the ratio 2:1 which contained particles of carbide alloy WC3M (with concentration 40%). The impregnation was carried out in an ultrasonic bath during 5 minutes. After that, the plate was sintered again at the temperature of 1380 °C in hydrogen medium. Then its surface was covered with multilayer coating of gas-vapour phase. As a result, we obtained a gradient plate which had a core made of WC20, a transition section not less than 250 mkm thick composed of a mixture of large grains of WC20 and grains WC3M and an even layer of finely-grained alloy WC3M 150 mkm thick. Cobalt concentration was variable across the layer and changed from 6% at the surface to 20% at the core of the plate.

But this method has some drawbacks: the technology of manufacturing a multilayer carbide alloy with gradient structure is rather complicated, long, energy consuming and requires advanced equipment. A multilayer carbide alloy can be obtained only on a flat surface. Filling a press mould by several feeders is difficult to proportion, which in practice prevents obtaining a plate with optimal combination of useful properties.

Nevertheless, there is a theoretical possibility to obtain gradient strength of carbide plates from viscous core to wear resistant surface by means of electrodischarge hardening by alloying. Until now depending on the alloying electrode material coating by means of electrodischarge hardening by alloying has been applied to increase several-fold resistance to wearing of metal-cutting, bench-work, medical tools, cold metal treatment dies, as well as to improve anti-frictional properties of wear surfaces, increase heat and corrosion resistance of metal parts, change conducting characteristics of contact elements. In other words, working characteristics of parts made of the most widely spread structural and tool steels were improved by means of covering the surface with a thin layer of coating with the help of electrodischarge hardening by alloying

It is useful to study ways of electrodischarge hardening by alloying of carbide alloy WC10KS surfaces in order to reinforce and restore them after use. Such application has not been described before.

Depending on technological purpose different coatings can be applied onto a surface:

a)blanket coating, b) mixed coating consisting of different materials, c) discrete coatings consisting of micro sections.

Within this research, a low-cobalt coating on the surface of high-cobalt carbide alloy WC10KS was formed by means of electrodischarge hardening by alloying. Carbide alloy WC6-OM was used as an alloying element. As a result, less strong core of the part can bear considerable shock loads and bending stresses and the harder surface has high wear resistance, which provides great durability of carbide alloy WC10KS.

Advantages of electrodischarge hardening by alloying consist in

1) availability of complex micrometallurgical processes;
2) extremely strong adherence of the applied coating layer with the surface;
3) simplicity of the process;
4) eliminating of contraction of the part being treated;
5) small size and low energy consumption of the equipment.

The essence of electrodischarge hardening by alloying is as the following: a spark discharge causes anode (electrode) erosion and the products of the erosion are transferred to the cathode (the part). In this process, the part is being covered with a layer with transformed composition and structure which depend on an electrode material the electrical mode of the equipment. High discharge temperature (5000 ÷ 10000 ^{0}C), short action ($10^{-6}...10^{-3}$ seconds), instant cooling of the heated areas bring us to the conclusion that ultrafast temperring takes place after electrodischarge hardening by alloying. Alongside with this process there is a contact or drop transfer of alloying elements from the hardening electrode onto the treated surface when they osculate and further diffusion spreading of these elements through the coating.

The work principle of the device for application of coatings is based on alternating discharges of charge-storage capacitors when a vibrating electrode touches the surface of the treated part during its reciprocating motion. Spark discharge of charge-storage capacitors cause electrode erosion, transfer diffusion of the electrode material into the surface layer of the treated part.

Surface hardening of carbide alloy WC10KS by means of electrodischarge hardening by alloying was performed with the help of portable device UR – 121 produced by Open stock company "Kontsern Podolsk". Treatment of the carbide alloy by an alloying electrode WC6-OM was carried out in two modes:

Mode 1 – 2 minutes per 1 cm^2 with electrode vibration frequency 100Hz, + 1 minute per 1cm^2 with electrode vibration frequency, set automatically by a magnetic system in the range from 140 to 500Hz;

Mode 2 – 1 minute per 1cm^2 with electrode vibration frequency, set automatically by a magnetic system in the range from 140 to 500Hz.

Alloy WC6-OM which consists of 91.9 WC + 6 % Co + 2 % TaC + 0.1 % VC has constant strength, doping agents VC and TaC help to obtaining small-grain structure by preventing growth of WC-phase grains. At the same time WC6-OM alloy wear resistance is as high as that of the most wear resistant alloy of WC–Co (B 253) group, but its strength is higher.

The technique of obtaining coating on a carbide alloy surface is the following: the surface of a carbide plate is treated with spirit to remove thoroughly all traces of oil and grease before electrodischarge hardening by alloying. After that, a graphite layer is applied onto the deoiled surface in order to prevent decarbonisation and as a consequence embrittling η_1- phase of double tungsten and cobalt carbide which is inadmissible for carbide alloy structure.

To study special characteristics of the alloy structure after the treatment and to determine the depth of the hardened surface zone optical microscope OLIMPUS – GX 50 and scanning electronic microscope «Philips SEM 515» were used.

Fig 1 shows microstructure of WC10KS alloy with hardened layers of WC6-OM after electrodischarge treatment at different modes. Badly etching layers of different thickness (5 ... 25 mkm) can be seen on the surface.

Scanning electronic microscopy proves gradient structure on WC10KS alloy: increased content of tungsten and reduced content of cobalt in the surface layer (Fig. 2).

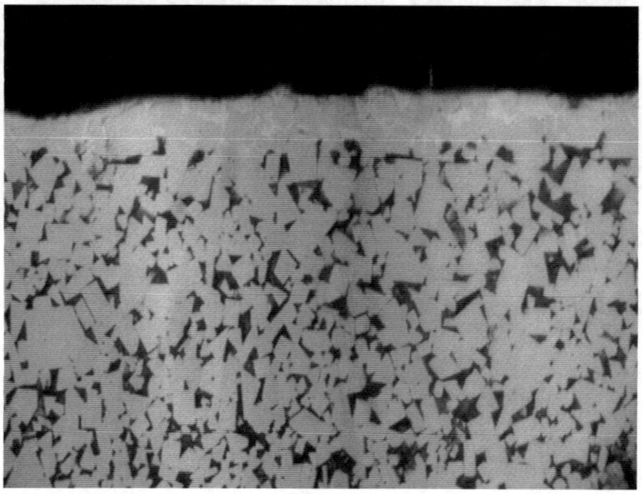

a

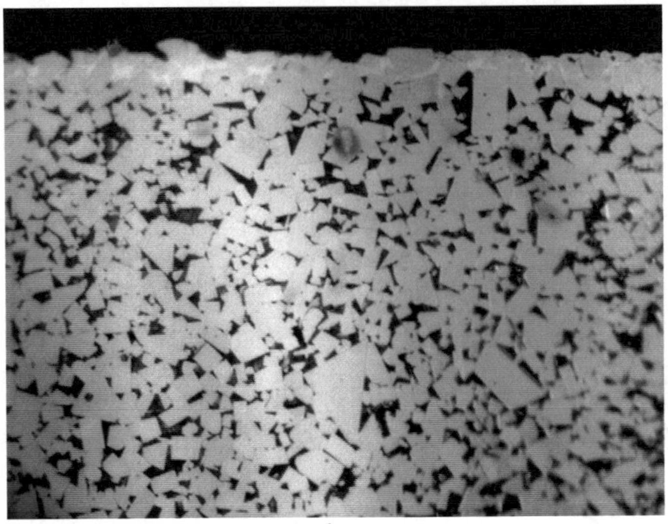

b

a – mode 1; b – mode 2

Figure 1. Microstructure of WC10KS alloy with WC6-OM coating applied at different modes, x 1000

X-ray research showed that treatment of carbide alloy surface by means of electrodischarge hardening by alloying causes changes in its phase composition (Fig.3).

Alongside with tungsten monocarbide WC and cobalt bonding which are found in the original carbide alloy, ditungsten monocarbide W_2C and carbide WC_{1-x} are formed in the surface layers of the parts treated at two modes. It is known that ditungsten monocarbide W_2C has greater hardness than that of WC (Fig. 3).

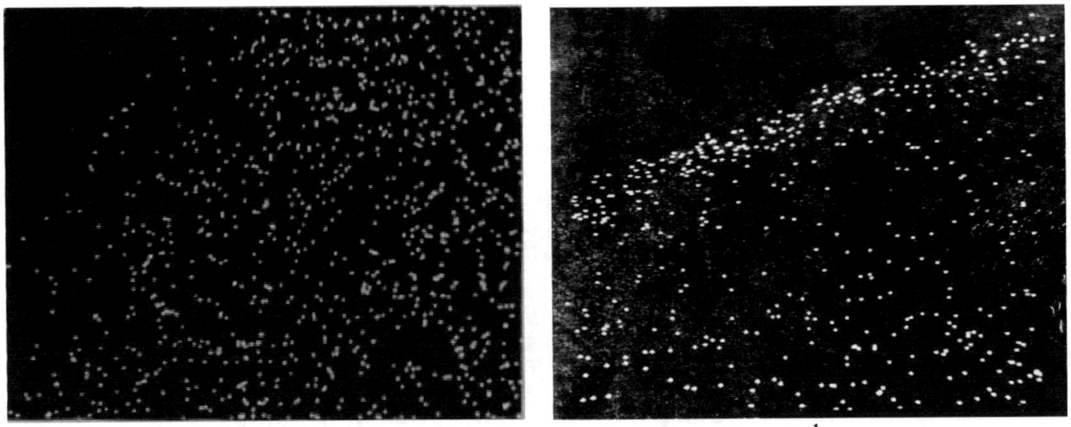

a b

Figure 2. Microstructure of WC10KS alloy with WC6-OM coating at X-rays: cobalt (a); tungsten (b) x 1000

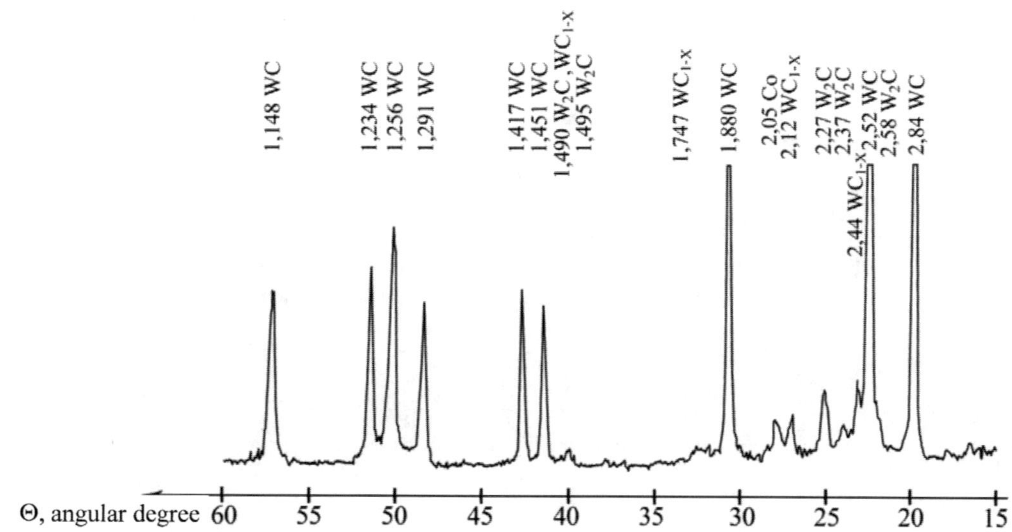

Figure 3. A fragment of X-ray diffraction pattern WC10KS alloy with WC6-OM coating

Changes in phase composition of the surface layer of the carbide alloy influenced nanohardness indices. Nanoindentation of the carbide alloy results in increasing nanohardness up to 22000 MPa (Fig. 4).

Wear resistance tests were carried out on «PC-Operated High Temperature Tribometer» at the room temperature. A degree of wearing was determined by means of measuring the depth and the area of the track made by an unmoving diamond indenter on a rotating sample under the load of 3H, with rotating speed 4000, linear speed 2.5 cm/s. For comparison similar tests were carried out with original parts.

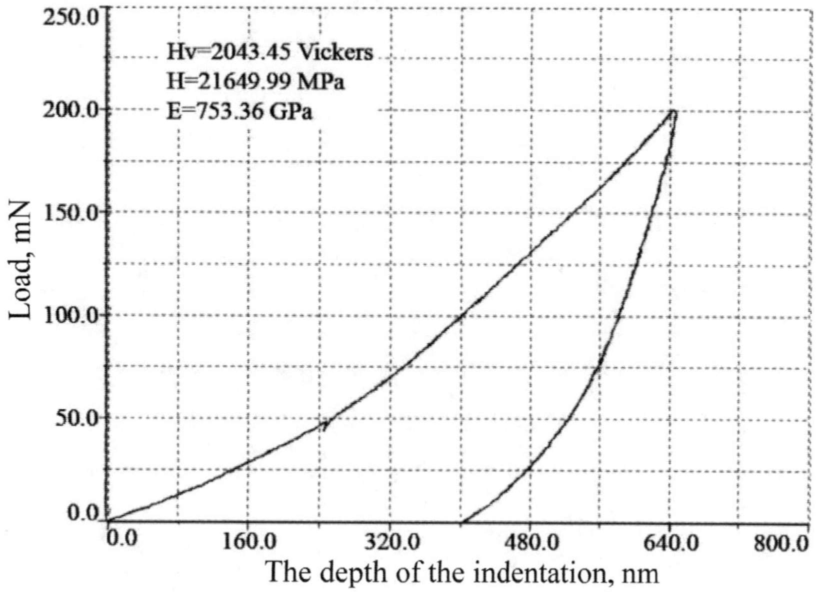

Figure 4. Loading-unloading curves of nanoindentation on WC10KS alloy with WC6-OM hardening layer

IOP Conf. Series: Materials Science and Engineering **91** (2015) 012019 doi:10.1088/1757-899X/91/1/012019

Wearing of samples with WC6-OM coating after electrodischarge treatment showed that the track depth on the coated sample is 10.8mkm while the original sample had the track of 58mkm. Track section area of wear out samples is 941mkm^2 and 12921mkm^2 correspondently. Friction coefficient (μ) of WC10KS alloy with WC6-OM coating applied by electrodischarge treatment is 0.23 which differs from the original sample for which $\mu = 0.41$.

As far as electrodischarge hardening by alloying always leads to increasing roughness of a surface it is important to study microgeometry of the surface after hardening of WC10KS. The results of the study are shown in the table in Fig. 6. In this research the microgeometry if the carbide alloy surface after treatment was determined by profilometry method at «Micro Measure 3D station». Hardening of the carbide alloy surface increases roughness of the surface but it remains within technical specifications with allowed surface finish characteristics $R_a = 2.5$mkm.

Table Microgeometry of WC10KS carbide alloy surface

Treatment	Roughness R_a, mkm
Original sample	1.32
After surface hardening	2.15

Electrodischarge hardening by alloying for application of coating on carbide alloy surface by proper electrodes allows producing carbide plates with gradient structure, e.g.WK3 – WK6 – WK8 – WK10 – WK15 – WK20. It is also possible to restore surfaces of different carbide alloys after use. It will increase service life of tools consisting of such alloys and save expensive materials (tungsten, cobalt). Thus developing tungsten carbide alloys with gradient structure can be considered as a new stage in carbide alloy production.

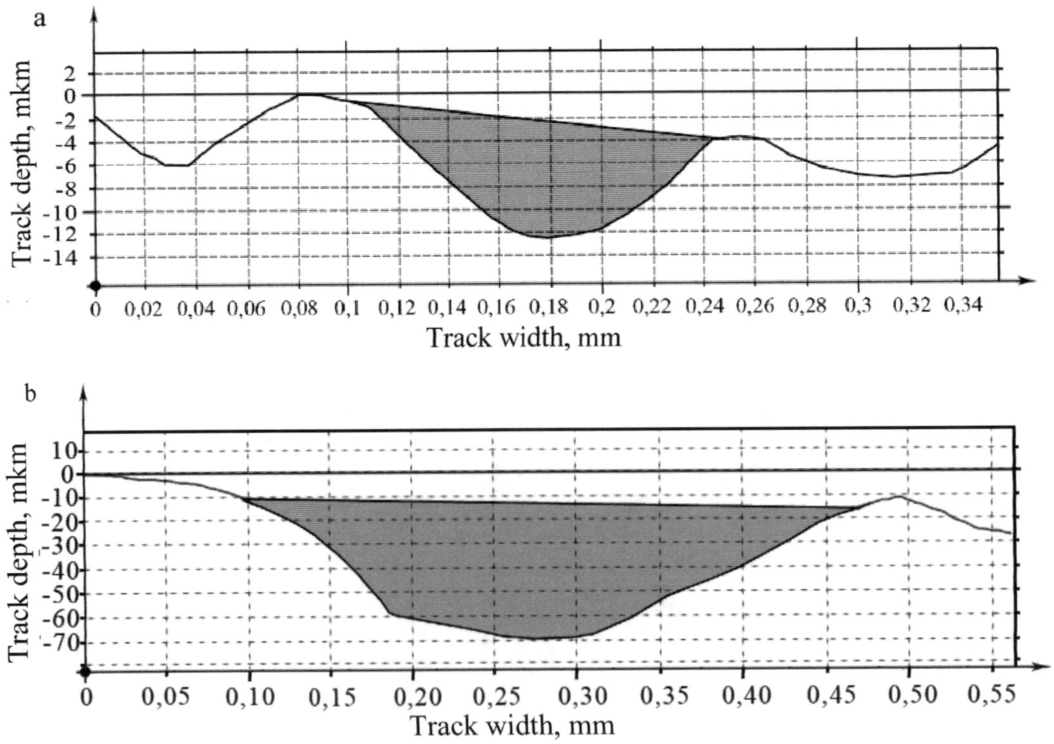

Figure 5. Track profiles and wear areas of samples with coatings (a), without coatings (b).

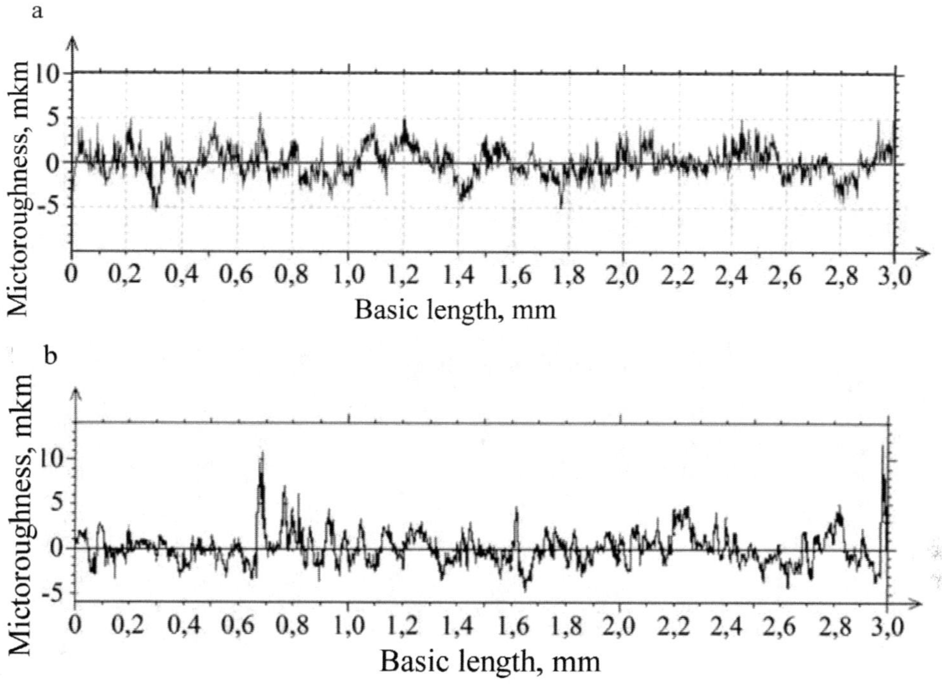

Figure 6. Microgeametry of WC10KS alloy in the original state (a), after electrodischarge treatment (b)

Conclusion

A carbide alloy with gradient structure consisting of viscous core made of WC10KS alloy and more wear resistant surface made of WC6-OM alloy is produced by means of electrodischarge hardening by alloying. Increasing wear resistance is caused by ditungsten monocarbide W_2C, which improves working characteristics of inserts made of carbide alloy based on tungstem monocarbide WC.

This research was carried out within a state order by Ministry of Education and Science of the Russian Federation № 11.153.2014/K

References

[1] Nozdrin I V Particle size of vanadium and chromium borides and carbides in a plasma flux / I V Nozdrin G V Galevskii L S Shiryaeva M A Terenteva *Steel in Translation* 2011 T 41 № 10 p 799-804

[2] Nozdrin I V Reactor for plasmometallurgical production of refractory borides and carbides / I V Nozdrin L S Shiryaeva G V Galevskii V V Rudneva *Steel in Translation* 2011 T 41 № 8 p 644-648

[3] Nozdrin I V Plasma synthesis and physicochemical certification of chromium nanocarbide / I V Nozdrin V V Rudneva *Steel in Translation* Vol 53 No 12 2012 p 797-801

[4] Oskolkova T N Pulse plasma treatment of the surface of alloy VK10KS / T N Oskolkova E A Budovskikh *Metal Science and Heat Treatment* 2012 vol 53 № 11–12 p 608 – 610

[5] Oskolkova T N Electric explosion alloying of the surface of hard alloy VK10KS with titanium and silicon carbide / T N Oskolkova E A Budovskikh *Metal Science and Heat Treatment* 2013 vol 55 № 1 p 96 – 99

[6] Oskolkova T N Features of structure formation of the surface layer in the course of electroexplosive alloying tungsten carbide hard alloy / T N Oskolkova E A Budovskikh V F *Goryushkin Non-Ferrous Metals* 2014 vol 55 № 2 p 196 – 200

IOP Conf. Series: Materials Science and Engineering **91** (2015) 012020 doi:10.1088/1757-899X/91/1/012020

Wear Resistant Coating on Tungsten Carbide Hard Alloy

T N Oskolkova

Russia, Novokuznetsk, ul. Kirova,42, tel. .(3843)74-87-63

E-mail: oskolkova@kuz.ru

Abstract. The article reveals new knowledge about the role of zirconium in the composition of (Ti, Zr)N ion-plasma coating applied on WC10KS alloy. It is determined that when zirconium is introduced into ion-plasma coating TiN (50%) wear resistance and adhesion strength grow, nanohardness increases by 23 % (up to 38500MPa), Yung's modulus rises by 67 %, friction coefficient reduces to $\mu = 0.07$ and performance characteristics of a carbide alloy improve.

Introduction

Development of many industries such as metalworking, machine building, mining, wood processing involves application of sintered tungsten carbide hard alloys which are widely used as tool materials. A promising trend of improving hard alloys is developing reinforcing technologies which provide better performance characteristics. Among these technologies there is quenching of carbide alloys in water solutions of polymers [1], modification of alloy surface with refractory carbides and borides and some others [2 – 4].

Increasing wear resistance without sacrificing viscosity provides durability of any tools which undergoe intensive loads during machining, pressing or rock drilling. In this context great attention is being paid to working out reinforcing technologies for tungsten carbide alloys in Russia and abroad. Using new methods of surface hardening with the help of concentrated energy flows is a priority trend in increasing durability of carbide alloys [5 – 7]. Russian industry produces WC-Co carbide plates with TiC and TiN ion-plasma coatings. These coatings are widely spread due to their satisfactory performance characteristics and adopted production technology. Nevertheless TiC and TiN have some drawbacks, low heat resistance being the main one. In this connection it can be useful to test introducing zirconium into the composition of TiN ion-plasma coating and to study its properties on WC10KS carbide alloy.

Results and discussion

For this research we used carbide alloy plates of WC10KS alloy produced by open stock company "Kirovograd Hard Alloys Plant" (Russia). The plates were covered with (Ti, Zr)N ion-plasma coating from separate cathodes of titanium and zirconium (50 % Ti + 50 % Zr). Nitrogen was used as a reaction gas. Two cathode made of titanium alloy were placed in front of each other in a chamber, and a cathode of zirconium alloy was placed between them.

The structure of the carbide alloy with ion-plazma coating was studied with the help of scanning electronic microscope «Philips XL-30».

Metallographic research of cross-sections (Fig. 1) showed that 15mkm thick (Ti, Zr)N ion-plasma coating applied on carbide plates WC10KS do not cause micro-cracks in the coating or on its border

IOP Conf. Series: Materials Science and Engineering **91** (2015) 012020 doi:10.1088/1757-899X/91/1/012020

with the main metal. Two phases of TiN and ZrN prove microlaminated structure of the coating (Fig. 2) which improves performance characteristics. In spite of the microlaminated structure of the coating its elements are uniformly distributed without microspores between layers. (Fig. 3).

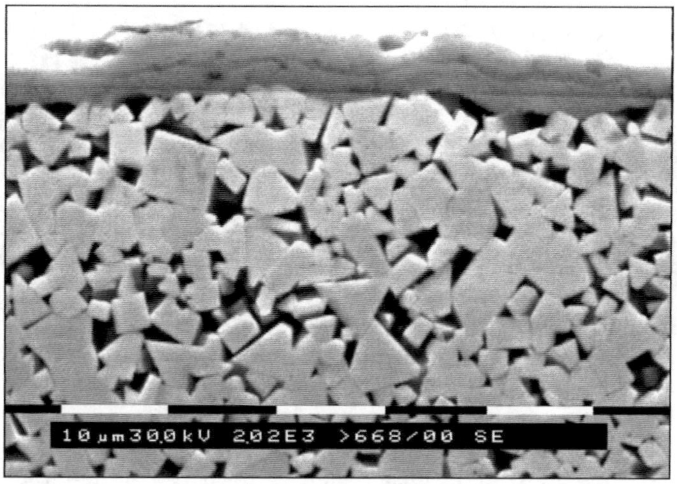

Figure 1. Microstructure of WC10KS alloy with ion-plasma coating (Ti, Zr)N

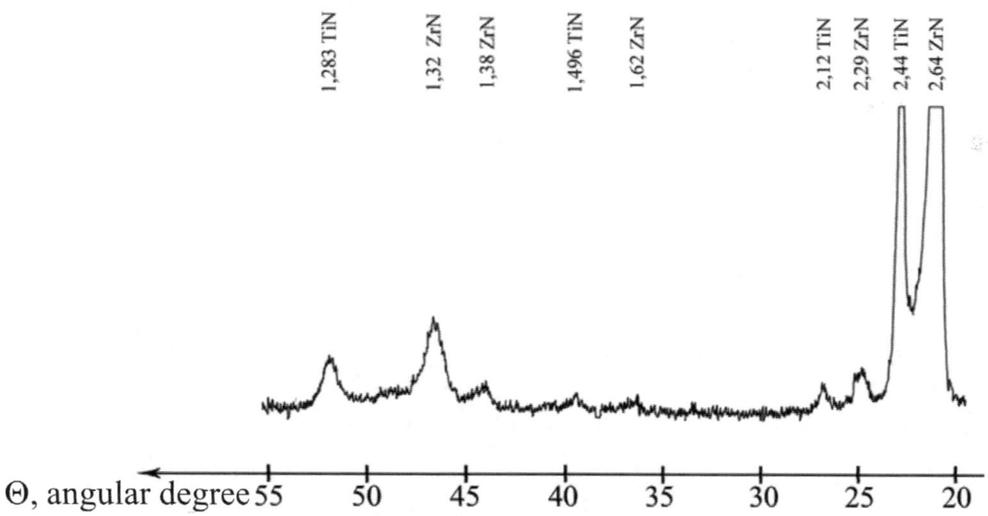

Figure 2. Fragment of diffraction pattern of WC10KS alloy with ion-plasma coating (Ti, Zr)N

To determine surface hardness of alloys with ion-plasma coating a method of nanoidentation (kinetic hardness) by means of «Nano Hardness Tester» was used, the results were processed with the help of Mountains Map Universal, 2.0.13 software.

The results of nanoidentation of a carbide alloy with (Ti, Zr)N ion-plasma coating proved that this coating is superhard with nanohardness 38500MPa (Fig. 4).

IOP Conf. Series: Materials Science and Engineering **91** (2015) 012020 doi:10.1088/1757-899X/91/1/012020

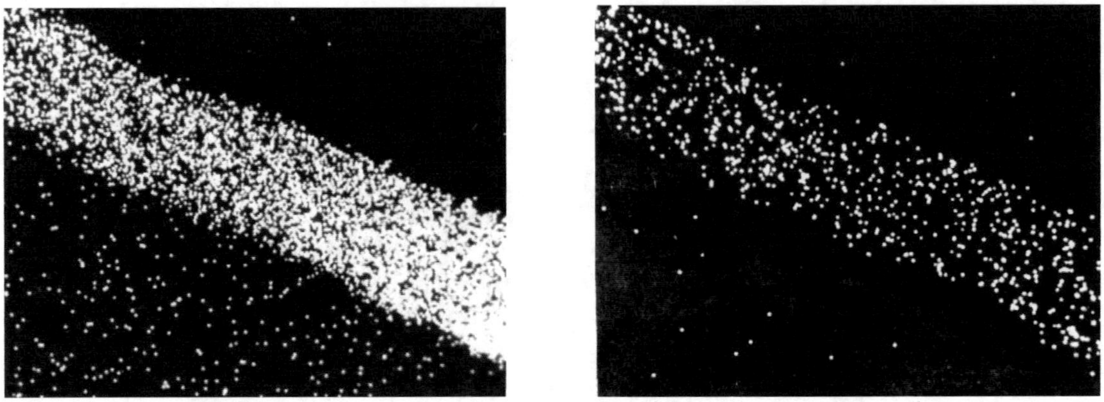

Figure 3. Microstructre of WC10KS alloy with ion-plasma coating (Ti, Zr)N in X-rays of titanium (a), zirconium (b); x 1000

Introduction of zirconium alloying element into the composition of TiN coating improves its mechanical properties: nanohardness increases by 22 -23%, Young's modulus grows by 66 -67% (See the Table). The growth of Young's modulus means increasing energy of atomic links and thus increasing the strength of the coating.

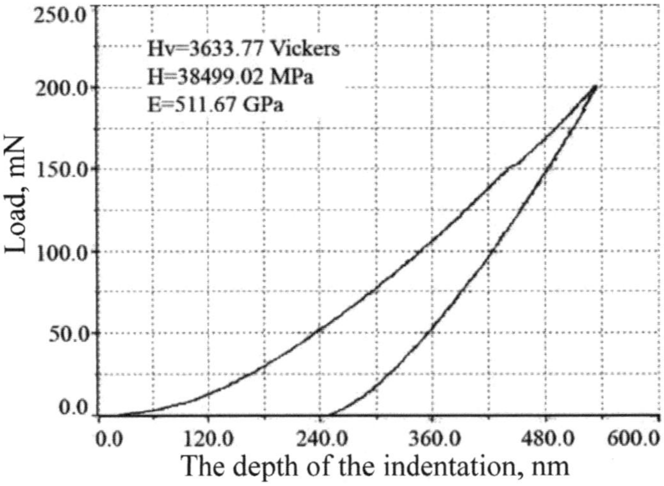

Figure 4 Loading-unloading curves of nanoidentation of (Ti, Zr)N ion-plasma coating on WC10KS alloy

Table Properties of wear resistant coatings

Properties	Coatings	
	TiN	(Ti, Zr)N
Nanohardness, MPa	31450	38499
Young's modulus, GPa	307.00	511.67

Tribological testings of coated samples in original sintered state at room temperature were carried out with the help of high-temperature tribometer programmed «PC-Operated High Temperature Tribometer». A degree of wearing was determined by means of measuring the depth and the area of the track made by an unmoving diamond indenter on a rotating sample under the load of 3H, linear speed 2.5 cm/s, with rotating speed 4000 for uncoated samples and 12000 for coated samples. The analysis was carried out with the help of Mountains Map Universal, 2.0.13 software.

Tribological testings of samples with ion-plasma coating showed that wear track depth on WC10KS alloys is 76.6nm and that on the original sample is 58mkm (Fig. 5).

Track cross-section area on worn out coated samples makes up 4.2 mkm^2 and on uncoated samples it is 12921mkm^2 (Fig. 6). Friction coefficient (μ) of WC10KS alloy with (Ti, Zr)N ion-plasma coating after treatment is fixed at 0.07 which differs from friction coefficient of the original sample equaling 0.41.

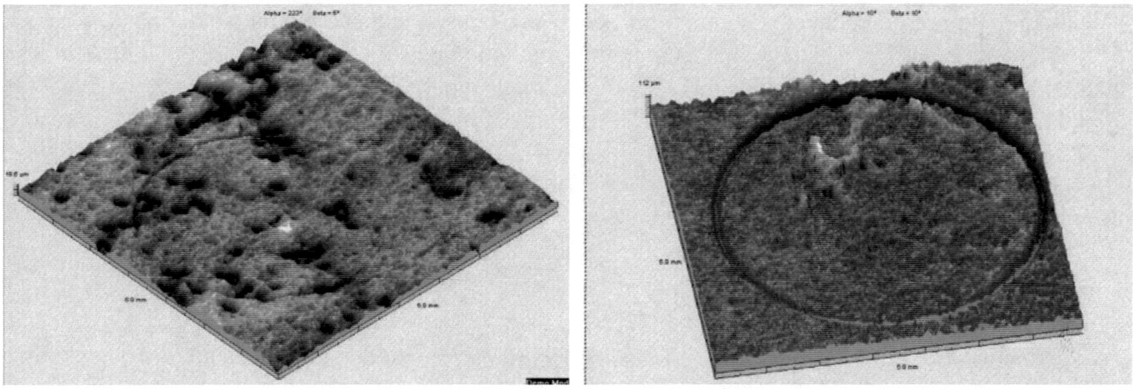

a b

a - WC10KS alloy with (Ti, Zr)N ion-plasma coating; b – original alloy
Figure 5. View of tracks after tribological tests

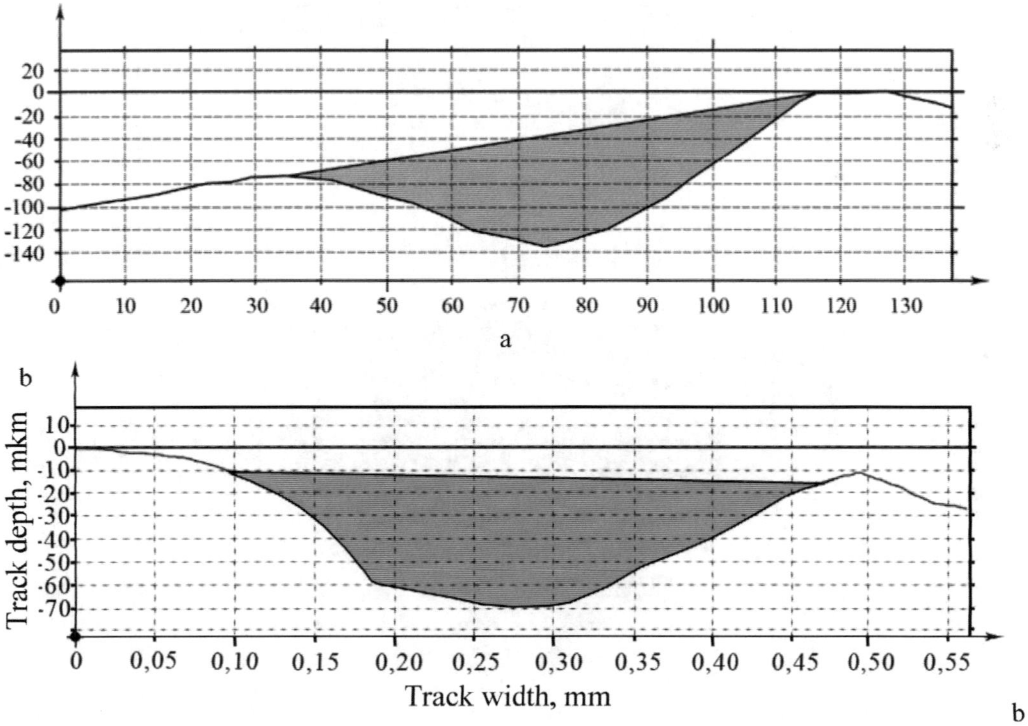

Figure 6. Profiles and areas (in grey color) of wear tracks cross-sections of (a) coated samples, (b) uncoated samples

High wear resistance of (Ti, Zr)N ion-plasma coating can be explained from a perspective of atomic concept by V.F. Moiseev which claims that nitrides of metals from the fourth group of Mendeleev's table provide less intensive wearing because chemical compounds with the highest atomic binding energy (maximal hardness and heat of atomization) are to provide the highest wear resistance.

Ion-plasma spraying is the finishing operation and it does not require any other perfections. After spraying a part can be used for intended purpose. Thus surface roughness is a very important characteristics of the part quality.

It is known that a tool with great roughness has short service life, that is why when a part is being designed an admissible value of surface microgeometry is set to provide the part durability. According to State Standard TU 48-19-367-92, the admissible value of surface finish is R_a = 2.5mkm. The surface microgeometry of a carbide alloy before and after applying (Ti, Zr)N ion-plasma coating was studied with the help of «Micro Measure 3D station». Profilometric research showed that the ion-plasma coating improves the samples surface. The surface roughness of an original sample is R_a = 1.32mkm. After applying the(Ti, Zr)N ion-plasma coating on WC10KS alloy the roughness makes up 0.97mkm (Fig. 7).

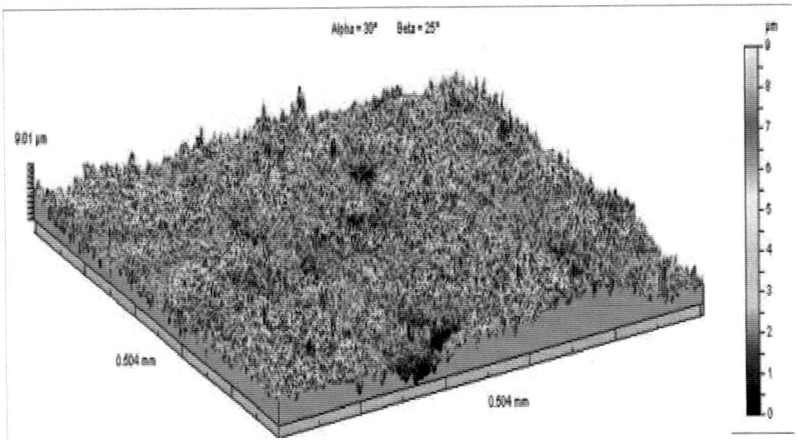

a

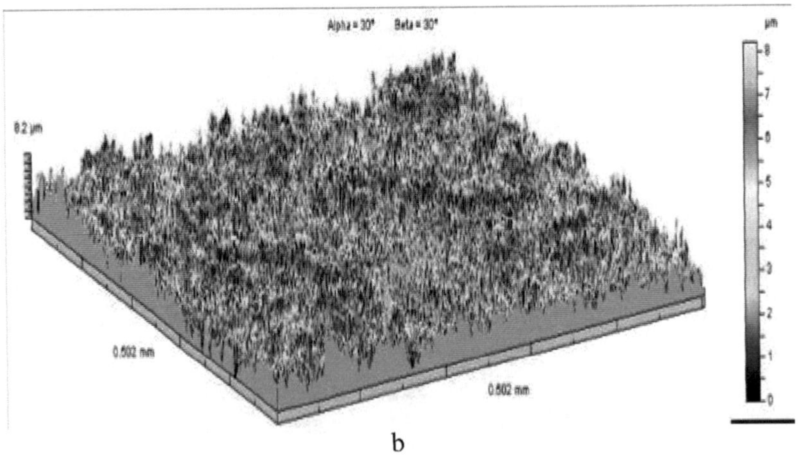

b

Figure 7. Microgeometry of WC10KS alloy with (Ti, Zr)N (a) coating; (b) original sample

The coating fully performs its functions when it is strongly adhesive to the main material. Adhesion considered as binding between of surface layers of dissimilar bodies and phases when they contact is

IOP Conf. Series: Materials Science and Engineering **91** (2015) 012020 doi:10.1088/1757-899X/91/1/012020

one of the main characteristics the obtained thin hard coatings. It characterizes the contact strength of the coating with the base material. It is suggested that it depends greatly on energy of a particle falling on the base plate: the higher is the energy, the greater is the adhesion. During ion-plasma spraying due to the impulse received from an ion knocked-on atoms cross a striking distance and settle down on the base plate. The energy of the atoms which approach the base plate is considerably higher than that during thermal vacuum spraying. It let us suggest that the adhesion of ion-plasma coatings with carbide alloy base plate will be satisfactory.

For this research adhesion tests of ion-plasma coating on WC10KS alloy were carried out with the help of Micro Scratch Tester using a method of scratching with diamond indenter which undergoes linear increasing loads.

To analyse thin hard coatings it is not enough just to increase the speed of load growth. That is why one of the possible solutions is using swing module which allows to prolong multifold the device track and to use high coefficient of normal strength growth. One of the advantages of oscillating sample action is a greater length of the scratch in comparison with a usual scretch. In the case of progressing loads it leads to smoother load growth between the minimal and maximal values.

Load strength at the moment when coating starts peeling off is an important characteristics of adhesive strength. It is called critical load strength and is determined with great accuracy with the help of acustic sensor, tangential load sensor, penetration depth and normal strength sensors, and build-in optical microscope (Fig. 8).

Within our research the adhesion tests of ion-plasma coating on WC10KS alloy were carried out at the following modes: diamond indenter speed – 7.97 mm/min at constantly growing load beginning with initial value of 0.1H to the final ones of 150.0H; signal level from acoustic emission sensor – 9; scratch length – 8mm.

The test showed that at normal loading of 22.3H (Fig. 8, a) a scratch is formed on the coating. As the load grows up to 46.8 H the scratch becomes wider (Fig. 8, b, c). When the load reaches 55.6 H (Fig. 8, d) a simultaneous leap of acoustic emission and fraction signals takes place. This value is critical because it concurs with the moment when the layer begins separating.

In a whole we can conclude that viscosity of (Ti, Zr) N ion-plasma coating with hard base metal is high. In fact high viscosity of (Ti, Zr) N ion-plasma coating with hard base metal is explained by microlamination which is characteristic for coatings produced by separate electrodes. A boundary between microlayers of the coating prevents crack growth and thus increases crack resistance.

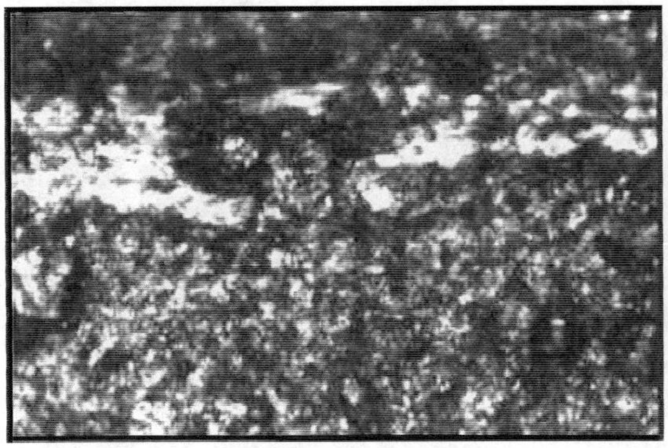
a

b

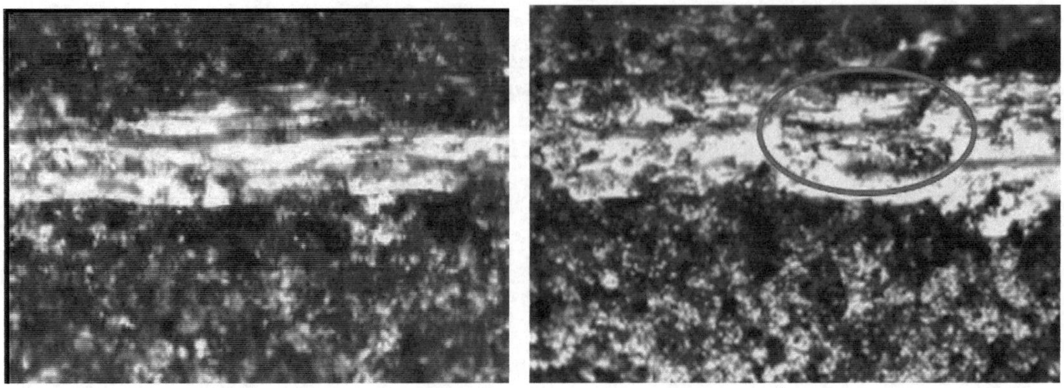

c d

a – 22.3 H; b – 40.3 H; c – 46.8 H; d – 56.6 H

Figure 8. Scratch with growing load on ion-plasma coating

Conclusion

Superhard 10 ÷ 15mkm thick (Ti, Zr)N ion-plasma coating with nanohardness 38500MPa characterized by high wear resistance and adhesion strength was obtained on the surface of carbide alloy WC10KS. It was determined that extra alloying of ion-plasma coating TiN with zirconium leads to increasing nanohardness by 22 ÷ 23 %, Yung's modulus by 66 ÷ 67 %. Growth of Tung modulus means increasing atom links and thus rising the strength of the coating

This research is done within a state order by Ministry of Education and Science of Russia № 11.153.2014/K

References

[1] Shorokhova O V Termovit-M Water-polymer Quenching Fluid / O V Shorokhova T N Oskolkova *Steel in Translation* 2011 T 41 № 4 p 291 – 293

[2] Nozdrin I V Particle size of vanadium and chromium borides and carbides in a plasma flux / I V Nozdrin G V Galevskii L S Shiryaeva M A Terenteva *Steel in Translation* 2011 T 41 № 10 p 799-804

[3] Nozdrin I V Reactor for plasmometallurgical production of refractory borides and carbides / I V Nozdrin L S Shiryaeva G V Galevskii V V Rudneva // *Steel in Translation* 2011 T 41 № 8 p 644-648

[4] Nozdrin I V Plasma synthesis and physicochemical certification of chromium nanocarbide / I V Nozdrin V V Rudneva *Steel in Translation* Vol 53 No 12 2012 p 797-801

[5] Oskolkova T N Pulse plasma treatment of the surface of alloy VK10KS / T N Oskolkova E A Budovskikh *Metal Science and Heat Treatment* 2012 vol 53 № 11–12 p 608 – 610

[6] Oskolkova T N Electric explosion alloying of the surface of hard alloy VK10KS with titanium and silicon carbide / T N Oskolkova E A Budovskikh *Metal Science and Heat Treatment* 2013 vol 55 № 1 p 96 – 99

[7] Oskolkova T N Features of structure formation of the surface layer in the course of electroexplosive alloying tungsten carbide hard alloy / T N Oskolkova E A Budovskikh V F Goryushkin *Non-Ferrous Metals* 2014 vol 55 № 2 p 196 – 200

IOP Conf. Series: Materials Science and Engineering **91** (2015) 012021 doi:10.1088/1757-899X/91/1/012021

Systems to Control Molten Metal Transfer in Arc Welding

A V Filonov[1], A V Kryukov[1] and D E Gusarov[1]
[1]Yurga Institute of Technology National Research Tomsk Polytechnic University
26 Leningradskaya st., Yurga, Kemerovo region, 652050, Russia,
ph. +7(38451)77765

E-mail: a.filonow@gmail.com

Abstract. The paper analyzes the systems used for controlling molten wire metal droplets during the arc welding process in shielding gases. The variations for implementing the relevant systems are given, with the positive and negative aspects of such implementation taken into account. Electrical systems are currently investigated to the fullest extent possible and implemented in different power sources for pulsed welding arc. Mechanical systems are represented by different types of feeders that provide the pulsed wire feeding process. The feed mechanisms driven by electric motors and electromagnets are analyzed. In addition to the mechanical and electrical systems, the examples of combined control systems are given.

1. Introduction

Mechanized and automatic consumable electrode arc welding is the basic technology to have a permanent joint. Gas-shielded metal arc welding is the most widely used in industries over the world.

The developed and offered on the market consumable electrode welding techniques are constantly improved with the purpose to obtain joints having an optimal balance between geometry and metal quality, reduce the costs of further processing works, and reduce the consumption of material and energy resources. A number of features pertaining to the gas-shielded welding process depend on a type of electrode metal transfer.

The transfer of wire electrode metal to a weld pool can normally be divided into three main stages. During the first stage a droplet of a specified size is formed at the electrode tip. Within the second stage the droplet's growth stops, and this droplet together with the electrode contact tip approaches to the weld pool. The third stage completes with the metal transfer to the pool when the bridge between the electrode tip and molten metal drop is destroyed.

There are several variations of metal transfer in the AW process using gas shields; the basic types are featured as follows:

- with naturally occurring short circuits;
- with continuous arcing and droplets or globs of metal transfer;
- with continuous arcing and spray or rotating spray transfer of metal.

The types of metal transfer as well as forces acting on a wire electrode metal while arcing are described in [1, 2]. Each type of metal transfer has both advantages and disadvantages.

2. Analysis

The droplet of a specified mass can be obtained using several possible processes. One of the promising trends in techniques to solve problems in terms of controlling droplets of molten wire metal is to apply

IOP Conf. Series: Materials Science and Engineering **91** (2015) 012021 doi:10.1088/1757-899X/91/1/012021

power pulse cycles to affect the welding process [3]. Currently, three systems have been developed to control the metal transfer:

- Electrical systems, where power sources generate energy pulses to act on the transfer process (Pulsed Power Welding);
- Mechanical systems, where pulsed wire feeding devices are used;
- Combined systems, where the combined effect of electrical and mechanical systems is received.

The first technique, when using an electrical system, is mostly fully investigated and implemented in different pulsed power supplies [4]. A wide range of ways in implementing the first technique makes possible to program almost any algorithms to change the energy parameters of welding arc [5, 6]. The feedback processed in similar devices also enables the design and development of adaptive control systems.

Complexity in designs and difficulty in finding technical solutions may refer to the disadvantages of those devices, and as a consequence they are more expensive. Their operation is limited and sometimes impossible in difficult conditions of high electromagnetic disturbances.

The second group when using a mechanical arrangement, according to the classification proposed in [7], represents devices that can make an influence on a filler material feed system (figure 1).

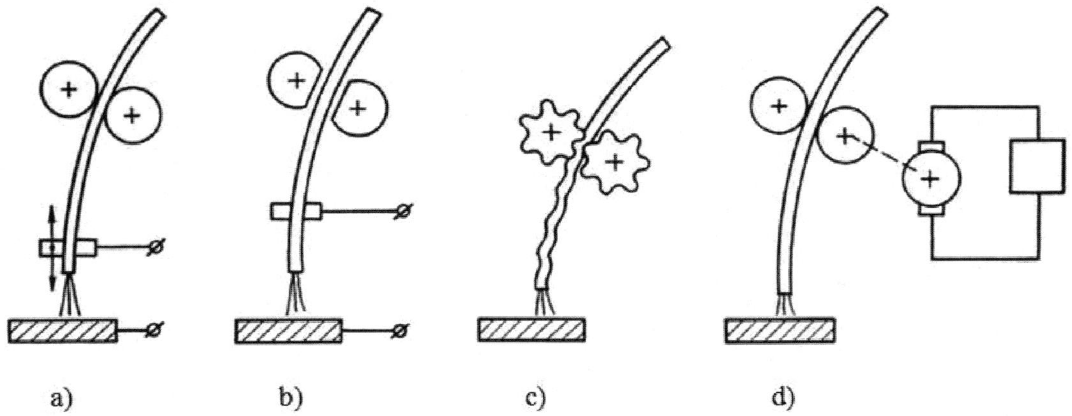

Figure 1. Devices to control the feeding process of filler material.

The devices acting on a feeding system of filler material are presented by mechanisms with the position of the current contact jaw be changeable (figure 1, a), mechanisms with non-circular (figure 1, b) and specifically profiled drive rollers (figure 1, c), and mechanisms fitted with a variable speed wire feed motor (figure 1, d). These mechanisms increase the stability of the welding process and improve the weld formation [8]. Describing this group, we can set the common shortcomings of these mechanisms are as follows: a narrow frequency range, complexity, and sometimes impossibility to correct operating modes during welding. On the other hand, such devices are less sensitive relating to utilized power sources and types of current, i.e. these techniques can be employed using standard power sources.

At the present time, the materials relating to this problem and presented in [7], have been further developed. Therefore, it seems necessary to update the information already available.

The currently available mechanisms of pulsed wire feed can be divided according to a number of features defining their characteristics.

One of the important criteria characterizing both continuous and pulsed feeders is a weld wire feed method. On the basis of this feature, the wire feed mechanisms can be divided into:

IOP Conf. Series: Materials Science and Engineering **91** (2015) 012021 doi:10.1088/1757-899X/91/1/012021

- a push-type feed system (a feeder is located in front of the drive rollers, its job is to push the electrode wire through the insulated conductor tube to a welding zone);
- a pull-type feed system (a feeder is located behind the drive rollers, its job is to pull the electrode wire from the insulated conductor tube).

Another classifying criterion is a type of electrode wire feed drive. According to this criterion, there are two main tendencies for designing and developing devices being used for pulsed wire feed:

- Devices driven by an electric motor (AC or DC, stepper motors);
- Devices driven by electromagnets.

We consider three conventional schemes used for feeding units based on electromagnets (figure 2) [9, 10].

One of the generalized variations for electromagnetic feeding units is shown in figure 2, a. A returnable member in these structures is made in the form of a spring [11] or in the form of elastic membranes [12, 13]. The disadvantage of these devices is irrational use of electromagnet power, which is used not only for wire feed but also for compressing the returnable member.

The device where the second electromagnet is used instead of the returnable member [12, 14] is shown in figure 2, b. Its disadvantages are a low rate of rise in the wire feed and instability of pulse intervals, hence, like the previous device, this mechanism requires the increased electromagnet power in order to overcome initial inertia, and the electromagnet used for the return action does not operates in full.

The device that does require much forces while returning a clamp and simultaneously compresses the spring (figure 2, c) [15] is accepted as the most successful technical solution, i.e. this device accumulates energy as long as the electromagnet armature retracts, and then this energy is translated into an initial point of wire feeding that leads to its rapid acceleration and enables the maximum use of the complete energy supplied to the feeder.

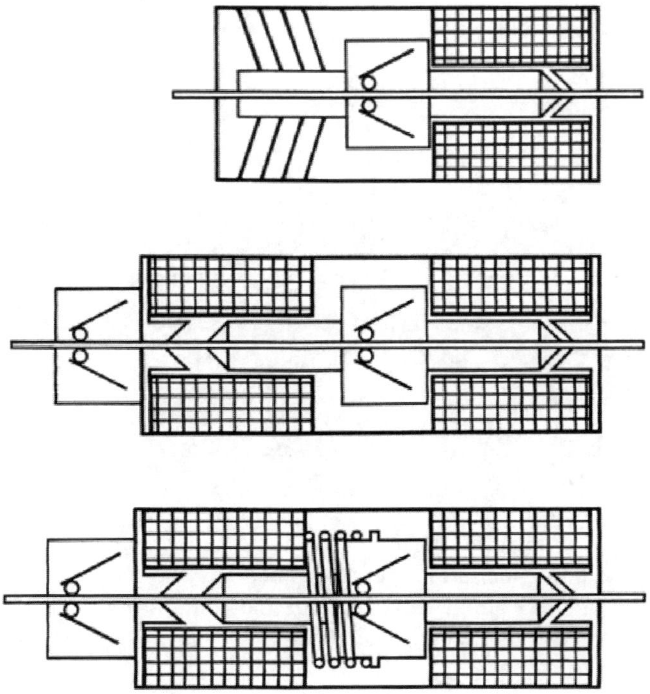

a)

b)

c)

Figure 2. Feeders based on electromagnets.

The feeding mechanisms driven by an electromotor, except for those presented in [7], can additionally be divided into:

1. Direct-drive mechanisms driven by an electromotor [16];
2. Mechanisms with an eccentric roller or profiled jaw;
3. Mechanisms with a programmable voltage supply motor [17];
4. Mechanisms with a movable clamp on the basis of a quasi-wave inverter (QWI) [18, 19];
5. Feeders with a stepping electromotor;
6. Direct-drive mechanisms with a valve electric drive [20].

The E.O. Paton Electric Welding Institute in Ukraine and the Fronius Company are the leading designers of pulse wire feeders.

According to the Ukrainian experts' opinion, mechanisms with a quasi-wave inverter are the most sophisticated all-purpose devices, featuring the improved capability to adjust pulse parameters, in which the feed roller is mounted directly on a motor shaft, and a microprocessor control system enables the programmable pulse rotation of the shaft [3]. One of the solutions offered by Ukrainian designers is the pulsed wire feed system using a special valve electric drive fitted with a digitally controlled unit for adjusting the shaft rotation characteristics [21, 22].

Fronius developed a new technology for the controlled metal transfer process branded the CMT (Cold Metal Transfer) process. During the short circuit, the wire metal moves rearward, the current flow stops and the droplet transfers to the weld pool without spattering [23, 24]. In fact, this technology is the one of those representing the third type of wire transfer control systems – the combined system.

The pulsed wire feed devices driven by the electric motors enabling the direct-drive pulsed feeding are research developments; however, their price exceeding that of conventional systems 1.2-1.5 times needs to be taken into account [25].

The analysis of the systems described in this paper has proved that wire feeding units show a great potential for further development. The pulsed wire feeding process finds its successful implementation in the scheme shown in figure 3.

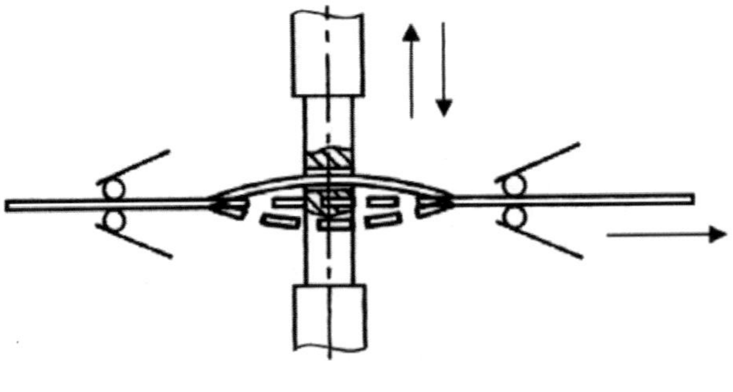

Figure 3. Schematic pulsed wire feed mechanism.

The operation of this mechanism can be described as follows: when moving "upward", the wire is pulled through the left clamp, at this time the right clamp is jammed, preventing the wire from passing. While moving to a "low position", the left clamp is jammed and the right clamp is open, enabling the wire feeding process. The reciprocating movement of the wire rod in the axial direction can be performed using one [26] or two [27] electromagnets, cams [28, 29, 30], or an electromotor driven crank-and-rod mechanism [31].

3. Conclusion

Thus, the wire metal transfer control process using various kinds of pulsed cycles is of great interest; this fact is evidenced by the development of modern welding technologies and machines, aiming at welds having specified properties. The most advanced types of feeders are electromagnetic devices with quasi-wave feeders and direct-drive devices with electric motors. The electromagnetic mechanisms have the advantages as they provide a wide range of control feed rates, relatively low feed forces, and high efficiency. The simple design, highly reliable mechanism and low mass-and-dimension parameters result in the advantages for the direct-drive devices. Therefore, these two types of mechanisms as well as their combination and the usage in combined systems are supposed as most promising relating to their further research and development.

References

[1] Potapievskiy A G 2007 *Gas Metal Arc Welding. Part 1. Active Gas Shielded Welding* (Kiev: Ekotekhnologiya).
[2] Potapievskiy A G, Saraev Yu N and Chinakhov D A 2012 *Gas Metal Arc Welding of Steels. Equipment and Technology of the Future: Monograph* (Tomsk: Publ. Tomsk Polytechnic University).
[3] Paton B E, Lebedev V A, Pichak V G and Poloskov S I 2009 *Development of Pulsed Wire Feed System for Welding and Deposit Welding* (Svarka i Diagnostika (Eng: Welding and Diagnostics) vol 3) pp 46–51.
[4] Lebedev V A 2010 *Development Tendencies for Controlling Metal Transfer in Mechanized Welding (Survey)* (Avtomaticheskaya Svarka (Eng: Automatic Welding) vol 10) pp 45–53.
[5] Kryukov A V, Zernin E A and Polishchuk V A 2013 *Modeling impulse wire feed welding in controlled gas shielding* (*Applied Mechanics and Materials* vol 379) pp 183–187.
[6] Krampit A G and Dmitrieva A V 2014 *Movement of liquid metal in a weld pool when welding with pulsed arc* (*Applied Mechanics and Materials* vol 682) pp 386–391.
[7] Shigayev T G 1983 *Techniques for Modulating Welding Current and Modulating Units* (Avtomaticheskaya Svarka (Eng: Automatic welding) vol 8) pp 51–55.
[8] Pavlov N V, Kryukov A V, Zernin E A and Polishchuk V A 2014 *Modelling of Welded Joint Geometry in Sheet Metal Welding* (*Applied Mechanics and Materials* vol 682) pp 443–447.
[9] Brunov O G 2007 *Mechanized Welding in Active Gas Shields with Pulsed Wire Feed: Scientific Publication* (Tomsk: Publ. Tomsk Polytechnic University).
[10] Brunov O G, Fedko V T and Slistin A P 1999 *Mechanisms for Pulsed Weld Wire Feed* (Technologiya Metallov (Eng: Technology of Metals) vol 11) pp 7–9.
[11] Voropai N M, Saveliev O N and Semergeev S S 1980 *Electromagnetic Mechanisms of Pulsed Welding Wire Feed* (Avtomaticheskaya Svarka (Eng: Automatic welding) vol 1) pp 46–49.
[12] Voropai N M 1998 *Principles for Designing Pulsed Welding Wire Feed* (Avtomaticheskaya Svarka (Eng: Automatic welding) vol 8) pp 19–25.
[13] Rodionov Yu A, Brunov O G and Lysenko A F 1997 *Mechanism of Pulsed Welding Wire Feed* Russian Patent 2090325.
[14] Fedko V T, Brunov O G and Lysenko A F 1998 *Mechanism of Pulsed Welding Wire Feed* Russian Patent 2104134.
[15] Brunov O G, Fedko V T, Knyazkov A F and Slistin A P 1999 *Mechanism of Pulsed Welding Wire Feed* Russian Patent 2136463.
[16] Koveshnikov S P, Belousov A N, Pavlov V F and Poloskov S I 1984 *Direct-Drive Mechanisms of Pulsed Welding Wire Feed* (Svarochnoe Proizvodstvo (Eng: Welding) vol 5) pp 32–34.
[17] Krasnoshapka V V, Kuznetsov V D and Skachkov I O 1993 *Using a DC Drive for Pulsed Welding Wire Feed* (Avtomaticheskaya Svarka (Eng: Automatic Welding) vol 9) pp 53–54.
[18] Lebedev V A 2003 *Peculiarities of Designing Mechanisms for Pulsed Welding Wire Feed in Welding Machines* (Avtomaticheskaya Svarka (Eng: Automatic Welding) vol 3) pp 48–52.
[19] Lebedev V A, Moshkin V F and Pichak V G 1996 *New Mechanisms for Pulsed Welding Wire Feed* (Avtomaticheskaya Svarka (Eng: Automatic Welding) vol 5) pp 39–44.
[20] Lebedev V A, Maksimov S Y, Pichak V G et al 2011 *New Mechanisms of Welding Wire and Filler Feed* (Svarochnoe Proizvodstvo (Eng: Welding) vol 5) pp 35–39.

IOP Conf. Series: Materials Science and Engineering **91** (2015) 012021 doi:10.1088/1757-899X/91/1/012021

[21] Lebedev V A 2014 *Features of Controlling GMAW with Pulsed Welding Wire Feed* (Svarka i Diagnostika (Eng: Welding and Diagnostics) vol 1) pp 15–18.

[22] Paton B E, Lebedev V A, Lengel I V and Poloskov S I 2013 *Using Mechanical Impulses to Control Automatic and Mechanized Gas Metal Arc Welding* (Svarka i Diagnostika (Eng: Welding and Diagnostics) vol 6) pp 16–20.

[23] Bondarenko V L 2004 *Arc Welding with Pulsed Welding Wire Feed – SMT process proposed by "Fronius"* (Avtomaticheskaya Svarka (Eng: Automatic welding) vol 12) pp 55–58.

[24] Lebedev V A 2012 *Transistor Power Source for Electric Arc Welding (Survey)* (Avtomaticheskaya Svarka (Eng: Automatic welding) vol 9) pp 34–40.

[25] Lebedev V A 2010 *Classification of Wire Feed Mechanisms in Welding Equipment* (Svarochnoe Proizvodstvo (Eng: Welding) vol 1) pp 31–37.

[26] Kirikov S O and Kim Yu Yu 2011 *A Method of Consumable Electrode Arc Welding and Welding Machine for Implementing This Method* Russian Patent 2474487.

[27] Brunov O G, Fedko V T and Vasiliev V I 1999 *Mechanism of Pulsed Welding Wire Feed* Russian Patent 2154560.

[28] Brunov O G, Fedko V T, Kryukov A V et al 2004 *Mechanism of Pulsed Welding Wire Feed* Russian Patent 2254969.

[29] Brunov O G, Fedko V T, Kryukov A V et al 2004 *Mechanism of Pulsed Welding Wire Feed* Russian Patent 2275998.

[30] Sednev V V, Brunov O G, Kryukov A V et al 2006 *Mechanism of Pulsed Welding Wire Feed* Russian Patent 2296654.

[31] Bulychev V V, Dubrovskiy V A and Golubina S A 2007 *Method of Vibrating Electrode Arc Welding and a Device for Wire Feed* Russian Patent 2301728.

IOP Conf. Series: Materials Science and Engineering **91** (2015) 012022 doi:10.1088/1757-899X/91/1/012022

Study of the Effect of Shrinkage Porosity on Strength Low Carbon Cast Steel

E Ol'khovik

[1] Admiral Makarov State University of Maritime and Inland Shipping, Saint-Petersburg, Dvinskaya str. 5/7, Russia

e-mail: olhovick@gmail.com

Abstract. Today there are many computer systems for modeling of the casting technology processes. All of them allow calculating the availability and distribution of the shrinkage porosity in the test casting, but this information allows only making changes in existing casting technology. In this paper you obtain the information about changes in the local and structural mechanical properties of the casting in the presence of its volume shrinkage porosity. Article presents the results of direct experimental studies of technological defects (shrinkage and gas porosity) impact on the mechanical properties of low carbon steel castings. Methods of investigation are also disclosed, including the methods for producing of molded samples obtained at different process conditions and the crystallization apparatus which is described for the measuring of the density of the samples. There are the mathematical relationship for the elastic modulus, yield stress, elongation and fatigue characteristics fracture cast steel with low carbon content in the presence of the volumetric shrinkage porosity.

1. Introduction

Shrinkage and porosity are always presented at steel castings which have difficult geometry and form. This known problem leads to uncertainty in the calculations of the structural strength of the cast steel. Porosity is uneven, that affects the overall reliability of the design. Moreover there are no accurate methods for its accounting in the technical strength. As a result, for reliability designers have used the safety factors higher than necessary that leads to heavy steel castings.

The experimental results for the steel plates of the castings are presented at Pellini's work [1]. Kubo and Pehlk in [2] proposed the mathematical model of the porosity formation process during solidification, based on the solution of the unsteady heat conduction. Paper [3] presents the multi-phase model that predicts melt pressure, feeding flow, and porosity formation and growth in the steel castings during solidification. Niyama, Uchida and Morikawa in [4] examined some castings and confirmed that the feeding gradient could be used to predict the formation of shrinkage porosity. Sigworth and Wang in [5] proposed the simple "geometric" criterion to describe, where there will be shrinkage porosity. Nastac and

IOP Conf. Series: Materials Science and Engineering **91** (2015) 012022 doi:10.1088/1757-899X/91/1/012022

Marsden in [6] looked at the solution of the multiphase modeling of macrosegregation and crystallization under various temperature conditions as the cause of the formation of the shrinkage porosity in large steel ingots.

Hardin and Beckermann in [7,8] proposed the methodology to predict the fatigue life of the components made of the cast steel with the shrinkage porosity. They checked data from X-ray tomography measurements of porosity in the castings. They also made combined experimental and computational study to investigate the effect of the central shrinkage porosity on the deformation properties of steel, studied in detail the damage and destruction of the cast steel under tensile tests. More recently Sutaria, Sharma and Ravi in [9] proposed the model that is based on the solution of the Stefan equation for solid and liquid phase with the terms of boundary interface. This solution allows to calculate the mass of the particles by tracking the optimal feed casting and solidification.

In article [10] we proposed the methods and the apparatus for the mechanical testing of the materials under conditions of three dimensional stress state. This procedure gives results better than conventional tensile test of specimens and can be used to assess the strength of porosity. In work [11] we developed the mathematical models to predict the formation of the dendritic structure of the cast steel in various technologic conditions, using the method of the phase field model. These data are also needed to evaluate the strength of the castings by the criterion of austenitic grain size and they are used by us in this work and in the future.

This paper presents the results of the experimental study about effect of the shrinkage porosity on the mechanical properties in the steel castings. There is also offered the analysis of problem of the steel casting parts failure due to the presence of shrinkage or gas porosity in the body casting. A lot of existing literary data show the presence of the characteristic defect in the steel castings under various conditions of their formation: gas or shrinkable porosity. Porosity in the steel casting is the central problem for the most foundries. Most designers are not sure how to account the effect of porosity in the cast steel parts. The casts study using the methods of partial or complete fragmentation on the templet and subsequent separate study do not guarantee the quality of the results [11,12] and they are quite expensive. Various industries use a combination of these strategies to try to establish reliable cost-effective designs. It is produced if there is a lack of liquid metal for the casting feeding during solidification, or as a result the dissolving of gases in the casting. Porosity as systemic phenomenon depends on how the shrinkage process is going. In the case of successful technology management the shrinkage porosity is minimized, on the contrary it becomes chaotic.

Porosity itself can be harmful for the fatigue durability of the casting, because the act that is manifested as stress concentration and cracks with subsequent formation of «kernel» leads to details destruction. Quantitative influence of porosity on fatigue resistance of the steel castings is not fully studied, and there is no good method for predicting of effect of the mechanical properties reducing in the presence of shrink or gas porosity in the casting.

2. Materials and method

Most of the major body parts have complex geometry or contoured shape and they are made of cast steel. This type of steel has good mechanical properties, including yield strength and elongation (plastic properties) and it is able to provide high resistance to elongation, low-cycle and cyclic loads, even in excess of the calculated parameters. A lot of the cast parts have a wall thickness which is varied from 15mm to 80 mm. This difference in thickness almost always leads to the formation of local parts of the casting, malnutrition and as a consequence affects the shrinkage porosity. Therefore, there were two objectives: determine the mechanical properties of the presence in the metal shrinkage porosity and explore the speed of the development of fatigue cracks in the different values of local porosity of cast metal. For better study the tests of some castings different geometry (thickness, length, shape) for real castings were conducted. In all cases the steel with low carbon content (C = 0.15 ÷ 0.25%) was studied. This steel crystallizes at the peritectic reaction and has difficulty for most foundries. For the experiments in this work several metal melts were performed, for which the control of the chemical composition was produced by optical emission method and the basic mechanical properties. All data are shown in Table 1.

Table 1. Chemical composition and mechanical properties of the experimental research of low-carbon steel

№	Mass, %						Mechanical properties		
(melting)	C,%	Mn,%	Si,%	Al,%	S,%	P,%	YS[MPa]	UTS[MPa]	Elong,%
1	0.17	0.3	0.25	0.09	0.015	0.025	196	390	24
	0.15	0.3	0.24	0.09	0.015	0.025	192	389	24
	0.15	0.25	0.23	0.08	0.013	0.022	194	390	25
2	0.18	0.32	0.25	0.08	0.014	0.025	201	390	23
	0.17	0.29	0.24	0.08	0.015	0.024	203	392	23
3	0.19	0.36	0.25	0.08	0.014	0.021	213	407	20
	0.20	0.33	0.25	0.08	0.014	0.024	215	408	21
4	0.23	0.75	0.23	0.09	0.013	0.023	231	438	19
	0.26	0.58	0.23	0.09	0.012	0.022	240	447	18
5	0.28	0.83	0.27	0.09	0.015	0.025	247	452	17
	0.25	0.61	0.26	0.08	0.015	0.025	242	450	17

All test samples for mechanical properties underwent the pre-heat treatment. Melting was carried out in the open induction furnace using slag. As a charge steel scrap metal and steel with the known chemical composition were used. These mechanical properties of the alloy (Table 1) were obtained by testing samples taken immediately during melting. Thus, all the samples can be divided into two parts on the strength of the group: Group 1 - melting №1,2,3; Group 2 - melting №4, 5.

There are many methods to determine the volume of porosity in the samples, each of them has certain advantages and disadvantages. The most commonly used method is x-rays, it is well-suited for determining the local zones affected by porosity, but it cannot give the accurate information about the volume of porosity. Porosity detection through non-destructive testing is a common method. Unfortunately, often the amount of porosity in the critical section can be so slight that it cannot be detected. Now there are no standards defining the critical values of porosity, but even if there is a quality system in a separate company, which determines the critical volume of porosity in casting body, it is likely that this approach will be subjective.

In experiments, it was important to limit impact force action on the part of sand mold. During solidification shrinkage also creates a force field in the casting – [12, 13]. We used the sand mold with minimal sealing that ensures its high ductility after the start of crystallization and during this process. Thus, solidification and shrinkage of the steel samples were conducted without force influence of foundry mold.

3. Experimental methods

In this paper, the method of hydrostatic weighing was used to determine the density of the cast steel samples and amount of shrinkage or gas porosity in the samples, thus, in the future, these samples will be used to determine the mechanical properties.

a) b)

Figure 1. The casted cone samples for studying of the mechanical properties of the metal (a); The scheme for the tenderloin of the samples to determine the density and strength (b)

For testing of research methods, on the first stage there was cast special steel cone sample (Fig.1(a)) of sufficient length without profits which scheme is shown in (Fig.1(b)). Two samples were cut for further study of determination of the mechanical properties and the density of the metal. The advantage of this scheme is the region where the upper sample feeds the lower one by liquid metal. Thus it can be initially assumed that lower molten metal sample has higher density, while the upper sample is struck by the shrinkage pores. The experimental castings were tapered samples and they had sufficient length and different cross-sectional area. To perform this work the series of the experimental castings, various lengths and diameters were casted. The general characteristics of the casting of cone geometry are represented in detail in Table 2, where L - length; d_1 - smallest diameter (bottom); d_2 - largest diameter (top).

Using a thermocouple the time of full curing was recorded in seconds (τ [sec]) for each sample. The total time of crystallization was the time of achievement of the solidus temperature at the top of the casting. About 40 different cast steel samples were made. All data about experimental castings are shown in Table 2. Additional samples for research were obtained from actual castings with a wall thickness from 15 to 50mm and the test block (ISO 4990:2003, par. 6.2.2.2) was casted from the casting and feeding. It is necessary to compare the overall results obtained from the sample of the cone's alloy and the test block. These results have approximately the same mode of solidification, which is made in the real castings.

Table 2. Geometric characteristics of experimental castings and time of their complete solidification

N	L[mm]	d_1[mm]	d_2[mm]	Volume [mm^3]•10^3	Square Lateral surface [mm^2]•`10^3	τ_{sol}[sec]
1	100	60	80	387.3	22	310
2	150	45	60	326.7	24.7	265
3	180	35	50	257.8	24	190
4	200	30	40	193.6	22	160
5	250	20	30	124.3	19,6	45
6	290	30	45	324.4	34.1	95
7	320	15	70	517	42.7	150
8	320	40	60	636.3	50.2	320
9	320	25	60	479.3	42.8	200
10	350	50	65	913.5	63.1	515
11	380	10	100	1103.7	65.6	610

The samples were cut from the experimental castings by machining, with the great control of geometrical accuracy. To perform hydrostatic weighing it is required more precise measurements and therefore specially designed installation shown in Figure 2 was created. To measure the strain gage specially made bar with four strain gauges is used which is connected to the Wheatstone bridge to maximize accuracy. In the processing of signal from the Wheatstone bridge 14-bit analog-to-digital converter (ADC) is used, which provides the measurement accuracy at least 0.025% from the sample weight. During the preparation the samples were heat treated at the temperature of 150-200°C for 10-15 minutes to eliminate the effect of factors associated with the possible fluid or other substances accumulation in the sample body.

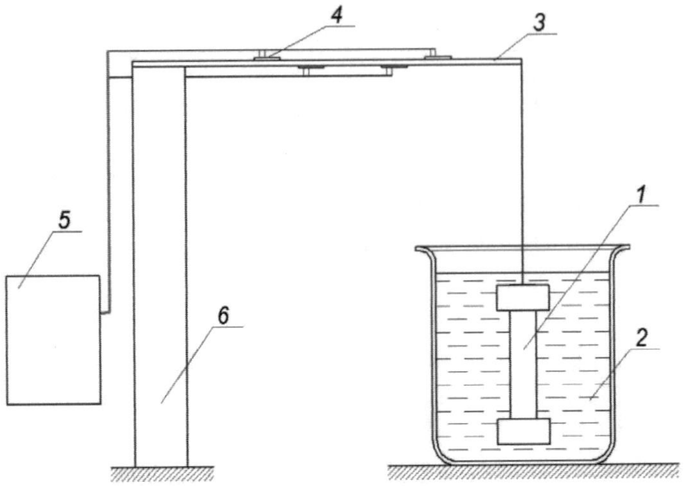

Figure 2. The schematic diagram of the laboratory apparatus for hydrostatic weighing: 1-Laboratory sample; 2-Distilled water; 3-Strain gauge beam; 4- strain gauges; 5-Measurement processing unit; 6- Pillar.

Each sample was analyzed sequentially: initially the sample is weighed into the open air and then in the distilled water. Previously, the geometric dimensions of the sample were measured and its volume was calculated precisely. Using the difference in the results of the sample weighing in the air and in the water the density was calculated. The density changing during sample immersion in the liquid indicates the presence of voids (porosity) in the sample, which have buoyancy force.

4. Results

The value of the porosity volume for all received (tested) samples was ranged from 0% to 10% from mass, however, the samples with the volume of shrinkage porosity more than 5% were found faulty and they have been excluded because for the real castings, such porosity volume is not valid. The study of mechanical properties of cast steel was performed for all samples in order to identify the change of deformation and strength characteristics depending on the amount of porosity.

The mechanical tests of all samples were carried out in accordance with ISO 6892, the preparation of the samples was carried out in accordance with ISO 377. The geometric dimensions of the samples were different, but the ratio of the length to the size of the working part of the samples cross-section was compiled in respect from 1/10 to 1/5. The mechanical tests were performed on the cylindrical fivefold and tenfold samples on the testing machines that provide the sufficient accuracy. According to the results of the mechanical tests the standard mechanical properties of the samples were determined, such as - the elastic modulus, yield strength, tensile strength and elongation.

The figure 3 presents the test results as a function of elasticity modulus from the volume of shrinkage porosity for the samples obtained from our melts №1 ÷ 5, and for the additional samples which were cut from various castings operating under load for 5-10 years. For more accurate analysis, the samples were divided into two groups according to the strength category.

It can be concluded that the porosity does not have serious influence on the modulus of elasticity, while in the case of excess amount of the porosity (more than 3%) the reduction of the elastic properties occurs steadily. It should be understood that the elastic modulus has an integral property, which has the total dependence for a large group of steel. Nevertheless, these data are necessary in the case of calculation the overall structural strength of the cast parts. This relationship (Fig.3) can be applied in the calculations where the finite element method is being used.

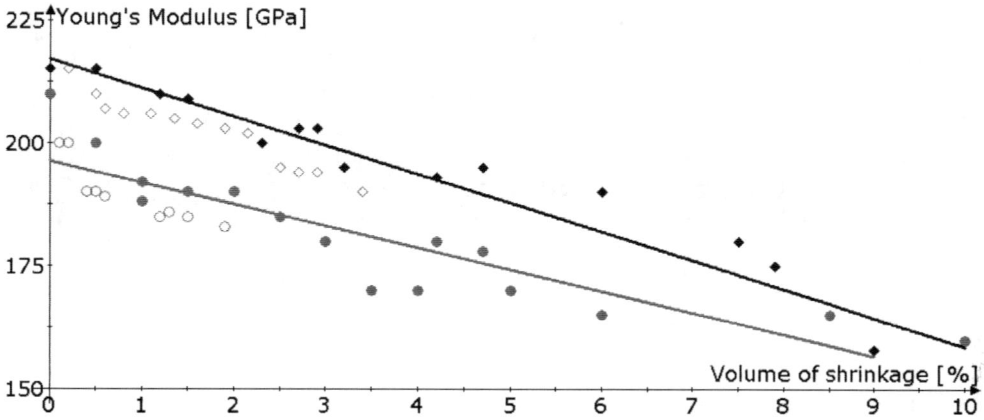

Figure 3. Experimental relationship of modulus of elasticity from the volume of shrinkage porosity

• Samples from melting №1,2,3; ♦ Samples from melting №4,5;

○,◊ Samples of castings which are operated under load for 5-10 years.

To calculate the local changes in the strength the mathematical relationship for the yield stress (YS) was developed – fig.4. Thus it is possible to construct the distribution of such important parameter as the yield stress for all castings, in case of shrinkage porosity.

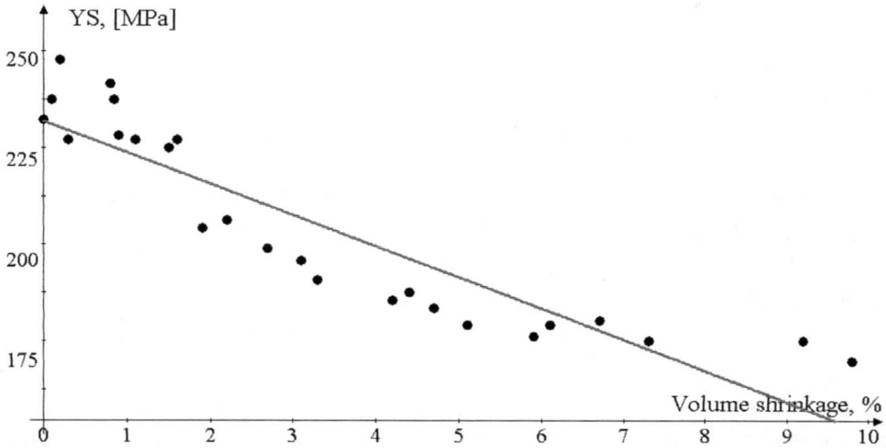

Figure 4. Experimental relationship of yield strength from volume shrinkage porosity

• Samples from melting №1-5

The greatest interest is the change in the plastic properties of the cast steel in the presence of porosity, because the steel castings are working in harsh conditions and cyclically varying loads that require high plastic properties - yield strength and elongation.

5. Conclusion

Currently castings designers don't have quantitative information about change of the mechanical properties of the cast steel directly into the casting. The elaboration of engineering principles for improving of the casting quality allows selecting the assurance factor using methods proposed in this paper. Most foundries conduct such experiments, but don't systematize the data influencing on the strength of the cast steel.

The proposed method for the structural analysis of the shrinkage porosity impact on the quality of steel castings has sufficient capacity and the qualitative methodology. Moreover, all obtained data can be used in the computer analysis of the casting technology and they allow determining the possible reduction in the structural strength of castings in the most loaded areas.

All developed techniques are being applied for production the steel castings of high strength and quality, for example, for the production of the shafting blanks [15], which surface will be processed in the future and for the bodies of ship's pipe fittings [16] or railway bogie[17,18], including some of its highly stressed parts.

6. Highlights

- The methods for obtaining of the cast steel samples (casting cone) that replicate the wide range of conditions of solidification process
- Experimental relationship of changing of mechanical properties of cast steel with low content of carbon with technologic defects as gas porosity and shrinkage
- All obtained data can be used to simulate the casting technology in order to prevent the formation of defects such as shrinkage porosity
- All obtained data can be used in the future to assess the structural strength of the steel castings in their long-term operation

References

[1] Pellini W S Trans Amer Foundrymen's Soc 61 (1953) p 61–80
[2] Kubo K Pehlke R D Mathematical modeling of porosity formation in solidification Metallurgical Transactions B 16 2 (1985) p 359-366
[3] Carlson K D Lin Z Hardin R Beckermann C Mazurkevich G & Schneider M (2003) Modeling of porosity formation and feeding flow in steel casting In Proceedings of the 56 SFSA *Technical and Operating Conference Paper* (No 4 4)
[4] Niyama E Uchida T Morikawa M Predicting shrinkage in large steel castings from temperature gradient calculations Int Cast Met J 6 2 (1981) p 16-22
[5] Sigworth G K Wang C Mechanisms of porosity formation during solidificationp A theoretical analysis *Metallurgical Transactions* B 24 2(1993) p 349-364
[6] Nastac L Marsden K Numerical modelling of macrosegregation and shrinkage in large diameter steel roll castingsp *A mould study International Journal of Cast Metals Research* 26 6 (2013) p 374-382
[7] Hardin R A Beckermann C Prediction of the fatigue life of cast steel containing shrinkage porosity *Metallurgical and Materials Transactions* A 40 3 (2009) p 581-597
[8] Hardin R A Beckermann C Effect of porosity on deformation damage and fracture of cast steel *Metallurgical and Materials Transactions Ap Physical Metallurgy and Materials Science* 44 12 (2013) p 5316-5332

[9] Mayur Sutaria Vinesh H Gada Atul Sharma B Ravi Computation of feed-paths for casting solidification using level-set-method *Journal of Materials Processing Technology* 212 6 (2012) p 1236-1249

[10] Ol'khovik O E and Ol'khovik E O Endurance of structural materials in the triaxial stress state Industrial Laboratory 63 7 (1997) p 423-428 (httpp//www scopus com/inward/record url?eid=2-s2 0-33748894154&partnerID=MN8TOARS)

[11] Ol'khovik E O Desnitskaya L V Prediction of structures of castings Izvestiya Vysshikh Uchebnykh Zavedenij Chernaya Metallurgiya 11 (2004) p 49-53 (httpp//www scopus com/inward/record url?eid=2-s2 0-10144260026&partnerID=MN8TOARS)

[12] Ol'khovik E O Desnitskii V V Molchanyuk R A Interaction between casting and mold during solidification *Steel in Translation* 37 5 (2007)p 422 – 424 (doip 10 3103/S0967091207050051)

[13] Desnitskii V V Ol'khovik E O Molchanyuk R A Investigations of deformative-strength properties of the molds materials during metal crystallization *Blanking productions in mechanical engineering* (Forging and stamping foundry and other productions) 1 (2007) p 16-19

[14] Ol'khovik E O Desnitskii V V Development of methods aided design foundry technology steel castings fitting oil and gas *Izvestiya Vysshikh Uchebnykh Zavedenij Chernaya Metallurgiya* 12 (2006) p 40-43

[15] Khmelevskaya V B Moseyko Y S Ol'khovik E O Reseach of hardening of ship propeller shaft by plasma spraying coverage with ultrasonic treatment *Vestnik gosudarstvennogo universiteta morskogo i rechnogo flota imeni admirala S O Makarova* 19 3 (2013) p 81-87

[16] Ol'khovik E O Reznik U A Development of laser welding technology of bellows assemblies for ship pipeline valves *Vestnik gosudarstvennogo universiteta morskogo i rechnogo flota imeni admirala S O Makarova* 25 3 (2014) p 119-122

[17] Ol'khovik E O Desnitskiy V V Research of influence of dimensional accuracy of casting of bogie sidebar of rail freight car on its strength *Vestnik Mashinostroeniya* 10 (2014) p 47-49

[18] Ol'khovik E O Development process design with effects of technology *quality Applied Mechanics and Materials Applied Mechanics and Materials* 770 (2015) p 419-423 (doip10 4028/www scientific net/AMM 770 419)

Prospects for Establishment of Technological Complexes in Machine Building Industry on The Basis of Electromechatronic Propulsion Systems

Yu M Osipov [1, 2, a], O Yu Osipov[2, b], V A Trifonov [1,c]

[1]Yurga Technological Institute (affiliate) of National Research Tomsk Polytechnic University, Russian Federation, 652055, Kemerovo region, Yurga, Leningradskaya str., 26
[2]National Research Tomsk State University, Russian Federation, 634050, Tomsk, Lenina ave., 36
[a] e- mail: umo1943@yandex.ru, [b]e-mail: ems2009@mail.ru,
[c] e-mail: v.trifonov@rambler.ru

Abstract. The authors consider prospects for technological complex establishment in machine building industry on the basis of electromechatronic propulsion systems for production of innovative products with different novelty levels: world, state, brunch, region, etc.

Nowadays there are many areas where people use robots and manipulation mechanisms. Among them there are both robot machines of Emergencies Ministry for saving people's lives (they are able to find a person under ruins, destroy a bomb, or check nuclear power plant after an accident), and robots for our daily life irrespectively of their specialization.

In comparison with strong gains in robotics intelligent control in technological complexes engineering achievements, which concerns creation of their basic elements such as final control elements of manipulator and robot movements, are far from perfect. This is the main component of robotic tools in any technological complex of machine building industry, where end-effector movements are done.

In the majority of Russian and foreign robot manipulators, in very many cases automated hand (lag) joints are single-axles with reduction electric drives, which are the source of noise, low speed, inaccuracy due to drive shafts, gear belts, gear trains, etc.

A robot construction needs till nine single-exle joints to move, for example, a hand (shoulder, fore arm, etc.) in a thee axes system. It creates big constructions, needs difficult control procedures of multicoordinate stage movements. To decrease influence of enumerated factors they decrease gearing burden, design rational robot configurations, certain contours, limit working space, etc.

It is proved by modern research and development situation in many companies: Fanuc Ltd., Yaskawa Electric Corp., Fuji Machine Mfg. Co., Toshiba Machines Co., Okuma Corp., MoriSeiki Co., Makino Milling Machines Co., Hitachi Seiki Co: Japan, Broetie-Automation GMBH, Germany, TSNIIPTK, Russia, etc.

Nowadays there are many interesting ideas and inventions that concern traverse actuators of robot moving parts, which are still in concepts or prototypes. They will be robotic tools tomorrow [1-6].

Today modern technological complexes of machine building industry on the basis of robots and manipulators represent development needs and social production improvement as a base for economical and military technical state power. They also represent necessity for research and assimilation of new alternative fields and spheres for people's activity.

The authors study electromechatronic robot and manipulator components, which are driven by electric drills, have operation and control sections. Progress in microprocessor technology sphere, electronics, and computing machinery has made possible to expand functions and quality of electric drill control sections, which are almost perfect now. There is no further development of control sections, robots and manipulators in general without qualitative development of operation parts, what lowers their quality and competitive ability [7, 8].

Very actual developments are those that are aimed at robot and manipulator operation parts creation with advanced control sections, which provide integrated move of every part of a robot set, for example, an arm (a shoulder, a fore arm and a hand), with movement characteristics that are similar to real man's movements (from soft, creeping speeds till abrupt movements adapted to an event) with assimilation of difficult and adequate movements. There is a new design tendency in mechatronics that is why constructional decisions become easier: they remove drive units, cam gears, conversion devices for transformation of roll motion into prismatic motion, etc. New magnetic materials on the basis of rare earth metals, new electrotechnical and constructional materials define electromechatronic development and provide usage of electric drill of private action in designing operation parts for technological complexes of machine building industry, consumer electronics, medicine, etc. It is possible to progress by synergetic joining nods of electrical machines with mechanical, electronic and computer components. It ensures production of whole new units – electromechatronic modules (EMD) and electromechatronic movement systems (EMMS) with intellectual control.

The authors propose to develop technological complexes of machine building industry on the basis of new functional element that is "active" driveline (ADL), which provides integrated movement of every part in a robot set, for example, an arm (a shoulder, a fore arm and a hand), with movement characteristics that are similar to real man's movements (from soft, creeping speeds till abrupt movements adapted to an event) with assimilation of difficult and adequate movements [9].

The results of Research and Development (see Table 1).

1. The authors formed the conception of intellectual scheme-constructive design – synergetic integration of electromechanical, electronic and computer control units, as a result of which people have innovations: industrial objects with different protection levels (world, state, branch, region, etc.) involved in commercial turn-round. Theory basis and principles for EMD and EMMS formation are done in the monograph "Multicoordinate electromechatronic movement systems" of the authors O.Yu. Osipov, Yu.M. Osipov, S.V. Shcherbinina:

- methodology for intellectual scheme-constructional design: in innovation processes realization of information conversion and energy in the system "person – tool – work object – environment" with the help of worker intellectual-creative resources on the basis of their educational knowledge, including their heuristic and creative thinking, metadata of design intellectual systems, synergetically integrated electromechanical, electronic and computer control units, as a result of which people have innovations: industrial objects with different protection levels (world, state, branch, region, etc.) involved in commercial turn-round;
- mathematical models and logic for interconnected space movements of joint groups on an "active" driveline (ADL), for example, for an assembly manipulator or an automated "arm" of a manipulator robot;
- systems of accumulator electric power supply and management of ADL voltage (24÷27) Volt with possibility to put some components of electric power supply and control in constructions of mobile robot sets, including robot arms (legs) and body;
- methods to calculate magnetic discontinuity inductive capacity and electromagnetic moving forces of ADL with power active core piece;

IOP Conf. Series: Materials Science and Engineering **91** (2015) 012023 doi:10.1088/1757-899X/91/1/012023

- methods for automated forecasting of EMD large, force-torque and speed characteristics to coordinate order production;
- automated technologies for display of inductor and EMD rotor core pieces;
- systems for automated documentation design for mobile robot sets with joints on the basis of ADL.

Table 1

R&D of electromechatronic movement systems

№	Research and Development of units with electromechatronic movement systems	Time periods (years)
1.	R&D «Development and making an experimental model of isolated generating plant with photovoltaic panels, automatically oriented to the Sun».	2010÷2011
2.	R&D «Development and making experimental prototypes of a high accuracy positioning system for X-ray microbarograph to study material patterns of organic and inorganic nature».	2010÷2011
3.	D «Development and making experimental prototypes of positioning mount to study antenna characteristics».	2011÷2012
4.	D «Development and making a device for nondestructive quality control of heating unit production and calibration».	2012÷2014
5.	R&D «Scientific-technical basis development of multicoordinate electromechatronic movement system formation for robotic tools on the basis of a functional unit that is an "active" driveline».	2013÷2014

2. The authors developed construction documentation for a range: arc EMD (Figure 1), line EMD (Figure 2), rotating EMD (Figure 3).

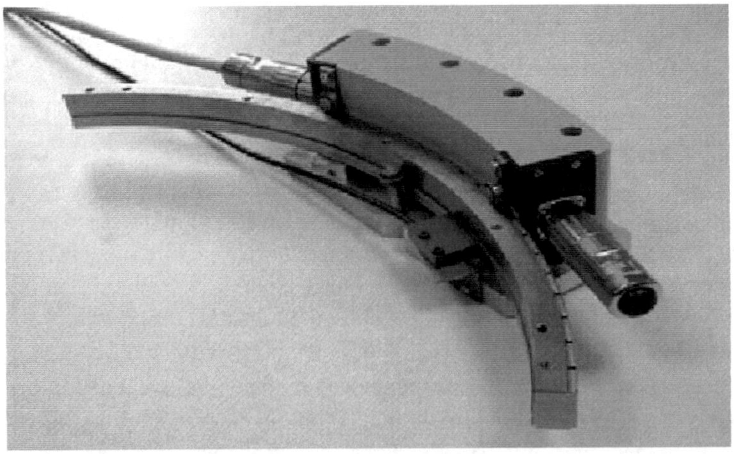

Figure 1. General view of arc electromechatronic module of movement

IOP Conf. Series: Materials Science and Engineering **91** (2015) 012023 doi:10.1088/1757-899X/91/1/012023

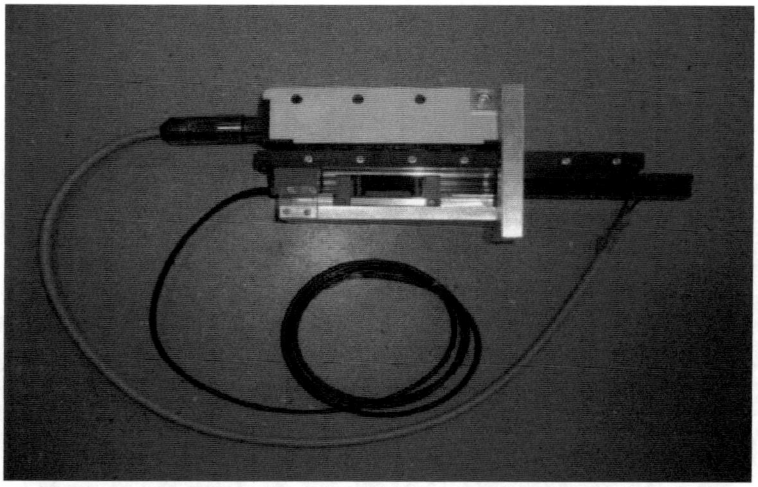

Figure 2. General view of line electromechatronic module of movement

Figure 3. General view of rotating electromechatronic module of movement

On the basis of developed element base the authors made experimental and small-lot prototypes of robotic sets and integrated in technological complexes of client machine building industry: positioning mount to test antennas of Radio Detection and Ranging equipment, device for nondestructive quality control of heating unit production, etc.

3. The authors presented a scheme-constructive conception of a new functional element, segment joints of robotic tools that is active driveline (ADL), which can help obtain advanced results in movement characteristics of robotic tools that are similar to real man's movements and provide integrated move of every part of a robot set, for example, an arm – from soft, creeping speeds till abrupt movements adapted to a real event, with assimilation of difficult and adequate movements especially in extreme and extraordinary situations.

The authors designed some constructive variants of ADL. For an experimental prototype they selected the variant that had design advantages: bigger rotary traverse of an output element in comparison (about 125h180 degree); possibility to put feedback probes and electromagnetic breaks without increasing a structure gauge; operational integrity of electrical wire of power supply and feedback by means of fixed arc inductors.

Main theories and principles for ADL formation are done in the article "Creation of extreme robotic sets on the basis of an active driveline" [9].

The authors made a joint of a robotic set on the basis of a new functional element that is an "active" driveline. The joint with radar antenna was tested in situ at a scanner assembly (Figure 4).

Figure 4. General view of a joint of a robotic set on the basis of a new functional element that is an "active" driveline with radar antenna

R&D in 2010÷2014 and developed scientific-technical work (scientific publications, monographs, RF patents, constructive-technical documentations, software support, etc.) allowed the authors to summarize experience and formulate some ways for establishment of technological complexes in machine building industry on the basis of electromechatronic movement systems for the next period till 2020 according to two R&D directions:

<u>Products to achieve economic growth of a country.</u>

1. Worktable manipulator of a laser beam machine for working spherical and curved surfaces, holes, etc., which has a working detail (or an article) with angle indexing in three rectangular axes.
2. Therapeutic training device with application of energy from impulse low-frequency electromagnetic field, thermal field and mechanical field to influence the zona of small pelvis and back bone with some manipulations (vibratory massage, medicine + infrared light).
3. X-ray 3D tomographic scanner to study substance structure of organic and inorganic nature, and for nondestructive product control.

<u>Products to provide safety and defense capacity of a country</u>.

1. Compact hybrid electromechatronic inertial navigation system for no-maintenance mobile objects.
2. Compact electromechatronic computer for mobile object position, for example, an unpiloted deep research vehicle.
3. A robot-android with remote intelligent control for rescue operations and so on.

Conclusion

1. During developing robotic tools of technological complexes for machine building industry scientists have to solve a compromise problem where the stress is made either on an operation or

control part. The level of techniques, which are used in formation of an electromechatronics operation or control part, plays a certain role in compromise decision selection.

2. Theory and methodology base of robotic systems with new type joints will have principles of construction integration and combination of electromechanical, electrical, electronic, and computer components, technologies for balancing moving parts and assembly of inductor cores and energy active rotors, optimization of magnetomotive construction force of passive and active core pieces and there components, windings, etc.

3. Active driveline application will allow obtaining new results in characteristics of robotic tool movements, which are similar to real man's movements (from soft, creeping speeds till abrupt movements adapted to a real event).

4. Originality of proposed ways for scientific problem solving shows practicability of theoretical and experimental researches, possibility to get results that have a right to be protected as inventions, utility models, production prototypes, application software, database, know how.

5. Importance and ranking of project results in a state safety related system with project results realization are extremely high: increasing of a state alert level till 2030, including antiterrorism protection, defense problem solving, fire fighting, space and World ocean exploration, problems of nuclear energetic and hazardous production facilities, medicine.

6. Technological development forecasting (project results will allow going the next step) is to design robotic tools with EDM with flexible controlled core pieces, which have high tactile sensitivity and sensor properties.

References

[1] Guerman-Galkin S G 1989 Some problems of electromechanotronic element classification News of Higher Educational Institutions. Electromechanics pp 11-15
[2] Kosykin Yu P 1991 Introduction to electromechanotronics St. Petersburg: Energoatonizdat, St. Petersburg office p 192
[3] I World (III Russian) conference in Electromechanotronics 1997 St. Petersburg: SETU p 335
[4] Krailuk A D, Komchenkov V I, Ivlev A A, Iuria A D 2009 Basic concepts of robotic development for military purposes till 2030 // Mechatronics, automation, management pp 10-15
[5] Vladimir V 2010 Vantsevich Education in Mechatronics // Editors: David Bradley, David W. Russell Mechatronics in Action Case Studies in Mechatronics – Applications and Education. — Springer-Verlag London Limited p 200
[6] Koskin Yu P and Samokhvalov D V 2013 About terminology and specialists training in electromechatronics // News of St. Petersburg SETU "LETI" pp 57-65
[7] Izotkina N Yu, Osipov Yu M, Syriamkin V I, Trifonov V A 2014 Commercialization of a position control system for X-ray microtomographs Topical issues of modern machine building industry Tomsk: TPU Publ pp 55-58
[8] Tceplit A, Grigoreva A, Osipov Yu 2014 Developing the model for assessing the competitiveness of innovative engineering products Applied Mechanics and Materials pp 623-630
[9] Osipov Yu M and Osipov O Y Creation of extreme robotic sets on the basis of an active driveline Works of 7 world symposium "Extreme robotic – robotic sets for work in dangerous environment (RISE-ER'2013)"

IOP Conf. Series: Materials Science and Engineering **91** (2015) 012024 doi:10.1088/1757-899X/91/1/012024

Patterns of folded structure formation in the maximum bending zone of [111] FCC single crystals

E A Alfyorova[1,2], D V Lychagin[1,3], A A. Popov[2]

[1] National Research Tomsk State University 634050, Russia, Tomsk, pr. Lenina, 36
[2] National Research Tomsk Polytechnic University 634050, Russia, Tomsk, pr. Lenina, 30
[3] Tomsk State University of Architecture and Building 634003, Russia, Tomsk, Solyanaya sq., 2

e-mail: dvl-tomsk@mail.ru , katerina525@mail.ru

Abstract. Formation of folded structure in the zone of [111] FCC single crystals maximum bending has been studied on the example of nickel under compression tests. Quasi-periodical behavior of misorientation change in the folded structure has been established following the direction to increasing Schmid factor for the acting slip systems. Change in misorientation inside and on the boundaries of the formed folding coincides with its accumulation in dislocation system: inside and on the boundaries of deformation bands.

1. Introduction

While pressure shaping of metallic materials folding is observed in bulk materials, as well as wrinkling of surface during presswork of irregular-shaped products made from thin sheets and tubes flexure.

Generally folds and wrinkles are located perpendicularly to compression stress direction and are presented as local instabilities releasing compression strength. Folds and wrinkles could be considered as functional defects decreasing structural properties of material under further static and dynamic loadings. For the other part, formation of controlled folds could be applied as the means of obtaining special purpose products possibly used in micro- and nano- units of different applications.

Folding can be observed under different ways of deformation; it could be seen differently and could be initiated both by slipping and by twinning. Taking as an example, folding was observed in steel deformed by twinning [1]. Gubernatorov V.V. considered formation of corrugated boundaries to be the reason of folding [2]. The given results were obtained while studying the surface of rolled 3% silicon iron. It was established that such boundary in separate zones created significant misorientation (around 30°). Panin A.V. studied folding under tension on the surface of EK-181 steel polycrystals [3]. Panin V.E., Panin A.V. pay special attention to the fact that deformation of surface layer can be considered as deformation of the separate structural level of deformation system [4]. Folding effect occur ahead the moving slip while surface smoothing [5]. Finite element modeling demonstrated that the folding zone does not differ by the significant increase in deformation rate, and maximum deformation follows the slip. Similar results have been described in [6], through the mentioned work is devoted to the process of metal cutting. Apart from the given above studies important research direction in metal treatment under pressure is modeling of folding under deformation (pressing) of sheet material, e.g. Hamdan M.N. [7]. Research results in this field enable to predict folding depending on the parameters of deformed system and deformation-inducing system.

Therefore studying folding in ductile metallic materials is of fundamental and applied interest for multilevel analysis of plastic strain and working out the pressure shaping modes.

The present research is aimed at studying the patterns of folding in FCC single crystals on the example of nickel with compression axis orientation [111].

2. Materials and method

The object of the present study is nickel single crystals with compression axis orientation [$\bar{1}$11] and lateral faces (110) and (112). Slipping is performed along crystal systems <110> {111}. Selection of this type of crystal-lattice orientation is defined by its demonstration of maximum instability of samples under compression tests and subsequently the intensively developing bending of boundaries. In the zone of maximum degree of boundaries bending the most intensive folding occurs. The mentioned zone is under study within the present work. The samples under study were deformed by 22% (0.22 in relative units), thus corresponding to transition to the stage III of strain hardening curves. According to data of transmission electron microscopy this deformation leads to formation of dislocation cell structure and disoriented dislocation substructures start to grow [8]. The deformation relief were examined using the SEM instrument Tescan Vega II LMU c EBSD- console nd optical microscope Leica DM 2500P. To study the surface morphology using other methods, such as in [9] for estimating fracture parameters using the acoustic emission method. Methodology of samples preparation, deformation and their further study are described in [10].

3. Experimental results

Samples tested under compression are always influenced by frontal friction. Friction value can be reduced by using different kinds of lubricants and binders. However according to calculations by finite element analysis method it was noted that in frontal-surrounding zone stress redistribution always occurs creating inhomogeneous stress field in the sample volume [11]. While activation of friction forces the scheme of primary stresses of uneven all-round compression is realized in this zone, and in the central part of crystal – uniaxial compression scheme. At the boundaries of zones with different schemes of primary stresses macro-bending of sample is observed and this zone is characterized by intensive folding. Structural study of folding process in this zone has been held.

The zone of folded structure formation has been studied by electron backscatter diffraction. Studies of misorientations accumulations have been carried out on perpendicular boundary with excluded deformation relief (boundary (110)). This allowed considering misorientation development as long as removing from folding on the surface of sample.

Figure 1a demonstrates crystal-lattice orientation of lateral faces and sample orientation with respect to microscope coordinate system. Dashed lines demonstrate form alteration under compression by 22%. Given are the zones with different schemes of primary stresses and disoriented zones. Location of folds is almost parallel to the sample face; they are well developed and have sharp boundaries. Apart from that, rough short slip traces intersecting the folds can be distinguished. Folds are shown in Figure 1b.

The data received during experimental studies prove development of reorientation of local zones inward the crystal. In the zone of maximum bending reorientation of boundary (110) zones from its initial direction (as shown by arrow from the center of straight pole figure, fig. 1d) can be observed. Reorientation process in folded structure is characterized by blurring of orientations. This process is realized by means of alternate arrangement of zones with orientation that is close to direction [120] and orientation located between directions [221]-[121] (blurring of central spot, Fig. 1c-d). At this time inside the misorientation bands some oscillation of misorientation close to the mentioned directions is observed. Similar results are noted in works of Gubernatorov V.V., Gervasieva I.V. and their colleagues [12-13]. They show that corrugation is caused by loss of stability of material layers under plastic strain in conditions of constrained shear. Resulting from that, alternating load occurring at meso-level lead to formation of volumes with regularly alternating misorientation with respect to the initial orientation. Authors note that corrugation is the consequence of gradient of the stress-strain state along the sample section and deformation constraint of material layers in the core zone. Corrugation in its turn influences the formation, structure and location of banded structures.

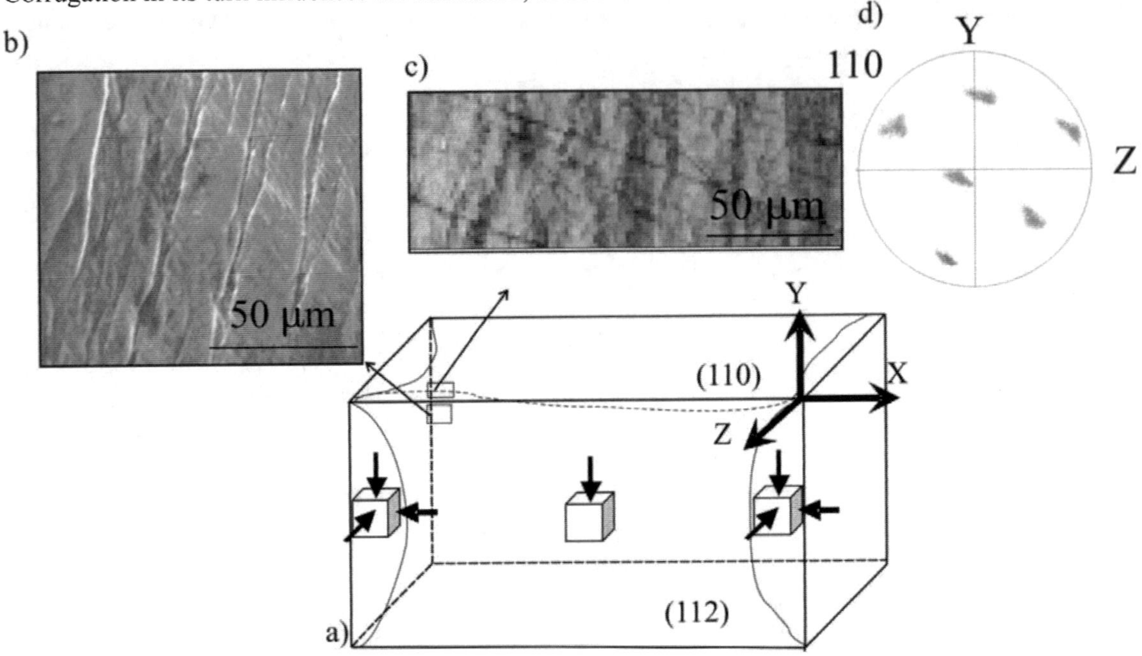

Figure 1. Crystal-lattice orientation and sample orientation with respect to the microscope coordinate system (a), folds morphology (b), reorientation in local zones (c), pole figure (d).

Figure 2a shows distribution of misorientation boundaries in the zone under consideration. It is seen that inside the reorganized bands the value of angles is 2°...8°, while along the boundaries of zones with different orientation it increases up to 30°. At this deformation degree the largest part is composed by boundaries with misorientation value 2°...3° (Fig. 2b). By analyzing alteration in the value of misorientation angles along the bands it was found that alteration has oscillating nature and varies within 3° for the zones oriented in direction [221]-[121] and within 7° for the zones that are oriented in direction [120]. Similar analysis across bands indicated compatible behavior and value of angle within 5°...6° in both cases.

When analyzing distribution of misorientation angles of the neighboring zones along the secant line in direction from point 1 to point 2 (secant line is given in Fig. 2a) oscillating behavior of alteration in value of angles is noted. In this case the values of angles that are higher than 10°...20° fall into the boundary of zones oriented in direction [120] and [221]-[121 (Fig. 3a). Misorientation accumulation with respect to the fixed point 1 (Fig. 3b) also occurs by means of periodical alteration of misorientation value. Along with that arises tendency to increase in misorientation value relatively to the initial point. This could be connected with macro-bending of the sample.

The maximum values of angles are observed with interval of 5 µm. Correlation with the behavior of dislocation structure [8] proves that such misorientation can be accumulated in misoriented dislocation cellular or micro-banded nickel substructure. Structural elements having the size of 0,5 µm at the marked higher step value misorientation of 5°...6° are accumulated on about 10...11 structural elements. When passing through the boundary of bands the angle value increases up to 15°...30°, i.e. boundaries with larger misorientation angles outline the local zones with less misorientation. Therefore the process of how misorientation zones occur from cellular structure can be seen. Similar mechanism has already been discussed before in the works of Khaterly M, Malin A.S. [14-15].

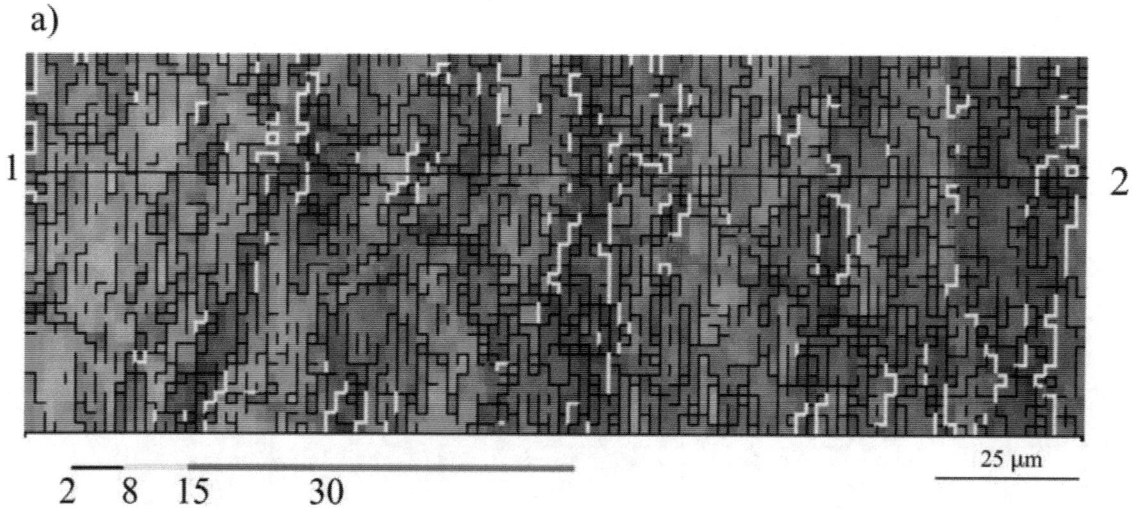

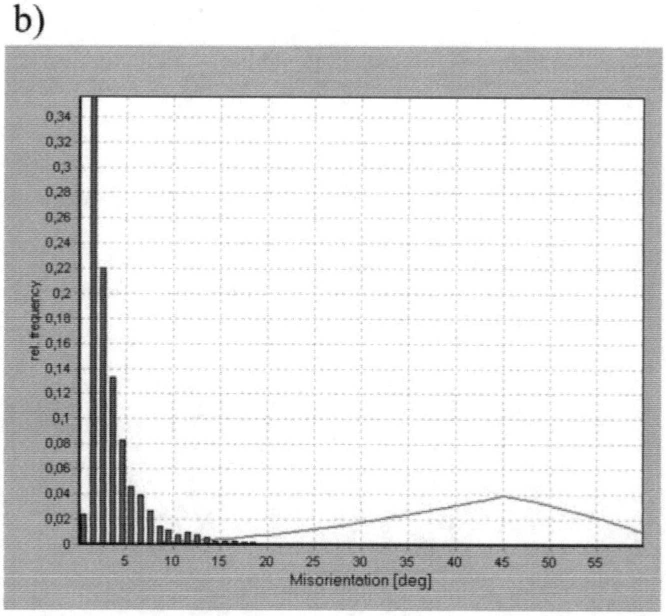

Figure 2. Distribution of misorientation boundaries (a), misorientation angles distribution bar chart (b).

The process of folded structure formation in single crystal in the zone of maximum bending can be presented as follows. With the deformation initiation six equally loaded slip planes with Schmid factor 0.27 start to act. However, only some of them are acting in local zones. Due to action of the particular shear systems local zones are characterized by bending of boundaries. In this case in order to accommodate shape alteration of boundaries the process of folding is being launched. Due to orientation alteration in folds, the acting slip systems are also changing. This phenomena observed in the experiment corresponds well with the common convictions on orientation alteration as a result of slipping in materials with fcc lattice and alterations in dislocation structure.

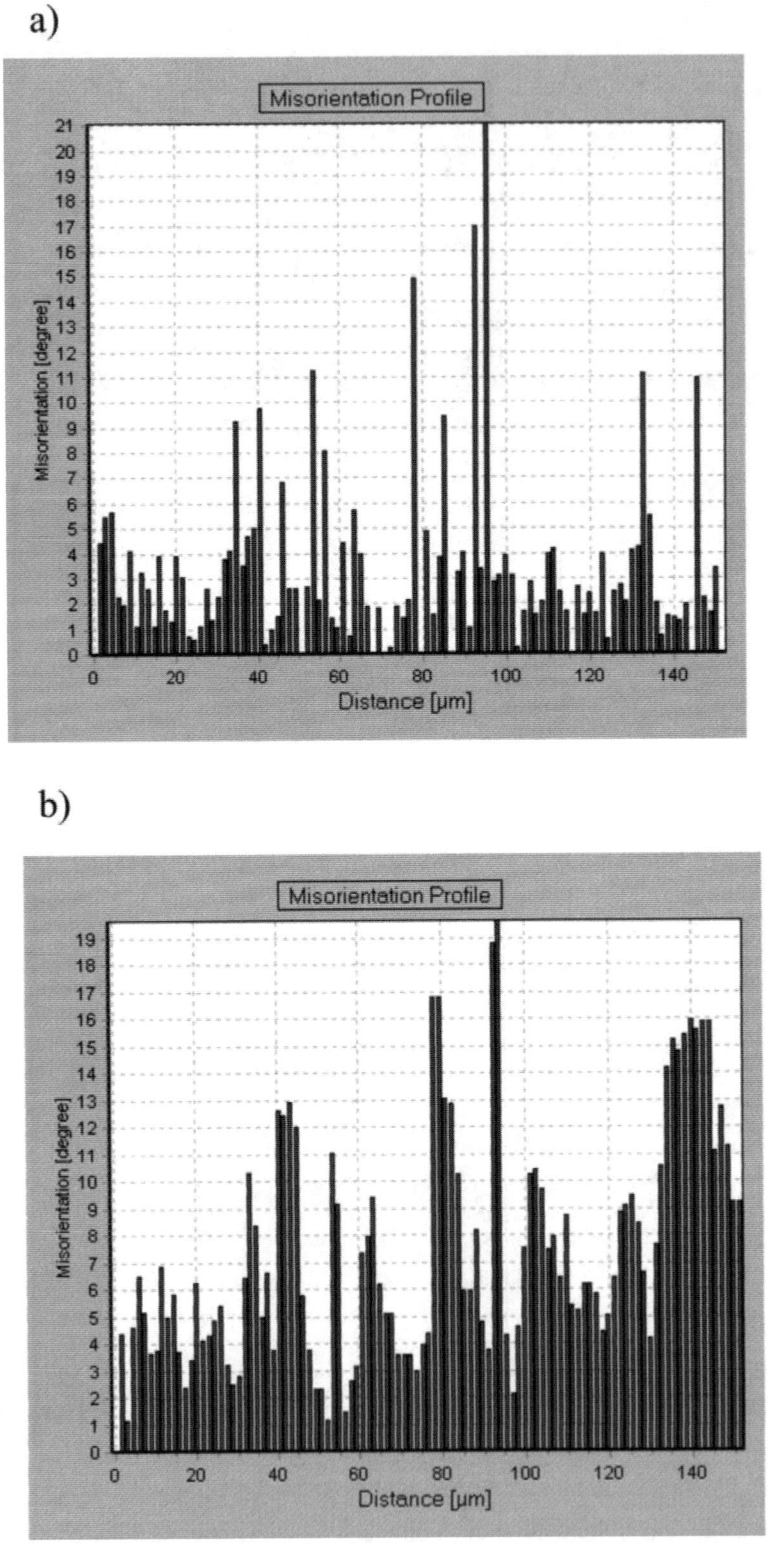

Figure 3. Distribution of misorientation angles along secant line (a), distribution of misorientation angles along secant line with respect to the fixed point 1 (b).

4. Conclusions

Zones of sample maximum bending are considered to be preferable zones of folding formation. The zone of maximum bending is the boundary of zones with different schemes of primary stresses.

Folding formation in the zone of macro-bending reflects periodic behavior of the spatial process of reorientation development. The value of period varies within 10…30 µm. Schmid factor increases in the zone of folded structure formation. Every period includes two zones, each of them having its preferable orientation, generally [221] and [120].

Alteration of orientation between neighboring zones correlates with the earlier obtained data of transmission electron microscopy.

The value of reorientation corresponds with reorientation in elements of dislocation structure inside deformation band and with significant alteration of orientation while transition from one deformation band to the other.

Acknowledgements

Investigations were carried out using the equipment center for collective use "Analytical Center Geochemistry of Natural Systems" Tomsk State University.

This studying (research grant No 8.1.76.2015) was supported by The Tomsk State University Academic D.I. Mendeleev Fund Program in 2015.

References

[1] Khoddam S, Beladi H, Hodgson P D and Zarei-Hanzaki A 2014 *Materials and Design* **60** 146
[2] Gubernatorov V V, Sycheva T S and Pyatygin A I 2003 *Phys. Met. Metallogr.* **96** **1** 28
[3] Panin A V et. al. 2011 *Phys. Mesomech.* **14** **4** 57
[4] Panin V E and Panin A V 2005 *Phys. Mesomech.* **8** **5-6** 7
[5] Narayan K, Sundaram, Yang Guo and Srinivasan Chandrasekar 2012 *Phys. Rev. Letters.* **109** **10** Article number 106001.
[6] Filippov A V and Gorbatenko V V 2014 *Applied Mechanics and Materials* **682** 525
[7] Hamdan M N, Al-Qaisia A A and Abdallah S 2012 *Int. J. of Modern Nonlinear Theory and Application.* **1** 55
[8] Starenchenko V A, Lychagin D V, Shaekhov R V and Kozlov E V 1999 *Russian Physics Journal* **7** 71
[9] Petrova V A, Bakanov A A and Walter A V 2014 *Applied Mechanics and Materials* **682** 431
[10] Lychagin D , Chumaevskiy A and Alferova E 2014 *Applied Mechanics and Materials* **682** 448
[11] Alferova E, Lychagin D and Chernyakov A 2014 *Applied Mechanics and Materials* **682** 485
[12] Gubernatorov V V, Sokolov B K and. Gervasyeva I V 1999 *Phys. Mesomech.* **2** **1-2** 157
[13] Gubernatorov V V, Sokolov B K, Sbitnev A K, Vladimirov L R and Gervasyeva I V 1999 *Textures and Microstructures* **32** **1-4** 41
[14] Hatherly M 1982 *Strength Met. and Alloys, Proc. 6th Int. Conf., Melbourne, Oxford* **2** 1181
[15] Hatherly M and Malin A S 1984 *Scr. Met.* **18** 449

Preferred Orientation Evolution of Olivine Grains as an Indicator of Change in the Deformation Mechanism

D.V. Lychagin[1,2], P.A. Tishin [1], A.S.Kulkov[1,3], A. I. Chernyshov[1] and E.A. Alfyorova[2]

[1] National Research Tomsk State University 634050, Russia, Tomsk, pr. Lenina, 36
[2] National Research Tomsk Polytechnic University 634050, Russia, Tomsk, pr. Lenina, 30
[3] Institute of Strength Physics and Materials Science, Siberian Branch of the Russian Academy of Sciences, 634021, Russia, Tomsk, 2/4 Akademicheskii Ave.

e-mail: tishin_pa@mail.ru, dvl-tomsk@mail.ru, 727@sibmail.com, aich@ggf.tsu.ru, katerina525@mail.ru

Abstract. The paper presents the results of investigations of deformed natural polycrystalline olivine. The relationship of the structure of polycrystalline olivine grains to three modal size distributions has been revealed. Grains of different size were observed to be strained at threshold temperatures of 950, 775, and 650°C. It has been demonstrated that the microstructure develops as the dislocation mechanism changes from diffusion creep to grain boundary sliding. The changes in deformation mechanisms promote the change in the preferred crystallographic orientations of olivine from type A to type D and then to type B. The relation of the transitions between different types of orientations to the conditions of deformation in the lower layers of the lithosphere at the plate boundaries is discussed.

1. Introduction

Olivine is considered an important indicator of the deformation regime of the lower layers of the lithosphere that carry information about the stress pattern of the crust–mantle interface. Information of this kind can serve to gain insight into the stress relaxation mechanisms that underlie seismic events [1, 2]. Much attention drawn to olivine is due to the fact that the olivine rock (dunite) contains 90–95% of olivine, which therefore can be considered to be responsible for the physicochemical and deformation processes occurring in the rock. Dunites are characterized by high plasticity, which occurs at temperatures ranging between 400 and 2000°C and pressures ranging between 0.3 and 20 GPa [3–6]. These parameters vary with rock occurrence depth and characterize the state of different layers of the lithosphere. Consequently, the deformational microstructures of dunite indicate the stress state in different layers of the earth crust and upper mantle. Therefore, to gain an understanding of the mechanisms of plastic deformation of olivine is of considerable scientific importance.

There are two main lines of research in this field associated with examination of experimentally or naturally deformed mono- and polycrystalline olivine by using optical methods, electron microscopy, and x-ray analysis [10, 14–16]. The first one is to study the structures that are formed in a material subject to deformation under known experimental conditions and the other is to examine the microstructure of naturally deformed olivine and reconstruct the stress pattern of the lithosphere [1, 2; 11, 12]. The most common techniques of reconstruction of plastic deformation mechanisms are based

IOP Conf. Series: Materials Science and Engineering **91** (2015) 012025 doi:10.1088/1757-899X/91/1/012025

on using functions of the preferred orientation of material in the active stress field [7–10, 13, 15]. Therefore, the present study was aimed to reveal the mechanisms by which the olivine grains of a polycrystalline aggregate had been naturally deformed by analyzing the aggregate morphology and orientation structure.

2. Materials and method

Samples of four structural types of naturally deformed olivine were selected as test objects. The selected structural types demonstrate different steps of deformation [7, 8, 15–17]. The morphology of polycrystalline olivine was examined by optical and scanning electron microscopy. The crystallographic orientation of grains was determined by the electron backscatter diffraction technique with the use of a Leica DM 2500P optical microscope attached to a Tescan Vega II LM scanning electron microscope. The chemical composition of the olivine and associated spinel was determined using an INCA Energy 300 energy dispersive x-ray spectrometer. The data obtained were statistically processed to identify the grain size distribution and the preferred crystal orientation (the source information was derived from electron micrographs and grain orientation maps).

3. Results

3.1. Phase and component composition. The test samples contained olivine (93–98.5 ± 3%) and spinel (1.5–7 ± 3%). The olivine was found to belong to a rhombic crystal system and have mmm symmetry. Its composition featured by a significant predominance of magnesium in A_2SiO_4 cations and by an iron content no more than 12%. The spinel belonged to a cubic crystal system and had m3m symmetry. Its composition was identified as a mixture of bivalent and trivalent $AO*B_2O_3$ oxides (here A stands for magnesium or iron and B for chromium or aluminum). In the test samples, a significant dominance of Cr_2O_3 (47–56 ± 0.2%) over Al_2O_3 (8–18 ± 0.3%) and of FeO (18–24 ± 0.2%) over MgO (6–9 ± 0.4%) was observed. The results obtained allowed us to determine the temperature of formation of the coexisting olivine and spinel. Using Fabri's phase equilibrium equations [18], the temperature range in which the material of the test samples was formed was estimated to be 650– 955°C.

3.2. Types of deformation microstructures. Structure analysis of the samples has revealed four types of structure in the polycrystalline olivine: protogranular, porphyroclastic, porphyrolastic, and mosaic. These structures were different in volume fraction (V) of grains of three modal sizes: coarse (more than 1 mm), medium (0.1–1 mm), and fine (less than 0.1 mm) (Table 1), indicating an increasing degree of deformation of the polycrystalline olivine.

Table 1. Characteristics of grains in the deformation microstructures and the temperatures of their formation.

Structure type	Grain type								
	coarse			medium			fine		
	D^{mod} [mm]	V [%]	T [°C]	D^{mod} [mm]	V [%]	T [°C]	D^{mod} [mm]	V [%]	T [°C]
Protogranular	5.5	81	955	0.75	19	795			
Porphyroclastic	3.0	44	945	0.65	42	780	0.06	14	655
Porphyrolastic	2.5	32	930	0.60	27	774	0.05	41	668
Mosaic				0.40	24	766	0.03	76	650

The protogranular type of structure featured a prevalence of the coarse fraction compared to the medium one. The medium size grains were localized mostly in triple joints of grains. This suggests that they had been formed as a result of recrystallization. The porphyroclastic structure was characterized by an increased fraction of medium grains and by the presence of fine grains, which where absent in the protogranular structure (see Table 1). It can be supposed that this effect was due to the evolution of primary recrystallization. The porphyrolastic type of structure featured nearly the same volume fractions of the three grain sizes. Besides, coarse grains showed a pronounced preferred orientation. The mosaic type of structure was distinguished by an obvious prevalence of fine grains over medium ones and by the absence of coarse grains.

Calculations of the phase equilibrium of olivine grains of different size at the sites adjoined to spinel [18] made it possible to estimate the temperature of formation of the polycrystalline structure types (see Table 1). Analysis has shown that the grains diminished in size as the temperature of their deformation decreases in the range 950–930°C for coarse grains, 795–765°C for mid-sized grains, and 670–650°C for fine grains.

3.3. Preferred orientation. The preferred orientation of grains for each type of olivine was determined by the EBSD technique. Stereographic projections of the basic crystallographic lines are given in Figure 1 where the elements of anisotropic structures, such as foliation and lineation, are marked. Foliation is associated with a plane similar to the rolling plane at pressure metal treatment and lineation with the direction of statistical pulling of grains, i.e. with the direction of maximal extension. For the protograinular structures, the maximum [100] axis orientation density corresponds to lineation. The [010] axes are arranged normal to foliation. The effect of orientation blurring over several maxima was due to the statistical contribution of coarse grains (figure 1a). Similar but more orderly orientations are typical for porphyroclast structures (figure 1b). This is A type of preferred orientations, which arise due to the activity of the [100] (010) slip system at high temperatures and low stresses [13]. The stereographic projections of the olivine crystallographic axes in porphyrolaste structures show a pronounced maximum of parallel linear [100] axes and girdle dispersion of [010] and [001] axes (figure 1c). The orientations of this kind corresponds to the D type and arise due to the activity of the [100] {0kl} slip system. The orientations of mosaic olivine are determined by the maximum concentration of [010] axes normal to the foliation and by the girdle distribution of [100] and [001] axes (figure 1d). In this case, the [001] axes form a weak maximum along lineation which allows the obtained pattern to be associated with B-type preferred orientations. This type of orientations is formed due to the activity of the [001] (010) slip system at low temperatures and high stresses [9, 13].

4. Discussion and Conclusions

The investigations performed have shown that the formation of olivine grains of three modal sizes in the test deformation system occurred at three threshold temperatures: 950, 775, and 650°C. Knowledge of the temperatures of formation and size of olivine grains makes it possible to use a Jung and Carato piezometer [19] to determine the differential stress. The obtained data were plotted on a map of deformation mechanisms (fig. 2). The dots corresponding to the three main sizes of grains are arranged in three areas on the graph. The dots corresponding to coarse grains are concentrated in the area of deformation by the dislocation creep mechanism, and the dots corresponding to fine grains are localized in the area of deformation by the diffusion creep mechanism. The dots associated with mid-size grains are on the boundary between the two areas. It was shown [12] that under these conditions, the deformation can occur by the mechanism of sliding along the grain boundaries. Consequently, the activity of this mechanism can be related to a separate area on the map of deformation mechanisms.

In view of that the grains of each of the isolated modal sizes make a certain volumetric fraction in a structure of particular type (see Table 1), we can suggest that its deformation occurred by successively changed or simultaneously operated deformation mechanisms. The dislocation creep mechanism prevailed in the formation of the protogranular polycrystal. A combination of two mechanisms – dislocation creep and grain boundary sliding – was characteristic of the porphyroclast type structure.

Diffusion creep was probably responsible for the formation of the fine-grain aggregate in the porphyrolaste type structures, and we believe that this mechanism prevailed in the formation of the mosaic polycrystalline structure.

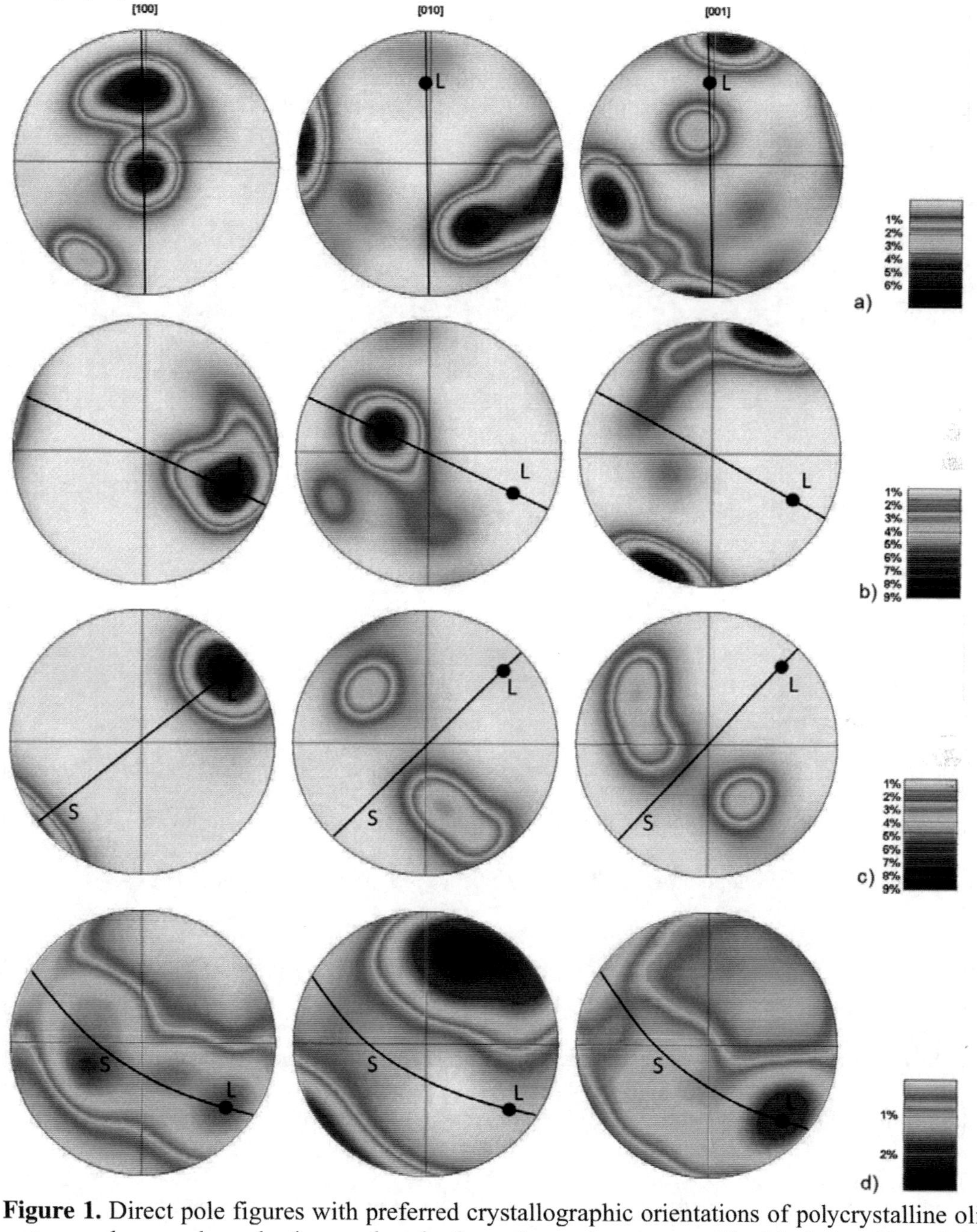

Figure 1. Direct pole figures with preferred crystallographic orientations of polycrystalline olivine of protogranular, porphyroclastic, porphyrolastic, and mosaic structure. The black line (S) and the dot (L) indicate the plane projections of foliation and lineation, respectively.

The features of the preferred orientations display their identity in protogranular and porphyroclast structures where they correspond to the A type [13]. The porphyrolaste structures are characterized by D-type orientations, which differ from A-type orientations by a higher degree of dispersion of [010] and [001] axes and by the occurrence of girdle dispersion (figure 1, a–d). The more significant misorientation of olivine grains in the porphyrolaste aggregates can be accounted for by different creep mechanisms. In our early studies [17], we have shown that the deformations of coarse grains are caused by consecutive activation of the (010)[100] → {0kl}[100] → {100}[001] slip system.

For mid-size grains, a combination of grain boundary sliding and translation by {0kl}[100], {110}[001], and (100)[010] slip systems is characteristic. Consistent activation of different slip systems in translation and grain boundary sliding promotes the development of rotation plasticity and re-orientation of grains. The deformation processes have the result that the fine-grain aggregate shows B-type preferred orientations, which are most pronounced in the mosaic polycrystal (figure 1, d).

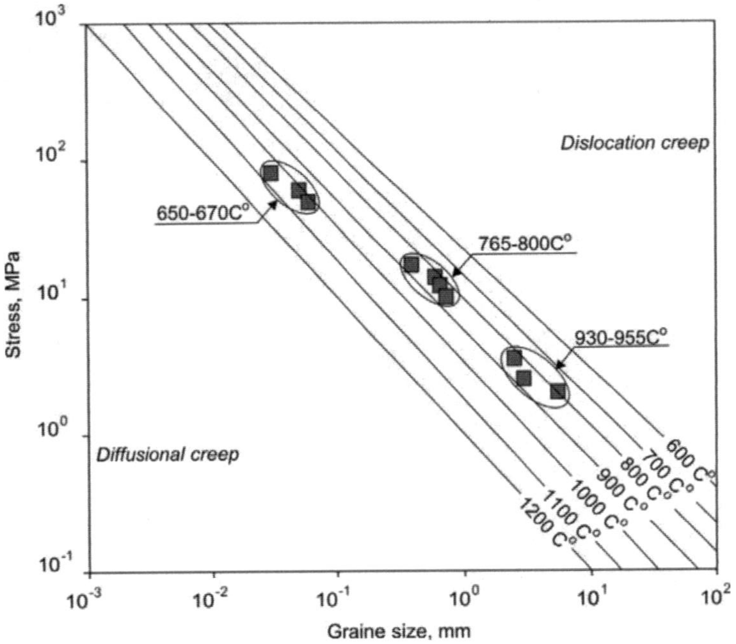

Figure 2. Map of deformation mechanisms of olivine showing the changes in grains size against the magnitude of differential stress according to [3] and the boundaries of the dislocation slide-to-diffusion transition in the temperature ranges 600–900°C [4–5] and 1000–1200°C [6].

B-type orientations occurring under the conditions of diffusion sliding are often mentioned in the literature [2, 9, 11, 12], as a rule, in the context of the transition from the C-type to the B-type orientation. A-type orientations are seldom and they bear witness to the deformation processes that took place during horizontal movements of the earth crust and mantle. C-type orientations are observed for subduction of the oceanic crust. B-type orientations are associated with tectonic acretion of the lower layers of the lithosphere at continental boundaries. The lack of C-type orientations in the test samples indicates the territorial features associated with the small angle of dipping of the oceanic plate under the continental plate.

Thus, the investigations performed revealed the deformation mechanisms, orientation types, and conditions of formation of the basic structural types of natural polycrystalline olivine.

Acknowledgements

This work has been supported by the Tomsk State University Academic D.I. Mendeleev Fund Program in 2015, Russia (research grant No 8.1.76.2015) and the Ministry of Education and Science of the Russian Federation (State Project No. 2282). The investigations were carried out using the equipment center for collective use “Analytical Center Geochemistry of Natural Systems” at Tomsk State University, Russia.

IOP Conf. Series: Materials Science and Engineering **91** (2015) 012025 doi:10.1088/1757-899X/91/1/012025

References

[1] Mainprice D, Nicolas A 1989 *J. Struct. Geol.* **11** 175

[2] Wheeler J 2009 *Geophys. J. Int.* **178** 1723

[3] Karato S, Paterson M S and Fitz J D 1986 *J. Geophys. Res.* **91** (1986) 8151

[4] Mei S, Kohlstedt D L 2000 *J. Geophys. Res.* **105** 21457

[5] Mei S, Kohlstedt D L 2000 *J. Geophys. Res.* **105** 21471

[6] Karato S, Jung H 2003 *Philos. Mag.* **83** 401

[7] Nicolas A, Poirier J P 1976 *Crystalline plasticity and solid state flow in metamorphic rocks* (New York: Wiley-Interscience)

[8] Goncharenko A I 1989 *Deformation and petrofabric evolution alpine-ultramafic rocks* (Tomsk: TSU Published, in Russia)

[9] Katayama I, Karato S 2006 *Phys. Earth Planet. Inter.* **157** 33

[10] Lee K, Jiang Z and Karato S 2002 *Tectonophysics* **351** 331

[11] Nagaya T, Wallis S R, Kobayashi H, Michibayashi K, Mizukami T, Seto Y, Miyake A and Matsumoto M 2014 *Earth Planet. Sci. Lett.* **387** 67

[12] Wang Y, Zhang J, Shi F 2013 *Earth Planet. Sci. Lett.* **376** 63

[13] Zhang S, Kararo S-I 1995 *Nature* **375** 774

[14] Leroux H, Libourel G, Lemelle L, Guyot F 2003 *Meteorit. Planet. Sci.* **38** 81

[15] Chernishov A I, Goncharenko A I, Gertner I F, Betcher O V 1997 *Petrostructure evolution ultramafic rocks* (Tomsk: TSU Published in Russia)

[16] Kulkov SN, Kulkov A S and Chernyshov A I 2010 *Physical Mesomechanics* **13** 83

[17] Kulkov A S, Chernishov A I, Lychagin D V, Tishin P A and Kulkov S N 2014 *AIP Proceedings* 323

[18] Fabries J. 1979 *Contributions to Mineralogy and Petrology* **69** 329

[19] Jung H and Karato S 2001 *Journal of Structural Geology* **23** 1337

Methods of noninvasive electrophysiological heart examination basing on solution of inverse problem of electrocardiography

M Grigoriev and L Babich

Tomsk Polytechnic University, Tomsk, Russia

E-mail: Mishatpu@mail.ru

Abstract. The article represents the main noninvasive methods of heart electrical activity examination, theoretical bases of solution of electrocardiography inverse problem, application of different methods of heart examination in clinical practice, and generalized achievements in this sphere in global experience.

1. Introduction

This Modern functional diagnostics provides the most variable instrumental methods of examination both invasive and noninvasive. The most widespread and available methods of heart examination is electrocardiography (ECG). In spite of prior use in cardiology it is also successfully used for examination of patients with diseases of lungs, kidneys, liver, endocrine glands, and blood system and also in pediatrics, geriatrics, oncology, sports medicine, etc. Using ECG it is possible to detect heart rate and as a result to identify any heart rhythms disturbances; to detect the disturbances of heart electrical conduction which may lead to decrease of its pumping ability and even to its complete cessation; to detect defects or damages of heart muscle caused by chronic or acute disease.

In spite of availability and informational content under real conditions ECG records are affected by internal and external disturbances which disrupt the given informational fragments and as a result cause additional problems in the process of generation of ECG analysis and interpretation computer systems. Even the solution of supposedly absolutely simple problem of ECG division into separate cardiac cycles (RR intervals) requires using quite difficult detection algorithms of QRS-complexes.

Voltage is significantly influenced by correct recording technique and also the distance from explorative electrode to current source. The size of ECG waves is inversely related to squared distance from electrode to current source. It means that the farther electrode is located from current source the less voltage of electrocardiogram complexes is. Therefore after removal of electrodes more than 12 sm from heart further change of voltage appears to be insignificant.

2. Methods and techniques

Electrocardiographic method also has essential limtations. Activity of definite myocardial subdivisions is insufficiently resembled in electrocardiographic signals recorded in direct leads. The example is the difficulty in ECG diagnosing of myocardial infarction of left ventricular posterobasal subdivisions. Besides according to principle of superposition in electrodynamics electrocardiogram is a sum of

IOP Conf. Series: Materials Science and Engineering **91** (2015) 012026 doi:10.1088/1757-899X/91/1/012026

electrical potentials occurring from the sources in multiple myocardial points. Since electrophysiological processes in different parts of cardiac muscle run simultaneously it is quite difficult to determine local electrical activity of myocardium using indirect ECG leads. For example human atrial repolarization wave under normal rhythm conditions is not apparent in ECG because it is "covered" by high-amplitude QRS complex resembling ventricular depolarization. The same limitations are also characteristic for the method of vector cardiography [1].

Wider opportunities are peculiar for the method of superficial electrocardiographic mapping of chest. The method involves simultaneous recording of multiple (from 40 to 250 and more) unipolar ECG leads from the surface of chest and construction of maps of electrical potential distribution on the chest surface by means of interpolation for each cardiocycle time interval. However the given method does not provide accurate determination of local electrical myocardial activity. If electrode is located on the surface of chest contributions to ECG signal from the nearest and the most distant relative to recording electrode of myocardium segment have approximately one order difference. For the electrode placed on heart surface this difference equals to three orders. Therefore, methods of invasive ECG recording are used aimed at maximum approach of electrodes to heart surface to determine local electrical activity of heart.

Transesophageal electrophysiological heart examination is based on the introduction of a probe with recording electrodes into the esophageal cavity. Esophagus in a definite region is quite closely adjacent to posterior wall of left atrium and posterior wall of left ventricle that is why intraesophageal ECG-signals selectively record activity of these heart compartments. Intraesophageal electrocardiography is used particularly for differential diagnostics of supraventricular and ventricular arrhythmias. However the given method allows determining local electrical activity of only certain heart structures.

Invasive electrophysiological heart examination based on direct recording of electrogram complexes from epicardial or endocardial heart surfaces is used for complex evaluation of electrophysiological heart processes and topical diagnostics of heart rhythm disturbances. The given methods can be applied to the "open heart" under thoracotomy conditions and also basing on interventional technologies of introduction of recording devices (catheters) into the heart cavities by transvessel access or into pericardium cavity by its transcutaneous puncture under fluoroscopic supervision.

Modern implementations of the given methods provide precise three-dimensional coordinates of recording electrodes by nonfluoroscopic methods and visualization of results in the form of isopotential and isochronic maps on the models of heart compartments by means of computer graphics. Computer models of heart compartments are built basing on multiple points of electrocardiogram recording with the known coordinates and also on the data of computed and magnetic resonance heart tomography.

This group includes methods of noncontact endocardial mapping based on introduction of "floating" balloon catheter into heart cavity, recording of electrocardiogram complex on its surface and reconstruction by calculation method according to received data of endocardial electrograms.

The drawback of the given methods which is overcome in this invention is their invasive character. The analogues of the invention are methods of reconstruction of electrograms in the inner points of the chest by calculating method according to the data of simultaneous recording of ECG complex on the chest surface. The given methods are based on the solution of inverse problem of electrocardiography. Inverse ECG problem mathematically is the task of harmonic extension of potential towards the sources that is Koch's task for Laplace equation. Calculating region in which Laplace equation is assumed, is the part of the chest limited by the external surface of heart, chest surface on which ECG recording is available and imaginary cross-sections of the chest at the level of diaphragm and clavicles.

On the part of chest surface where ECG recording is available means of electrical potential received as a result of ECG-mapping and also statement of zero equation of normal potential derivative are assumed. These data represent the statement of Koch's task.

IOP Conf. Series: Materials Science and Engineering **91** (2015) 012026 doi:10.1088/1757-899X/91/1/012026

Koch's task involves finding the potential of electrical field in the given region and its trace on heart surface and cross-sections of the chest for the potential in calculating region to satisfy Laplace equation and on the surface of torso where ECG recording is available Koch's statement.

One of the methods of electrocardiography inverse problem solution is the method of reconstruction of electrical field on "quasiepicardium" which is conventional spherical surface surrounding heart. Mathematically the method is based on the presentation of potential of heart electrical field in the form of harmonic polynomial (solid spherical harmonic) ratios of which from statement of equality (or minimum root-mean-square deviation) of means of polynomial and ECG-signal in the points of its recording taking into account zero equation of normal derivative of potential are located on the chest surface. Polynomial of not more than 4 level is used for stabilization of the solution. The method has a significant drawback: reducing sphere radius that is on drawing near the surface of "quasiepicardium" to the real heart surface the accuracy of potential reconstruction abruptly falls. On drawing surface of "quasiepicardium" near the chest surface, resolution of the method in terms of determination of local electrical myocardium activity reduces.

To solve boundary problems for Laplace equation methods of integral equations of potential theory in English-language literature more known as boundary element methods are widely used. The given methods particularly provide the presentation of heart and torso surfaces in the form of polygonal surfaces that is the division of boundary surfaces into multiple triangular elements.

The closest to the represented method (prototype) is the method of Noninvasive Electrocardiographic Mapping (Noninvasive Electrocardiographic Imaging, ECGI) where the superficial mapping is performed using 224 unipolar electrodes placed on a special waistcoat which patients wear during the examination. Surfaces of torso and heart are determined on the basis of computed or MR-tomography of the chest. Algorithm of reconstruction is based on the solution of electrocardiography inverse problem by the method of boundary elements. Heart and torso surfaces are represented in the form of polygonal surfaces. System of matrix-vector equations which is by the way of elementary transformations amounts to simultaneous linear algebraic equations is used for solution of inverse ECG problem. The method was used for determination of localization of accessory conduction pathways in WPW symptomatic syndrome, ectopic sources in ventricular extrasystole and tachycardia, reconstruction of dynamics of myocardial activation in atrial flutter.

A significant drawback of the analyzed method is the use of chest model with fixed factor of specific electrical conduction. Specific electrical conduction of different organs and tissues of the chest significantly differs. Floating factor of electrical conduction of biological tissues significantly influences electrical field of the heart in the chest that is confirmed by the data of experimental examinations. The difference between electrical conduction of lungs and surrounding soft tissues (4-5 times) plays the greatest role. Potentials of cardiac electrical field of model sources calculated for homogenous and inhomogenous models of the chest differ by 15%-20%. Therefore, ignoring of electrical inhomogeneity of the chest tissues leads to large deviations in reconstruction of cardiac electrical field [1]. To solve the inverse ECG problem method of regularization by A.N.Tikhonov [2] is used. It fundamentally includes the following claims:

1. Heart condition determines its electrical activity.
2. The amount, position and choice of heart model points assume maximum resolution for examination of electrical processes taking place in the heart.
3. Heart model points of the patient are elementary heart dipoles each of which in the system of heart coordinates has location, direction and behavior in time of its value, and define electrical activity of a patient.
4. According to electrocardiographic presentation potential value generated by the heart and recorded on the torso of a patient (direct task of electrocardiography) is determined from the formula including the value of electrical potential in j-point of standard lead ($j = 1, \ldots, 12$); average specific electrical resistance of torso and proportionality factor based on the characteristics of electrical activity of the region on the surface of patient's heart model surface [3].

IOP Conf. Series: Materials Science and Engineering **91** (2015) 012026 doi:10.1088/1757-899X/91/1/012026

In the developed system for noninvasive supervision of patient's heart state determination of electrical heart activity is performed by means of construction of linear algebraic equation system in which the number of unknowns equals to the number of points of patient's heart model. For this purpose the surface of patient's heart model is "divided" into adjacent Sm noncrossing regions. The number of possible different m regions equals to the number of electrically active i heart points. The given solution of electrocardiography inverse problem allows multifold increasing of accuracy in location of myocardium damages. If in the process of standard electrocardiographic examination it is possible to determine electrical activity of only 8 regions of heart surface then in this approach the determinant is the number of electrically active points in patient's heart model. Such methods of signal processing based on averaging of ECG cycles distortion in phase space of coordinates allow increasing efficiency of assessment of reference cardiac cycle [3].

The solution of electrocardiography inverse problem was firstly introduced already in 70s of the last century (B.Taccardi, R.Barr, R.Plonsey). The first efficient algorithm of electrocardiography inverse problem solution was developed in 1981 by V.V. Shakin. The first clinical tests of noninvasive electrophysiological method based on the solution of ECG inverse problem were performed in1985-87 at Scientific Center of Cardiovascular surgery named after A.N. Bakulev (L.A. Bokeria, V.V. Shakin, G.V. Mirskiy, A.Sh. Revishvili, I.P. Polyakova). Results of clinical tests showed potential perspective of this method, however the level of computing and medical equipment development at that time did not allow its full introducing into clinical practice. For the first time all stages of noninvasive epicardial mapping method was realized by research team headed by prof. W.Rudy (USA) who in 2004 suggested method variant called by the authors as Noninvasive Electrocardiographic Imaging which provides besides superficial ECG mapping computer tomography or MRI of the chest and heart. In 2006 at the premises of tachyarrhythmia surgical treatment department at Scientific Center of Cardiovascular surgery named after A.N. Bakulev under supervision of A.Sh. Revishvili modern software and hardware appliance for noninvasive electrophysiological heart examination was developed; it was based on the solution of electrocardiography inverse problem. Calculative noninvasive activation mapping included several stages:

1. Performance of multichannel electrocardiogram recording from the chest surface in 80 or 240 unipolar leads using different systems of superficial ECG-mapping.
2. Patients with already placed superficial electrodes underwent helical computed tomography of the chest with intravenous contrast. Different helices pitch distance was used: 5-7mm for the whole chest scanning and 3 mm helices pitch distance for heart region scanning.
3. Computed tomography data determined the boundaries of the chest surfaces, epicardial and endocardial heart surfaces, and helped to build realistic tree-D voxel models of torso and heart [4].

Such projects are also very important due to development of interventional and surgical methods of heart rhythm disturbances treatment. For example noninvasive activation heart mapping allows performing electrophysiological and topical diagnostics of heart rhythm disturbances with accuracy which was earlier achieved only on the basis of direct electrocardiographic measurements on the myocardial surface under conditions of surgical intervention [2].

The main advantages of the method are:

- noninvasive character;
- simultaneous mapping of all four heart chambers;
- simultaneous mapping on endocardium and epicardium of heart;
- mapping for one cardiac cycle;
- mapping of arrhythmias characterized by aperiodical electrophysiological processes.
- The main spheres of use of noninvasive activated mapping:
- preoperative diagnostics of hard disturbances of cardiac rhythm before surgical or interventional treatment;
- postoperative supervision, diagnostics of recurrent and incisional arrhythmias after surgical or interventional treatment;

IOP Conf. Series: Materials Science and Engineering **91** (2015) 012026 doi:10.1088/1757-899X/91/1/012026

- electrophysiological and topical diagnostics of rhythm disturbances characterized by aperiodical or rarely recurrent electrophysiological processes;
- electrophysiological heart examination in patients having contraindications for heart catheterization.
- cardiac resynchronizing therapy.

Noninvasive activation mapping performed during preoperative period reduces the time of invasive electrophysiological examination, duration of surgical intervention and intraoperational fluoroscopy, and significantly reduces absorbed radiation doses for a patient and medical staff.

During postoperative period noninvasive activation mapping provides an opportunity to control the efficiency of performed catheter ablation without repeated invasive interventions, to determine recurrences of rhythm disturbances in proper time and to determine topical characteristic and electrophysiological nature of incisional.

The opportunity of mapping of all four heart chambers per one cardiac cycle allows performing detailed diagnostics of cardiac arrhythmias having unstable character or nonperiodic course of electrophysiological processes. This group involves up to 30% of all rhythm disturbances: from rare actopic extrasystole to atrial fibrillation. Simultaneous mapping on epicardium and endocardium allows determining the depth of the source of pathologic myocardium excitation relative to epicardium and endocardium. This information is important for the choice of tactics of surgical (interventional) treatment of Wolff–Parkinson–White (WPW) and ectopic arrhythmias [2].

Thus progress in this sphere goes on. Over the last years effective algorithms of electrocardiography inverse problems solution were created [5-7] and all necessary technical means have been already widely use in clinical practice for a long time (Computed tomography and electrocardiotopography with multichannel ECG recording on the body surface). The development of special hardware and software appliances for electrophysiological heart examination based on solution of electrocardiography inverse problems goes on. Many tens of computer electrocardiographic databases with hundreds and thousands of well annotated ECG in standard, orthogonal and other low-channel leads and also large libraries of intracardiac electrograms were created. Many of them are essentially computer implementations of atlases which are predecessors of databases. Some databases such as for example "The Common Standards for Electrocardiography (CSE) Database" [7] and "The Physikalisch-Technische Bundesanstalt (PTB) Diagnostic ECG Database" played significant part in development of modern computed electrocardiography. All algorithms of ECG preprocessing, filtration isoline, cardiocycle marking, automatic diagnostics, etc.; which now are involved in electrocardiograph software were developed and verified with the help of this databases [8, 9].

Databases of tree-D medical images resulted from the use of tomographic of other intrascopy methods were created and are actively developed. For example as part of "The Cardiac Atlas Project (CAP)" big database containing results of noninvasive intrascopic examinations of normal and pathological hearts aimed at clinical research and educational purposes was created and is constantly extended.

However there are no databases containing all necessaries elements for verifications and improvement of methods and algorithms applied for solution of electrocardiography inverse problem. In internet it is possible to find only certain cases containing simultaneously multichannel ECG measured on the chest surface and new medical description of torso and heart shape.

Synchronized base combining data of simultaneously performed electrophysiological and tomographic clinical investigations could become foundation for improvement of methods and algorithms used for solution of inverse electrocardiographic problem and for their detailed experimental check. Moreover such database would be also useful for solution of a variety of other problems restricting development of noninvasive electrocardiotopography. At many scientific centers intensive complex examinations of heart activity mechanisms are held among them: modeling of structural and functional relations at different levels (biomolecules, myocytes, myocardial tissues, the whole heart and the whole body) and modeling of electrophysiological, mechanical and biophysical processes (metabolism and blood circulation). The important and useful result of this work is program packages with open code for multiscale modeling of electrophysiological processes such as SCIRun, CHASTE (Cancer, Heart and Soft Tissue Environment) and CARP (Cardiac Arrhythmia Research Package).

At the clinic several different systems of electroanatomic mapping which allow performing mapping of activation process on endocardium with the help of catheters are used. In the process of simultaneous mapping (recording) of electrical potential on the body surface and potential reconstruction on the heart surface there is an opportunity to compare received results with real measurement.

The first verification method of location of initial myocardial activation area is comparison of results of inverse electrocardiographic problem solution with data received in the process of invasive catheter operations, search and elimination of ectopic excitation sources. The search of octopic sources suggests direct measuring of electrical potentials on epicardium during open heart operations or endocardium in the process of invasive catheter electoanatomic heart examination. In addition to the above before operative treatment noninvasive electrocardiotopographic examination is performed (multielectrode superficial electrocardiogram is recorded) and also tomography (CT or MRI) for detailed description of torsoanatomy and accurate position of superficial electrodes is performed. These data are necessary for solution of inverse electrocardiological problem and determination of wave excitation code. After performing the operation of ectopic source removal coordinates of identified ectopic excitation sources are compared with the results of restoration of excitation wave front pathway received while solving inverse problem [8].

The second verification method is simultaneous performance of electrocardiotopographic examination (records of superficial electrocardiograms) and stimulation (cardiac stimulation by method of periodic electric stimulation of certain points of excitable myocardium with the help of cardiostimulatar electrodes or using special active catheter electrode in the process of invasive catheter electroanatomical examination). Accurate position of stimulating electrodes in the heart is registered with the help of tomographic examination (in case of permanent cardio stimulator use) or in the process of invasive catheter electroanatomical heart examination as for example using CARTO XP [8].

3. Conclusion

Based on the above to solve a direct problem (calculation of potential on the body surface according to set parameters of electrical generator) and inverse problem (noninvasive reconstruction of heart electrical field according to ECG on the body surface) the following data are needed:

- Simultaneously recorded electrocardiosignals in multiple points on the chest surface
- Synchronized with electrocardiosignals numeric tree-D model of the chest and internal organs of a patient
- Position of points on the chest surface in which electrodes were located
- Synchronized with electrocardiosignals numeric tree-D model of the heart and its internal chambers
- Data about electrical conduction of internal organs, myocardium and endocavitary blood
- Parameters of myocardium biophysical model as excitation environment
- Position of stimulation points and electrical stimulation parameters

Therefore numeral solution of electrocardiography inverse problem with anatomical accuracy providing possibility of faultless surgical intervention is complex task including development, improvement and providing with necessary speed of algorithms operation at all stages: starting with voxel models of density distribution of tissues resulted from CT or MRI, localization of electrodes on the body surface to visualization of physiological processes in myocardium and formation of reasonable diagnostic conclusion.

A big number of noninvasive mapping systems such as ИРМ-7100 of FUKUDA DENSHI company (Japan) and CARDIAC -112.2 system 2PA company (Czech Republic) has recently appeared abroad. These systems are designed in the form of permanently installed appliances with the

IOP Conf. Series: Materials Science and Engineering **91** (2015) 012026 doi:10.1088/1757-899X/91/1/012026

patient connected to them by means of many wires. In addition there is an insistent necessity to study the patient's organism in different types of activity and also while physical exercises. Taken into account these circumstances nowadays diagnostic multichannel electrocardiographic systems with telemetry connection channel are being developed.

In general informational content and as a result diagnostic value of above mentioned methods is different and significance of each of them is relative. Nevertheless analyzed approaches to development of informational technologies of ECG processing for solution of the problem of dangerous heart diseases determination allow performing timely diagnostics and providing appropriate treatment.

4. Acknowledgement
The research was financially supported by the Federal Target Program "Research and development on priority directions of scientific-technological complex of Russia in 2014-2020 years», project No. 14.578.21.0032 "Development of an experimental model of the hardware-software complex for noninvasive recording of the heart micropotentials in a broad band without filtering and averaging in real time in order to detect the early signs of sudden cardiac death".

References
[1] Information on http://www findpatent ru/patent/241/2417051 html
[2] Revishvili A Sh Kalinin A V Kalinin V V Labartkava E Z Alexandrovna S A Spiridonova E I Sopov O V Noninvasive diagnostics and results of cardiac arrhythmia interventional treatment using new system of noninvasive body surface mapping "Amicard 01K" Annals of Arrhythmology 3 (2012) pp 39–47
[3] Mashchenko T G Trubchanova N V Analysis of cardiosignals processing methods Collection of studies "Vestnik NTU "KPI": *Automatics and professional equipment* 20 (2010) pp 115–123
[4] Fetisova E A Results of use of a new method of calculating activation mapping for topical diagnosis of WPW syndrome *Author's abstract by Candidate of Medicine Moscow* 2011 p 24
[5] Denisov A M Zaharov E V Kalinin A V Kalinin V V Using the method of regularization by Tikhonov for solution of electrocardiography inverse problem *Reporter of Moscow State University Series 15 Calculus mathematics and cybernetics* 2 (2008) pp 5–10
[6] Denisov A M Zaharov E V Kalinin A V Kalinin V V Numerical methods of solutions of some heart electrophysiology inverse problems *Differential equations* 7 (2009) pp 1014–1022
[7] Denisov A M Zaharov E V Kalinin A V Kalinin V V Numerical solution of electrocardiography inverse problem for media with piecewise-constant conductivity *Journal of Calculus Mathematics and Mathematical Physics* 7 (2010) pp 1233–1239
[8] Aidu E A -I Trunov V G Principles of database formation for verification of noninvasive electrocardiotopography systems in silico and in vivo Informational processes 12 (2012) pp 438–446
[9] Bokeria L A Revishvili A Sh Kalinin V V lyadzgina O S Fetisova E A Simonyan G Yu Noninvasive endocardial mapping of heart ventricles in terms of solution of electrocardiography inverse problem *Reporter of Arrhythmology* 57 (2009) pp 24–28

IOP Conf. Series: Materials Science and Engineering **91** (2015) 012027 doi:10.1088/1757-899X/91/1/012027

Results of research on causes of the defects of laser marking articles

V V Markov[1, a], L I Lebedeva[1], N V Kanatnikov[2]

[1]State University – Education-Science-Production Complex, Branch of Karachev, 120, Lunacharsky St., Karachev, Bryansk Region, Russian Federation, 242500

[2]State University – Education-Science-Production Complex, 29, Naugorskoe Shosse, Oryol,Russian Federation, 302020

[a]E-mail: pms35vm@yandex.ru

Abstract. The article describes technological aspects of the marking process using solid-state fibre laser. It presents the analysis of the technological modes and errors of the precise laser marking and establishes low density of the laser system as one of the causes of the poor quality of a laser seam. The article reviews the use of the following material laser marking is performed on - plastic with fine metal and conductive surface coating.

Introduction

The laser technologies are widely use in machine-building and device-building. Laser technological equipment is successful apply for the cutting of materials, grinding of instruments, laser soldering and welding, as well as for the marking of products. Precision laser marking is differing from others methods of marking the symbols a high productive, clear, miniature and wear-firmness of imprints.

A technology of laser marking possesses the several advantages [1]:

– high economic effectively and small prime cost of marking operation even by making articles the small batches;

– small prime cost operations of check by taking and periodic tests;

– high speed produce of inscriptions and pictures is ensure a better ability to handle of equipment and a abnormally high of coefficient its using as compared with other methods of marking;

– maximum spare method by any materials and products, absence any mechanical influences, which are can destroy a surface of article;

– high-contrast inscriptions are ensures a excellent they reading, in that number machine finishing (automatically reading);

– combination the reliability, convenience reading and defend of information (the inscriptions impossible remove or change, has fine defense from any attempts of falsification);

– high stability with influence of temperature, erasing, alkali's, acids, solvents and others aggressive substances, abnormally high a humidity of air and sea salt mist;

– possibility the making of inscriptions a enough great cubic capacity on the small grounds surfaces of articles and possibility the making of additional inscriptions.

IOP Conf. Series: Materials Science and Engineering **91** (2015) 012027 doi:10.1088/1757-899X/91/1/012027

An inculcation of laser marking is not demand great expenditures; the articles are marked on the compact plant with period repaying about five months.

That way, laser marking of articles is be the same perspective method a making of inscriptions (as compared with marking of paint, casting, fording, electro-erosion processing). Actual problem a widening of possibilities a laser marking of articles is be analysis she technological regimes and errors, as well as ground peculiarity she application in the conditions, which are tie with risk a destruction of article.

A aim this work is be analysis of quality a laser joint, technological regimes and errors of laser processing, as well as experimental ground a fit laser technology by making of inscriptions and pictures on the thin-film covers without they damage.

Description of system laser marking and she program ensuring

In Russian Federation make the several kind of technological equipment for the laser marking. A one is kind of they equipment – a system precision laser marking SPLM «MiniMARKER-M10/20» («Laser Centre» Ltd, Saint-Petersburg). System is serve for the making of images in the contour and raster (filling) kinds on the surfaces of metal, plastic, rubber, foil and others articles (figure 1) [2].

Figure 1. Versions of designs a system precision laser marking SPLM «MiniMARKER-M10/20»

Main technical characteristics of system «MiniMARKER-M10/20»:

1) dimensions of area of processing: 100 x 100 mm;
2) speed a transference of laser ray: 0 – 2.5 m/s;
3) type of making images: contour, raster (filling), text, graphic;
4) dimensions of making symbols: 0.25 – 100 mm;
5) breadth of line with automatic burn through: 0.05 – 3 mm;
6) materials, which are may be marking: metals and alloys, protective covers of metal surfaces, plastics, rubbers, foils;
7) type of laser: impulse, fibrous;
8) maximum power of laser: 10 W;
9) control of laser: over personal computer (Windows 2000, XP, Vista, type interface: USB);
10) conditions of exploitation: temperature from +15 to +35 °C, relative humidity to 80% without droplets of moisture.

A system precision laser marking SPLM «MiniMARKER-M10/20» is consisting of: block of optical transference a ray (BOTR) with optical fibro; working table with bracket for fixing a BOTR and transference they along vertical axis (OZ); block of electro-feeding and control; personal computer (from Pentium III/450 MHz, operative memory from 256 Mb, monitor from 15", video adapter from 800x600 points, interface USB 1.0 or 2.0, Windows 2000 Professional SP3, Windows XP SP1 или SP2, Vista, Internet Explorer 6.0 or higher).

A system precision laser marking SPLM «MiniMARKER-M10/20» is has original program ensuring – a program complex «SinMark TM». Basis possibilities of this complex [2]:

1) monitoring and testing of laser equipment;

2) programming of technological regimes;

3) possibility of inclusion a automatic regime of work a laser equipment in composition of technological line;

4) regimes of many-passages marking, execution the putting program;

5) vector and raster regimes of laser marking;

6) regime of marking the images wide and capacious line;

7) import of graphic information from format BITMAP and PCX, as well as vector information in the format HPGL/2 (PLT) and AutoCAD;

8) quick creation of program for the marking text and small graphic images, support print «True Type», availability the vector prints;

9) automatic change number and batch marking articles;

10) by marking of articles is indicate the real time and date;

11) has possibility generation touch-code all main types;

12) control the additional systems of transference.

Program complex «SinMark TM» is ensure a preparation of program laser marking, control of laser work station in the processing, launch the testing programs.

Technological itinerary and regimes the laser marking of articles

On the undertaking Open Joint-Stock Company «Karachevsky Zavod «Electrodetal» (Russian Federation, Bryansk region, Karachev) a system precision laser marking SPLM «MiniMARKER-M10/20» is use for the making of inscriptions and trade marks on the main production this undertaking – the rectangular electrical connectors. The marking are expose to corpuses of electrical connectors, which are making of the plastics and metal, as well as the plastics corpuses of electrical connectors with thin-film and conducting electricity cover.

The technological itinerary of precision laser marking is including the several actions:

1) preparation a working place. It is necessary is including an electro-feeding of personal computer and laser equipment;

2) start a program for the marking of articles, which is create of before-hand and contain the main technological regimes of precision laser marking;

3) check a correct of preparation, which is will be make;

4) definition a focus distance, which is necessary for the laser marking. By this necessary rationality dispose of article on the working table and checking with help of mould of focus distance from objective a laser radiator to surfaces of article;

5) optimization the regimes of marking. By this necessary is selection the optimal regimes on the quality of making symbols;

6) making of symbols on the surfaces of article. By this necessary is make marking symbols way pressing on the pedal of laser equipment or with help a manipulator «mouse» of computer;

7) check. By this necessary is make a visual check quality of marking a surface of article and put a ready article in the package.

Fixing the technological regimes of marking are take place by include a program, which is will be create a before-hand. A program of control is maintaining several technological regimes:

1) frequency the impulses of laser radiation: 1 – 10 kHz;

2) speed transference of ray: 50 – 300 mm/s;

3) breadth a track of laser ray (laser joint): 0.05 – 3 mm;

4) power a laser radiation: 20 – 100 % from maximum power a laser radiation of equipment;

5) period a burn through of inscription: 5 – 10 s;

6) coordination's a burn through of inscription about working table: include users of program by making of working draught a inscription.

IOP Conf. Series: Materials Science and Engineering **91** (2015) 012027 doi:10.1088/1757-899X/91/1/012027

Technological regimes of laser marking are be distinguished: a high degree of adaptation with peculiarities the surfaces of articles; a possibility of automation a process of marking; high mechanical and physics-chemistry properties of inscriptions. Thank to this dignities, a laser marking one may a recommend as preferable method the making of inscriptions on the articles, which are has complicated forms and special properties. A flexible technology of laser marking is allow by demand of consumers is add to composition of transference additional information, which is not including in the compulsory block of dates [3].

Analyses the defects of laser marking

An actual problem of precision laser marking is appearing a quality of joint. A quality of joint – it is absence the defluxions of dimensions a joint with direction, which is perpendicularity a motion of laser ray. For the appraisal a quality of laser marking joint be making experimental researches the symbols of marking with help a scientific microscope AKSTOCKOP2 MAT (equipment a «Centre of collective use measuring and testing equipments» (CCU), which is appear a structurally subdivision of Federal State Budget Educational Institution High Professional Education «State University – Education-Science-Industrial Complex»). Results the observation of marking inscriptions to scientific microscope AKSTOCKOP2 MAT are show on the figure 2 and 3.

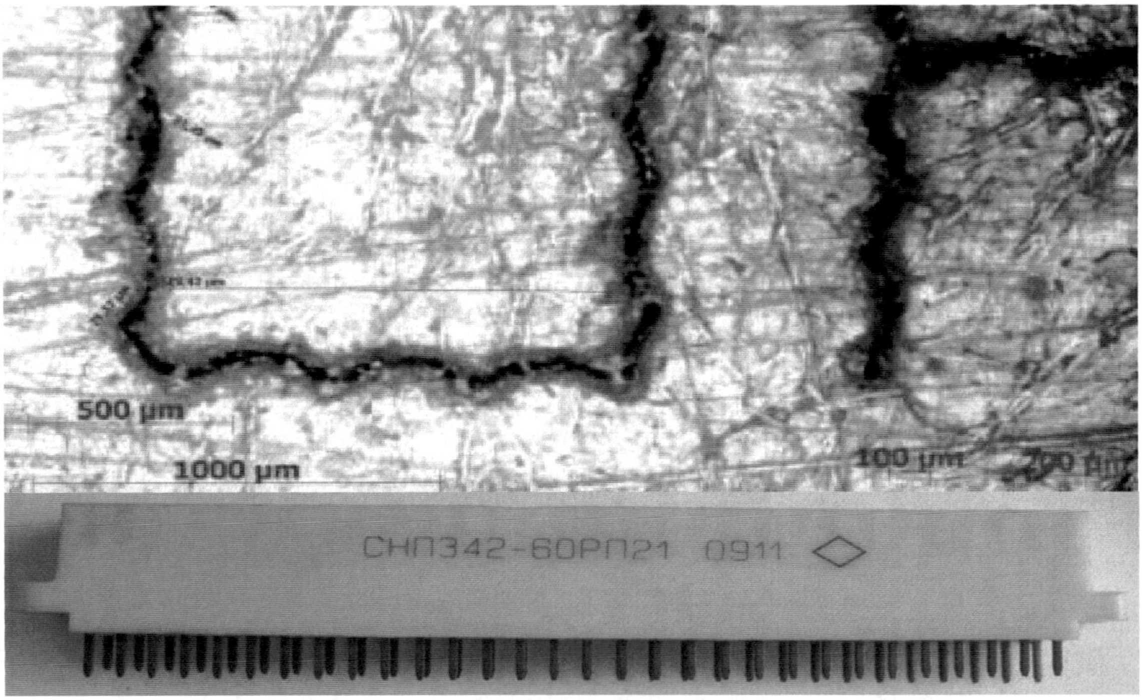

Figure 2. Appearance an electric plug of electrical connector SNP342
and increasing image of marking inscription

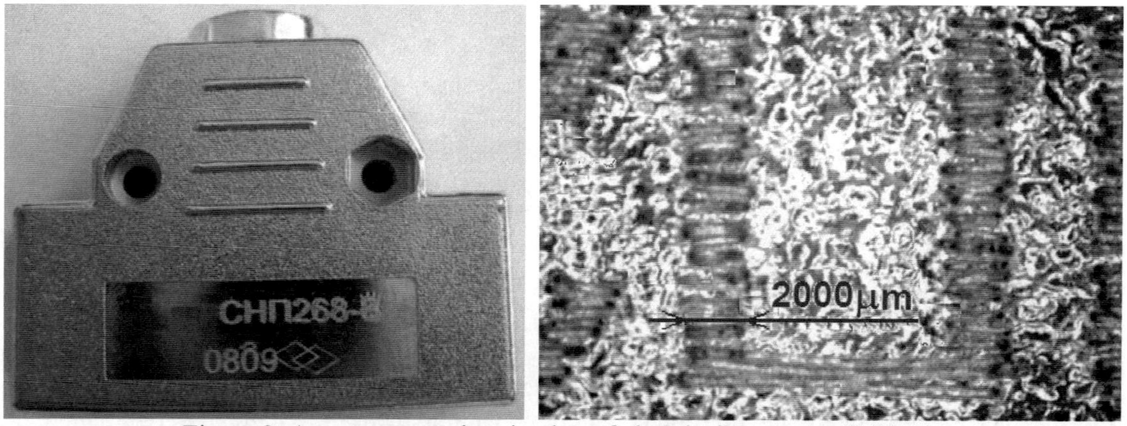

Figure 3. Appearance a electric plug of electrical connector SNP268
and increasing image of marking inscription

By photographs is one can see that laser joint is go of zigzags, and besides an unevenly of laser joint is show as to longitudinal (figure 2), so and to diametrical (figure 3) directions. By results measuring dimensions an increasing image of marking inscription (with use scientific microscope AKSTOCKOP2 MAT) is definite a step of vibration laser joint to longitudinal and diametrical directions (figure 4). Are receiving following significances (is receiving on the results of measuring a step of vibrations by electronically scales of scientific microscope AKSTOCKOP2 MAT):

1) to longitudinal direction a step of vibration laser joint (L_1) is put together:

$$L_1 = 300 \ \mu\text{m}\,; \quad (1)$$

2) to diametrical direction a step of vibration laser joint (L_2) is put together:

$$L_2 = 120 \ \mu\text{m}\,. \quad (2)$$

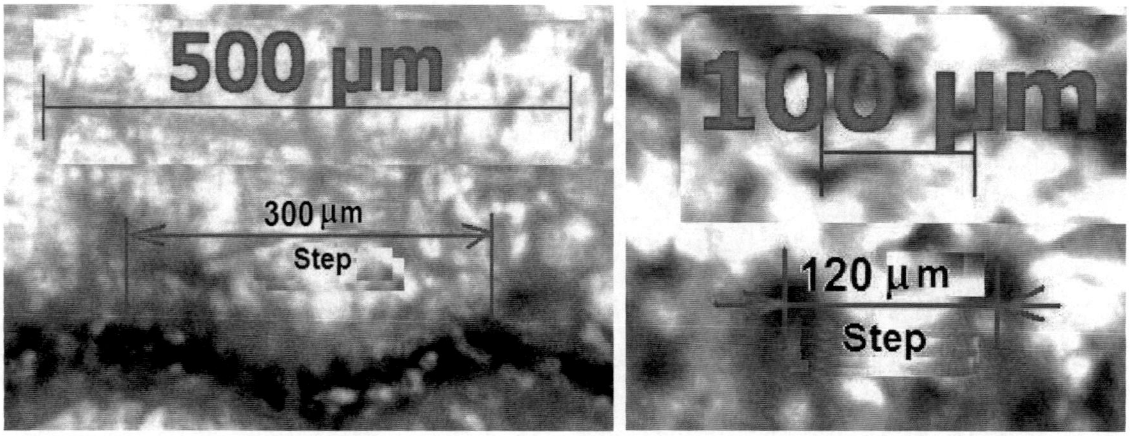

a) b)

Figure 4. Vibration a laser joint to longitudinal (a) and diametrical (b) directions

Results of measuring vibration of laser joint are show, what period of vibration to longitudinal and diametrical directions is different. The possible causes uneven of laser joint are considered.

It is known speed (υ) a burn through of inscription:

$$\upsilon = 300 \text{ mm/s}\,. \quad (3)$$

Then one may is period, behind which a laser ray is accomplish the vibrations to longitudinal and diametrical directions, that is period of vibrations of inscription to longitudinal (T_1) and diametrical (T_2) directions:

$$T_1 = \frac{L_1}{\upsilon} = \frac{0,3}{300} = 1 \cdot 10^{-3} \ s; \quad (4)$$

$$T_2 = \frac{L_2}{\upsilon} = \frac{0,12}{300} = 4 \cdot 10^{-4} \ s. \quad (5)$$

It is known, what frequency (f) and period (T) of vibration are connect a formula:

$$f = \frac{1}{T}. \quad (6)$$

If knows the periods vibration of inscription to longitudinal (T_1) and diametrical (T_2) directions, one may calculate a frequency of vibration and compare she with frequency of vibration a laser radiation:

$$f_1 = \frac{1}{T_1} = \frac{1}{1 \cdot 10^{-3}} = 1 \cdot 10^3 \ \text{Hz} \quad \text{and} \quad f_2 = \frac{1}{T_2} = \frac{1}{4 \cdot 10^{-4}} = 2,5 \cdot 10^3 \ \text{Hz}. \quad (7)$$

A frequency of vibration a laser radiation is equally: $f = 10$ kHz. Well then, she one may creation of vibration a laser joint with period $T = 0.1$ ms and step of vibration $L = 30$ µm. It is to four times less a step longitudinal vibration of inscription.

Consequently, a frequency of laser radiation is not being cause the unevenness of inscription. One may, that low exactness of laser joint is evoke a vibration of non-standard device for the marking. This way, main condition a security of quality laser marking is appear a hardness of technological system «equipment-device-instrument-detail» (system EDID) of laser equipment [4].

Peculiarities of laser marking articles with thin-film covering

A one of actual problem a laser marking is appear a fixing of regimes making symbols on the surfaces of miniature articles, which has a thin-film covering. An example one may serve the corpuses of electrical connectors SNP268 (Russian Federation), which are appear analog the connectors international series «D-Sub» (figure 3).

Designing peculiarity this corpuses are appear an enough complication form, which is receive of casting from thermoplastic polymer materials. Afterwards the corpuses are cover two-coating metal covering. By this a intact of thin metal cover, which is consist a level of copper (30 – 40 µm) and a level of nickel (40 – 60 µm), is not be break. Results the observation of cut marking inscriptions to scientific microscope AKSTOCKOP2 MAT are show on the figure 5.

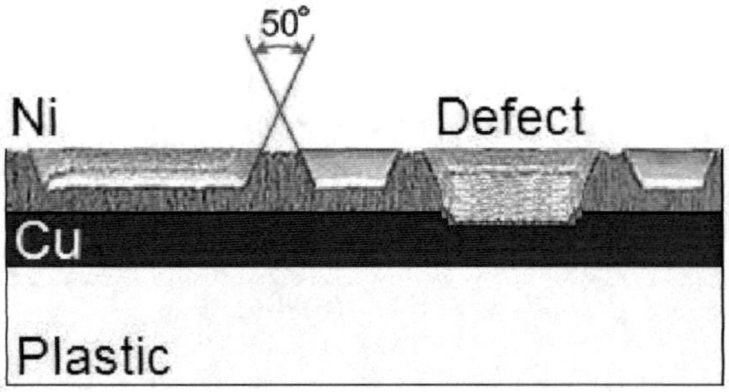

a)

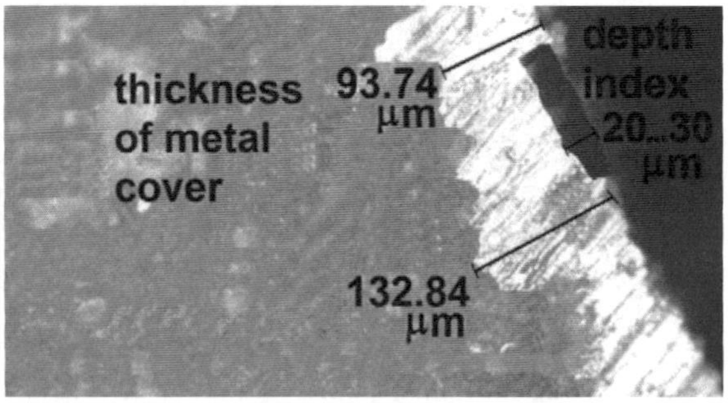

b)

Figure 5. Rule making a marking on the thin copper-nickel cover (a) and they actual condition (b)

Are allows the insignificant (pointing) defects in the difficult-accessible sections surface of article. Such defects are arising by making a cover of chemical method, when electrolyte is not penetrating to the thin openings and ditches. A maximal risk – this defects one may not right away, but afterwards some time, by keeping and transporting, and to the process exploitation of article. Therefore selection optimal technological regimes of laser marking surfaces articles with thin-film metal covering appear is main condition security a quality of article. By selection this regimes it is necessary a careful determine probable part of defective articles in the batch. A laser marking is appear a final technological operation, therefore defects by marking one may strongly increase a volume of defective articles in the batch [5].

Observation a cut of surface marking corpus electrical connector SNP268 to scientific microscope AKSTOCKOP2 MAT is show, what real thickness metal cover is equally 90 – 150 μm. Depth a symbol, which is making a laser marking, is equally 20 – 30 μm (figure 5, b).

This way, by observe the fixing regimes of processing, a technology of laser marking is ensure a double stock by thickness of cover. Therefore it is technology one may successful use for the marking of articles with thin cover of surfaces.

Conclusions

Laser technological equipments for the precision marking of articles are one may ensure a high quality inscriptions. Allowing is unevenness with step to 30 μm. A quality laser joint on the inscriptions one may strongly aggravate a use insufficiency of devices by fixing articles and laser radiator. This way, main condition ensures a quality of inscriptions is appearing a hardness of technological system «equipment-device-instrument-detail» (system EDID) of laser equipment.

By observe the fixing regimes of processing, a technology of laser marking is ensuring a double stock by thickness of cover. Therefore it is technology one may successful use for the marking of articles with thin cover of surfaces.

Executed researches are carrying out on order of undertaking Open Joint-Stock Company «Karachevsky Zavod «Electrodetal» – is main undertaking of Russia Federation in field of manufacturing the rectangular electrical connectors for the radio electronics and computer equipments. The results of these researches are passed to manufacture and use for the confirmation of faculty a undertaking is carry out the demands of customers with quality of connectors.

Acknowledgments

Authors considers it his pleasant duty is express of gratitude the technical specialists of «Centre of collective use measuring and testing equipments» (CCU), which is appear a structurally subdivision of Federal State Budget Educational Institution High Professional Education «State University – Education-Science-Production Complex»: Alexander A. Simakov, Anatoly V. Tarasov, Michail I. Daniluk. We are gratitude for the useful participation, a great technical help by carry out the experiments, a moral encouragement and the valuable advices.

References

[1] Afonkin M, Larionova E, Gornyi S 2010 *Analysis of Laser Engraving Complexes Technological Capabilities* (Photonics Vol. 5) pp. 20-24.

[2] Liu J Yang L 2010. *Fiber lasers: move over, solid-state lasers – Femtosecond fiber lasers are here* (LaserFocusWorld Vol. 4) pp. 101-106.

[3] Shaymaradov R 2011 *CO2 Laser: Flexible, Reliable and Approved Device* (Photonics Vol. 5) pp. 73-80.

[4] Petronic C Milosavljevic A Milovanovic D 2011. *Influence of picosecond laser pulses on the microstructure of austenitic materials* (Journal of Russian Laser Research Vol. 32. N. 6) pp. 564-570.

[5] Ristic S Polic-Radovanovic S 2010. *Ruby laser beam interaction with ceramic and copper artifacts* (Journal of Russian Laser Research Vol. 31. N. 4) pp. 380-386.

[6] Katunin A D Markov V V Vetchinnikov D A 2012 *Technological regimes and errors of precision laser marker the articles* (Fundamental and applied problems of technical and technology. – Vol. 3-3 N 293) pp. 74-81.

IOP Conf. Series: Materials Science and Engineering **91** (2015) 012028 doi:10.1088/1757-899X/91/1/012028

Thermocapillary model of formation of surface nanostructure in metals at electron beam treatment

V D Sarychev, S A Nevskii, S V Konovalov, I A Komissarova, E V Chermushkina

Siberian State Industrial University, Novokuznetsk, 42 Kirov Street, 654007, Russia

E-mail: nevskiy.sergei@yandex.ru

Abstract. The paper presents the thermocapillary model for the formation of nanostructures in surface layers of materials. It is based on Navier-Stokes hydrodynamic equations, the thermal conductivity equation, the state equation and the boundary conditions. A search for the solutions in the form of a progressive wave has been carried out. The dispersion equation has been obtained and analyzed. The dependence of the instability increment on wavelength has been built. The values of the critical wavelength, at which thermocapillary instability for iron and titanium comes, have been obtained. Value comparison of the calculated wavelengths with cell sizes of crystallization has showed a satisfactory agreement.

1. Introduction

Earlier in the papers [1 – 3], it has been found out that the fatigue life of steels of different structural classes irradiated by low-energy high-current electron beams from 10 to 40 J/cm^2 increases in 2 – 5 times compared with the non-irradiated samples. One of the reasons for this increase, for example for pearlitic steel, is considered to be, firstly, the melting of the surface layer, leading to the formation of columnar structures. The transverse dimensions of the columns change insignificantly with the increase of the beam energy [3] ranging from 0.3 to 0.6 μm, and the longitudinal dimensions that characterize the thickness of the melted layer, undergo a significant increase. Secondly, there is a hardening of the surface layer due to the formation of a martensitic structure; thirdly, the formation in the crystallization cells of nanoscale martensite and, fourthly, the graphite precipitation at the cell borders.

The electron beam treatment of titanium alloys [4 – 6] also leads to the formation of a gradient nanostructures and strengthening of their surface layers. This leads to the increased wear resistance. Thermal effect modeling [6] of the electron beam on titanium, carburized by electric explosion has showed that carbon particles of nanometer range are dissolved in about 10 μs. This suggests that in the process of the combined treatment a graphitized carbon fiber should be used, in which the diameter of microfibrils is about of tens of nanometers.

The impact of the electron beams on the silumin has been studied in [7 – 9]. In [7] it has been found out that low-energy high-current electron beams contribute to the dissolution of large silicon particles in the layer with the thickness of from 40 to 55 μm with the formation of cellular-dendritic structure. Silumin hardening is due to the formation of substitution solid solution, as well as the dispersion of the structural components. In [8, 9] it has been shown that electron beams contribute to the formation in the surface layers of a nanocrystalline structure, which also contributes to its hardening.

The formation models of gradient structures under the influence of concentrated energy fluxes have been considered in the works [10 – 15]. The main formation mechanism of nanoscale structures with this effect is the hydrodynamic instability of a shear flow of the molten layer relative to solid supports. In the papers [10, 11] it has been found out that the nanostructure formation in the surface layer of the material, when exposed to a heterogeneous plasma flows of electric explosion of conductors, is due to Kelvin-Helmholtz instability, which occurs due to the reversal of the plasma front and the emergence of the parallel flow field of plasma and molten metal. The dependence analysis of the instability increment on a wavelength [11] has showed that it has two maximums in nano- and microrange, and microwave mode is formed by the interaction of the layers of a perfect liquid with a surface tension at the border, but nanowave mode is formed due to viscosity. Another formation mechanism of periodic surface structures is that in the irradiated material by reason of the inhomogeneous heating along the depth thermocapillary and thermo-gravitational convections occur [12 – 15]. If the layer is thin enough, the thermocapillary convection mode is dominated; it is caused by the heterogeneity of the surface tension [12]. Flows of this type are characterized not only by the movement of a liquid along the free surface, but also the deformation of the surface [13]. In these works the approximation of incompressible liquid is used. The development of these models is presented in the papers [14, 15] for beams with a power density of $\geq 10^6$ W/cm^2. At such values of power density, as such works show, the approximation of incompressible liquid is not enough; that is why it is necessary to use a wide-range state level.

There are also works devoted to the modeling of thermocapillary liquid flows on the surface of the material [16 – 19]. The paper [16] gives the numerical investigation of the flow in a double-layer medium with the neglecting of gravity force influence. It has been established that the introduction of viscous encapsulant leads to the intensity reduction of thermocapillary flow in an encapsulated layer. Interface deformation is small if it passes between the liquid and solid. The increase of viscous encapsulant leads to the formation of pressure gradient in this layer and interface deformation, which differs qualitatively from the observed liquid on a free surface at the lack of this layer. The structure of the encapsulated layer and the interface deformation depends on the viscosity and the thickness. In [17] the influence of the moving local heat source on the flow structure in a thin liquid layer on a horizontal padding has been theoretically analyzed. A two-dimensional problem in the approximation of the layer has been considered. For the case of the small value of Reynold's number in an attendant coordinate system a steady-state equation has been received; it describes the deformation of a liquid film. The received equation is applicable, particular, for the conditions of the horizontal padding. In the paper [18] the mathematical model of crystallization process has been offered and substantiated; it reflects experimentally established properties of structure formation in the instability zone. Its essence is reduced to the design of a mathematical object including mathematical Kahn-Hillard models [20], heat-and-mass transfer as Stephen task generalization [21]. The results of numerical experiments have shown that in a isotropic surface tension a banded heterogeneity is formed. It on the strength of plane front instability and the development of the recrystallization processes undergoes the deformation. The similar situation is observed in the eutectic crystallization process, when one of the phases disintegrates on the separate fine cells [22]. Further development of the banded structure is the result of the origin of surface waves, which are similar to Marangoni surface. The works [19, 23, 24] are devoted to the research of the long wave instability in the task on thermocapillary convection in a horizontal layer with a free deforming border and a solid bottom. In these works the dispersed equations have been received, from which the equations to define critical values of the parameters for all three main types of stability loss have been obtained. With the fixed values of frequency and amplitude of the oscillations neutral curves of monotonic and oscillatory instability in the form of the dependence of Marangoni number on wave number have been built. The areas of parametric resonances, corresponding to synchronous and subharmonic modes have been defined. A frequency rate, at which a yield to the high-frequency asymptotics occurs, has been found. It has been shown that the longitudinal vibrations have no effect on convective instability and the transverse ones lead to stabilization of the interface.

Thus, it should be concluded that the analysis of the dispersion relations of thermocapillary waves in the nanoscale range has not been practically conducted. Therefore, the aim of this work is to obtain and analyze the dispersion relations in the nanoscale range of wavelengths.

2. Another section of your paper

To solve the given problem we will consider a viscous heat-conducting incompressible fluid, which has the layer thickness h on the free surface $z = \eta(x, y, t)$, and absorbs heat. At the values of thermal flow ~ 10^5 W/cm^2, used in the experiments [1 – 9], the approximation of an incompressible fluid should be considered justified [25]. After the electron beam impact in the liquid layer the temperature profile $T = T_0(z)$ is set. This temperature is called undisturbed. If the wave vector of the perturbation is directed in the plane *XOY*, then they depend on coordinates x, y and the time according to the law $\exp(\omega t - i(mx + ly))$. We shall only consider the short-wave perturbations. The behavior of such waves is determined by the temperature variations in the surface layers, where $|z| \le \kappa^{-1}$, κ^{-1} – the depth of wave attenuations. In this case, the gradient of the unperturbed temperature along the z-axis (along the layer depth) will be considered constant.

Write the linearized Navier-Stokes equations:

$$\frac{\partial u}{\partial t} = -\frac{\partial p}{\rho \partial x} + \nu\left(\frac{\partial^2 u}{\partial x^2} + \frac{\partial^2 u}{\partial y^2} + \frac{\partial^2 u}{\partial z^2}\right), \frac{\partial \mathrm{v}}{\partial t} = -\frac{\partial p}{\rho \partial y} + \nu\left(\frac{\partial^2 \mathrm{v}}{\partial x^2} + \frac{\partial^2 \mathrm{v}}{\partial y^2} + \frac{\partial^2 \mathrm{v}}{\partial z^2}\right),$$
$$\frac{\partial w}{\partial t} = -\frac{\partial p}{\rho \partial z} + \nu\left(\frac{\partial^2 w}{\partial x^2} + \frac{\partial^2 w}{\partial y^2} + \frac{\partial^2 w}{\partial z^2}\right), \frac{\partial u}{\partial x} + \frac{\partial \mathrm{v}}{\partial y} + \frac{\partial w}{\partial x} = 0, \quad (1)$$
$$\frac{\partial T}{\partial t} + w\frac{d T_0}{d z} = \chi\left(\frac{\partial^2 T}{\partial x^2} + \frac{\partial^2 T}{\partial y^2} + \frac{\partial^2 T}{\partial z^2}\right).$$

Boundary conditions on the melt surface $z = 0$ have the form of:

$$-p + 2\nu\rho\frac{\partial w}{\partial y} = \sigma\left(\frac{\partial^2 \eta}{\partial x^2} + \frac{\partial^2 \eta}{\partial y^2}\right), \rho\nu\left(\frac{\partial u}{\partial y} + \frac{\partial \mathrm{v}}{\partial x}\right) = \frac{\partial \sigma}{\partial x}, \rho\nu\left(\frac{\partial \mathrm{v}}{\partial z} + \frac{\partial w}{\partial y}\right) = \frac{\partial \sigma}{\partial y}$$
$$\frac{\partial T}{\partial z} = 0, \frac{\partial \eta}{\partial t} = w, \quad (2)$$

Adherence conditions and the absence of temperature perturbations on the boundary «melt – solid» $z = -h$ will have the following form:

$$u = \mathrm{v} = w = 0, T=0. \quad (3)$$

Here u, y, w are the components of the perturbed flow velocity along the surface (along the axes x and y) and perpendicular to it; $\rho, \nu, \chi, \sigma(T)$ – density, kinematic viscosity, thermal diffusivity and coefficient of surface tension of the melt; p – pressure in the fluid layer; T – temperature perturbation $T_0(z)$. The value perturbations, which are included in (1) and (2) have the following form:

$$u(x,y,z,t) = U(z)\exp(\omega t - i(mx + ly)), v(x,y,z,t) = V(z)\exp(\omega t - i(mx + ly)),$$
$$w(x,y,z,t) = W(z)\exp(\omega t - i(mx + ly)), \quad (4)$$
$$T(x,y,z,t) = T(z)\exp(\omega t - i(mx + ly)), \eta(x,y,t) = \eta_0 \exp(\omega t - i(mx + ly))$$

where ω – perturbation frequency; $k = (m,l)$ – wave vector, i – imaginary unit. Substitution (4) into (1) – (3) subject to $\frac{\partial\sigma}{\partial x} = \frac{d\sigma}{dT}\left(\frac{dT_0}{dz}\frac{\partial\eta}{\partial x} + \frac{\partial T}{\partial x}\right)$, $\frac{\partial\sigma}{\partial y} = \frac{d\sigma}{dT}\left(\frac{dT_0}{dz}\frac{\partial\eta}{\partial y} + \frac{\partial T}{\partial y}\right)$ allows obtaining the system of the ordinary differential equations:

$$P(z) = i\frac{\rho\nu}{k^2}\left(W'''(z) - k_1^2 W'(z)\right)$$
$$W''(z) - k_1^2 W'(z) - \frac{1}{\rho\nu}P'(z) = 0 \tag{5}$$
$$T''(z) - k_2^2 T(z) - G_0\frac{W(z)}{\chi} = 0$$

where $k_1^2 = \frac{\omega}{\nu} + k^2$, $k_2^2 = \frac{\omega}{\chi} + k^2$, $G_0 = \frac{dT_0}{dz}$. The equation system (5) is reduced to one equation, containing $W(z)$

$$W^{IV}(z) - (k^2 + k_1^2)W''(z) + k^2 k_1^2 W(z) = 0 \tag{6}$$

We use (2) – (4) to get boundary conditions

$$\begin{aligned} &-P(0) + 2\rho\nu W'(0) = -\eta_0 k^2 \sigma_0 \\ &\rho\nu\left(U'(0) - imW(0)\right) = -im\sigma_T \Sigma_0 \\ &\rho\nu\left(V'(0) - ilW(0)\right) = -il\sigma_T \Sigma_0 \\ &T'(0) = 0, W(0) = \eta_0\omega, W'(-h) = W(-h) = T(-h) = 0 \end{aligned} \tag{7}$$

where $\sigma_T = \frac{d\sigma}{dT} = const$, $\Sigma_0 = G_0\eta_0 + T(0)$. After the transformation (7) will have the form of:

$$\begin{aligned} &W'''(0) - (k_1^2 + 2k^2)W'(0) - k^3\omega_c^2 W(0)/\omega\omega_\nu = 0 \\ &W''(0) - k^2 W(0) - k^2\omega_T^2 (W(0) + (\omega/G_0)T(0))/\omega\omega_\nu = 0 \\ &T'(0) = 0, W'(-h) = W(-h) = T(-h) = 0 \end{aligned} \tag{8}$$

where $\omega_c^2 = k^3\sigma_0/\rho$, $\omega_T^2 = k^2\sigma_T G_0/\rho$, $\omega = \nu k^2$.

3. Results and discussion

The solution of (6) which satisfies the conditions at the boundary of «melt – solid», will be written in the form:

$$W(\text{z}) = W_1(\text{z}) + W_2(\text{z}). \tag{9}$$

where

$$\begin{aligned} &W_1(\text{z}) = A_1 k_1 \text{sh}\,(k\,(z+h)) + A_2 \text{ch}\,(k\,(z+h)), \\ &W_2(\text{z}) = -A_1 k\,\text{sh}\,(k_1\,(z+h)) - A_2 \text{ch}(k_1\,(z+h)). \end{aligned} \tag{10}$$

Then the boundary value problem for the temperature equation has the form:

$$T''(z) - k_2^2 T(z) - G_0\frac{W(z)}{\chi} = 0,\ T'(0) = 0,\ T(-h) = 0. \tag{11}$$

Its solution is

$$\mathrm{T(z)} = \frac{G_0}{\chi}\left(\frac{W_1(z)}{k^2-k_2^2}+\frac{W_2(z)}{k_1^2-k_2^2}-S_1\frac{\mathrm{sh}(k_2(z+h))}{\mathrm{ch}(k_2h)}-S_2\frac{\mathrm{ch}(k_2z)}{\mathrm{ch}(k_2h)}\right), \quad (12)$$

where $S_1 = \frac{1}{k_2}\left(\frac{W_1'(0)}{k^2-k_2^2}+\frac{W_2'(0)}{k_1^2-k_2^2}\right)$, $S_2 = W_1(-h)\left(\frac{1}{k^2-k_2^2}-\frac{1}{k_1^2-k_2^2}\right)$. Subject to the relations of $k^2-k_2^2 = -\frac{\omega}{\chi}$; $k_1^2-k_2^2 = \frac{\omega(1-\varepsilon)}{\chi\varepsilon}$, $\varepsilon = \frac{\nu}{\chi}$ – Prandtl number. Convert S_1 and S_2 to the form

$$S_1 = \frac{\chi}{\omega k_2}\left(-W_1'(0)+\frac{\varepsilon}{(1-\varepsilon)}W_2'(0)\right),\ S_2 = -\frac{\chi}{\omega(1-\varepsilon)}A_2 \quad (13)$$

Then (11) will have the form:

$$\mathrm{T(z)}=\frac{G_0}{\omega}\left(-W_1(z)+\frac{\varepsilon}{(1-\varepsilon)}W_2(z)+\left(W_1'(0)-\frac{\varepsilon}{(1-\varepsilon)}W_2'(0)\right)\frac{\mathrm{sh}(k_2(z+h))}{k_2\mathrm{ch}(k_2h)}-\frac{\mathrm{ch}(k_2z)}{(1-\varepsilon)\mathrm{ch}(k_2h)}A_2\right) \quad (14)$$

at z=0

$$\mathrm{T}(0) = \frac{G_0}{\omega}\left(-W_1(0)+\frac{\varepsilon}{(1-\varepsilon)}W_2(0)+\left(W_1'(0)-\frac{\varepsilon}{(1-\varepsilon)}W_2'(0)\right)\frac{\mathrm{th}(k_2h)}{k_2}-\frac{A_2}{(1-\varepsilon)\mathrm{ch}(k_2h)}\right) \quad (15)$$

Subject to (14) convert the system (8) to the form

$$\omega\omega_\nu\left((k_1^2+k^2)W_1'(0)+2k^2W_2'(0)\right)+k^3\omega_c^2\left(W_1(0)+W_2(0)\right)=0$$

$$2k^2\omega\omega_\nu W_1(0)+\left(\omega\omega_\nu(k^2+k_1^2)-\omega_T^2k^2/(1-\varepsilon)\right)W_2(0)+\frac{\omega_T^2k^2A_2}{(1-\varepsilon)\mathrm{ch}(k_2h)}-$$

$$-\omega_T^2k^2\left(W_1'(0)-\frac{\varepsilon}{(1-\varepsilon)}W_2'(0)\right)\frac{\mathrm{th}(k_2h)}{k_2}=0 \quad (16)$$

$$2k^2\omega\omega_\nu W_1(0)+\left(\omega\omega_\nu(k^2+k_1^2)-\omega_T^2k^2/(1-\varepsilon)\right)W_2(0)+\frac{\omega_T^2k^2A_2}{(1-\varepsilon)\mathrm{ch}(k_2h)}-$$

$$-\omega_T^2k^2\left(W_1'(0)-\frac{\varepsilon}{(1-\varepsilon)}W_2'(0)\right)\frac{\mathrm{th}(k_2h)}{k_2}=0$$

Evaluate, coming into (15) values $A_2, W_2(0), W_2'(0)$ through $W_1(0), W_1'(0)$, as a result one gets

$$2k^2\omega\omega_\nu W_1(0)+\left(\omega\omega_\nu(k^2+k_1^2)-\omega_T^2k^2/(1-\varepsilon)\right)\left(b_{11}\mathrm{W}_1(0)+b_{12}\mathrm{W}_1'(0)\right)+$$

$$+\frac{\omega_T^2k^2\mathrm{ch}(kh)}{(1-\varepsilon)\mathrm{ch}(k_2h)}\left(W_1(0)-W_1'(0)\,\mathrm{th}(kh)/k\right)-$$

$$-\omega_T^2k^2\left(W_1'(0)-\frac{\varepsilon}{(1-\varepsilon)}\left(b_{21}\mathrm{W}_1(0)+b_{22}\mathrm{W}_1'(0)\right)\right)\frac{\mathrm{th}(k_2h)}{k_2}=0,$$

$$\left(\omega\omega_v\left(2k^2+(k^2+k_1^2)b_{11}\right)+\frac{\omega_T^2k^2}{(1-\varepsilon)}\left(\frac{\mathrm{ch}(kh)}{\mathrm{ch}(k_2h)}+\frac{b_{21}\varepsilon\,\mathrm{th}(k_2h)}{k_2}-b_{11}\right)\right)W_1(0)+$$
$$\left(\omega\omega_v(k^2+k_1^2)b_{12}\right)-\frac{\omega_T^2k^2}{(1-\varepsilon)}\left(b_{12}+\frac{\mathrm{sh}(kh)}{\mathrm{ch}(k_2h)k}+((1-\varepsilon)-\varepsilon b_{22})\frac{\mathrm{th}(k_2h)}{k_2}\right)W_1'(0)=0 \quad (17)$$

where

$$\begin{aligned} b_{11}&=\left(k\sinh(k_1h)\sinh(kh)-k_1\cosh(k_1h)\cosh(kh)\right)/k_1,\\ b_{12}&=\left(k_1\cosh(k_1h)\sinh(kh)-k\sinh(k_1h)\cosh(kh)\right)/(kk_1),\\ b_{21}&=\left(k\cosh(k_1h)\sinh(kh)-k_1\sinh(k_1h)\cosh(kh)\right),\\ b_{22}&=\left(k_1\sinh(k_1h)\sinh(kh)-k\cosh(k_1h)\cosh(kh)\right)/k. \end{aligned} \quad (18)$$

Substitution (18) into (17) leads to the system

$$A_{11}\mathrm{W}_1(0)+A_{12}\mathrm{W}_1'(0)=0;\ \ (A_{21}+B_1)\mathrm{W}_1(0)+(A_{22}+B_1)\mathrm{W}_1'(0)=0. \quad (19)$$

Where

$$A_{11}=2k^2\omega\omega_v b_{21}+k^3\omega_c^2(1+b_{11}),\ A_{12}=\omega\omega_v(2k^2b_{22}+k_1^2+k^2)+k^3\omega_c^2b_{12},$$
$$A_{21}=\omega\omega_v\left(2k^2+b_{11}(k^2+k_1^2)\right),\quad A_{22}=\omega\omega_v(k^2+k_1^2)b_{12},$$
$$B_1=\frac{\omega_T^2k^2}{(1-\varepsilon)}\left(\frac{\mathrm{ch}(kh)}{\mathrm{ch}(k_2h)}+\frac{b_{21}\varepsilon\,\mathrm{th}(k_2h)}{k_2}-b_{11}\right), \quad (20)$$
$$B_2=-\frac{\omega_T^2k}{(1-\varepsilon)}\left(kb_{12}+\frac{\mathrm{sh}(kh)}{\mathrm{ch}(k_2h)}+((1-\varepsilon)-\varepsilon b_{22})\frac{k\mathrm{th}(k_2h)}{k_2}\right).$$

Equating the determinant of the system (17) to zero we get the dispersion equation for the finite layer

$$\Delta=\Delta_1+\Delta_2,\Delta_1=A_{11}A_{22}-A_{12}A_{21},\ \Delta_2=A_{11}B_2-A_{12}B_1. \quad (21)$$

Consider the case of short waves $kh >> 1$, then the approximation of a semi-infinite layer is fair. As a result we will get the dispersion equation for such layer:

$$\begin{aligned}&\omega^2\left((\omega+2\nu k^2)^2-4k_1k^2\nu^2+\omega_c^2\right)-\\ &-\omega_T^2\left((1+\delta(1-k_1/k_2))(\omega^2+2k^2\nu\omega+\omega_c^2)-k/k_2(2kk_1\nu\omega+\omega_c^2)\right)=0.\end{aligned} \quad (22)$$

where $\delta=\dfrac{\varepsilon}{1-\varepsilon}$. It can be reduced to an algebraic equation of the 16-th degree. This equation depends on three dimensionless parameters: the Prandtl number and two squares of frequency relations a^2, b^2 ($a=\omega_c/\omega_v, b=\omega_T/\omega_v$). For numerical calculations we consider that $k_0 \approx 10^7 m^{-1}$, $h \approx 10^{-5} m$, $G_0 \approx (T_b - T_m)/h \approx 10^8 K/m$, T_b, T_m – the melting and boiling temperatures. Using these values, one will calculate the key parameters for iron and titanium by the formulas

$$a_0^2=\frac{\sigma_0}{\nu^2\rho k_0},\ b_0^2=\frac{\sigma_T G_0}{\nu^2k_0^2} \quad (23)$$

These calculations are shown in the table, from which it is visible what can be used as a first approximation $\varepsilon \approx 0$ (the Prandtl number is equal to zero). Then the algebraic equation splits into two equations

$$(z^3 + z^2 + 3z - 1)(z^2 - 1)^2 (z - 1) - b^2 (z^2 - 1)(z + 1)^2 + a^2 (z^2 - 1)^2 - 2a^2 b^2 = 0$$
$$(z^3 + z^2 + 3z - 1)(z^2 - 1) - b^2 (z - 1) + a^2 (z + 1) = 0 \quad (24)$$

Here it is denoted $z = \frac{k_1}{k}, a^2 = a_0^2 \frac{k_0}{k}, b^2 = b_0^2 \frac{k_0^2}{k^2}$.

We select only such roots, in which the real part is greater than zero (Re(z)>0). On the detected values of the roots of $z = x + iy$ we find the expression for the complex frequency

$$\omega = \omega_v (z^2 - 1) = \omega_v (x^2 - y^2 - 1) + 2i\omega_v xy \quad (25)$$

Decrement is calculated by the formular

$$\alpha = \mathrm{Re}(\omega_v (z^2 - 1)) = \omega_v (x^2 - y^2 - 1) \quad (26)$$

The mentioned procedure allows determining the dependence of the decrement on the wave number k. Figures 1 a and b show the dependences $\alpha(k/k_0)$ for iron and titanium.

Table 1. Physical parameters

Parameter	Iron	Titanium
Density, $\rho, 10^3$ kg/m^3	7.87-7.02	4.5-4.1
Surface tension coefficient , σ, N/m	1.872	1.650
$d\sigma/dT$, 10^{-3}N/K·m	-0.49	-0.26
Coefficient of viscosity ν, 10^{-7} m^2/s at 100-2000°C	4.0	13
coefficient of thermal conductivity , χ, 10^{-7}m^2/s	150	40
Melting temperature, °C	1536	1682
Boiling temperature, °C,	2862	3260
Prandtl number Pr= ν/χ	0.02	0.3
$\omega_v = \nu k_0^2$, μs^{-1}	40	130
Nondimensional capillary number, a_0	12.2	4.7
Nondimensional thermocapillary number, b_0	55.3	12.4

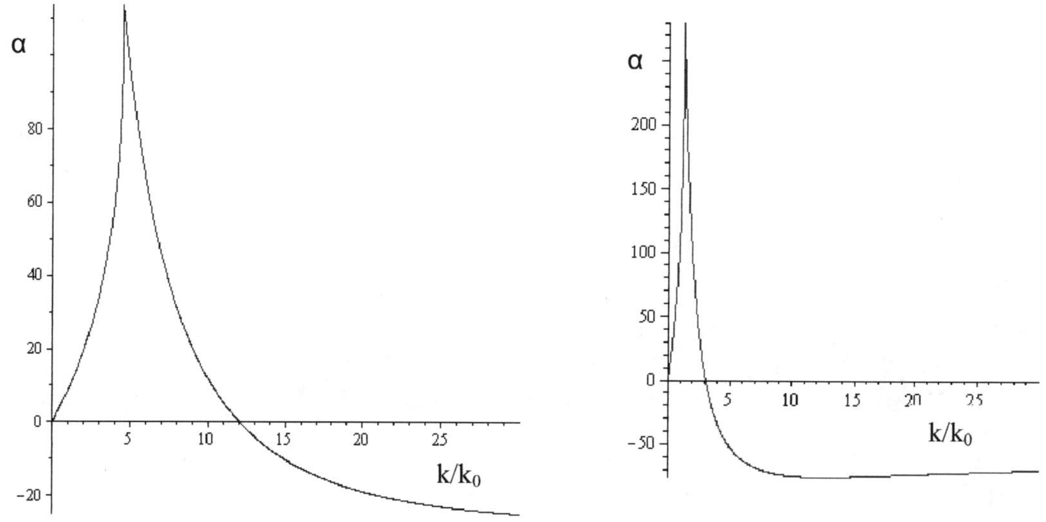

Figure 1. Dependencies of instability decrement on wave number for iron (a) and titanium (b).

From the analysis of the represented Fig. 1. it can be seen that the critical value of the nondimensional wave number for iron is equal to $k = 5$. This corresponds to the critical wavelength λ = 125 nm. As mentioned in the Introduction, the experimental values of cell sizes of crystallization are approximately of 0.3 μm [3], which is in three times more than the calculated wavelength. This discrepancy can be explained by the fact that the experiment has been carried out with steel containing 0.76 % of carbon. Such small additives have an effect on the magnitude of the surface tension of liquid iron [26].

For titanium (Figure 1 b) the critical wavelength is 419 nm; that does not contradict the experimental values [4 –6].

4. Conclusions

Thermocapillary model of the impact of low-energy high-current electron beams on metals has been developed. The dispersion equation, which is given by the algebraic equation of the 16-th degree, and in the approximation of short waves splits into two equations, has been obtained.
For iron and titanium the critical wavelength, at which thermocapillary instability comes, has been defined.

5. Acknowledgements

The work is fulfilled at the financial support of RFBR (projects №14-08-00506a and 15-08-03411a), grant of the President of the Russian Federation for state support of young Russian Scientists – Doctors of Sciences (project MD-2920.2015.8) and state task № 3.1496.2014/K.

References

[1] A reference

[2] Ivanov Y F, Kolubaeva Y A, Bessonov D A, Vorobev S V, Gromov V E, Tsellermaer V Y, Ivanov K V 2013 *J. Surf. Investig. X-ray, Synchrotron Neutron Tech.* **7** 90

[3] Gromov V E, Sizov V V, Vorob'ev S V, Konovalov S V, Ivanov Y F 2013 *J. Surf. Investig. X-ray, Synchrotron Neutron Tech.* **7** 94

[4] Grishunin V A, Gromov V E, Konovalov S V, Ivanov Y F, Teresov A D 2013 *J. Surf. Investig. X-ray, Synchrotron Neutron Tech.* **7** 990

[5] Raikov S V, Sosnin K V, Ivanov Yu F, Budovskikh E A, Gromov V E 2014 *Russ. Phys. J.*

57 252.
[6] Potekaev A I, Klopotov A A, Ivanov Yu F, Volokitin O G 2013 *Russ. Phys. J.* **56** 914.
[7] Ivanov Yu F, Budovskikh E A, Gromov V E et all 2012 *The structure, phase composition and properties of surface layers titanium alloys after electroexplosive alloying and electron-beam treatment* (Novokuznetsk, Publishing House of the SPC).
[8] Ivanov Yu F, Petrikova E A, Cherenda N N, Teresov A D 2014 *Adv. Mater. Res.* **872** 162
[9] Song R G, Zhang K, Chen G N 2003 *Vacuum* **69** 513
[10] Gao B, Hao Y, Zhuang W F, Tu G F, Shi W X, Li S W, Hao S Z, Dong C, Li M C 2011 *Phys. Procedia* **18** 187
[11] Sarychev V D, Vashchuk E S, Budovskikh E A, Gromov V E 2010 *Tech. Phys. Let.* **36** 656.
[12] Granovskii A Yu, Sarychev V D, Gromov V E 2013 *Tech. Phys.* **58**, 1544.
[13] Urpin V A, Yakovlev D G 1989 *Sov. Phys. Tech. Phys.* **59** 19
[14] Volkov N B, Leyvi A Ya, Mayer A E, Talala K A, Yalovets A P 2006 *AIP Conf. Proceed.* **849** 268
[15] Volkov N B, Leivi A Y, Talala K A, Yalovets A P 2010 *Tech. Phys.* **55** 484
[16] Astashinski V M, Leyvi A Y, Uglov V V, Cherenda N N, Yalovets A P 2014 *J. Surf. Investig. X-ray, Synchrotron Neutron Tech.* **8** 519.
[17] Nivedita R. Gupta, Hossein Haj-Hariri, Ali Borhan 2006 *J. Col. and Inter. Sci.* **293** 158
[18] Lee P D, Quested P N, McLean M 1998 *Phil. Trans. R. Soc. Lond. A* **356** 1027
[19] Lukashov E A, Radkevich E V, Yakovlev N N 2011 *J. Math. Sci.* **179** 491
[20] Zen'kovskaya S M 2007 *J. Appl. Math. and Mech.***71** 755
[20] Golovin A A, Davis S H, Nepomnyashchy A A 1998 *Phys. D.* **118** 202
[21] Visintin V 1996 *Models of Phase Transitions* (Boston: Birkhauser).
[22] Lacey A A, Tayler A B 1983 *IMA J. Appl. Math.* **30** 303
[23] Zen'kovskaya S M, Novosyadlyi V A 2009 *J. Appl. Math. and Mech.* **73** 271
[24] Zen'kovskaya S M, Novosyadlyi V A 2011 *J. Appl. Math. and Mech.* **69** 277
[25] Kirichenko N A 2009 *Quantum Electronics* **39** 442
[26] ASM Metals HandBook - Materials Characterization **10** (ASM International, 2002).

IOP Conf. Series: Materials Science and Engineering **91** (2015) 012029 doi:10.1088/1757-899X/91/1/012029

Fatigue life of silumin irradiated by high intensity pulsed electron beam

S V Konovalov[1,2], K V Alsaraeva[1], V E Gromov[1] and Yu F Ivanov[3,4]

[1]Siberian State Industrial University, 42, Kirov str., Novokuznetsk, 654007, Russia

[2]Wenzhou University, Wenzhou, Zhejiang Province, 325035, China

[3]Institute of High Current Electronics Siberian Branch of Russian Academy of Science, 2/1, Akademicheskii av., 30, Lenin str., Tomsk, 634021, Russia

[4]National Research Tomsk State University, Tomsk, 634050, Russia

E-mail: konovserg@gmail.com

Abstract. The electron-beam processing of silumin, leading to the evolution of structure-phase state of its surface is carried out. It has been shown that this alloy is a multiphase material and contains, except an aluminum-based phase, the particles of intermetallic compounds of Al-Si-Fe-Mn. It is shown that electron beam treatment of the eutectic silumin surface increases the fatigue service life more than in 3.5 times. The analysis of structure-phase states modification of silumin subjected to electron beam treatment with the following fatigue loading up to the failure is carried out by methods of optical and scanning electron diffraction microscopy. Analysis of the surface layer structure revealed the sources of nucleation of submicrocracks. It is revealed that the large silicon plates located on the surface and in the subsurface layer are the most dangerous stress concentrators.

1. Introduction

Fatigue failure of parts is one of the most commonly occurring causes of equipment, machinery, vehicles and structure failure. Therefore the problem of preventing the fatigue failures (service life increase) of critical parts is a very important one, especially in the branches of industry where the occurrence of emergencies leads to disastrous consequences. As a rule, fatigue cracks are initiated in the surface layer of the part. Therefore, the state of the surface layer affects significantly the fatigue life of the material.

Currently, in various branches of industry aluminum alloys are becoming increasingly popular. The most common of them is an aluminum alloy with silicon – silumin. This is due to its relatively low cost and relatively low specific gravity. However, the relatively low strength properties of silumin narrow significantly its scope of application. The most important method of improving their mechanical properties is the modification [1, 2].

The effective method of the specified modification and as a consequence, fatigue life increase is the processing of material surface by high intensity electron beam of submillisecond effect duration allowing to change the surface layer structure tens of micrometers thick transforming it into multi-modal structure-phase state and essentially without changing the structure-phase state of the alloy basic volume [3-5].

In papers [6-8] it was shown that surface treatment of stainless steels with high intensive pulsed electron beam is accompanied by increase (more than by 3.5 times) in fatigue life of these materials. It is obvious that deformation effect taking place under fatigue tests is favorable for structural transformation of material.

The aim of this work is to analyze the regularities of structure modification silumin by high-intensity pulsed electron beam, subjected to high-cycle fatigue testings up to the fracture.

2. Materials and research methods

Al-Si alloy (silumin) in an as-cast condition has been used as the material of the research. Pulsed melting of the silumin surface layer has been carried out by high-intensity pulsed electron beam on the installation «SOLO». Processing modes: electron energy is 18 keV, energy density of electron beam is (10–25) J/cm^2, effect pulse duration is 50 and 150 μs, the number of effect pulses is 1, 3, 5; pulse recurrence frequency is 0.3 s^{-1}; irradiation has been carried out in argon at a residual pressure of 0.02 Pa. These modes have been provided to heat the surface layer of a thickness of 50 μm up to melting temperature and quenching from the melt at speeds up to 10^6 K/s.

Fatigue tests, as in [6-8], have been performed on a special installation according to the scheme of a circular asymmetric cantilevered bending. The samples have had the shape of a parallelepiped with dimensions of 8x14x145 mm. The simulation of cracks has been carried out by a cut in the form of a semicircle with a radius of 10 mm. Test temperature is 300 K, the frequency of sample loading with curve is 15 Hz, the load is 10 MPa. At least 5 samples have been tested for each irradiation mode.

Investigation of the structure and phase composition of the alloy has been carried out by the methods of optical (etched thin section) and scanning electron microscopy (SEM). The elemental analysis has been carried out by the methods of electron micro-X-ray spectral analysis.

3. Results and their discussions

Figure 1 provides the image of the structure of the aluminum-silicon alloy in an initial state, obtained by the methods of scanning electron microscopy in inversely reflected electrons. It is evident that the alloy is a multiphase material. The selection of the second phases has predominantly a platelet form, and may differ significantly in intensity. The main volume of the inclusions of the second phase has a low intensity and is slightly different in intensity from the matrix image. It is known that the inclusions formed by the atoms of higher atomic weight, reflect the incident flux of electrons more intensively. Therefore, they have a lighter contrast in the images of the structure obtained in the inversely reflected electrons [9]. This fact allows concluding that the inclusions detected in the alloy differ significantly in elemental composition. In addition to the plates, there are also formations in the form of polyhedrons and selections of irregular shape in the alloy.

The main elements of the researched alloy are aluminum and silicon, which concentration in average is equal to 10.5 mass %. For elemental analysis three types of particles have been selected. They are differ in the intensity of the image in the inversely reflected electrons and morphology (Fig. 1, c). Analyzing the results presented in Table 1, it can be noted that in the investigated alloy, along with aluminum and silicon, there are iron and manganese, distributed extremely unevenly along the volume of the material. The matrix (spectrum 1), within the sensitivity of the analyzer, is formed by aluminum. Particles, having relatively low contrast in the images of the patterns in the inversely reflected electrons, are generated by the atoms of silicon (spectrum 2). Particles with brighter phase contrast contain atoms of aluminum, silicon, iron and manganese (spectrum 3 and spectrum 4).

The morphology of the particles is determined by the relative concentration of atoms of iron and silicon in them. Namely, in the particles in the form of "skeletons" (spectrum 3) the concentration ratio is Al/Fe/Si = 7/4/1; in the particles of a plate-like morphology (spectrum 4) the concentration ratio is Al/Fe/Si = 3/2/1. Following the results, represented in [1, 10], it can be supposed that the particles of a plate-like morphology (spectrum 4) are β-phase; its structure is defined as $Al_5(Fe, Mn)Si$; the particles of a skeletal form (spectrum 3) are α-phase; its structure is defined as $Al_8(Fe, Mn)_2Si$. Silicon and intermitallide particles have plate-like and skeletal forms. Therefore, they can be a source of cracks in mechanical testing of the material.

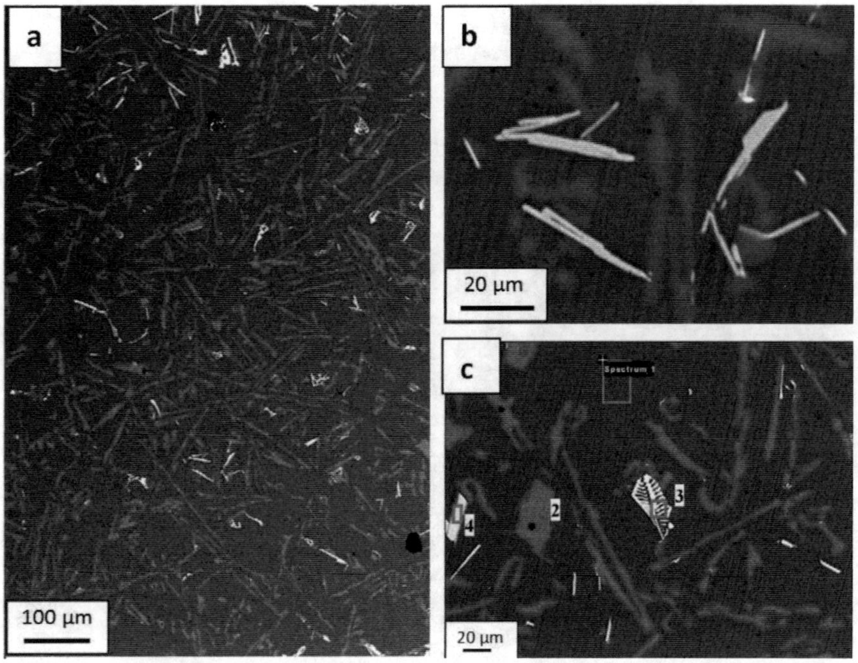

Figure 1. The structure of aluminum-silicon alloy (initial state). (a, b) – SEM analysis of the surface in inversely reflected electrons, (c) – surface image, obtained by micro-X-ray spectral analysis; on (c) the areas, which are chosen for the material elemental analysis, are indicated.

Table 1. Results of the micro-X-ray spectral analysis of the section of aluminum-silicon alloy surface represented on Fig. 1.

Element, mass %	Number of the section			
	1	2	3	4
Al	99	2.0	57	54
Si	1.0	98	8	15
Mn	0.0	0.0	1	1
Fe	0.0	0.0	34	30

In the fatigue tests of silumin samples a non-monotonic dependence of the number of cycles up to fracture on the irradiation mode by high-intensity pulsed electron beam has been received [11]. In the initial state silumin samples withstood $N=1.3 \cdot 10^5$ the number of cycles up to the fracture. The greatest increase in fatigue life (~3.5 times) has been provided by the irradiation mode No. 2 with the parameters of irradiation – 20 J/cm^2; 150 µs; 0.3 s^{-1}; 5 pulse.

It is evident that the fatigue life of silumin is determined primarily by the structure of the surface layer, modified by electron-beam processing. For structural studies of the irradiation surface of silumin the samples have been selected; they have showed minimal (for mode No. 1 with the parameters of irradiation – 15 J/cm^2; 150 µs; 0.3 s^{-1}; 3 pulse) and maximum (for mode No. 2) fatigue life.

The irradiation of silumin surface with pulsed electron beam depending on the energy density of the electron beam is accompanied either by melting the surface of the sample (Fig. 2, a, b), or the melting of the surface layer of the material of a certain thickness (from one to tens of micrometers) (Fig. 2, c, d). In the first case, the fatigue life in some cases has been below the fatigue life of the initial samples, and in the second one it has exceeded the fatigue life of the initial material in more than 3.5 times.

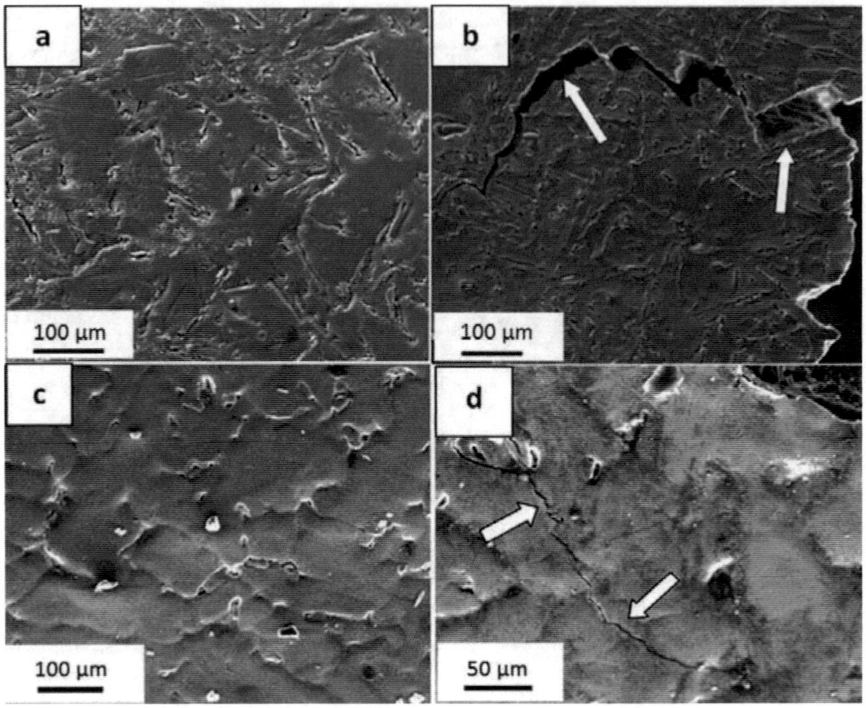

Figure 2. Electron microscope image of silumin irradiation surface structure, processed by electron beam according to mode No. 1 (a, b) and No. 2 (c, d). The pointers on (b) and (d) indicate the cracks, formed as a result of fatigue tests

The irradiation of silumin surface for mode No. 1 leads to the partial melting of the excess silicon inclusions (Fig. 2, a). In the surface layer numerous micropores along the boundary between the plate/matrix and microcracks are formed; they are located in the silicon plates, which weaken the material. On the image of fatigue fracture surface of the silumin sample (Fig. 3) it can be clearly seen that the fatigue crack is formed on the sample surface (Fig. 3, a). The reasons for the formation of fatigue cracks are coarse inclusions of silicon (Fig. 3, b), which are stress concentrators. As a result, the fatigue tests lead to the fracture of plates and the formation of long microcracks (Fig. 2, b).

During silumin irradiation with high-intensity electron beam in mode No. 2 the structure of the surface layer by morphological characteristic differs significantly from the structure of the original sample (Fig. 1) and the sample irradiated in the mode of the surface melting (Fig. 2, a, b). On the irradiation surface the homogeneous structure of a grain type is formed (Fig. 4, a-c), and the thickness of the molten layer varies in the range up to 20 μm (Fig. 4, d).

Resulting from high-speed crystallization a multimodal structure at the macro level is represented by the grains on the basis of aluminum, whose dimensions range from 30 to 50 μm. There are silicon particles on the grain boundaries; their transverse dimensions do not exceed 20 μm (Fig. 4, b). Mesolevel of the modified layer consists of two-phase (silicon and solid solution on the basis of aluminum) crystallization cells detected on the surface of a fatigue fracture. The cell crystallization sizes vary in the range of 100 nm up to 250 nm (Fig. 4, e). This reflects the submicrocrystalline structure of the near-surface layer. It is important that the stress concentrators, which can be a source of fracture of silumin samples at this mode of irradiation, on the edge of a fracture are not revealed.

Apparently, the concentrators, which caused the fracture of the sample, are situated below the surface, most likely at the interface of the liquid and solid phases.

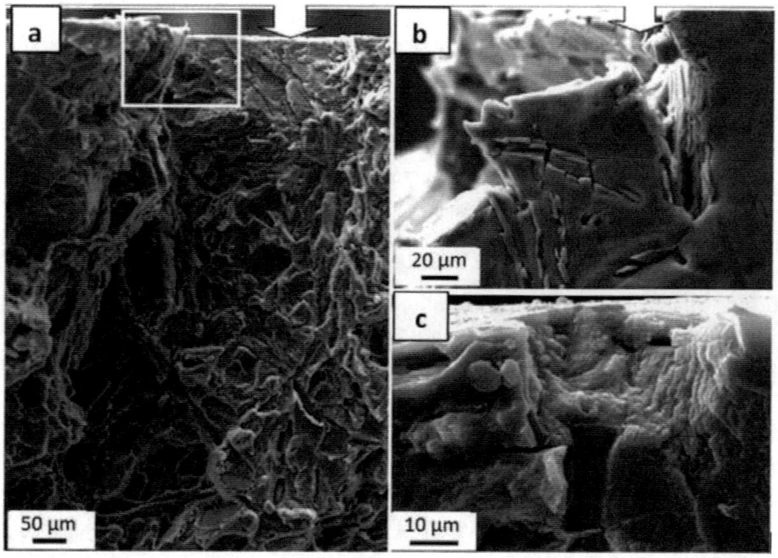

Figure 3. Electron microscope image of the fracture surface structure (a, b) and irradiation surface (b) of silumin, processed by electron beam according to mode No. 1. On (a) the frame singles out a fatigue crack formation range; the pointers indicate: on (a) – irradiation surface, on (b) – place of origin of fatigue crack

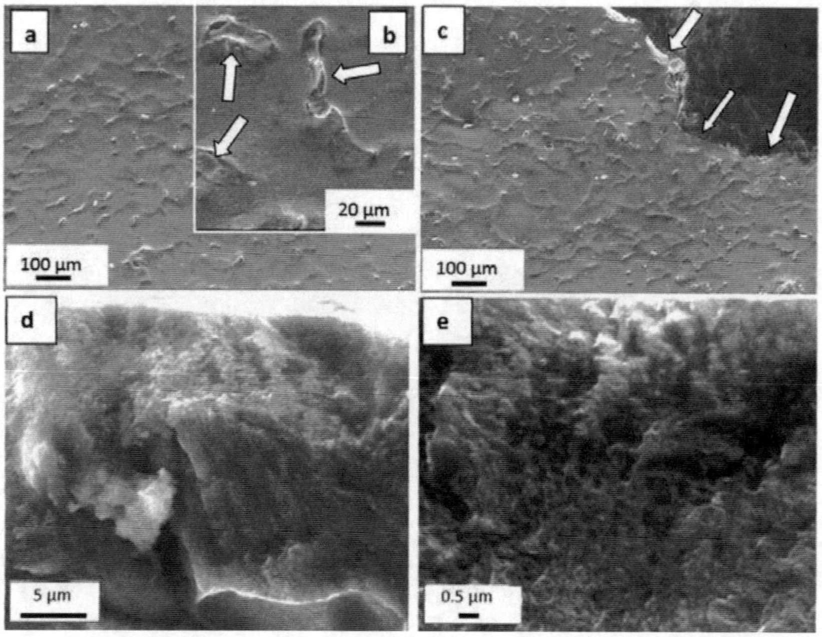

Figure 4. Electron microscope image of irradiation surface structure (a-c) and silumin fracture surface (d, e), processed by electron beam according to mode No. 2. The pointers on (b) indicate silicon particles; on (c) – edge of the sample fatigue fracture

The formation of multi-level structural-phase state determines the development of the damping properties in the surface layer of the modified silumin with respect to the main material by mechanical and thermal external influences. This prevents the premature birth and spreading of fragile microcracks from the surface into the bulk of the material. These cracks can lead to the formation of main cracks and to the fracture of the basic material. Thus, the formation of a submicro- and nanoscale multiphase structure at the irradiation of silumin according to mode No. 2 is also the defining reason assisting a multiple increase of its fatigue life.

4. Conclusion

Processing of silumin by high intensive electron beam has been carried out. It has been established that the fatigue tests of silumin subjected to irradiation with high intensity pulsed electron beam resulting in the increase in fatigue life by more than 3.5 times. Investigations of structure and surface modified layer destruction of silumin subjected to high-cycle fatigue tests to fracture have been carried out by methods of optical and scanning electron microscopy.

It has been shown that this alloy is a multiphase material and contains, except an aluminum-based phase, the particles of intermetallic compounds of Al-Si-Fe-Mn. These particles have a plate or skeletal form and therefore, can be a source of cracks in the material mechanical testing. The researches of an irradiation surface structure and the surface of fatigue failure of silumin in an initial state and after modification states with intense pulsed electron beam according to different regimes have been carried out.

It has been established that the main reason of multiple increase of silumin fatigue life, irradiated by a pulsed electron beam, is the formation of the submicro- and nanoscale multiphase structure in a modified surface layer.

This work has been supported by a grant of the President of the Russian Federation for state support of young Russian scientists – Doctors of Sciences (project MD-2920.2015.8) and state task № 3.1496.2014/K.

References

[1] Zolotorevskiy V S and Belov N A 2005 *Metal science of casting aluminum alloys*, (Moscow, Publishing House MISiS).
[2] Borowski J and Bartkowiak K 2010 *Physics Procedia* **5** 449-456.
[3] Hao S Z, Qin Y, Mei X X, Gao B, Zuo J X, Guan Q F, Dong C and Zhang Q Y 2007 *Surf. and Coat. Technol.* **201** 8588–8595.
[4] Hao Y, Gao B, Tu G F, Li S W, Hao S Z and Dong C 2011 *Applied Surface Science* **257** 3913–3919.
[5] Hao Y, Gao B, Tu G F, Cao H, Hao S Z and Dong C 2012 *Applied Surface Science* **258** 2052–2056.
[6] Ivanov Yu F, Koval N N, Gorbunov S V, Vorobyov S V, Konovalov S V and Gromov V E 2011 *Rus. Phys. J.* **54** 575-583.
[7] Gromov V E, Ivanov Yu F, Sizov V V, Vorob'ev S V and Konovalov S V 2013 *J. of Surf. Investigation. X-ray, Synchrotron and Neutron Techniques* **7** 94-98.
[8] Gromov V E, Ivanov Yu F, Vorobiev S V and Konovalov S V 2015 *Fatigue of steels modified by high intensity electron beams*, (Cambridge, Cambridge International Science Publishing Ltd).
[9] Krishtall M M, Yasnikov I S and Polunin V I 2009 *Scanning electron microscopy and X-ray spectral microanalysis in the examples of practical application,* (Moscow, Technosfera).
[10] Belov H A 2009 *Phase construction of aluminum alloys,* (Moscow, Publishing House MISiS).
[11] Ivanov Yu F, Alsaraeva K V, Petrikova E A, Teresov A D, Gromov V E and Konovalov S V 2015 *IOP Conf. Series: Materials Science and Engineering* **81** 012011.

IOP Conf. Series: Materials Science and Engineering **91** (2015) 012030 doi:10.1088/1757-899X/91/1/012030

Investigation of defect copper substructure disrupted in creep condition under the action of magnetic field

S V Konovalov[1,2], **N G Yaropolova**[1], **D V Zaguyliaev**[1], **V E Gromov**[1], **Yu F Ivanov**[3,4] **and I A Komissarova**[1]

[1]Siberian State Industrial University, 42, Kirov str., Novokuznetsk, 654007, Russia

[2]Wenzhou University, Wenzhou, Zhejiang Province, 325035, China

[3]Institute of High Current Electronics Siberian Branch of Russian Academy of Science, 2/1, Akademicheskii av., Tomsk, 634021, Russia

[4]National Research Tomsk State University, 30, Lenin str., Tomsk, 634050, Russia

E-mail: konovserg@gmail.com

Abstract. The defect substructure of M00b copper samples loaded up to disruption in creep condition both under and without the action of 0.35 T magnetic field is investigated in the paper. Material near the disrupted surface and at certain distances from it received the serious study. It has been ascertained that when copper disrupting without magnetic action on creep process the main type of dislocation substructure is the cellular one irrespectively of the distance to disruption surface. As the result of magnetic field influence on creep process the main type of dislocation substructure is replaced by the stripe-like one. The distinctive quantitative characteristics of dislocation substructures are observed. Moreover, a gradient behavior of the number of stress raisers has been revealed when moving away from disruption surface both in deformation conditions with and without a magnetic field.

1. Introduction

The progress in engineering and technologies facilitates development of high-power electric power plants forming magnetic and electric fields, which constantly effect on the materials around them. On the other hand, the most of essential products, components of the machines and structures are maintained in conditions of constant stress (creep), so strict requirements are specified for them in terms of operational characteristics. However, such factors as external power effects and mechanical stress acting simultaneously can significantly reduce an operational time of essential products and units causing their premature failure [1-6]. Extended operational time of metals and alloys determines the advances in improvement of metal products reliability. One of the factors responsible for physical and mechanical properties of materials is considered to be their structure evolving in conditions of stress. Therefore, one needs exact information on material structure near the point which caused its complete disruption. For the fore-going reasons the work focuses on investigation of magnetic field effect on the gradient of copper defect substructure subject to creep tests near the zone of disruption.

2. Materials and research methods

Polycrystalline M00b copper was used as material of research. Flat samples with 150×5×0.46 mm^3 working sections were applied in experiments. Before creep tests material structure was balanced via recrystallization annealing for two hours at the temperature 700 ^{0}C succeeded by chilling in water necessary for removal of annealing cinder. The tests in conditions of creep were carried out under constant tension stress σ = 130 MPa and the temperature 25 ^{0}C up to disruption. The first half of samples was deformed in conditions of external magnetic field with induction 0.35 T, the second half – in normal conditions.

To record the deformation of creep a modernized testing machine for research into the processes of metal plastic deformation was used. The testing machine operates in coordination with the software intended for recording the data on sample elongation in regard to the time of experiment with a possibility to analyze creep phases and approximate creep curves [7].

As the source of a magnetic field we used a permanent electromagnet which can adjust magnetic field induction. Magnetic field induction was monitored by milli-teslameter TPU with the accuracy up to 10^{-3} T.

Dislocation substructure of samples was researched via transmission electron diffraction microscopy (TEM) of thin foils by electron microscope EM-125. The foils were made of material taken near the zone of disruption. Images of thin structure of material were used for classification of morphological characteristics of structure [8, 9].

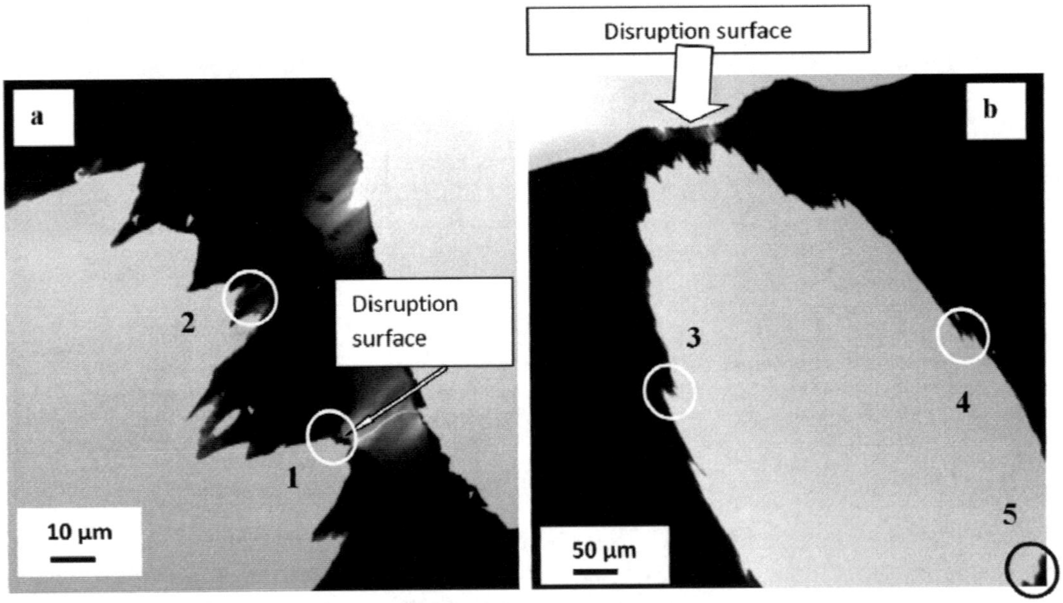

Figure 1. General view (b) and sub-surface part (a) of ion thinned foil (the sample is disrupted by creep in the magnetic field). As ovals TEM analyzed foil zones are marked; (a) – 1 – disruption surface; 2 – the layer over the distance ~20 µm from the disruption surface; (b) – layers, located at the depths 3 – ~265 µm; 4 – ~325 µm; 5 – ~610 µm.

3. Results and their discussions

3.1. Layer by layer analysis of dislocation substructures emerging in copper in conditions of creep

The experiments demonstrated that in the layer bordering with disruption surface a stripe-like substructure is forming, the azimuthal component of full disorientation angle in a stripe-like

IOP Conf. Series: Materials Science and Engineering **91** (2015) 012030 doi:10.1088/1757-899X/91/1/012030

substructure is up to $\Delta\alpha$ = 5.7 degrees. In stripes a dislocation substructure occurs in the form of chaotically distributed dislocations with scalar density ~3.5x10^{10} cm^{-2}. Moreover, a lot of bend contours were revealed.

The experiments ascertained that irrespectively of the distance to disruption surface a cellular dislocation substructure prevails – is the main one (Fig. 2b). On grain boundaries a stripe-like substructure occurs (Fig. 2a).

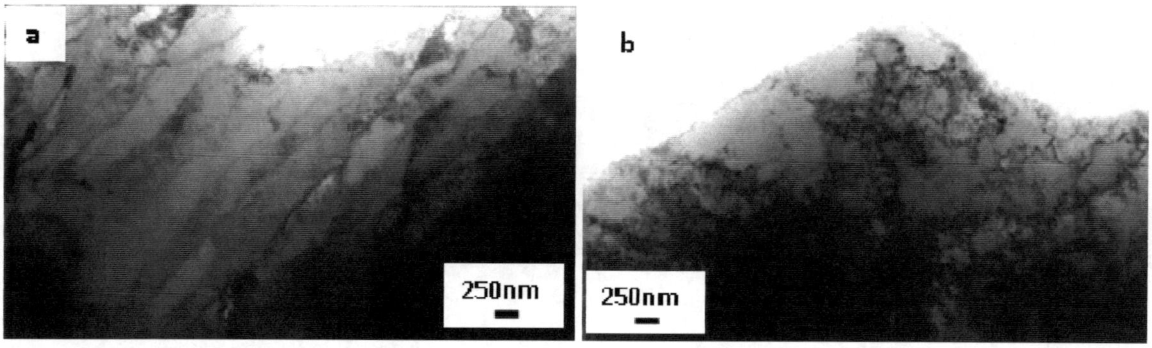

Figure 2. Electron microscopical image of dislocation substructure of the layer ~85 μm away from copper disruption surface

In the layer ~85 μm away from disruption surface the scalar density of dislocations in the stripe-like substructure is ~2.2x10^{10} cm^{-2}; in the cellular one - ~2.6x10^{10} cm^{-2}. In the layer ~325 μm away from disruption surface the scalar density of dislocations in the stripe-like substructure is ~2.7x10^{10} cm^{-2}; in the cellular one - ~2.9x10^{10} cm^{-2}. In the layer ~495 μm away from disruption surface the scalar density of dislocations in the cellular substructure is ~2.5x10^{10} cm^{-2}; in the stripe-like one - ~2.3x10^{10} cm^{-2}. In the layer ~635 μm away from disruption surface the main type of dislocation substructure is the cellular one (Fig. 3a). On grain boundaries the stripe-like substructure occurs (Fig. 3b).

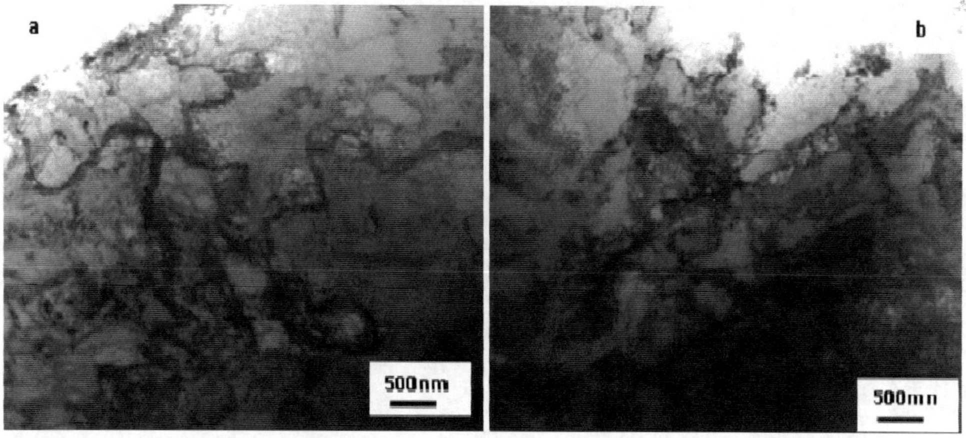

Figure 3. Electron microscopical image of dislocation substructure of the layer ~635 μm away from disruption surface of copper

Moreover, grains with dislocation substructure in form of a grid and chaotically distributed dislocations are found. In cells and stripes there is a dislocation substructure in form of chaotically distributed dislocations. The scalar density of dislocations in cells is ~2.7x10^{10} cm^{-2}; the scalar density of dislocations in stripes is ~$2.5x10^{10}$ cm^{-2}; the scalar density of grid-like dislocations in grains is ~$6.0x10^{10}$ cm^{-2}.

Copper structure analyzed in terms of quantity demonstrated changes in contours density relatively to the distance to disruption surface. The analysis of results shown in Fig. 4 demonstrates the gradient character of changing number of stress concentrators in copper. The layer bordering with the surface of disruption is in the most stressed condition.

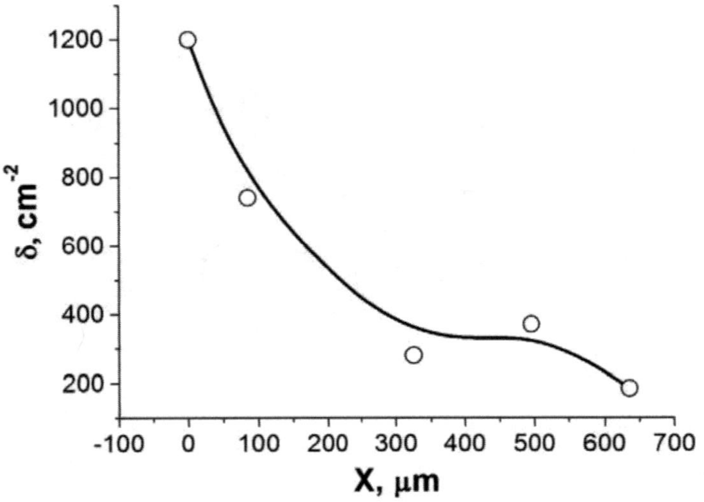

Figure 4. The dependence of bend contours density of extinction (density of stress concentrators) on the distance to copper disruption surface

3.2. Layer-by-layer analysis of dislocation substructures emerging in copper in conditions of magnetic field creep

A lot of bend extinction contours characterize electron-microscopical images of copper structure in disruption zone (the density of contours ~$2.9x10^3$ cm^{-2}), it confirms curvature and torsion of material crystal lattice.

The analysis of electron-microscopical images of a substructure forming in ~2 μm layer of copper after its disruption in condition of magnetic creep demonstrates that stripe-like and sub-grain substructures are forming in this layer in conditions of creep (the sphere of sub-grains is marked with a circle). Quasi circular arrangement of reflexes is typical for micro-electron diffraction patterns of this structure (Fig. 5 b), it indicates a nano-crystal condition of the substructure, and secondly, a high level of disorientation in these crystals. Taken measurements, indeed, showed that sub-grain dimensions vary from 35 to 85 nm. A dislocation substructure in form of chaotically distributed dislocations with the scalar density ~$3.85x10^{10}$ cm^{-2} is observed in stripes.

The experiments stated that irrespectively of the distance to disruption surface a stripe-like substructure is the main type of dislocation substructure.

For copper structure of the layer, which is ~6 μm away from disruption surface a dislocation substructure in form of chaotically distributed dislocations is typical. The scalar density of dislocations in stripe-like substructure is ~$4.1x10^{10}$ cm^{-2}; in grains ~$2.4x10^{10}$ cm^{-2}. The density of bend extinction contours in this material is ~$0.67x10^3$ cm^{-2}.

In stripe-like copper substructure in the layer, which is ~20 μm away from disruption surface there is a dislocation substructure in form of chaotically distributed dislocations, grids are infrequent. The scalar density of dislocations is ~$3.3x10^{10}$ cm^{-2}. The density of bend extinction contours in this material is ~$0.54x10^3$ cm^{-2}.

In stripe-like copper substructure in the layer, which is ~20 μm away from disruption surface there is a dislocation substructure in form of chaotically distributed dislocations, grids are infrequent. The scalar density of dislocations is ~$3.3x10^{10}$ cm^{-2}. The density of bend extinction contours in this material is ~$0.54x10^3$ cm^{-2}.

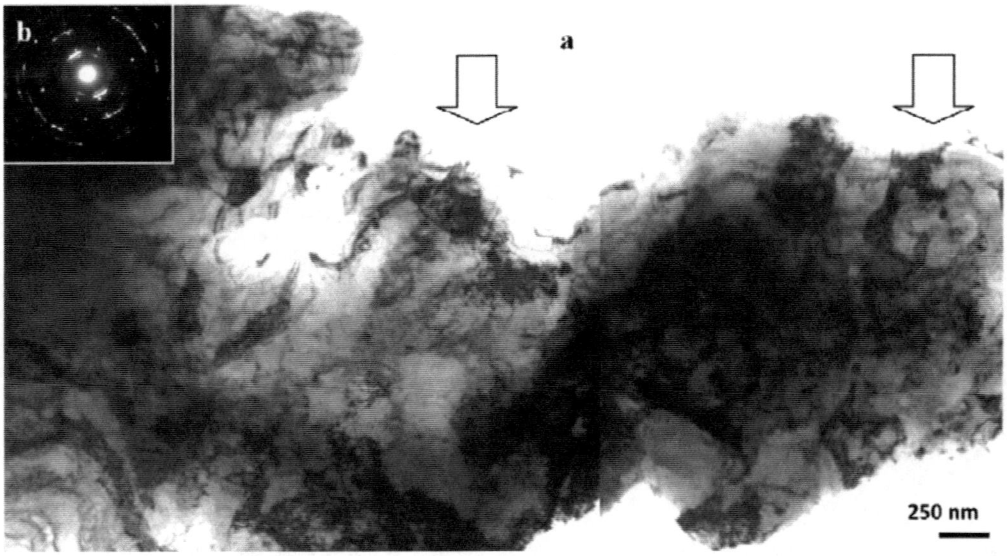

Figure 5. Electron-microscopical image of copper surface layer disrupted in conditions of creep in magnetic field; a – light-colored fields; b – micro-electron diffraction pattern. The arrows show disruption surface.

In the layer ~265 μm away from disruption surface in stripe-like substructure there is a dislocation substructure in form of chaotically distributed dislocations, and grids are less frequent; the scalar density of dislocations is ~2.8×10^{10} cm^{-2}. The scalar density of dislocations in grains without stripe-like substructure is ~3.3×10^{10} cm^{-2}. The density of bend extinction contours in this material is ~0.62×10^{3} cm^{-2}.

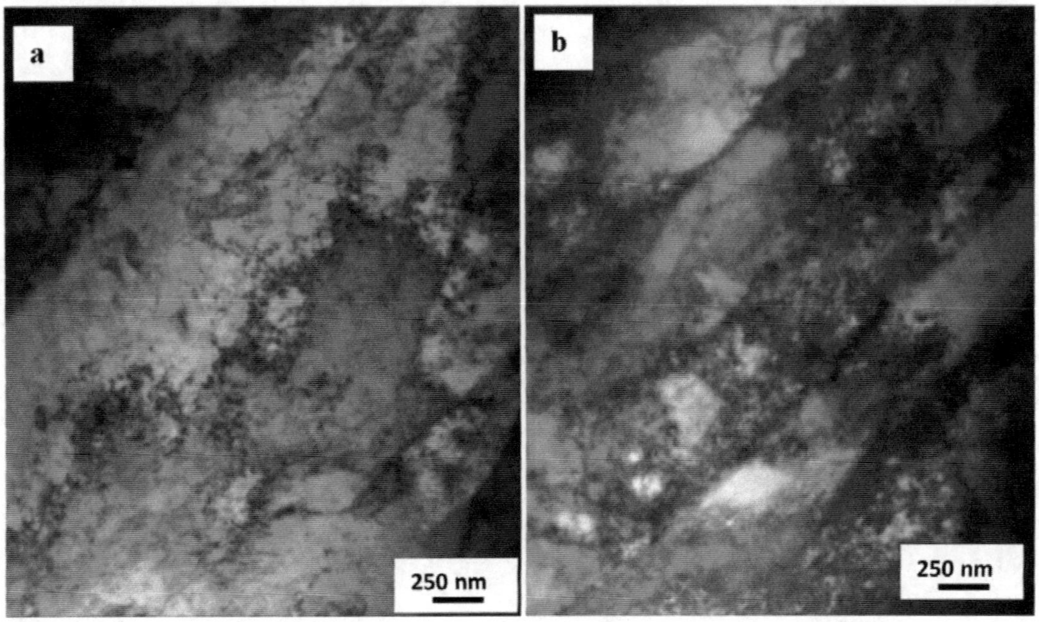

Figure 6. Electron-microscopical image of the layer structure, which is ~325 μm away from disrupted copper surface in conditions of magnetic creep

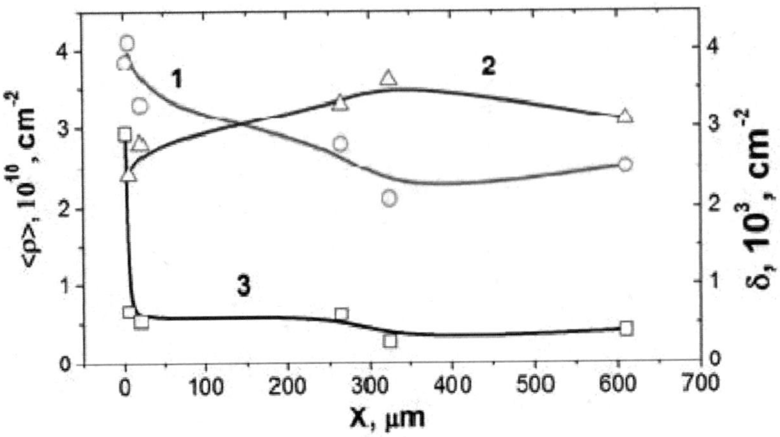

Figure 7. The dependence of scalar density of dislocations <ρ> in the stripe-like substructure (curve 1) and in grains (curve 2), densities of bend contours of extinction (curve 3) on the distance to copper disruption surface (creep in conditions of magnetic field)

For the layer of material ~325 μm away from disruption surface the same two types of dislocation substructure as in the above mentioned layers are typical, however, in grains with chaotic dislocation structure broken sub-grain boundaries are often observed (Fig. 6). In stripe-like substructure there is a dislocation substructure as chaotically distributed dislocations, grids are less frequent (Figure 7); the scalar density of dislocations is $\sim 2.1 \times 10^{10}$ cm^{-2}. The scalar density of dislocations in grains without a stripe-like substructure is $\sim 3.6 \times 10^{10}$ cm^{-2}. The density of bend extinction contours in this material is $\sim 0.28 \times 10^3$ cm^{-2}.

In a stripe-like copper substructure in the layer ~610 μm away from disruption surface the scalar density of dislocations is $\sim 2.5 \times 10^{10}$ cm^{-2}. The scalar density of dislocations in grains without stripe-like substructure is $\sim 3.1 \times 10^{10}$ cm^{-2}. The density of bend extinction contours in this material is $\sim 0.42 \times 10^3$ cm^{-2}.

It has been stated that in the layer bordering with disruption surface a stripe-like dislocation substructure is mainly forming. It has been revealed that the scalar density of dislocations naturally decreases as it is moved away from disruption surface (Fig. 7, curve 1). The second type of copper structure includes grains with chaotically distributed dislocations or grid dislocation substructure. In this case the scalar density of dislocation slightly increases (within measurement errors) as moving away from disruption surface (Fig. 7, curve 2). In copper disruption zone there are internal stress fields, indicated by bend extinction contours seen on electron-microscopical images of thin foil structure. The density of bend extinction contours reduces as moving away from disruption surface (Fig. 7, curve 3).

4. Conclusion

It has been determined that when copper disrupting without magnetic action on creep process the main type of dislocation substructure is a cellular one irrespectively of the distance to disruption surface. On grain boundaries a stripe-like substructure is revealed.

The distinctive quantitative characteristics of dislocation substructures have been revealed, as well as a gradient nature of changing number of stress concentrators when moving away from disruption surface irrespectively of magnetic action.

Under magnetic action on creep the main type of dislocation substructure is a stripe-like one, irrespectively of the distance to disruption surface. In some cases grains with dislocation chaos structure, as well cellular and grid substructures are revealed.

The reducing character of extinction bend contours density has been determined when moving away from disruption surface in conditions of magnetic creep.

Acknowledgement

The work is performed with financial support of Russian Foundation for Basic Research (projects No 14-08-00506a, 14-08-00686a), Grant of the President of the Russian Federation for state support of young Russian scientists – Doctors of Science (project MD-2920.2015.8) and state assignment No 3.1496.2014/K.

References

[1] Zagulyaev D V, Konovalov S V, Litvinenko N G , Komissarova I A and Gromov V E 2013 *Tsvetnye Met.* **4** 74-77.
[2] Golovin Yu I 2004 *Phys. Solid State* **46** 789-824.
[3] Kurek E I, Kurek I G, Oleinich-Lysyuk A V and Raranskii N D 2014 *Phys. Solid State* **56** 1597-1604.
[4] Lei Y, Zhang Z, Nie J and Chen X 2008 *Chinese J. of Materials Research* **22** 420-424
[5] Chen X, Shen Z, Wang J, Chen J, Lei Y and Huang Q 2012 *International J. of Advanced Manufacturing Technology* **60** 537-544.
[6] Al'shits V I, Darinskaya E V, Koldaeva M V and Petrzhik E A 2008 *JETP Letters* **88** 428-434.
[7] Konovalov S V, Danilov V I, Zuev L B, Filip'ev R A and Gromov V E 2007 *Phys. Solid State* **49** 1457–1459.
[8] Hirsch P B, Howie A, Nicholson R B, Pashley D W and Whelan M J 1977 *Electron microscopy of thin crystals* (Melbourne: Krieger Publishing Co.).
[9] Koneva N A and Kozlov E V 2004 *Materials Science and Engineering* **A 387–389** 64-66.

IOP Conf. Series: Materials Science and Engineering **91** (2015) 012031 doi:10.1088/1757-899X/91/1/012031

Influence of layer-by-layer laser sintering conditions on the quality of sintered surface layer of products

N A Saprykina[1,a], A A Saprykin[1], I F Borovikov[2,b], Y P Sharkeev[3,c]

[1]Yurga Institute of Technology of National Research Tomsk Polytechnic University Affiliate, Tomsk, Russia
[2]The Moscow State Technical University of N.E. Bauman, Moscow, Russia
[3]Institute of Strength Physics and Materials Science of SB RAS
The study was performed by a grant from the Russian Science Foundation (project №15-19-00191).

E-mail: [a] saprikina@tpu.ru, [b] bif1986@mail.ru, [c] sharkeev@ispms.tsc.ru

Abstract. The influence of technological modes of sintering: the displacement velocity of laser beam V, laser power P, scanning step S and preheating temperature of powder material t on the quality of sintered surface layer of aluminum powder PA-4, copper powder PMS-1 and cobalt-chromium-molybdenum powder DSK-F75 were studied.

1. Introduction

Nowadays the most innovative direction in modern industry is additive technologies allowing manufacturing complex functional products from powder materials based on 3D-CAD model [1]. These technologies vary from the other by the way of obtaining ready components – not by removal of material layer from rough materials but by using layer-by-layer building-up of materials at the same time obtaining the required shape and size of product [2, 3]. Using this technology brings reduction in raw materials consumption by 75 %. The main problems of using considered technologies in Russia are connected with application of imported equipment and materials [4, 5, 6]. Production by means of layer-by-layer laser sintering from domestic powder materials requires conducting experimental studies to define sintering modes and find out their influence on the quality of sintered surface layer.

2. Result and Discussion

The present work describes influence of technological modes of sintering: the displacement velocity of laser beam V, laser power P, scanning step S and preheating temperature of powder material t on the quality of sintered surface layer of domestic powder materials with different melting temperature. Experimental studies were held on technology laser complex consisting of ytterbium fiber laser LK-100-V, three-axis table, computer numerical control (CNC) and original program software [7].

Powder materials of different size and different melting temperature were selected for processing to analyze behavior of powder materials under laser sintering.

- Aluminum powder PA-4 is manufactured from primary aluminum by means of pulverization. Percentage of active aluminum: 98 %. Powder melting temperature: 660 °C. Bulk density: 0.96 g/sm³. Powder size: 0.07 mm. At the present moment aluminum and its alloys are used in many fields of industry and engineering. Primarily aluminum and its compounds find application in aviation and automobile industries. Aircraft structures, engines, blocks, cylinder heads, crankcases, pumps and other components are manufactured from aluminum and its alloys.

- Copper stabilized powder PMS-1, with copper content of 99.5 %. Range of powder melting temperature: 1030-1070 °C. Bulk density: 1.25-1.9 g/sm³. Powder size: 0.07 mm. Powder is used in powder metallurgy for production of sintered products, and in instrument engineering [8].
- Cobalt-chromium-molybdenum powder DSK-F75, with cobalt content of 66.4 %, chromium of 28 %, molybdenum of 3 %. Range of powder melting temperature: 1350-1450 °C, bulk density: 8.4 g/sm³. Powder size: 0.1 mm. Powder is used in machinery engineering for manufacturing components operating under high temperatures.

In order to provide quality of sintered surface layer from powder materials with different range of melting temperature, alteration in the level of deformations and thickness of sintering depending on sintering modes were studied [9].

Samples of sintered single layer (20 mm long, 5 and 10 mm wide) were obtained during experimental studies. The areas of technological modes of layer-by-layer laser sintering were distinguished resulting from exploratory studies. For obtaining sample from powder PA-4, with some mechanical strength the rational mode is: P=(10-20) W, V=(1000-3000) mm/min, S=(0.1-0.2) mm, t=(26-200) °C. For copper powder PMS-1 P=(14-30) W, V=(200-3000) mm/min, S= (0.1-0.3) mm, t=(26-200)°C. Rational mode for powder DSK-F75 is: P=(10-20) W, V=(100-300) mm/min, S= (0.1-0.15) mm, t=(26-200)°C. Power of laser radiation is varied depending on the melting temperature of powder material and temperature conductivity coefficient, particles shape and size. Hard-melting powder material DSK-F75 (powder size 0.1 mm) is marked by increase in power of laser radiation and decrease in beam displacement velocity. Geometric state of the sintered surface was carried out by specially developed methodology with applying instrumental digital microscope [10].

Figures 1-6 demonstrate influence of power of laser radiation on the quality of sintered surface layer. Lack of power while sample sintering leads to their diffusion; power excess, on the contrary, leads to deformation, appearance of longitudinal and transversal cracks and powder inflammation.

Figures 1-2 shows samples made from aluminum powder PA-4. Power alteration: from 10 to 20 W, at the constant velocity V=1000 mm/min, sintering temperature t=26 °C and scanning step S=0.2 mm result in increase in thickness of sintered layer from 0.765 to 1.55 mm. Sample presented at Figures 1-2 has low mechanical strength and diffuses when contacting with it.

Figure 1. Appearance of the sintered surface PA-4 (x2), Sintering modes V=1000 mm/min, t=26 °C, S=0.2 mm, P=10 W.

Figure 2. Appearance of the sintered surface PA-4 (x2), Sintering modes V=1000 mm/min, t=26 °C, S=0.2 mm, P=20 W.

Figures 3-4 demonstrates influence of power on appearance of the sintered surface from copper powder PMS-1. Power alteration from 15 to 30 W while V=200 mm/min, t=200 °C, S=0.3 mm leads to alteration of Rz from 475 to 975 μm. Samples resulted to be strong, but with formation of longitudinal and transverse cracks. As the power increases, cracks start to grow due to increase in thermal stresses and high thermal conductivity of powder PMS-1.

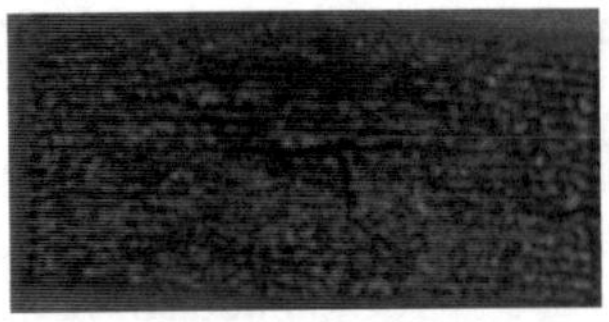

Figure 3. Appearance of the sintered surface of PMS-1 (x2), Sintering modes V=200 mm/min, t=200 °C, S=0.3 mm, P=15 W.

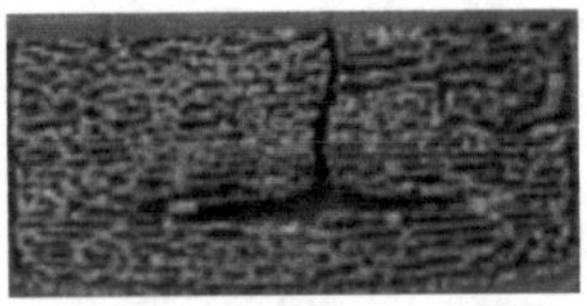

Figure 4. Appearance of the sintered surface of PMS-1 (x2), Sintering modes V=200 mm/min, t=200 °C, S=0.3 mm, P=30 W.

Figures 5-6 demonstrates influence of power on the appearance of sintered surface DSK-F75. While sintering this powder coagulation was observed. Coagulation is combining of small particles of dispersed systems into larger ones under the action of adhesion power. Power alteration from 10 to 20 W, while V=300 mm/min, t=26 °C, S=0.1 mm leads to increased coarseness of surface layer from 425 to 625 μm, diameter of coagulated particles from 175 to 325 μm and thickness of sintered layer from 0.65 to 1.0 mm. Sample presented in Fig. 3a is not characterized by large mechanical strength.

Figure 5. Appearance of the sintered surface DSK-F75 (x2), sintering modes V=300 mm/min, t=26 °C, S=0.1 mm P=10 W.

Figure 6. Appearance of the sintered surface DSK-F75 (x2), sintering modes V=300 mm/min, t=26 °C, S=0.1 mm P=20 W.

Figures 7-16 show photographs of appearance of the studied powder materials depending on the displacement velocity of laser beam. Figures 7-8 shows photographs of appearance alteration of the sintered surface of powder PA-4 depending on the displacement velocity of laser beam. Increasing displacement velocity of laser beam from 1000 to 3000 mm/min, while P=20 W, t=26 °C, S=0.2 mm has lead to decrease in the thickness of sintered layer from 1.55 to 1.33 mm.

Figure 7. Appearance of the sintered surface PA-4 (x2), sintering modes P=20 W, t=26 °C, S=0.2 mm, V=1000 mm/min.

Figure 8. Appearance of the sintered surface PA-4 (x2), sintering modes P=20 W, t=26 °C, S=0.2 mm, V=3000 mm/min.

Figures 9-10 shows samples from copper powder PMS-1, obtained while P=15 W, t=200 °C, S=0.3 mm. When V=200 mm/min defects along and across cracks formation appear on the sample. When V=3000 mm/min, powder does not have time to sinter. When velocity increases the thickness of sintered layer decreases from 1.700 to 0.7 mm. Defects are conditioned by high heat conductivity of powder material, and sharp temperature gradient during and after sintering. Sample given in Figures 10 has low mechanical strength and diffuses when contacting with it.

Figure 9. Appearance of sintered surface PMS-1 (x2), sintering modes P=30 W, t=200 °C, S=0.3 mm, V=200 mm/min.

Figure 10. Appearance of sintered surface PMS-1 (x2), sintering modes P=30 W, t=200 °C, S=0.3 mm, V=3000 mm/min.

Sintering while V=200 mm/min and P=15 W, t=200 °C, S=0.1 mm leads to combustion of powder material (Figures 11-12). Defects placed along and across the cracks formation are observed. While velocity of 3000 mm/min the obtained sample has high strength, and it has no defects. The thickness of sintered layer decreases with velocity increase from 1.0 to 0.7 mm, coarseness 590 to 225 μm.

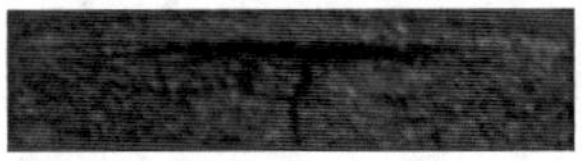

Figure 11. Appearance of the sintered surface PMS-1 (x2), sintering modes V=200 mm/min и P=30 W, t=200 °C, S=0.1 mm, V=200 mm/min.

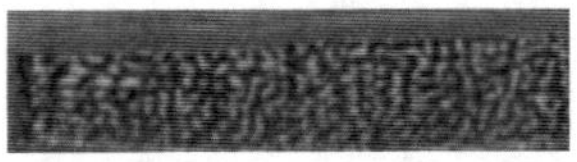

Figure 12. Appearance of the sintered surface PMS-1 (x2), sintering modes V=200 mm/min и P=30 W, t=200 °C, S=0.1 mm, V=3000 mm/min.

Samples obtained while modes P=22 W, t=114 °C, S=0.2 mm, when V=200 mm/min and 3000 mm/min form strong sintered surface (Figures 13-14). Increase of velocity leads to alteration of sintered layer thickness from 0.9 to 0.41 mm, Rz from 930 to 550 μm. Sample given in Figures 13-14 has structure of molten metal and is black in color due to formation of copper oxide.

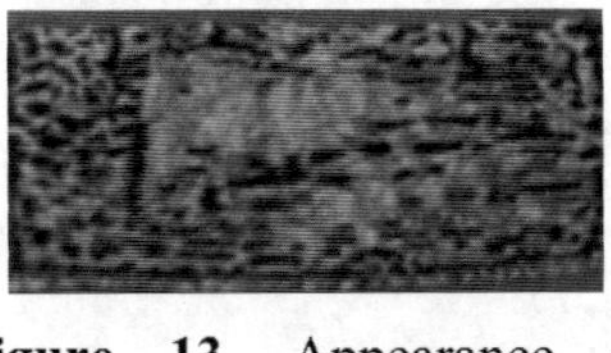

Figure 13. Appearance of sintered surface PMS-1 (x2), sintering modes P=22 W, t=114 °C, S=0.2 mm, V=200 mm/min.

Figure 14. Appearance of sintered surface PMS-1 (x2), sintering modes P=22 W, t=114 °C, S=0.2 mm, V=3000 mm/min.

While velocity increase from 100 to 300 mm/min, when P=10 W, t=26 °C, S=0.1 mm, coarseness of surface decreases from 560 to 425 μm, thickness of sintered layer from 0.88 to 0.65 mm (Figures 15-16).

Significant influence of laser beam on the quality of surface layer should be taken into account when setting displacement velocity of laser beam. Velocity increase results in decrease of layer thickness and coarseness; in some cases it is large for formation of sintered surface. Velocity decrease in some cases results in combustion of powder material and appearing of defects due to overheating higher than boiling point.

Figure 15. Appearance of sintered surface DSK-F75 (x2), sintering modes P=10 W, t=26 °C, S=0.1 mm, V=100 mm/min.

Figure 16. Appearance of sintered surface DSK-F75 (x2), sintering modes P=10 W, t=26 °C, S=0.1 mm, V=300 mm/min.

Figures 17-22 present comparison of appearance of sintered surface PA-4 when changing the scanning step. By increase of the scanning step from 0.1 to 0.2 mm, while sintering modes P=10 W, t=26 °C, V=1000 mm/min (Fig. 9) it results in increased thickness of sintered layer from 0.74 to 0.765 mm.

Figure 17. Appearance of sintered surface PA-4 (x2), sintering modes P=10 W, t=26 °C, V=1000 mm/min, S=0.1 mm.

Figure 18. Appearance of sintered surface PA-4 (x2), sintering modes P=10 W, t=26 °C, V=1000 mm/min, S=0.2 mm.

Figures 19-20 shows sintered surface with some strength (scanning step S=0.1 mm and P=15 W, t=200 °C, V=3000 mm/min). Changing scanning step from 0.1 to 0.3 mm decreases thickness of sintered layer from 0.7 to 0.66 mm, increasing Rz from 225 to 425 μm.

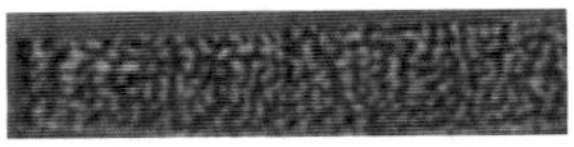

Figure 19. Appearance of sintered surface PMS-1 (x2), sintering modes P=30 W, t=200 °C, V=3000 mm/min, S=0.1 mm.

Figure 20. Appearance of sintered surface PMS-1 (x2), sintering modes P=30 W, t=200 °C, V=3000 mm/min S=0.15 mm.

While increasing scanning step from 0.1 to 0.15 mm, sintering modes P=10 W, t=26 °C, V=300 mm/min, Figures 21-22 shows decrease of surface coarseness from 425 to 300 μm, thickness of sintered layer decreases from 0.65 to 0.4 mm, diameter of coagulated particles decreases from 175 to 150 μm.

Scanning step influences quality of surface layer not that significantly. Its influence on powder materials differ. During sintering of powder PA-4 increase of scanning step results in increase of thickness of sintered layer. During sintering of cobalt-chromium-molybdenum composition increase of scanning step results in decrease of thickness of sintered layer, Rz and diameter of coagulated particles. During sintering of copper powder PMS-1 thickness of sintered layer decreases, coarseness increases.

Figure 21. Appearance of sintered surface DSK-F75 (x2), sintering modes P=10 W, t=26 °C, V=300 mm/min, S=0.1 mm.

Figure 22. Appearance of sintered surface DSK-F75 (x2), sintering modes P=10 W, t=26 °C, V=300 mm/min S=0.15 mm.

Figures 23-26 demonstrate comparison of appearance of sintered surface DSK-F75 according to preheat temperature of powder material. Preheat temperature of powder material being increased from 26 to 200 °C, while sintering modes P=10 W, S=0.15 mm, V=300 mm/min, coarseness of surface decreases from 300 to 280 μm, thickness of sintered layer increases from 0.4 to 0.65 mm, diameter of coagulated particles decreases from 150 to 115 μm.

Figure 23. Appearance of sintered surface DSK-F75 (x2), sintering modes P=10 W, S=0.15 mm, V=300 mm/min, t=26 °C.

Figure 24. Appearance of sintered surface DSK-F75 (x2), sintering modes P=10 W, S=0.15 mm, V=300 mm/min, t=200 °C.

Comparison of appearance of sintered surface PMS-1 according to preheating temperature of powder material is given in Figures 25, 26. Increased preheating temperature of powder material from 26 to 200 °C leads to decrease of defects size, while the following sintering modes P=30 W, S=0.3 mm, V=200 mm/min. Thickness of sintered layer increases from 1.4 to 1.67 mm.

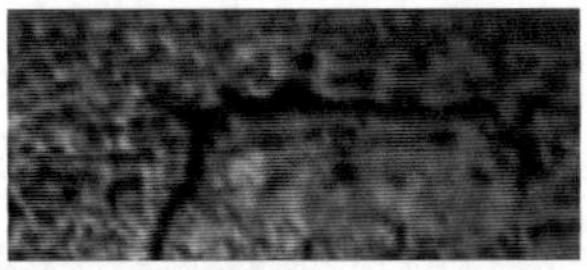

Figure 25. Appearance of sintered surface PMS-1 (x2), sintering modes P=30 W, S=0.3 mm, V=200 mm/min, t=26 °C.

Figure 26. Appearance of sintered surface PMS-1 (x2), sintering modes P=30 W, S=0.3 mm, V=200 mm/min, t=200 °C.

Conclusion

Studies that have been carried out enable to conclude that power has significant influence on the quality of sintered surface layer. Inappropriate setting of power can result in either diffusing from contact or deforming, or combustion of powder so that the process goes out of control. Another significant factor is influence of displacement velocity of laser beam on the quality of surface layer. Increase of displacement velocity results in decrease of thickness and coarseness of sintered layer, in some cases it is insufficient for surface formation. Velocity decrease in some cases leads to combustion of powder material and appearing of defects. Scanning step influences the quality of sintered surface layer insignificantly. During sintering of cobalt-chromium-molybdenum composition increase of scanning step results in decrease of sintered layer thickness, Rz and diameter of coagulated particles. Alteration of preheating temperature of powder material leads to insignificant increase of sintered layer thickness, decrease diameter of coagulated particles and coarseness and increase of strength properties. The present study will be useful for setting the sintering modes for new powder compositions.

References

[1] Otto A, Combinations of laser material processing and metal forming Photonica 5 2007 p 2 – 6
[2] Panchenko V Ya , Laser technologies of machining: contemporary problems of fundamental research and applied designs, monograph M Phismatlit 2009
[3] Proskokov A V, Petrushin S I, Process clip formation with a single conditional shear surface, Applied Mechanics and Materials 682 2014 p 304-312
[4] Gubaidulina R H, Petrushin S I, Galeeva A A, Selecting an Economical Variant of the Manufacturing Method of Engineering Product Fabrication under Current Conditions Applied Mechanics and Materials 682 2014 p 613-616.
[5] Arkhipov P V, Yanyushkin A S, Lobanov D V, The effect of diamond tool performance capability on the quality of processed surface Applied Mechanics and Materials 379 2013 p 124-130
[6] Chinakhov D A, Dependence of Silicon and Manganese Content in the Weld Metal on the Welding Current and Method of Gas Shielding Applied Mechanics and Materials 756 2015 p 92-96
[7] Saprykin A A, Saprykina N A, Improvement of surface layer formation technology for articles produced by layer-by-layer laser sintering, Applied Mechanics and Materials 379 2013 p 56-59
[8] Lychagin D V, Tarasov S Yu, Chumaevskii A V, Alfyorova E A, Macrosegmentation and strain hardening stages in copper single crystals under compression International Journal of Plasticity. 69 2015 p 36-53
[9] Babakova E V, Gradoboev A V, Saprykin A A, Ibragimov E A, Yakovlev V I, Sobachkin A V, Comparison of Activation Technologies Powder ECP-1 for the Synthesis of Products Using SLS Applied Mechanics and Materials 756 2015 p 220-224
[10] Saprykina N A, Saprykin A A, Shigaev D A, Investigation of the factors affecting the quality of the surface obtained by laser sintering Metal 4 2011 p 78-82
[11] Saprykina N A, Saprykin A A, Matrunchik M S, Formation of Surface Layer of Cobalt Chrome Molybdenum Powder Products with Differentiation of Laser Sintering Modes Applied Mechanics and Materials 682 2014 p 294-298

IOP Conf. Series: Materials Science and Engineering **91** (2015) 012032 doi:10.1088/1757-899X/91/1/012032

Reduction of longitudinal ultrasonic oscillations in $NaClO_3$ crystals

E G Soboleva[1,a], A L Igisheva[2,a] and T B Krit[3,b]

[1,2]Yurga Institute of Technology Tomsk Polytechnic University,
Leningradskaya st., 26, 652055, Yurga, Russia
[3]Moscow State University,
Leninskie Gory 1/2, 119991, Moscow, Russia

[a]e-mail: sobolevaeno@mail.ru, [b] timofey@acs366.phys.msu.ru

Abstract. The authors study reduction of longitudinal ultrasonic oscillations (internal friction Q^{-1} and logarithmic decrement δ) in $NaClO_3$ crystals within the temperature range of 78 – 525 K and relative amplitudes of oscillatory deformation of 10^{-7} – 10^{-5} according to the method of compound piezoelectric vibrator at the frequencies of ≈ 10 ^{5} Hz. Q^{-1} maximums under the temperatures of 110 K, 240 K, 260 K, ≈ 400 K and its fast growth under the temperature over 500 K are revealed as well as athermic peak of δ in the region of small deformation amplitudes (near 10^{-7}).

1. Introduction

Elastic waves in solid bodies are characterized by reduction and propagation velocity. Measurements of these parameters allow studying the impact produced upon the elastic waves by any solid properties associated with the lattice of the solid. At present time over a dozen of various types of interactions, influencing elastic waves propagation, have been estimated [1, 2]. If in the process of dynamic testing of an elastic body, as, for example, with ultrasound application, the changes of mechanical stresses and deformations are not co-phased, then, in the inelastic body some amount of the input energy is always spent not on the deformation itself but on other various processes in the lattice and finally turns into (internal friction causing reduction of elastic vibrations). As the measure of internal friction we take the loss tangent tgφ (φ – angle of deformation lagging stress in phase); value Q^{-1}, reciprocal to mechanical Q-factor of the system executing forced oscillations; logarithmic decrement of decaying oscillations δ and decay ratio α_B of the elastic wave with the length λ, propagating in the solid body. The value of internal friction depends upon the frequency and the amplitude of applied stress and the innumerated measures are related to each other with the ratio [3]:

$$\mathrm{tg}\varphi = Q^{-1} = \delta/\pi = \alpha_B\lambda/\pi. \quad (1)$$

As a rule, various mechanisms causing decay of vibrations are divided into relaxation, resonance and hysteresis [4]. Under the periodic deformation of the solid body all mechanisms reveal themselves to some extent but their contribution to the process of vibration decay is different. When the frequency of oscillations changes from infrasound to hypersound we obtain a number of peaks (maximums) on the curve of internal friction of the real crystal. This kind of Q^{-1} curve with maximums, by analogy with the theory of light, is sometimes called the "mechanical" absorption spectrum. It is impossible to create a method which would allow measuring the internal friction continuously in the interval of frequencies from 10^{-4} to 10^{11} Hz. That is why the experiments were carried out with the application of a number of methods where the frequencies overlapped each other. The experimental difficulties of Q^{-1} measuring in a wide interval of frequencies are often compensated by an indirect technique, namely – by obtaining the dependence of the internal friction upon the temperature. Curve Q^{-1} (T) is often called "thermal spectrum" for short [4].

At the present moment it is estimated that in ion monocrystals for the frequency range ~ 10^5 Hz the basic contribution to the internal friction is made by dislocation mechanisms (for example, decay determined by the thermoelastic effect, for metals makes about 50% of general decay, and for semiconductors and non-conductors its contribution to general decay does not exceed several per cent [1]). Internal friction can be determined by two different phenomena: inelasticity and plastic deformation of the system. Inelasticity is characterized by amplitude-independent parameters of a part of deformation energy input into the system. Plastic deformation, as distinguished from inelasticity, is a function of amplitude of deformation of the studied macroscopic system.

Systematic study of internal friction in the kilohertz range of frequencies in the crystals of A_1B_7 type (alkali halide crystals with ionic bond) was completed by S.P. Nikanorov and B.K. Kardashev [5]. In the course of theoretical analysis of experimental data the given authors studied the role of Peierls barrier and the role of fields of force of point defects of different origin under thermally activated oscillating motion of dislocations near the equilibrium state. According to the results of measurements of the amplitude-dependent absorption of ultrasonic oscillations the authors unambiguously proved that in the stress field of the ultrasonic waves the dislocations dissipate the energy of oscillations at the same obstacles which play an important role in formation of the stress level of the flow. It is of interest to obtain the data on internal friction in crystals with significant covalent component of the interatomic bond. The crystal of sodium chlorate ($NaClO_3$) is one of such objects.

2. Theory

In the non-conducting crystals for one of extreme cases of interaction in the region $\omega\tau \ll 1$ (ω – circular frequency, τ – time of thermal relaxation of phonons) absorption of ultrasound is determined mainly by phonon-phonon interaction according to Akhiyeser mechanism [6]. This mechanism works as follows. When the acoustic wavelength is considerably larger than the free length of thermal phonons ($\omega\tau \ll 1$), the sound wave causes macroscopic deformation which leads to changing frequency ω_I (k) of the given oscillation mode. In its turn, change of frequency leads to the change of the function of phonon energy distribution. Sound wave dissipation is related to the fact that elastic modulus which the system of thermal phonons contributes to, relaxes, with the absorption ratio described by the formula:

$$\alpha = \frac{\Delta c \tau \omega^2}{2\rho \upsilon^3}, \tag{2}$$

where Δc – difference of non-relaxed and relaxed elastic moduli, ρ – density, υ – sound velocity.

It is possible to think, as Mason did in his time [6], that phonon contribution into the non-relaxed modulus is determined by changing thermal phonons energy due to changing the boundary Debye frequency in the field of ultrasonic wave deformation. Then, in an isolated situation of longitudinal wave we obtain the following expression for the non-relaxed modulus

$$c_{11}^{нр} = c_{11} + 3\sum_i \varepsilon_i (\gamma_i^{11})^2, \tag{3}$$

where c_{11} – elastic modulus in harmonic approximation, ε_i – thermal energy of i-th phonon branch. Grüneisen ratio γ_i^{jk} for the i-th phonon branch in Debye approximation is calculated according to Bragger's formula [6]

$$-\gamma_i^{jk} = p_j p_k + \frac{\ell_p \ell_q}{c}(c_{jkpq} + p_r p_s c_{jkprqs}), \tag{4}$$

where p_j – direction cosine of displacement vector, ℓ_p – direction cosine of wavevector (indices p, q, r, s are related to phonons, j, k – to the sound wave), c_{jkpq} – rigidity second-order constants, c_{jkprqs} – rigidity third-order constants, c –elastic modulus actual for phonons.

In the process of equilibration between various phonon branches the elastic modulus relaxes to its equilibrium value which equals

$$c_{11}^{p} = c_{11} + \overline{\gamma}^2 c_V T, \tag{5}$$

where $\overline{\gamma}$ - Grüneisen constant, T – temperature, c_V – heat capacity of a volume unit. Absorption constant according to (2) under the condition of equality $\Delta c = c_{11}^{\text{нр}} - c_{11}^{p}$ is transformed to

$$\alpha = \frac{\left[3\sum_i \varepsilon_i \left(\gamma_i^{11}\right)^2 - \overline{\gamma}^2 c_V T\right]\omega^2\tau}{2\rho\upsilon^3}. \tag{6}$$

In work [7] Mason formula (6) was analyzed in respect to the calculation of phonon absorption of the ultrasound in the crystals of alkali metals halogenides. To coordinate the calculated and experimental values of α in this case the necessity arises to consider the defect of Δc modulus in quasi-harmonic approximation supposing that the amount of phonons N_i of every type of oscillations remains constant until relaxation and internal energy change is determined only by the changes of ion oscillation frequencies. The difference between the non-relaxed and relaxed elastic moduli appeared to equal

$$\Delta c = T\sum_i c_i \gamma_i^{jk}\gamma_i^{\ell m} - \frac{T}{c_V}\sum_i c_i \gamma_i^{jk}\sum_i c_i \gamma_i^{\ell m}. \tag{7}$$

3. Calculation ratios

Study of inelastic properties of $NaClO_3$ monocrystal was completed in two variants. Amplitude-independent internal friction within the temperature interval 77 – 525 K at the frequencies of ≈ 100 kHz was studied by the “half-width” of the resonance curve

$$Q^{-1} = \frac{\Delta f}{f_p}, \tag{8}$$

where Δf – interval of frequencies between the resonance curve points, corresponding oscillation amplitudes equaling to 0.707 of the maximum amplitude, and f_p – resonance frequency. Description of the unit and details of the experimental method for this case are provided in monograph [8].

Amplitude-dependent internal friction was studied under the same frequencies and discrete temperatures of 300, 400 and 500 K with application of an electric acustic unit with load resistance of R = 1,2 kilohm, similar to that described in [5]. Logarithmic decrement δ was taken as a measure of internal friction, and it was found as

$$\delta = \left(\frac{2bd_{12}}{s_{22}}\right)^2 \frac{R_э}{f_p m_b}, \quad R_э = R\left(\frac{U}{U_R} - 1\right), \tag{9}$$

and amplitude of relative oscillative deformation of the sample was calculated according to the formula

$$\varepsilon = \frac{s_{22}}{2bd_{12}\ell_0 f_p} \cdot \frac{U_R}{R}. \tag{10}$$

Here b, d_{12}, s_{22} and f_p – accordingly thickness, piezoelectric modulus, compliancy constant and resonance frequency of piezoelectric quartz, m_b – mass of vibrator (piezoelectric quartz + sample), R_e – equivalent resistance of vibrator at the moment of resonance, U and U_R accordingly – decrease in current at the section "piezoelectric quartz – resistance R" and on the resistance at the moment of resonance, ℓ_0 – length of the sample.

$NaClO_3$ monocrystals were grown from aqueous liquids of a CP preparation by slow evaporation under room temperatures. The best samples had perfect cubic form and their size reached 25 × 15 × 10 mm^3. Samples in the form of cores were cut out according to the wet thread technique and their length along the crystallographic direction <100> corresponded to the condition of resonance excitation of the double compound vibrator at the basic frequency of 100 kHz for the longitudinal standing ultrasonic wave. No new dislocations were introduced into the samples before the measurements started. Maximum amplitude of the relative oscillative deformation in the crest of the standing wave ε changed within two orders (from 10^{-7} to 10^{-5}). Internal friction measurements were completed with an average error of about 10%.

4. Results and their discussion

Thermal dependence of internal friction in $NaClO_3$ samples practically in the whole region of solid phase existence is presented in Figure 1. Here we also present the changes of Young modulus in direction <100> of sodium chlorate monocrystal. The fact that at once catches the researcher's eye that the mechanical spectrum of $NaClO_3$ sample contains plenty of internal friction maximums (five completely resolved ones under the temperatures 110, 225, 260, 400 and 450 K) and an ascending branch of Q^{-1} maximum under T > 500 K. Function $E_{<100>}$, first of all, is considerably non-linear and, second, it has a number of anomalies. The obtained mechanical spectrum is even more surprising as under similar conditions in terms of temperature and frequency Q^{-1} (T) and $E_{<100>}$ (T) in NaCl crystal have a common linear form – with the growth of the sample temperature internal friction and elastic modulus change monotonously (Q^{-1} increases, $E_{<100>}$ decreases).

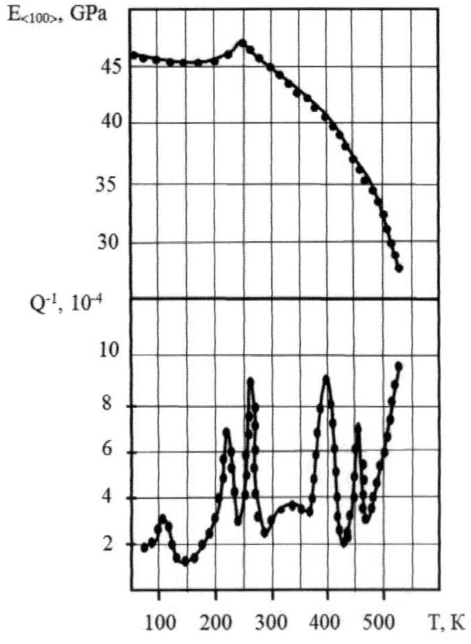

Figure 1. Thermal changes of internal friction and Young modulus of $NaClO_3$ monocrystal in the cube plane.

The observed peculiarities in the thermal behavior of internal friction and elastic modulus in sodium chlorate crystal are obviously related to the order-disorder processes in its lattice. In simplified form the interpretation of results can be presented as follows. When the crystal of $NaClO_3$ is heated, its anion sublattice, consisting of pyramidic groups of ClO_3^- with atom of chlorine in the vertex and atoms of oxygen in the base, is destabilized first. Near the temperature of 110 K in the local areas of anion sublattice atoms of chlorine start transitions between two equivalent sites and, by convention, this can be considered the trigger mechanism to the so called isostructural (isomorphous) phase transition in the crystal of sodium chlorate under low temperatures. Maximum Q^{-1} is followed by changing the sign of elastic modulus thermal dependence – $E_{<100>}$ increases in the interval of 170-260 K. A certain culmination of the described process is achieved under the temperatures of 225 – 260 K when the intensity of chlorine atoms transitions becomes enough to impart the quasi-molecular group ClO_3 additional degree of freedom – hindered rotation (libration). Separate anions randomly ("in a Brownian way") change their orientation for two directions. An indirect proof to this is the fact that under the temperature of 260 K $NaClO_3$ crystal becomes an auxetic (its Poisson's ratios $\sigma_{<100>}$ and $\sigma_{<110,001>}$ become negative). As the temperature of the crystal grows more and more anions become involved into the process of reorientation which finally leads (T ≈ 400 K) to cation sublattice destabilization. Activation energy of cation vacancy is 0.46 electron-volt and energy of Frenkel pair formation 1.98 electron-volt [8]. Q^{-1} peak under T = 400 K in $NaClO_3$, this way, is associated with elastic relaxation in the lattice of this crystal due to sodium ions "jumping" between the sites. When the temperature approaches 500 K ($T_{melting}$= 536 K) the crystal becomes strongly ionized (chlorate ions separate from sodium ions) and, as Curie point in $NaClO_3$ is taken over the melting temperature (T_c = 593 K), then, it is supposed that spontaneous polarization can be caused only by dipoles appearing due to free rotation of ClO_3 group.

Our studies of internal friction in $NaClO_3$ monocrystal under the temperature of T>300 K correlate to anomalous behavior of many physical properties of the given material in this temperature region (Fig. 2) [8].

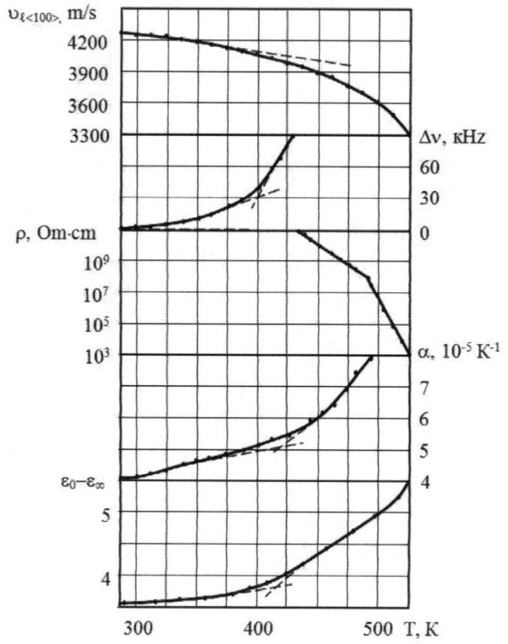

Figure 2. High-temperature changes of longitudinal waves velocity $\upsilon_{\ell<100>}$, frequency displacement of nuclear quadruple resonance ^{35}Cl $\Delta\nu$, specific electrical resistance ρ, thermal expansion coefficient α and difference of low- (ε_0) and high-frequency (ε_∞) transmittivities of $NaClO_3$ monocrystal.

An unexpected and surprising result was obtained in the experiments on amplitude dependence of internal friction in the studied object (Fig. 3). We mean peak δ under remarkably small amplitudes of oscillative deformation (near 10^{-7}). The plastic properties of $NaClO_3$ crystal in the ultrasound field of such low intensity are not quite understandable. Results presented in Fig. 3 clearly demonstrate that the view of δ (ε) curves significantly depends upon the temperature of the samples. The most marked peak of internal friction corresponds to the temperature of sodium chlorate sample of 400 K under which many non-linear properties start (Fig. 2). Let us notice that on dynamic curves of deformation for some materials which have absolutely different chemical nature than our object of study, for example, metals [2, 5, 9], we observe similar "sharp plasticity point". As an example in Fig. 4 we present a curve of steel deformation in static and dynamic modes taken from the monograph of Bulgarian authors [9]. That's what they say about the considered phenomenon on page 99 of their book: "... The dynamic curve σ_x (ε_x) often has a peculiar "tooth" explained by some structural characteristics of material..." In the piezoelectric (sodium chlorate) and in the ferromagnetic (steel) identical "structural characteristics" caused by such insignificant mechanical effects of elastic pulse advancing, indicate a single mechanism of quasi-microplasticity effect characteristic for various solids being significantly different in nature and physical and mechanical properties [10, 11].

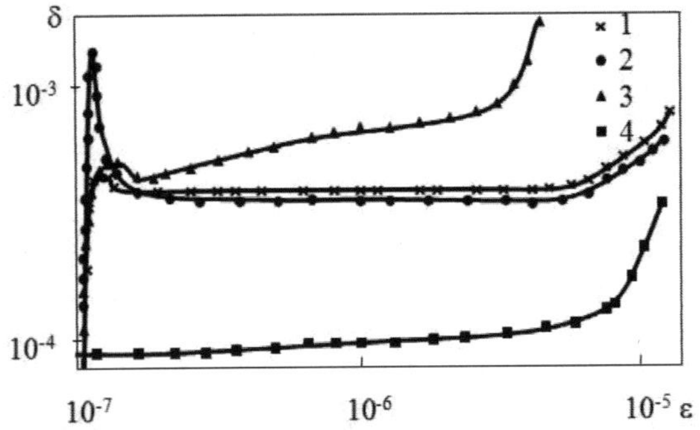

Figure 3. Amplitude dependence of internal friction of $NaClO_3$ (1, 2, 3) and NaCl (4) crystals. 1, 4 – 300 K, 2 – 400 K, 3 – 500 K.

The developed microplasticity of $NaClO_3$ under the temperature of 400 K is associated with the beginning of the given crystal transition into the superionic state due to cation sublattice "melting". Elastic wave modulates the depths of potential wells of quasi-free sodium ions in the conducting sublattice leading to the redistribution of Na^+ due to thermally activated motion. This phenomenon is the phenomenon of inelastic relaxation [12].

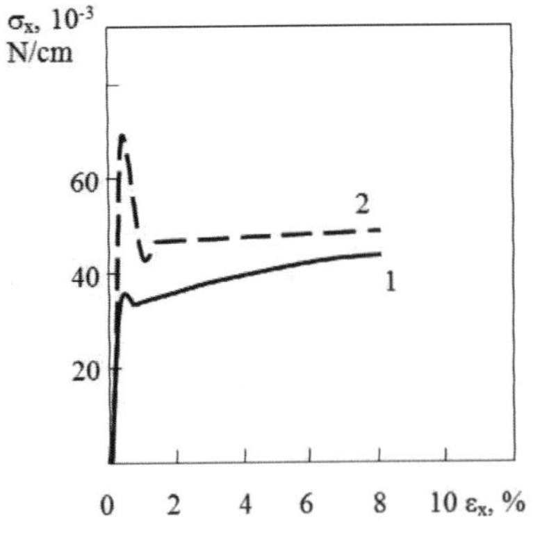

Figure. 4. 1 – static curve for steel; 2 – dynamic curve for steel [8].

The test experiment with NaCl crystal (curve 4 in Fig. 3) demonstrates the typical view of function δ (ε) with amplitude dependence of internal friction under $\varepsilon \geq 10^{-5}$ determined by motion of dislocations. In $NaClO_3$ crystal under T = 500 K amplitude independent component of internal friction is practically absent (curve 3 Fig. 3).

5. Findings

1. It is established that mechanical spectrum of $NaClO_3$ sample contains maximum of internal friction (five completely solved ones under the temperatures of 110, 225, 260, 400 and 450 K) and an ascending branch of Q^{-1} maximum under T > 500 K.
2. A marked peak of internal friction is revealed under T = 400 K when many non-linear properties of sodium chlorate start.
3. The study revealed that the test experiment with NaCl crystal demonstrates the typical view of δ (ε) function with amplitude dependence of internal friction under $\varepsilon \geq 10^{-5}$ determined by motion of dislocations. In $NaClO_3$ crystal under T = 500 K amplitude independent component of internal friction is practically absent.

References

[1] Truell R, Elbaum C, Chick B 1972 *Ultrasonic methods in solid physics* Moskow 307p
[2] Klimenov V, Borozna V, Zaitsev K 2009 *METAL 2009 - 18th International Conference on Metallurgy and Materials* 273-278
[3] Kuchin V A, Ulyanov V L 1986 *Elastic and inelastic properties of crystals* Moskow 136p
[4] Postnikov V S 1974 *Internal friction in metals* Moscow 352p
[5] Nikanorov S P, Kardashev B K 1985 *Elasticity and dislocation inelasticity of crystals* Moskow 250p
[6] Nozdrev V F, Fedorishchenko N V 1974 *Molecular acoustics* Moskow 288p
[7] Kovalenok R V, Merkulov L G 1972 *Physics of the Solid State* **14** 340-344
[8] Belomestnykh V N, Soboleva E G 2009 *Acoustic, elastic and inelastic properties of sodium halogenate crystals* Tomsk, Publishing TPU 276p
[9] Kolarov D, Boltov A, Boncheva N 1979 *Mechanics of plastic medium* Moskow, Publishing World 302p
[10] Alfyorova E A, Lychagin D V, Chernyakov A A 2014 *Applied Mechanics and Materials* **682** 485-490
[11] Alfyorova E A, Lychagin D V 2013 *Applied Mechanics and Materials* **379** 66-70
[12] Aliyev A E, Saidov A A, Khabibullayev P K, Shinder I I 1995 *Acoustic Journal* **42** 322 – 333

Estimation of Temperature Conductivity Coefficient Impact upon Fatigue Damage of Material

V Bibik[1, a], **A Galeeva**[2, b]
[1]Associate professor, [2]student
Yurga Institute of Technology, TPU affiliate
Russia, 652055, Yurga, Leningradskaya street 26

e-mail: [a]bibik@tpu.ru, bnastena.soldatova.93@mail.ru

Abstract. In the paper we consider the peculiarities of adhesive wear of cutting tools. Simulation of heat flows in the cutting zone showed that, as thermal conduction and heat conductivity of tool material grow, the heat flows from the front and back surfaces to tool holder will increase and so, the temperature of the contact areas of the tool will lower. When estimating the adhesive wear rate of cemented-carbide tool under the cutting rates corresponding to the cutting temperature of up to 900 °C, it is necessary to take the fatigue character of adhesive wear into consideration. The process of accumulation and development of fatigue damage is associated with micro- and macroplastic flowing of material, which is determined by the processes of initiation, motion, generation, and elimination of line defects – dislocations. Density of dislocations grows with increase of the loading cycles amount and increase of load amplitude. Growth of dislocations density leads to loosening of material, formation of micro- and macrocracks. The heat capacity of material grows as the loosening continues. In the given paper the authors prove theoretically that temperature conductivity coefficient which is associated with heat capacity of material, decreases as fatigue wear grows.

1. Introduction

One of the main factors affecting durability of cutting tools is temperature in the contact area [1, 2]. The thermal and physical characteristics of tool and machined material will influence the temperature in the cutting area [3-6]. Simulation of heat flows in the cutting area showed that, as thermal conduction and heat conductivity of tool material grow, the heat flows from the front and back surfaces to tool holder will increase and so, the temperature of the contact areas of the tool will lower [3]. The diagram of heat flows in the cutting tool is presented in Figure 1.

Analysis of cutting tool use under factory conditions showed that cemented-carbide tool is quite often used under the cutting rates which correspond to the temperatures of up to 900°C in the cutting area, i.e. under the cutting rates which are lower than the rate corresponding to the inflexion point of the curve showing the dependence of wear upon cutting rate [7]. At this segment of the curve the durability of the tool increases as the cutting temperature grows Figure 2. It is commonly accepted [8, 9] that this temperature region corresponds to adhesive wear of the tool. According to the current opinion [8, 10-12], adhesive wear is based upon the phenomenon of adhesive bond – developing strong temporary bonds between the contacting surfaces resulting from their mutual deformation under high temperature. The ability of materials for adhesive wear grows under recrystallization temperature. Plastic deformation is localized in the areas of actual contact of chip or machined surface with the tool where the protecting oxide and adsorbed films are destroyed, thus, the conditions for physical contact and formation of chemical bonds are created. As a result, we observe adhesive bonding of machined and tool material [13, 14]. It should be mentioned that formation and destruction of seizure bridges occurs with a certain periodicity. Destruction takes place in the surface or subsurface layer of less strong (machined) material. At the same time the crystal lattice in the surface layers of the tool is also damaged, as repeated seizure and breaking of adhesive bonds cause cyclic stress of the surface layer of tool material which is characterized by increased fragility (in comparison to the machined material) [13, 14]. Cyclic loading affecting the contact edges of the tool leads to formation of fatigue cracks resulting in separation of wear particles [7].

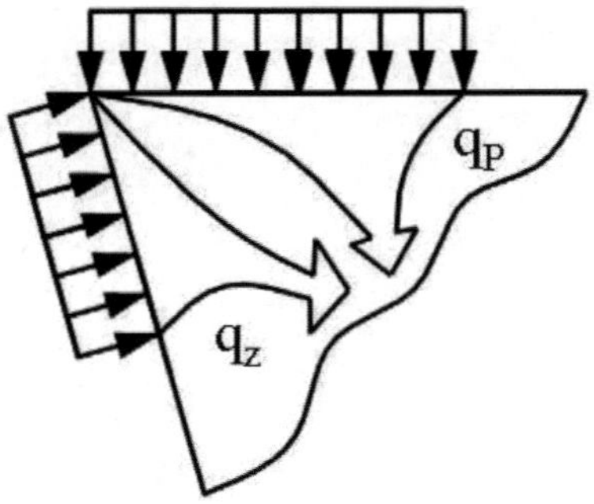

Figure 1. Cutting tool heat flows

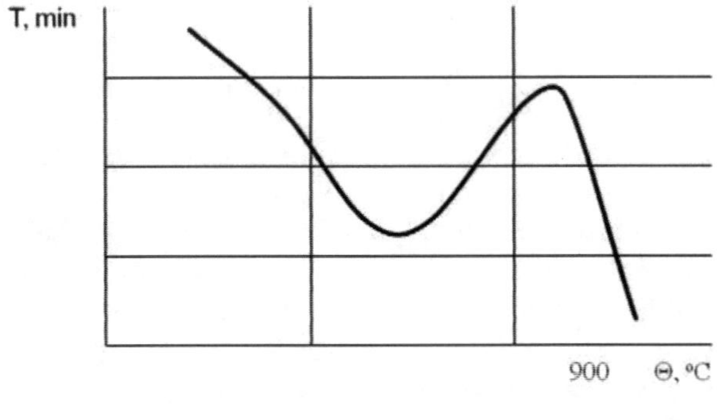

Figure 2. The curve of the temperature-tool life

The fatigue nature of adhesive wear was experimentally revealed in many works, such as [15, 16]. That is why estimation of wear rate of cemented-carbide tool under the temperatures below 900°C requires taking the fatigue nature of adhesive wear into account.

Results and Discussion

According to many researchers, the process of accumulation and development of fatigue damage is associated with micro- and macroplastic flowing of material, which is determined by the processes of initiation, motion, generation and elimination of line defects – dislocations [17-20]. Density of dislocations grows with increase of the loading cycles amount and increase of load amplitude. Growth of dislocations density leads to loosening of material, formation of micro- and macrocracks.
Temperature conductivity coefficient [21]

IOP Conf. Series: Materials Science and Engineering **91** (2015) 012033 doi:10.1088/1757-899X/91/1/012033

$$a = \frac{\lambda}{cp} \quad (1)$$

where c - specific thermal capacity;

ρ - density,

characterizes the rate of temperature distribution in space.

Basically, the temperature conductivity coefficient allows judging of the processes of accumulation and development of fatigue damage but reference literature does not provide any articulate information on this problem. Nevertheless, there is a number of data indicative of such dependence.

For metals, temperature conductivity is determined by the total of electron and phonon contribution (flat waves of deformation) [22, 23]

$$\lambda = \lambda_e + \lambda_p \quad (2)$$

The phonon component λ_p of heat-transfer resistance determined by dissipation (disappearing of one (or two) phonon and appearing of two (or one) phonons) at the dislocations, is directly associated with viscous drag of dislocations by phonons. We observed great changes of λ_p according to density of dislocations of LiF crystal [23] see Figure 3. Also in work [24] the authors describe decrease of thermal conduction for lead with addition of bismuth in comparison to pure metal.

In work [25] the authors note decrease of thermal conduction coefficient of KCl with addition of KNO_2 in comparison to pure metal as well as decrease of thermal conduction under deformation of KCl and CaF_2.

The electron component of thermal conduction λ_e is directly proportional to electrical conductivity [23, 26], i.e. they speak about correlation between these parameters.

Reference literature provides information on relation between electrical conductivity, level of peak voltage and number of load cycles for some metals (C1010, C1020, Armco iron) [19]. As an example in Figure 4 we provide one of such diagrams for C1010 under the number of load cycles $N=1{,}56\cdot10^6$ at every level of peak voltage σ.

Decrease of thermal conduction with growth of load cycles is associated with gradual accumulation of discontinuity in the metal, the tendency for higher rate of electrical conductivity decrease are more noticeable when submicroscopic cracks develop to become microcracks. Immediately before destruction of the sample electrical conductivity falls most abruptly [19].

Thus, the information in reference literature on relation between electrical conductivity coefficient and thermal conduction coefficient λ, correlating with it, and generation and accumulation of dislocations, allows making a conclusion that, as fatigue grows thermal conduction of material in the center of damage decreases.

As for temperature conductivity coefficient we did not find any direct information conserning its relation with fatigue in reference literature.

In the process of fatigue damage accumulation the crystal lattice also accumulates defects and loosening of material, resulting from it, occurs [18, 19]. Thermal capacity of material (c) increases in the course of material loosening [21]. That is why it is possible to suppose that temperature conductivity coefficient which is associated with thermal capacity of material (1), to a greater extent (than λ) will decrease as fatigue damage grows.

One of important parameters of dislocation structure is density of dislocations ρ'. Density of dislocations is total length of dislocation lines in the unit of volume of materials and it is determined by the following correlation [23]

$$\rho' = \frac{1}{A\Lambda}, \quad (3)$$

where Λ - mean free path of phonons;

A - scattering cross-section of phonons per unit of dislocation length.

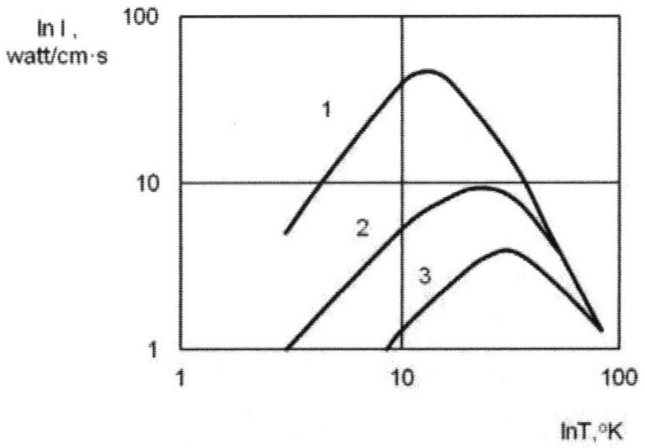

Figure 3. Dependence of temperature conductivity of LiF crystal upon absolute temperature under various degree of deformation:1 – after keeping during 2 hours under 830° C (ρ' =3·104 cm-2), 2 – after deformation of 2,4% under room temperature , (ρ' =1,8·107 cm-2),3 – after deformation of 4% under 180° C (ρ' =4,6·107 cm-2)

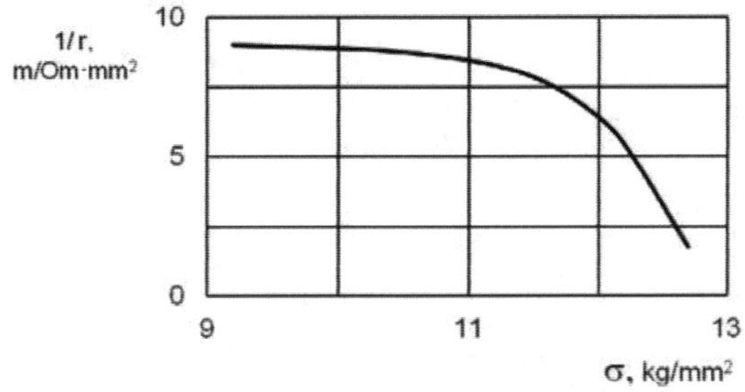

Figure 4. Influence of cyclic stress amplitude upon electric conductivity change

Scattering cross-section of phonons per unit of dislocation length can be determined as [23]

$$A = \frac{\gamma^2 b^2}{S} 2\pi\nu, \tag{4}$$

where S - sound velocity;
b - Burgers vector;
γ - Gruneisen constant;
ν - oscillation frequency of dislocations.

Gruneisen constant is determined by the expression [27]

$$\gamma = \frac{\alpha E a_0^3}{C_a}, \tag{5}$$

where a_0 - atomic size;
E - elasticity modulus;
C_a - atomic heat capacity;
α - coefficient of linear expansion.

free path of phonons can be found from equation from Debye equation [28]

$$\lambda = \frac{1}{3} cS\Lambda\rho, \tag{6}$$

where λ - thermal conduction;
c - specific heat capacity;
ρ - density.

Thus

$$\Lambda = \frac{3\lambda}{cS\rho}. \tag{7}$$

Thermal conduction is expressed through temperature conductivity (a) [21]

$$\lambda = c\rho a. \tag{8}$$

After combining (8) and (7) we obtain

$$\Lambda = \frac{3a}{S}. \tag{9}$$

From equation (9) it follows that the free path of phonons is directly proportional to temperature conductivity of material which is proved by the works of A.A. Kusov [29].

Thus, using (9) and (4) in (3) we determine dislocation density as function of temperature conductivity

$$\rho' = f(a) \tag{10}$$

or

$$\rho' = \frac{S^2}{6\pi a\gamma^2 b^2 v}, \tag{11}$$

Taking

$$\frac{S^2}{6\pi\gamma^2 b^2 v} = m, \tag{12}$$

we obtain the formula to calculate the density of dislocations

$$\rho' = \frac{m}{a} \tag{13}$$

Conclusion

When estimating the rate of adhesive wear of cemented-carbide cutting tool under the cutting speeds corresponding to the cutting temperature of up to 900 C, it is necessary to take the fatigue nature of adhesive wear into account. Accumulation of fatigue damage causes change of thermal and physical characteristics of material. Analysis of formula (13) allows saying that, under accumulation of fatigue damage, i.e. under growth of dislocation density, temperature conductivity coefficient decreases, thus, proving the suggested hypothesis that the temperature conductivity coefficient which is associated with thermal capacity of material, decreases due to the growth of fatigue damage. So, temperature conductivity coefficient can be considered a structurally-sensitive characteristics of material and used as information parameter for forecasting the durability of cemented-carbide cutting tools.

References

[1] Makarov A. D. Optimizacija processov rezanija. - M.: Mashinostroenie, 1976.
[2] Poletika M. F. Teorija obrabotki metallov rezaniem. Ch.1 - Tomsk: Izd.TPI, 1974.
[3] Reznikov A. N. Teplofizika rezanija. - M.: Mashinostroenie, 1969.
[4] Uteshev M. H. Razrabotka nauchnyh osnov rascheta prochnosti rezhushhej chasti instrumenta po kontaktnym naprjazhenijam s cel'ju povyshenija ego rabotosposobnosti. Diss. dokt. tehn.

nauk. - Tjumen', 1995.
[5] Bibik V., Petrova E. Methods of Forecasting Wear Resistance of Cutting Tools Based on the Properties of Their Materials. Applied Mechanics and Materials Vol. 682 (2014) 491-494. Trans Tech Publication, Switzerland.
[6] Ovcharenko V.E., Mokhovikov A.A., Ignatev A.S. Influence of surface nanostructure on the life of cermet in metal cutting // Steel in translation. 2013. №6. p. 348-350.
[7] Poletika M. F. Teorija obrabotki metallov rezaniem. Ch.2 - Tomsk: Izd.TPI, 1975.
[8] Loladze T. N. Iznos rezhushhego instrumenta. - M.: Izd-vo nauchno-tehnicheskoj literatury, 1958.
[9] Razvitie nauki o rezanii metallov / Pod red V. F. Bobrova - M.: Mashinostroenie, 1967.
[10] Avakov A. A. Fizicheskie osnovy teorii stojkosti rezhushhih instrumentov. - M.: Mashgiz, 1960.
[11]. Loladze T. N. Prochnost' i iznosostojkost' rezhushhego instrumenta. - M.: Mashinostroenie, 1982.
[12] Pushnyh V. A. Issledovanie adgezionnogo iznosa tverdosplavnyh instrumentov pri obrabotke titanovyh splavov: Diss ... kand. tehn. nauk. - M., 1979.
[13] Bel'skij S. E., Tofpenec R. L. Strukturnye faktory jekspluatacionnoj stojkosti rezhushhego instrumenta. - Minsk: Nauka i tehnika, 1984.
[14] Zhilin V. A. Subatomnyj mehanizm iznosa rezhushhego instrumenta. - Rostov: Izd. Rostovskogo universiteta, 1973.
[15] Egorov S. V. Kontaktnaja prochnost' rabochih poverhnostej instrumenta kak faktor, opredeljajushhij ego rezhushhie svojstva // Prochnost' rezhushhego instrumenta. - Har'kov, (1967) 57-75.
[16] Kuklin L. G. Issledovanie ustalosti tverdogo splava // Izv. vuzov. Mashinostroenie. 7 (1963) 66-71.
[17] Akulov N. S. Dislokacii i plastichnost'. - Minsk: Izd-vo AN BSSR, 1963.
[18] Ivanova V. S., Terent'ev V. F. Priroda ustalosti metallov. - M.: Metallurgija, 1975.
[19]. Ivanova V. S. Ustalostnoe razrushenie metallov. - M.: Izd-vo po chernoj i cvetnoj metallurgii, 1963.
[20] Mehanika razrushenija i prochnost' materialov: Spravochnik. - Kiev: Izd. Naukova dumka, 1990. - T.4.
[21] Lykov A. V. Teorija teploprovodnosti. - M.: Vysshaja shkola, 1967.
[22] Frenkel' Ja. I. Vvedenie v teoriju metallov. - L.: Nauka, 1972.
[23] Fridel' Zh. Dislokacii. Per s angl. - M.: Mir, 1967.
[24] Scattering of Phonons and Electrons by Imperfections in a Metal //The Philosophical magazine. J. of Theoret. Exper. and Appl. Phys. 8 (1956) - vol.1. Eighth Series 718-721.
[25] Oskotskij V. S., Smirnov I. A. Defekty v kristallah i teploprovodnost'. - L.: Nauka, 1972.
[26] Jakubov F. Ja. Jenergeticheskie sootnoshenija processa mehanicheskoj obrabotki materialov. - Tashkent: Izd. Fan, 1985.
[27]. Zhurkov S. N. K voprosu o fizicheskoj osnove prochnosti // Fizika tverdogo tela. 22 (1980) 3344-3349.
[28] Teploprovodnost' tverdyh tel: Spravochnik / A. S. Ohotin, R. P. Borovikov i dr. - M.: Jenergoatomizdat, 1984.
[29] Kusov A. A. Fononnaja model' razrushenija nagruzhennoj atomnoj cepochki // Fizika tverdogo tela. 21 (1979) 3095-3099.

IOP Conf. Series: Materials Science and Engineering **91** (2015) 012034 doi:10.1088/1757-899X/91/1/012034

Simulation of thermal fields using different types of wide burnishing

N M Bobrovskij [1,a], P A Melnikov [2,b], S N Grigoriev [3,c], I N Bobrovskij [4,d]

[1,2,4] 14, Belorusskaya st. ,Togliatti, Russia
[3] 1,Vadkovskij per., Moscow, Russia

e-mail: [a]Bobrnm@yandex.ru, [b]Topavel@mail.ru, [c]Rector@stankin.ru, [d]Bobri@yandex.ru

Abstract: The article features simulation of heat build-up in the tool while treatment using different types wide burnishing. The understanding of conditions of heat build-up and heat distribution and the quantitative estimation on the finishing stages of part's treatment is always challenging, because there is a possibility of negative and nonreciprocal effect onto the development of performance properties of treated parts. This issue gains particular importance during treatment, including burnishing using no means of lubrication.

Introduction

The understanding of conditions of heat build-up and heat distribution and the quantitative estimation on the finishing stages of part's treatment is always critical, for minimization of negative and nonreciprocal effect of the heat onto the development of performance properties of treated parts, which is relevant for machining process, including wide burnishing. The reference literature does not provide analytical behavior for wide burnishing process, enabling temperature calculation in any area of machining of both the tool and the blank. The use of analytical methods enables us to get the generalized solution, relevant for simple shaped bodies (a plate, a cylinder, a ball) keeping some simplification in mind [1]. 2 and 3 dimensional tasks, however appears a difficulty integrating differential equation of thermal conductivity on unambiguous conditions, which corresponds to thermal processes in technological systems [2, 3]. This article features results of simulation. Two fundamental entries, affecting the overall image while wide burnishing: temperature of the tool and temperature by using different types of burnishing.

The heat distribution within the tool during wide burnishing, by self-positioning tool

The existing models for estimation of heat build-up while wide burnishing with feed (or "burnishing") are not usable for wide burnishing owing to following features of wide burnishing: various type of integrity within these technologies – in tens of times more efforts by wide burnishing; absence of linear feed (only transverse direction); increased machining speed (machining cycle within 1-3 rotation of part); shapes of tools [4, 5, 6]. The authors have received details of dependence in order to estimate the contact surface temperature in the process of wide burnishing without use of cooling taking into account the thermo-physical characteristics of materials, speed of process, number of loading cycles and technical scheme of machining used, the use of which enables decrease of temperature.

As initial expression for heat distribution simulation onto the surface of the tool for wide burnishing let us use the known expression of isolated continuous source of q_{max}[7, 8, 9] intensity, which is found within infinite body:

$$\Theta(x,y,z,\tau)=\frac{q_{max}}{4\pi\lambda R}\cdot\left(1-erf\left[\frac{R}{\sqrt{4\omega\tau}}\right]\right) \quad (1)$$

where $R=\sqrt{(x_u-x)^2+(y_u-y)^2+(z_{vu}-z)^2}$.

Considering the accepted shape of body (half-space), and taking into account the principle of source reflection, while transition to final expression let us multiply it by 2.

Since the source is decided to be found on the surface of the part, let us accept $y_u = 0$.

We should carry out integration of initial mathematical expression into necessary coordinates. We should bear in mind the shape and dimensions of thermal source, the heat-intensity distribution behavior on the contact surface, schematization of the part's shape etc.

By integration on x_u and z_u axils on the contact surface, then we get information for case of equal distribution of heat-source density:

$$\Theta(x,y,z,\tau)=\frac{q_{max}}{2\pi\lambda}\cdot T_2(x,y,z,\tau) \quad (2)$$

$$T_2(x,y,z,\tau)=\iint\frac{\left(1-erf\left[\sqrt{\frac{(x_u-x)^2+(z_u-z)^2+y}{4\omega\tau}}\right]\right)}{\sqrt{(x_u-x)+(z_u-z)+y^2}}\cdot dx_u\cdot dz_u \quad (3)$$

For normal one-direction distribution of heat-source, behavior (2), (3) is reconstituted in the following way:

$$T_2'(x,y,z,\tau)=\iint\exp\left[-k\left(\frac{2(x_u-0{,}5l)}{l}\right)^2\right]\times$$
$$\times\frac{\left(1-erf\left[\sqrt{\frac{(x_u-x)^2+(z_u-z)^2+y^2}{4\omega\tau}}\right]\right)}{\sqrt{(x-x_u)^2+(z-z_u)^2+y^2}}dx_u\cdot dz_u\cdot \quad (4)$$

The temperature while using different technological types of wide burnishing

Let use study the process of wide burnishing and its technological aspects from the point of view number and disposition of sources and heat sinks in the technological scheme.

When using the process of wide burnishing it's possible to use different technological schemes of process (Figure 1):

1. Scheme A represents traditional process scheme of burnishing using single tool, effected by force P.
2. Scheme B contains 3 and more tools, asymmetrically proportionally displaced on the on the part's surface.
3. Scheme C contains 2, 4 and more tools, symmetrically and proportionally displaced.
4. Scheme D contains 2 and more tools, proportionally symmetrically/asymmetrically displaced.

The use of each scheme alternates in definite way the thermal pattern in the surface layers of treated material, consequently its physically-mechanical features [10].

With regard to above mentioned, we should consider, the by wide burnishing treatment same area of the surface might be effected by several loading cycles within several revolutions of the part or by use of several tool schemes [11].

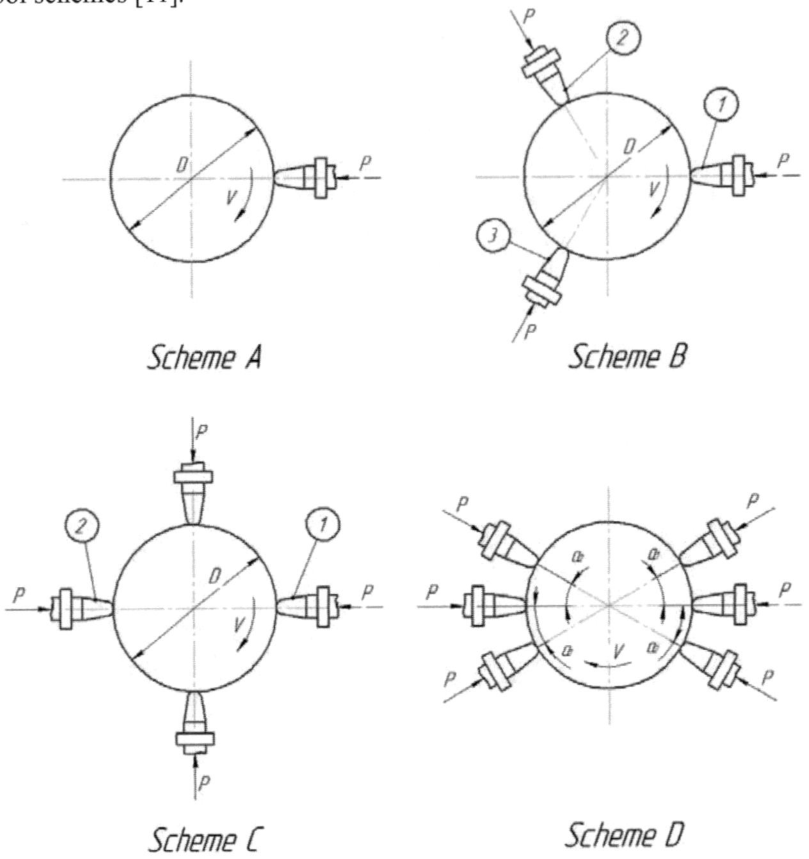

Figure 1. Basic technological schemes of wide burnishing process

Let us reconstitute expression (2) and (3), with regard to mentioned factors. Regard to the number and displacement of the tools shall be effected through estimation of distance x and composition of temperatures of each tool. Taking into account, that estimation of temperature is carried out in coordinates relative to the tool (thermal source).

For Scheme B:

$$\Theta_{\Sigma}(x,y,z)=\Theta(x,y,z)+\Theta\left(x+\frac{\pi D}{3},y,z\right)+\Theta\left(x+\frac{2\pi D}{3},y,z\right), \tag{5}$$

Where $\Theta_{\Sigma}(x,y,z)$ - temperature of single thermal source, calculated using one of the formulas (2), (3).

For Scheme C:

$$\Theta_{\Sigma}(x,y,z)=\Theta(x,y,z)+\Theta\left(x+\frac{\pi D}{2c},y,z\right), \tag{6}$$

With the use of 4 tools, the dependence behavior attributes the following shape:

$$\Theta_{\Sigma}(x,y,z)=\Theta(x,y,z)+\Theta\left(x+\frac{\pi D}{4},y,z\right)+\Theta\left(x+\frac{\pi D}{2},y,z\right)+\Theta\left(x+\frac{3\pi D}{4},y,z\right), \quad (7)$$

For Scheme D:

$$\Theta_{\Sigma}(x,y,z)=\Theta(x,y,z)+...+\Theta_i(x+l_{i,},y,z)+...+\Theta_n(x+l_n,y,z) \quad (8)$$

Where l_n – is distance of part's circle periphery to the specific tool.

The consideration of several loading cycles is done through accumulation of heat from each loading cycle:

$$\Theta_{\Sigma N}(x,y,z)=\Theta_{\Sigma 1}(x,y,z)+\Theta_{\Sigma 2}(x+\pi D,y,z)+...+\Theta_{\Sigma R}(x+\pi D(n-1),y,z) \quad (9)$$

Results and Discussion

Below is the image of results of estimation on above formula (Figure 2-4).

The thermal distribution by width of indentor shows, that the level of temperature is proportional by width of treated surface, which leads to equability of physical-mechanical features of the final surface layer. It's also visible that character of distribution barely depends of speed of burnishing, which enables us to use wider range of speed of burnishing to choose the needed temperature (Figure 2).

The results of estimation depending on burnishing speed of treated material are shown. The analysis of estimation results shows, that the temperature on the surface layer of the part by use of technological scheme of single tool burnishing increasing not significantly (Figure 3).

The use of multi-tool scheme, does not significantly affect the level of temperature of part's surface, however enables us to reduce lead-time of wide burnishing process, increasing the performance. The main effect onto the features of the surface layers would be done caused by maximum temperature level in contact area directly next to indentor, where the basic physical-mechanical processes take place during burnishing (Figure 4).

It has been established that, despite of the differences of technological conditions of burnishing by cylindrical indentor and wide self-positioning tool the maximum temperature of contact area in both cases is on the same level, averagely, does not exceed limit of 300 °C.

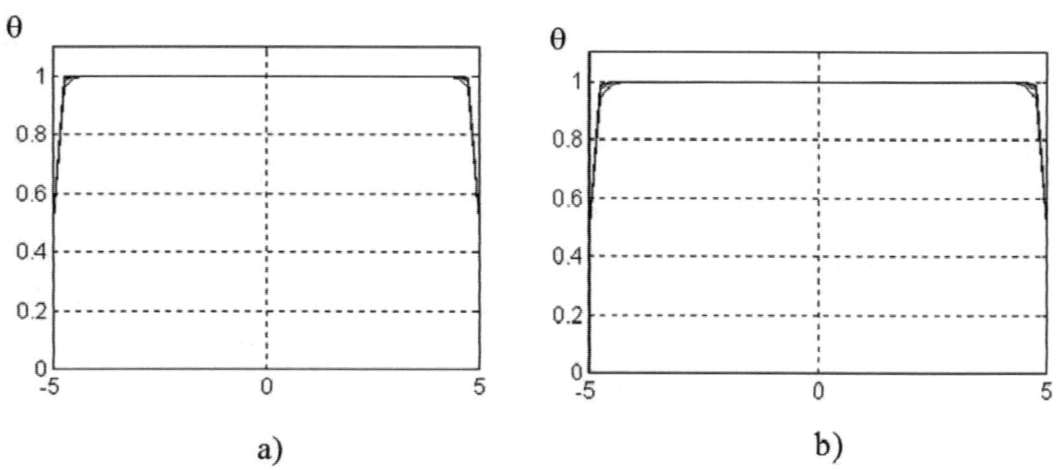

Figure 2. Temperature distribution by width of burnihing: a) steel 40; b) cast iron; θ =Θ(z)/Θmax – relative temperature in dimensionless form

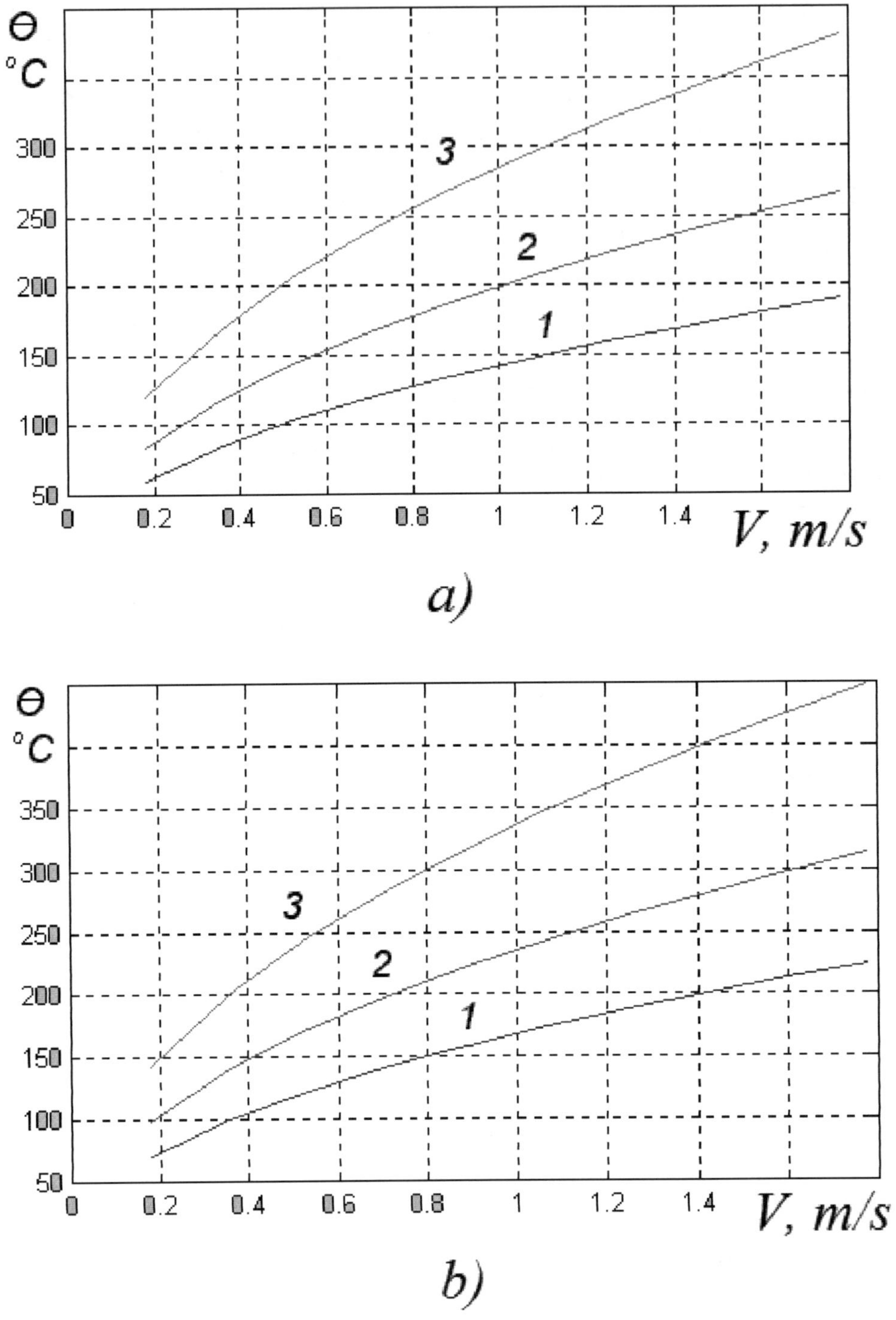

Figure 3. Dependance of average temperature on contact area by wide burnishing on speed
a) - steel 40; б) – cast iron; 1 - P=40 N/mm; 2 - P=56 N/mm; 3 – 80 N/mm

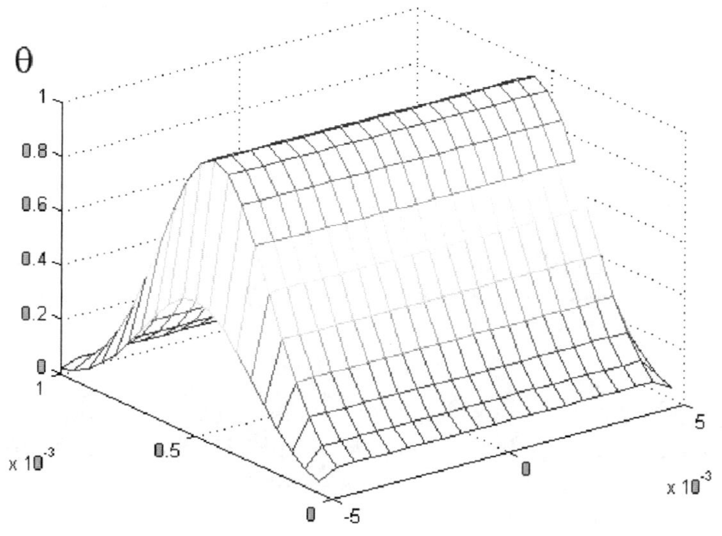

a)

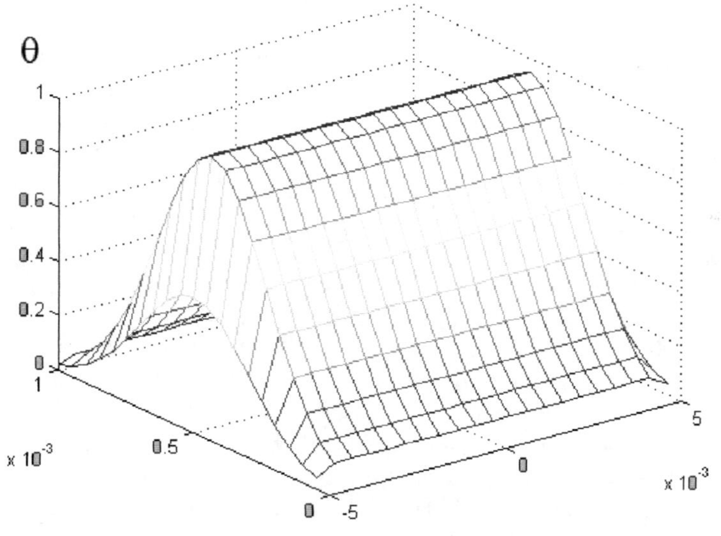

b)

Figure 4.Heat distribution on contact surface: a) steel 40; b) cast iron; θ =Θ(z)/Θmax – relative temperature in dimensionless form

Conclusion

1. We have established dependencies necessary to define contact temperature in the process of wide burnishing without cooling with regard to thermo-physical characteristics, speed of process, quantity of loading cycles and technological scheme used, the use of which enables us to minimize the heating.

2. The temperature significantly depends on speed and force of burnishing: together with increase of speed and pressure onto the surface of the part the average temperature of contact area by

wide burnishing may increase up to 8 – 9 times. But even in the most intense loading conditions the temperature of contact surface does not exceed 400 °C by wide burnishing of steel 40 and 450 °C – by treatment of high-tensile cast iron 75-50-03.

3. It has been established that, despite of the difference of technological conditions of wide burnishing by indentor and wide self-positioning tool the maximum heating temperature of contact area in both cases is on the same level, averagely, not exceeding limit of 300 °C.

4. The heat pattern of wide burnishing enables us to develop technological limitations to treatment process from the point of view of two factors. First point is, in order to prevent possible negative structurally-phase transformations in the surface layer of burnished part with the following lowering of its quality and operational reliability. And the second point is, in order to reduce tool-ware rate, especially using no lubrication.

References

[1] Grigor'ev S N Myachenkov V I Kuzin V V Automated thermal-strength calculations of ceramic cutting plates *Russian Engineering Research* 31 (2011) 1060-1066 Information on http://www scopus com/inward/record url?eid=2-s2 0-83055191231&partnerID=40&md5=ac9e4119ebc830ffab25f55adde9a968

[2] Kuzin V V Grigoriev S N Fedorov M Y Role of the thermal factor in the wear mechanism of ceramic tools Part 2: *Microlevel Journal of Friction and Wear* 36 (2015) 40-44 Information on http://www scopus com/inward/record url?eid=2-s2 0-84924358898&partnerID=40&md5=000a9ebdbd2961408c3bc2991678da69

[3] Grigoriev S N Vereschaka A A Vereschaka A S Sladkov D V Nano-scale multi-layered coatings for cutting tools generated using assisted filtered cathodic-vacuum-ARC deposition (AFCVAD) *Applied Mechanics and Materials* 327 (2013) 1454-1459

[4] Grigoriev S N Fominski V Y Romanov R I Volosova M A Gnedovets A G Experimental study and modeling of laser plasma ion implantation for WSEX/57FE interface modification *Applied Surface Science* 276 (2013) 242-248

[5] Grigoriev S N Metel A S Fedorov S V Modification of the structure and properties of high-speed steel by combined vacuum-plasma treatment *Metal Science and Heat Treatment* 54 (2012) 8-12 Information on http://www scopus com/inward/record url?eid=2-s2 0-84861686621&partnerID=40&md5=d49b3b305157e65e2a5a7dd128c89b1d

[6] Grigor'ev S N Myachenkov V I Kuzin V V Automated thermal-strength calculations of ceramic cutting plates *Russian Engineering Research* 31 (2011) 1060-1066 Information on http://www scopus com/inward/record url?eid=2-s2 0-83055191231&partnerID=40&md5=ac9e4119ebc830ffab25f55adde9a968

[7] Reznikov A N Thermophysics of cutting Mashinostroenie Moscow 1969

[8] Vorontsov A L Sultan-Zade N M Albagachiev A Y Savkin A I Development of a new theory of thermal cutting processes 12 General mathematical apparatus of the thermophysics of cutting (Part 1) *Russian Engineering Research* 30 (2010) 1241-1247 Information on http://dx doi org/10 3103/S1068798X10120142

[9] Vorontsov A L Sultan -Zade N M Albagachiev A Y Savkin A I Development of a new theory of thermal cutting processes: 17 Determining the temperature fields and contact temperatures in cutting (Part 3) *Russian Engineering Research* 31 (2011) 471-478 Information on http://dx doi org/10 3103/S1068798X1105025X

[10] Zakharov O V Stability of forced contact closure in centerless grinding on fixed supports *Russian Engineering Research* Vol 31 Issue 10 (2011) 1007-1009

[11] Vereshchaka A S Grigor'ev S N Kim V A Mokritskii B Y Kirichek A V Design of an effective manufacturing process for tool materials *Russian Engineering Research* 34 (2014) 516-521 Information on http://www scopus com/inward/record url?eid=2-s2 0-84906826462&partnerID=40&md5=e955a86c2277e24a6aa35a783c8356d6

Aspects of thermal field by wide burnishing

N M Bobrovskij [1,a], P A Melnikov [2,b], S N Grigoriev [3,c], I N Bobrovskij [4,d]

[1,2,4] 14, Belorusskaya st. ,Togliatti, Russia
[3] 1,Vadkovskij per., Moscow, Russia

e-mail: [a]Bobrnm@yandex.ru, [b]Topavel@mail.ru, [c]Rector@stankin.ru, [d]Bobri@yandex.ru

Abstract: This research features 3D model of development of thermal fields by wide burnishing. By development of 3D model the source of heat build-up is represented as additive aggregate of connecting by width of spot heat sources, operating on the surface of the tool.

Introduction

An important aspect affecting the quality of the part's surface layer, is the roughness and performance of the tool during operation by using methods of surface-plastic deformation is the temperature of the tool and part, including contact area between them.

A model of thermal pattern of the process of smoothing without lubricants was created, allowing estimating the temperature effects in the contact area between the treated surface and the burnishing tool. Determination of the thermal pattern allows to develop the technological limitations of the process in terms of two factors:

- possible structural transformations in the surface layer after burnishing, which can affect the quality performance of the machined surface;
- wear rate of the tool and the ability to predict tool life at the stage of the process planning that plays an important role in the burnishing technology stability in mass production.

Research of Reznikov A.N. features a block diagram of the total heat transfer during burnishing [1] (Fig. 1), separate components of the solids are represented as separate components (burnishing tool and part) involved in the process system. The author mentioned "*in order to rationally organize the burnishing process, providing the most efficient performance and creation of conditions for normal operation of diamond, it's necessary to define the temperatures, appearing on the contact surface of the later*". In accordance with the block diagram establish theoretical calculation to determine the temperature dependence over a wide burnishing, including contact area. For different types of processing SPD methods heat is released to a certain extent. Therefore, the source of heat in the developed model is three-dimensional. However, by schematization of the problem, an assumption accepted that the sources operating in the area of contact of indentor with the blank are two-dimensional. Now, let us consider the laws of heat flow density distribution in the volume of heat area.

IOP Conf. Series: Materials Science and Engineering **91** (2015) 012035 doi:10.1088/1757-899X/91/1/012035

Thermal field model while burnishing by a wide self-adapting tool

As the results of solving the problem of Hertz are shown on the contact of two cylinders of limited length with parallel axes, the contact patch shape is a rectangular. Therefore, at small depths t, characteristic for wide burnishing let us represent heat sources as two-dimensional rectangular shape b*l, located on the surface of the blank (moving source J1 and J2; and stationary source on the surface of the tool J3 and J4, see Fig. 1).

The density distribution of these sources can initially rendered as equal (Fig. 2a), although linear or normal distribution is more accurate (Fig. 2b, c).

Let us more precisely consider the problem of the distribution of the intensity of the heat on the contact area of the indentor with the part. For the burnishing process the issue described in details in [2]. The author put forward a proposal for a normal distribution. It is confirmed by the fact that the nature of the thermal intensity corresponds to a change of the degree of strain hardening (change of micro hardness), which almost corresponds to the normal distribution [3, 4].

Therefore, to clarify the calculation of the temperature by a wide burnishing procedure we accept normal distribution of temperature on the x-axis in the direction of the tool (Fig. 2 c):

$$q(x) = q_{\max} \cdot \exp\left[-k \cdot \left(\frac{2(x-0{,}5l)}{l} \right)^2 \right], \tag{1}$$

where k - coefficient characterizing the concentration of heat flow in the focus of deformation in the direction of flow; $q_{\max}$ - maximum heat flux in the center of the contact area.

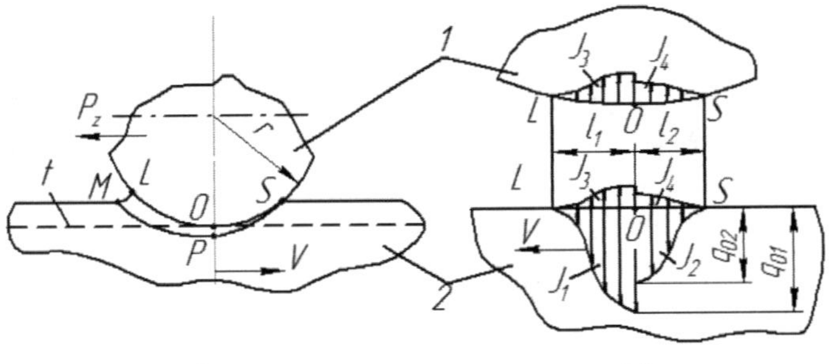

a) b)

Figure 1. The process of plastic deformation of the surface (a) and the heat block diagram (b) [1]

These more complex distributions, as will be shown below, are slightly complicate mathematical description of thermal processes in a wide burnishing. Y.I. Baratz [2] estimates the value of k as:

$$k = \pi \cdot l / \Delta \tag{2}$$

where Δ – the thickness of the plastically deformed layer of machined material.

For the actual burnishing process conditions factor is concentrated within a range of from 5 to 6 - when the processing of high-hardened steel (HRC 62 ... 64); 3 ... 4 - when processing steel hardness of 40 ... 50 HRC and not more than 2.5 ... 3 - carbon steel processing in the state of delivery and non-ferrous metals [2].

To solve the problem we schematize the shape of solids existing in the process system.

Tools (burnishing tool) schematized as half-space, on the surface of which acts rectangular flat heat source dimensions $b * l$.

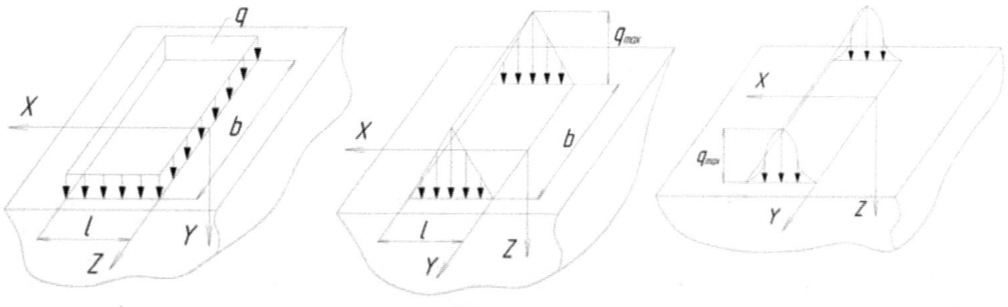

a) b) c)

Figure 2. Density distribution Law of the heat source during a wide burnishing: a) uniform; b) linear; c) normally distributed

A part, due to the small size of the source in the l direction of movement in relation to the size of the items, as the half schematized as a half space, onto surface of which a flat rectangular heat source size $b * l$ is moving at a speed (Fig. 2).

Such a transition at schematization of the shape of the part is considered in [2]. In the present research we investigated the amendments made to the calculation of the temperature at such transitions. The need for such amendments in the calculation of the temperature in a wide burnishing discussed in more detail below.

However, it is necessary to make a comment. Subsequently, the calculation should take into account the overall temperature of the accumulation of heat in the tool during the N cycles of processing time. This condition is taken into account through the calculation of non-stationary thermal pattern in the tool [5, 6].

Considering the boundary conditions on the surfaces of solids, not occupied by heat sources or sinks. As a first approximation we take these surfaces as adiabatic, provided the process of smoothing without cooling cutting fluid. If necessary to take into account the effect of the lubricant should introduce additional heat sinks into considered structural diagram.

Consider the definition of the maximum heat flux q at the contact surface with the heat source power Q_u.

For equal distribution law (Fig. 2, a):

$$q_{\max} = \frac{Q_u}{b \times l}. \tag{3}$$

For a linear distribution law (Fig. 2, b):

$$q_{\max} = \frac{2Q_u}{b \times l}. \tag{4}$$

For a normal distribution of heat source (Fig. 2, c):

$$Q_u = b \cdot q_{\max} \int_0^l \exp\left[-k\left(\frac{2 \cdot (x - 0{,}5 \cdot l)}{l} \right)^2 \right] \cdot dx \tag{5}$$

For convenience of integration displace the origin along the x axis to the point $x = \frac{l}{2}$ and introduce the notation $R = \frac{l}{2}$.

Formula (5) will take on appearance of:

IOP Conf. Series: Materials Science and Engineering **91** (2015) 012035 doi:10.1088/1757-899X/91/1/012035

$$Q_u = b \cdot q_{\max} \int_{-R}^{R} \exp\left[-k\left(\frac{x}{R}\right)^2 \right] \cdot dx \tag{6}$$

Let us integrate and transform the formula (6) with respect to $q_{\max}$.

$$q_{\max} = \frac{Q_u}{b \cdot \frac{\sqrt{\pi} R}{\sqrt{k}} erf\left[\sqrt{k}\right]} \tag{7}$$

Definition of basic dependencies for calculating temperatures

To derive the mathematical laws apply the method of heat sources [1].

The total capacity of the heat source is described by

$$Q = P_Z \cdot \upsilon \tag{8}$$

Where υ- indentor movement speed; P_z - the main component of the force during machining.

Expressing thermal source power through the normal force P_N and friction coefficient f, let us put down (8) in the form of:

$$Q = P_Z \cdot \upsilon = P_N \cdot f \cdot \upsilon \tag{9}$$

As noted in [7], while plastic contact (burnishing) the coefficient of friction includes f_{def} deflection f and adhesive components f_{adh}:

$$f = f_{def} + f_{adh} \tag{10}$$

An adhesive component [7] is negligible for most materials f_{adz} = 0.01 ... 0.05, and depends on the material and surface roughness, not the machining modes.

The deformation component can be determined by the approximate formula:

$$f_{def} = m \cdot \sqrt{\frac{h_b}{R}}, \tag{11}$$

where h_b - the depth of the burnishing tool penetration; R - radius of the operating part; m - the coefficient depending on the parameter of support surfaces (at by burnishing m = 0.2 ... 0.55).

The research features [8] the experimental dependence of the friction coefficient f of micro hardness at burnishing different steels, which shows that f is found in the range of values of f = 0.04 ... 0.08. There are also some general estimates of the magnitude of the coefficient of friction at different burnishing materials f = 0.03 ... 0.12.

In the solution of the obtained problem the expression for the coefficient b^*, which describes the share of heat going into the tool:

$$b^* = \frac{1}{1 + 2 \cdot \sqrt{2} \cdot \frac{\lambda_q}{\lambda_n} \cdot \sqrt{\frac{V \cdot L}{\omega_n}}}, \tag{12}$$

The research features [9] a rather more general dependence of β_f, taking into account the ratio of q and heat, leaving a tool with respect to heat, Q_u obtained by parts surface:

$$\beta_f = \frac{Q_u}{Q_n} = \frac{\lambda_u}{\lambda_n} \cdot \sqrt{\frac{a_n}{a_u}} \tag{13}$$

Thus, the proportion of heat going into the tool and the part can be determined according to the formula, respectively:

IOP Conf. Series: Materials Science and Engineering **91** (2015) 012035 doi:10.1088/1757-899X/91/1/012035

$$Q_u = b^* \cdot Q; \quad Q_n = (1-b^*) \cdot Q \tag{14}$$

or

$$Q_u = \frac{Q \cdot \beta_b}{1+\beta_b}; \quad Q_n = \frac{Q}{1+\beta_f} \tag{15}$$

Calculation of the temperature in the tool using heat sources approach

As an initial expression let us attribute known expression for the isolated continuous source intensity $q_{\max}$, located in unlimited body:

$$\Theta(x,y,z) = \frac{q_{\max}}{4\pi\lambda R} \cdot \left(1 - erf\left[\frac{R}{\sqrt{4\omega}}\right]\right) \tag{16}$$

Where λ – heat-conduction coefficient, J/(m·s·°C); ω – temperature conductivity coefficient, m²/s; $R = \sqrt{(x_u - x)^2 + (y_u - y)^2 + (z_u - z)^2}$.

Considering adopted at schematization body shape (half-space), taking into account the principle of sources reflection, the transition to the final expression multiply it by 2.

Since the source adopted located on the part's surface, we attribute $y_u = 0$.

It is necessary to carry out the integration of the original mathematical expression for the corresponding coordinates. This requires taking into account the shape and size of the heat source, source reflection principle, the nature of the intensity distribution on the heating contact surface schematization body shape, etc [10, 11]. Since the source adopted located on the surface tool, attribute $y_u = 0$.

Integrating over x_u and z_u within the contact area, we obtain for the case of uniform density distribution of the heat source:

$$\Theta(x,y,z) = \frac{q_{\max}}{2\pi\lambda} \cdot T_2(x,y,z) \tag{17}$$

$$T_2(x,y,z) = \iint \frac{\left(1 - erf\left[\sqrt{\frac{(x_u - x)^2 + (z_u - z)^2 + y}{4\omega}}\right]\right)}{\sqrt{(x_u - x) + (z_u - z) + y^2}} \cdot dx_u \cdot dz_u \quad 8)$$

In case of the normal distribution in one direction of the heat source, depending on (17, 18) are transformed as follows:

$$T_2'(x,y,z) = \iint \exp\left[-k\left(\frac{2(x_u - 0{,}5l)}{l}\right)^2\right] \times \\ \times \frac{\left(1 - erf\left[\sqrt{\frac{(x_u - x)^2 + (z_u - z)^2 + y^2}{4\omega}}\right]\right)}{\sqrt{(x - x_u)^2 + (z - z_u)^2 + y^2}} dx_u \cdot dz_u \cdot \tag{19}$$

Results and Discussion

Figures 3-4 present the results of calculations using formulas (17-19). From these graphs we can see that the maximum temperature is reached by the center line of contact between the burnishing tools and the machined surface. As by classic burnishing [12], the magnitude of temperatures that occur during the normal speed burnishing (5 ... 25 m / min), do not exceed the critical temperatures by which would be structural changes in the surface after burnishing [13, 14, 15] that has no negative impact on the surface quality indicators.

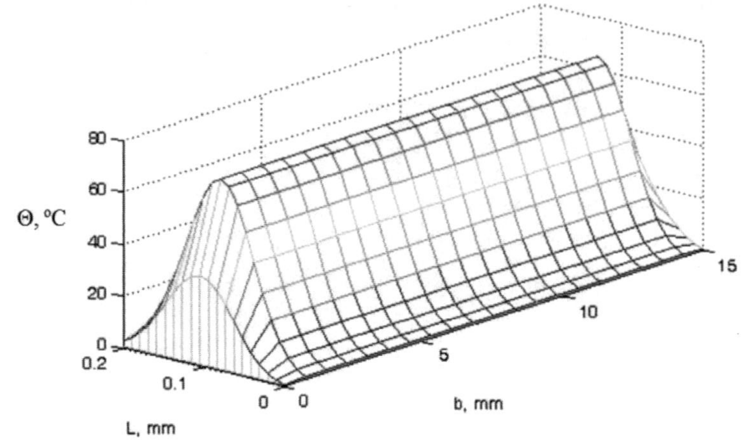

Figure 3. The temperature distribution on the surface by burnishing with a force F = 8000 N, and υ = velocity of 10 m / min, high-duty cast iron 75-50-03 (HRC 45)

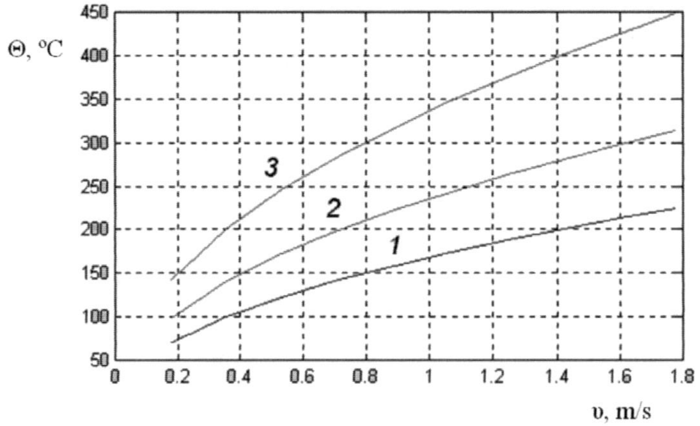

Figure 4. Dependence of the average temperature in the contact area by the wide burnishing high-duty cast iron 75-50-03 on speed: 1- P = 40 N/mm; 2 - F = 56 N/mm; 3 - 80 N/mm

Conclusion

1. Understanding of conditions of heat build-up and thermal distribution and their quantitative evaluation on the final stages of treatment of the parts is always challenging, because there is a possibility of negative and nonreciprocal effect onto the development of performance properties of treated parts. This issue gains particular importance during treatment, including burnishing using no means of lubrication

2. The calculation data, received with the help of this 3D model of heat fields development could be further used by prognostication of wear process of the burnishing tool, which is relevant by treatment using no means of lubrication.

IOP Conf. Series: Materials Science and Engineering **91** (2015) 012035 doi:10.1088/1757-899X/91/1/012035

References

[1] Reznikov A N Reznikov L A Thermal processes in technological systems Mashinostroenie Moscow 1990

[2] Baratz Y I Finishing metals by plastic working (Thermal physics and quality) *Publishing house of Saratov University Press Saratov* 1982

[3] Kuzin V V Grigoriev S N Fedorov M Y Role of the thermal factor in the wear mechanism of ceramic tools Part 2: *Microlevel Journal of Friction and Wear* 36 (2015) 40-44 Information on http://www scopus com/inward/record url?eid=2-s2 0-84924358898&partnerID=40&md5=000a9ebdbd2961408c3bc2991678da69

[4] Grigoriev S N Fominski V Y Romanov R I Volosova M A Gnedovets A G Experimental study and modeling of laser plasma ion implantation for WSEX/*57FE interface modification Applied Surface Science* 276 (2013) 242-248

[5] Vorontsov A L Sultan-Zade N M Albagachiev A Y Savkin A I Development of a new theory of thermal cutting processes 12 General mathematical apparatus of the thermophysics of cutting (Part 1) *Russian Engineering Research* 30 (2010) 1241-1247 Information on http://dx doi org/10 3103/S1068798X10120142

[6] Zakharov O V Stability of forced contact closure in centerless grinding on fixed supports *Russian Engineering Research* Vol 31 Issue 10 (2011) 1007-1009

[7] Reznikov A N Basis of calculation of thermal processes in technological systems KuAI Kuibyshev 1986

[8] Papshev D D Hardening of parts by ball-burnishing Mashinostroenie Moscow 1968 – M : *Machine Building* 1968

[9] Baratz Y I Shaposhnik R K Solids of revolution processing by surface plastic deformation *Finishing machine parts* (1985) 102-107

[10] Grigor'ev S N Myachenkov V I Kuzin V V Automated thermal-strength calculations of ceramic cutting plates *Russian Engineering Research* 31 (2011) 1060-1066 Information on http://www scopus com/inward/record url?eid=2-s2 0-83055191231&partnerID=40&md5=ac9e4119ebc830ffab25f55adde9a968

[11] Grigoriev S N Vereschaka A A Vereschaka A S Sladkov D V Nano-scale multi-layered coatings for cutting tools generated using assisted filtered cathodic-vacuum-ARC deposition (AFCVAD) *Applied Mechanics and Materials* 327 (2013) 1454-1459

[12] Grigoriev S N Metel A S Fedorov S V Modification of the structure and properties of high-speed steel by combined vacuum-plasma treatment *Metal Science and Heat Treatment* 54 (2012) 8-12 Information on http://www scopus com/inward/record url?eid=2-s2 0-84861686621&partnerID=40&md5=d49b3b305157e65e2a5a7dd128c89b1d

[13] Grigor'ev S N Myachenkov V I Kuzin V V Automated thermal-strength calculations of ceramic cutting plates *Russian Engineering Research* 31 (2011) 1060-1066 Information on http://www scopus com/inward/record url?eid=2-s2 0-83055191231&partnerID=40&md5=ac9e4119ebc830ffab25f55adde9a968

[14] Vorontsov A L Sultan-Zade N M Albagachiev A Y Savkin A I Development of a new theory of thermal cutting processes: 17 Determining the temperature fields and contact temperatures in cutting (Part 3) *Russian Engineering Research* 31 (2011) 471-478 Information on http://dx doi org/10 3103/S1068798X1105025X

[15] Vereshchaka A S Grigor'ev S N Kim V A Mokritskii B Y Kirichek A V Design of an effective manufacturing process for tool materials *Russian Engineering Research* 34 (2014) 516-521 Information on http://www scopus com/inward/record url?eid=2-s2 0-84906826462&partnerID=40&md5=e955a86c2277e24a6aa35a783c8356d6

IOP Conf. Series: Materials Science and Engineering **91** (2015) 012036 doi:10.1088/1757-899X/91/1/012036

Influence of Cooling Rate in High-Temperature Area on Hardening of Deposited High-Cutting Chrome-Tungsten Metal

N N Malushin [1], D V Valuev[2], A V Valueva [3], A Serikbol[4], I F Borovikov[5]
[1] 654007 Kemerovo region, Novokuznetsk, Kirova str. 42 Siberian State Industrial University,
[2,3,4] 652050, Kemerovo region, Yurga, Leningradskaya str.26 Yurga Technological Institute branch of Tomsk Polytechnic University
[5] The Moscow State Technical University of N.E. Bauman, Moscow, 5, 2nd Baumanskaya
E-mail: valuev@tpu.ru

Abstract: The authors study the influence of cooling rate in high-temperature area for thermal cycle of high-cutting chrome-tungsten metal weld deposit on the processes of carbide phase merging and austenite grain growth for the purpose of providing high hardness of deposited metal (HRC 64–66).

Introduction

Hot-working steels like P18, P6M5, P9, P2M8, 3X2B8, 3X2B4Ф, X10B14, etc., which meet almost all requirements for deposited metal to harden metallurgical equipment parts, have as good service characteristics so unsatisfactory weldability. Usually to prevent cold cracks formation traditional technology of weld deposit uses preparatory and additional high-temperature heating (T. heat. = 400-700°C) and retarded piece cooling. During this process malleable austenite products of breakdown are formed and they have low hardness and enduring quality that causes difficult heat treating. During the process of bimetal piece heat treating it is not always possible to make use of all high-alloyed metal characteristics and to ensure its maximum hardness.

In developed weld deposit ways for hot-working steels with good hardness in order to prevent cold crack formation it is supposed to use a kinetic malleability effect ("extramalleability"). A special feature of proposed ways for weld deposit is application of preparatory and additional low-temperature heating (T. heat. = 230 –280 °C).

We studied influence of a kinetic malleability effect on a formation character of temporary and retained stress when cooling patterns made of deposited metal of P18-type steel and 12X17-type steel in a hammered state with the help of the microscopy heat installation IMASH–5C–69. We made the patterns, which have the form of a two-sided "shovel" with a throat area in 9 mm^2 and a working section length in 46 mm, out of indicated steels in an annealed state. General view and a cutaway pattern scheme from deposited metal are presented in Figure 1.

Kinetic malleability effect was evaluated according to temporary stresses decreasing when cooling steady patterns in the interval of martensitic transformation. Steadiness was ensured by electromechanical drive of a stressing installation system. Patterns were heated in vacuum $5 \cdot 10^{-5}$ mm Hg by means of heat, which came forth in the process of current flow. The temperature of heating was 800 C, which was lower the point Ac_1, and 1200 °C, which ensured high degree of carbide phase merging in austenite. After heating and holding the patterns were held rigidly for saving their lengths on following cooling. Conditions of heating and cooling imitated a welding thermal cycle. Fixed length caused proper increase of pattern elastic plastic deformation under cooling. Stain-gauge indicator correct to ± 1.5 % persistently transduced pattern stress in the process of cooling. Platinum took temperature: platinum-rhodium couple, welded to the center of pattern. Kinetics of temporary stress formation in fixed patterns depending on temperature was fixated on PDP-4-type two-dimensional potentiometer.

For test material we selected rapid steel P18, harden on martensite, and chromous rust-resisting steel 12X17, referred to ferrite class. Constitution of studied steels is shown in Table 1.

The steel 12X17 does not experience $\gamma \leftrightarrow \alpha$-transformation in contrast to rapid steel, hardened even in the air. The data of Figure 2 shows that when cooling patterns made of the steel P18 from the temperature 1200 °C one can see a kinetic malleability effect on a phase transformation, which is shown up in decrease of temporary stresses in the interval of martensitic temperatures.

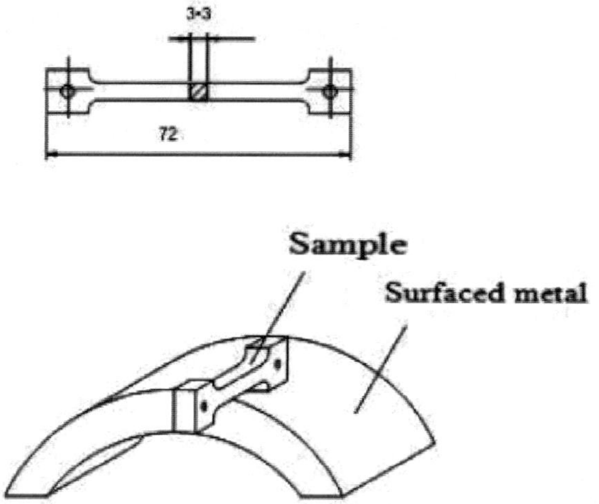

Figure 1. General view of a pattern for high-temperature researches and its cutaway pattern scheme from deposited metal

Stress maximum on the curve of temporary stress formation in steel is about 240 °C, which is quite close to the start point for martensitic transformation of the steel P18 on its hardening from the temperature T_{st} = 1200 °C, and minimum 100 °C corresponds to such temperatures, laying higher than the temperature of martensitic transformation end ($M_K^{cond} > M_K$), when transformation degree is quite high, and with the help of strong "frame" formation out of numerous martensitic crystals flow limit increases, and alloy malleability decreases.

Table 1. Constitution of the steels P18 and 12X17

Steel class	Steel type	Constitution, in %					
		C	Si	Mn	Cr	W	V
M	P18	0.74	0.39	0.27	4.09	18.5	1.25
F	12X17	0.1	0.65	0.7	17.5		

The data of Figure 2 shows that in the interval of martenistic transformation temporary stresses have become a third-double fewer in comparison with stresses that are in austenite area.

During further cooling lower the temperature M_k^{cond} stresses increase till 120-180 MPa by means of certain increase of martensite quantity. A kinetic malleability effect is not shown up when cooling patterns made of the steel 12X17 because of lack of transformation phase.

Along with increasing temperature stresses persistently increase and reach the levels far exceed retained stress in the steel PI8, which experienced martenistic transformation. When we heat patterns till 800 °C, that is lower than A_{C1} temperature because of lack of a phase transformation both in ferrite and hardened metal of PI8-type steel, and when we cool them a kinetic malleability effect is observed, stresses persistently increase. Research results show that in high-cutting chrome-tungsten metal it is possible to observe an extra malleability effect at the moment of martenistic transformation.

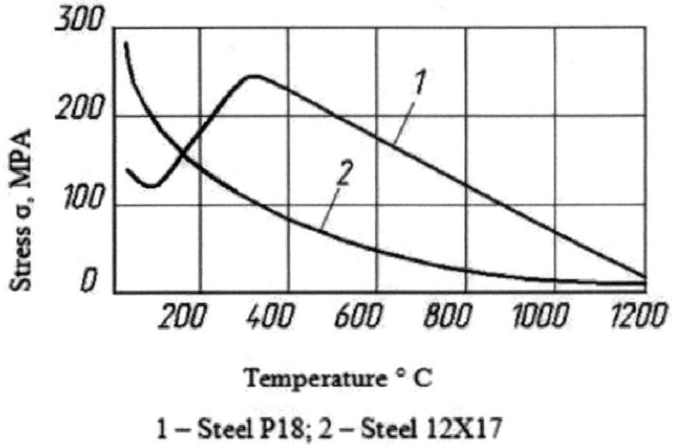

Figure 2. Temporary stress formation in patterns made of the steel PI8 and 12X17

To get deposited steel almost without a tendency to crack formation one should level temporary stresses during a deposit process with the help of their partial relaxation by means of kinetic malleability effect activity at the moment of martenistic or bainitic transformation. Special feature of a proposed thermal deposit cycle is in its three stages. The first one ensures heat time limit and cooling rate increase in the areas with high temperatures, prevents from grain growth and austenite decay with break-even low-strength structure formation. It could be realized by means of high-concentrated heating source application (for example, plasma deposit) and accompanied by cooling. The second stage of a thermal cycle allows deposited metal to be in an austenite state during making all the layers in a deposit process. It is achieved by heating with $T_{heat} = M_{st} + (50 – 100\ °C)$. To get deposited metal almost without a tendency to crack formation temporary stress level should be compensated in a deposited process during the third stage of a thermal cycle by means of temporary T_{heat} decreasing lower than M_{st} temperature. Here temporary stresses decrease with the help of partial relaxation at the moment of martenistic or bainitic transformation. It allows getting deposited metal in a hardened state with a low level of retained stresses [1].

The basis of multilayer plasma deposit method is abidance by a certain thermal cycle [2]. In the context of forming roll plasma deposit with the help of hot-working rapid steels a thermal deposit cycle should have the following characteristics:

a) heating rate should be at its maximum in the interval of the temperatures 750 °C – T_{heat};

b) the duration of stay under temperatures higher the point A_1 should be as small as possible and for rapid P18-type steel it should not exceed 60 sec;

c) cooling rate in the interval $T_{heat} – M_{st} + (50 – 100)$ °C should be higher than critical, that is 3 °C/sec, what allows stopping a decay process of high-alloyed austenite;

d) heating temperature during a deposit process should not be lower than $M_{st} + (50 – 100)$ °C, that is 230–280°C for P18 steel.

Recommended thermal cycle for plasma deposit of forming rolls is done in Figure 3 [2, 3]. This thermal cycle prevents from crack formation and ensures high hardness of deposited metal along with not difficult deposit technology. Higher cooling rate in the areas with high temperatures prevents from grain growth and austenite decay with break-even low-strength structure formation. Higher cooling rate influences almost only those characteristics of hardened and tempered deposited chrome-tungsten metal that are connected with carbide merging increase and increase of austenite grain, what in its turn influences hardness of deposited metal.

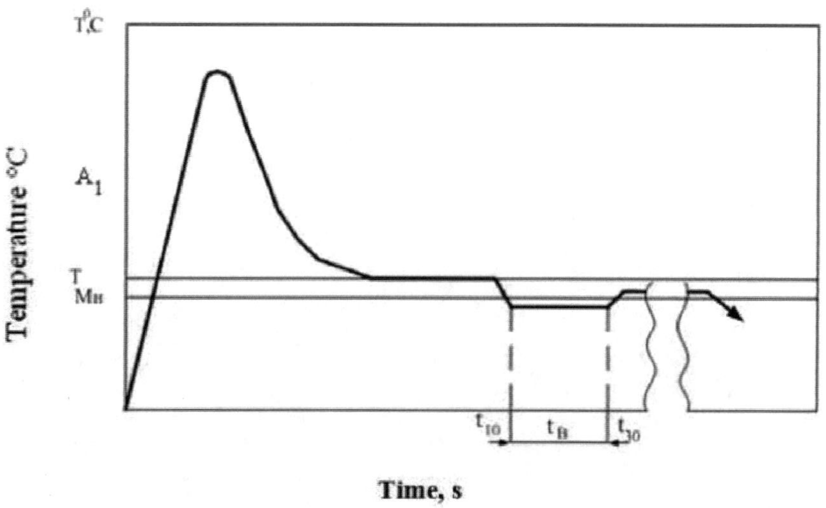

Figure 3. Recommended thermal cycle for plasma deposit of forming rolls

Research objective is to study cooling rate influence in areas with high temperatures of thermal cycle for high-cutting chrome-tungsten metal deposit on carbide phase merging processes and austenite grain growth to ensure high hardness of deposited metal.

In the proposed method of multilayer deposit we recommend to apply higher cooling rate in a high temperature area of a thermal cycle because it is aimed at high-alloyed austenite preservation during layer deposit process. Higher cooling rate application should prevent from carbide precipitation in temperatures of minimum austenite stability and from unstable pearlite formation. Austenite stability in chrome-tungsten steels is quite high. It is especially high in rapid steels, where diffusion transformation in the point higher than M_{st} usually is not observed. Less alloyed die steels (of 3X2B8-type steel) have less austenite stability in 750–550 °C temperature areas. To define allowable cooling rate when quite alloyed austenite is preserved we studied cooling rate influence in a high temperature area (higher than the temperature of 550 °C) on chrome-tungsten metal hardening [1]. We heated patterns made of P18 and 3X2B8 steel in a vacuum vessel of a microscopy heat installation "ALA–TOO" till the temperature of 1200 °C, which allows obtaining quite alloyed austenite. After 10 min holding we cooled the patterns at different rate. To consider cooling rate influence on metal hardening we measured pattern hardness at room temperature.

Obtained data is presented in Figure 4: cooling rate of rapid steels, when quite alloyed austenite is preserved as well as high hardness, is much lower than analogical critical rate for 3X2B8 die steel. HRC 64–66 hardness is preserved by P18 steel at cooling rate more than 6°C/sec, and 3X2B8 steel has critical cooling rate at 60°C/sec. Consequently to harden metal during deposit process it is necessary to cool it at a rate that is higher than critical one. Therefore more alloyed steels with stable austenite are more preferable for multilayer deposit because of their low critical hardening rate.

IOP Conf. Series: Materials Science and Engineering **91** (2015) 012036 doi:10.1088/1757-899X/91/1/012036

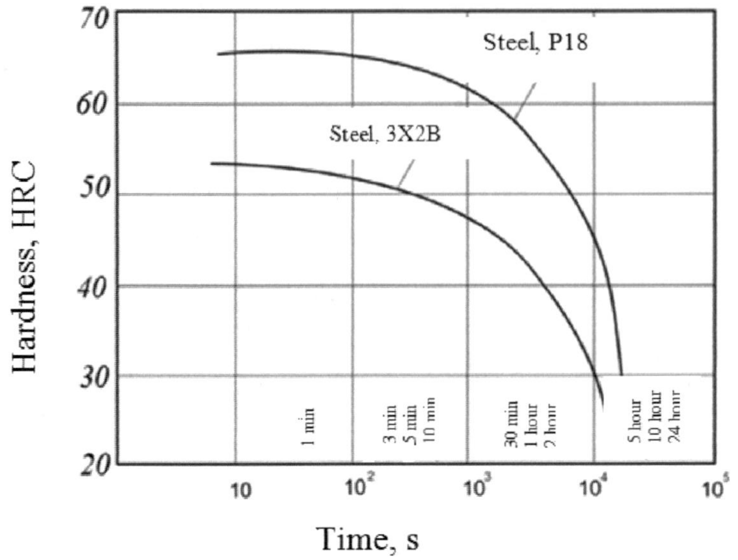

Figure 4. Cooling rate influence in high a temperature area on hardness of hardened steels

Application of P18-type rapid steel alloyed by aluminium and nitrogen as deposited metal favorably affects preservation of smaller austenite grains, preparation quite alloyed mixed crystals, which provide high mechanical properties of deposited metal.

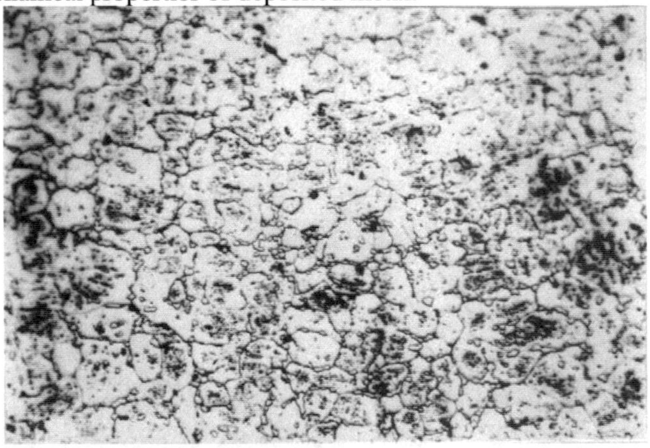

Figure 5. Structure of the deposited metal P18UN × 340

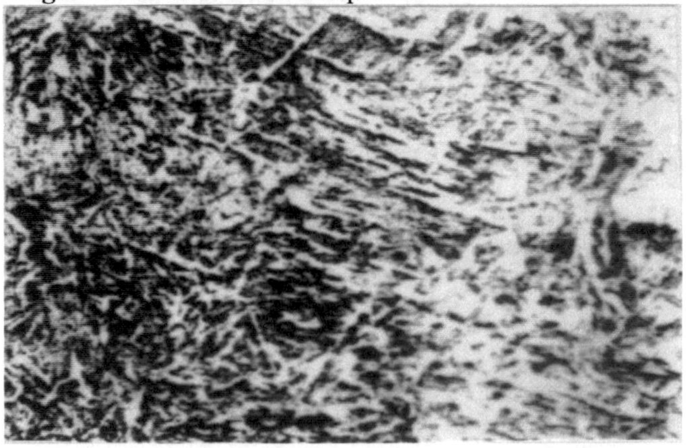

Figure 6. Structure of the deposited metal 3X2B8 × 340

To provide such properties it is unnecessary to hold deposited chrome-tungsten metal in high-temperature area for more than 60 sec and cool it in this area at rate that is lower than critical one. Р18ЮН and 3X2B8 metal structures are shown in Figures 5 and 6.

Our research devoted to influence of a thermal deposit cycle on hardening of chrome-tungsten deposited metal absolutely confirms basic process principles of tool steel hardening, which are given in the works [4-10]. Central tenets of the theory about tool chrome-tungsten steel thermal processing are acceptable for analysis of deposited metal hardening processes.

Therefore the work demonstrates that the first stage of a proposed deposit thermal cycle for chrome-tungsten metal provides heating time limit and higher cooling rate in high temperatures areas, prevents from grain growth and austenite decay with break-even low-strength structure formation. It could be realized by means of high-concentrated heating source application (for example, plasma deposit) and accompanied by cooling.

Conclusion
It is found that increased cooling rate in high temperature areas and holding time under this temperature influence greatly the processes of carbide phase coagulation and merging, austenite alloy degree and its grain size in chrome-tungsten deposited metal of high hardness. To provide maximum hardness of chrome-tungsten deposited metal it is necessary to limit heating time and ensure increased cooling rate in high temperature areas. For these purposes it is proposed to use high-concentrated heating sources (plasma deposit) and accompanied cooling of a deposited detail.

References
[1] N N Malushin D V Valuev Quality ensuring of components for their whole life cycle by plasma surfacing with hot-work steels Tomsk polytechnic University Publishing 2013
[2] Malushin N N T G Vostretsova Regulated thermal cycle for plasma deposit of forming rolls by hot-working steels *J Blank production in machine building industry* 13 (2014) 14–17
[3] N N Malushin V L Osetkovskiy I V Osetkovskiy Deposit of metallurgical equipment details with low temperature heating by hot-working steels *J Blank production in machine building industry* 10 (2014) 6–10
[4] Gueller Yu A Tool steels Metallurgy (1975) 584 p
[6] Chinakhov D A Calculation of gas-dynamic impact of the active shielding gas on the electrode metal drop in gas jet shielded welding J Applied Mechanics and Materials 379 (2013) 188-194
[7] Chinakhov D A Zuev A V Filimonenko A G Gas-dynamic impact of a shielding gas jet on the drop transfer when welding with a consumable electrode J Advanced Materials Research 1040 (2014) 850-853
[8] Chinakhov D A Il'yaschenko D P Gotovschik Y M Investigating the influence of the power supply type upon the properties of the weld joints under manual arc welding J Advanced Materials Research 1040 (2014) 837-844
[9] Chinakhov D A Gas dynamic control of properties of welded joints from high strength alloyed steels *J China Welding* Vol 23 No 3 (2014) 27-31
[10] Babakova E V Gradoboev A V Saprykin A A Ibragimov E A Yakovlev V I Sobachkin A V Comparison of activation technologies powder ECP-1 for the synthesis of products using SLS *J Applied Mechanics and Materials* 756 (2015) 220-224
[11] Valuev D V Serikbol A Stress state and its definition in hardfacing heat-resistant steel cold rolling mill rollers *J Advanced Materials Research* 1040 (2014) 703-707

IOP Conf. Series: Materials Science and Engineering **91** (2015) 012037 doi:10.1088/1757-899X/91/1/012037

Oxidation Processes in Blowing Steel With Inert Gas into the Ladle

R A Gizatulin[1], D V Valuev[2], V A Trifonov[3], A V Valueva[4], A Serikbol[5]
[1,2,3,4,5] 652050, Kemerovo region, Yurga, Leningradskaya str.26 Yurga Technological Institute branch of Tomsk Polytechnic University

E-mail: valuev@tpu.ru

Abstract. This work reports the possible development of oxidative processes in a metal when treating the melt in the ladle under intensive stirring with an inert gas. The industrial data have been received, confirming the possibility of reducing the concentration of silicon and aluminum in the metal, as well as changing the slag chemical composition with the bath blowing with the inert gas through the top submerged lance.

Introduction

When an inert gas is blown into steel, the oxidation process of silicon, manganese and chromium, with adding into a ladle to enhance its chemical composition, is possible to occur. Regardless of the source of oxygen, the element in a melt that is oxidized first has the highest deoxidizing capacity (affinity for oxygen) at a given concentration and temperature. In Table 1, the values of equilibrium oxygen concentrations are compared, with various contents of manganese, silicon and chromium at a temperature of 1580° C in the melt (Formula [1]). It also shows the oxygen content is balanced with aluminum. According to Table 1, with the concentration of aluminum being approximately 0.001%, the oxygen content is so small in the melt that manganese and chromium in structural steels can not react with oxygen coming from different sources. Aluminum binds to oxygen. The specified above content of aluminum is enough to prevent oxidation of silicon, if the silicon concentration does not exceed 0.3 - 0.4%. At higher concentrations of silicon, this aluminum content can not protect it against oxidation. In practice, the vast majority of melts contains more than 0.01% aluminum in the molten metal after tapping. At such aluminum concentrations, silicon is not oxidized despite any concentrations in the metal. The oxygen injected to the melt only oxidizes aluminum.

In cases of the melts blown with argon and their chemical composition not adjusted in the ladle at all, the oxidation process of silicon, manganese and chromium does not occur; contents of these elements in the ladle analysis and grade analysis samples were similar (with the observed deviations within the chemical tolerance limits). On the other hand, when argon is blown, the oxidation of aluminum occurs and its concentration in the steel decreases depending on the blowing time (τ) and its content in the ladle sample (Al_l)

$$\Delta Al = 0{,}2[Al]_l + 0{,}0012\tau - 0{,}04; \qquad R_{MH} = 0{,}6. \tag{1}$$

According to this equation, when the content of aluminum in the ladle sample is 0.03%, its melting loss during the 6-minute blowing makes 0.009%.

Table 1. Equilibrium concentrations of oxygen dissolved in iron at various concentrations of manganese, chromium, silicon and aluminum at 1580°C

Element	Equilibrium concentration of oxygen with elements, $\times 10^3$, % by weight with the concentration of elements in iron, % by weight											
	0.001	0.01	0.02	0.05	0.1	0.2	0.3	0.4	0.6	0.8	1.0	1.6
Mn								97	66	50	40	27
Cr								57	47	42	37	25
Si			30	17	13	10	8	7	5	5	5	5
Al		0.8	0.7	0.4								

We assume the two main sources of oxygen to the metal: from ladle slag and atmosphere. In the case of the single-slag process, the metal and slag are de-oxidized during the tapping process when ferrosilicon, silicon manganese and aluminum are basically added. Table 2 summarizes the chemical compositions of slag after tapping and blowing with argon; it is observed, first, the concentration of ferrous and manganese oxides in the slag after tapping is relatively small - 5% maximum, with an average content of 2.6% and 2.1% accordingly in the investigated material mass, and, secondly, during the tapping process, the de-oxidation of slag is not able to complete. During the following argon blowing, a decrease in concentrations of both components is observed. It is proportional to the sum of these oxides in the slag before blowing: the higher their initial total concentration, the greater the reduction (Figure 1). They are apparently reduced with aluminum.

Table 2. Summarized data on the chemical composition of slag in the ladle after tapping and blowing

	Content, wt %							
	CaO	SiO_2	MgO	FeO	MnO	Al_2O_3	S	$B=CaO/SiO_2$
After tapping	38.5-53.2 45.4	16.7-27.6 22.3	6.1-19.9 13.9	1.3-5.3 2.61	0.9-4.7 2.12	5.2-12.5 9.7	0.11-0.61 0.35	1.5-3.0 2.09
After blowing	37.2-53.4 44.9	17.7-28.0 23.1	6.5-22.1 14.4	0.7-2.8 1.66	0.53-3.95 1.31	5.7-14.6 11.0	0.15-0.71 0.38	1.5-3.0 2.01

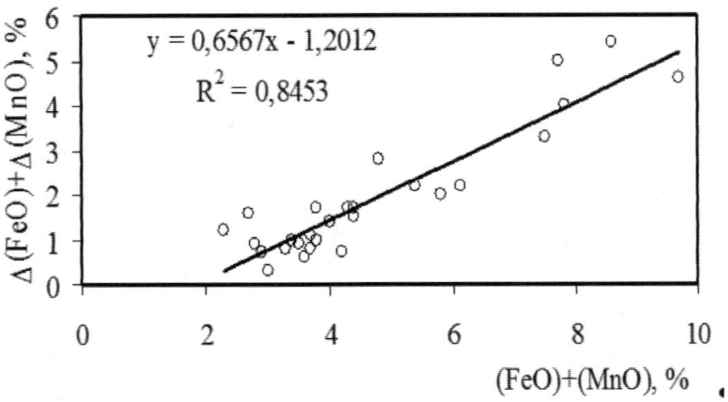

Figure 1. Reduction of ferrous and manganese oxides in slag when blowing with argon

The amount of the oxidized aluminum while the blowing process (Δ [Al]) can be determined relative to changes in the content of ferrous and manganese oxides in the slag:

$3\,(FeO) + 2\,[Al] = (Al_2O_3) + 3\,[Fe]$,

$3\,(MnO) + 2[Al] = (Al_2O_3) + 3\,[Mn]$,

$$\Delta[\mathrm{Al}]=\frac{2\cdot 27\cdot \mathrm{S}(\Delta\mathrm{RO})\mathrm{M}}{3\cdot 72\cdot 10^4},\ \%, \qquad (2)$$

where 27 is the atomic mass of aluminum; 72 is the molecular mass of ferrous oxide. When aluminum is oxidized with manganese oxide, the molecular mass of manganese oxide should be inserted into the formula (71). To simplify the calculations, there is no need to make such replacement due to negligibility of the error occurred.

S is the mass of slag, t; Δ (RO) = Δ (FeO) + Δ (MnO) is the total decreasing in concentration of ferrous and manganese oxides, %; M is the mass of a melt, t. In further calculations we take M = 110 t.

In calculations based on this formula, it becomes difficult to define an amount of slag relative to a particular melt; the amount may range from 20 - 30 to 50 - 60 kg/t.

An attempt has been made to compare the expected "melting loss" of aluminum due to ferrous and manganese oxides with the actual loss observed for specific values ΔRO in the melts. For this purpose, the values Δ[Al] are derived from ΔRO at a slag mass of 2t, 3t and 5t, which are indicated by the lines in Figure 2. The points represent the actual values of ΔAl depending on the actual ΔRO. As it can be seen, the actual melting losses of aluminum in most melts are similar to those of with the slag mass ranging 3 - 5 t. Considering the slag layer thickness in the ladle, such amounts of slag are quite possible. When the slag density ρ = 3t/m^3 and ladle diameter D = 3 m, the slag layer thickness is 0.23 m at a slag mass of 5t, and 0.14 m at 3 t. The value of aluminum melting loss while oxidizing with slag can vary from 0.005 to 0.05%.

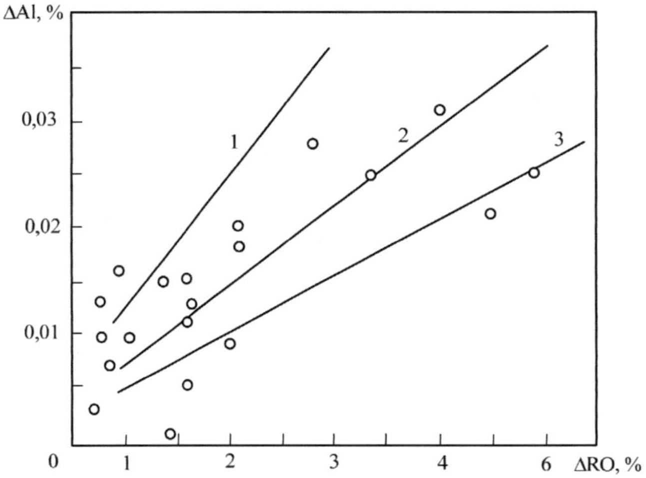

Figure 2. Aluminum loss depending on the total concentration of ferrous and manganese oxides in the slag

Another source of oxygen is the atmosphere. The amount of oxygen entering a metal from the air can be determined using the Temkin equation [2], with the activation energy E = 117000 J/mol. For calculating we have assumed that oxygen passes into the metal through the slag-free surface near the place where argon bubbles reach the surface forming a 1 m diameter. It turns out that the 6-minute blowing into the melt can result in absorption of 1.633 kg oxygen that is sufficient for the oxidation of 0.0018% aluminum in a 100-ton ladle. 0.0027% of aluminum can be oxidized under the similar conditions with the 10-minute blowing. Hence, the main source of oxygen that enters the metal while blowing with argon is slag. A portion of oxygen entering the melt from the air is considerably small.

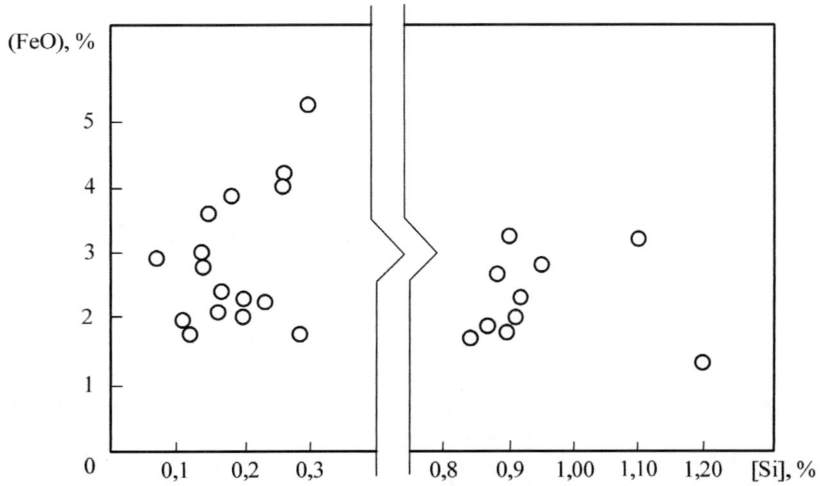

Figure 3. Influence of a content of silicon in the metal on the slag oxidation

Figure 3 gives the comparison between the content of ferrous oxide in the slag and silicon in the metal after tapping. As one might anticipate, the degree of slag oxidation should be higher when the content of silicon is low in the ladle analysis. Figure 3 gives no evidence to this: in cases of the experimental melts, the (FeO) content in slag practically does not depend on the concentration of silicon in metal. No evidence is given referring to aluminum (Figure 4).

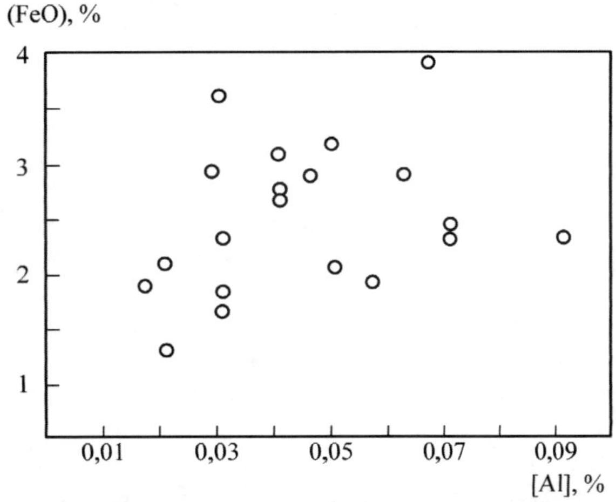

Figure 4. Influence of the aluminum content on the degree of slag oxidation

A higher concentration of aluminum in the ladle analysis sample does not prove sufficient de-oxidation of slag. This proves that at tapping, firstly, a considerable portion of aluminum is absorbed by metal, and second, the metal is not completely stirred with slag [3-15].

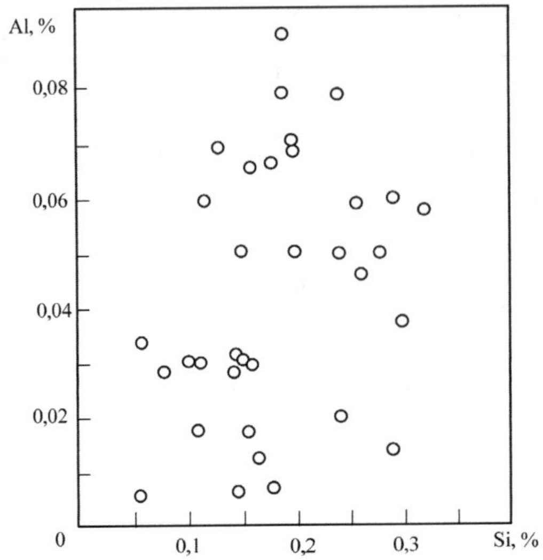

Figure 5. Relationship between the concentrations of aluminum and silicon in the ladle sample

As a result, the metal - slag system does not reach equilibrium as evidenced by the lack of relationship between the contents of silicon and aluminum in the ladle samples (Figure 5).

Conclusion

We have investigated the behavior of silicon, manganese and chromium added to the ladle to obtain the required chemical composition of steel, followed by blowing with an inert gas. It has been proven that absorbing elements mainly depends on the degree of de-oxidation in slag and metal, and first of all, the contents of FeO and MnO in slag and aluminum in metal.

References

[1] G Wagner Thermodynamics of alloys Metallurgiya (1957) -179 p

[2] D A Chinakhov Calculation of gas-dynamic impact of the active shielding gas on the electrode metal drop in gas jet shielded welding *J Applied Mechanics and Materials* 379 (2013) 188-194

[3] D A Chinakhov A V Zuev A G Filimonenko Gas-dynamic impact of a shielding gas get on the drop transfer when welding with a consumable electrode *J Advanced Materials Research* 1040 (2014) 850-853

[4] Valuev D V Serikbol A Stress state and its definition in hardfacing heat-resistant steel cold rolling mill rollers *J Advanced Materials Research* 1040 (2014) 703-707

[5] Gizatulin R A Valuev D V Secondary Metallurgical Treatment Saarbruecken : LAP LAMBERT Academic Publishing 2013 pp 380

[6] Gizatulin R A Valuev D V Edesheva Ch V Technology of combined processing of steel at the Release of Calcium and Aluminium *J Modern Problems of Science and Education* 2 (2014) 1 - http://www science-education ru/116-12782

[7] Gizatulin R A Valuev D V Serikbol A Improving the Process Parameters Regenerative Processes at Steel Processing Ladle Furnace *J Modern Problems of Science and Education* – 3 (2014) 1 http://www science-education ru/117-12800

[8] Gizatulin R A Secondary technology and ladle treatment processes in metallurgy – Tomsk: Publ TPU 2008 - 198 p

[9] Gizatulin R A Nokhrina O I Kozyrev N A Steel microalloying with nitrogen by ladle bottom and top submerged lance blowing *J Proceedings of Higher School Ferrous Metallurgy* 6 (2010) 30–33

[10] Malushin N N Valuev D V Valueva A V Serikbol A Kinetic study of the effect of plasticity and its role in stress relaxation in the weld speed steel during the martensitic transformation *J Applied Mechanics and Materials* 682 (2014) 58-63

[11] Chinakhov D A Il'yaschenko D P Gotovschik Y M MAW productivity development and reduction of its harmful effect on human organisms *J Applied Mechanics and Materials* 682 (2014) 122-126

[12] Il'yaschenko D P Chinakhov D A Gotovschik Y M Influence of Power Supply Energy Characteristics upon the Stability of MMA Process *J Applied Mechanics and Materials* 756 (2015) 97-100

[13] Malushin N N Valuev D V Valueva A V Serikbol A Kinetic study of the effect of plasticity and its role in stress relaxation in the weld speed steel during the martensitic transformation *J Applied Mechanics and Materials* 682 (2014) 58-63

[14] Chinakhov D A Agrenich E P Computer simulation of thermo-mechanical processes at fusion welding of alloyed steels *Materials Science Forum* 575-578 (2008) 833-836

[15] Malushin N N Valuev D V Measuring stress condition in hard faced work rolls for cold rolling *J Applied Mechanics and Materials* 379 (2013) 71-76

Extraction of Chromium from Carbon Ferrochromium Residual Wastes

P P Lazarevskiy [1, a], R A Gizatulin [2,b], Yu E Romanenko [1,c], D V Valuev [2, d], A V Valueva [2, e], A Serikbol [2, f],
[1] 654007 Kemerovo region, Novokuznetsk, Kirova str. 42 Siberian State Industrial University, Russia
[2] 652050, Kemerovo region, Yurga, Leningradskaya str.26 Russia
Yurga Technological Institute branch of Tomsk Polytechnic University
[a] email: lazura@ya.ru, [d] email: valuevden@rambler.ru,
[e] email: valueva@tpu.ru, [f] email: aikosha94s@mail.ru

Abstract. This work reports the problem of processing residual wastes after producing carbon ferrochrome by recycling dust using a hydrometallurgical method with the purpose of extracting the basic component - chromium. The X-ray diffraction analysis results, chemical and granulometric compositions of dust from the carbon ferrochrome production are given. The method for the production of chemical-enrichment concentrate (CEC) by processing ferrous dust is described, with obtaining a middling product - sodium mono-chromate with its further reduction to chromium hydroxide, followed by autoclave leaching, and resulting in the production of chemically enriched chrome concentrate. The plant used for autoclave leaching and filtering is schematically depicted. The smelting process of metallic chromium using the ladle aluminothermic method is described.

Introduction

Today, metallurgical enterprises pay much attention to both the rational use of both internal material returns and industrial waste recycling. In recent years, a top-priority has been given to the disposal of ferroalloy production wastes and their maximum return to production processes that can result, in its turn, in saving material and energy resources and reducing production costs [1, 2].

One of the technogenic waste types after the production of high-carbon ferrochrome is dust (ferrous dust). This ferrous dust is formed in the production process of ferroalloys and is collected by a gas cleaning system fitted in the ore smelting furnace. When ferrochrome is melted in the open furnace, the volume of dust entrained makes 50 - 60 g/m^3, in some cases up to 100 g/m^3. The specific dust yield per 1 ton of ferrochrome for an open furnace is 25 - 75 kg/t [3]. The dust after the ferrochrome smelting has the following chemical composition in %: Cr_2O_3 22.5 – 43,6; SiO_2 10.0 – 15.5; CaO 0.2 – 0.4%; MgO 25.2 – 33.1; Al_2O_3 3.2 – 5.4; FeO 4.1 – 6.0; C 5.4 – 6.2; S 1.0 – 1.2.

A granulometric composition of dust depends on many factors and can vary greatly. The distribution of dust particles per fractions is given in Table 1.

Table 1. Distribution of dust particles per fractions.

Particle size, micron	>200	200 – 100	100 – 60	60 – 20	20 – 10	10 – 1	<1
Content, wt. %	34.5	12.3	19.0	25.0	7.5	1.65	0.05

Given the considerable amount of ferrous dust and high content of chromium oxide in it, recycling is an important task in modern technological scheme of ferrochrome production.

According to the X-ray diffraction phase analysis, chromium in ferrous dust is present in the form of a complex compound (Mg, Fe) $CrAl_2O_4$, and this requires the application of a hydrometallurgical process to extract the main component, including the treatment of material with acids or alkalis solutions and transfer of the extracted component into a solution with its further extraction.

The process for obtaining chemical enrichment chromium concentrate consists of two stages:

1) Ferrous dust is processed to convert chromium from the complex compound into sodium mono-chromate and to receive a concentrated solution;

2) Sodium mono- chromate is reduced to chromium hydroxide, with further production of CEC.

The technology used in producing sodium mono- chromate is based on chrome ore enrichment and involves the oxidation heat treatment of a charge consisting of ferrous dust and soda ash (Na_2CO_3 not less than 98%) [4].

The component ratio is calculated based on the stoichiometric reaction:

$$2Cr_2O_3 + 4Na_2CO_3 + 3O_2 = 4Na_2CrO_4 + 4CO_2 \quad (1)$$

The charge components thoroughly mixed in a drum mixer were subjected to the oxidation heat treatment at 1150 - 1250 °C for three hours in a muffle furnace. After cooling, the resulting sinter was ground to a fraction of less than 50 microns and X-Ray analyzed. The results of the X-ray diffraction phase analysis given in Figure 1 indicate that sodium mono- chromate was formed due to chromium contained in the dust. In addition, during the oxidative sintering the interaction of ferrous dust impurities (SiO_2, Al_2O_3, Fe_2O_3) with sodium results in forming silicates, sodium aluminates and ferrites.

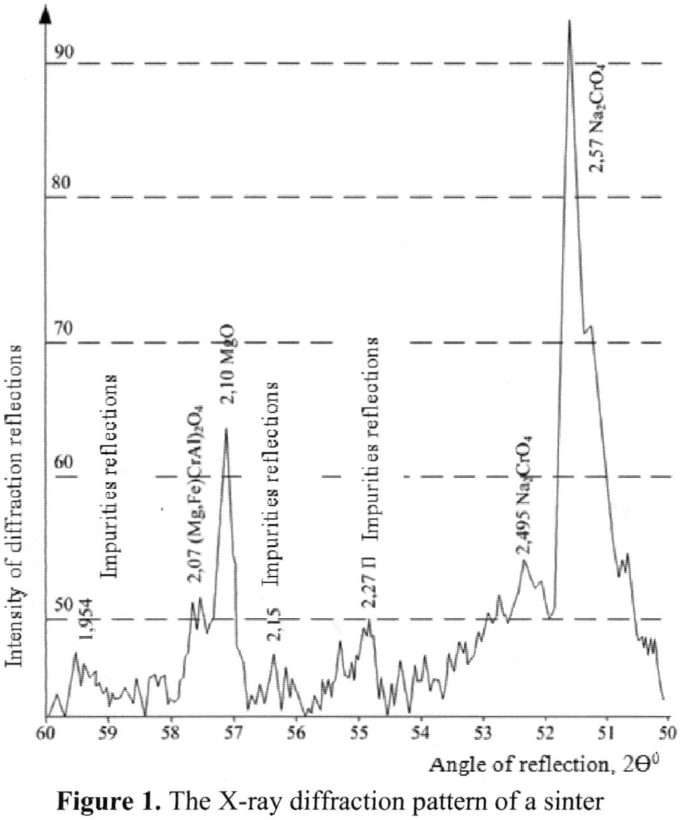

Figure 1. The X-ray diffraction pattern of a sinter

To extract the sodium mono-chromate from the sinter, the leaching process in water with a solid - liquid volumetric ratio of 1:1 was used [4]. The leaching was conducted in an autoclave at 70 - 90°C for one hour. In this process, the pulp consisting of the sodium mono- chromate solution and sludge was formed. Further separating the pulp in a Nutsche filter resulted in the formation of the sodium mono- chromate solution with a chromium concentration of 130 - 150 g /l and sludge with the following chemical composition, %: 0.9 – 1.2 Cr_2O_3; 7.4 – 8.5 FeO; 68.3 – 69.4 MgO 15.9 – 16.4 Al_2O_3. The content of impurities in the sodium mono- chromate solution does not exceed 30 mg/l.

During the second stage in producing CEC, sodium mono- chromate was reduced to chromium hydroxide. The process is based on the autoclave leaching of the solution at a temperature of 140 - 150 ° C and a pressure of 4 - 6 kg/cm^2 for one hour, with adding elemental sulfur according to the reaction in [6]:

$$4Na_2CrO_4+6S+(2n+1)\cdot H_2O \rightarrow 2Cr_2O_3\cdot nH_2O\downarrow+3Na_2S_2O_3+2NaOH. \quad (2)$$

Figure 2 displays the schematic equipment used.

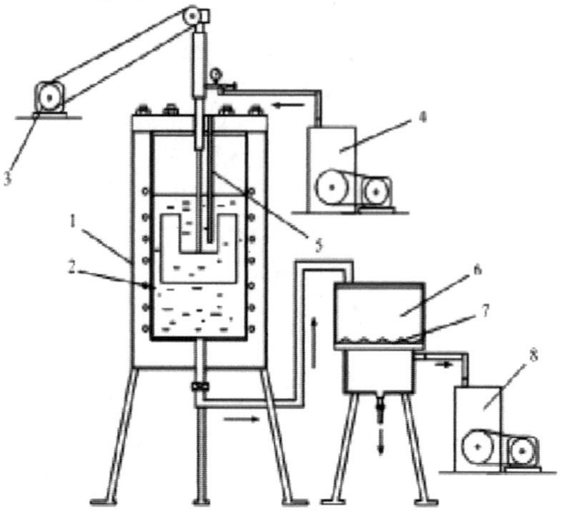

Figure 2. An autoclave with the mechanical mixing and Nutsch filter: 1 - reactor with paddle mixer (a rotation speed of 15 - 20 rev/min); 2 - sodium mono- chromate solution Na2CrO4; 3 – electric driver; 4 - compressor; 5 - thermocouple; 6 - Nutsch (suction) filter; 7 - filtering cotton cloth; 8 - vacuum pump

After an hour, the contents were pumped to the Nutsche filter and washed with running water; on the surface of the filtering cloth there appeared a bluish green precipitate (chromium hydroxide), the result of further heating at a temperature of 600°C was a gray-green powder with the following chemical composition, %: Cr_2O_3 95,8; FeO 3,9 [7-10]. The results of the X-ray diffraction pattern in analyzing CEC are shown in Figure 3.

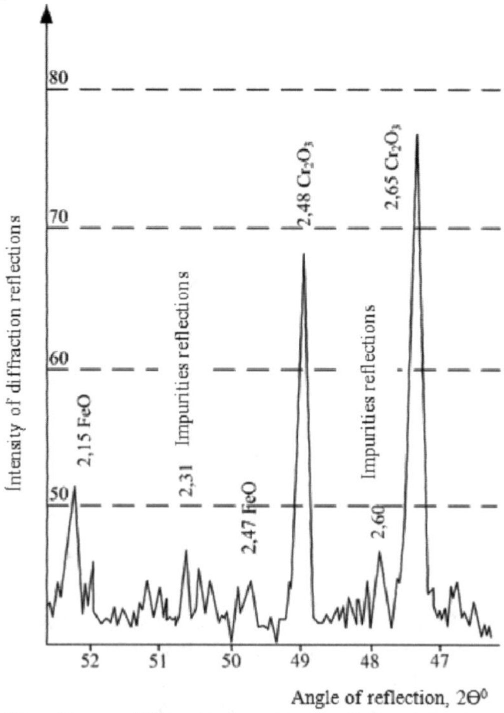

Figure 3. The X-ray diffraction pattern of chromic concentrate

The hydrometallurgical processing of the dust removed in the production of carbon ferrochrome enables us to obtain chemically enriched chromic concentrate, which can be used for the production of metallic chromium. The technological chart for the production of chromic concentrate using ferrous dust is shown in Figure 4.

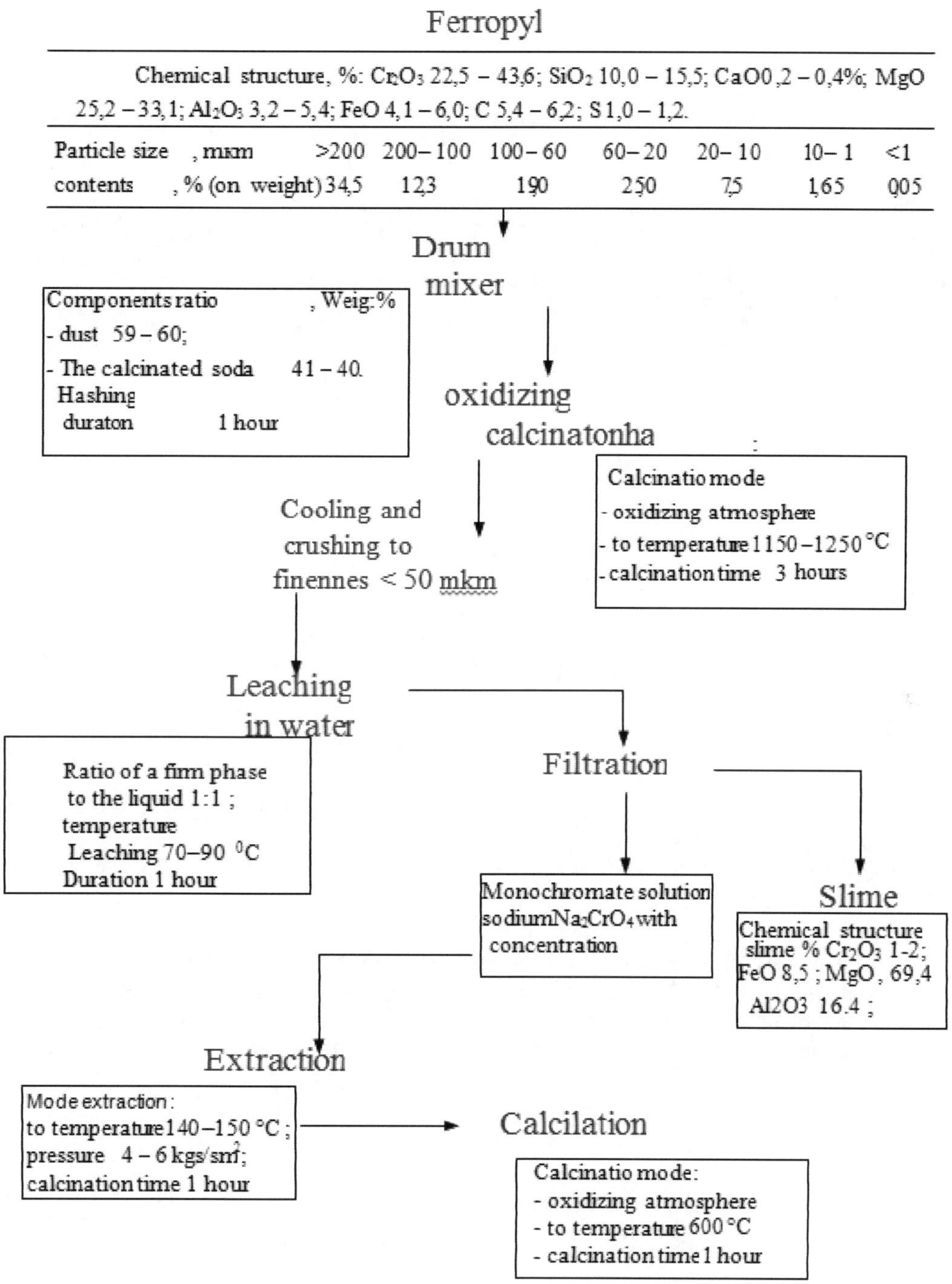

Figure 4. Block flow chart for extracting the chemical-enrichment chromium concentrate from ferrous dust

Metallic chromium was smelted using the out-of-furnace aluminothermic reaction. Charge materials used were chromic CEC (Cr_2O_3 95.8%; FeO 3.9%), primary aluminum powder, sodium nitrate (> 99.0% $NaNO_3$) and fluorspar. The fraction of the charge materials was less than 1 mm. Magnesium swarf and potassium nitrate in a proportion of 4:1 were used as an ignition mixture. Smelting was conducted using the plant shown in Figure 5. It consists of a graphite crucible 1, where a weighted charge undergoes deep melting; a device to ignite the charge by heating a nichrome coil 2 connecting two electrodes 3; a pallet 4, on which the crucible with the weight and a mechanism for moving the electrodes with the spiral are placed. The electrodes are fixed to the roof, covering the crucible during the operation.

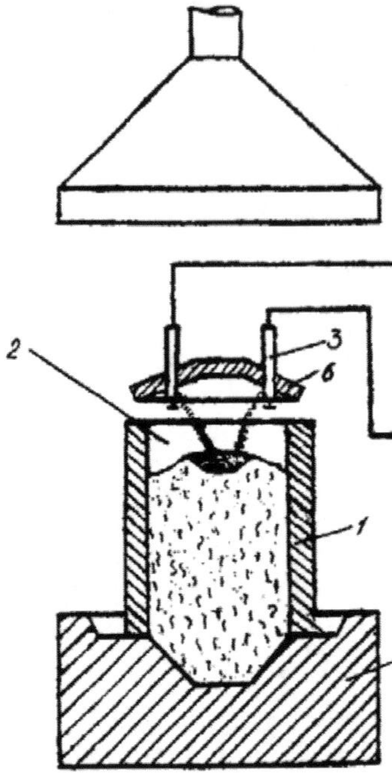

Figure 5. Schematic melting furnace: 1 - crucible; 2 - spiral; 3 - electrode; 4 - pallet; 5 - transformer; 6 - roof

Using a wheel, the device to ignite the charge was raised to its highest position; the crucible was then placed on the pallet and filled with the charge.

The ignition device was powered by transformer 5. Magnesium swarf and potassium nitrate in a proportion of 4:1 were used as an ignition mixture. A hollow was made in the center of the charge, and the hollow was then filled with the ignition mixture. After preparing the plant for work, current was supplied to bring the spiral to red heat, and the roof was lowered using the wheel till contact of the spiral with the ignition mixture. Magnesium inflamed and ignited the charge. After igniting the charge, current was turned off, then the process run spontaneously due to the heat of exothermic reactions.

The basic aluminothermic reactions used in producing ferrochromium are shown in Table 2.

Table. 2 - Basic aluminothermic reactions in ferrochrome production

Reaction	Oxygen content in oxide, %	Metal content in oxide, %	Aluminum content per 1kg oxide, kg	Thermal effect of reaction at 2227 °C, kJ	Thermal effect per 1 kg oxide, kJ	Specific heat of reaction at 2227 °C, kJ/kg
$Cr_2O_3 + 2Al + 2Cr + Al_2O_3$	31.6	68.4	0.36	- 394000	2590	1913
$3FeO + 2Al + 3Fe + Al_2O_3$	22/2	77.8	0.25	- 848250	3920	3140
$6NaNO_3 + 10Al = 3Na_2O + 3N_2 + 5Al_2O_3$	56.5	Nitrogen content 16.5	0.53	- 6871400	13470	8820

The process is spontaneous because of the heat generated due to the exothermic reactions for a period of 20 - 25 minutes.

The resulting metal and slag were analyzed using the X-ray diffraction; the results are given in Figure 6. The slag multiplicity is 2.4.

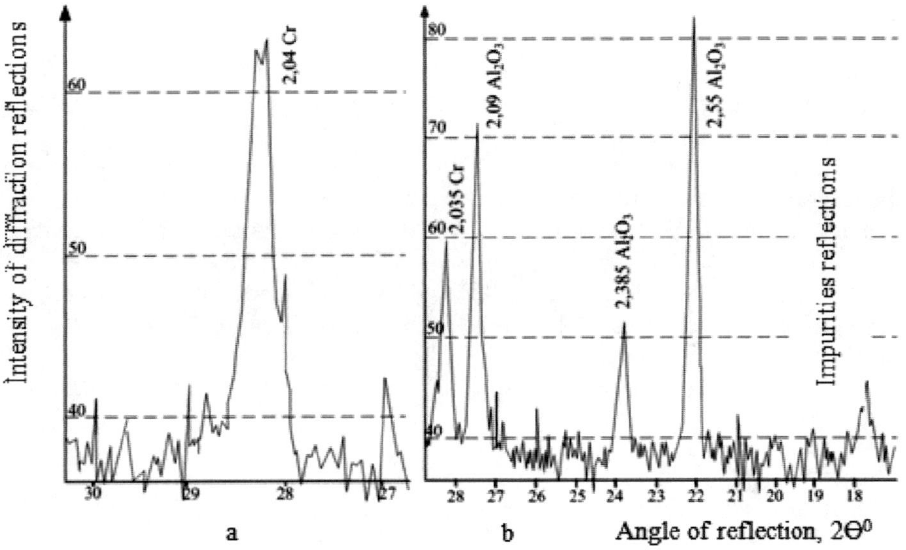

Figure 6. The X-ray diffraction pattern of metallic chromium (a) and slag (b)

The results of the X-ray diffraction analysis show that the metallic fraction of the reduced products is presented in the form of chromium. Chemical composition, %: 98.0 – 98.5 Cr, 0.4 – 0.5 Si, 0.5 – 0.7 Al, 0.6 – 0.8 Fe, 0.03 – 0.04 C, 0.02 – 0.03 S, 0.02 – 0.03 P. The alloy corresponds to GOST-5905-2004.

Conclusion

The X-ray diffraction analysis of slag has revealed that the slag contains metallic chromium. This is due to the fact that the main component of the slag is high-melting aluminum oxide that causes the metallic shot to tangle in the slag. In this connection, it is required to reduce a temperature for melting the slag and increase its flowability in liquid state.

On grounds of the results obtained in the experiments, it follows that chromium concentrate produced by recycling the dust from the ferrochrome production can be used for the production of standard metallic chromium.

Thus, the technologies developed for the purpose of rational utilizing dispersed materials, containing a target element in the smelted ferroalloy, are cost-effective and environmentally sound.

References

[1] Malushin N N Khaibulina N E Multiple use of raw materials in the industry Chelyabinsk South Ural Publishing House 1986

[2] Bagryantsev G I Thermal deactivation and disposal of industrial and domestic wastes Municipal and industrial wastes: methods of disposal and recycling – analytical surveys Novosibirsk: Series Ecology 1995

[2] Il'yaschenko D P Chinakhov D A Gotovschik Y M Influence of power supply energy characteristics upon the stability of MMA process *J Applied Mechanics and Materials* **756** (2015) pp 97-100

[3] Russian Patent RU (11) 2281250 (13) S2 MPK C01G 37/14 (2006 01) A method for producing sodium mono- chromate / V Ya Pivvuev; Appl 2004 03 02; Publ 2006 08 10

[4] Russian Patent RU (11) 2281250 (13) S2 MPK C01G 37/14 (2006 01) A method for producing sodium mono- chromate / V Ya Pivvuev; Appl 2004 03 02; Publ 2006 08 10

[5] Nokhrina O I Producing low-phosphorous concentrate from ore and ferro-manganese nodules *J News of higher schools - Ferrous Metallurgy* 8 (2000) pp 40 – 45

[6] Chinakhov D A Gas dynamic control of properties of welded joints from high strength Alloyed steels *J China Welding* Vol 23 No 3 (2014) pp 27-31

[7] P P Lazarevskiy Resource-saving production of chromium oxide by utilizing technological wastes Proceedings of the All-Russian scientific-practical conference "Metallurgy: technology management innovation quality" - Novokuznetsk: SibGIU 2012 pp 125 – 127

[8] Iljyashchenko D P Chinakhov D A Improvement of health and safety characteristics of shielded metal arc welding with inverter power source *J The 7th International Forum on Strategic Technology IFOST2012* September 17-21 2012 Tomsk Polytechnic University Vol I pp 19-21

[9] Valuev D V Serikbol A Stress state and its definition in hardfacing heat-resistant steel cold rolling mill rollers *J Advanced Materials Research* 1040 (2014) pp 703-707

[10] Malushin N N Valuev D V Valueva A V Serikbol A Testing of metallurgical equipment on contact strength and wear resistance *J Applied Mechanics and Materials* 682 (2014) pp 53-57

IOP Conf. Series: Materials Science and Engineering **91** (2015) 012039 doi:10.1088/1757-899X/91/1/012039

Optimizing Technological Parameters of the Reduction Processes in Treating Steels in a Ladle Furnace

R A Gizatulin [1], D V Valuev [2], A V Valueva [3], A Serikbol [4], I F Borovikov[5]
[1,2,3,4] National Research Tomsk Polytechnic University,Tomsk, Lenin Avenue 30, Russia
[5] The Moscow State Technical University of N.E. Bauman, Moscow, 5, 2nd Bauman-skaya

E-mail: valuev@tpu.ru

Abstract. This work reports the possible development of reduction processes when treating the molten metal and slag using a ladle furnace under conditions of intensive stirring with an inert gas. The industrial data have been received, confirming the possibility of decreasing the concentration of ferrous and manganese oxides in the slag and stabilizing the contents of manganese and silicon in the metal.

Introduction

In modern metallurgy, much attention is paid to the problem of reducing the consumption of deoxidizing agents and alloying materials by developing efficient techniques employed for their adding to steel, with simultaneous improving steel properties by reducing concentrations of harmful impurities, as well as increasing a penetration coefficient for effective use of alloying elements. One of the solutions to this problem is increasing the extraction of an alloying element from slag by intensifying reduction processes while treating melts in modern ladle furnaces equipped with various systems for metal and slag mixing.

With the diffusion reaction to reduce metal oxides, a diffusion flux is determined by the known equation:

$$\Pi = \beta(C - C^*), \qquad (1)$$

Where β is the mass transfer coefficient, m/s; C, C* are the concentration of substance in volume and at the boundary interface, mol/m^3.

According to the surface renewal theory, the mass transfer coefficient β is defined as follows:

$$\beta = \sqrt{\frac{4DU}{\pi l}}, \qquad (2)$$

Where D is the diffusion coefficient, m^2/s; U is the rate of convective flow, m/s; l is the characteristic dimension of convective flow, m.

1. Отношение скоростей потоков металла и шлака обратно пропорционально динамической вязкости

When calculating a mass transfer coefficient, two assumptions have been made:

1. The ratio of metal and slag flow rates is inversely proportional to the dynamic viscosity

$$\frac{U_{Me}}{U_{slag}} = \frac{\eta_{slag}}{\eta_{Me}}, \qquad (3)$$

Where η_{Me}, $\eta_{ш}$ are the dynamic viscosity of metal and slag, Pa·s.

2. The ratio of characteristic dimensions for metal and slag convective flows is directly proportional to the relative heights of metal (h_{Me}) and slag (h_{slag})

$$\frac{l_{Me}}{l_{slag}} = \frac{h_{Me}}{h_{slag}} = \frac{m_{Me}\rho_{slag}}{m_{slag}\rho_{Me}}, \tag{4}$$

Where m_{Me} and $m_{ш}$ are the masses of metal and slag, kg; ρ_{Me} and ρ_{slag} are the densities of metal and slag, kg/m^3.

Due to these assumptions, such important conditions as viscosity of slag and its quantity affecting the process behavior can be taken into account.

The slag viscosity (η_{slag}) is defined by equations constraining the basicity and viscosity of slag. The constraint equations are derived from the data given by the author in [2]. Slag becomes heterogeneous when its basicity exceeds 2.5 causing constrained diffusion processes. The dependencies for 1500 and 1600 ° C temperatures are received (Figure 1).

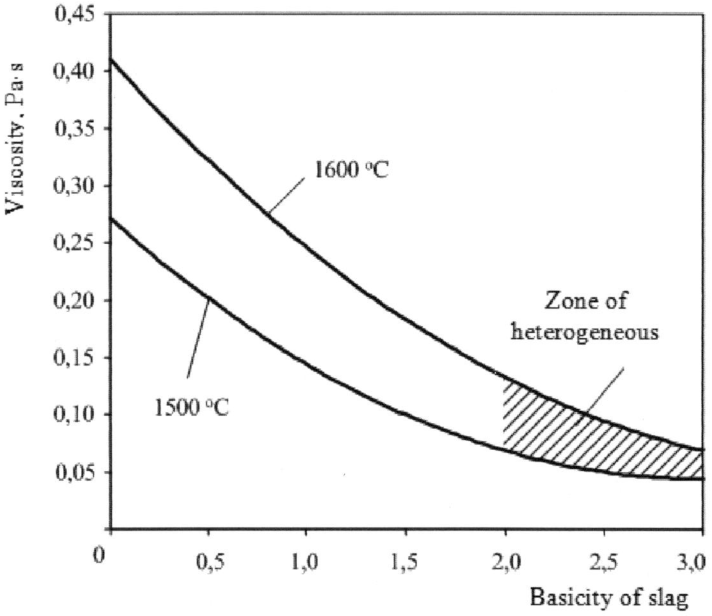

Figure 1. The dependency of slag viscosity on the slag basicity

$$\eta_{ш} = 0.0062B^2 - 0,1069B + 0.511, \tag{5}$$
$$\eta_{ш} = 0.00664B^2 - 0,0893B + 0.3543, \tag{6}$$

Where B is the basicity of slag.

The oxidation rate for a reducing agent can be calculated according to the equations applicable to manganese and silicon, when the reducing agent is added to the surface of the molten slag, for the cases of limiting the process by diffusion of the reducing element to the metal/melt interface and oxides of interacting components in the molten slag:

$$V_{Si} = -\frac{d[\%Si]}{d\tau} = \beta_{[Si]}\{[\%Si] - [Si\%]*\}\frac{S}{V_{slag}}, \tag{7}$$

$$V_{Si} = \beta_{(SiO_2)}\frac{28}{60}\{(\%SiO_2) - (SiO_2\%)*\}\frac{S}{V_{slag}}, \tag{8}$$

$$V_{Mn} = \beta_{(MnO)}\frac{28}{142}\{(\%MnO) - (MnO\%)*\}\frac{S}{V_{slag}}, \tag{9}$$

where $\beta_{[Si]}, \beta_{(SiO_2)}, \beta_{(Mno)}$ are the mass transfer coefficients, m/s; (% SiO2), (% MnO) are the concentrations of SiO2 and MnO in the slag, respectively, % by; [% Si] is the concentration of silicon in ferrosilicium, % by wt; (% SiO2) *, (% MnO) * are the equilibrium concentrations of SiO2 and MnO in the slag, respectively, % by wt; [Si] is the concentration of silicon in ferrosilicon, % by wt; [Si] * is the equilibrium concentration of silicon, % by wt; S is the area of phase interface, m^2; Vslg is the volume of slag m^3.

On the assumption that the system is balanced at the phase interface boundary, the equilibrium values of silicon, manganese, SiO2 and MnO concentrations are calculated using equilibrium constants.

This system of equations is solved together with the equilibrium constant expression using the method of successive approximations.

The equation below calculates a period required to transfer ferrosilicon drops

$$\tau = \frac{h_{slag}}{\omega}, \tag{10}$$

where $h_ш$ is the height of molten slag, m; ω is the velocity of moving particles, m/s.

The Hadamard–Rybczynski equation is used to estimate a velocity of moving the liquid drops through the molted slag [3]

$$V_ж = \frac{2}{3} g \frac{\rho_{slag} - \rho_{Me}}{\eta_{slag}} r^2 \frac{\eta_{slag} + \eta_{Me}}{2\eta_{slag} + 3\eta_{Me}}, \tag{11}$$

where η_M is the dynamic viscosity of a molten ferrosilicon drop, Pa·s; ηш is the dynamic viscosity of molten manganese oxide, Pa·s; r is the radius of metal particles, m; g is the free fall acceleration, m/s; ρ_M is the drop density, kg/m^3; $\rho_ш$ is the slag density, kg/ m^3.

Post reduction of oxides to metal in the molten slag occurs at the boundary interface of the steel - oxide melt. In this case, the oxidation rate of silicon can be described by the equation

$$V_{Si} = -\frac{d[\%Si]}{d\tau} = \beta_{[Si]} \{[\%Si] - [Si]^*\} \frac{S}{V_{slag}}, \tag{12}$$

where β[Si] is the mass transfer coefficient for silicon, m/s; [% Si] is the concentration of silicon in steel, % by wt; [Si] * is the equilibrium concentration of silicon, % by wt.

The freshly burnt limestone is added to bind the resulting SiO_2.

The dissolution rate of limestone dispersed particles is determined by the convection diffusion equation [4, 5]

$$\frac{d(CaO)}{d\tau} = \beta_{(CaO)} \{(CaO) - (CaO)\} \frac{S}{V}, \tag{13}$$

where $(CaO)_{нас}$ and (CaO) are the concentrations of calcium oxide in the slag near the surface of limestone lumps (the concentration of saturation) and in the bulk of slag, respectively; S is the total surface area of limestone lumps, m^2; V is the volume of slag m^3.

The total surface area of spherical particles can be expressed by the equation [5]

$$S = \frac{6M}{d\rho}, \tag{14}$$

where M is the total mass of particles, kg; $\rho_{тв}$ is the density of particles, kg/m^3; d is the particle diameter, m.

For modeling a mixing process the pattern of successively located baths is applied [4]. Varying the concentration of substance C in a time for the successive baths is expressed by the equation

$$\frac{\partial C_i}{\partial t} = \frac{Q}{V_m} (\rho_{i-1} C_{i-1} - \rho_i C_i) \bar{\beta}_i, \tag{15}$$

where ρ_i and ρ_{i-1} are the density of metal in the i-th and i-1 successive bath, kg/m^3; Q is the metal flow between the baths, m^3/s; Vm is the bath volume, m^3; $\bar{\beta}_i$ is the mass transfer coefficient of sub-

IOP Conf. Series: Materials Science and Engineering **91** (2015) 012039 doi:10.1088/1757-899X/91/1/012039

stance through the metal-slag boundary layer, $\bar{\beta}_i = \beta_i \cdot P$ (where P is the value to numerically characterize a power of melt mixing), m/s.

A schematic flow chart showing the model can be divided into several parts, including calculating the following indicators: a period of time for interacting a deoxidizer and oxides in the slag; an amount of the deoxidizer used for the reduction of metal oxides in the slag; a composition of metal in bulk and the molten oxide; activity of components of the metal and slag phases; a rate of reactions between deoxidizers and oxides in the slag; mass transfer coefficients of interacting components in the metal and slag; a mass of slag and metal phases; a dissolution rate of limestone in the slag; compositions of metal and slag phases; slag basicity.

Below are the values for a specified calculation error relative to various parameters required for calculations (Table 1).

Table 1. Specified calculation errors for parameters available from experiments

Parameter	Specified calculation error
Metal mass, kg	10
Slag mass, kg	10
Manganese content in metal, %	0.001
Silicon content in metal, %	0.001
Manganese content in slag, %	1.0
Silicon content in slag, %	1.0
Ferrum content in slag, %	0.1
Basicity of slag	0.1
Mass of additives, kg	1.0

A prerequisite for constructing the mathematical model is the application of the law of conservation of mass, which is taken as a basis for instantaneous material balances.

The initial data to start working in the 'Advisor Wizard' system, are:

- Metal state parameters: a mass at a point in time τ_1, kg; temperature, °C; composition, % by wt;
- Slag state parameters: a mass of oxidative slag at a point in time τ_1, kg; basicity, units; composition, % by wt;
- Chemical composition of coke;
- Ferrosilicon parameters: chemical composition, % by wt; mass, kg;
- Parameters of lining: a mass of lining washed out for a melt, kg; composition, % by wt;
- Parameters of limestone: a mass of lime, kg; time required for complete dissolution of lime in the slag min; composition, % by wt;
- Parameters of a ladle- furnace, a diameter of bath at the slag belt level, m; bath depth, m;
- Increments in calculations

Figure 2 depicts the schematic structure of the model

IOP Conf. Series: Materials Science and Engineering **91** (2015) 012039 doi:10.1088/1757-899X/91/1/012039

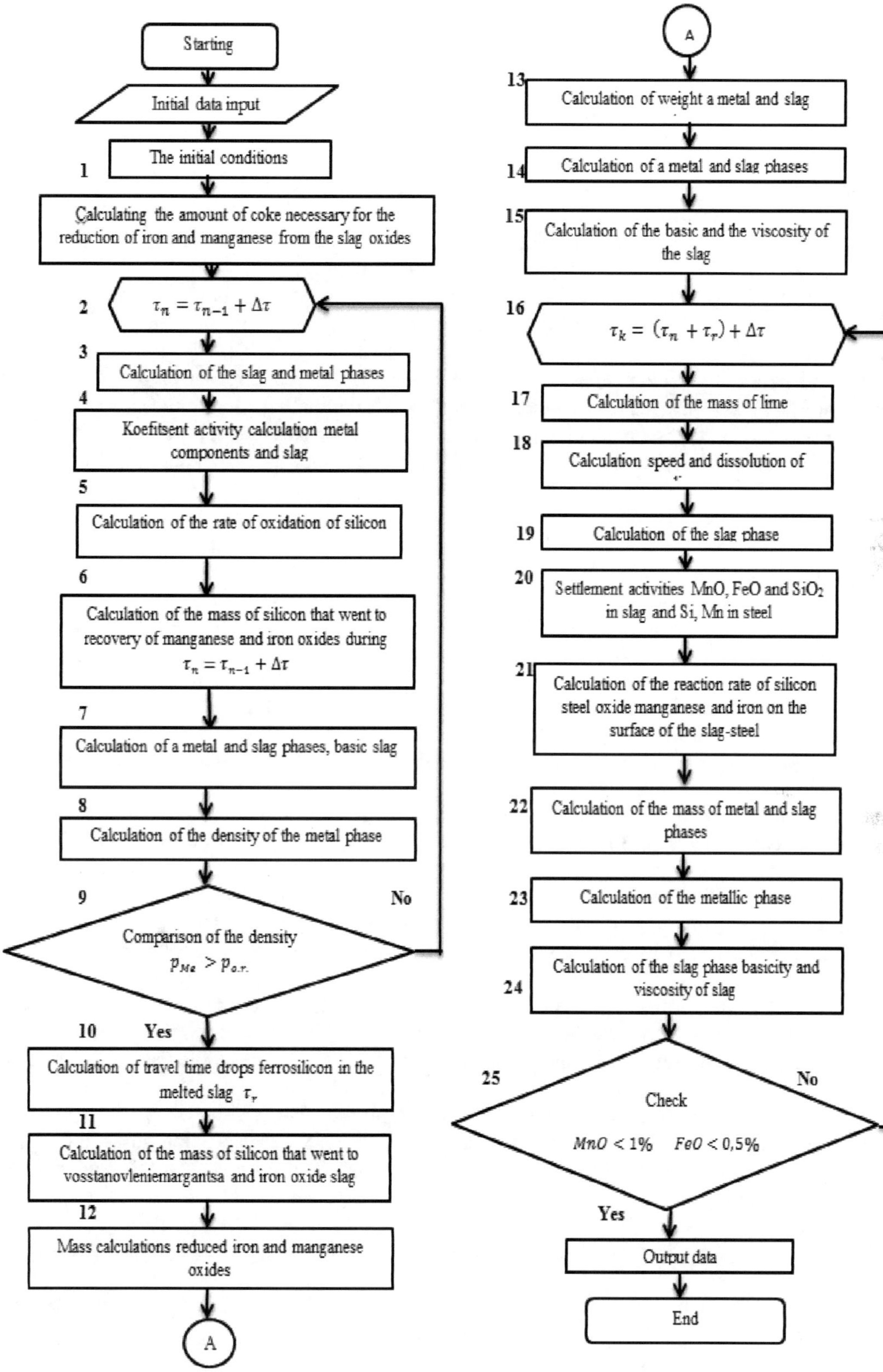

Figure 2. Block flow chart of the mathematical model

IOP Conf. Series: Materials Science and Engineering **91** (2015) 012039 doi:10.1088/1757-899X/91/1/012039

Calculations are performed in the following way: knowing the metal and slag compositions, the compositions of these phases can be calculated after adding coke. These compositions are accepted as initial at a point in time τ_1.

After adding ferrosilicon the metal and slag compositions can be determined at a point in time $\tau_2 = \tau_1 + \Delta\tau$, where $\Delta\tau$ is the increment in calculations (block 1, 2, 3).

In block 4, the activity coefficients of components in the metallic and slag phases [7] are calculated using the method proposed by the authors in [6]. In block 5, the oxidation rate of silicon in ferrosilicon is calculated using equation (12). Block 6 gives the calculation for an amount of silicon required for the reduction of oxides in the slag. In block 7, the compositions in the metallic and slag phases are determined; block 8 is for calculating a density of the metallic phase.

Block 9 - the density of the metallic phase and the density of the molten slag are compared. With the condition $\rho_{Me} > \rho_{o.p.}$, we proceed to calculate the time of the motion of a metal drop in the oxide melt (block 10).

Block 11 is used to calculate the mass of silicon used for the reduction of oxides in the slag to form metals and the mass of the reduced element (block 12). The metal and slag masses are then determined by solving a system of three balance equations: the total material balance, the balance of the reduced element, the balance of silicon (block 13), with further calculating the metal and slag compositions (block 14) and the basicity and viscosity of slag (block 15).

In blocks 17, 18, the mass, rate and dissolution time for limestone are calculated.

Then, the activity coefficients of components of the metallic and slag phases and the composition of the slag phase are calculated in blocks 19, 20. Activities of components of the metallic phase are determined using parameters of first-order interactions, and activities of components of the slag phase are determined using the method developed by the authors [7], with calculating the reaction rate for reducing oxides dissolved in the steel.

After block 20, the reaction rates are calculated used for the reduction of slag oxides to metals with silicon dissolved in the steel (block 21). Further, the masses and compositions of the metallic and slag phases are determined (block 22, 23, 24).

Calculations are made as long as the content of a metal in oxides in the slag phase is less than an estimated value of calculation accuracy as to this component.

To prove the model adequacy, a number of experimental melts were made using a ladle-furnace at Novokuznetsk Iron and Steel Plant. The melting process and oxidation time were observed according to the plant's production process instruction. After pouring the molten steel from the furnace, the ladle with steel was moved to the 'ladle-furnace' unit, where an amount of slag in the ladle was measured. Then, coke breeze in quantity of 50 - 100 kg was added onto the slag surface, with the holding time 10 - 20 min. After that, ferrosilicon FS65, FS75 was added in amounts to ensure the silicon content in the steel closer to the upper limit. Simultaneously with the silicon mixing, limestone was added in an amount to obtain the free running slag. The additives were introduced while stirring the steel with an inert gas. The treatment time was 40-70 min. In the course of experiments, samples of the metal and slag were taken to monitor the chemical composition of the steel, with the exact timing of all technological operations.

The initial data on each melt were entered into the program 'Advisor Wizard' for calculating. The results of changes in the content of manganese and silicon in reference points in time in the samples taken to identify melts, and the content of silicon and manganese estimated for the same points in time are shown in Table 2.

The residual content is 0.01 - 0.04% for manganese, and 0.01 - 0.08% for silicon. The relative error is 2 - 4% for manganese and 6 - 8% for silicon; the errors are within acceptable limits (less than 10%).

Table 2. Change in the contents of manganese and silicon during the melting process, Steel 25G2S, %

No. melt	No.* sample	Experimental melts		Estimated data		Residual	
		Mn	Si	Mn	Si	Mn	Si
1	1	0.98	0.85	1.01	0.90	0.02	0.05
	2	1.11	0.75	1.15	0.70	0.04	0.04
	3	1.35	0.69	1.36	0.74	0.01	0.05

No. melt	No.* sample	Experimental melts		Estimated data		Residual	
		Mn	Si	Mn	Si	Mn	Si
	4	1.45	0.66	1.46	0.70	0.01	0.04
	5	1.47	0.64	1.46	0.70	0.01	0.06
2	1	0.94	0.86	0.98	0.92	0.04	0.06
	2	1.07	0.77	1.12	0.80	0.05	0.03
	3	1.31	0.63	1.33	0.70	0.02	0.06
	4	1.39	0.61	1.41	0.68	0.02	0.07
	5	1.42	0.60	1.41	0.68	0.02	0.08
3	1	1.22	1.10	1.18	0.72	0.04	0.03
	2	1.29	0.89	1.25	0.82	0.04	0.07
	3	1.37	0.79	1.33	0.77	0.02	0.02
	4	1.47	0.70	1.48	0.75	0.01	0.05
	5	1.47	0.70	1.48	0.75	0.01	0.05
4	1	0.91	0.89	0.94	0.90	0.03	0.01
	2	1.10	0.81	1.12	0.82	0.02	0.01
	3	1.34	0.78	1.35	0.75	0.01	0.02
	4	1.37	0.64	1.42	0.72	0.05	0.08
	5	1.42	0.64	1.42	0.72	0.01	0.08

* - samples 1, 2, 3, 4, 5 are refinery samples taken in a 10, 20, 30, 40 and 60 minute interval after adding ferrosilicon.

Conclusion

Thus, it can be claimed that the model adequately exhibits the main features of the process under investigation and allows you to predict the reduction process of oxides to metals and evaluate the slag oxidation.

References

[1] Popel S I Sotnikov A I Boronenko V N Theory of metallurgical processes Metallurgy 1986
[2] Medzhibozhsky M Y Fundamentals of thermodynamics and kinetics of steelmaking processes Donetsk Vishcha shk Golovn Publishing House 1986
[3] Grigoryan V A Belyanchikov L N Stomakhin A Y Theoretical Principles of Electric Steelmaking Metallurgy 1979 pp 115 – 120
[4] Levich V T Physicochemical hydrodynamics Publishing House of the USSR Academy of Sciences 1952
[5] Chinakhov D A Study of thermal cycle and cooling rate of steel 30ХГСА single-pass weld joints J Applied Mechanics and Materials 52-54 (2011) 442-447
[6] Il'yaschenko D P Chinakhov D A Gotovschik Y M Calculation of the heat content of the electrode metal droplet wshen applying power supplies for manual arc welding with different volt-ampere characteristic *J Applied Mechanics and Materials* 756 (2015) 101-104
[7] Gizatulin R A Valuev D V Combined treatment of liquid metal with calcium and aluminum *J Metal Technology* 10 (2012) 3-11
[8] Chinakhov D A Zuev A V Filimonenko A G Gas-dynamic impact of a shielding gas jet on the drop transfer when welding with a consumable electrode *J Advanced Materials Research* 1040 (2014) 850-853
[9] Malushin N N Valuev D V Measuring stress condition in hard faced work rolls for cold rolling *J Applied Mechanics and Materials* 379 (2013) 71-76
[10] Gizatulin R A Valuev D V Edesheva Ch V Technology of combined treatment at tapping metals with calcium and aluminum *J Modern Problems of Science and Education* 2 (2014) 1-6
[11] Valuev D V Serikbol A Stress state and its definition in hardfacing heat-resistant steel cold rolling mill rollers *J Advanced Materials Research* 1040 (2014) 703-707

Ultrasonic Surface Treatment of Titanium Alloys. The Submicrocrystalline State

V A Klimenov[1,2], V A Vlasov[1,2], V Y Borozna[1], A A Klopotov[1,3]
[1]Tomsk State University of Architecture and Building, 2, Solyanaya Sq., Tomsk, 634003, Russia
[2]National Research Tomsk Polytechnic University, 30, Lenin Ave., Tomsk, 634050, Russia
[3]National Research Tomsk State University, 36, Lenin Ave., Tomsk, 634050, Russia

E-mail: klimenov@tpu.ru, vik@tsuab.ru, klopotovaa@tsuab.ru,

Abstract. The paper presents the results of the research on improvement of physical-and-mechanical properties of titanium alloys VT1-0 and VT6 by modification of surfaces using ultrasonic treatment, and a comprehensive study of the microstructure and mechanical properties of modified surface layers. It has been established that exposure to ultrasonic treatment leads to formation in the surface layer of a structure with an average size of elements 50 – 100 nm, depending on the brand of titanium alloy.

1. Introduction

Currently, in the literature it is shown that refinement of a grain structure up to submicro and nano-crystalline states in metals and alloys leads to significant changes in physical-and-mechanical properties (increase in the yield point and the stress limit, improvement of tribological characteristics, and other) [1]. For example, in papers [2,3] a monotonous increase in the microhardness with an increase in the dispersion of the grain structure has been shown on the example of titanium alloys. In this connection, on the surface of metals it is possible to form submicro- and nanocrystalline states. Such surfaces posess high performance properties. One of the most promising methods in this area is the ultrasonic surface treatment [4]. Titanium-based alloys are very promising in this area.
The aim of the work is to carry out structural studies of the effect of ultrasonic treatment on physical-and-mechanical properties and on the microstructure of titanium alloys VT1-0 and VT6.

2. Materials and Research Methods

Titanium-based alloys VT 1-0 (α- phase) and VT6 ($\alpha + \beta$) were used in the study. Samples of alloys VT 1-0 and VT6 were incubated for 1 hour at temperatures of 680 °C and 750 °C, respectively, followed by air cooling.

Ultrasonic surface treatment of samples was carried out using an ultrasonic processing kit UZTK-02. The kit consists of an ultrasonic generator UZG-02/22 and a tool which includes: a magnetostrictive transducer PMS-063 and an indenter. A special ultrasonic impactor tool was installed into a toolholder of the machine (Fig. 1). Technical characteristics of the ultrasonic equipment: capacity – 0.25 kW; ultrasonic transducer - magnetostrictive; operating frequency – 22 kHz; inverter cooling - air; cooling air flow rate – 30 m at a pressure of not less than 0.3 MPa.

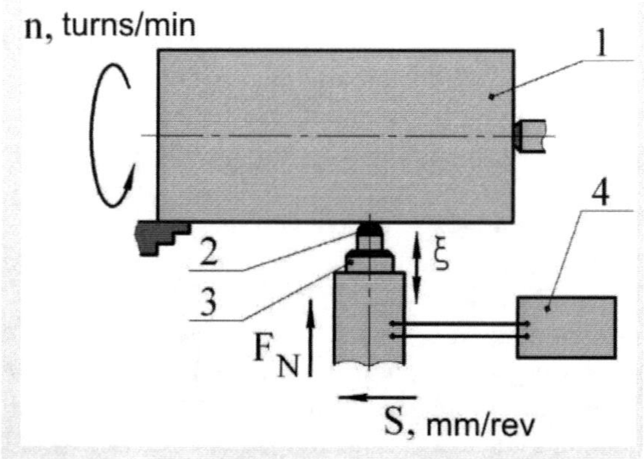

Figure 1. The scheme of ultrasonic surface treatment: 1 – treated workpiece, 2 – ball made of hard alloy, 3 – magnetostrictive transducer, 4 – ultrasonic generator

The study of the morphology and the surface roughness of titanium samples, as well as the measurement of the cross-section area of the track, were carried out using a profilometric complex "MICRO MEASURE 3D station" and a non-contact optical profilometer NewView 7300.

Surface hardening of titanium samples was evaluated by an increase in microhardness values using the microhardness tester PMT-3 with a load on the indenter of 20 grams.
Roughness parameters were measured using the profilograph software allowing to register the obtained measurements and to carry out their statistical processing with acquisition of various data on the surface of the studied sample: Ra, Rz, Rq, Rp, Rv, Rt, Rsk, Rru, RTp, RHTp.

The metallographic study of the surface structure of titanium samples was carried out using an electronic inverted universal metallographic microscope Axiovert 200 MAT / M MAT (Carl Zeiss). Electron microscopic studies of the structure of titanium alloys were carried out using an electron microscope JEM-200FX.

Prior to ultrasonic exposure the titanium samples were mechanically treated on a lathe under the modes shown in the table 1. For formation of an inclination angle of lateral surfaces of ridges the parameters of lathe turning were varied: longitudinal feed of the slide S which setts the distance between separate ridges of the forming surface microrelief and the angle at the tip of the cutter.

Table 1. Modes of mechanical treatment of samples

Alloy brand	N of the sample	Cutting speed V, m/min	Longitudinal feed of the cutter S, mm/rev	Angle at the tip of the cutter ε, degrees
VT1-0	1	55	0,1	45
	2	55	0,1	60
	3	55	0,1	90
	4	55	0,2	45
	5	55	0,2	60
	6	55	0,2	90
VT6	1	55	0,1	45
	2	55	0,1	60
	3	55	0,1	90
	4	55	0,2	45
	5	55	0,2	60
	6	55	0,2	90

Six samples of each brand of titanium alloys with different surface microgeometry were prepeared for the study of the effect of preliminary surface roughness of titanium alloys on the surface hardening process during ultrasonic treatment. Samples had a cylindrical shape with a diameter of 40 mm and a height of 20 mm. The outer cylindrical portion of the sample were subjected to treatment. Surface treatment of titanium alloys was carried out using different cutters depending on the brand of titanium (VT1-0 samples were treated using cutters R6M5; samples VT6 using the alloy VK6).

IOP Conf. Series: Materials Science and Engineering **91** (2015) 012040 doi:10.1088/1757-899X/91/1/012040

The microgeometry of the surface formed as a result of mechanical treatment depends on the geometry of the tool and the feed, as well as on other factors significantly altering the theoretical profile. The latter include: elastic and plastic deformations, friction on the rear surface of the tool, outgrowth, vibrations, and changes in the contour of the blade during its wear. However, in this study the factors influencing the process of microgeometry formation generated by the cutting process and the factors that depend on the equipment and the quality of tools were excluded from consideration. Only factors determining the geometric formation of the residual ridge were taken into account.

In this regard, the variable parameters during lathe treatment were the longitudinal feed of the slide S, which setts the distance between separate ridges of the formed surface microrelief, and the angle 8 - the angle at the tip of the cutter which forms an inclination of lateral surfaces of ridges. It should be noted that treatment of titanium alloy VT1-0 was carried out using a cutter with a cutting part made of high-speed steel R6M5. However, during treatment of titanium alloy VT6 the tip of the abovementioned cutter broke down immediately, not allowing to perform the intended treatment. In this connection, for treatment of the alloy VT6 the cutter was replaced with a cutter with a cutting part made of hard alloy of the brand VK6.

3. Results and Discussion

The indenter under the static and dynamic force generated by an oscillating system plastically deforms the surface layer of the workpiece preliminary treated by cutting. Surface treatment of the workpiece was carried out using a hard-alloy indenter oscillating with an ultrasonic frequency. In the area of the local contact of the indenter with the treated surface a plastic deformation zone occurs which shifts together with the indenter. All samples were treated under the same mode in order to eliminate the influence of modes of the ultrasonic treatment on the structure of the forming hardening layer.

The process of development of plastic deformation during the collision of a single indenter with an elastic-plastic material with known physical-and-mechanical properties and roughness consists of three stages. At the first stage the load is concentrated on ridges of microroughnesses, ultralocal stresses cross the elasticity line, and plastic deformation of the material begins from crumpling of microasperity tops. A continuous contact area is absent.

At the second stage with an increase in the load the plastic deformation captures groups of ridges, at the same time a mass formation of contact microareas occurs. These microareas, in turn, can be continuous or divided into separate areas depending on the overall surface microgeometry. In some areas plastic deformation may be combined with elastic deformation of the material depending on the shape of asperities, sizes, combination in their location, and, finally, depending on elastic and plastic properties of the material. The third stage is characterized by the beginning of deformation of the microprofile base, at the same time a continuous contact zone of the indenter with the surface occurs, capable of bearing the load without significant plastic deformations.

After treatment with an ultrasonic tool the surface of all samples appeared to be a homogeneous smooth microrelief with the roughness value Rz of 3.5÷5.5 micron, almost no different from each other. The carried out detailed study of the profilogram of the sample made of titanium alloy of the brand VT6 (sample number 2, Table 1) after ultrasonic treatment proves the suggestion that the greatest effect of ultrasonic treatment is achieved on roughness ridges, and in the underlying areas it is significantly smaller. Thus, Fig. 2 clearly shows residual depressions of roughness ridges obtained during mechanical treatment which did not undergo any changes during the ultrasonic treatment. All this indicates that a deformation zone of the first kind occurs, which due to the low penetration of the tool into the detail leads only to partial smoothing of the roughnesses. Contact of the tool with the workpiece in this case has an intermittent nature. The mean microhardness value of titanium alloy VT1-0 is 2400 MPa [4]. Microhardness measurements of titanium alloy VT1-0 samples which had undergone ultrasonic treatment have shown that there is s significant spread in its mean value in the range of values from 3200 MPa up to 4000 MPa. This spread depends on the shape and the size of the roughness relief. A similar situation occurs when measuring the microhardness of titanium alloy VT6 samples.

The analysis of profilograms of titanium alloy VT1-0 samples has shown that a regular microrelief of the surface with the roughness value from Rz = 5 microns up to Rz = 100 microns occurs as a result of mechanical treatment, consisting of successively located ridges and roots of a certain height at a constant pitch (Fig. 2). The regularity of the profile is of great importance in formation of hardened surfaces because it provides the consistency of properties of the treated surface throughout the entire area of the contact.

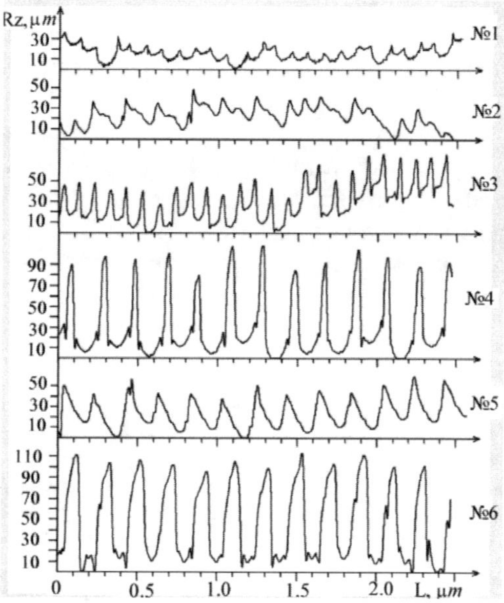

Figure 2. Surface profilograms of samples after mechanical treatment of the alloy VT 1-0

The analysis of profilograms of titanium alloy VT1-0 samples has shown that a regular microrelief of the surface with the roughness value from Rz = 5 microns up to Rz = 100 microns occurs as a result of mechanical treatment, consisting of successively located ridges and roots of a certain height at a constant pitch (Fig. 3). The regularity of the profile is of great importance in formation of hardened surfaces because it provides the consistency of properties of the treated surface throughout the entire area of the contact.

It has been established that the microrelief of the surface of titanium alloy VT6 samples obtained under same modes of mechanicacl treatment has a considerably lower surface roughness as compared to samples of titanium alloy VT 1-0. This can be explained by the fact that with a decrease in the plasticity the metal deforms less, the build-up reduces, vibrations reduce and, as a result, the cleaniness level of the treated surface increases.

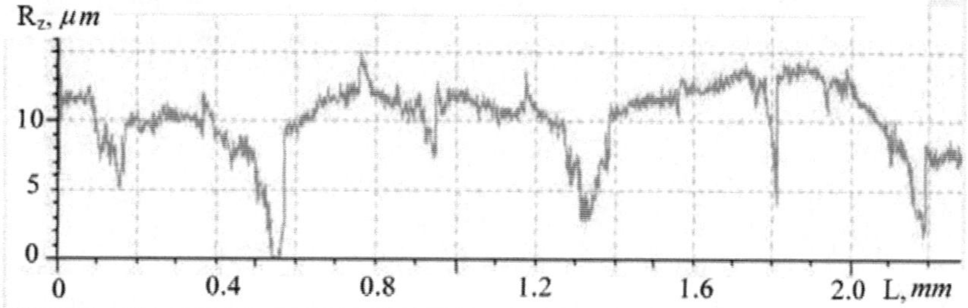

Figure 3. The surface profilogram of the sample of titanium alloy VT6 after ultrasonic treatment

Fig. 4 presents the results of microhardness studies Hμ of titanium alloys from the initial surface roughness. The analysis of these dependencies taking into account a considerable spread of values Hμ

IOP Conf. Series: Materials Science and Engineering **91** (2015) 012040 doi:10.1088/1757-899X/91/1/012040

allows to draw a conclusion that an increase in the surface roughness results in an increase in the microhardness. This leads to the fact that samples with the highest initial roughness have the greatest increase in surface microhardness values.

To study the microhardness distribution along the depth of the sample the transversal metallographic sections were prepeared. Samples of each brand of the titanium alloy with the highest microhardness were subjected to the study.

The analysis of the obtained data for titanium alloy VT 1-0 has allowed to establish that ultrasonic treatment of the surface layer leads to a gradient increase in microhardness values from the initial 2400 MPa up to 4000 MPa on its surface. The value of the transitional zone has ammounted to nearly 120 microns (Fig. 5).

The study of the sample of titanium alloy VT6 has revealed that ultrasonic modification of the surface layer leads to a gradient increase in microhardness values from the initial 4000 MPa up to 5400 MPa on its surface. Despite the fact that during treatment of titanium alloy VT6 the hardening effect from ultrasonic treatment is somewhat smaller and amounts to 1400 MPa the value of the transitional zone, wherein a monotonous decrease in the microhardness is observed, is considerably smaller and is about 40 microns.

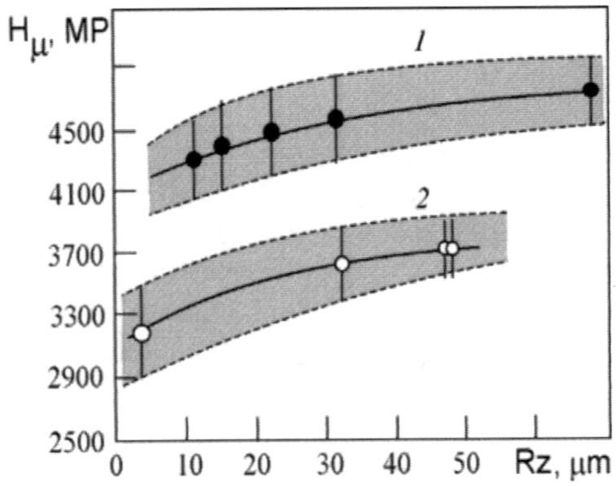

Figure 4. The dependence of the microhardness on the roughnes of alloys VT6 (curve 1) and VT1-0 (curve 2). Gray color indicates the area of changes in the microhardness taking into account the measurement error

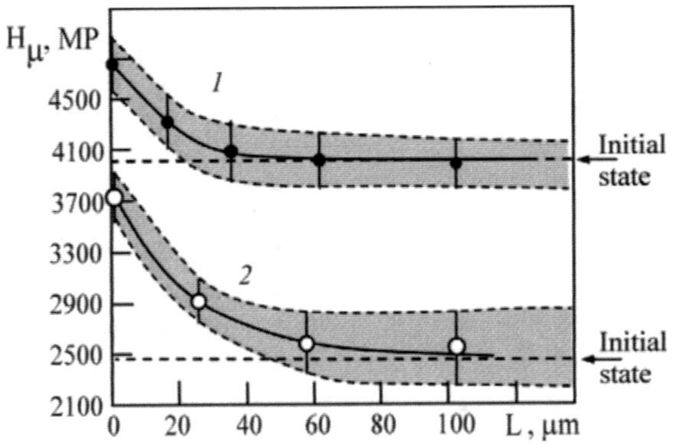

Figure 5. Distribution of the microhardness along the depth from the sonicated surface: 1 – VT6; 2 – VT1-0. Gray color indicates the area of changes in the microhardness taking into account the measurement error

Metallographic studies of the structure of the surface layer of titanium alloy VT1-0 after mechanical treatment have shown that mechanical treatment of the surface leads to formation of a defect structure. The micrograph (Fig. 6) shows that grains are elongated in the direction of the cutting tool movement. Here, grains repeat the contour of the roughness ridge formed by the cutting tool.

Studies of the structure of the surface layer of titanium alloy VT1-0 after ultrasonic treatment have shown its strong refinement with the texture, also directed towards the longitudinal feed of the indenter movement. The value of this surface layer and the refined structure is about 100 microns. As a result of the study of the structure of the surface layer of titanium alloy VT1-0 after sonication it has been found that the surface layer of the sample has a profoundly refined structure with a texture also directed towards the longitudinal feed of the indenter movement. This indicates that the structure has undergone significant changes as a result of the ultrasonic treatment. The degree of refinement of the grain structure is quite substantial, as evidenced by the photomicrographs (Fig. 6).

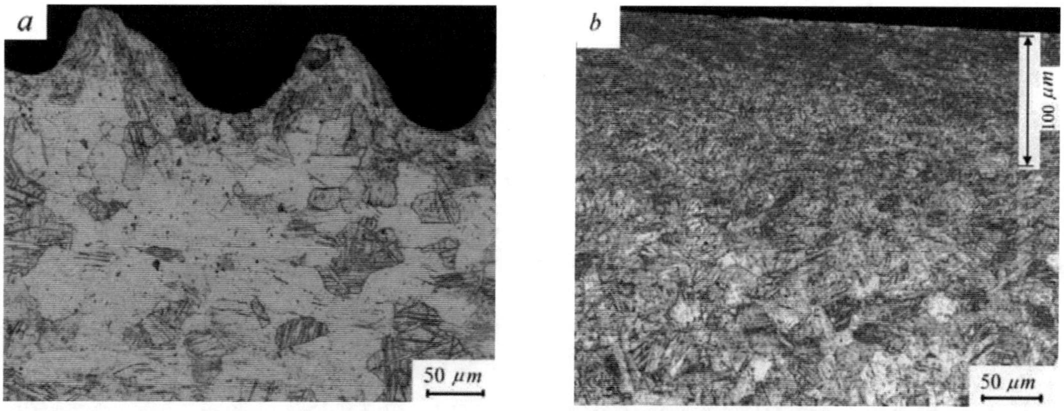

Figure 6. The metallographic image of the surface of titanium alloy VT1-0 after mechanical treatment (a) and after the subsequent ultrasonic treatment (b)

Electron-microscopic studies of samples made of titanium alloy VT1-0 with the highest microhardness have allowed to establish that sonicated surfaces have a heterogeneous microstructure in terms of the volume. The average size of structure elements on the surface of the sample are about 100 nm. At a distance from the surface the average size of structure elements increases up to 200 nm with a further increase up to a regular coarse-grained state.

Similar electron-microscopic studies of sonicated samples of titanium alloy VT6 with the highest microhardness have shown that the microstructure is also heterogeneous in terms of the volume of the near-surface region. The average size of structure elements on the surface of the sample on average is equal to 50 nm. At a distance from the surface the average size of structure elements increases up to 120 nm with a subsequent increase up to a regular coarse-grained state.

Conclusion

1. The existence of the dependence between the initial surface roughness of titanium alloys and the microhardness after ultrasonic treatment has been established. The microhardness of titanium alloys after ultrasonic treatment increases with an increase in the initial surface roughness.

2. It has been shown that an increase in the initial surface roughness of titanium alloys up to 80 microns allows, as a result of ultrasonic treatment, to form in the surface layer a structure with an average size of elements of 100 nm for titanium alloy VT1-0 and 50 nm for titanium alloy VT6.

The paper includes the results obtained in the course of the project implementation of the Ministry of Education and Science of the Russian Federation (number 8.1.42.2015), as a part of the programme "D.I. Mendeleev Science Foundation of Tomsk State University" in 2015, and the state task in the field of scientific research (project number 11.351.2014/K).

References

[1] Valiev R Z and Alexandrov I V 2007 *Bulk nanostructured metallic materials* (Moscow: ICC Akademkniga)

[2] Beliy A B Kukareko V A Kononov A G Kopylov V I Sharkeev Yu P 2008 *Friction and wear.* **29** **6** 571-577.

[3] Lebedev VA Streltsova I P 2006 *Strengthening technologies and coatings.* **7** 7-12.

[4] Klimenov V A Kovaievskaya Zn G Borozna V Yu Zeming Sun Qifang Zhu, 2009 *Rare Metals,* **28** 195-198

IOP Conf. Series: Materials Science and Engineering **91** (2015) 012041 doi:10.1088/1757-899X/91/1/012041

Forecasting of operational indicators of grinding tools with the controlled form and orientation of abrasive grains

V A Korotkov[1, 2, a], **E M Minkin**[2, b]

[1]National Research Tomsk Polytechnic University,
Tomsk, 30 Lenin Avenue, 634050, Russia
[2]Kuzbass State Technical University named after T. F. Gorbachev,
Kemerovo, 28 Vesennjaja Street, 650000, Russia

[a]e-mail: korotkov-va@mail.ru, [b]e-mail: jonm1@mail.ru

Abstract. The interconnection of the abrasive grain front angle parameter with the form, orientation and wear out parameters is investigated. The form of the abrasive grains was estimated by means of form coefficient which represents the relation of diameters of the spheres described around contours of grains, to diameters of the spheres entered in them. The spatial orientation angle of the abrasive grains was defined between main (i.e. the biggest) axis of the grains and the cutting plane. It is established that, depending on an orientation angle at increase in a form coefficient of the abrasive grains can be either an increase or a decrease in the values of their front angles. In most cases, with an increase in a form coefficient of the oriented grinding grains (at orientation angles Θ=10°÷125°) the growth of their front angles is fixed. At tangential orientation of grains (Θ=0°) and at the close directions of orientation (Θ=135°÷180°) the return picture is observed. Also established that the longer the abrasive grain wears along the main axis and located in the tool body, the larger is its front angle. Besides that, the front angles of the abrasive grains reach the maximum positive values at orientation angles Θ=22.5°÷45°.Dependence of tension in grains during the work with parameters of their form, orientation and depth of embedment in the bundle is investigated. It was found that for all orientation angles of grains their tension significantly increases with an increase in their form coefficient. Besides that it is confirmed that the deeper the grain is in the bundle, the lower the tension is there. Also found that tension is minimal when the grains are tangential orientated. Further on increase the option of the grains in the direction of action of the cutting force follows. Such option of orientation is the most rational both from the point of view of minimization of tension, and for ensuring rational sizes of front angles of the abrasive grains. The rational angle of grain orientation for cutting disc are in range Θ=65°÷75°.
The joint analysis of dependences of front angles and tension in grains from factors of their form and orientation allows the prediction of operating characteristics of grinding tools at the design stage for specific tasks and processing conditions.

1. Introduction

The effectiveness of the grinding process is largely predetermined by the efficiency of the abrasive grains of which grinding tools are manufactured [1, 2, 3]. The most successful work of grains, with other things being equal (brand and granularity of an abrasive, type of the abrasive tool, hardness, structure and bundle type of the tool, the grinding conditions), is possible when grains possess a favorable geometry for the chosen processing case. The geometry of the cutting part of grain is defined by two factors – a form and orientation in a body of the tool [4, 5, 6]. These factors in the process of production of abrasive tools are often remained uncontrolled and uninvolved.

So, bulk of grains from which manufacture the abrasive tools, has a form changing in the wide range from isometric to lamellar versions. Such dispersion of a form of grains is caused by specifics of standard technologies of their production which are based on crushing of abrasive ingots and the subsequent division of abrasive particles according to the sizes. At the same time, the grain form factor remains almost uncontrolled and uncontrollable. In world practice the attempts to streamline the form of grains in the abrasive tools by introduction of the special technologies directed on creation or allocation of fractions of grains with an identical form are known. Manufacturing techniques of spherical grains and grains of other special forms, technology of ovalization of grains in mills with pig-iron spheres, manufacturing techniques of abrasives in which thanks to the accelerated cooling of the abrasive melt and emergence on ingots a grid of micro cracks receive grains of needle and lamellar forms, etc. are known [1, 3, 7, 8, 9]. Along with the form parameter, other factor which actively influences geometry and overall performance of grains is their orientation in a body of the tool [3, 10, 11]. This is exemplified by the use of abrasive belts and skins produced using electrostatic field exposure. This effect makes it possible to orient the grain spikes perpendicular to the base tape [12]. Industrial and positive experience of creation of abrasive tools of other types with application of technologies on management of orientation of grains isn't known yet. Still more it belongs to production of abrasive tools at which the form and orientation of the abrasive grains would purposefully be regulated at the same time. The creation of such tools makes it possible to further improve the efficiency of grinding processes [13].

In this regard the problem of creation and application of new abrasive tools made of abrasive grains with controlled form and orientation is actual. Such tools allow to maximize the usage of cutting capabilities of each individual grain. The solution of this problem requires carrying out researches which would allow to predict the operational indicators of grinding tools depending on factors of a form and spatial orientation of abrasive grains.

2. Methodology of researches

For forecasting of operational characterizations of grinding tools at a design stage it is expedient to investigate the geometry of grains of which they are made, and to establish dependence of sizes of front angles of grains on factors of their form and orientation. After that it is necessary to analyze the scheme of operation of the grinding tool and to establish the value of the components of the cutting force acting on the grinding grains. Using the established sizes of components of the cutting force, it is necessary to establish dependence of the maximum tension in the abrasive grains on the factors of their form and orientation. The joint analysis of the received dependences of front angles and tension in grains from the factors of their form and orientation allows the design of grinding tools for specific tasks and processing conditions, based on the fact that values of front angles of grains have a direct impact on the cutting ability of tools, on sizes of deformations of the processed material, effective power and temperature of cutting, and the level of the maximum tension in the grains affect the wear of tools [14, 15].

3. Results and Discussion

Researches were conducted on the example of cutting discs with the characteristic Type 41 230×4×32 A 46 C BF 80 m/s, which performs cutting of workpieces (21.3×2.8 pipe steel 1010 (U.S. standard AISI)) at a speed of V=80 m/s and with constant workpiece clamping force to the disc F=32 N (P_y=32 N) [16, 17]. In the course of researches it was established dependences of the front angles of oriented abrasive grains A 46 on a variety of their form, taking into account a wear factor. Also the interrelation of a form and orientation of grains with sizes of tension arising in them during the work was established. The form of grinding grains was estimated by means of specially developed technique and the computer program [3, 18] by using the concept of the form coefficient (K_f) equal to the relation of diameters of the spheres described around contours of grains, to diameters of the spheres entered in them.

For an assessment of front angles of the oriented abrasive grains the special program complex allowing to turn images of abrasive grains (received by means of a microscope and the digital camera) under the demanded corner concerning the cutting plane and to define a front angle in all points of a contour is developed [19, 20].

The received results were tabulated with the subsequent construction of three-dimensional plots of dependence of front angles of abrasive grains on their form coefficient (K_f) and the spatial orientation angle (Θ), measured between main (i.e. the biggest) axis of the grains (L) and the cutting plane (figure 1).

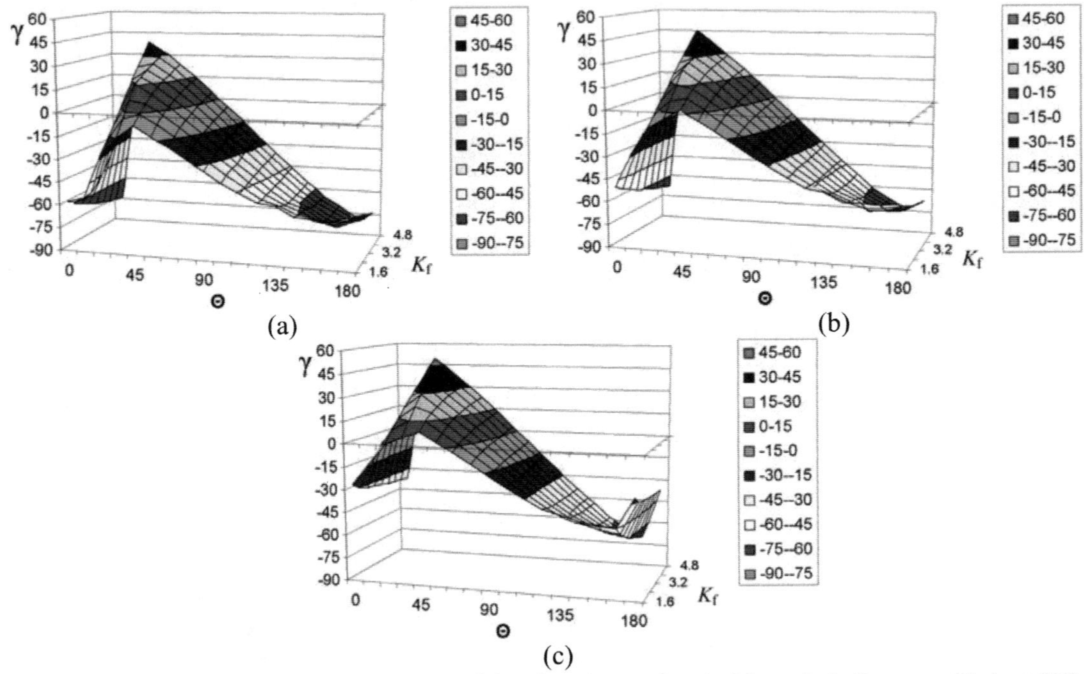

Figure 1. Dependences of front angles (γ) of the abrasive grains A 46 on their form coefficient (K_f) and the spatial orientation angle (Θ) at hypothetical wear to $1/3L$ (a), to $1/2L$ (b) and to $2/3L$ (c)

Results of researches allow to note that, regardless of the form coefficient, the more long the abrasive grain works, i.e. wears out along the main axis and is in a tool body, the more sharply (on average) there is its front angle. Thus the factor of orientation of grains concerning to the cutting plane very significantly influences sizes of front angles. So, for example, at K_f=1,6 front angles of abrasive grains can vary within -60° ÷ -6° (wear to $1/3L$) and -44° ÷ +9° (wear to $2/3L$).

Front angles of abrasive grains at K_f=4,8 depending on an orientation angle Θ change in the range of -77 ° ÷ +39 ° (wear to $1/3L$) and -69 ° ÷ +49 ° (wear to $2/3L$). Thus, the same grain which is differently oriented concerning the cutting plane is characterized by a considerable difference of sizes of front angles. It reaches 53° ÷ 54° at isometric abrasive grains (K_f=1,6) and 116° ÷ 118° at needle and lamellar grains (K_f=4,8) that essentially distinguishes geometry of grinding tools and for example cutters, drills and mills at which the cutting wedge has invariable geometry and is sharpened to within degree shares. The findings also show that the front angles of the abrasive grains reach the maximum positive values at orientation angles Θ=22.5° ÷ 45°.

Furthermore, depending on the angle of orientation with increasing grain form factor can be either an increase or a decrease in the values of their front angles. Thus, in most cases, with an increase in the form factor of the oriented abrasive grains (in the range of orientation angles Θ=10° ÷ 125°) their front angles increase. In tangentially oriented grains (Θ=0°) and grains with similar orientation direction (Θ=135° ÷ 180°) the opposite picture is observed.

For research the values of tension in abrasive grains at various options of their orientation in a tool body the components of cutting force acting on the grains A 46 during the work of cutting discs were defined previously. For this purpose in processing of preparations, among other indicators, the effective power of cutting was defined. Proceeding from it the tangential component of cutting force was established (P_z=8.46 N) [3].

After that the sections of a working surface of cutting discs by the digital camera in the mode of macro shooting were photographed and calculation of amount of grains on surface unit of area in the graphic Paint editor was made. It is established that, on average, on a working surface of the researched discs the number of grains makes N_s=2.68 of grains/mm^2. Proceeding from this value, as well as the contact area of the cutting disc with the workpiece to be cut (S=22.4 mm^2), the amount of the abrasive grains which are at the same time participating in work (N_s=60 grains) and the sizes of forces operating on single abrasive grain is established (P_y=0.533 N; P_z=0.141 N). The obtained data allowed to make calculation of tension in models of abrasive grains of brand A 46 with various form coefficients and options of spatial orientation in a tool body by the program SolidWorks. Models of grains with various form coefficients are constructed on the basis of radially oriented grains given according to front angles received by means of special computer programs [19, 20].

Software settings for calculation of tension in grains: linear elastic isotropic model; material - corundum; strength at stretching $8.5 \cdot 10^7$ N/m^2; strength at compression of $7.6 \cdot 10^8$ N/m^2; module of elasticity of $3.74 \cdot 10^{11}$ N/m^2; Poisson's coefficient is 0.22; mass density is 4025 kg/m^3; module of shift of $1.5 \cdot 10^{11}$ N/m^2; coefficient of thermal expansion $8 \cdot 10^{-6}$ 1/K. Grid type: a grid on a solid body on the basis of curvature; 4 points of the Jacobian; the maximum element size 0.00968809 mm; the minimum element size 0.00322933 mm. Criterion of durability – the criterion of the maximum normal tension used for fragile materials.

The forces operating on real single abrasive grain (P_y=0.533 N; P_z=0.141 N) is applied to models of grains with different variants of their orientation and depth of setting in a bundle (figure 2 (a), (b)). It was found that depending on concrete option of orientation of models of grains, the maximum tension arises on their various locations along their boundary seal in a bundle.

For a more complete picture of the state of stress models of grinding grain at all variants of their orientation were recorded and analyzed the data in three specific areas on the border seal in the bunch: zone on the left, central zone and zone on the right (figure 2).

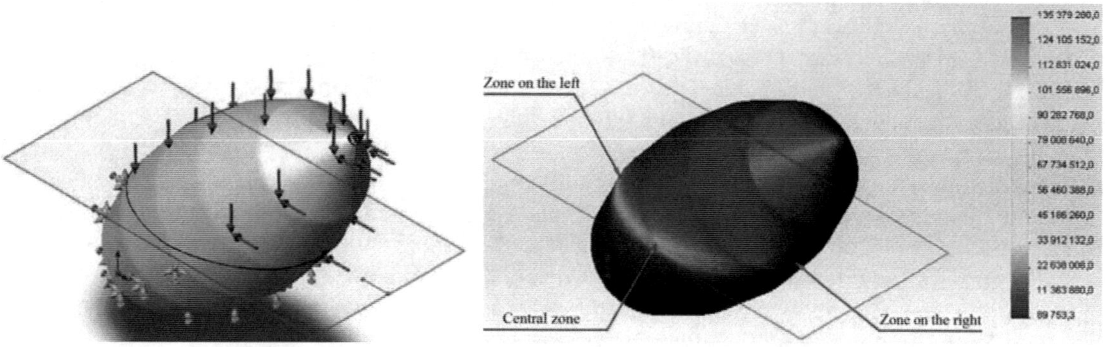

(a) (b)

Figure 2. The scheme of loading (a) and distribution of tension along the border in the bunch (b) in the model of grain A 46 with K_f=2.4 and the orientation angle Θ=45°, that is fixed in a bundle to 1/2L

The results of evaluation of tension in the model of abrasive grain with K_f=2.4 (average form coefficient of grains in the studied fraction) at various angles of orientation are presented in figure 3.

Given graphics (figure 3) indicate that the smallest tension arises at tangential orientation of grain. Further, in the range of grain orientation Θ=11°15′ to Θ=45° there is a significant increase in tension, with the maximum values achieved in the "zone on the right" under the console of the protruding portion of grain. In the following range from Θ=45° to Θ=67°30′ there is a decrease in the maximum tension to values, approximate to option of tangential orientation.

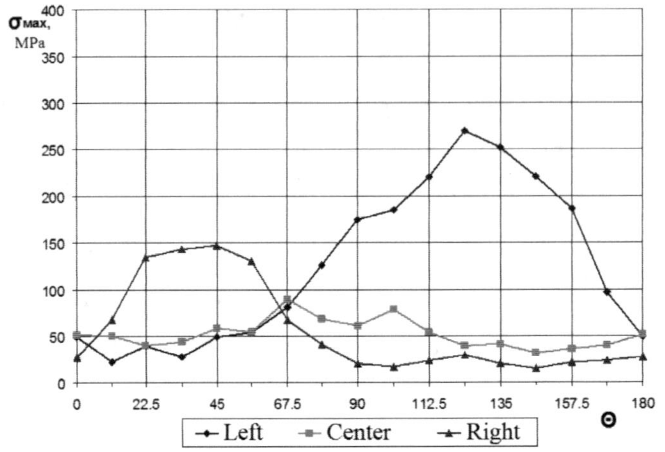

Figure 3. Tension in the model of abrasive grain A 46 with K_f=2.4 that is fixed in a bundle to 1/2*L* on border in a bundle at various corners of orientation

Thus, it is obvious that decrease in the maximum tension at Θ=67°30′ results from uniform distribution of tension in a grain body (at the left, on - to the center and on the right) along boundary seal in a bundle. Such distribution of tension is reached because the grain orientation angle Θ=67°30′ is close to the direction of the cutting force *P*.

In this case the components of the cutting force P_y=0.533 N, P_z=0,141 N, i.e. have a ratio 3,78:1 and the vector of cutting force *P* coincides with the maximal axis of grain at its orientation with Θ=75°12′. Calculation and the subsequent comparison of tension for orientation angles Θ=67°30′ and Θ=75°12′ show that more rational to reduce the maximum tension is the angle Θ=67°30′, which is explained by the peculiarities of termination of grain in the bundle.

With further increase in the orientation angle in the range of Θ=67°30′ to Θ=123°45′ there is a significant increase in tension to the largest observed values. The maximum loaded area here is "zone on the left" under the console of the acting part of grain. The obtained data also show that in all cases the more rational is the grain orientation by top towards to the direction of the component of the cutting force P_z. It finds an explanation that at orientation of grain towards to P_z the vector of cutting force *P* is near the main axis of grain. At a grain arrangement by its top from the P_z direction, the cutting force *P* works under big corners in relation to the main axis of grain, creating the maximum tension in "a zone at the left".

Further researches on establishment of rational angles of orientation of abrasive grains taking into account kinds of their form (K_f=1.6; K_f=2.4; K_f=3.2; K_f=4.0; K_f=4.8) and seal depths in a bundle revealed similar nature of tension distribution (figure 4 (a), (b), (c)).

The obtained data (figure 4) show that for all orientation angles of grains at increase in their form coefficient the maximum tension significantly increases. So, in the most rational angle of orientation of grains of an order Θ=67°30′ upon transition from K_f=1.6 to K_f =4.8 (seal in a bundle to 1/2*L*) the maximum tension increases by 4.2 times (from σ_{Max}=79,8 MPa to σ_{Max}=335,6 MPa). At the least rational orientation angle near Θ=123°75′ upon transition from K_f=1.6 to K_f =4.8 the maximum tension increases in the area by 7,7 times (from σ_{Max}=187,7 MPa to σ_{Max}=1443,9 MPa).

The revealed dependence correlates with results of tests according to wear and grinding coefficient of cutting discs with a controlled form of grains [16, 17] and allows to confirm them theoretically.

IOP Conf. Series: Materials Science and Engineering **91** (2015) 012041 doi:10.1088/1757-899X/91/1/012041

Besides, it is established that with reduction of depth of seal of grain in a bundle the maximum tension in grains also noticeably increases.

So, for various form coefficients of grains, upon transition from seal in a bundle to 2/3*L* to the sealing of up to 1/2*L* the maximum tension in grains (for various angles of orientation on average), increases in 1,03÷1,25 times, and upon transition from seal in a bundle to 1/2*L* to the sealing of up to 1/3*L* growth makes 1,24÷1,77 times. Thus, the effect of planting depth in the bundle increases with increasing of the grain form coefficient.

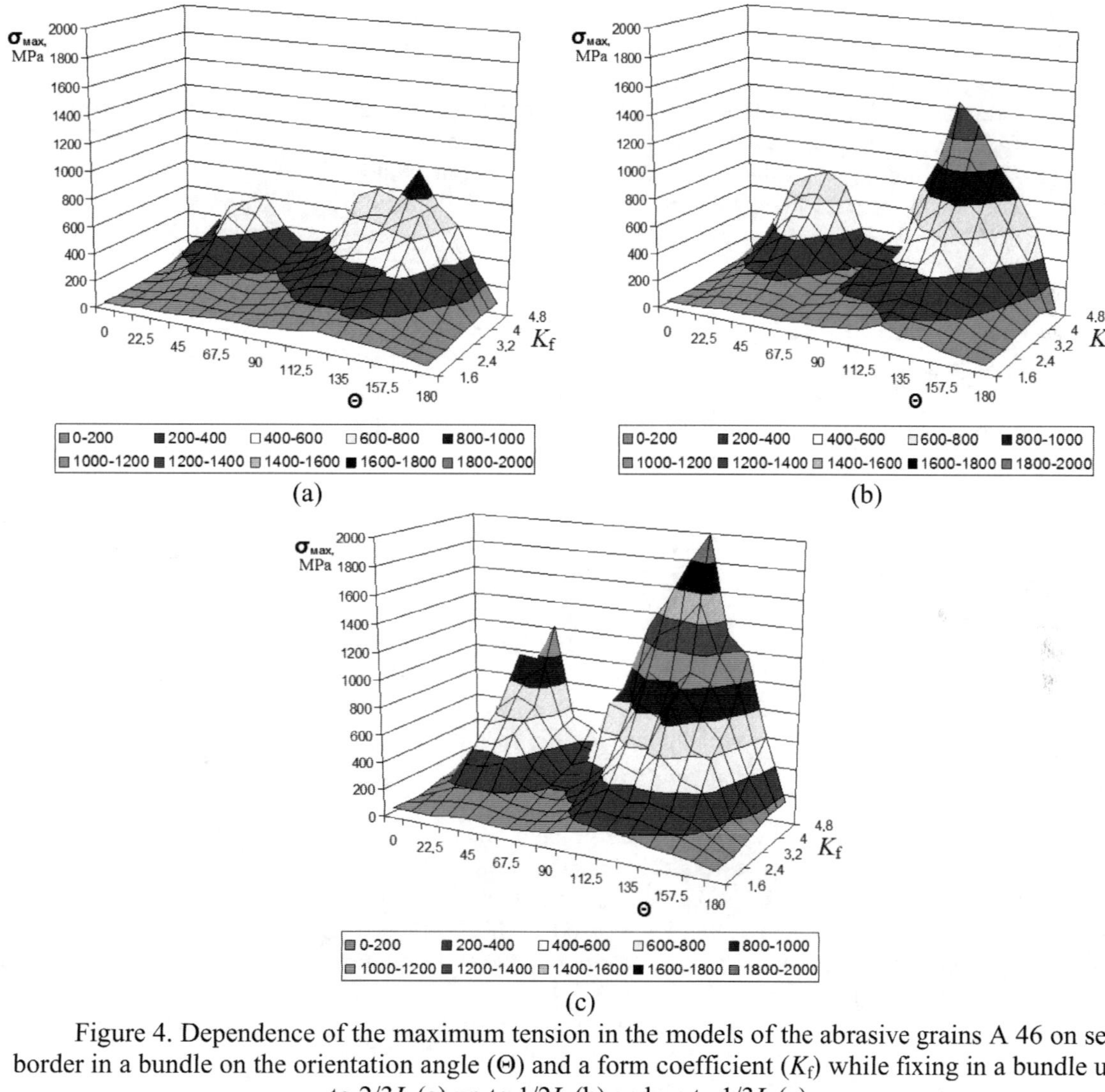

Figure 4. Dependence of the maximum tension in the models of the abrasive grains A 46 on seal border in a bundle on the orientation angle (Θ) and a form coefficient (K_f) while fixing in a bundle up to 2/3*L* (a) up to 1/2*L* (b) and up to 1/3*L* (c)

4. Conclusion

The joint analysis of the evaluation results of the front angles and tension in the oriented abrasive grains with a controlled form allowed to formulate the following conclusions:

- To achieve the maximum positive values of the front corners of the grains and the maximum work productivity of grinding tools should be used needle and lamellar grains (with high values of K_f) and orient them in the tools under orientation angles in the range Θ=22.5° ÷ 45°;

- To ensure maximum durability of grinding tools, with other things being equal, it is expedient to use abrasive grains of an isometric form (with the minimum K_f values), and also to focus grinding grains in the tangential direction (Θ=0°);
- Grinding tools with the increased work productivity and at the same time with a sufficiently high durability can be produced by orienting the abrasive grains under angles, close to the direction of the cutting force vector *P*;
- For the grinding wheels made of corundum abrasive grains and used for processing of structural steel at which operation the radial component of cutting force P_y in 3÷5 time bigger than the tangential component P_z, the rational angles of grain orientation are in range Θ=65°÷75°.

References

[1] Korotkov A N, Increase of operational opportunities of grinding tools, *Publishing House of Kuzbass State Technical University*, Kemerovo, 2005.
[2] Zaitsev A G, Influence of the diamond grains form on the wear resistance of grinding wheel for processing of firm alloys, *J. Sc. Bulletin of machine building*. 2 (1975) pp 76–77.
[3] Korotkov V A, Increase of operational opportunities of cutting discs, *Machine building*, Moscow, 2009.
[4] Korotkov A N, Operational properties of abrasive materials, *Publishing house of Krasnoyarsk University*, Krasnoyarsk, 1992.
[5] Granovsky G I, Granovsky V G, Cutting of metals, High School, Moscow, 1985.
[6] Vakser D B, Ways to increase productivity in grinding, *Publishing House of Moscow State Technical University named after N.E. Bauman*, Moscow, 2001.
[7] Tyrkov V N, Abrasive materials and instruments. Catalogue, All-Russian Research Institute of Information and Feasibility Studies in Mechanical Engineering and Robotics, Moscow, 1986.
[8] Pliss D A, Anakhin V A, Monakhov V N, Vibration separators, *Bowels of the earth*, Moscow, 1991.
[9] Korotkov V A, Technology and equipment for producing of the experimental cutting discs *Modern techniques and technologies (MTT – 2006) : Proceedings of the 12-th international scientific and practical conference of students, postgraduates and young scientists*, March 27–31, 2006. – Tomsk, Russia, 2006. pp 177–179.
[10] Lyukshin V S, Increase of operability of grinding tapes by use of grains with a controlled form and orientation, *Abstract of the thesis of Candidate of Technical Sciences*, Kemerovo, 2007.
[11] Korotkov V A, Russian Federation Patent 2,369,474. (2009).
[12] Kovalchuk A M, Fundamentals of design and manufacturing technology of abrasive and diamond tools, *Mechanical Engineering*, Moscow, 1984.
[13] Korotkov V A, Ways analysis operational characterization increase of the polishing wheels // Modern techniques and technologies (MTT 2003) : *Proceeding of the 9-th international scientific and practical conference of students, post-graduates and young scientists*, April 7–11, 2003. – Tomsk, Russia, 2003. pp 102–103.
[14] Korotkov V A, Minkin E M, Research of geometrical and strength characteristics of the oriented abrasive grains. , *J. Sc Processing of metals*. 3 (2012) pp 42-47.
[15] Petrushin S I, Proskokov A V, Theory of not free cutting of materials. Calculation of the intense deformed state in a cutting zone, J. Sc. *Bulletin of machine building*. 2 (2010) pp 34-40.
[16] Korotkov V A, Research of operational characterizations of cutting discs with oriented abrasive grains *Applied Mechanics and Materials* : Trans Tech Publications, doi : 10.4028/www.scientiffic.net/AMM.682.224 – Switzerland, 2014, Vol. 682(2014) pp 224–230.
[17] Korotkov V A, Research of the exploitation abilities of the cutting discs, which consist of the abrasive grains with controlled orientation *Modern Techniques and technologies* (MTT 2007) : Proceeding of the 13-th international scientific and practical conference of students,

post-graduates and young scientists, March 26–30, 2007. – Tomsk, Russia, 2007. pp 49–51.
[18] Korotkov V A, Rilov G M, Russian Federation Certificate on official registration of the computer program 2006,613,051 (2006).
[19] Korotkov V A, Minkin E M, Russian Federation Certificate on official registration of the computer program 2011,614,263 (2011).
[20] Korotkov V A, Minkin E M, Russian Federation Certificate on official registration of the computer program 2011,615,114 (2011).

Entering the operative correction machining processes CNC

R Yu Nekrasov[1], A I Starikov[1] and A A Lasukov[2]
[1]Tyumen State Oil and Gas University, 38, Volodarskogo street, Tyumen, 625000, Russia
[2]Yurga Institute of Technology, TPU affiliate 26, Leningradskaya street, Yurga, 652050, Russia

E-mail: syncler@mail.ru

Abstract. The article describes the solution to the problem of compensation of errors occurring during machining on CNC machines. We propose a method of mathematical modeling of processes diagnostics and control of technological equipment. The results of the diagnosis of the CNC machine, as well as the mathematical model describing the dependence of the positioning error of the executive bodies of operating component of cutting force P_Z, in the range of movement OX.

Introduction.

The problem of improving the accuracy of process equipment associated with the cutting force caused by the deformation is solved for many years, but it is still not completely solved.

Errors, inaccuracies predefined positioning of the cutting tool in the operating space of the machine are determined or eliminated by CNC systems, where the bodies of stepper drives constantly monitor the position of the working elements of the technological system. However, attempts to solve the problem of occurrence of errors associated with strains did not find their wide application. So the use of adaptive control systems invented B.S. Balakshin not widely used due their improper conditions of modern market economy. After all, in order to stabilize the individual parameters of the cutting process had understate modes, despite the fact that she did not meet the stabilization of optimal performance.

Path to the maximum optimization of cutting process was found only in the 21st century. Scientists from the Tyumen State Oil and Gas University have proposed to introduce corrections in the work process equipment directly in the processing, based on operative diagnostic [1, 2, 3]. So for the regulation of the cutting conditions, it was decided to base, the defined directly in the process of machining, deformation coefficient of the shear layer, and to enter corrections to compensate for deformation in the process system data on power strain on the servo drive of the machine.

Figure 1 shows a diagram of the transformation commands required linear movement into control signals supplied to drive the technological systems that provide real displacement of technological system elements. In modern CNC systems installed better interpolators [4, 5], which provide an estimate on several coordinates and allow you to enter correction to compensate for the errors of single steps and accumulated leadscrew errors. However, the verification of accuracy of positioning is done periodically, and the error is determined by the entire length of the movement of the body of the machine.

Massive use of operative diagnostics and correction is not possible without the of the universalization of process, which in turn requires the use of mathematical modeling techniques [6, 7].

So for process modeling, it was decided to use the language of mathematical logic, the possibility of having a clear description of the patterns of interaction of elements of technological system.

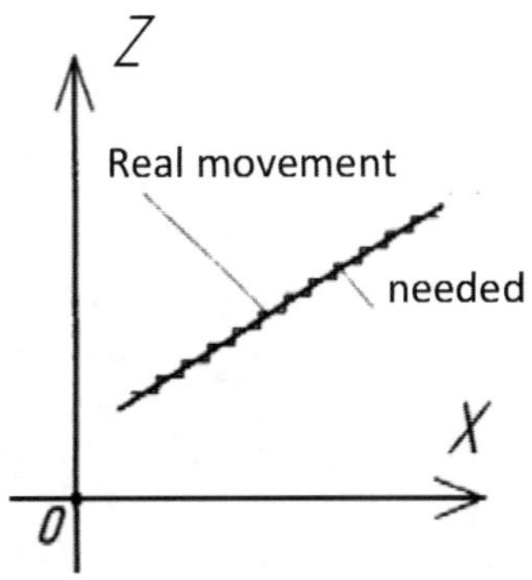

Figure 1. Linear interpolation of movement

To derive a mathematical model, by the language of predicate logic, necessary to draw the alphabet of its operands.

The operands may include elements of the technological system and thereof parts. In addition, alphabet must include all occurring in the process of diagnosis and management of processes and their parameters.

Table 1. The alphabet of operands of mathematical models of processes of diagnostics and management

			Technological system elements
A			Machine tool
	A1		Bed
		$A1_1$	Guide
	A2		Headstock
		$A2_1$	Spindle
	A3		Caliper
		$A3_1$	Instrumental head
	A4		Tailstock
	A5		Power elements
		$A5_1$	Longitudinal feed drive
		$A5_2$	Cross feed drive
		$A5_3$	Maine drive
	A6		Table
		$A6_1$	Rails
B			Instrument
	Bn		Instrumental element
C			Workpiece
	Cn		Workpiece element
D			Load device
	D1		Simulating the cutting force P
	D2		Simulating the component of cutting force P_x
	D3		Simulating the component of cutting force P_z
E			Measuring device
	E1		Measuring device probe

	E2		Instrumental measuring device
	E3		Measuring ruler
	E4		Cutting layer deformation meter
F			Auxiliary tool
	F1		Holder
	F2		Front center
	F3		Rear center
G			Information system (CNC)
	G1		Diagnostic data registration system
	G2		Correction calculation system
	G3		Lathe management system
			Processes
L			Diagnostic system
	L1		Technological system cutting force deformation process
		$L1_1$	Cutting force workpiece deformation process
		$L1_2$	Caliper deformation diagnostic process
		$L1_3$	Instrumental head shift diagnostic process
		$L1_4$	Temperature deformation diagnostic process
		$L1_5$	Straightforwardness rails deviations process
		$L1_6$	Cutting layer deformation diagnostic process
		$L1_7$	Feed drive load characteristics diagnostic process
M			Machining process
	M1		Process parameters
		$M1_1$	Cutting modes
		$M1_2$	Cutting tool movement trajectory
N			Diagnostic data registration process
O			Correction calculation process
	O1		Correction vectors according to the operational diagnosis force parameters
	O2		Correction vectors according to of operative diagnostics of temperature settings
	O2		Correction vectors according to preliminary diagnostic
	O3		Cutting modes corrector
P			Corrections entering process
			Measurement parameters
S			Diagnostic parameters
	Δ^{TC}		Lathe technological system deformation
		ΔX^{3}	X-axis workpiece deviation
		ΔZ^{3}	Z-axis workpiece deviation
		ΔX^{A3}	X-axis caliper cutting force deviation
		ΔZ^{A3}	Z-axis caliper cutting force deviation
		ΔX^{A31}	X-axis instrumental head deviation
		ΔZ^{A31}	Z-axis instrumental head deviation
		ΔR	Cutting tool wear
		V^{z}	Longitudinal feed drive stress
		V^{x}	Cross feed drive stress
		ξ	Shear layer deformation coefficient

For example, to simulate the process of diagnostics deviations contour cutting tool [8, 9, 10], the operands are used, indicating the location of the operation, the object and the subject of the process, as well as the conditions and the output function describing the result.

1) Model diagnostic tool wear:

$$A\exists\left(B \underset{\varphi_1-\varphi_2}{\overset{1^\circ}{\exists}} \Delta_R \wedge E2\right) \in \left(f:\varphi \to \Delta_R\right) \Rightarrow G1\exists G \tag{1}$$

Reading as: For A machine tool in there instrument B, where there are deviations from the original contour Δ_R in the range of φ_1 - φ_2 with steps 1° when measuring device E2, described by the function *f* with range φ and domain Δ_R of the recorded data acquisition system diagnostics *G1*, of the CNC G

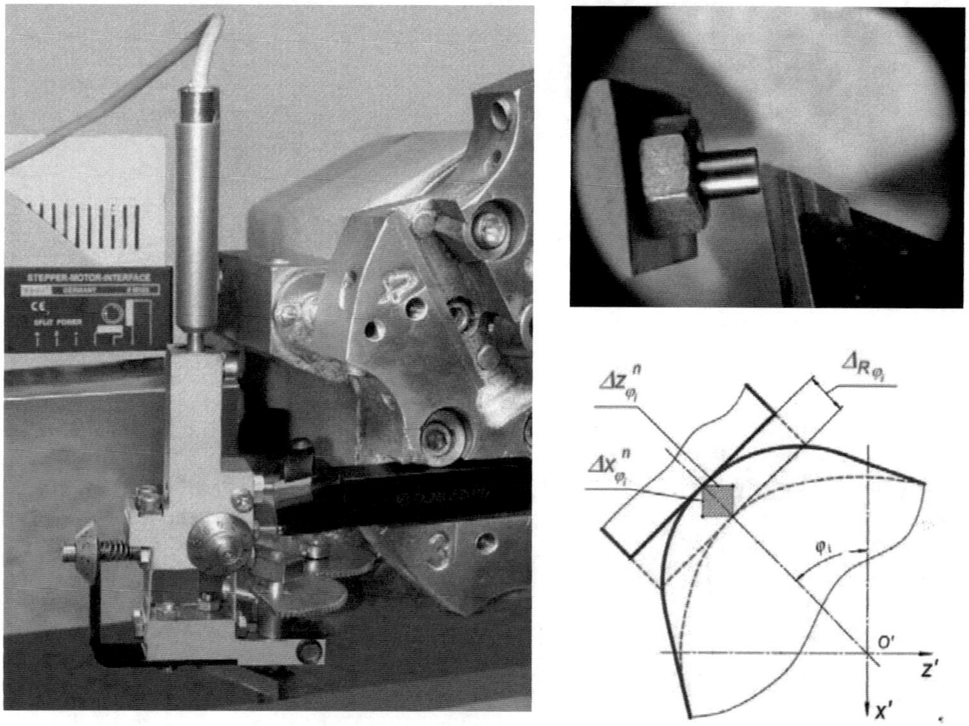

Figure 2. Tool wear diagnostic

Thus, the mathematical model of the process diagnostics of deformations in the technological system has the form:

$$A\exists\left(\underset{\pm OZ}{\exists} A3 \underset{\pm OX}{\exists} A3_1 \wedge D3\exists\Delta X^{A3_1} \wedge E1\right) \in \left(f:OZ \to \Delta X^{A3_1}; P_X\right) \Rightarrow G1 \tag{2}$$

Reading as: For *A* machine tool exists, having the ability to move along the axis OZ, caliper *A3* and installed on it, which has the ability to move along the axis OX toolholder *$A3_1$*, measuring arising under the influence of the load device *D3* positioning error ΔX^{A3} determined by reacting with measuring device *E1*, described by the function fc range of values OZ and domain and P_X, recordable data recording system diagnostics *G1*.

For diagnostic results generated depending on the direction of movement and direction of the load elements of the technological system (figure 3) were constructed three-dimensional graphs of variation in the positioning of the load carriage operation range shown in Figure 4.

As a result of the approximation of diagnostic data we obtain the dependence of deviations positioning along the OZ axis of the component P_X of the cutting force over the entire range OZ machine coordinate. So to determine the surgical correction is enough to know only the forces acting at a particular time on the elements of the machine.

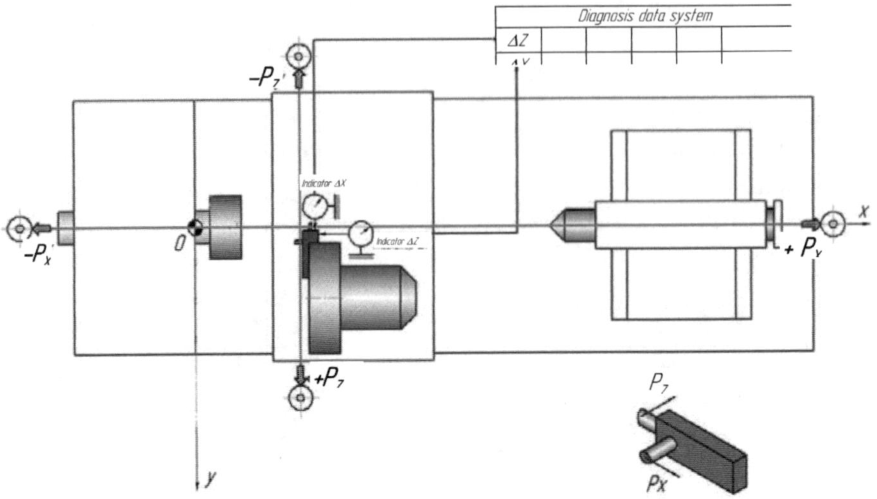

Figure 3. Technological equipment deformations diagnostic

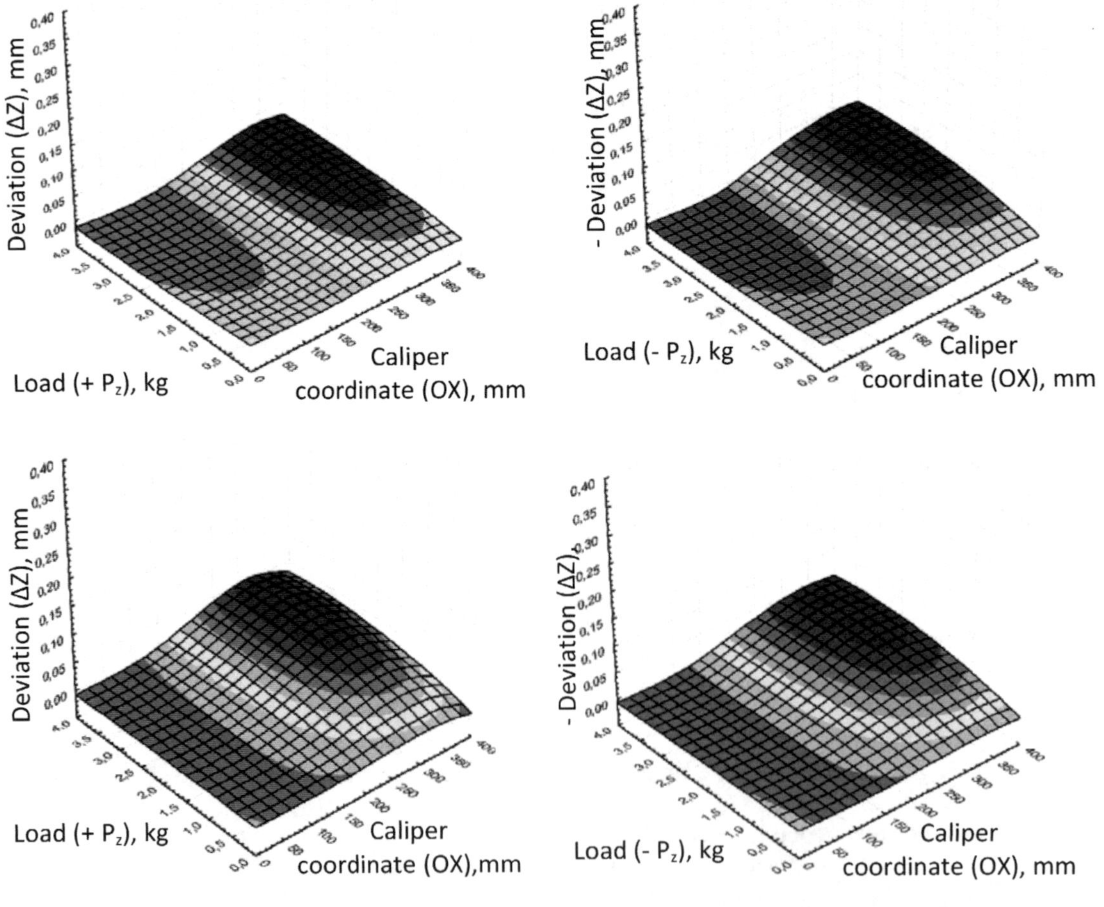

Figure 4. The dependence of deviations from the positioning of the load caliper in the operating range

After converting the surface by smoothing the quadratic polynomials we obtain the second degree, respectively, with positive and negative loading and forward and backward movement of the caliper.

Mathematical models of deviations with positive and negative loading and forward and backward movement of the caliper

$$\Delta Z=-0{,}0027-2{,}9192\text{-}5*X-0{,}0039*P_z+1{,}5938\text{-}7*X^2+3{,}4677\text{-}5*X*P_z-0{,}0005*P_z^2$$

$$\Delta Z=-0{,}0048-3{,}1299\text{-}5*X-0{,}0008*(-P_z)+2{,}1588\text{-}7*X^2+4{,}4859\text{-}5*X*(-P_z)-0{,}0013*P_z^2$$

$$\Delta Z=-0{,}0134+0{,}0001*X+0{,}0087*P_z-1{,}7195\text{-}7*X^2+3{,}8369\text{-}5*X*P_z-0{,}0028*P_z^2$$

$$\Delta Z=-0{,}0069+1{,}6564\text{-}5*X+0{,}0067*(-P_z)+8{,}5116\text{-}8*X^2+4{,}2091\text{-}5*X*(-P_z)-0{,}0025*P_z^2$$

Supplied directly to the interpolator move command can take into account the necessary corrective feedback, calculated on the basis of the application submitted by mathematical models. That is counted identified errors of technological system.

In addition to the data obtained as a result of diagnostics, it is assumed the input error correction eliminates determined by operative diagnosis and regulatory regimes in order to optimize the cutting process.

Summary

The method of simulation processes using mathematical logic and the proposed method of surgical correction of the input to the control program of the CNC machine directly to the interpolator - tested and have been known for many years. However, to realize the full possibilities offered by these methods appeared only with the invention of new ways to diagnose the cutting process. Applying the proposed methods can also be offset and error caused by other factors

References

[1] Putilova U S, Nekrasov Yu I and Lasukov A A 2014 Loading of the manufacturing systems elements in the process of unsteady mode cutting and the models of their arrangement deviations *Applied Mechanics and Materials* Vol. 682 pp 192-195.

[2] Nekrasov R Yu, Soloviev I V and Putilova U S 2015 Laser interferometry method of stress determination in loaded cutting tool parts *Applied Mechanics and Materials* Vol. 698 pp 537-541.

[3] Nekrasov R Yu, Putilova U S, Starikov A I and Kharitonov D A 2014 Diagnostics of cut-layer deformation and rational tool loading in numerically contolled lathes *Russian engineering research* vol. 34 No. 12 pp 826-828.

[4] Tetia R, Jemielniakb K, O'Donnellc G and Dornfeldd D 2010 Advanced monitoring of machining operations *CIRP Annals - Manufacturing Technology*, Vol. 59 Issue 2 pp. 717–739.

[5] Schneidera Ulrich, Olofssonb Björn, Sörnmob Olof and Drusta Manuel Anders Robertssonb, Martin Hägelea and Rolf Johanssonb 2014 Integrated approach to robotic machining with macro/micro-actuation *Robotics and Computer-Integrated Manufacturing*, vol. 30 Issue 6 pp. 636–647.

[6] Kadira Aini Abdul and Xua Xun 2011 Towards High-Fidelity Machining Simulation *Journal of Manufacturing Systems* vol. 30 Issue 3 pp. 175–186.

[7] Shalamov V G 2007 Mathematical simulation when cutting metals: study guide Chelyabinsk. SUrSU press.

[8] Nekrasov R Yu, Starikov A A and Korchuganov M.A. 2014 Performance assessment of carbide tooling under thermal and loading conditions *Applied Mechanics and Materials* Vol. 682 pp 414-417.

[9] Petrova V A, Bakanov A A and Walter A V 2014 Crack Resistance of Weld Seals, Hardening Methods and Composite Tools Quality *Applied Mechanics and Materials* vol. 682 pp. 431–437.

[10] Ovcharenko V E 2012 Evolution of the structure of plasma metal-ceramic coating under pulsed electron-beam treatment *Inorganic Materials: Applied Research* vol.3 pp. 210-215

IOP Conf. Series: Materials Science and Engineering **91** (2015) 012043 doi:10.1088/1757-899X/91/1/012043

Reliability Modelling while Designing and Improving Working Members of Agricultural Implements

V.I. Myalenko, D.E., professor

5, Markovtseva Street, Kemerovo, 650056, Russia

e-mail: nir@ksai.ru

Abstract. The author suggests the method of reliability prediction in making new constructions of working members of agricultural implements by using the solid diagrams of their real load.

The efficiency of using agricultural implements is evaluated as a rule by their maintenance reliability. Forced stoppage of agricultural units and failures of particular assemblies trouble agrotechnical operation rhythm and reduce final productivity. The most frequent cause of agricultural implements failure is the necessity of replacement of rapidly wearing parts of working members. Wear mechanism is due to the presence of abrasive particles in the natural soil consistency the cultivation of which results in galling (fretting) from the soil-cutting parts surface.

The friction force between the soil and the parts of working members in the Amontons – Coulomb relation $F = N \cdot \mu$ determines the magnitude and the rate of wear, where N means normal pressure and μ means friction ratio. Normal pressure magnitudes on distinct segments of soil-cutting parts differ which determines various wear rate.

In general the initial problem of modelling reliability of a soil-cutting part comes to the determination of estimated nature of their loading in the conditions of real maintenance. It is important to know the normal pressure magnitudes on distinct segments of a soil-cutting part. The efforts to find hydromechanical analogy for the determination of cutting resistance are familiar but they cannot be a reliable determination basis [1]. At the same time the dynamical soil deformation taking place in the practice of agrotechnical operations performance that can be predicted in calculating the mode of deformation rates also depends on initial physical properties of a plot [2]. Thus, bottom pressure testing on different segments of a soil-cutting part by an experimental approach is a rather complex engineering problem.

The objective of this research is to investigate the display technique of the load diagram of the surface of a soil-cutting part according to a priori determined general force load characteristics and in further reliability prediction.

It is not difficult to determine the total external load by an experimental approach with the help of familiar devices for application of solid dynamometry of tillage implements (certificate of authorship 1744527, patent 2498245).

As a result of solid dynamometry, as a rule, essential spatial load bearing characteristics acting from the soil on a given part or a working member are defined. Load bearing characteristics in most cases represent the projections of the resultant vector and principal moment on proper coordinate axes [3].

IOP Conf. Series: Materials Science and Engineering **91** (2015) 012043 doi:10.1088/1757-899X/91/1/012043

Figure 1 shows the outlines of the soil-cutting part $ABCD$ with the coordinate positions of the points $A(x_1, y_1, z_1)$; $B(x_2, y_2, z_2)$; $C(x_3, y_3, z_3)$; $D(x_4, y_4, z_4)$; where $\vec{F}$ means the resultant vector and $\vec{M}$ means the principal moment.

The resultant vector and principal moment magnitudes are defined by the following formula (Eq.1):

$$F = \sqrt{F_x^2 + F_y^2 + F_z^2} \ ; \ M = \sqrt{M_x^2 + M_y^2 + M_z^2}\,, \qquad (1)$$

where $F_x, F_y, F_z, M_x, M_y, M_z$ mean the projections of the resultant vector and principal moment on the coordinate axis.

The resultant vector of external resistance forces is an invariant quantity and doesn't depend on reduction centre, and the angle between the resultant vector and principal moment can change subject to the chosen reduction centre. In the event that the direction of the resultant vector and principal moment coincide "the wrench" takes place, in other words, there is a force and a couple of forces acting in the plane perpendicular to the force.

The wrench axis is a line in the space that is defined by the formula (Eq.2):

$$\frac{M_x - (y \cdot F_z - z \cdot F_y)}{F \cdot x} = \frac{M_y - (z \cdot F_x - x \cdot F_z)}{F \cdot y} = \frac{M_z - (x \cdot F_x - y \cdot F_z)}{F \cdot z}\,, \qquad (2)$$

where x, y, z are current coordinates.

Since the vectors $\overrightarrow{AB} = (x_2 - x_1, y_2 - y_1, z_2 - z_1)$; $\overrightarrow{AD} = (x_4 - x_1, y_4 - y_1, z_4 - z_1)$; $\overrightarrow{AM} = (x - x_1, y - y_1, z - z_1)$ are complanar, we can work out the equation of plane (Eq.3):

$$\begin{vmatrix} x - x_1 & y - y_1 & z - z_1 \\ x_2 - x_1 & y_2 - y_1 & z_2 - z_1 \\ x_4 - x_1 & y_4 - y_1 & z_4 - z_1 \end{vmatrix} = 0. \qquad (3)$$

The solution of the set of equations (2) and (3) allows to define the unknown x, y, z (Eq.4)

$$\begin{cases} \dfrac{M_x - (y \cdot F_z - z \cdot F_y)}{F \cdot x} = \dfrac{M_y - (z \cdot F_x - x \cdot F_z)}{F \cdot y} \\ \dfrac{M_x - (y \cdot F_z - z \cdot F_y)}{F \cdot x} = \dfrac{M_z - (x \cdot F_x - y \cdot F_z)}{F \cdot z} \\ \begin{vmatrix} x - x_1 & y - y_1 & z - z_1 \\ x_2 - x_1 & y_2 - y_1 & z_2 - z_1 \\ x_4 - x_1 & y_4 - y_1 & z_4 - z_1 \end{vmatrix} = 0 \end{cases} \qquad (4)$$

Within the framework of the given model it is assumed that the current coordinates x, y, z determine the position of the point of intersection of the wrench axis with the soil-cutting part surface. Received values of the current coordinates determine the position of the point

$n(x_0, y_0, z_0)$ (Fig. 1), which can be considered with the reasonable approximation as a certain centre of pressure of the force whose modulus is equal to the resultant vector magnitude F, the value of a quantity M can be neglected as a source of pressure for it can be considered as influencing rather the path of motion of abrasive particles on the soil-cutting part surface.

In the figure 1 the position of the soil-cutting part is chosen so that the coordinate axis oz is parallel to the resultant vector $\vec{F}$ of the external resistance forces.

The key condition in modelling a solid or distributed load diagram acting from the soil on the soil-cutting part is to be the following equality (Eq.5):

$$\vec{F} = \sum_{i=1}^{i=n} p \cdot S_i \cdot h_i \tag{5}$$

where p, S_i, h_i are the unit values of the specific pressure, the pressure surface and of the load diagram height in the place of the load action (Fig. 2). Let us imagine conditionally that the soil-cutting part A, B, C, D being under the influence of the force equal to the resultant vector rests only on three points $A(x_1, y_1, z_1)$; $\mathrm{D}(x_4, y_4, z_4)$; $\mathrm{E}(x_5, y_5, z_5)$ located on the sides of the part. Responses R_A, R_D and R_E are defined from the equilibrium condition (Eq.6):

$$\begin{cases} \sum_{npoz} = R_A + R_D + R_E - F = 0 \\ \sum_{momx} = F \cdot y_0 - R_A \cdot y_1 - R_D \cdot y_4 - R_E \cdot y_5 = 0 \\ \sum_{momy} = F \cdot x_0 - R_A \cdot x_1 - R_D \cdot x_4 - R_E \cdot x_5 = 0 \end{cases} \tag{6}$$

Solving the set of equations (6) we will have (Eq.7):

$$\begin{aligned} R_A &= F\frac{(y_5 - y_0)(x_4 - x_0) - (x_4 - y_0)(x_5 - x_0)}{(y_5 - y_1)(x_4 - x_1) - (y_4 - y_1)(x_5 - x_1)}; \\ R_E &= F\frac{(y_0 - y_1)(x_4 - x_1) - (y_4 - y_1)(x_0 - x_1)}{(y_5 - y_1)(x_4 - x_1) - (y_4 - y_1)(x_5 - x_1)}; \\ R_D &= F\frac{(y_5 - y_1)(x_0 - x_1) - (y_2 - y_1)(x_5 - x_1)}{(y_5 - y_1)(x_4 - x_1) - (y_4 - y_1)(x_5 - x_1)}; \end{aligned} \tag{7}$$

Let us change the point E by the points B and C and determine the responses in these points on the basis of the equilibrium condition R_E, R_B and R_C (Eq.8):

$$\begin{cases} \sum_{ox} = R_B + R_C + R_E = 0 \\ \sum_{mox} = R_B y_2 + R_C y_3 + R_E y_5 = 0 \end{cases} \tag{8}$$

$$R_C = R_E \frac{y_2 - y_5}{y_3 - y_2}\ ;\ R_B = R_E \frac{y_5 - y_3}{y_3 - y_2}$$

In the model under study the vectors $\vec{R}_A, \vec{R}_B, \vec{R}_C, \vec{R}_D$ are defined, therefore, the extreme points of these vectors are the points $a\,(x_1', y_1', z_1')$; $b\,(x_2', y_2', z_2')$; $c\,(x_3', y_3', z_3')$; $d\,(x_4', y_4', z_4')$. We write down the equation of the plane in which the vectors $\overrightarrow{ab} = (x_2' - x_1', y_2' - y_1', z_2' - z_1')$ and $\overrightarrow{ad} = (x_4' - x_1', y_4' - y_1', z_4' - z_1')$ are located (Eq.9).

$$\begin{vmatrix} x - x_1' & y - y_1' & z - z_1' \\ x_2' - x_1' & y_2' - y_1' & z_2' - z_1' \\ x_4' - x_1' & y_4' - y_1' & z_4' - z_1' \end{vmatrix} = 0 \tag{9}$$

Thus, the received outlines of the solid figure A, B, C, D, a, b, c, d represent the required volume of the diagram of external force load of the soil-cutting part. Comparing the vectors R_A, R_B, R_C and R_D according to the absolute values and defining minimal of them, for example $|R_C| < |R_A|$, $|R_B|$, $|R_D|$ we draw the plane parallel to $ABCD$ through the point C and obtain the points c, a', b', a corresponding to the limits of this plane. The total volume of the load diagram V includes the volume of the prism $ABCDa'b'cd'$ $-V_1$ and the volumes of two pyramids with vertices in the point c and with bases $a'b'ab$ $-V_2$ and $a'ad'd$ – V_3 (Fig.1).

Consequently (Eq.10):

$$V = V_1 + V_2 + V_3 \quad \text{or} \quad V = \sum_{i=1}^{n} S_i \cdot h_i \tag{10}$$

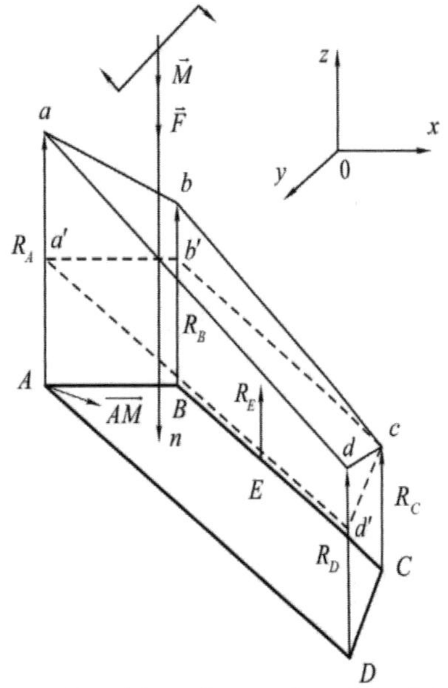

Figure 1 Position of wrench and load bearing characteristics on the soil-cutting part:
$ABCD$ are the outlines of the soil-cutting part;
$\vec{F}$, $\overline{M}$ are the resultant vector and principal moment of the wrench.

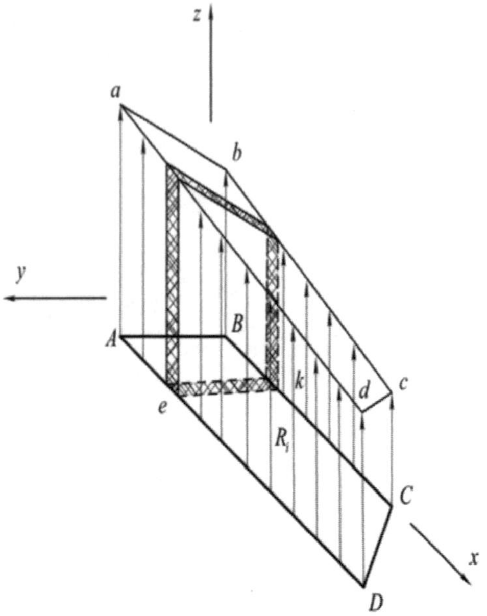

Figure 2 General view of the solid load diagram:
ek is the element of section of the soil-cutting part

While estimating the compliance of the derived diagram of the solid load with the real maintenance conditions we should mention that the given load diagram has some averaged values and doesn't take into consideration the variability of operation by the value and the direction of the resultant vector and principal moment. To obtain more accurate calculations one can consider the variability of characteristics of the solid load diagram.

Multiplying the right and left sides of the equality (10) by the specific volume weight γ we have (Eq.11):

$$V \cdot \gamma = (\sum_{i=1}^{n} S_i \cdot h_i) \cdot \gamma \,, \tag{11}$$

the value $V \cdot \gamma$ which is equal to the value of the force F, that is $V \cdot \gamma = F$, then the required quantity γ is defined by the formula:

$$\gamma = \frac{F}{V} \,;$$

For the practical use of the derived load diagram let us consider a certain element of the section ek of the soil-cutting part separately (Fig.2). Consider the element of the section ek in the rectangular coordinate system (Fig.3). The figure 3 illustrates R_i (load forces). The expected wear rate (fretting) $\Delta\tau$ from the surface of the part (Fig.3b) is known to be proportional to the normal pressure.

$$\Delta\tau = f(k \cdot R_i \cdot \mu) \quad , \tag{12}$$

where $\Delta\tau$ means the wear rate,
k means the factor allowing for departure R_i of from normal,
μ is the friction ratio.

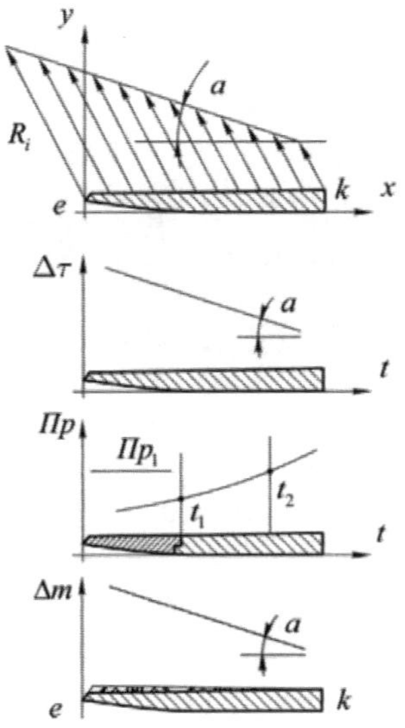

Figure 3. Nature of external load and the rate of abrasive wear of the soil-cutting part: the R_i are the load forces; $\Delta\tau$ is the wear rate Πp is the limiting state of the soil-cutting part; Δm is the thickness of carbide phase

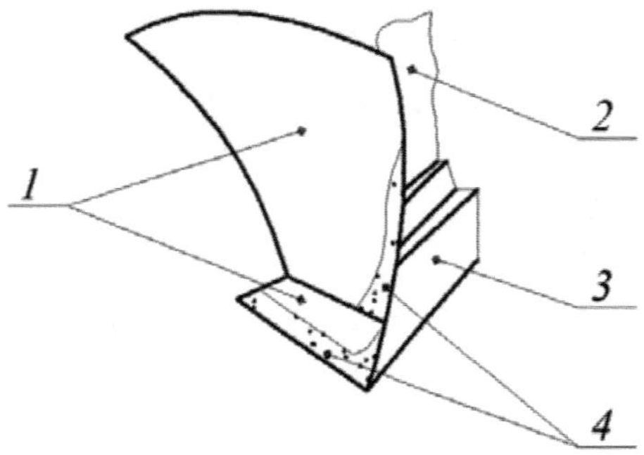

Figure 4 The working member of the agricultural implement –the plough body: 1 – the front face of working member, 2 – the shank, 3 – the landside, 4 – the outlines with additional work hardening

Change in R_i and $\Delta\tau$ lengthwise the section of the soil-cutting part follows the equal law. Indeed, in the relationship (12) k and μ are constants, therefore, the wear rate is proportional to the response of load forces. At the same time Πp, the limiting state of the soil-cutting part followed by the failure of the agricultural machinery, is achieved after the instant of time t_1 (Fig.3c). In the event of availability of a split construction of the soil-cutting part in case of its replacement at the instant of time t_1 the regular limiting state takes place at the instant of time t_2.

While designing and improving the working members of agricultural implements the results of this research can be carried out in a different way. For that purpose the additional work hardenings of separate areas of the front face of the working member of an agricultural implement can be applied. It is reasonable to carry out the work hardenings by applying abrasion-resistant alloy. At present we know the empirical equations derived in vitro characterizing the wear rate of working members of ploughs [4]. Unfortunately, they are valid only for particular plots and therefore can not extend to other soil types. Welding alloys are good protection from abrasive wear and represent high-chrome cast iron alloy or alloys with different content of carbide phase [5]. It is reasonable to choose the thickness Δm of carbide phase application to the section of the soil-cutting part according to the load diagram or expected wear rate (12). Figure 3d shows change in thickness of carbide phase layer along the length of the section of the soil-cutting part. Figure 4 illustrates the working member of an agricultural implement – the plough body and the outlines of additional work hardening segments.

Conclusion

Therefore, while designing and improving the working members of agricultural implements with specified reliability life time it is essential to have the real load diagram derived by using solid dynamometry. Then, focusing on material costs we should choose a constructive way of designing a split soil-cutting part with a changeable wearing segment or an integral part with application of hardening layer to wearing segments.

References

[1] Gadzhiev G M Kuznetsova Yu A Onuchin E M Eshtukov I V Bessonova E S The analysis of the processes of soil-cutting on the basis of hydromechanical analogies *The Siberian Bulletin of Agricultural Science* № 6 2013 – pp 59-66

[2] Zolotarevskaya D I Change in rheological properties and density of soddy-podzolic soil in dynamic load *Soil science* №3 2010 pp 313-323

[3] Myalenko V I Methods of experimental determination of load bearing characteristics of working members of implements *Novosibirsk State University* Novosibirsk 1991

[4] Sahu R K Raheman H Draught prediction of Agricultural Implements using Reference Tillage Tools in Sandy Clay Loam Soil *Biosystems Engineering* №2 2006 – pp 275-284

[5] Bhakata A K Mishraa A K Mishrab N S Jhaa S Metallurgical life cycle assessment through prediction of wear for agricultural grade steel *Wear* 3 4 (2004) pp 338-346

IOP Conf. Series: Materials Science and Engineering **91** (2015) 012044 doi:10.1088/1757-899X/91/1/012044

Manganese concentrate usage in steelmaking

O I Nokhrina[1], I D Rozhihina[2]

[1] DSc, prof., Yurga Institute of Technology, Yurga, Russia
[2] DSc, prof.,Siberian State Industrial University, Novokuznetsk, Russia

E-mail: kafamsf@sibsiu.ru

Abstract. The results of the research process of producing metalized products by solid-phase reduction of iron using solid carbonaceous reducing agents. Thermodynamic modeling was carried out on the model of the unit the Fe-C-O and system with iron ore and coal. As a result of modeling the thermodynamic boundary reducing, oxidizing, and transition areas and the value of the ratio of carbon and oxygen in the system. Simulation of real systems carried out with the gas phase obtained in the pyrolys of coal. The simulation results allow to determine the optimal cost of coal required for complete reduction of iron ore from a given composition. The kinetics of the processes of solid-phase reduction of iron using coal of various technological brands.

Abstract. The paper describes experiments on effects of metal deoxidizer composition, component proportion, pelletizing mixture, particle size distribution of basic materials and flux on manganese recovering from oxides under direct melting.

Introduction.

Results of varied briquette usage for direct-ladle steel melting has shown that manganese recovery from briquettes is increased to 1,5-fold more than recovery from conventional ferroalloys, taking into account some losses due to melting and in-steel inclusions. Direct-alloyed steel satisfies standard specifications of gas and nonmetallic content and mechanical properties as well.

Analysis of modern steelmaking technology shows that costs of deoxidizing and alloying and the cost of consumed energy significantly affect the cost of the product; while the cost of energy consumption affects not only directly but also indirectly, through the price of ferroalloys. Besides, when alloying steel with manganese, Mn heavy losses are inevitable: up to 20 % of manganese is lost during extraction, 20 – 25 % – during ore concentration, 20 – 25 % – during smelting of ferroalloys, up to 25 % – during steel alloying, i.e. total recovery of manganese does not exceed 50 %. One way to improve the efficiency of deoxidizing and alloying of steel is modification of manufacturing of alloying agents, in some cases it is the use of direct steel alloying in the furnace or the ladle, using ores and concentrates [1 – 6].

Manganese can be reduced by aluminum, silicon or carbon. The simplest version of ladle steel deoxidizing and alloying is the use of melting briquettes composed of oxide materials and reducing agents [7]. Completeness of the reaction in the briquette depends on the size of the contact surface of starting materials, fragmentation and concentration of major components, temperature, rate of reaction of products disengagement and forming of low-melting fluid slag at a given temperature.

Impact of various factors on metallothermic reduction of manganese with silicon has been examined in laboratory conditions. Conventional silicon-manganese alloy and unconventional alloys with different silicon and aluminum contents were used as reducing agents. Lime and dolomite were used as fluxes.

Use of aluminum-containing alloy as a reducing agent can improve the reduction of manganese. Table 1 shows the results of manganese reduction with various compositions of the reducing metal.

Briquettes were made from manganese ore, metal-reducing agent and flux. The briquettes were loaded into a graphite crucible which was put into a Tammann furnace and heated to 1500 °C. After a five-minute heating alloy and slag were poured and analyzed.

A fusible slag is formed when reducing with both aluminum and silicon in predetermined proportions; that increases the reaction rate in briquettes, allows more complete reduction of the major component and cuts flux consumption. Presence of 15 – 20 % of Al_2O_3 in the slag contributes to high speed of the process and good separation between the metal and the slag.

Mn-Si-Fe alloy-composition with a manganese content of 15 – 30 % and silicon content of 40 – 45 % used as a reducing agent (see Table 2) allows increasing of manganese-to-silicon ratio by 1.5 – 2-fold in the end metal. This enables the use of such briquettes for ladle deoxidizing and alloying of carbon steels, low-alloyed and alloyed steels of various grades.

Table 1. Results of manganese reduction with various reducing agents.

Indicators	FS45 (44 % Si, 1,36 % Al)	FS65 (66,2 % Si, 1,7 % Al)	FS75 (75,65 % Si, 2,1 % Al)	Alloy composition Al-Mn-Si-Fe (6,5 % Al, 22,1 % Mn, 32,47 % Si)
Briquettes composition, %				
Ferro-Silicium	50	42	41,5	42
Ore	35	42	41,5	35
CHPP ash	5	5	5	5
Dolomite	10	11	12	11
Alloy Composition,%				
Mn	36,30	38,29	38,21	46,95
Si	17,43	28,34	34,65	18,29
Mn Removing ,%	84,9	80,4	75,6	94,0
MnO content in the slag,%	10,9	12,2	12,9	6,0

Table 2. Results of manganese recovery from briquettes.

	Standard FS45	FS45Mn25
Briquette composition, mass.%:		
FerroSilicium	50,0	50,0
Roasted ore	35,0	35,0
Hearth slag	5,0	5,0
Dolomite	10,0	10,0
Metal Yield,%		
Consumption of FS 45	133	130
Alloy composition:%		
Mn	32,3	52,7
Si	22,1	24,1

	Standard FS45	FS45Mn25
P	0,09	0,12
Recovering:		
Mn	84,9	90,6
Si	65,3	69,6

Ratio between roasted concentrate consumption and reducing agent consumption should be (0.7 – 0.8) / 1.0 in briquettes. With such a ratio between components, manganese recovery is 90 – 95 %; recovery from the briquette concentrate is 85 – 90 %. The manganese content in the slag, which is formed in the briquette, is reduced to 7 – 10 % and below, with silicon residual content of 15 – 20 % in metal droplets. Slag multiplicity in the briquettes, after process completion, is only 0.70 – 0.75 when compared with the slag multiplicity of 2.0 – 2.5 at conventional low-carbon ferromanganese smelting.

For increased recovery of manganese from carbonate manganese ore, a product of manganese-carbonate ore thermal treatment was proposed as the oxidizer obtained by longer roasting of the ore in oxidizing environment.

Chemical composition of the product of carbonate ore thermal treatment is shown in Table 3, and its phase composition – in Table 4.

Table 3. Chemical composition of the product of carbonate ore thermal treatment.

Manganese Carbonate ore	Yield after roasting,%	Chemical composition, %							
		Mn	MnO	CaO	MgO	SiO_2	Al_2O_3	P_2O_5	Fe_2O_3
Usinskaya 26 % Mn	72,50	35,86	28,36	13,24	4,83	17,93	1,38	0,51	7,60
Usinskaya 30 % Mn	70,27	42,69	33,77	14,94	1,42	11,38	1,42	0,53	6,70

Table 4. Phase composition of the product of carbonate ore thermal treatment.

Manganese Carbonate ore	Phase composition %		
	$(Ca, Mg)Mn_2O_4$ + $(Ca, Mg) Fe_2O_4$	$Mn_2O_3 + Fe_2O_3$	Silica and other oxides
Usinskaya 26 % Mn	75,0	5,5	19,5
Usinskaya 30 % Mn	64,2	23,0	12,8

The best results are obtained by roasting of manganese carbonate ore, containing 26 – 31 % of Mn, 8 – 11 % of calcium oxide and 1 – 3 % of MgO, 2-7% of Fe_2O_3, 8-17% of SiO_2, in oxidizing environment.

The product of thermal treatment, which can be used for production of water-resistant exothermic briquettes for direct steel alloying, should be roasted in oxidizing environment for 1 hr at 850 – 950 °C. Cooling to 500 – 600 °C should be performed in oxidizing environment. Fusible manganese oxides which accelerate formation of ferrites and calcium manganites are formed in the ore.

Thus, the product of carbonate ore thermal treatment does not contain free bases and is presented by following, compounds, mass. %:

$(Ca, Mg) (Mn, Fe)_2O_4$ 55 – 75;
$Mn_2O_3 + Fe_2O_3$ 5 – 25;
Silica and other oxides 10 – 30.

The study shows that the product of thermal treatment of carbonate ore in the oxidizing environment is not hygroscopic. Moisture content does not increase during storage for 3 – 4 weeks in air. X-ray phase study shows that all lime and magnesia in the products are fixed isomorphous mixtures of ferrites and manganites of Ca and Mg.

Briquettes, having compositions as shown in Table 5, are manufactured from the material obtained and the reducing metal (Mn 30,13 %; Si 30 %; Al 7 %). Results of manganese reduction from the briquettes are given in Table 6.

Table 5. Composition of briquettes.

Components	Compositions, mass %				
	1	2	3	4	5
Alloy of Aluminum, manganese, silicon and iron	56,5	47,5	43,0	60,0	40,0
Product of carbonate manganese ore thermal processing	39,5	47,5	51,5	36,0	45,5
Liquid glass	4,0	5,0	5,5	4,0	5,5

During sintering of the briquettes of the proposed composition, manganese reduction from aluminum oxide and silicon occurs at high speed as well as inter-reacting of Al_2O_3 and SiO_2 which are resulted from the reaction.

$Mn_2O_3 + Al = 2Mn + Al_2O_3$

$Fe_2O_3 + Al = 2Fe + Al_2O_3$

$Mn_2O_3 + 3/2Si = 2Mn + 3/2\ SiO_2$.

Table 6. Results of manganese reduction from products of thermal treatment.

Characteristics	Compositions of exothermic briquettes				
	1	2	3	4	5
Alloy compositions, %					
Mn	49,52	52,37	52,9	The metal is not completely separated from the slag	54,9
Si	19,11	17,11	17,42		13,4
Al	0,22	0,24	0,20		0,20
Slag compositions, %					
MnO	4,36	6,68	6,03	The metal is not completely separated from the slag	15,0
SiO_2	43,0	43,78	43,62		47,0
Al_2O_3	24,16	20,28	20,93		not found
CaO	19,7	17,7	21,2		
Slag multiplicity	0,48	0,52	0,53	0,6	0,7
Recovery of Manganese to alloy, %	87,2	82,97	80,14	75,0	70,0

At the first process stage, when aluminum is being oxidized, silica ore is an active flux; whereby products of thermal treatment of high-grade ore (Mn 30 – 35 %; SiO_2 8 – 10 %; CaO 8 – 11 %; MgO 1 – 3 %) and of relatively low-grade ore (Mn 25 – 27 %; $SiO_2$13 – 17 %; CaO 6 – 10 %; MgO 1 – 3 %; Fe_2O_3 3 – 7 %) can be successfully used.

In laboratory conditions, the main technological parameters of mixtures for direct steel alloying were investigated as well as their impact on volume of manganese reduction from oxides. These parameters include value of component mixing, particle size distribution (PSD) of oxidant, reductant and flux, type of coupling agent and a mixture pelletizing technology.

It is established experimentally that grinding of ore fines together with reductant powder ensures total contact between the particles of metal reducing agent and ore. This increases (by 5 %

approximately) the amount of recovered manganese in comparison with briquettes made by conventional techniques. Besides, in briquettes with the charge subjected to co-grinding, recovery of silicon increased by about 3 %. Investigation of effect of starting material PSD on rates of manganese reduction from oxides by direct alloying showed that quantity of particles larger than 1 mm should be limited in mixtures.

It is established experimentally that during pelletizing of mixtures for ladle direct steel alloying with manganese, lignosulfonate concentrates and liquid glass can be used as coupling agents. Particularly effective is the use of boric oxide as a coupling agent. In this case, even briquettes with lime content retain strength in air for two weeks. When utilizing TPP ash as a flux (8.88 % Al_2O_3, 23.98 % SiO_2, 0.56 % TiO_2, 45.85 % CaO, 4.98 % MgO, 6.32 % FeO, 8.18 % Fe_2O_3, 1,82 % other impurities) it is practical to use water for moistening the mixture. Oxides of CaO, SiO_2, Al_2O_3 presenting in the ash, react with water to form stable compounds: $2CaO{\cdot}SiO_2{\cdot}4H_2O$, $3CaO{\cdot}Al_2O_3{\cdot}6H_2O$, $SiO_2{\cdot}2H_2O$ having astringent properties. It is found that pelletizing of mixtures for direct steel alloying can be performed by pelletizing or briquetting at a disk granulator; wherein, pelletizing method does not significantly affect the parameters of manganese reduction from oxides. The briquettes and pellets have a sufficient strength for transportation.

The technology of ladle deoxidizing and alloying of steel with manganese using mixtures of different composition passed industrial tests in 25-ton electric arc furnaces. Mixtures of the following compositions were used:

I – friable alloy [8] Al–Mn–Si–Fe–C (7 % Al, 25.5 % Si, 27 % Mn) – 39.2 %, manganese ore – 39.2 %, dolomite – 19.6 %, a coupling agent – 2 %.

II – FS45 – 42 %, manganese ore – 42 %, dolomite – 12 %, a coupling agent – 4 %.

III – FS75 – 17.4 %, manganese ore – 43 %, dolomite – 11.5 %, CHP ash – 23.2 %, water – 4.9 %. Steel grades St40, St45, St20 (briquettes, composition I and II) and St45L, St35L (composition III, pellets) were smelted by direct alloying. Technological characteristics of certain melts, using briquettes of composition I, are shown in Table 7.

The mixtures were fed into the ladle during tapping. The temperature of the steel before the tapping was 1883 – 1893°K, that is close to the upper limit recommended by technological instruction. Briquettes dissolved completely at the period of tapping. Use of briquettes and pellets containing manganese oxide materials gave rise to manganese recovery by 78 – 88 % compared with 70.5 % at conventional smelting.

It is known that, when melting ferromanganese and silicomanganese, about 80 % of manganese is recovered from the ore; and the coefficient of manganese total recovery does not exceed 60 %. Obviously, the above-described technology increases total recovery of manganese and eliminates the use of conventional manganese ferroalloys which contain 1 % of manganese. The use of oxide manganese materials for ladle deoxidizing and alloying of steel reduces the cost of deoxidizing by 276 rubles per ton for the steel with 0.35 – 0.65 % manganese content.

Table 7. Technological characteristics of some experimental melts

Characteristics	conventional (average for 10 melts)	melt № and steel grade						
		1	2	3	4	5	6	7
		St35	St45	St35	St40	St40	St40	St40
Before deoxidizing, %								
C	0,26	0,37	0,44	0,37	0,39	0,41	0,38	0,38
Mn	0,19	0,17	0,18	0,16	0,22	0,22	0,23	0,23
In finished steel, %								
C	0,35	0,35	0,35	0,36	0,42	0,44	0,38	0,39
Mn	0,63	0,70	0,68	0,68	0,68	0,69	0,70	0,68
Si	0,22	0,21	0,25	0,27	0,22	0,22	0,24	0,22
S	0,030	0,034	0,032	0,031	0,032	0,032	0,031	0,030
P	0,019	0,017	0,014	0,018	0,014	0,022	0,018	0,018

IOP Conf. Series: Materials Science and Engineering **91** (2015) 012044 doi:10.1088/1757-899X/91/1/012044

Consumption of deoxidizer, kg								
To furnace: FMn70	123	-	-	-	-	-	-	-
SMn17	150	-	-	-	-	-	-	-
FS65	-							
FS25	-	150	150	150	100	100	100	100
To ladle: FS65	150	70	60	60	60	60	60	60
Briquettes	-	800	750	750	700	700	700	700
Melt weight, t	30,4	31	30,4	30,4	31	31	30,4	30,4
Recovery of Mn, %	70,5	78,08	77,05	75,5	77,5	79,1	77,6	75,5
Total recovery of Mn %	57,5	74,64	73,65	76,57	74,1	75,6	74,1	72,2

Quality tests of steel, produced by direct alloying with oxide manganese-containing materials in the ladle have shown that macrostructure, gas content, nonmetallic inclusions and mechanical properties of the resulting metal satisfy all requirements and are identical to the steel produced by the conventional technique (see Figures 1 and 2).

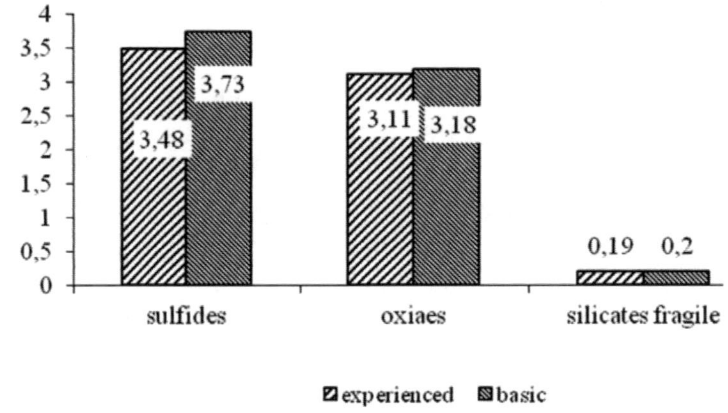

Figure 1 – Pollution of steel 40 with nonmetallic inclusions

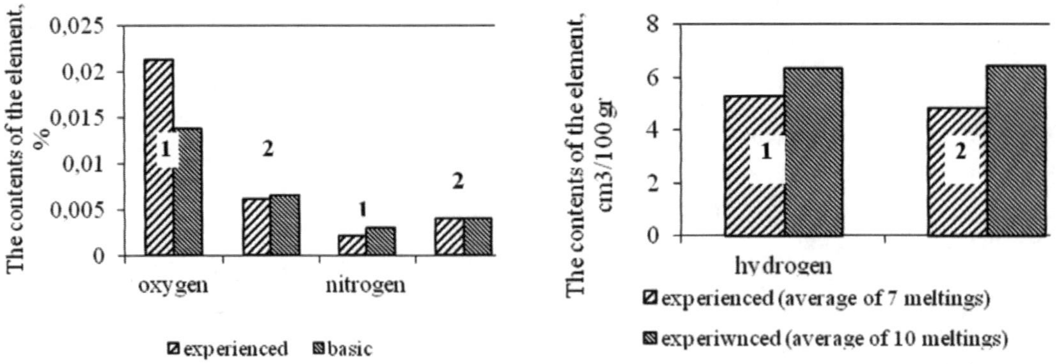

1 – sample of metal from furnace before tapping; 2 – sample of metal from ladle after tapping

Figure 2 – Content of gases in steel

IOP Conf. Series: Materials Science and Engineering **91** (2015) 012044 doi:10.1088/1757-899X/91/1/012044

Findings

1. Mixtures for direct alloying, based on the Al–Mn–Si–Fe–C friable alloy and on the conventional ferrosilicium, were tested in industrial environments.
2. In direct steel deoxidizing and alloying with briquettes based on Al–Mn–Si alloy, manganese is recovered with high stability. Total recovery of manganese from briquettes exceeds the recovery using conventional ferroalloys by almost 1.5 times (taking into account all the collateral manufacturing losses).
3. The stable recovery of manganese is approximately 90 % in direct steel deoxidizing and alloying with briquettes and pellets in the ladle, using exhaust system dust as a reducing agent in fractionating of conventional ferrosilicium FS45 and FS75.
4. The metal from experimental melts satisfies standard specifications of macrostructure, gas content, nonmetallic inclusions and mechanical properties.

References

[1] Bobkova O S 1991 Moscow. *On mechanism of oxide material melting and recovery of metals from oxide melts. Stee №* **23** 78
[2] Nosov Y N, Kamshukov V P and Sokolov V V 2004 Moscow. *Direct ladle alloying of steel with manganese sinter during converter tapping. Steel № 5.* **35** 85
[3] Nokhrina O I and Dmitrienko V I 2004 Moscow. *Reduction of manganese from its oxides by silicon in direct furnace alloying of steel. Steel № 6.* **34** 82
[4] Nokhrina O I, Komshukov V P and Dmitrienko V I 2004 Moscow. *Developing technology for direct alloying of steel with manganese in electric-arc furnace. Metallurgist № 5.* 264
[5] Protopopov E V, Shakirov K M, Ayzatulov R S, Fokin K.S. 1997 Moscow. *Thermodynamic fundamentals of metal converting with elements of direct reduction. Proceedings of universities. Iron and steel № 8.* **13** 86
[6] Gasik M I, Liakishev N P end Velichko B F 1999 Dnepropetrovsk. *Direct alloying of steel in conditions of mineral resource and energy base of ferrous metallurgy. Actual problems and prospects of electro-metallurgy.* **48** 125
[7] Nakonechny A Y, Romanenko V I end Zaitsev A Y 1994 Moscow. *Efficiency of steel direct alloying with manganese.* Steel *№ 1.* **17** 82
[8] Gasik M I, Myachin V G end Polyakov O I 1990 Moscow. Phase *composition of friable complex alloys of Mn-Fe-Si-Al-C-P. Proceedings of universities. Iron and steel № 1.* **103** 120

IOP Conf. Series: Materials Science and Engineering **91** (2015) 012045 doi:10.1088/1757-899X/91/1/012045

The use of coal in a solid phase reduction of iron oxide

O I Nokhrina[1], I D Rozhihina[2], I E Hodosov[3]

[1] DSc, prof., Yurga Institute of Technology, Yurga, Russia
[2] DSc, prof.,Siberian State Industrial University, Novokuznetsk, Russia
[3] Assistant, Siberian State Industrial University, Novokuznetsk, Russia

E-mail: kafamsf@sibsiu.ru

Abstract. The results of the research process of producing metalized products by solid-phase reduction of iron using solid carbonaceous reducing agents. Thermodynamic modeling was carried out on the model of the unit the Fe-C-O and system with iron ore and coal. As a result of modeling the thermodynamic boundary reducing, oxidizing, and transition areas and the value of the ratio of carbon and oxygen in the system. Simulation of real systems carried out with the gas phase obtained in the pyrolys of coal. The simulation results allow to determine the optimal cost of coal required for complete reduction of iron ore from a given composition. The kinetics of the processes of solid-phase reduction of iron using coal of various technological brands.

Introduction.

Promising raw material for the production of high-quality steel is a so called direct reduced iron, which is obtained by outside the blast furnace redistribution of iron-containing materials. Production technology metalized products developed and improved with 60s pro-nineteenth century, and to date the volume of production of DRI reached 76 million tons per year. In Russia and in the world DRI generally prepared using as reducing agent of the converted gas. Granular iron ore is fed into the shaft type furnace, where temperatures not exceeding the melting temperature materials, reduction of iron oxides occurs. The use of natural gas leads to a rise in the cost of iron and limits the development of these technologies in areas where natural gas is scarce. Reducing the consumption of natural gas is possible when replacing coal. Currently, up to 8 % of the world's iron produced by direct reduction is carried out with the use of coal. There are various technological proposals for the use of coal metalized products; there is no single technology, which received wide commercial distribution.

Processes solid reduction of iron oxide with iron ore as a solid carbonaceous reductant coal are not unambiguous assessment, due to multi-component system, and the occurrence of simultaneous reduction reactions involving solid carbon and gaseous products. Keep in mind that in the process of solid reduction reactions simultaneously develop both direct and indirect reduction of iron oxide ores as-is useful [1].

1. Actual job is to research and develop energy-efficient technology of complex processing of iron-ore and coal to produce metalized materials suitable for use in steelmaking critical applications [2].

In conducting research as iron-containing materials used iron ore (iron ore deposit Tashtagolsky) and iron ore concentrate produced by ore beneficiation. As solid carbonaceous reducing agents used nut coke and coals of different technological marks (lignite brand – B2; long-flame coal brand – D;

Coal weakly caking brand – SS; coal skinny mark – MOT). The compositions of the materials used and reducing technical analysis are shown in Tables 1–3.

Table 1. The phase composition of the starting iron ore.

Lot Of:	hematite (Fe_2O_3)
Attendant:	magnetite ($FeO \cdot Fe_2O_3$), goethite (α–FeOOH), quartz (SiO_2), kaolinite ($Al_4[Si_4O_{10}](OH)_8$)
Lil Bit:	iron chloride, feldspar

Table 2. The chemical composition of iron-containing materials.

The chemical composition of the iron ore, % mass							
$Fe_{overall.}$	SiO_2	Al_2O_3	P_2O_5	CaO	MgO	S	moisture
52.3	19.15	3.31	0.072	0.94	0.76	0.42	5.78
Chemical composition of iron ore concentrate, % mass							
61.2	7.92	2.41	0.02	1.96	2.31	0.41	6.27

Table 3. Technical Analysis reducing.

Reducing agent	Characteristic			
	The content [C] in the working masses,%	A^d (ash),%	V^{daf} (volatile),%	W^r (moisture),%
Coal –B2	49.1	7.83	46.7	24.7
Coal – D	55.6	5.64	43.5	10.4
Coal – SS	70.2	6.41	24.3	6.7
Coal – TO	76.3	8.34	15.2	5.1
Coke	84.4	12.3	1.3	1.2

The studies included thermodynamic modeling of solid-phase reduction of iron oxide iron ores using software system "Terra", developed at the Moscow State Technical University. N.E. Bauman [3].

The process of reduction of iron oxides, iron ore according to the principle Baykova takes place in steps:

$Fe_2O_3 \rightarrow Fe_3O_4 \rightarrow FeO \rightarrow Fe$ at $T \geq 843$ K.

The use of coal as reducing processes associated with the development of the direct and indirect restoration. Direct reduction of iron oxide with solid carbon is in accordance with the reactions:

$3Fe_2O_3 + C = 2Fe_3O_4 + \{CO\} - 129{,}07$ MJ; (1)

$Fe_3O_4 + C = 3FeO + \{CO\} - 187{,}28$ MJ; (2)

$FeO + C = Fe + \{CO\} - 152{,}67$ MJ. (3)

According to [1] in the presence of a solid carbonaceous reducing the interaction between carbon and iron oxides is carried out with the assistance of the gas phase which is recovered by a carbon gasification reaction:

$CO_2 + C = 2CO - 166{,}32$ MJ. (4)

The thermal decomposition of coal is accompanied by the formation of gaseous products that are involved in the reduction of iron oxides according to reactions:

$3Fe_2O_3 + \{CO\} = 2Fe_3O_4 + \{CO_2\} + 37{,}25$ MJ; (5)

$Fe_3O_4 + \{CO\} = 3FeO + \{CO_2\} - 20{,}96$ MJ; (6)

$FeO + \{CO\} = Fe + \{CO_2\} + 13{,}65$ MJ; (7)

$3Fe_2O_3 + H_2 = 2Fe_3O_4 + \{H_2O\} - 4{,}2$ MJ; (8)

$Fe_3O_4 + H_2 = 3FeO + \{H_2O\} - 62{,}41$ MJ; (9)

$FeO + H_2 = Fe + \{H_2O\} - 27{,}80$ MJ. (10)

2. Thermodynamic modeling includes research on the model of the elementary system and study the recovery of iron ore from the oxides of iron ore using coal of various technological brands.

Model systems elemental composition element wise Fe-C-O, formed by setting the composition of the starting mixture in a 1 mol Fe_2O_3 oxide and the number of moles of carbon and oxygen, respectively are varied parameters n and m (Fe_2O_3+nC+mO). In the simulation process is carried out the following tasks: determination of the temperature and the amount of carbon to ensure full recovery of iron; determining the optimum values of the index a* (measure a*=m/n – number ratio of oxygen to carbon in the system). Figures 1 shows the dependence of the absolute content of the gas components (a) and the condensed phase (b) of the parameter a*.

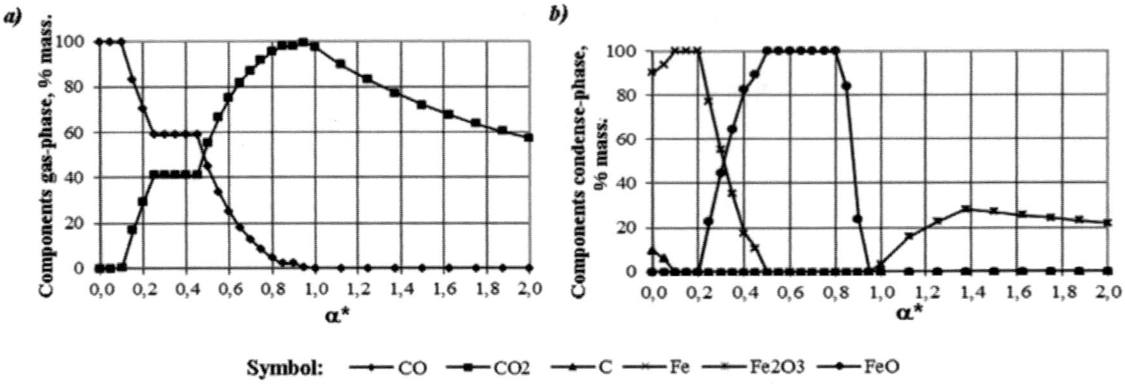

a – the composition of the gas phase depending on the values of a*;
b – the components of the condensed phase, depending on the value of a*

Figures 1. Thermodynamic modeling in the Fe-C-O.

The analysis of the calculation results showed the presence of three areas. Restorative region exists for values of a* < 0,2. The transition region is characterized by a decrease in the concentration of the reduced iron to extinction and increasing the FeO content in the condensed phase. Oxidative area begins at a* = 0,5, when in the condensed phase completely disappears iron. When a* = 0,95 in the gas phase completely disappears from, and in a condensed FeO. Thus, as a result of research of redox conditions in the system Fe-CO reducing identified boundaries, transitional and oxidizing areas and corresponding values a*. The best conditions for iron reduction corresponds to the value of the index a < 0,2 [4].

Thermodynamic modeling of solid-phase reduction of iron oxide ore using coal of various grades of technology was carried out taking into account the element, reducing carbon oxide compounds (Table 4, 5) and the composition of the gas produced by heating the reducing agents (Table 6) [5 – 7].

Table 4. The elemental composition of reducing.

Reducing agent	Elemental analysis for fuel mass, %				
	C^{daf}	H^{daf}	N^{daf}	O^{daf}	S^{daf}
Coal – 2B	70.41	3.88	0.98	22.58	1.15
Coal – D	72.02	6.12	1.74	20.12	0.34
Coal – SS	85.30	4.78	2.16	6.52	0.14
Coal – TO	90.79	3.96	2.93	2.16	0.31
Coke	96.97	0.56	1.42	0.56	0.51

Table 5. The oxide composition of the ash reducing, %.

Reducing agent	Fe_2O_3	Al_2O_3	CaO	MgO	P_2O_5	SiO_2	S
Coal – 2B	10.30	5.40	45.30	5.90	0.20	26.90	0.68
Coal – D	8.17	25.60	9.70	3.70	1.80	50.00	0.61
Coal – SS	17.46	26.55	1.2	1.58	0.5	51.1	0.02
Coal – TO	16.4	25.01	5.38	1.66	-	45.2	0.06
Coke	13.12	23.72	5.15	1.76	0.69	47.9	0.02

Table 6. Composition of reducing gas when heated to 1173 K.

Reducing agent	The gas composition, %					
	CO_2	CO	H_2	CH_4	O_2	N_2
Coal – 2B	10.21	5.42	44.30	5.90	0.20	-
Coal – D	8.17	25.60	50.24	3.70	1.80	-
Coal – SS	3.54	7.15	56.35	3.1	0.5	2.45
Coal – TO	1.35	8.26	67.3	2.35	0.5	4.2
Coke	13.12	23.72	5.15	1.76	0.69	-

The results of process simulation solid reduced iron using coal as a reductant possible to determine the dependence of the degree of reduction of iron and the gas phase flow of coal from each processing mark. Figures 2 shows the dependence of the volume and composition of the gas phase formed as a result of the evolution of volatile components by heating long-flame coal. The degree of recovery and composition of the gas phase from the flow of long-flame coal in the reduction of 100 kg of ore are shown in Figures 2b.

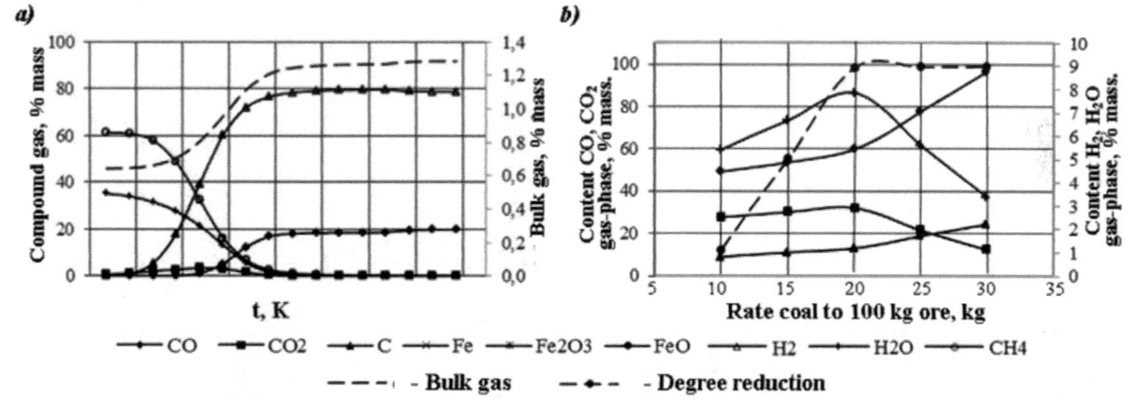

a – the dependence of the volume and composition of gas phase temperature;
b – the degree of recovery and gas phase composition of long-flame coal consumption in the reduction of 100 kg of ore

Figures 2. Thermodynamic modeling of solid-phase reduction of iron oxides long-flame coal.

3. The study included high temperature laboratory experiments. The resulting simulation reducing cost values corresponding to the highest degree of reduction, were used in the manufacture of ore-coal compositions. Pre components subjected to crushing and desired fractions followed by classifying, then in predetermined proportions of ore and reducing agent composition were averaged and subjected to briquetting. These briquettes were placed in a graphite crucible and covered with a lid. The crucible with braces injected into the hot zone resistance furnace heated to a temperature of the experiment and kept at isothermal holding time. For the experiments was chosen isothermal holding temperature 1173 K at the holding time of 90 min. In the resulting experiments metalized materials by chemical analysis

determined the content Fe_{total}, Fe_{met}, by calculation determines the degree of metallization – φ_{met}, according to the formula:

$\varphi_{met} = Fe_{met}/Fe_{total} \cdot 100$,

where Fe_{met} – metallic iron content,%;

Fe_{total} – total iron content (amount of oxidized and metallic iron) %.

The highest degree of metallization (close to 97 %) was obtained by applying the ore-coal briquettes, which corresponds to an amount of reducing agent, the resulting thermodynamic modeling (Figure 2b).

Influence of physico-chemical properties of the carbonaceous reducing processes was studied by solid state reduction of kinetic studies [8]. Experiments were performed using a batch weighing method in a resistance furnace at temperatures 1273, 1373, 1473 K, and the isothermal hold duration 90 min. The calculation of the degree of recovery in the construction of the kinetic curves was carried out taking into account the loss of mass of sample corrected for moisture content in a mixture of volatile and fly away. The criterion considered 100% recovery linkage weight loss corresponding to the total carbon monoxide formation reactions – 1, 2, 3. The experimental results obtained by using as a reducing coke nut, caking, long-flame coal and lignite with different isothermal soak temperatures are shown in Figure 3.

From these dependences that after 60 minutes of isothermal heating at a temperature of 1473 K have the greatest degree of reduction in briquette made up using as a reducing long-flame coal. The resulting material contains Fe_{met} – 83%, with φ_{met} – 96 %. As a result, the isothermal hold at a temperature of 1373 K for 60 min. the greatest degree of recovery in briquettes made up using as a reducing agent of brown coal. The resulting material contains Fe_{met} – 67 %, with φ_{met} – 97 %. As a result, the isothermal hold at 1273 K for 60 minutes in the highest degree of reduction observed for briquettes made up using as reducing lignite. The resulting material contains Fe_{met} – 50 %, with φ_{met} – 97 %. Fig. 3 shows that the intensity of the processes of solid-phase reduction of iron oxides, iron ore depends not only on the material composition used in iron and carbon components, but also on their physical and chemical properties.

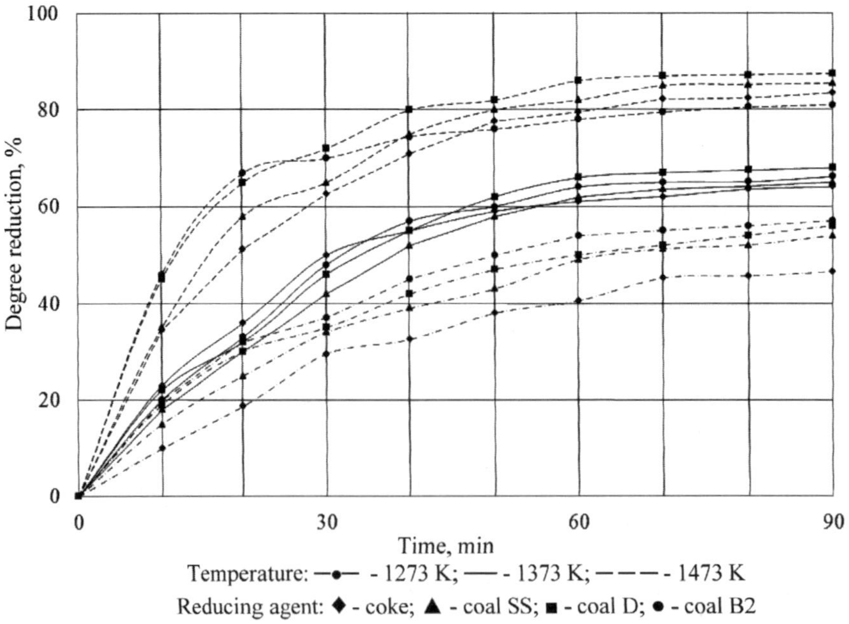

Figures 3. Dependence of the degree of reduction of the isothermal holding time.

Study of the processes of thermal decomposition was carried out by reducing periodic weighing sample placed in a resistance furnace at a constant heating furnace from 300 to 1200 K. The experiments used a fraction of the corresponding coal used in the manufacture of ore-coal briquettes. To create conditions of oxygen deficiency in the furnace supplied with a neutral gas, argon flow rate of 50 ml/min. Weight loss of sample recorded at intervals of 50 K. The processed results of the experiments are shown in Figures 4.

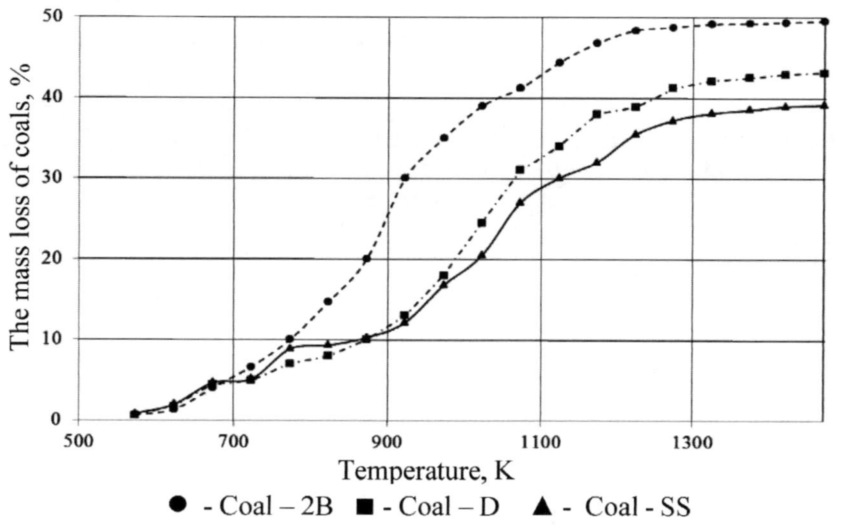

Figures 4. Dependence of the mass loss of coals of different technological brands of temperature.

From the figure it is shown that thermal decomposition process different grades of coal proceeds differently. Thermal decomposition of brown coal begins at temperatures lower than the thermal decomposition of the stone. Amount of gaseous products evolved during the thermal decomposition of brown coal, significantly more than the expansion of long-flame and caking. The process of thermal decomposition of coal caking occurs less intense compared to other brands of coal under consideration [9].

Conclusions.

As a result of the research, a thermodynamic model of the process of obtaining the metalized products by solid-phase reduction of iron oxides from iron ore using coal as a reductant. Application of the model allows to determine the optimal cost of coal of different brands in the preparation process of metalized products with a given composition and degree of metallization. In the course of the research also obtained data on the influence of physical and chemical properties of different carbonaceous reducing agents on the intensity of the recovery processes. The data obtained can be used in the development of co-processing techniques iron ore and coal in order to obtain high quality metalized materials.

References

[1] Yusfin Y S 2007 Moscow. *Metallurgy of iron* 464
[2] Hodosov I E and Nokhrina O I 2014 Novokuznetsk. *Preparation of pure iron by vnedomennoy processing of iron ore and coal. Kuzbass Innovation Convention "Kuzbass: education, science, innovation"* **379** 520
[3] Sinyarev G B, Vatolin N A and Cowards Y G 1982 Moscow. *The use of computers for thermodynamic calculations me-metallurgical processes.* 32 47

[4] Rybenko I A 2000 Novokuznetsk. *Development of methodology and calculation system technology options continuous semi-tion of metal in aggregates jet-emulsion type: dis. cand. tehn. Sciences: 05.16.02: 28.03.00 protected: approved. 14.06.00.* 165

[5] Mizin V and Serov G 1976 Moscow. *Carbonaceous reducing agents for ferroalloys*. 272.

[6] Stumpf G G, Ryzhkov Y A, Shalamanov V A and Petrrov A I 1994 Moscow. *Physical and technical properties of rocks and coals of Kuznetsk Basin: Directory*. 447

[7] Gasik M I 1988 Moscow. *Theory and technology of production of ferroalloys*. 784

[8] Amdur A M, Potapov A M, Raznitsina A L and Lhamsupen M 2012 Moscow. *Reduction kinetics of iron ore concentrate carbon. Proceedings of the higher educational institutions. Iron and steel № 8*. 17 56

[9] Rozhihina I D, Romanenko U E, Lazarevskiy P P, Hodosov I E 2014 Novokuznetsk. *Alternative carbon material in the recovery processes for the production of manganese, chromium alloys and metallized iron. In Proceedings of the XVIII All-Russian scientific-practical conference "Metallurgy: technology, management, innovation, quality"*. 128 420

Processing of energy materials in electromagnetic field

A P Rodzevich[1,a], L V Kuzmina[2,b], E G Gazenaur[2,b], V I Krasheninin[2,b]

[1]Yurga Institute of Technology (affiliate) of Tomsk Polytechnic University, Russia
[2]Kemerovo State University, Russia

e-mail: [a]nimez@tpu.ru, [b]specproc@kemsu.ru

Abstract. This paper presents the research results of complex impact of mechanical stress and electromagnetic field on the defect structure of energy materials. As the object of research quite a typical energy material – silver azide was chosen, being a model in chemistry of solids. According to the experiments co-effect of magnetic field and mechanical stress in silver azide crystals furthers multiplication, stopper breakaway, shift of dislocations, and generation of superlattice dislocations – micro-cracks. A method of mechanical and electric strengthening has been developed and involves changing the density of dislocations in whiskers.

Introduction

The control over physical and mechanical properties of materials by different methods of treatment effecting on their defect structure has been a burning issue so far.

For semiconducting and energy materials this urgency is caused by probable intensifying chemical processes, as well as by improvement of some functional properties, for instance, stability and reactivity.

This paper provides the research results of complex influence of mechanical stress and electromagnetic field on the defect structure of energy materials.

As the object of research quite a typical energy material – silver azide was chosen, being a model in chemistry of solids.

The defect structure of silver azide has been well studied. Silver azide is defect according to Frankel with mainly labial interstitial cations of silver (Ag^+). The surface of silver azide crystals is positively charged, and the surficial zone is enriched by negatively charged cation vacancies (V_k^-). The information on qualitative and quantitative composition of impurities of positive metal ions is available: Cu^{2+}, Fe^{3+}, Al^{3+}, Bi^{3+}, Pb^{2+}, Ca^{2+}, Si^{4+}, Ti^{2+}, Mg^{2+} with concentration $3\cdot10^{-5}\div10^{-4}$ mole percent; edge dislocations have electric charge ($\sim10^{-16}$ C) and magnetic moment (5×10^{-21} $A\cdot m^2$) [1-3].

A decomposition reaction in an anion sub-lattice of silver azide is thought to occur when localizing of two holes on a cation vacancy. The concentration of anion and cation vacancies is determined by impurity concentration in silver azide and is to depend on doubly-charged cations concentration.

Therefore, intensifying mechanical influence is expected while electromagnetic field is effecting on the defect structure, which is of considerable importance for physical and chemical properties of silver azide. The defects are particularly important for forming reaction zones in crystals [3].

Experimental methods

The objects of research are silver azide whiskers (AgN_3), grown according to well-known methods by F.I. Ivanov, the average dimensions are $10\times0.1\times0.03$ mm^3 [4].

IOP Conf. Series: Materials Science and Engineering **91** (2015) 012046 doi:10.1088/1757-899X/91/1/012046

For sample preparation optically transparent and perfectly facetted crystals were selected as shown in Fig. 1.

To carry out experiments the samples were prepared in planar geometry which makes it possible to indicate a gaseous product, emitting during decomposition, and record topography of its distribution: both faces of crystals were БФ–6 – glued on mica plate preliminary defatted with alcohol.

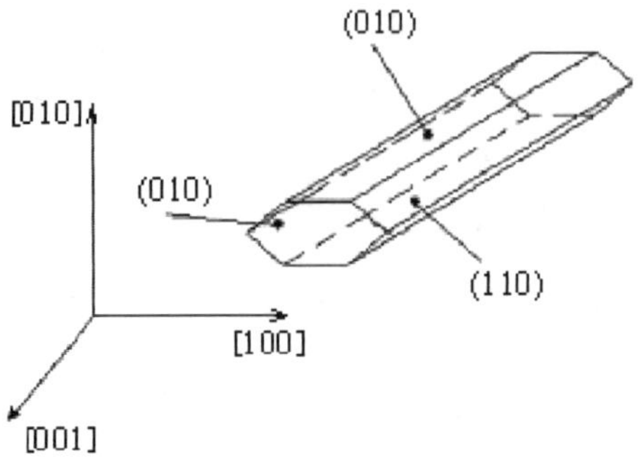

Figure 1. Indices of crystal faces and directions in silver azide whisker.

Mechanical stress in samples was developed by uniaxial compression on the axis [010] in a specially made acrylic resin cell (Fig. 2.a). Samples were stressed in magnetic and electric fields in a non-magnetic cell in the conditions of uniaxial compression generated by constant stress (Fig. 2. b).

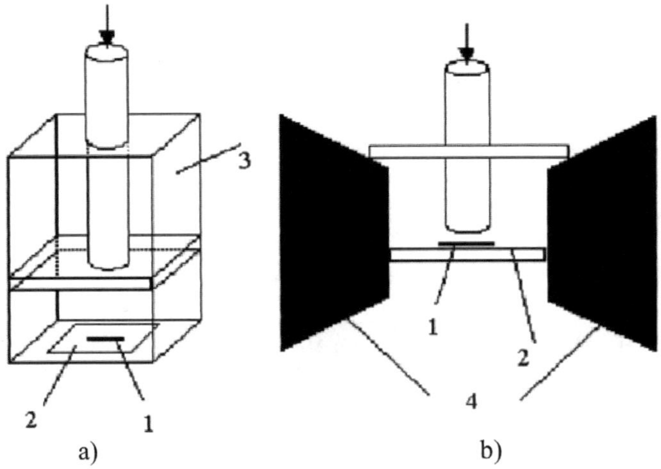

Figure 2. Plan of experiments: a – mechanically stressed samples; b – mechanically stressed samples in constant magnetic or electric fields: 1 – crystal; 2 – mica plate; 3 – acrylic resin cell; 4 – electromagnet poles or copper electrodes.

To provide bend deformation of a sample a dielectric bar with a conic edged base was put into an inter-electrode space.

Local pressure in both cases was changed within the range $5 \cdot 10^5$ H/m^2 - $5 \cdot 10^7$ H/m^2. Electrodes were energized by a source of direct current (Б5-44, Б 5-50). Electrodes – crystal space was minimal and didn't exceed 50 μm (external voltage decreases completely on the crystal provided that air clearance has such dimensions). Non-contact electric field intensity was 10÷40 W/m, that one of a contact electric field – 30 kW/m (gallium was used as a contact).

Magnetic field was formed by electromagnet (ЭМ-1) with adjustable intensity up to 0.3 T.

Magnetic induction was measured with a magnetic induction indicator Ш1-8 or tesla-meter (sensitivity 10^{-5} T).

A supersonic wave in samples was excited by the method of complex vibrator involving the magnetostrictive converter with the proper frequency 20 kHz. The intensity of exited waves didn't exceed ≈1.5 W/cm^2. The samples were placed on the top of the magnetostrictive converter with the plane (010).

Gaseous products of decomposition were tested volumetrically [3-5]. Over a period of time after energy deposition the sample was placed into a cell with water solution of sodium thiosulfate. Dissolution was observed by a microscope in transmitted red light, and diameter and space coordinates of bubbles of emitted gaseous product were registered.

For dissolving whiskers and further calculation of emitted gas we used a standard cell for measuring optical constants of liquids with the space between walls up to 10^{-3} m. The solvent was held in the horizontally placed cell by surface tension forces. When measuring the volume of gas bubble it was assumed that gas pressure inside the bubble was equal to air pressure.

The volume of gas emitted when crystals dissolving was attributed to face area where gas emitting was observed.

We know that reaction regions in silver azide, i.e. in crystal zone where gaseous products of dissolution emit (Fig. 3), are similar to places of dislocation rising, therefore, it is necessary to reveal them.

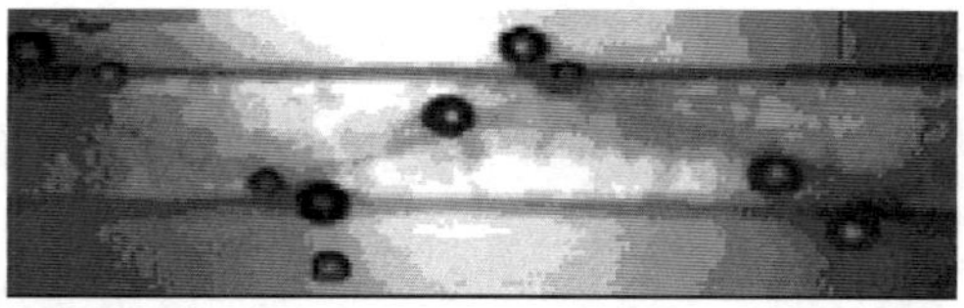

Figure 3. Dislocation structure of silver azide revealed volumetrically.

The dislocation structure was tested by the method of etching pits. Contrast etching pits were formed when etching of crystals AgN_3 in 10 % water solution of sodium thiosulfate. The crystal glued on both faces was put in solution $Na_2S_2O_3$ for 2 – 3 seconds, and then it was rinsed in distilled water or alcohol. This procedure was repeated in order to assure that the density of etching figures (pits) didn't increase, but their dimensions and depth grow slightly, then etching pits can be caused by dislocation rising.

The density of dislocations was determined as relation of the number of etching pits to the area of crystal surface.

For each point of experimental curves at least 10 samples were taken. The results of experiments were processed by computer with the software Microsoft Excel.

Results and Discussion

The effect of different types of stresses (mechanical stress, electric field) on the defect structure (dislocations and impurity) has been already tested in silver azide crystals [1,6,7].

The kinetic dependencies of accumulating dislocations have been obtained for mechanical stress ($5 \cdot 10^5$ - $5 \cdot 10^7$ N/m^2), according to them it has been stated that in 7-9 seconds of exposure the sample loses its mechanical integrity and is subject to brittle fracture.

As for the co-effect of magnetic field (0.3 T) and mechanical stress, the kinetics of accumulating dislocations regarding the time of exposure is a complex dependence, depicted in Fig. 4. In these experiments preliminary deformed samples were used, the input density of edge dislocations was $4\cdot 10^3$ cm^{-2}.

Mechanical stress influencing the crystal equaled to $5\cdot10^5$ N/m^2 – an optimal value for reliable recording the accumulation of dislocations, provided that mechanical stress exceeds this value the sample is subject to brittle fracture.

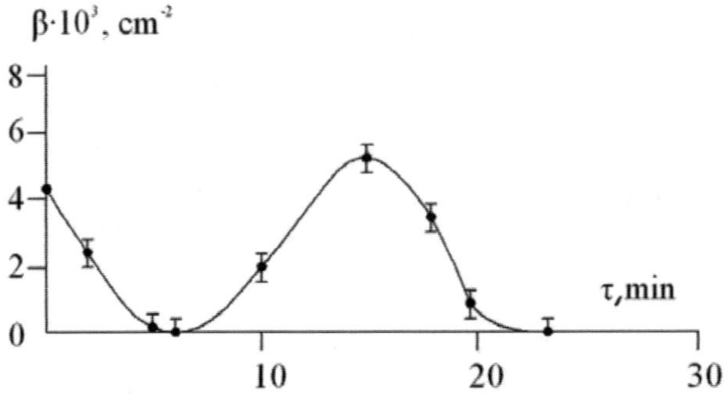

Figure 4. The dependence of dislocations density in silver azide crystals on the time of co-effect of mechanical stress ($5\cdot10^5$ N/m^2) and magnetic field (0.3 T).

Fig. 4 demonstrates that increasing time of exposure results in a cyclically changing density of dislocations, in 30 minutes of exposure the density of dislocations falls up to 0. However, while dislocations structure in silver azide crystals changes when being influenced mechanically, in magnetic field a slow dissolution is seen: gaseous products – gas bubbles are emitting (Fig. 5, curve 2). If the period of exposure is longer the samples are subject to brittle fracture without explosion. For comparison dissolution in magnetic field (0.3 T) without strength failure is depicted in the same Figure (Fig. 5, curve 1.).

It is worth noting that mechanical stress of samples, placed between the poles of electromagnet, is attended by directed movement of edge dislocations (it is determined according to displacing etching pits).

Let us discuss the obtained results of experiments.

Brittle fracture of samples recorded over a certain period of exposure can probably comprise 2 phases – micro-crack nucleation and its propagation. The cause of micro-crack nucleation is probably development of a super-dislocation which is formed as the result of integrating dislocations near the stopper. Meanwhile, the pressure of stopped dislocations exceeds several times the pressure of their free movement. This rising pressure can improve the strength of a crystal and at the stopper place there is a crack occures being the sink for newly forming dislocations. Any accumulation of point defects can be considered as a stopper.

In magnetic field one observe the decreasing energy of dislocations – paramegnetic impurity centers interaction, it facilitates the separation of dislocation from other obstakles. In non-magnetic materials dislocations move in constant magnetic field and without external effect due to the slow relaxation of dislocation structure in the field of internal stresses.

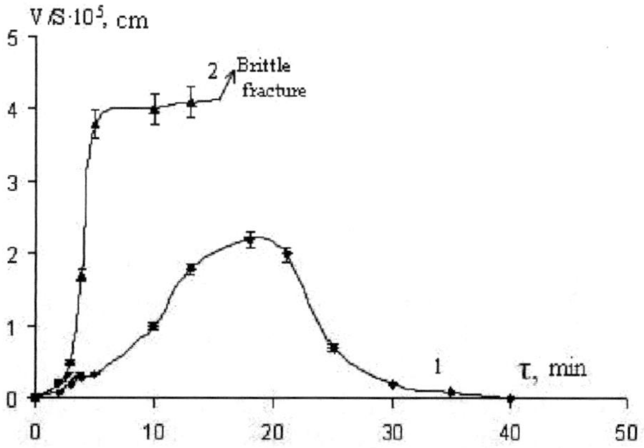

Figure 5. The dependence of retained gas in silver azide crystals on the time of exposure: 1 – magnetic field (0.3 T); 2 – co-effect of magnetic field and mechanical stress.

The external force (mechanical) influencing on the sample simultaneously with magnetic field accelerates the relaxation of dislocation structure that increases the path of dislocations when the sample is effected both by magnetic field and mechanical stress.

As we observe directed movement of negatively charged edge dislocations, possible in electric field only, an attempt was made to find a non-contact electric field, which provides no movement of dislocations recorded as changing etching pits. This external non-contact electric field, which compensates the developed by magnetic field internal one, was found, it doesn't contradict to the earlier discovered magnetoelectric effect in silver azide crystals [4].

Similar cyclic change in dislocations density is also observed when ultrasonic processing. In terms of specific character of the experiment initially dislocation-free crystals were used (Fig. 6). In this case the time of transformation of dislocation structure is far shorter, i. d. over 5 minutes of exposure 2 maximums of dislocations density are recorded, then the sample fails.

After samples are processed ultrasonically and the reaction of decomposition by a contact electric field is initiated (30 kW/m), it was recorded that in the zone of minimal dislocations the samples exploded after 3 minutes of being energized and in the zone of maximal dislocations – after 9 minutes. It demonstrates that the reaction of slow decomposition dominates principally over the critical fracture in conditions of significant concentration of dislocations in the sample.

Strengthening near-surface layer of the sample after ultrasonic processing is considered to be one of the reasons of this effect.

As the number of dislocation increases a longer effect of electric field up to sample explosion is required, whereas the samples expose earlier in conditions of decreasing density of dislocations.

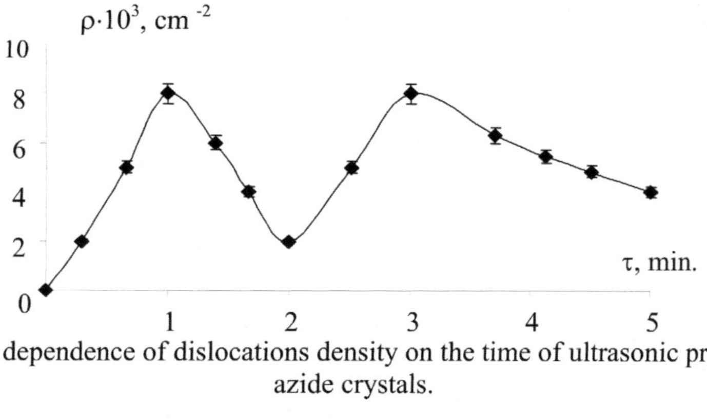

Figure 6. The dependence of dislocations density on the time of ultrasonic processing in silver azide crystals.

Various results can be probably explained by the fact that the increasing number of dislocations causes the dispersion of charge carriers on them and complicates the approach of these carriers to reaction zones. Meanwhile, slow decomposition of basic charge carriers – reagents of chemical reaction – is recorded and followed by explosion when the concentration of pits gets critical. As the number of dislocations decreases an inverse process takes place.

Conclusions

In conclusion it is to say that ultrasonic processing and mechanical stress of silver azide crystals causes generation and a wide-range change in the density of dislocations, constant magnetic field effect facilitates the separation of dislocations from accumulated defects, and electric field (internal field of polarization or the non-contact internal one) is the reason of dislocations output.

A co-effect of magnetic field and mechanical stress in silver azide crystals furthers generation, stopper breakaway, shift of dislocations, and generation of superlattice dislocations.

Therefore, a method of mechanical and electric strengthening is proposed and based on changing dislocations in whiskers.

References

[1] Rodzevich A P Kuzmina L V Gazenaur E G and Krasheninin V I Plasticity and decomposition of whiskers on electric-induced deformation / *AIP Conf Proc* 1623 (2014) 519-522

[2] Babakova E V Gradoboev A V Saprykin A A Ibragimov E A Yakovlev V I Sobachkin A V Comparison of Activation Technologies Powder ECP-1 for the Synthesis of Products Using SLS *Applied Mechanics and Materials* 756 (2015) 220-224

[3] Krasheninin V I Kuzmina L V Ivaschenco V E Decomposition of crystalline asides of heavy metals in constant magnetic field *Materials Research Innovations* 5 (2002) 219

[4] Krasheninin V I Kuz'mina L V Zakharov V Yu Stalinin A Yu Silver azide decomposition in an electric field *Chem Phys Reports* 4 (1995) 529-538

[5] Rodzevich A P Gazenaur E G Kuz'mina L V and Krasheninin V I The Polarization of Silver Azide in Electric Field *Advanced Materials Research* 1040 (2014) 744–747

[6] Malushin N N Valuev D V Valueva A V Serikbol A Kinetic study of the effect of plasticity and its role in stress relaxation in the weld speed steel during the martensitic transformation J Applied Mechanics and Materials 682 (2014) 58-63

[7] Zakharov V Yu Krasheninin V I Kuz'mina L V Zakharov Yu A The control of solid phase decomposition of silver azide by noncontact electric field *Solid State Ionics* 101-103 (1997) 161–164

Cutting capacity and strength of single grinding grains

V S Lyukshin*, A V Barsuk, R R Fazleev

Yurga Institute of Technology, TPU affiliate, 26, Leningradskaya St., Yurga, 652050, Russia

E-mail: *lwsfoa@rambler.ru

Abstract. The results received at research of the cutting ability and durability of single grinding grains are presented in article. In particular influence of such parameters as a form and a corner of orientation of grinding grain was considered.

Grinding process due to its particular nature is accompanied by removal of a large number of extra-fine cutting chips; significant amount of grinding grains are involved in running of the grinding belt. Removal of cutting chips of very small size occurs in time period measured by split seconds; therefore it makes it difficult to directly observe the ongoing processes.

For this reason characteristic properties of cutting by grains process, cutting capacity of grains and strength are commonly evaluated by indirect methods.

Research of cutting capacity of single grains.

Considering existing methods to define cutting capacity of grinding grains the most wide spread among them are:

– method of the special disc abrasion by the studied abrasive material [1];
– cutting by the single abrasive grain fixed in the rotating disc [2];
– scratching by the single grain of the processed material at low velocities of relatively linear motion [3];
– wearing of materials inside abrasive bulk [4];
– method of defining cutting capacity of grinding grains fixed by centrifugal forces [5];
– wearing in gas flow of abrasive [6].

Abrasive materials are intended for grinding, and therefore, in order to define their cutting properties it is reasonable to simulate the real grinding conditions. Grinding conditions are marked by the following attributes [7]:

1. Velocity of relative interaction of abrasive and the processed material reaches dozens of meters per second;
2. Processing is carried out by fixed abrasive (grains are connected by bunch);
3. Working surface of grinding layer is pressed to the processed material with certain forcing;
4. Multiple micro-cutting occurs by means of large amount of separate abrasive units.

Analyzing the existing methods for defining cutting capacity of grinding grains from the point of their accordance to the above listed attributes it can be concluded that method based on cutting by single grinding grain fixed in the rotating disc simulates adequately well the real grinding conditions.

For this reason this method has been taken as a basis to evaluate cutting capacity of grinding grains.

IOP Conf. Series: Materials Science and Engineering **91** (2015) 012047 doi:10.1088/1757-899X/91/1/012047

Experimental studies to define cutting capacity of single grinding grains were conducted at surface grinding machine, model 3G71 by the following cutting modes: velocity of disc with fixed grinding grain V_d=30 m/s; velocity of traverse motion V_{trav}=0.19 m/min; cutting depth t=0.04 mm; cut counts N=20, cutting scheme – plain two-axis vector grinding.

During experimental testing the grinding grain under study was fixed by bunch in a hoop under certain angle γ. The hoop was then fixed on the disc, and disc was set on the work spindle. Weight of the processed sample was preliminary measured on weight scales and it was fixed on magnetic holding plate afterwards. Cutting depth and supply were set by means of cutter control devices. Cutting depth was controlled by dial indicator. The sample was taken out and weighed after twenty runs. Then the experiment was repeated.

The samples selected for the experimental testing included: grains of aluminum-oxide abrasive, grade 13A125, isometric ($K_f \approx 1.14$), transient ($K_f \approx 1.56$) and plate-like form ($K_f \approx 2.27$).

The form of grinding grain was evaluated quantitatively by means of form coefficient K_f [15-17]. This parameter was defined as ratio of surface around grain projection circle ($S_{a.c.}$) to the square of grain projection ($S_{g.p.}$):

$$K_f = \frac{S_{a.c.}}{S_{g.p.}}. \tag{1}$$

Orientation angle of grinding grain (γ) was changed both in cutting direction from 0° to +90°, and in the opposite side from 90° to 180° through each 15°.

Rectangular bars LxBxH = 30x20x10 mm from material St3 (HB111) were used as the processed samples.

Before experiment each sample underwent grinding and finish-grinding of the processed surface. Weight of removed material was measured by weight scale ADV-200 with weighing accuracy 0.0005 g.

Cutting capacity of single grinding grains (ΔM) was defined as difference between weight of processed material sample before and after the experiment.

Experimental results are given in Figure 1.

According to the graphs it can be stated that:

1. form and angle of grains orientation affect their cutting capacity;
2. grains of plate-like form ($K_f \approx 2.27$) have better cutting capacity, while grains of isometric form ($K_f \approx 1.14$) – have less effective cutting capacity;
3. orientation angle of abrasive grains, providing better cutting capacity is:
   - for plate-like form ($K_f \approx 2.27$) – from 60° to 90°;
   - for transient form ($K_f \approx 1.56$) – from 60° to 90°;
   - for isometric form ($K_f \approx 1.14$) – from 75° to 90°.

Thus the given results enable to conclude that it is reasonable to use grinding grains of plate-like form with orientation angle of γ=75° in order to increase cutting capacity of grinding discs and belts.

<u>Research of strength of single grinding grains.</u>

The existing methods used to define strength of grinding grains are as follows:

1. MU2-036-057-82 – method of determination of grinding grain destruction [7];
2. MU2-036-105-85 – method of defining strength of single grain of grinding materials [7];
3. Method of defining brittleness of grinding grains in ball mill [8];
4. Method of cutting by single grain of V-shaped processed sample during flat grinding [9, 10];
5. Method of destruction by uniaxial compression on the facilities of roll crusher-type [8];
6. Method of strength tests of grinding grains by impact of special hammer head [11];

7. Method of strength tests of grinding grains by falling weight from different height [12];
8. Method of strength tests of grinding grains by oscillating weight [13].

Taking into account the real stress state of grains during operation in grinding tool, the indicated methods to different extent have a number of disadvantages.

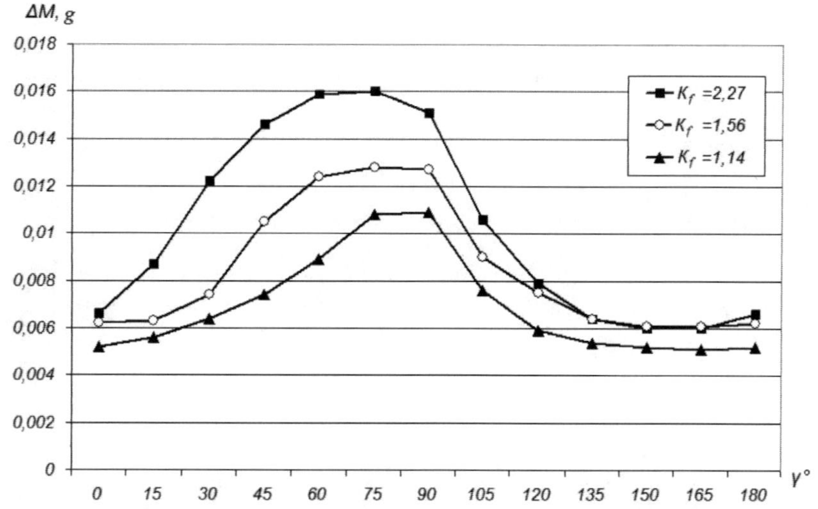

Figure 1. Influence of form and orientation angle of grinding grains on their cutting capacity.

Knowing that grinding grain operates under periodic short-term loads, methods based on dynamic loading will better simulate the real loading conditions. Method of strength tests of grinding grains by oscillating weight refers to one of such methods. However it is rather problematic to define influence of slope angle of grinding grain on its impact strength by the instruments used within this method. Keeping that in mind the present work implies using the instrument including apart from standard knots structural elements dealing with slope of grinding grain and controlling its height [14].

To perform strength tests grains of aluminum-oxide abrasives were used, grade 13A with grain number 125, 100, 80, and aluminum-oxide abrasives of grade NK F24 (Germany) with grain number 80. Grains were of isometric ($K_f \approx 1.14$), transient ($K_f \approx 1.56$) and plate-like form ($K_f \approx 2.27$).

Slope angle of grain (γ) in regards to horizontal plane was accepted as 90°, 75° and 60° (within the impact direction).Thus it enabled to evaluate the tendency of grain strength alteration depending on its orientation relative to the basis.

Impact strength of grinding grain (a) was defined according to the following formula [13]:

$$a = \frac{U}{S}, \tag{2}$$

where U – impulse force;

S – area of fracture surface of grinding grain.

Impulse force equals:

$$U = m \cdot (V_{xx} - V_{px}), \tag{3}$$

where V_{xx} and V_{px} – velocity of non-load and operating run;

m – hammer head weight.

Velocities of non-load and operating run were calculated according to the following formulas:

$$V_{xx} = 2 \cdot \sqrt{g \cdot l} \cdot \frac{\sin \alpha_{xx}}{2}, \quad (4)$$

$$V_{px} = 2 \cdot \sqrt{g \cdot l} \cdot \frac{\sin \alpha_{px}}{2}, \quad (5)$$

where g – free fall acceleration;

α_{xx} and α_{px} – angle of deviation of balance from the equilibrium position during non-load and operating run.

Experimental results are given in Figures 2-5.

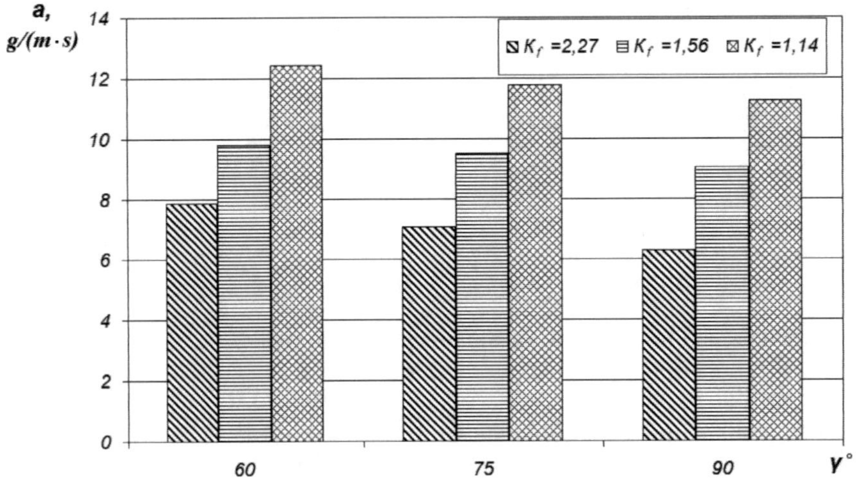

Figure 2. Influence of form and orientation of single grinding grains 13A125 on their impact strength.

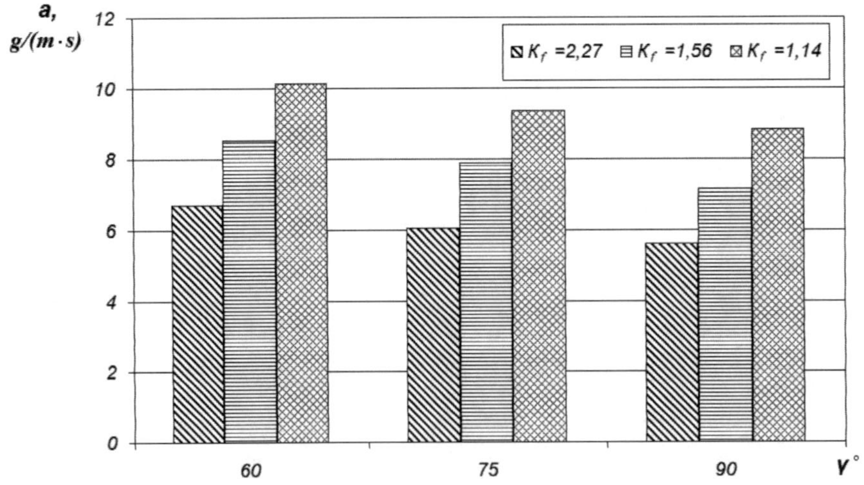

Figure 3. Influence of form and orientation of single grinding grains 13A100 on their impact strength.

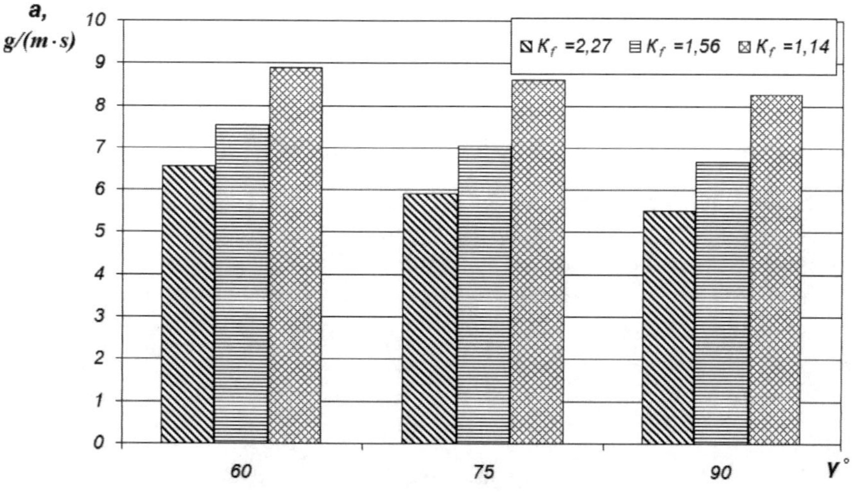

Figure 4. Influence of form and orientation of single grinding grains 13A80 on their impact strength.

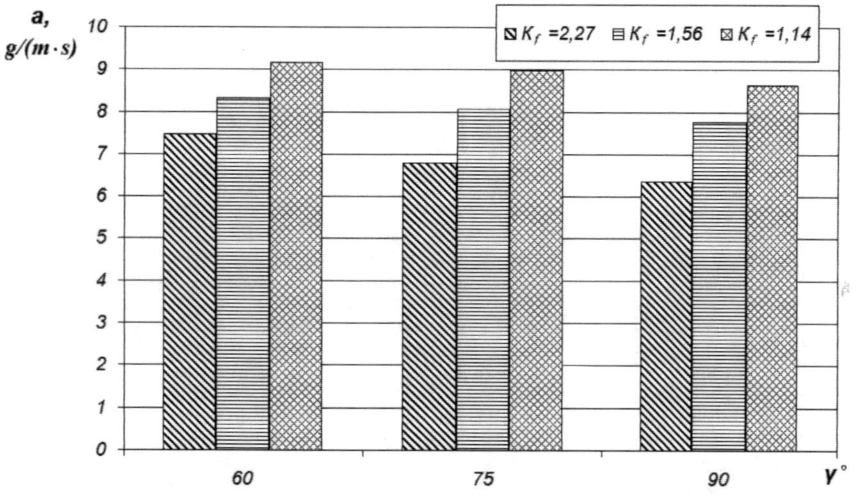

Figure 5. Influence of form and orientation of single grinding grains NK F24 on their impact strength.

The obtained results demonstrate as follows:

1. Form and orientation of grains influence significantly the value of impact strength;
2. All other conditions being equal, isometric grains are characterized by larger strength. Their strength is 1.2 times higher than of transient grains and 1.35 times higher than of plate-like grains;
3. In case of angle of slope alteration from 90° to 60° (within impact direction) grain strength increases:
   - for grains 13A125 – by 7÷18%;
   - for grains 13A100 – by 10÷16%;
   - for grains 13A80 – by 7÷18%;
   - for grains NK F24 – by 5÷17%.
4. Apart from form and orientation parameters grain strength is also affected by size and grade of abrasive.

The obtained data was analyzed by the program software “STATISTICA 6.0”. On the basis of this software connection between impact strength (a) and parameters of single grinding grain (K_f, γ) given as:

$$\ln a = a_0 + a_1 \cdot \ln K_f + a_2 \cdot \ln \gamma . \qquad (6)$$

Calculation of regression coefficients was carried out by least square method. Fisher’s ratio test was used to validate the accuracy of the obtained models.

Parameters of statistical model reflecting the influence of abrasive grade (13A or NK), forms of grinding grain, orientation angle and size of grinding grain on impact strength are given in Table.

In Table a_0 – free term of model;

K_f – form coefficient of grinding grain;

γ – angle of slope of grinding grain.

Table. Parameters of statistical models.

Abrasive material grade	Model terms	Coefficients of statistical model
13A125	a_0	3,802989
	$\ln K_f$	-0,752999
	$\ln\gamma$	- 0,290076
13A100	a_0	4,235779
	$\ln K_f$	-0,713419
	$\ln\gamma$	- 0,299496
13A80	a_0	3,463996
	$\ln K_f$	-0,517542
	$\ln\gamma$	- 0,296567
NK F24	a_0	3,225691
	$\ln K_f$	-0,382461
	$\ln\gamma$	- 0,232846

It can be seen from the models that with decrease of the grain size degree of influence of form on its impact strength decreases. The given result is explained by the fact that with decrease of grain size its cross-section area with form alteration will be changed to the less extent. Influence of grinding grain orientation on strength almost does not depend on grain size. Analysis of model for grain NK F24 shows that influence of form and orientation angle of this grinding grain on impact strength is significantly lower than of grain 13A80.

Conclusions

1. Form and orientation angle affect the cutting capacity of grinding grains;
2. Grains of plate-like form ($K_f \approx 2.27$) intentionally-oriented on the surface of basis have better cutting capacity, while grains of isometric form ($K_f \approx 1,14$) have less capacity;
3. Orientation angle of grinding grains when cutting capacity is the most efficient is:
   – for plate-like form ($K_f \approx 2.27$) – from 60° to 90°;
   – for transient form ($K_f \approx 1.56$) – from 60° to 90°;
   – for isometric form ($K_f \approx 1.14$) – from 75° to 90°.
4. Form and orientation of grains significantly affect the value of impact strength;

IOP Conf. Series: Materials Science and Engineering **91** (2015) 012047 doi:10.1088/1757-899X/91/1/012047

5. All other conditions being equal, isometric grains are characterized by larger strength. Their strength is 1.2 times larger than of transient grains and 1.35 times larger than of plate-like grains;
6. In case of angle of slope alteration from 90° to 60° (within impact direction) grain strength increases:
   - for grains 13A125 – by 7÷18%;
   - for grains 13A100 – by 10÷16%;
   - for grains 13A80 – by 7÷18%;
   - for grains NK F24 – by 5÷17%.

Results of statistical analysis correlate with and prove conducted model experiment. It is particularly noted that together with increasing K_f and γ (from 60° to 90°) impact strength of single grinding grains decreases.

References

[1] Yascheritsyn P I, Zaitsev A G 1972 *Enhancing quality of grinding surfaces and cutting properties of abrasive-diamond instrument* (Minsk: Science and Engineering)
[2] Kharchenko I V, Sayutin G I, Bogomolov N I 1972 *Abrasives* **4** 14–17
[3] Goepfert G I Williams J L 1959 *Mechanical Engineering* **4** 69–73
[4] Tretiyakov I P, Korotkov A N 1979 *Progressive methods of abrasive, diamond processing in machinery engineering* 93–94
[5] Korotkov A N 1990 *Grinding of components by abrasive mass* (Naberezhnye Chelny: Science for Industry: Proceedings)
[6] Kremen Z I, Masarsky M L Guzel V Z 1979 *Machines and instruments* **6** 25–26
[7] Korotkov A N 1992 *Operational properties of abrasive materials* (Publishing Office of Krasnoyarsk University)
[8] Breker D 1974 *Designing and technology of machine engineering* **4** 160–165
[9] Loladze T N, Bokuchava G V 1967 *Wearing of diamonds and diamond discs* (Moscow: Machine Engineering)
[10] Davydova G E. 1973 *Studying strength properties of abrasives and diamond and their interaction with the processed materials* (Tbilisi)
[11] Karpov A B 1973 *Studying interaction of grain and bunch of grinding instruments under dynamic loading* (Moscow)
[12] Kascheev V N 1970 *Abrasive destruction of solid bodies* (Moscow: Nauka)
[13] Murdasov A V 1975 *Abrasives* **9** 14–18
[14] Patent 38505 Russia, MPK^7 G 01 N 3/00 Facility for defining impact strength of grinding grains Korotkov A N, Romanenko A M, Lyukshin V S, Mineshev M A No.2004104702 applied 17.02.04 published 20.06.04, Cert. No.17
[15] Lyukshin V S 2014 *Applied Mechanics and Materials* **682** 148–153
[16] Korotkov A, Korotkova L, Gubaidulina R 2014 *Applied Mechanics and Materials* **682** 469–473
[17] Korotkov V A, Petrushin S I 2014 *Applied Mechanics and Materials* **682** 224-230

IOP Conf. Series: Materials Science and Engineering **91** (2015) 012048 doi:10.1088/1757-899X/91/1/012048

On the Problem of Wear Resistant Coatings Separation From Tools and Machine Elements

S I Petrushin[1], R H Gubaidulina[1,2], S V Gruby[2,3] and A V Likholat[1,4]

[1]Yurga Institute of Technology, TPU affiliate
Kemerovo Region, Yurga, 26, Leningradskaya RUS
[2]Moscow State Technical University by N.E. Bauman
Moscow, 5, 2nd Baumanskaya RUS

E-mail[1]: psi@tpu.ru, E-mail[2,4]: victory_28@mail.ru, E-mail[3]: gru@bmstu.ru

Abstract. The article considers separation of wear resistant coatings of tool and engineering materials which arises both during coating fabrication and use of the product. The cause of this phenomenon is assumed to be related to thermal residual stresses generating on the coating-substrate border. These stresses have been analyzed and methods are provided to calculate it after produced composite material is cooled down from the temperature of coating synthesis to the ambient temperature. A no-fracture condition has been stated in relation to coating-substrate thicknesses, temperature differences and physical and mechanical properties of combined materials. The issue of intermediate layer incorporation with pre-set parameters has been discussed. A co-effect of thermal residual and functional stresses on the strength of the boundary layer has been considered when heating, tension and compression of a product with wear resistant coating. Conclusions have been made, as well as recommendations to improve fracture strength of products with thin wear resistant coatings.

1.Introduction

Wear resistant coatings layered on the surfaces of tool and engineering materials are widely applied in mechanical engineering in order to improve frictional surface layer life along with retention of strength characteristics of the product. For instance, leading companies [1, 2] coat high-speed steel and hard alloy cutting edges with variously combined thin layers of refractory material carbides and nitrides and aluminum oxide (TiC, TiN, Al_2O_3 etc.). Here, the total thickness of coating doesn't exceed 10 – 20 μm. If this value is exceeded a coating gets brittle and can separate both under action of functional stresses and when coating layers [3,4]. A similar phenomenon can also appear when surfacing, spraying, chrome or nickel plating and during some other processes with a distinguished border between layers (there is no significant diffuse zone). When breaking away interlaminar cracks arise on coating – base material (substrate) border. We consider a general approach to solve the problem of fracture strength of coating in terms of thermal residual stresses analyzed and stated in papers [5-8].

2 Results and Discussion

Calculation of thermal residual stress in laminated composite materials.

A product with a coated surface is a two-layer composite material where a surface layer is much thinner than a substrate. The strength of multi-layer composite materials depends not only on ultimate strength of layer materials but on interlaminar strength. It is often the strength of two adjacent layers that restricts integrity of a composite. It is caused by product heating up to a significant temperature ($T_И$) when composite material manufacturing. After sintering, surfacing, coating, layer-by-layer synthesis a composite is cooled down up to the ambient temperature (T_O). Herewith, thermal residual stresses arise on the border of two adjacent layers because of different linear expansion coefficients. If it exceeds the ultimate strength of any layer material cold interlaminar cracks appear and a composite material gets improper for this product.

If thermal residual stresses are smaller than the ultimate strength of layer materials monolithic nature of a composite is assured, however, it can reduce strength properties of a product. The layer of material with a bigger coefficient of linear expansion tends to decrease its dimensions more than the layer with a smaller coefficient of linear expansion. As the result, the first layer will be stretched by the second one and it will be pressed by the first one. In material with a bigger coefficient of linear expansion thermal tension residual stresses arises, and in material with a smaller coefficient of linear expansion there are compressive thermal residual stresses [4].

Let us consider a two-layer composite material synthesized at temperature difference $\Delta T = T_И - T_O$ and subject to thermal residual stresses only (Fig.1). The coefficient of linear expansion of material 1 is α_1, its coefficient of elasticity is E_1 and Poisson's ratio is μ_1, those of material 2 are α_2, E_2 and μ_2, respectively. We consider first the case, when $\alpha_1 \succ \alpha_2$. As the thickness of a single layer in a composite product is far smaller then its length and width, the state of two-dimensional stresses will arise in it when cooling down, Hook's law for it is as follows [9].

For layer 1 (tension) - $\sigma = \dfrac{\varepsilon E_1}{1 - \mu_1}$; (1)

For layer 2 (compression) - $\sigma = \varepsilon E_2 (1 - \mu_2)$. (2)

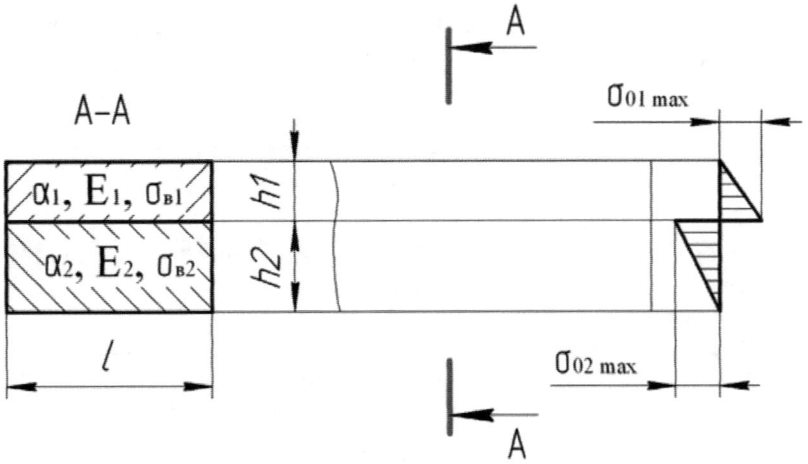

Figure 1. A diagram of two-layer composite material subject to thermal residual stresses only ($\alpha_1 \succ \alpha_2$).

For lack of external stresses maximum thermal residual stresses $\sigma_O^{\max}$ will be on the border of layers. Thermal residual stresses on the upper surface of layer 1 and on the lower surface of layer 2 equal to zero, and inside the layers they are distributed linearly (see Fig.1). Let us denote the cross-section area of layer 1 as $F_1 = bh_1$, and that of layer 2 as $F_2 = bh_2$, where b is the width of layers, h_1 and h_2 their

IOP Conf. Series: Materials Science and Engineering **91** (2015) 012048 doi:10.1088/1757-899X/91/1/012048

thicknesses, the condition of equilibrium of forces provided that composite material on the border is uniform is as follows

$$\frac{1}{2}\sigma_{O1}^{\max} b h_1 = \frac{1}{2}\sigma_{O2}^{\max} b h_2$$

Hence,

$$\sigma_{O2}^{\max} = \sigma_{O1}^{\max}\frac{h_1}{h_2} \quad (3)$$

The absence of thermal cracks on border of layers in a composite material is to be assured by the condition of deformations equality [6]. In layer 1 deformation caused by thermal cooling down $\varepsilon = \alpha\Delta T$ is opposite in direction to residual stresses caused by thermal residual stresses, in layer 2 these deformations are added. Therefore, in terms of equations (1) and (2) we have:

$$\alpha_1\Delta T - \frac{\sigma_{O1}^{\max}(1-\mu_1)}{E_1} = \alpha_2\Delta T + \frac{\sigma_{O2}^{\max}}{E_2(1-\mu_2)} \quad (4)$$

For the case under consideration ($\alpha_1 \succ \alpha_2$) maximum tension stresses in layer 1 are the most disadvantageous conditions in terms of interlaminar cracks arising in a composite material. It is calculated according to (4) and (3) as follows:

$$\sigma_{O1}^{\max} = \frac{\Delta T(\alpha_1 - \alpha_2)}{\dfrac{1-\mu_1}{E_1} + \dfrac{1}{E_2(1-\mu_2)}\dfrac{h_1}{h_2}} \quad (5)$$

We denote the ultimate tension strength of layer 1 material as σ_{b1}, and the safety factor as $k \succ 1$ in view of deviated real physical and mechanical parameters of layer materials in a composite from the designed parameters ($k_{\min} \approx 1{,}5$). Then the condition of fracture strength in a two-layer composite material (see Fig.1), cooled down by temperature difference ΔT and not subject to operating thermal and functional stresses, is written as follows:

$$\frac{\sigma_{b1}}{k} \geq \frac{\Delta T(\alpha_1 - \alpha_2)}{\dfrac{1-\mu_1}{E_1} + \dfrac{1}{E_2(1-\mu_2)}\dfrac{h_1}{h_2}} \quad (6)$$

It is worth noting that the condition (6) enables the first phase of designing a composite and a coated composite aimed at assuring their uniformity when manufacturing. For this purpose we express layer thicknesses relation in terms of equation (6):

$$\frac{h_1}{h_2} \geq \left[E_1\Delta T(\alpha_1 - \alpha_2)k - (1-\mu_1)\sigma_{b1}\right]\frac{E_2(1-\mu_2)}{E_1\sigma_{b1}} \quad (7)$$

Inequation (7) sets the limit for layer thicknesses relation in a two-layer composite, after it's exceeded interlaminar thermal cracks will arise in the lay.

Similarly to the case, when $\alpha_1 \prec \alpha_2$, the following condition of fracture strength is obtained:

$$\frac{h_1}{h_2} \leq \frac{\sigma_{b2}E_2}{\left[\Delta T(\alpha_2 - \alpha_1)kE_2 - (1-\mu_2)\sigma_{b2}\right]E_1(1-\mu_1)} \quad (8)$$

where σ_{b2} - ultimate strength of the second layer material

When deducing the inequation (8) we took into consideration that after cooling down layer 1 will be pressed and layer 2 – stretched.

Breaking away after coating.

In terms of mentioned above information we consider the issue of coatings breaking away in respect to technologies of soldering and surfacing. Equations (7) and (8) preset the condition of a coating non-breaking away from the substrate unless a composite is exposed to an external stress. Figure 2 provides an example of fracture strength borders calculated according to (9) and (10) for a two-layer composite consisting of steel 60 ($\sigma_b = 1500$ MPa; $\alpha = 11{,}5 \cdot 10^{-6}\, 1/{}^{o}C$; $\mu = 0{,}3$; $E = 218500$ MPa) and tungsten-cobalt hard alloy BK8 ($\sigma_b = 1590$ MPa; $\alpha = 5{,}1 \cdot 10^{-6}\, 1/{}^{o}C$; $\mu = 0{,}23$; $E = 598400$ MPa). It demonstrates that the relation of layer thicknesses depends significantly on operating temperature of a composite manufacturing. There will be no thermal cracks when heating less $500^{o}C$, it conforms to conditions of low-temperature soft soldering. If the temperature of the process is higher (hard soldering, powder sintering, surfacing), the relation of thicknesses of composite layers are to be taken into consideration. Figure 2 demonstrates that when $\alpha_1 \succ \alpha_2$ (steel coated hard alloy substrate) the thickness of the upper layer of a composite increases in direct proportion to the rise of operating temperature, and when $\alpha_1 \prec \alpha_2$ (hard alloy coated steel) it decreases according to a hyperbola dependence.

Therefore, coated composites can have thermal residual stresses inside their structure before they are used; the value of these stresses depends on the operating temperature of composite manufacturing and on the difference of thermal linear expansion coefficients of layer materials. One should take it into consideration when designing coated products because functional stresses being algebraically added to thermal residual stresses can cause either coating separation or increase in the strength of coating-substrate cohesion.

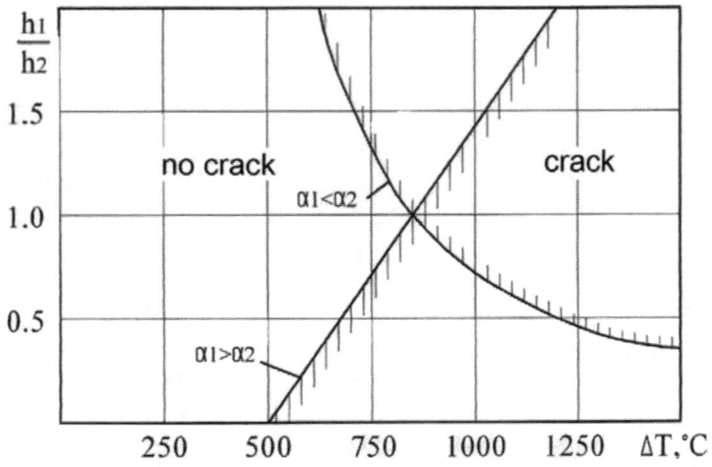

Figure 2. Fracture strength borders for a two-layer composite.

It is worth mentioning that after a two-layer composite is cooled down thermal residual stresses arise in the layer of steel in both cases, and thermal compression residual stresses grow in hard alloy layer, that is, we have a preliminary stressed material.

Layered composites whose coating and substrate are made of the same material have been studied so far. Provided that thickness relations resulted from conditions (7) and (8) are improper for composite functioning, the purpose of designing can be achieved by introducing between the layers an interlaminar layer being a mixture of source materials (two-layer coating). The simplest way to produce composites of this kind is powder metallurgy. In terms of mixture laws [10], physical and mechanical properties of an interlaminar layer are determined as follows:

- ultimate strength

$$\sigma_{bc} = \sigma_{b1}\upsilon_1 + \sigma_{b2}\upsilon_2 = \sigma_{b2} - \upsilon_1(\sigma_{b2} - \sigma_{b1});$$

- coefficient of linear expansion

$$\alpha_c = \alpha_1\upsilon_1 + \alpha_2\upsilon_2 = \alpha_2 - \upsilon_1(\alpha_2 - \alpha_1);$$

- Poisson's ratio

$$\mu_c = \mu_1\upsilon_1 + \mu_2\upsilon_2 = \mu_2 - \upsilon_1(\mu_2 - \mu_1);$$

- modulus of elasticity

$$E_c = E_1\upsilon_1 + E_2\upsilon_2 = E_2 + \upsilon_1(E_1 - E_2)$$

where $\upsilon_1 = V_1/V_c$ - volume concentration of material 1 in the mixture;
$\upsilon_2 = V_2/V_c$ - volume concentration of material 2 in the mixture, whereas $\upsilon_1 + \upsilon_1 = 1$

Let us substitute these values for the criterion of fracture strength (7). Then, a condition of no cold thermal cracks between the upper layer of coating and mixture is as follows:

$$\frac{h_1}{h_c} \geq [E_1\Delta T(\alpha_1 - \alpha_c)k - (1-\mu_1)\sigma_{b1}]\frac{E_c(1-\mu_c)}{E_1\sigma_{bc}} \tag{9}$$

The same condition for the mixture and substrate material:

$$\frac{h_c}{h_2} = [E_c\Delta T(\alpha_c - \alpha_2)k - (1-\mu_c)\sigma_{bc}]\frac{E_2(1-\mu_2)}{E_c\sigma_{bc}} \tag{10}$$

If a grade composite is produced according to the principle $\alpha_1 \prec \alpha_C \prec \alpha_2$ we have the similar conditions of fracture strength assuring obtained on the basis of equation (8)

$$\frac{h_1}{h_c} \leq \frac{\sigma_{bc}E_c}{[\Delta T(\alpha_c - \alpha_1)kE_c - (1-\mu_c)\sigma_{bc}]E_1(1-\mu_1)}: \tag{11}$$

$$\frac{h_c}{h_2} \leq \frac{\sigma_{b2}E_2}{[\Delta T(\alpha_2 - \alpha_c)kE_2 - (1-\mu_2)\sigma_{b2}]E_c(1-\mu_c)}. \tag{12}$$

Therefore, a monolithic nature of grade composite with interlaminar layers of multilayer coating can be obtained by varying volume concentrations of adjacent materials in the mixture with equations (9) – (12).

Functionally stressed breaking away

If a coated product is used under thermal functional stresses the relation of coating and substrate thicknesses can be selected on the ground of formulae (7) and (8), where ΔT is the difference of coating synthesis temperature and functional temperature. If coating is stressed by forces, for instance, friction force, the problems of strength of materials are to be resolved according to the loading conditions.

Let us consider a two-layer composite with b-wide rectangular cross-section synthesized at temperature difference $\Delta T = T_И - T_O$ and subject both to tension force P and thermal residual stresses. Top h_1-thick material has a coefficient of linear expansion α_1, modulus of elasticity E_1 and Poisson's ratio μ_1 while bottom h_2-thick material has parameters: α_2, E_2 and μ_2, whereas $\alpha_1 \prec \alpha_2$. As the result of force P action there is tension stress in the bar

$$\sigma_p = \frac{P}{(h_1 + h_2)\cdot b} \tag{13}$$

For the case under consideration according to formula (7) maximum thermal residual stresses can be calculated on the upper surface of the bottom layer and according to formula (8) – on the lower surface of the top layer in a two-layered composite. If this residual stress is added to functional tension stress in terms of (13), the total stresses on the border of layers will equal to

- for layer 1 - $$\sigma_{\Sigma 1}=\frac{\Delta T(\alpha_1-\alpha_2)}{\frac{1}{E_1(1-\mu_1)}-\frac{1-\mu_2}{E_2}\cdot\frac{h_1}{h_2}}+\frac{P}{(h_1+h_2)\cdot b};\quad (14)$$

- for layer 2 - $$\sigma_{\Sigma 2}=\frac{\Delta T(\alpha_2-\alpha_1)}{\frac{1-\mu_2}{E_2}-\frac{1}{E_1(1-\mu_1)}\cdot\frac{h_2}{h_1}}+\frac{P}{(h_1+h_2)\cdot b}.\quad (15)$$

In formula (14) the first summand will be negative ($\alpha_1 \prec \alpha_2$), in (15) – positive. Therefore, under action of a force compressive thermal residual stresses in coating will be compensated through tension stresses caused by an external force. On the contrary, in layer 2 (substrate) these stresses will sum up and probably cause a crack on the border of the layers. Then, the condition of fracture strength of a two-layer coated composite cooled down by the temperature difference ΔT and stressed by tension force P is as follows

$$\frac{\sigma_{b2}}{k}\geq\frac{\Delta T(\alpha_2-\alpha_1)}{\frac{1-\mu_2}{E_2}-\frac{1}{E_1(1-\mu_1)}\cdot\frac{h_2}{h_1}}+\frac{P}{(h_1+h_2)\cdot b}\quad (16)$$

If a coated composite is compressed the equation (14) is critical and on its ground the condition of fracture strength can be deduced which is similar to equation (16).

Summary

1. When designing wear resistant coated products a probability of coating separation is to take into consideration. It is caused by thermal residual stresses arising between coating and substrate when they are technologically heated and an obtained composite is cooled down.
2. For coatings with a distinct border with the product thermal residual stresses depend on physical and mechanical properties of coating and substrate materials (coefficient of linear expansion, modulus of elasticity, Poisson's ratio, and ultimate strength), temperature of product heating when synthesizing, and material thicknesses relation.
3. Obtained no-cracks conditions make it possible to match the thickness of coating with its production technology even at the stage of designing composite materials.
4. The problem is resolved to calculate fracture strength of thermally and functionally stressed layered composite materials, and depends on the type of stress coated products are subject to (tension, compression, bending etc.)

References

[1] Tyalina L N, Minaev A M and Pruchkin V A 2011*Tambov GOU TGTU* 80
[2] Bouznik B M, Fomin V M and Alkhimov A P 2005 *Novosibirsk Publishing House of SB RAS* 260.
[3] Vereshchaka A S 1993 *Machinebuilding.* 336
[4] Petrushin S I, Saprykin A A and Dureev V V 2014 *Tomsk TPU Publishing* 205
[5] Saprykina N A and Saprykin A A 2013 *Appl. Mechan. and Mater* **379** 56I
[6] Petrushin S I 2013 *Appl. Mechan. and Mater* **379** 100
[7] Saprykina N A and Saprykin A A, 2014 *Appl. Mechan. and Mater*I **1040** 808
[8] Petrova. V A, Bakanov A A and Walter A V 2014 *Appl. Mechan. and Mater.* **682** 431
[9] Mechanical engineering. Encyclopedic reference book 1948 Engineering calculations in mechanical engineering *Mashgiz* 457
[10] Bazhenov S L, Berlin A A, Kul'kov A A, Oshmyan V G 2010 *Publishing house «Intellect»* 352

IOP Conf. Series: Materials Science and Engineering **91** (2015) 012049 doi:10.1088/1757-899X/91/1/012049

Influence of boric anhydride upon the physical and chemical properties of ferrosilicon slag

A M Yefimets, E P Tesleva, A V Solovyan
Yurga institute of technology of National Research Tomsk polytechnic university
Yurga, Kemerovskaya obl., Russia

E-mail: tesleva@mail.ru

Abstract. The authors study the influence of boric anhydride upon the physical and chemical properties of slag in the manufacture of ferrosilicon. It is established that adding boric anhydride to the slag changes its refractory quality and its viscosity and eases pouring slag and metal. Slags with optimal composition and properties are described.

1. Introduction

Influence of slag upon the process of ferrosilicon production significantly increases when chip included into the fusion material is partly replaced by ferriferous oxide materials, for example, mixed ore pellets or ores with high content of iron. In this case the amount of slag in relation to metal achieves 20%. In the furnace considerable part of metal formation processes involves the slag phase and that is why so much attention is drawn to studying its properties [1-6].

In work [7] the influence of fluorides adding upon the cinematics of interaction between CaO-SiO_2-Al_2O_3 melts and silicon carbide (SiC) was studied. At the same time it is evident that the environmental hazard of fluorides calls forth the necessity of selecting other intensifiers. Borates are of interest in this context.

2. Theoretical basis

The most important characteristic of slags is their phase composition as it allows explaining many peculiarities of their physical properties. Most metallurgic slags are silicate, i.e they contain this or that amount of silice (SiO_2). The system CaO-SiO_2-Al_2O_3, is basic to describe the slag formation processes in the manufacture of ferrosilicon. The given system was studied many times to specify the boundaries of phase fields of various compounds. Reference literature presents methods of analytical description of phase equilibrium in complex systems when the amount of the phases formed is calculated according to mathematical formulas relating the phase composition to the chemical one [8, 9]. If boric anhydride is introduced into the fusion material, the slag formed can be described by the four-component system CaO-SiO_2-Al_2O_3-B_2O_3. The diagram of its phase composition is suggested in work [10]. To build the diagram of phase composition the authors used thermodynamically-diagrammatical method applying Gibbs energy minimization principle and regularities of construction of phasing diagrams. Application of the given method allowed dividing the four-component system CaO-SiO_2-Al_2O_3-B_2O_3 into four three-component subsystems including 18 elementary tetrahedrons of coexisting phases. Table 1 presents the consolidated characteristics of 15 main components of the given system.

Figure 1 provides the data on relative volume of tetrahedrons containing the studied phases called the probability of phase existence in the system of CaO-SiO_2-Al_2O_3-B_2O_3. The largest volume in this system is taken by anorthite (CAS_2), boric anhydride (B_2O_3) and silica. As all of them are acid it must be advantageous for ferrosilicon production which is completed with application of acid slags.

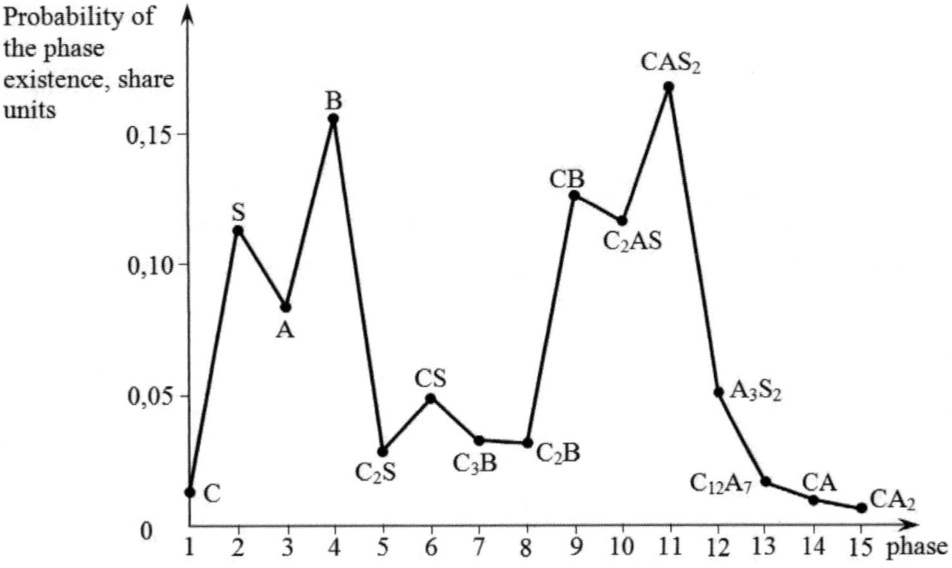

Figure 1. Probability of the phase existence in the system CaO-SiO_2-Al_2O_3-B_2O_3.

Table 2 presents specific compositions of slags from FS65 ferrosilicon production which are the result of not only transfer of some unreduced fusion mixture components into the oxide melt but also of fluxing stone addition to regulate the slag properties. It is easy to notice significant variety of slag composition resulting from insufficient optimality of slag fusion practice. Content of silica in the slags varies from 30 (slag 8) to 82.9% (slag 10), calcium oxide – from 6.32 (slag 7) to 32.69% (slag 2) and aluminium earth – from 13.47 (slag 10) to 44.02% (slag 5). Ratio CaO/SiO_2 which characterizes the acid-base properties of slags is within 0.04-1.25.

On the base of phase composition diagram of CaO-SiO_2-Al_2O_3-B_2O_3 system a mathematical model was developed allowing estimation of phase composition of slag on the base of its chemical composition. The calculation data (Table 2) show that production slags are crystalized in completely different areas of phase diagram where such phase components as anorthite, helenite (C_2AS), mullite (A_3S_2), corundum (Al_2O_3), cristobalite (S) differ markedly in terms of their physical and chemical characteristics giving the slags the same differing properties which does not aid stable operation of the furnace. Some slags of stably low activity are in the phase triangle cristobalite-anorthite-mullite and even being heated up to 1750-1850°C have high viscosity, are badly poured from the furnace due to the carcass type of anion component structure of the first two phases.

Very rarely we can notice the cases of slag in the region of crystallization of helenite-anorthite-corundum. But we can clearly see the intention of technologists to fall into the optimal area of phase triangle anorithe-helenite-pseudowollastonite (CS) recommended by some researchers [11] where the slags are grouped at the confluence of anorithe and helenite. The composition of slag recommended in work [12] as optimal for ferrosilicon production is close to real ferroalloy slag 6 (Table 2) containing 37.39 % SiO_2, 36.13 % Al_2O_3 and 26.48 % CaO. As phase components it includes 70.31 % of anorthite, 1.73 % of pseudowollastonite and 27.96 % of helenite.

Table 1. The consolidated characteristics of phases of system $CaO-SiO_2-Al_2O_3-B_2O_3$

№	Phase	Composition, mas %				$t_{melt.}$, °C	Polytopes, containing the phase	
		C	S	A	B		amount	volume
1	C	100	0	0	0	2625	1	0.0528
2	S	0	100	0	0	1713	3	0.4591
3	A	0	0	100	0	2050	5	0.3434
4	B	0	0	0	100	450	6	0.6298
5	C_2S	65.1	34.9	0	0	2130	5	0.1155
6	CS	48.3	51.7	0	0	1544	4	0.1984
7	C_3B	70.6	0	0	29.4	1490	7	0.1343
8	C_2B	61.5	0	0	38.5	1310	5	0.1262
9	CB	44.4	0	0	55.6	1160	7	0.5076
10	C_2AS	40.9	21.9	37.2	0	1513	12	0.4706
11	CAS_2	20.2	43.2	36.6	0	1553	7	0.6742
12	A_3S_2		28.2	71.8	0	1910	2	0.2020
13	$C_{12}A_7$	48.65	0	51.5	0	1455	2	0.0661
14	CA	35.5	0	64.5	0	1605	3	0.0359
15	CA_2	21.6	0	78.4	0	1750	2	0.0230

Table 2. Chemical and phase composition of slags of FS65 alloy production.

№	Chemical, %			$\frac{CaO}{Al_2O_3}$	$\frac{CaO}{SiO_2}$	Phase, %					
	CaO	SiO_2	Al_2O_3			C_2AS	CS	CAS_2	S	A_3S_2	A
1	24.38	43.27	32.35	0.75	0.56	4.34	11.66	84.00			
2	32.69	30.35	36.96	0.88	1.08	60.31		39.69			
3	8.95	67.83	23.22	0.39	0.13			44.30	45.94	9.76	
4	22.21	42.82	34.97	0.64	0.52	3.58	4.49	91.93			
5	18.75	37.23	44.02	0.43	0.50	4.39		83.95			11.66
6	26.48	37.39	36.13	0.73	0.71	27.96	1.73	70.31			
7	6.32	61.93	31.74	0.20	0.10			31.28	40.44	28.28	
8	37.64	30.00	32.36	1.16	1.25	67.02	12.67	20.31			
9	30.69	35.19	34.12	0.90	0.87	40.56	7.43	52.01			
10	3.63	82.90	13.47	0.27	0.04			18.97	72.43	9.60	
11	29.30	32.80	37.90	0.77	0.89	45.53		52.82			1.62

Results and discussion

We applied the developed mathematical model to calculate the possible change of slag phase composition in case of adding boric anhydride into them. The results of calculation are presented in the graphical form in Figure 2. The calculation showed that, as the concentration of B_2O_3 in slag increases, concentration of other substances reduces – that of helenite, anorthite, pseudowollastonite; at the same time formation of $CaO{\cdot}B_2O_3$ (CB), $CaO{\cdot}2B_2O_3$ (CB_2) and free silica (S) is observed being the result of typical for the given conditions reaction $CaO{\cdot}SiO_2 + B_2O_3 = CaO{\cdot}B_2O_3 + SiO_2$. Formation of low-melting calcium borates in slag (Table 1) opens up an opportunity of regulating its refractory quality and viscosity and free silica allows better reduction of silicon. To check the given calculations experimentally we completed a special study.

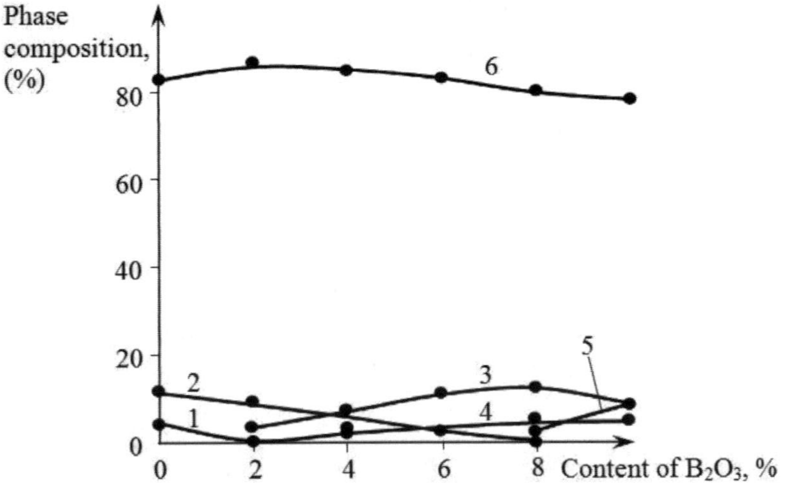

Figure 2. Influence of boric anhydride upon the phase composition of slag in ferrosilicon production: 1 – C_2AS, 2 – CS, 3 – CB, 4 – S, 5 – CB_2, 6 – CAS_2.

Viscosity of slag was estimated with electrovibrating Stengelmeyer viscosimeter in molybdenum melting pots with inner diameter of 20 and height of 35 mm. Diameter of spindle was 2.5 mm and depth of its immersion into the liquid melt was 10±0.5 mm. The temperature of slag was read by the thermal couple PR 30/6 which measuring junction was led to the bottom of the melting pot through a special sinking. Viscosity was measured in Tamman vertical tube furnace. Microscrew hoist was used to immerse the spindle to suitable depth.

The research showed that production slag which compositions are presented in Table 2 demonstrate unfavorable temperature dependence of viscosity. Slag 3 is highly viscous both in the furnace and when tapping even under considerable overheating. Under 1600-1650°C its viscosity is over 10 Pa·s (100 poise) which makes its tapping difficult. It is determined by such phases as tridymite (45.94%) and anorthite (44.30%) having carcass structure of anion component. Slag 8 is too base (CaO/SiO_2 = 1.25), silicon is not reduced well from it, it is fused within a narrow temperature range, with unstable characteristic. It is situated in the region of high-melting helenite (67.02 %) and is not typical for FeSi production as well as slag 3 (CaO/SiO_2 = 0.13). Presence of slags with such composition and characteristics means that the process is not well proven.

The properties of another group of slags (№№ 1,2,6 и 9) are close to each other. They are suitable for normal working conditions. But slags 2 and 9 are characterized by slightly increased basicity and higher viscosity. Their characteristics can be improved by adding boric anhydride. Bor-containing slag retains high mobility which guarantees its easy tapping from the furnace and clean runner and slag pots. Slag 6 has the best characteristics of all production slags (Table 2). It has the lowest viscosity and is stable in terms of this characteristic within a wide temperature range. Slag 6 is received after adding limestone to the fusion material as in the other case it would be acid and situated deeper in the region of anorthite crystallization.

To improve its characteristics we added a varying amount of borate ore (but not the increased amount of limestone) to basic slag 5. The compositions of the studied slags are presented in Table 3.

Slags 12, 13, 14 were received by adding 5, 10 and 15 % of borate ore and slags 15, 16 and 17 – by adding 1, 2, 4 and 6% of B_2O_3 to slag 5. With addition of borate ore into slags, together with increase of B_2O_3 content ratio of CaO to Al_2O_3 and SiO_2 also grows. That's why borate ore can be regarded as complex flux. So, 1% of B_2O_3 in the slag due to borate ore addition, ratio of CaO to Al_2O_3 and SiO_2 increases by 0.07-0.08 units. It has a significant influence upon slag viscosity. Thus, under the temperature of 1500°C the initial slag has viscosity of 5.3 Pa·s. Addition of 10% borate ore into it leads to viscosity decrease up to 3.2 Pa·s and content of B_2O_3 in it is 0.76%. To produce slag with optimal composition and properties it is necessary to maintain the content of B_2O_3 in it of about 1-2% by adding borate ore. The experimental data demonstrate that in this case the crystallization temperature of slags will be within 1500-1550°C and it is enough to overheat them up to 1600-1650°C to achieve viscosity of 1.0 Pa ·s which is optimal for tapping from the furnace.

Table 3. Compositions of experimental slags

№	Proportion of ingredients, %						
	CaO	SiO_2	Al_2O_3	B_2O_3	MgO	CaO/Al_2O_3	CaO/SiO_2
5	18.75	37.23	44.02	-	-	0.43	0.50
12	19.50	36.70	43.00	0.39	0.38	0.45	0.53
13	20.15	36.26	42.10	0.76	0.73	0.48	0.56
14	20.80	35.82	41.21	1.11	1.07	0.50	0.58
15	18.57	36.86	43.58	0.99	-	0.43	0.50
16	18.39	36.50	42.33	1.96	-	0.43	0.50
17	18.03	35.80	42.33	3.85	-	0.43	0.50
18	17.69	35.13	41.53	5.66	-	0.43	0.50

Formation of free-running slag in the presence of boric anhydride creates favorable kinematic conditions for reduction of silicon into metal due to surface renewal resulting from washing the products (SiC) of interaction between the reducing agent and the silica. Silicon carbide can react not only with silica with formation of silicon and carbon monoxides, but also with B_2O_3 in slag with boron carbide and silicon monoxide formation leading to silicon

IOP Conf. Series: Materials Science and Engineering **91** (2015) 012049 doi:10.1088/1757-899X/91/1/012049

reduction and preventing excessive carbide forming especially when producing high-grade ferrosilicon.

Conclusion

1. When we introduce boric anhydride (B_2O_3) into production slags their composition includes low-melting calcium borates allowing changing their refractory quality and viscosity to optimize the process and ease taping of cast products.

2. Adding boric anhydride to the slags during ferrosilicon production promote process acceleration of silicon carbide destruction and reduces the diffusion barrier of silica reduction by silicon carbide.

3. B_2O_3 as a stronger acid oxide displaces SiO_2 in the compounds which increases its activity and more complete transfer into metal.

4. It is recommended to establish the borate ore consumption in the amount of receiving 1-2% of B_2O_3 in the slag increasing it when carbide content in the furnace increases and in case of tight furnace operation due to excessive SiO_2 in the slag.

References

[1] Selivanov V N Dyuldina E V Lozovsky Ye P Korotin A V 2013 *News of higher educational institutions:Ferrous metallurgy* **5** 7 (in Russian)
[2] Pavlovich L B Medvedskaya O O 2010 *News of higher educational institutions: Ferrous metallurgy* **10** 3 (in Russian)
[3] Rozhikhina I D Nokhrina O I 2007 *Steel in Translation* **37** 660
[4] Pavlov A V Ostrovsky D Ya Glavatskikh Yu V 2011 *News of higher educational institutions: Ferrous metallurgy* **5** 20 (in Russian)
[5] Nokhrina O I Proshunin I E Rozhikhina I D 2009 *Steel in Translation* **39** 314
[6] Nokhrina O I Rozhikhina I D Dmitrienko V I Golodova M A Efimenko Yu A 2014 Steel in Translation **44** 99
[7] Khrushchev M S 1989 *Ferroalloy production* 21 (in Russian)
[8] Heath D L 1957J Amer Ceram Soc **40** 50
[9] Palatnik L S 1954 *DAS USSR* **95** 1227 (in Russian)
[10] Nazarov B K Akberdin A A 1986 *Journal of Academy of Sciences USSR Metals* **6** 61 (in Russian)
[11] Akberdin A A Kireyeva G M Kim A S 1996 *KIMS* **3** 27 (in Russian)
[12] Kolyvanov S Yu Maksimov Yu S 1983 *New in ferrosilicon production technology* 50 (in Russian)

IOP Conf. Series: Materials Science and Engineering **91** (2015) 012050 doi:10.1088/1757-899X/91/1/012050

Increasing the Operational Stability of Railway Core by Modifying

S N Fedoseev, Z M Mukhtar

Yurga Institute of Technology, National Research
Tomsk Polytechnic University Affiliate
Kemerovo region, Yurga, Leningradskaya str. 26, RUS

E-mail: sfedoseev@tpu.ru

Abstract. At the present time in conditions of high intensity use of the railways of great importance paid to longer service life the railroad tracks, elements turnout in particular crosspieces. The main element of the railway track turnouts are carrying out a change of direction of movement of trains. Many translations have exhausted their time and are obsolete. From is technical condition turnout's largely dependent reliability of rail transport.
Turnouts are the most complicated structure in the railway. Individual elements of the railway line – the spider who perceives high impact loads. Crossings should have sufficient strength and wear resistance because undergo complex dynamic loads. High dynamic impact and wear areas are also experiencing the antennae, which are located near the cores.
The basic technological aspects to improve the characteristics steel by modifying. The technology of producing high-manganese steel modified special modifier to improve the structural strength of receiving the item. As a result of a series of experimental heats received the samples that were subjected to mechanical testing. Structural studies have shown improvements steel microstructure as well as its mechanical properties.

Introduction

Successful decision facing rail problems in a timely manner and quality needs of the national economy and population in traffic due to the intensification of the transportation process based on advanced technologies and improve the reliability of hardware.

The introduction of the railway rolling stock of a new generation, characterized by an increase in consumer – performance and environmental qualities, is impossible without solving to improve the maintenance of track, including rails, fasteners, rail base, ballast, turnouts.

A special place in the road economy occupy crossing railway tracks, are quite complex in construction, operating in more difficult conditions than the main path, and which are responsible and expensive elements of the permanent way.

Turnouts are time-tested technical devices. Simultaneously, due to the increase in load due to increased train speed and weight, they become a weak point in the operational activities and limit the capacity of railways.

Currently is West-Siberian railway in operation for about 11% of defective turnouts. In the main ways is a 9.8% defective frog. Defects switch ties more than 11%. On the road, not enough attention is paid to conduct work to extend the life of turnouts and their elements. Complexity of conducting of traveling facilities in Siberia caused by the harsh climate, high traffic volumes in and axial loads.

All this determined special urgency and urgency of improving the operational reliability of switches, their components and elements, an improved system of management of the switch.

And increase the requirements for automation and remote control devices in rail transport, track superstructure, which should provide a high degree of safety of trains. Link automation devices and track superstructure are track circuits, which have made it possible to ensure the safety, maximum throughput and the carrying capacity of the land and plants. Unfortunately, in the current track circuits on the railway network situation is far from satisfactory. Over 40% of device failures falls on the automation track circuits. In this case, statistics show that 50% of the track circuit failure occurs due to a malfunction of the insulating joints. Therefore, as a responsible member of the system, ensuring the safety of trains, track circuit failures on the number of standing in one of the first places.

In this regard, great importance attaches to measures aimed at improving the reliability of existing track circuits, in particular, an increase in the probability of failure-free operation of the insulating joints by improving their quality and to determine their optimal reserve to rebuild failed and preventive replacements.

One of the most widely used wear-resistant materials for the manufacture of switches is manganese austenitic steel serving for the manufacture of parts, wear is accompanied by strikes and heavy loads.

Improving the structure and properties of high-manganese steel wear, which wear resistance, resistance to high static and dynamic loads are much superior to most known materials used for the manufacture of switches, and at the same time the material is highly sensitive to the melting conditions, a slight change in the content of C, Si, S, carbide and other alloying elements by conventional means - is a complex production problems.

Traditional ways of improving the properties of materials due to their composition is almost exhausted. In general, the properties of materials depend on the structural construction material at the micro level. In metallurgy it is known that steel and alloys with fine-grained structure has several advantages structural and technological properties to steels and alloys with a coarse texture. One of the promising areas produce alloys with fine-grained structure is a modification of insoluble ultrafine powders.

These studies are of interest both from the practical and theoretical points of view, systematization of knowledge in the production of casting alloys with given technological and structural properties and structure.

Results and Discussion

Effect of modification depends not only on the amount of administered modifier, but also on the conditions of subsequent cooling rate and heat treatment of the alloy. Inoculation efficiency is dependent on the activity of the additives, oxidation, gas saturation, metal contamination, its physic-chemical properties, the melt temperature, the duration of its exposure to the ladle, etc. Therefore, the development of steels and alloys, new grades should undertake a systematic study to the accumulation of reliable data to determine the optimal concentration of processing aids. The chemical composition of different brands of modifiers is presented in Table 1.

Furnace steel processing modifiers performed in order to improve its quality indicators and the following tasks:

- increasing fluidity;
- crushing of cast structure;
- reducing nonmetallic inclusions in steel;
- decrease in the propensity to crack,
- improving cold resistance, corrosion resistance and other service characteristics.

Table 1 – Chemical composition of the different brands of standard modifiers

Brand modifier	The content of the basic chemical elements, mass. %								
	Si	Ca	Ba	Al	Mg	Ti	Mn	Zr	Fe
SIBAR®12	60.0-70.0	<2.0	11.0-14.0	<2.0	–	–	–		rest
SIBAR®22	45.0-60.0	<3.0	20.0-30.0	≤3.0	–	–	–		rest
INSTEEL®1.2	40.0-50.0	12.0-15.0	12.0-15.0	≤2.0	1.0-1.5	–	–		rest
INSTEEL®4.1	40.0-50.0	8.0-10.0	7.0-10.0	1.0-3.0	1.0-1.5	8.0-12.0	–	–	rest
INSTEEL®6	30.0-50.0	8.0-16.0	7.0-12.0	7.0-12.0	–	–	–	–	rest
Z-GRAPH®	60.0-70.0	–	2.5-3.0	–	–	–	6.0-7.0	6.0-7.0	rest
ZIRCALLOY®	70.0-75.0	1.5-2.5	–	<2.5	–	–	–	1.0-2.0	rest

The use of modifiers causes binding of sulfur, phosphorus, harmful gases and non-ferrous metal impurities, size reduction, spheroid zing and reduction in the remaining amount of metal inclusions, grain boundaries purification crystallizing the melt. Non-metallic inclusions become favorable morphology.

We use the modifier is different from the standard modifiers. Modifier used consists of ultra- and nanopowders zirconium, titanium, niobium, hafnium, vanadium, tantalum, copper, and aluminum; used as a reductant Na_3AlF_6.

In this paper we investigated a series of samples cores of manganese steels produced by conventional manufacturing technology JSC "Novosibirsk Switch Plant" and with the use of a modifier based on ultrafine particles of metal oxides and cryolite. Modification of high-manganese steel production was carried out in conditions of "Novosibirsk Switch Plant" in the bucket with the addition of a modifier in the amount of 4 kg per ton of melt.

The purpose was to study the effect of the modifier on the basis of ultrafine metal oxide and cryolite on the chemical and phase composition, microstructure and mechanical properties of high-manganese steel cores.

In this paper, we obtain samples of the modified special steel modifier investigated its physical and mechanical properties that affect the structural strength of the material. Chemical composition of the investigated steel is given in Table 2.

Table 2 – Chemical composition of the investigated cores

Sample	Mass fraction of elements, %				
	C	Mn	Si	P	S
				no more	
normal core	1.00 – 1.30	11.50 – 16.50	0.30 – 0.90	0.09	0.02
not modified core	1.20	14.70	0.52	0.03	0.001
modified core	1.20	14.10	0.47	0.03	0.001

X-ray diffraction studies were performed on the X-ray diffract meter Shimadzu XRD 6000 ("Nano-Center" TPU, Tomsk) at a voltage of 40 kV and anode current of 30 mA, using CuKα-radiation (λ = 1.5418 Å), in increments of 2 °/min. Used to decrypt the database catalogs JCPDS.

Found that the test samples are matrix austenite having dissolved therein carbon – phase (Fe, C) (№ cards 31-0619, Fig. 1). Changing the relative peaks in samples is due to preferred orientation of the grains.

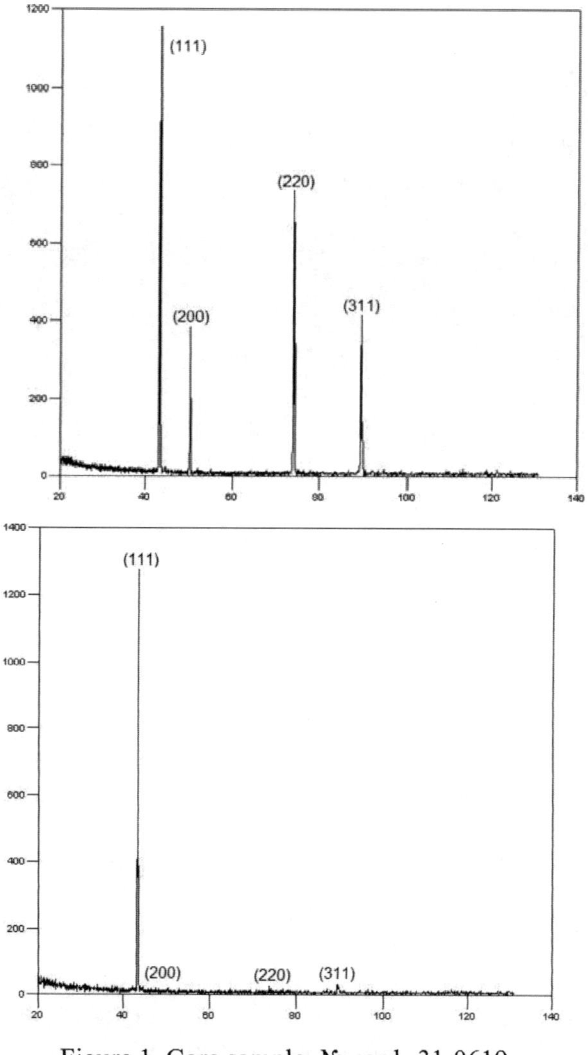

Figure 1. Core sample, № cards 31-0619

At cross-sectional samples are observed blowholes. The microstructure of the samples studied is an austenite inclusion along the boundaries and within the grains.

Impact modifier quality and structural characteristics of the samples were evaluated cores austenite grain size, number, shape and distribution of nonmetallic inclusions in the grain boundaries and inside the grains, the presence of the carbides.

Structure of unmodified and modified samples cores is austenite with non-metallic inclusions on the boundaries and within grains (Fig. 2). Studies have shown that after the introduction of the modifier changed austenite grain size and changed the character of the distribution of non-metallic inclusions (Fig. 2). The grain size in the unmodified sample is 85 μm. After modifying the grain size decreased to 15 μm.

In the unmodified samples in the horizontal cross-sectional at a distance of 10-15 mm from the bottom plane of the tide in all samples are observed globular carbides along the boundaries and within grains. Also in the modified sample non-metallic inclusions are observed along the boundaries and within the grains.

Study of microstructure and local chemical composition was performed with an electron microscope Vega II LMU integrated with the X-ray energy dispersive microanalysis Oxford INCA Energy 350 (Regional Center for collective use of scientific equipment, Tomsk).

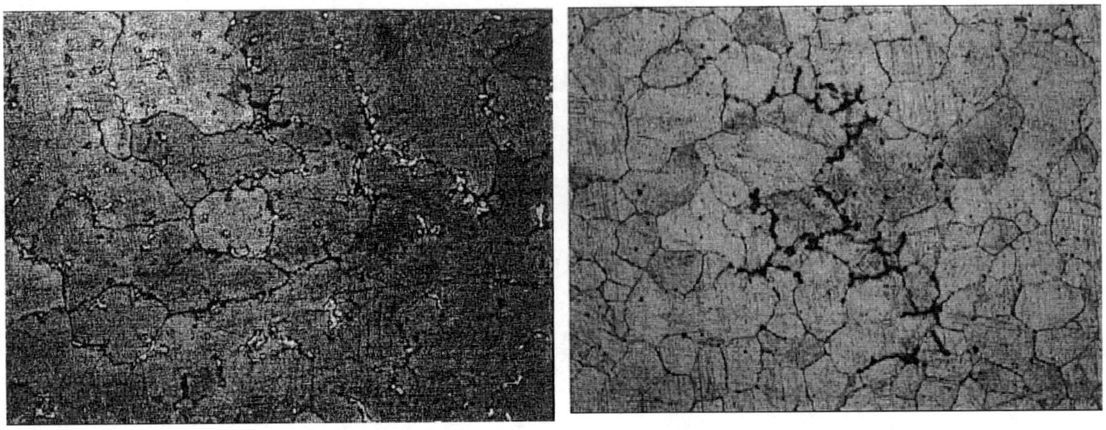

a) without the addition of modifier b) after adding the modifier

Figure 2. Optical metallographic increase ×70

Research samples without modifier showed that their microstructure is austenite with non-metallic inclusions on grain boundaries and inside (Fig. 3). In the not modified sample contains excessive carbides which reduce the strength and toughness of the steel.

Morphological characteristics and chemical composition of inclusions was investigated using scanning electron microscopy. Figure 3 shows that the non-metallic inclusions have a size of ~ 55 microns, and along the grain boundaries, they are placed in a thin layer between the grains.

Figure 4 shows the microstructure of the modified samples cores. The structure of the cores as well as in the unmodified sample is austenite with nonmetallic inclusions along the grain boundaries. It is seen that after modification significantly reduced the number of nonmetallic inclusions. In composition it is clear that non-metallic inclusions are mainly carbides, whose number has dropped significantly.

In the modified samples no carbides, blowholes, a decrease in the number and size of nonmetallic inclusions on the boundaries and within the grains, and visually observed to decrease grain size.

Modification of molten steel significantly improved homogeneity of the structure of steel, which in turn contributes to its density and it leads to improved isotropy, i.e. promotes uniform distribution of stress under load, reduces cracking, chipping and pore formation.

IOP Conf. Series: Materials Science and Engineering **91** (2015) 012050 doi:10.1088/1757-899X/91/1/012050

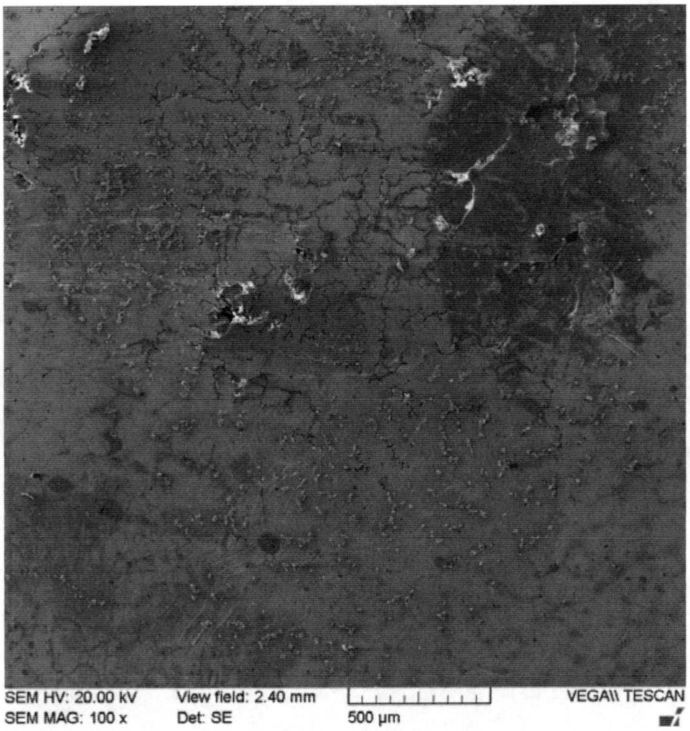

Figure 3. The microstructure of unmodified core

Figure 4. The microstructure of modified core

Calculating the size of nonmetallic inclusions showed that for samples without modifying the data sizes of inclusions are in the range from 18 microns to 145 microns, and the sample after introducing the modifier type inclusions sizes are in the range from 5 microns to 15 microns, i.e. observed a significant reduction in the size of non-metallic inclusions, which leads to a reduction in the range of sizes of inclusions and improves the density of the steel structure.

According to the results of mechanical testing of the modified and unmodified samples cores shown in Table 3, it is possible to judge small improvements in strength characteristics of steel.

Table 3 – Mechanical properties of cores

Sample	Tensile strength, kgf/mm^2	Tensile fluidity, kgf/mm^2	Specific elongation, %	Relative narrowing, %
not modified core	86,7	48	30,7	24,4
modified core	89,5	48	38,0	29,6

Conclusion

As a result, the modification does not change the basic chemical composition of the steel, but reduces the number and size of nonmetallic inclusions in the grain boundaries and also a decrease in grain size. Points to the improved quality obtained of metal. Introduction modifier significantly changes the mechanical properties of the samples, namely improving the service characteristics of the cores.

References

[1] Cherepanov A N, Poluboyarov V A, Kalinin A P, Korotaeva Z A 2000 J. Materials **10** 45-53
[2] Malushin N N, Valuev D V, Valueva A V, Serikbol A 2014 J. Applied Mechanics and Materials **682** 58-63
[3] Turcan D A, Korzunin J K, Roschupkin V I 2010 J. Omsk Scientific Bulletin **2** 113-115
[4] Davydov N G 1979 Manganese steels (Moscow)
[5] Volynova T F 1988 Metallurgy (Moscow)
[6] Fedoseev S N, Lychagin D V, Sharafutdinova A S 2014 J. Advanced Materials Research **1040** 236-240
[7] Kalinin V T, Hrychikov V E, Krivosheev V A 2004 J. Metallurgical and Mining Industry **6** 38-42
[8] Rodzevich A P, Gazenaur E G, Belokurov G M 2014 J. Applied Mechanics and Materials **682** 206-209.
[9] Mulyavko N M 2001 J. Proceedings of the Chelyabinsk Scientific Center **4** 28-30.
[10] Makarov S V, Sapozhkov S B 2014 J. World Applied Sciences Journal **29(6)** 720-723.

Research of influence of electric conditions of the combined electro-diamond machining on quality of grinding of hard alloys

A S Yanyushkin [1, a], D V Lobanov [1, b], P V Arkhipov [1, c]

1665709, Russia, Irkutsk region, Bratsk, Makarenko str. 40, Bratsk State University

e-mail: [a]yanyushkin@brstu.ru, [b]mf_nauka@brstu.ru, [c]pavded@yandex.ru

Abstract. Improvement of production engineering of finishing machining of the hard alloys providing set parameters of quality and exactitude of products is an actual problem in the conditions of the modern industry. To reach objectives it is possible only at the expense of the solution of a complex of questions of fundamental and applied nature. Basic researches are directed on identification of main factors and the reasons of inefficient use of diamond grinding wheel at finishing machining of products from high-strength hard alloys materials. The applied party of research consists in development of recommendations about implementation of offered solutions under production conditions. In paper the problems connected with effective machining of modern high-strength materials are observed. The reasons of limited application of diamond grinding wheels on a metallic binder are noted when machining hard alloys. The carried researches allowed to establish advantages of a method of the combined electro-diamond machining of hard alloys and to define the rational modes providing high quality of products.

Introduction

Effective processing of products from hard alloys is an actual problem in the conditions of modern engineering manufacture [1, 4, 6, 7]. Methods of abrasive processing of hard alloys applied at the operations not always allow to provide the set parameters of exactitude and quality [2, 3, 5]. The widespread abrasive tools used for processing of hard alloys, in most cases do not allow to decide posed problems, and use of the productive tool from super hard abrasive materials is limited, in an aspect of the problems related with restoring of their cutting properties [5, 9, 11, 13]. Heightening of performance of application of synthetic abrasive materials at manufacture of products from hard alloys and high-strength materials probably at the expense of the solution of a complex of fundamental, designer and technological problems.

The assaying of a condition of a problem of processing of modern high-strength hard alloys materials allowed to mark prospects of the physicochemical and combined methods of processing [3, 9, 11, 13, 19]. The combined methods allow to combine in one process some methods in a different combination that leads to heightening of performance of processing. Authors [4, 9, 11] it is marked that widespread aspects of the combined processing of hard alloys is electrochemical grinding which execute a electrically conductive diamond wheel on the metallic binder, the being cathode, grinding with a continuous electrochemical straightening of a wheel, and also the combined electro-diamond grinding. However, accurate and uniform recommendations about a choice of aspects and conditions

IOP Conf. Series: Materials Science and Engineering **91** (2015) 012051 doi:10.1088/1757-899X/91/1/012051

of the combined processing of hard alloys it is not worked out.

In this connection considerable interest represents research of features of influence of electric conditions of the combined electro-diamond processing on quality of grinding of hard alloys.

Material and research methods

In this research quality of the treated surface of hard alloys was evaluated at the combined method of electrochemical grinding with a simultaneous continuous straightening of a wheel. For estimate parameter of quality the roughness of the treated surface is selected [2, 3, 5].

The essence of a method consists in anode dissolution of products of grinding which are organized on a wheel surface that ensures functioning of a wheel in a self-sharpening condition, and a simultaneous electrochemical softening of a working area of a treated hard alloy. The method provides fixed cutting properties of a wheel and on the other hand under the influence of anode and chemical processes the specific is augmented removal a preform material, at the expense of it performance raises and expenditure of diamond wheels decreases. The technology is implemented using the developed techniques and special equipment designed [6, 8, 10, 12, 14-18].

The assaying of methods of electro-diamond grinding allowed determining the following factors influencing process of machining: current density of a straightening of j_{st} [A/cm^2], current density the etching of j_{et} [A/cm^2], rate of cutting of V [m/s], depth of cutting of t [mm/d.s] and stride of S [m/min]. In the conditions of the same instrument registration of process of machining these factors rather fully reflect technique of electro-diamond grinding.

For holding of research the design of experiments necessary for minimizing of amount of experiences in a stroke of definition of a mathematical sample piece of examined process for the purpose of abbreviation of material inputs and time is executed. In this case research of change of conditions and parameters of the combined machining from electric treatment schedules of hard alloys were carried out: current density of a straightening of a wheel of j_{st} [A/cm^2] and current density the etching of a detail of j_{et} [A/cm^2]. Thus, amount of factors of k = 2, amount of experiences – 4.

Treatment schedules are selected according to the recommendations guaranteeing excellence of a surface, developed earlier [11, 19]. According to recommendations current density of a straightening of a wheel should be no more j_{st} = 0.1…0.4 [A/cm^2], current density the etching of a detail of – j_{et} = 10…40 [A/cm^2], stride – S = 1.0…2.0 m/min, transversal feeding – t = 0.01…0.04 mm/d.s, rate of cutting – V = 15…40 m/s.

As an output parameter the roughness of the treated surface of Ra [μm] was examined. The condition of the treated surfaces of products from hard alloys was inspected with use of modern engineering of the visual and structural assaying.

Materials for holding of research selected hard alloys of several brands of WC3M (WC-97%, Co-3%, M - fine-grained), WC8 (WC-92%, Co-8%), WC15 (WC-85%, Co-15), TN20 (TiC-79%, Ni-15%, Mo-6%), in an aspect of their wide application on manufacture as structural and instrumentals materials.

In research diamond wheels on a on the metallic binder of brand M1 - 01 and M2 – 01 speckle 100/80 and 80/63 were used.

As an electrolyte the solute of salts was applied: 0,5 % of Na_2CO_3 and 1,0 % of NaCl in water. According to observations Instruments for the measurement of conductivity KSL-101, conductance of an electrolyte made 26 Cm/cm, the contents of salts – 20 g/l.

Results

By results of research of an alloy of WC3M dependence of a roughness of Ra [μm] of the treated surface on electric treatment schedules of j_{st} [A/cm^2] and j_{et} [A/cm^2] looks like:

$$Ra=0.2157+0.0174\cdot j_{st}-0.00065\cdot j_{et}-0.0106\cdot j_{st}\, j_{et}$$

The surface of a response of the examined parameter is presented in figure 1.

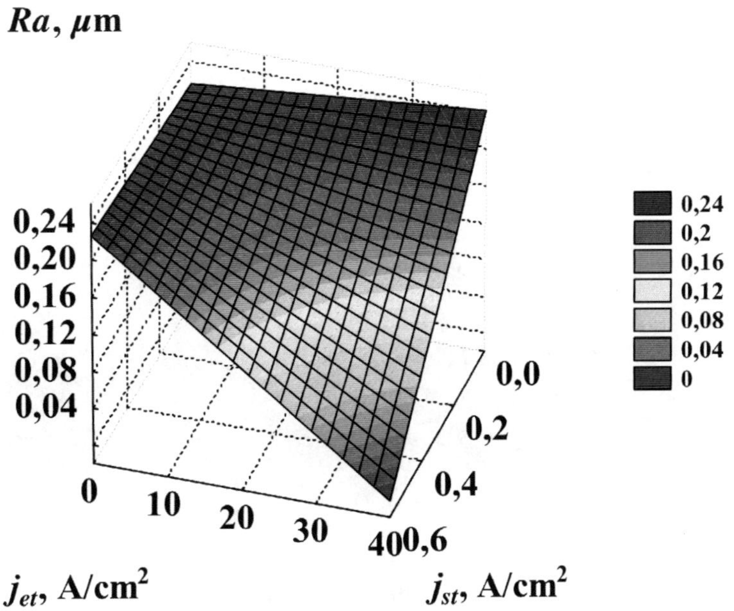

Figure 1. Dependence of a surface roughness of an alloy of WC3M from electric conditions of the combined electro-diamond grinding

Machining of a hard alloy of WC8 allowed to receive dependence of a roughness of Ra [μm] of the treated surface on electric treatment schedules of j_{st} [A/cm^2] and j_{et} [A/cm^2] the following aspect:

$$R_a=0.3286+0.1376 \cdot j_{st}-0.00568\, j_{et}-0.0017\, j_{st}\, j_{et}$$

Response surface on the results of experimental studies are presented in Figure 2.

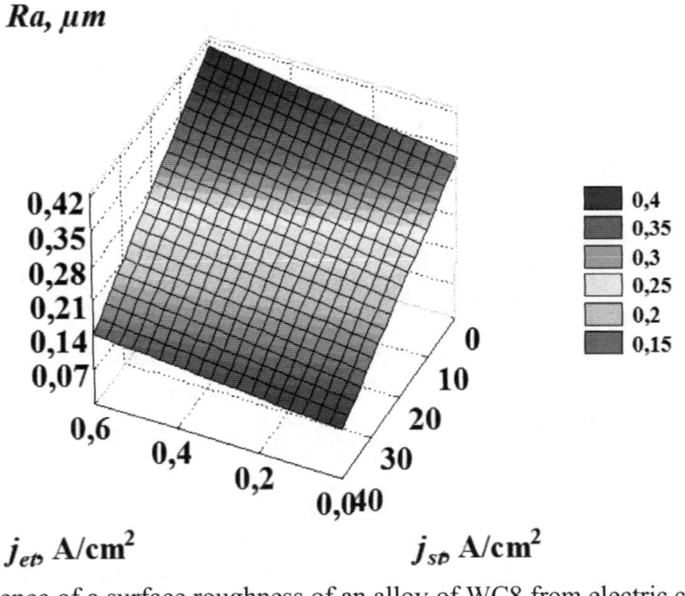

Figure 2. Dependence of a surface roughness of an alloy of WC8 from electric conditions of the combined electro-diamond grinding

By results of research of an alloy of WC15 dependence of a roughness of Ra [μm] of the treated surface on electric treatment schedules of j_{st} [A/cm^2] and j_{et} [A/cm^2] looks like:

$$R_a=0.456+0.1098\,j_{st}-0.0083\,j_{et}-0.0069\,j_{st}\,j_{et}$$

The surface of a response of the examined parameter is presented in Figure 3.

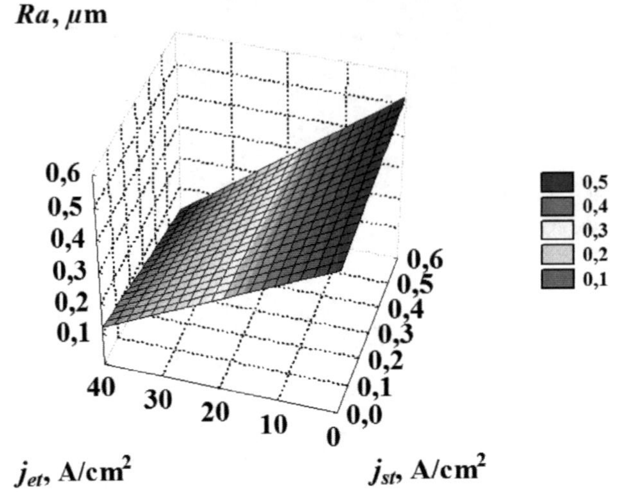

Figure 3. Dependence of a surface roughness of an alloy of WC15 from electric conditions of the combined electro-diamond grinding

Machining of a hard alloy of TN20 allowed to receive dependence of a roughness of Ra [μm] of the treated surface on electric treatment schedules of j_{st} [A/cm^2] and j_{et} [A/cm^2] the following aspect:

$$R_a=0.2685+0.0618\,j_{st}-0.0018\,j_{et+}0.0054\,j_{st}\,j_{et}$$

Response surface on the results of experimental studies are presented in Figure 4.

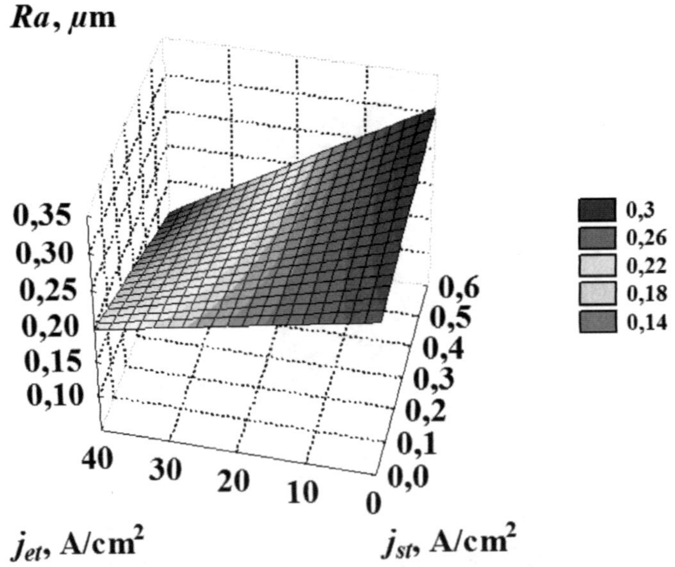

Figure 4. Dependence of a surface roughness of an alloy of TN20 from electric conditions of the combined electro-diamond grinding

IOP Conf. Series: Materials Science and Engineering **91** (2015) 012051 doi:10.1088/1757-899X/91/1/012051

Discussion

Thus, experimentally it is confirmed that application of a method of the combined electro-diamond machining of products from hard alloys, at the expense of a combination of anode dissolution of a worked stock and an electrochemical straightening of a diamond wheel, is reached a minimum roughness of the treated surfaces of products which on the average for 60 % is less, than at other methods to diamond and abrasive machining of such materials.

The obtained data allowed to define the rational electric conditions of the combined machining ensuring excellence of the treated surface of hard alloys of j_{st} = 0.2…0.3 A/cm^2 and j_{et} = 20…30 A/cm^2 which can be recommended for a semi-finishing and finish machining. As rational mechanical treatment schedules conditions are accepted: stride S = 1.5 m/min, transversal feeding of t = 0.02 mm/d.s, rate of cutting of V = 35 m/s.

The proposed combined method provides high performance and quality of the machined surface and is recommended for rational and effective processing of products from hard alloys and hard materials.

Reference

[1] Mokritsky B Y Burkov A A Choice of the effective metal-cutting instrument for turning machining of hard alloys of the lowest group *Technology of machine building* 12 (2011) pp 18-20

[2] Suslov A G Bishutin S G Mathematical sample piece of a roughness of the grinding surface Reference book *The engineering journal with the application* 8 (2004) p 17

[3] Vasiliev A S Gryazev V M Characterization of the surface layer of parts processed in different ways *Fundamental and applied problems of engineering and technology* 2-5 (292) (2012) pp 5-9

[4] Lobanov D V Yanyushkin A S Rychkov D A Petrov N P Optimal organization of tools for machining composites *Russian Engineering Research* 2 (2011) pp 156-157

[5] Lobanov D V Yanyushkin A S Influence of sharpening on the quality of hard-alloy tools for the cutting of composites *Russian Engineering Research* 3 (2011) pp 236-239

[6] Arkhipov P V Yanyushkin A S Lobanov D V Petrushin S I The effect of diamond tool performance capability on the quality of processed surface *Applied Mechanics and Materials* 379 (2013) pp 124-130

[7] Yuanyushkin A S Rychkov D A Lobanov D V Surface quality of the fiberglass composite material after milling *Applied Mechanics and Materials* 682 (2014) pp 183-187

[8] Yanyushkin A S Lobanov D V Arlhipov P V et all Cathode for grinding wheel trueing in combined electric powered diamond wheel processing Patent RF № 2446039 (2010)

[9] Lobanov D V Yanyushkin A S Influence of methods of sharpening quality hard alloy tools for machining composite materials *Messenger of machine industry* 3 (2011) pp 50-53

[10] Yanyushkin A S Lobanov D V Arlhipov P V The loss of cutting ability of diamond wheels on metal binder for grinding composite materials *Messenger the Siberian State Aerospace University Academician MF Reshetnev* 1 (2013) pp 178-183

[11] Yanyushkin A S Lobanov D V Kovalevsky S V Heightening in quality of preparation of hard alloy tools *Systems Methods Technology* 1 (2009) pp 95-98

[12] Arkhipov P V Lobanov D V Yanyushkin A S Perfecting equipment under processes of the combined machining *Messenger of the Tajik engineering university* 2 (2013) pp 32-37

[13] Arkhipov P V Balykov A V Deacons A A Yerenkov O J Ivanov V P Kalita E G Lipatova A B Lobanov D V Medvedeva O I Morozova A V Schmidt I V Yanyushkin A S Effective technologies of mechanical machining of non-metallic materials Moscow 2014

[14] Yanyushkin A S Rychkov D A Lobanov D V et all Abrasive wheel for electrochemical grinding with a perpendicular arrangement of current-carrying insertions Patent RF № 144707 (2014)

[15] Yanyushkin A S Rychkov D A Lobanov D V et all Abrasive wheel for electrochemical grinding with a parallel arrangement of current-carrying insertions Patent RF № 145108 (2014)

[16] Yanyushkin A S Rychkov D A Lobanov D V et all Abrasive wheel for electrochemical grinding with a slanting arrangement of current-carrying insertions Patent RF № 144708 (2014)
[17] Yanyushkin A S Medvedev O I Yakimov S A Arkhipov P V About mechanism destruction grain of diamond in process of electro-diamond grinding *Systems Methods Technology* 1 (2009) pp 34-36
[18] Arlhipov P V Yanyushkin A S Kovalevski S V On the nature of diamond grinding wheel glazing *Works of Bratsk state university* 2 (2008) pp 169-174
[19] Yanyushkin A S Technology of electro-diamond sharpening of cutting tools and methods of its realization Stary Oskol 2013

IOP Conf. Series: Materials Science and Engineering **91** (2015) 012052 doi:10.1088/1757-899X/91/1/012052

Finite Element Model of Trenchless Pipe Laying

P V Burkov[1,2], **S P Burkova**[3], **A N Kravchenko**[4]
[1] Professor, Tomsk State University of Architecture and Building, 2, Solyanaya Sq., Tomsk, Russia
[2] Professor, National Research Tomsk Polytechnic University, 30, Lenin Ave., Tomsk, Russia
[3] Doctor, National Research Tomsk Polytechnic University, 30, Lenin Ave., Tomsk, Russia
[4] Student, National Research Tomsk Polytechnic University, 30, Lenin Ave., Tomsk, Russia

E-mail: burkovpv@mail.ru

Abstract. The paper focuses on the stress and strain state of the underground main pipeline section using Autodesk Inventor software.

1. Introduction

Pipeline transfer is currently faces the problems of industrial safety, in particular, the safety of underwater lines. These problems are connected with difficulties of the current state control and the low possibility of timely identification and elimination of causes of pipeline breaks. Underwater main pipelines are the most critical sections of petroleum pipeline construction although their share in its total volume is comparatively small. In case of possible environmental incidents, the safety of underwater lines should meet strict requirements even at minor failures of petroleum pipelines [1]. The problem of the reliability improvement of pipelines is relevant at both design stage and construction and operation stages. The effect on petroleum pipeline due to external loads resulting in irreversible processes can be prevented by ensuring its adequate behavior. This can be achieved by the construction modeling allowing for both operational characteristics accepted in standards and regulations, and external effects on petroleum pipelines.

The trenchless pipe-laying technique is widely used alongside with traditional techniques of pipe-laying. This technique is currently the most cost-effective way of pipeline construction. Another factor that provides its wider use is the progressive development of new construction technologies by design organizations. Today, it is quite obvious that trenchless technology has no alternative to solve the problem of pipeline laying and reinstatement using the specific equipment. Presently, there are basically three possible ways of pipe laying below the water. The oldest traditional way includes not only a great deal of earthmoving operations, manpower, and long-term construction works but, first of all, an adverse environmental effect and creation of uncomfortable conditions in cities. The second way is tunneling or microtunneling with constructing of launch and reception shafts. This method is characterized by the accuracy control of tunneling paths.

The third way is horizontal directional drilling. It is a controlled trenchless laying of underground pipes using specialized drilling rigs. Lately, horizontal directional drilling is being widely used allowing to reduce time and labour consumption especially in constructing underground pipes below

IOP Conf. Series: Materials Science and Engineering **91** (2015) 012052 doi:10.1088/1757-899X/91/1/012052

natural barriers, such as rivers, ravines, lakes, woodlands, quick grounds and in urban conditions, namely: railways, highways, parks, etc. Horizontal directional drilling is more acceptable for landscape and ecological balance preservation as compared to other ways of pipe laying. It eliminates the man-induced impact on flora and fauna, stream-bank erosion, benthal deposits, and minimizes the negative influence on living conditions in the area of pipeline laying and reinstatement works. One of the main problems that must be solved for the infield pipeline design, is the construction of the linear pipeline portion. As the pipeline routing is often laid together with other linear communications, it is necessary to take all possible measures to preserve its integrity. In crossing automobile roads, the pipelines are usually cased in protective steel tubes [2].

Let us consider Aleksandrovskoe – Anzhero-Sudzhensk main pipeline which crosses 98 water barriers on its way. The territory includes the undeveloped, mixed-forested land along Aleksandrovskoe – Anzhero-Sudzhensk main pipeline at its 208th km distance. The broad, even, and peaty floodplain is represented by oblong alternating low hummock ridges and bogs covered with medium thick undergrowth of ledum and osier and birch. Undergrowth with sedge meadows are observed along the river bed. The bogginess of the river bed is 20%, forest coverage is 90%. Within the area of underwater line, the river sand bed is tortuous and one-armed. In dry weather period the width and the depth of the river bed is 10-15 and 1,5-2 m, respectively.

The financial and economic aspects are also important in horizontal directional drilling. Estimated cost of pipeline laying can be reduced due to the reduction of work execution terms, supplementary employee costs, and heavy excavation equipment. While designing underground structures, the stress and strain state should be calculated for soil adjacent to the pipeline and that located at a certain depth from the surface. To solve this problem, in most cases it is necessary to solve the elastic boundary value problem using the finite element method (FEM). The advantages of this method are minimum requirements for the source information and optimum forms of results.

2. Results and discussion

The aim of this work is to construct the finite element model of trenchless underwater pipe lining within the peaty floodplain area.

The strain-stress state analysis of pipeline structures based on methods of mechanics of materials and the structural theory does not allow the adequate and accurate analysis of the fuel and energy facility strength, and in some cases, it can show an incorrect pattern of stress and strain state of the structure. The structural analysis using FEM is, in fact, the worldwide standard of strength and other kinds of analysis. This is because FEM is a universal method allowing to simulate different structures having different material properties. Results obtained by FEM make possible to detect pre-emergency areas including those still having no defects, and take whatever necessary measures to increase the reliability of the pipeline system [3-8]. The longitudinal section of oil gathering pipeline was constructed based on the accepted design solutions, in which the point of crossing with the automobile road was indicated as shown in Figure 1. This section of the pipeline routing is laid in a protective steel casing having 45 m length and 426 mm in diameter. Figure 2 shows the schematic layout of the pipeline across the automobile road.

IOP Conf. Series: Materials Science and Engineering **91** (2015) 012052 doi:10.1088/1757-899X/91/1/012052

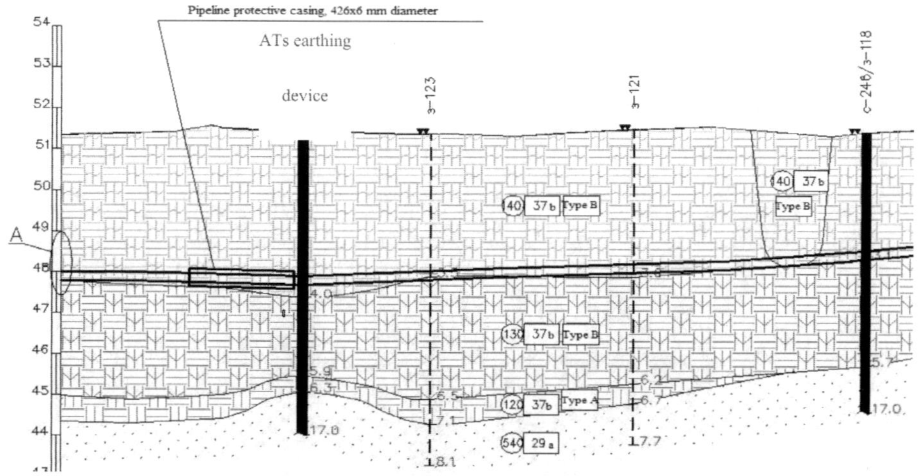

Figure 1. Cross-sectional plan view of the pipeline portion

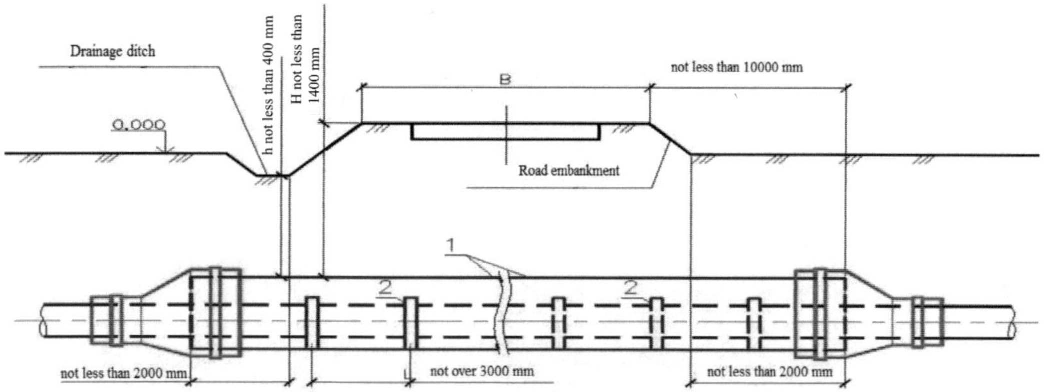

Figure 2. Schematic layout of the pipeline across the automobile road

The stress and stain state of the pipeline in protective steel casing was analyzed by the finite element method using Autodesk Inventor software. Once the solid model of the pipeline was invented, the stress and strain state analysis is conducted accounting for all loads that have been calculated at the design stage. The solid model of the pipeline is illustrated in Figure 3.

Figure 3. Solid model of the pipeline in protective steel casing

IOP Conf. Series: Materials Science and Engineering **91** (2015) 012052 doi:10.1088/1757-899X/91/1/012052

Autodesk Inventor is 3D mechanical CAD design software developed by the Company Autodesk, USA, for creating 3D digital prototypes of industrial products. Inventor's simulation tools let users create a complete design cycle and design documents.

A simulation model developed in Inventor is shown in Figure 4 and represents the interaction between the cone and soil. The cone is subjected to 10 MPa pressure. The model assumes that there are no loads acting during winter and the temperature effect on the pipeline. The Drucker–Prager pressure-dependent model was used to simulate the stress and strain state of soil.

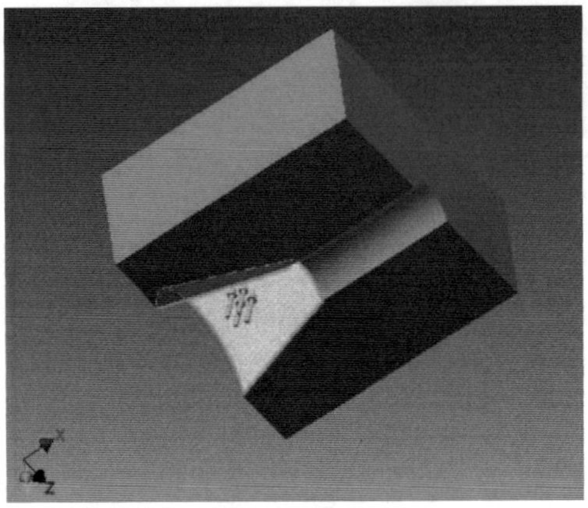

Figure 4. A simulation model in Inventor: the interaction between the cone and soil

Figures 5 -9 illustrate the results of FEM analysis of soil stresses and movements.

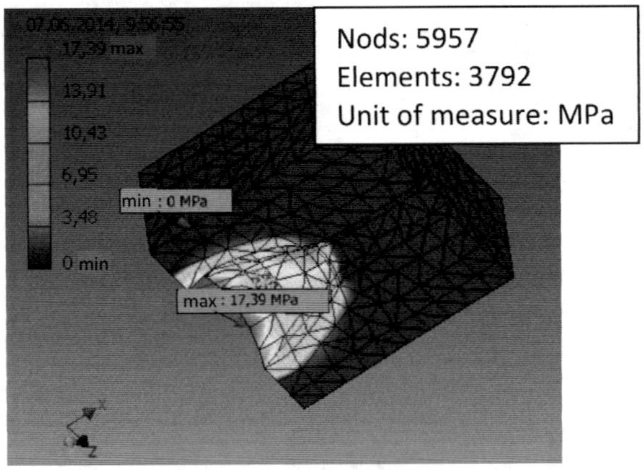

Figure 5. Von Mises stresses in soil

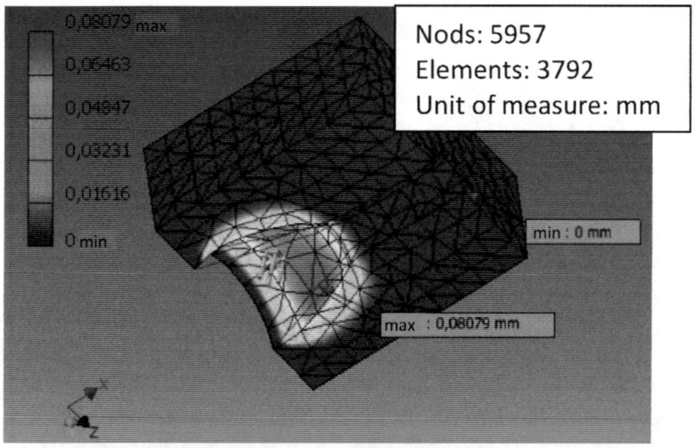

Figure 6. Total soil movements

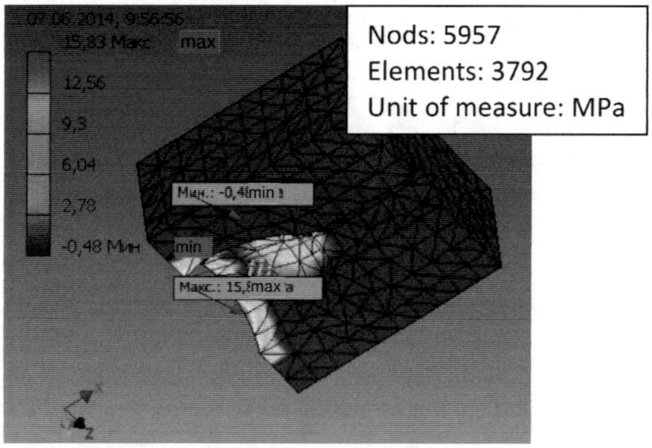

Figure 7. The first main stress

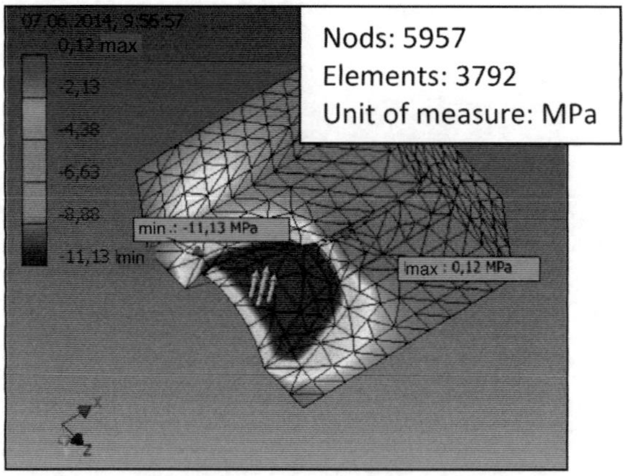

Figure 8. The third main stress

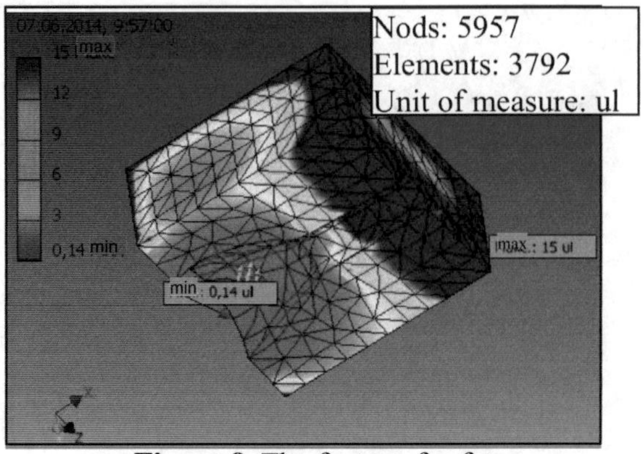

Figure 9. The factor of safety

Theoretically calculated pull forces can be provided by the up-to-date drilling rigs of the TiDrill type having pull force of 4,400 kN. Thus, the suggested model gives the adequate results. The distribution of total soil stresses and movements obtained herein are shown in Figures 10 and 11.

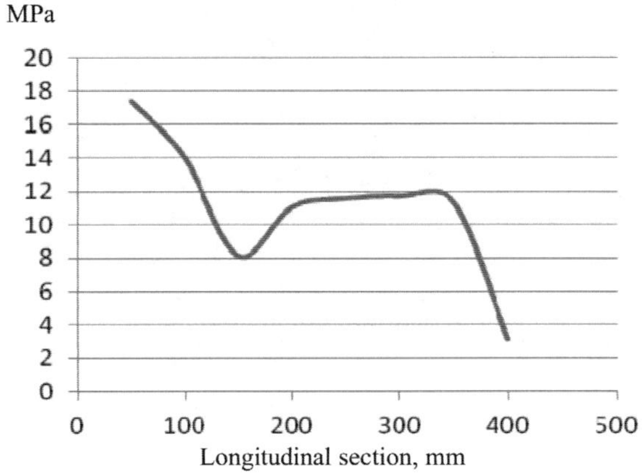

Figure 10. Von Mises longitudinal stress distribution

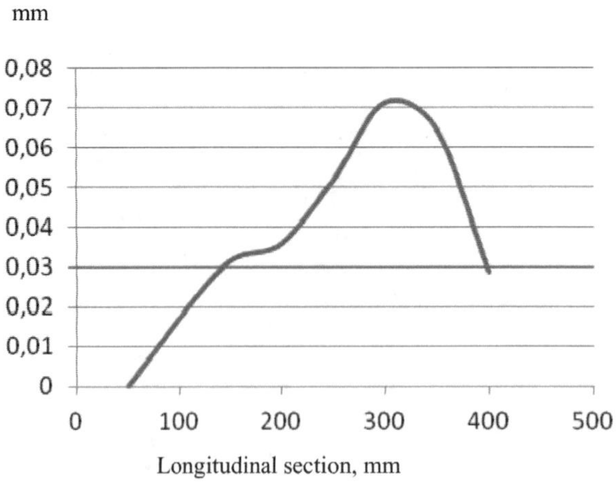

Figure 11. Total soil longitudinal movements.

3. Conclusions.

The FEM model was constructed to detect the stress and strain state of soil around the petroleum pipeline. The total soil movements vary by length and significantly depend on distributed loads applied to the cone.

4. References

[1] Shammazov AM Mugallimov FM and Nefedova NF 2000 Podvodnye perekhody magistral'nykh nefteprovodov (Ufa: Nedra-Biznestseptr Publ) p 237
[2] Mustafin F M Gas oil pipe technology Nedra Moscow 2007
[3] Burkov PV Burkova SP Timofeev VYu Ashcheulova AA and Klyus OV 2013 *Vest KuzSTU* 6 pp 77-79
[4] Burkov P V Chernyavsky D Y Burkova S P Konan A Simulation of pipeline in the area of the underwater crossing *IOP Conference Series*: E and Env Sc 21 (2014) 1-5
[5] Burkov P V Kalmykova K G Burkova S P Do T T Research of stress-deformed state of main gas-pipeline section in loose soil settlement *IOP Conference Series*: E and Env Sc 21(2014) pp 5-7
[6] Aleshin V V Kobyakov V A Seleznev V B Strength analysis of industrial pipelines in ANSYS El J *SAPR i grafika* 7(2007) pp 57-61
[7] Semin E E Tarasenko A A The use of software packages for technical diagnostics and repair of vertical steel containers *Trubopr Trans : teor i prak* Moscow 2 (2006) pp 5-8
[8] Sil'nitskii P F Tarasenko M A Tarasenko A A Design of foundation ring of reservoir with defects *Izv Vuz Neft' i gaz* 5(2011) pp 76–78

Analysis of the influence of warming on the quality of soldered instruments

V V Konovodov[1], A V Valentov[2], I S Kukhar[2]

[1] Novosibirsk State Agricultural University
147 Nikitin St, 630039 Novosibirsk, Russia; phone: (383) 267-35-07
[2] Yurga institute of technology (branch) of Tomsk polytechnic university
26 Leningradskaya St, 652050 Yurga, Russia; phone: (38451)-6-05-37

e-mail: valentov@mail.ru

Abstract. The study is focused on the analysis of the influence of short-term exposure to high temperature on tungstenless hard alloy metal plates while soldering iron-carbon. According to the results of the research with macrohardness tester the macrohardness deviation amounted to ±3%.

During the production process of turning cutters with tungstenless hard alloy metal plates soldered on iron-carbon there were micro- and macrocracks registered in the hard alloy metal plates. In authors' opinion, the reasons for this could include several curcumstances:

- being warmed with high temperature the structure of a hard alloy is changed, more specifically, there is a possibility of the change of microhardness through phase structural change while being exposed to high-temperature;
- disturbance of instrument soldering process conditions;
- unreasonable choice of the correlation of thicknesses of the soldering alloy and tungstenless hard alloy metal plate;
- defects that occurred during the production of the tungstenless hard alloy metal plate.

Timeliness of the research

Currently Russian enterprises in the area of repair and rebuild of parts of agricultural machines are in the phase of intensive development. According to the National information service [www.pulset.ru], in 2007 there were in Russia 249 companies of various forms of ownership which were occupied with rebuild of machine parts, while in 2012 they were 474. In general, during the period of five years the number of enterprises acting in the area of repair and rebuild of agricultural machines has increased almost twofold. This is connected with ageing of the population of machines, large delivery times of foreign spare parts and, as a consequence, increased down time of the machines, which negatively affects the profit of the households, especially taking into account isolated location of some subjects of the Russian Federation from the areas of concentration of spare parts and components production of agricultural machines.

The procedure of parts repair constitutes a subsequence of building-up operations followed by mechanical processing. Over 75% of parts reconditioning methods of agricultural machines, as of their structure, include various kinds of welding, while it is possible to rebuild over 90% of component spectrum. When the welding is made with a high quality, at the concluding phase of reconditioning —

IOP Conf. Series: Materials Science and Engineering **91** (2015) 012053 doi:10.1088/1757-899X/91/1/012053

the mechanical processing — the precision and the quality of rebuilt surfaces, as well as the general part of material and labor costs has to be defined for the rebuilding of parts of agricultural machines.

The factors that decrease the precision and the quality of the surface include a relatively high hardness of the deposited surface, irregular chemical and physical composition and irregular allowance for mechanical processing. To reduce the negative effect on the processing of the parts of agricultural machines a special tool is used which must have a very firm, stable cutting head, which is able to carry alternating loads by a relatively high cutting speed.

Popular instruments, including authorial, do not meet the requirements of hardness, stableness and wearing out. Using the wolfram-free hard alloy metal might help to solve this issue, though mechanical mounting of such plates due to the low hardness causes frequent failure of hard alloy metal plates because of the influence of buff load. Fabrication of soldered tools on the basis of wolfram-free hard alloy plates with use of traditional welding alloy is impossible because of the low wetting angle (<300) of the alloy on the surface of wolfram-free hard alloys. This results in the increased tool expenditure, reduced machine utilization and high original cost of processing and reconditioning. The need for high quality tools at the enterprises of repair and rebuild of the parts of agricultural machines only in Kemerovo region amounts to 100 thousand items annually by precalculated lifetime 45 minutes.

Currently there are several ways of solving the targeted problem. One of this ways involves using high-endurance iron-carbon alloy for production of turning tools with wolfram-free hard alloy metal plates. Thereby large reserve of performance gain of lathing of parts of agricultural machines rebuilt by welding by means of increasing of effectiveness and stableness of welded tool.

Methods of research, authenticity and validity of the results:

Experimental research has been completed in laboratory conditions according to the schedule of single-factor experiment. In order to define the quality of the soldered junction of the tool holder — iron-carbon alloy — wolfram-free hard alloy metal — metallographic method was used. The rack resistance of the tool was calculated by the method of S.I. Petrushin. In order to define the authenticity and repetition of the results following methods were used which were confirmed by theoretical research: Student method, Mann-Whitney method and method of rank correlation by Spearman. The results of the research were used in practice.

Scientific novelty:

1. The composition of the iron-carbon alloy has been established for welding of wolfram-free hard alloy metals with increased contact angle.
2. The strains have been determined which occur during tool welding, as well as new method of crack resistance calculation has been suggested.
3. The effectiveness has been defined of using of turning tools with wolfram-free hard alloy metal plates, which are welded with iron-carbon alloy while lathe turning surfaced coats of the parts of agricultural machines.

Practical implications:

1. The developed alloy composition provided a secure junction of the tool holder with wolfram-free hard alloy metal plates.
2. The suggested method of forecasting of imposed and operating stress while welding the tools helped to increase the production quality by 15% due to the reduction of defect ratio.
3. The suggested lathe turning tool with wolfram-free hard alloy metal plates raised the productivity by 30-40% with increase of stableness thrice-fivefold depending on technological processing modes.

Utilization:

The results of the research have been utilized by way of pilot batch of tools and released for implementation into repair and technical enterprises of Kemerovo Oblast.

IOP Conf. Series: Materials Science and Engineering **91** (2015) 012053 doi:10.1088/1757-899X/91/1/012053

Objects and methods of research

Now we analyze the reasons why the micro- and macrocracks occur in the tungstenless hard alloy metal plate soldered on iron-carbon.

Results of the research

The ability of the instrument to execute its functions in many ways depends on the real damages that occur during its production. The most dangerous defects are pores, contaminating impurities, structural irregularity, inhomogeneous granularity, micro- and macrocracks. The research objective is the influence of warming on the quality of cutters soldered on iron-carbon alloy.

Turning cutters with a tungstenless hard alloy metal plate were produced with iron-carbon soldering alloy. The parts ready for soldering were covered with a 3–4 mm layer of flux in order to avoid corrosion. The completely soldering-ready construction was put into the power coil of the HFC oven for 50 seconds. During this time the soldering alloy is partially melted and solidly bond with instrument holder and tungstenless hard alloy. Soldering temperature equals 1,160–1,200°C, however during such short period of time the structure of hard alloy is not changed, which means that physical and mechanical features of the hard alloy are preserved. The produced instrument from the power coil was put into sand for smooth cool down. High cool down speed provokes macrocracks in the hard alloy along the soldering line.

The warming temperature for FeC-soldering, with corresponding composure of the alloy, amounts to 1,160–1,200°C, taking into account possible technologic overwarming (30–50°C) in comparison with the start of alloy melting temperature (≈ 1,147°C). According to the literature data [1,2,9], as a result of high-temperature influence during instrument soldering hard alloy tends to durability reduction. The authors [3,4,5] however in most cases do not indicate the duration of the influence and reinforce their conclusions with the data obtained during long-lasting (several hours) exposure to oven heating.

This issue was analyzed during the research of microhardness with a microhardness tester of turning cutters produced with the given technology. In all samples the interface of hard alloy and soldering alloy corresponded with the coordinate x=0. The hardness was measured in Vickers units upon the diagonals of impression under applied force of 2,94 N and 1,96 N for cutters with KHT16 and TH20 plates respectively. The results of the research are presented on figures 1, 2.

Moreover, the influence of a short-term warming wa analyzed on the macrohardness of tungstenless hard alloy plates KHT16 and TH20, the results of which are presented in table 1.

Table 1: The influence of high-temperature HFC warming (warming up to 1,200°C, holding 25–30 s) on the microhardness of hard alloys

Alloy brand	Microhardness, MPa						Hardness change, %
	before warming			after warming			
	H_o min	H_o max	H_o	H_o min	H_o max	H_o	
KHT16	14080	19320	16480	12840	18320	17300	+ 5
TH20	15840	18650	17300	15240	18440	16870	– 2,5

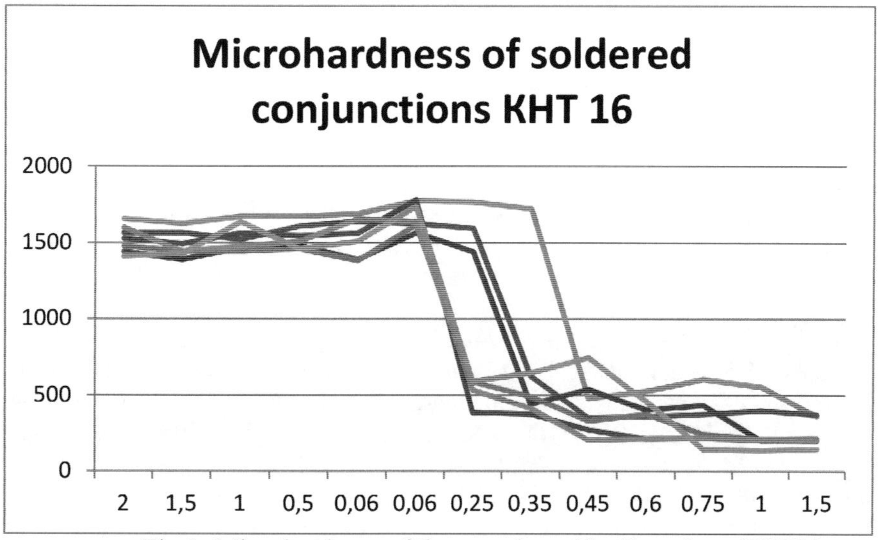

Fig.1: Microhardness of the samples with alloy plates KHT16

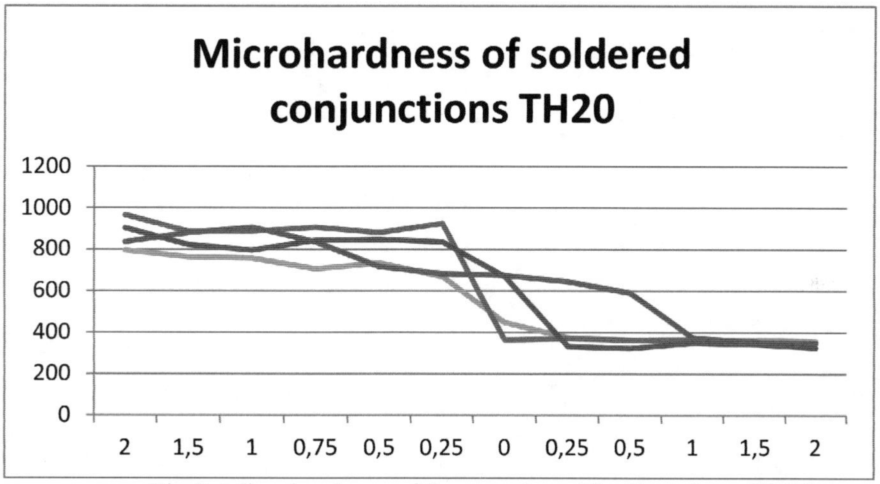

Fig.2: Microhardness of the samples with alloy plates TH 20

In the publication [7,8] a forecasting method is suggested for residual and operating stress in the tungstenless hard alloy plates soldered on iron-carbon alloy. Based on the results of the suggested measures the quality of the instruments has been enhanced (no micro- or macrocracks) by 60%.

Conclusions

As a result of the conducted research we can draw following conclusions:

- the microhardnes of the tungstenless hard alloy plates by short-term (40–50 sec) high-temperature (1,100–1,200°C) warming changed within the range of ±3% which is a statistically non-significant factor;

- with usage of the forecasting method of absence of residual and operating stress described in publication the quality of the instrument is enhanced by 60%;

- while soldering turning cutters the process conditions of time (40–50 seconds) and temperature (1,100–1,200°C) should be followed.

IOP Conf. Series: Materials Science and Engineering **91** (2015) 012053 doi:10.1088/1757-899X/91/1/012053

References

[1] N.A. Klochko. The basics of soldering technology and heat treatment of the hard alloy instruments. – Moscow; "Metallurgiya", 1981. - 200 p.

[2] I.N. Ivanov. The opportunities of the productivity enhancement of the soldered instruments with hard alloys. / Soldering in machine-engineering. - Omsk: TSP NTO "Mashprom", 1980.

[3] V.V. Konovodov, Y.V. Kalloyda, A.A. Malyshko, S.U. Glazachev, Z.A. Luzyanina. Concerning the hardness of the hard alloy cutting instruments / Analysis of modern agricultural problems.: Synopsis of the reports of the research and practice conference of the scientists of Novosibirsk state agricultural university and Humboldt university – Novosibirsk, 1995.

[4] M.G. Loshak. Hardness and durability of hard alloy. - Kiev: Nauk. dumka, 1984.- 328 p.

[5] N.A. Klochko, A.K. Khersonsky. The research of the influence of heat treatment on the condition and characteristics of sintered hard alloy products. / Hard alloys and instruments: Digest of scientific articles /VTIITS - Moscow: "Metallurgiya" 1986. - p. 56–60.

[6] O.Y. Retyunsky. Method of making performs for composite reversible cutting plates // Applied Mechanics and Materials. - 2013 - Vol. 379. - p. 24-27.

[7] A.V. Valentov, V.V. Konovodov, E.V. Agafonova. Forecasting residual and operating stress in soldering cutting tools with tungsten-free alloy inserts // Applied Mechanics and Materials. - 2013 - Vol. 379. - p. 28-31.

[8] A.I. Afonasof, A.A. Lasukov. Elementary chip formation in metal cutting //Russian Engineering research, 2014, Vol.34, No.3, hh. 152-155

[9] R.V. Chernukhin. Reliability of the steering gear of truck vehicles// Applied Mechanics and Materials. - 2013 - Vol. 379. - p. 36-42.

IOP Conf. Series: Materials Science and Engineering **91** (2015) 012054 doi:10.1088/1757-899X/91/1/012054

Influence of thick-walled cylinders length on the residual stresses generated during the single-cycle mandrelling

V F Skvortsov[1a], A O Boznak[1b], A B Kim[1c] and A Yu Arlyapov[1d]

[1]Tomsk Polytechnic University, 30 Lenin Avenue, Tomsk, 634050, Russia

E-mail: [a]TMRI@tpu.ru, [b]Aleksey@tpu.ru, [c]Bogowhich@tpu.ru, [d]Arlyapov@tpu.ru

Abstract. The paper presents methodology and results of experimental study by Sachs method the effect of length of mandrelled thick-walled cylinders made of steel grade 50 (0.5% C) with 5 mm diameter hole, outer diameter of 15 and 25 mm on residual stresses. Based on the review of studies, it was noted that this effect may be both due to uneven process of mandrelling along the cylinder length and due to zero axial residual stresses on its faces. It was found that the length of cylinders has the strongest influence on axial residual stresses. With allowances between 0.9 and 7.1% and length reduction from 40 to 10 mm, the largest absolute value of axial residual stresses decreases from 210 to 50 MPa. It was noted that when the outer diameter of cylinders is 15 mm their length significantly influence on both hoop and radial residual stresses. With the above-mentioned decrease in cylinders length, absolute values of hoop and radial residual stresses in the region adjacent to the hole, depending on the mandrelling allowance, increase (from –135 to –205 MPa and from –45 to 55 MPa respectively, when allowance is 7.1%), decrease (from –315 to –235 MPa and from –135 to –95 MPa respectively, when allowance is 0.9%) and remain almost unchanged (when allowance is 3.4%). Impact of length on these stresses is weak when the outer diameter of cylinders is 25 mm.

1. Introduction

Mandrelling is an effective method of processing holes in parts such as hollow cylinders [1, 2]. It is used for holes with diameters d=1...150 mm and relative depth L/d≤100.

Mandrelling is accompanied by generation of considerable residual stresses in hollow cylinders [1, 3, 4]. (Under these stresses hereinafter, as usual, we will understand their average values along the length of the cylinder.) Residual stresses are mainly determined by allowance and number of mandrelling cycles, mechanical properties of the cylinder material and wall thickness ratio, which is characterized by the ratio D/d (D – is outside diameter of the cylinder). When the wall thickness ratio D/d≤3, both compressive and tensile hoop residual stresses can be formed in the zone adjacent to the hole. Impact of hollow cylinder mandrelling configuration (tension or compression) on residual stresses is weak [5]. The absolute value of residual stresses in the mandrelled hollow cylinders can be close to the offset yield strength $\sigma_{0.2}$ of the part material. It was established that in order to generate favorable compressive hoop residual stresses in the hole surface layer that enhance performance characteristics of the cylinders, the last mandrelling cycle must be performed with small allowance (about 0.01d).

Residual stresses in the mandrelled thick-walled cylinders can be significantly influenced by length of the cylinders. With reduction of the length of cylinders, mandrelling process becomes less uniform, as the length of entrance and exit regions, in which axial metal flow predominates, accounts for higher share of the cylinder length. Thereby, we can expect all components of the residual stresses in the mandrelled cylinders to depend on the cylinders length. Furthermore, since the axial residual stresses always vanish on the faces of the cylinder, the absolute value of these residual stresses decreases as the length of the cylinder decreases. Hoop residual stresses slightly change too. In addition, as the review of the abovementioned studies showed, influence of the hollow cylinder length on the residual stress formed after mandrelling is not studied experimentally. This, to a certain extent, limits application of mandrelling for manufacture of hollow cylinders of different lengths.

The purpose of the study is to investigate the influence of thick-walled cylinders length on residual stresses generated in single-cycle mandrelling experimentally.

2. Experimental procedure

Experiments were performed on samples made of steel grade 50 (0.5% C content, $\sigma_{0.2}\approx470$ MPa, HB 2170...2290 MPa) with hole diameter of d=5 mm. The outer diameter D of the samples was 15 mm (D/d=3) and 25 mm (D/d=5). The length L of the sample was equal to 10 and 40 mm. Holes in the samples were twist drilled on a lathe and then reamed with hand reamers. Mandrelling in compression scheme was performed on the universal testing machine UME-10TM using special equipment [2]. Single-point carbide mandrels (carbide grade VK8, 8% Co and 92% WC) with cones angle equal to 6° and width of the cylindrical margin equal to 3 mm were used for mandrelling. Relative allowances a/d were 0.9%, 3.4%, 7.1%. Oil fluid MR-7 was used as a lubricant. Mandrelling speed was equal to 0.008 m/s. After mandrelling, in order to ensure required accuracy, external surface and faces of the samples were subjected to fine grinding. Three samples were used in each experiment.

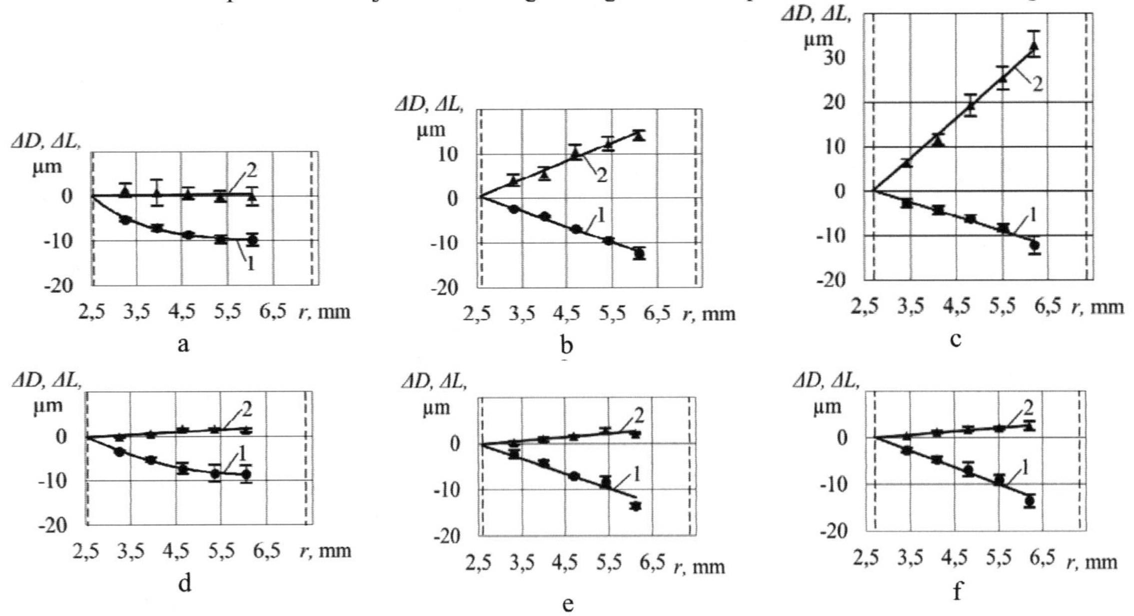

Figure 1. Dependence of change in outer diameter (1) and length (2) of the mandrelled samples with D=15 mm on the internal surface radius: *a* – a/d=0.9%, L=40 mm; *b* – a/d=3.4%, L=40 mm; *c* – a/d=7.1%, L=40 mm; *d* – a/d=0.9%, L=10 mm; *e* – a/d=3.4%, L=10 mm; *f* – a/d=7.1%, L=10 mm. Dotted lines show internal and external surfaces of the samples

Residual stresses were found by Sachs method using I.A. Birger's formulas [6]. In accordance with this method, 0.7 mm thick layers of metal were incrementally removed from the inner surface of the samples on a CNC wire-EDM machine model DK 7725 (PRC), resulted changes in external diameter and length were measured. External surface diameters were measured on an ultra-optimeter (Carl Zeiss Jena, Germany) with an accuracy of readings equal to 0.0002 mm. Length of the samples was measured with a mikrokator (accuracy of readings equal to 0.0005 mm) fixed in a stand. External diameters of 40 mm long samples were measured in three cross-sections located in the middle and at a distance of 2 mm from the sample faces. External diameters of 10 mm long samples were measured only in two cross-sections located at a distance of 2 mm from the sample faces. Diameter of the sample external surface was taken as the average of the measured values. Length of the sample was considered equal to the average of the two extreme values of the measured lengths.

Figure 1 shows an example of relationships between the change of outer diameter ΔD and length ΔL of the samples (D=15 mm) and the radius of the internal surface (95 per cent confidence intervals are given near the average values). These and other similar dependences (for samples with D=25 mm) were approximated using Statistica 8 with straight lines and second-degree polynomials, which equations were then used to calculate residual stresses. Young's modulus $E=2\cdot10^5$ MPa, Poisson's ratio $\mu=0.3$ were used. Calculations were performed using Microsoft Excel 2013.

3. Results and discussion

Distribution of hoop σ_θ, radial σ_r and axial σ_z residual stresses along the radius r of the mandrelled samples (stress curves) are shown in figure 2 (D=15 mm), and figure 3 (D=25 mm). It can be seen that the hoop residual stresses are compressive in the zone adjacent to the hole and tensile in the zone adjacent to external surface in all the cases studied. Radial residual stresses on the hole surface and external surface are equal to zero, and are compressive in the rest of the part material. Distribution of axial residual stresses along the radius of the sample is more complex. For samples with D=15 mm and length L=40 mm, in the region adjacent to the hole, axial residual stresses are compressive when allowance is 0.9% (figure 2a), and tensile when allowance is 3.4% and 7.1% (figures 2b and 2c).

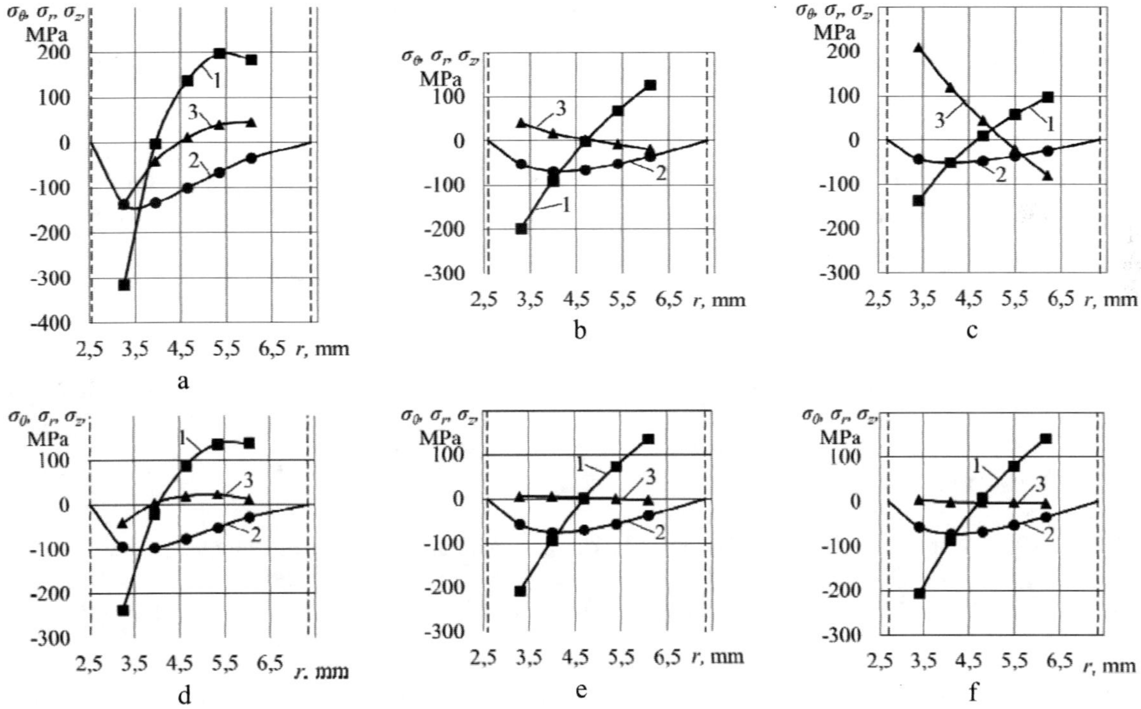

Figure 2. Curves for hoop (1), radial (2) and axial (3) residual stresses in the mandrelled samples with D=15 mm: a – a/d=0.9%, L=40 mm; b – a/d=3.4%, L=40 mm; c – a/d=7.1%, L=40 mm; d – a/d=0.9%, L=10 mm; e – a/d=3.4%, L=10 mm; f – a/d=7.1%, L=10 mm.
Dotted lines show internal and external surfaces of the samples

For samples with D=25 mm, axial residual stresses are compressive in this region for all allowances used (figure 3). As can be seen from figures 2 and 3, axial residual stresses near the hole are balanced by axial residual stresses of opposite sign in the outer surface layer. The largest in absolute value stresses are the hoop residual stresses, which are close to the offset yield strength $\sigma_{0.2}$ of the cylinders material for samples with D=15 mm (figure 3).

Studies have shown that length of the samples has the strongest influence on the axial residual stresses. With the length reduction from 40 to 10 mm of the samples of both wall-thickness ratios, the largest absolute value of axial residual stresses decreases to the values smaller than 50 MPa (figures 2 and 3). Thus, for samples with D=15 mm, mandrelled with 0.9% allowance, these stresses near the hole vary from –135 MPa (figure 2a) to –40 MPa (figure 2d), and for samples mandrelled with 7.1% allowance – from 210 MPa (figure 2c) to values almost equal to zero (figure 2f). In samples with D=25 mm, subjected to mandrelling with 3.4 and 7.1% allowances, the axial residual stresses near the hole vary from –190 MPa (figure 3a) to –50 MPa (figure 3c) and from –75 MPa (figure 3b) to –10 MPa (figure 3d), respectively.

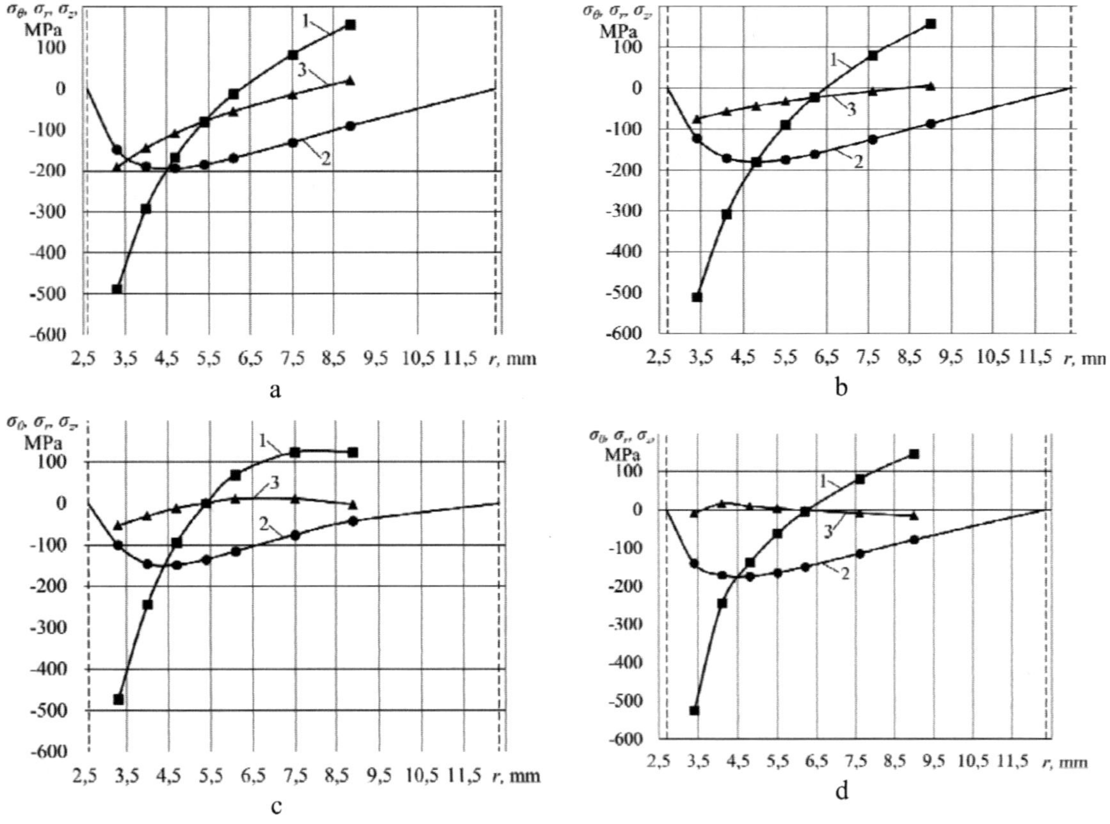

Figure 3. Curves for hoop (1), radial (2) and axial (3) residual stresses in the mandrelled samples with D=25 mm: a – a/d=3.4%, L=40 mm; b – a/d=7.1%, L=40 mm; c – a/d=3.4%, L=10 mm; d – a/d=7.1%, L=10 mm. Dotted lines show internal and external surfaces of the samples

When the outer diameter of samples D=15 mm, their length has a noticeable effect on both the hoop and radial residual stresses (figure 2). With the samples' length reduction from 40 to 10 mm, hoop and radial residual stresses may decrease, increase or remain almost unchanged depending on mandrelling allowances. Indeed, hoop and radial residual stresses near the hole reduce in magnitude from –315 MPa to –235 MPa and from –135 MPa to –95 MPa respectively when mandrelling allowance is 0.9% (figures 2a and 2d), these stresses remain almost unchanged when mandrelling allowance is 3.4% (figures 2b and 2e), increase in absolute values from –135 MPa, to –205 MPa and from –45 MPa to –55 MPa respectively when mandrelling allowance is 7.1% (figures 2c and 2f).

Impact of length on hoop and radial residual stresses is weak when the outer diameter of cylinders is 25 mm.

4. Conclusion

1. Length of mandrelled hollow cylinders has the strongest influence on axial residual stresses. With reduction of length of cylinders made of steel grade 50 (hole diameter is 5 mm, outer diameter is 15 mm and 25 mm, mandrelling allowances vary from 0.9 to 7.1%) from 40 to 10 mm, the largest absolute value of axial residual stresses decreases from 210 to 50 MPa.
2. If the outer diameter of cylinders is 15 mm their length significantly influences on both hoop and radial residual stresses. With the above-mentioned decrease in cylinders length, absolute values of hoop and radial residual stresses in the region adjacent to the hole, depending on the mandrelling allowance, increase (from –135 to –205 MPa and from –45 to 55 MPa respectively, when allowance is 7.1%), decrease (from –315 to –235 MPa and from –135 to –95 MPa respectively, when allowance is 0.9%) and remain almost unchanged (when allowance is 3.4%). Impact of length on these stresses is weak when the outer diameter of cylinders is 25 mm.

References

[1] Proskuryakov Yu G The Technology of Hardening Sizing and Forming Working of Metals Mashinostroenie Moscow 1971

[2] Skvortsov V F Arlyapov A Yu Okhotin I S Kim A B Mandrelling with High Allowances as a Method of Finishing Treatment of Small Diameter Deep Holes *7th International Forum on Strategic Technology (IFOST): Proceedings* Vol 2 (2012) pp 336-340

[3] Chunmei Zhao Hongyan Hu Yefei Zhou Yukui Gao Xunjun Ren Qingxiang Yang Experimental and Numerical Investigation of Residual Stresses Around Cold Extrusion Hole of Ultrahigh Strength Steel *Materials & Design* (2013) pp 78-84

[4] Maximov J T Duncheva G V Ganev N Bakalova T N The Benefit from an Adequate Finite Element Simulation of the Cold Hole Expansion Process *Engineering Failure Analysis* (2009) pp 503-511

[5] Skvortsov V F Arlyapov A Yu Boznak A O Kim A B Residual Stresses in Compression and Tension Mandrelling of Thick-Walled Cylinders *Mechanical Engineering Automation and Control Systems (MEACS)* (2014) pp 1-5

[6] Birger I A Residual stresses Mashgiz Moscow 1963

Stress and Strain State Analysis of Defective Pipeline Portion

P.V. Burkov[1,2], S.P. Burkova[3], S.A. Knaub[4]
[1] Professor, Tomsk State University of Architecture and Building, 2, Solyanaya Sq., Tomsk, Russia
[2] Professor, National Research Tomsk Polytechnic University, 30, Lenin Ave., Tomsk, Russia
[3] Doctor, National Research Tomsk Polytechnic University, 30, Lenin Ave., Tomsk, Russia
[4] Student, National Research Tomsk Polytechnic University, 30, Lenin Ave., Tomsk, Russia

E-mail: burkovpv@mail.ru

Abstract. The paper presents computer simulation results of the pipeline having defects in a welded joint. Autodesk Inventor software is used for simulation of the stress and strain state of the pipeline. Places of the possible failure and stress concentrators are predicted on the defective portion of the pipeline.

Introduction

After the construction of the main pipeline, the process of its long-term operation begins. A safe operation of a commercial construction built even in compliance with a perfect engineering design (with the use of advanced CAD systems) is impossible without its periodical inspection, technical diagnostics, estimation of its actual state, timely repair, and replacement of worn-out parts.
Unfortunately, the actual state of commercial constructions in operation is analyzed after the technical diagnostics often not only without the use of computer simulation techniques, but also without any appropriate engineering estimates at all. In these cases, they use tables with results of model tests created dozens years ago or, at most, the simplest semiempirical formulas for calculation of one or another defect allowance.

A non-destructive testing (NDT) of industrial pipelines included in the main oil transportation objects, often poses a problem of their technical diagnostics and further operation. Thus, such defect as displacement was discovered in the weld-affected zone of the pipeline. The static strength and the allowable lifecycle (service life) of welded joints with defects are calculated in accordance with the Standard 'Main pipelines. Circular, longitudinal, and spiral welds with defects and pipes with ripples. As preliminary estimates have shown, there is no a combination of external service loads capable to induce the limit state in the pipeline portion. It was decided to support by calculations the possibility of further operation of the pipeline portion with defect, and replace it at the next major repair.

The aim of this paper is to define the possibility of further operation of the pipeline by means of ascertaining the values of internal forces applied to the defective portion. The values obtained for maximum stresses and theoretical values of the pipeline material strength are to be compared. The stress values and the pattern in the defective portion are suggested to determine by the simulation model of stress and strain state of the pipeline with defect and the finite element method. Autodesk

Inventor, 3D mechanical CAD software allowing to design and study the stress and strain state of different pipelines is used for this simulation. In works [1-10], the authors describe the individuality and complexity of this software.

2. Results and discussion

The internal changes in metal lead to its fracture and are called a limit stress state. The structural reliability is provided by the comparison of maximum stresses which appear at the most dangerous point and admissible limit values of the given material. The limit stress state of a structure is just the limit beyond which its operation is impermissible. The structural reliability is higher when the level of actual internal stresses is farther from the limit state. Autodesk Inventor software allows the strain-stress computation of the pipeline wall and obtaining output data in the form of three values of primary stresses which are presented by roots of a cubic equation defined by the stress vector components.

The strain-stress computation is based on the calculation methodology of allowable elastic stresses. The stress and strain state of the displacement type defect located on a cylindrical pipe casing is determined by a three-dimensional design model subjected to combined service loads. Since the aim of this work is to study stresses occurring within the defective portion of the pipeline, the design model can be reduced to its defective portion. The internal excessive stress is the service load applied to the pipeline. The dead load of the pipeline portion, wind and snow/ice loads also should be taken into account. In order to provide a static determinability of the model, the boundary conditions include a pin-edge fixing of the pipeline portion to prevent linear motions. Schematic view of defect locations on the pipeline portion N 82140 is given below.

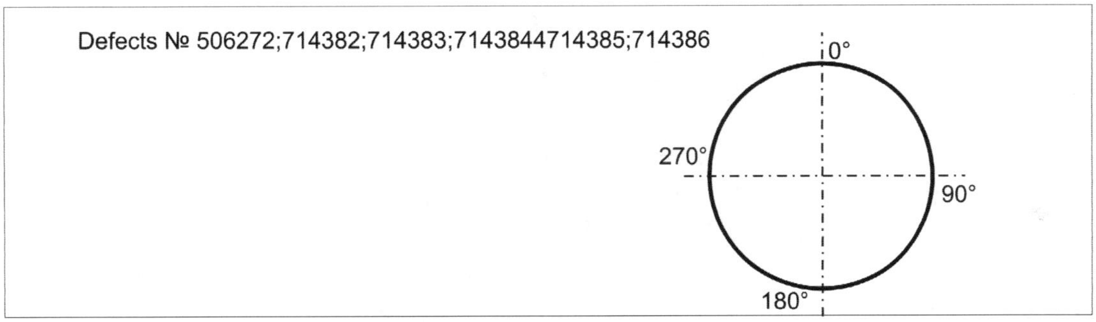

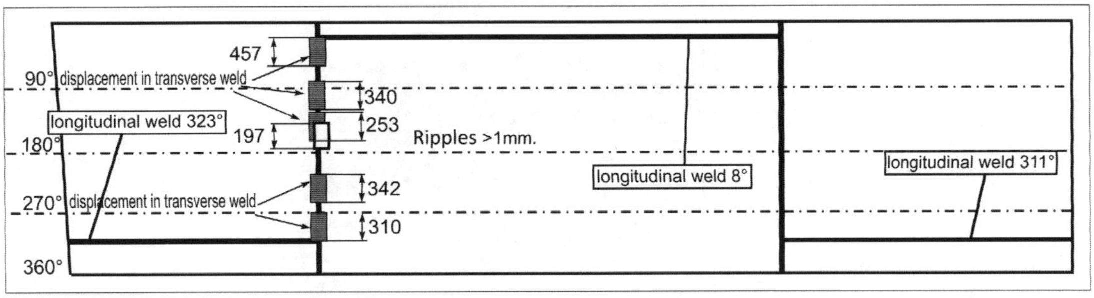

3. Defective portion simulation model

The length of the pipeline casing is selected such that to exclude the possible effect of end connections on the defect area. The displacement type defect is simulated in conformity with the measurement results obtained during technical diagnostics (Fig. 1a,b,c). Loads distributed across the casing area are applied to its defect area. Welded joints are neglected. The finite element model contains five casing objects unified with each other by the bonded contact, i.e. the complete dependence of displacements is provided in all assemblies. Cartesian system is used for this model. The finite element mesh reproduces the surface curvature and is superimposed automatically. The minimum mesh size was obtained at mash condensation within the defect area and equals to 0,6 mm. Mesh size of 10 mm was accepted for all other structural elements. The non-linear model simulation was carried out using the iteration technique. The values of the equivalent stresses are distributed in the sectional area of casing under the maximum load, i.e. in its outer surface. The graphic imaging of stress field distribution in the pipeline wall is depicted in Figures 2 - 6.

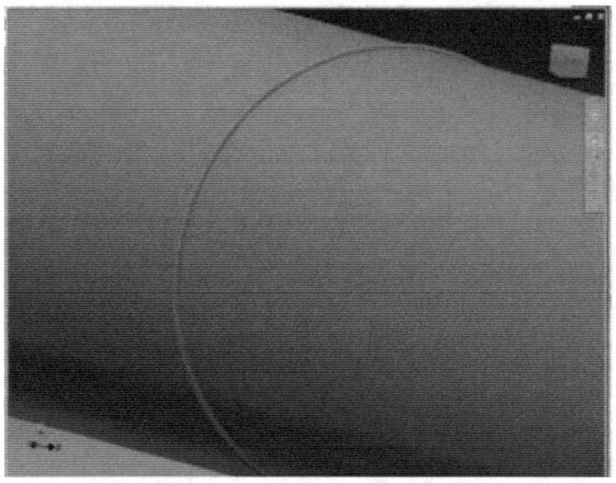

Figure 1a. Pipe the weld.

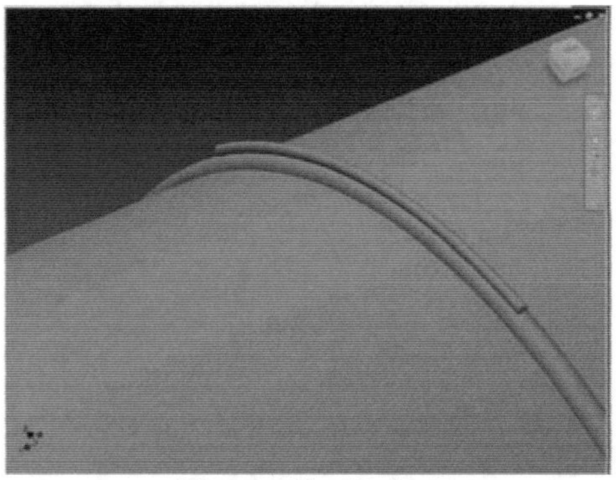

Figure 1b. Defect

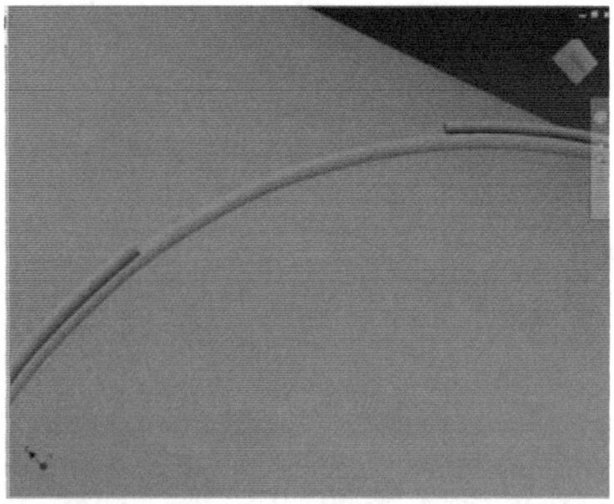

Figure 1c. Finite element model of pipeline with the displacement type defect.

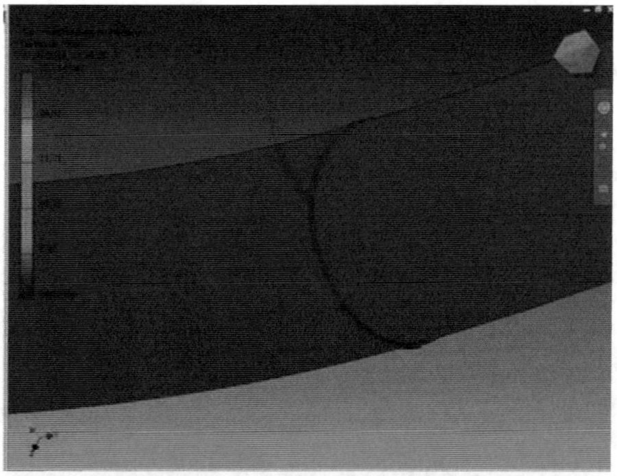

Figure 2. Equivalent stress distribution in the sectional area of casing

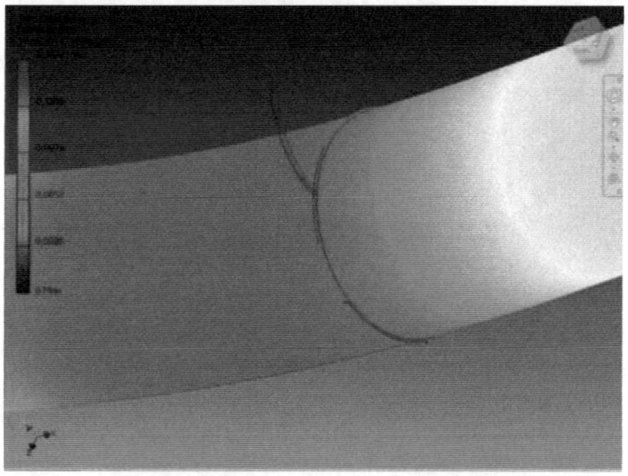

Figure 3. Bending in defect area

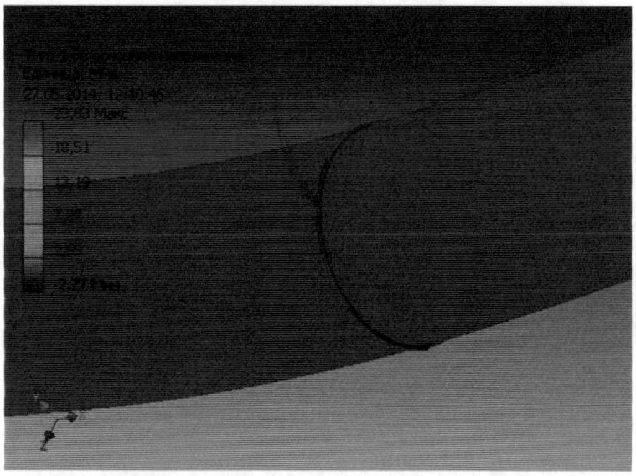

Figure 4. The first main stress

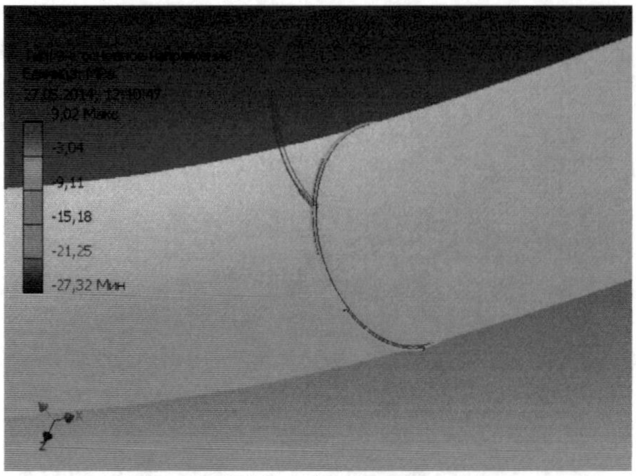

Figure 5. The third main stress

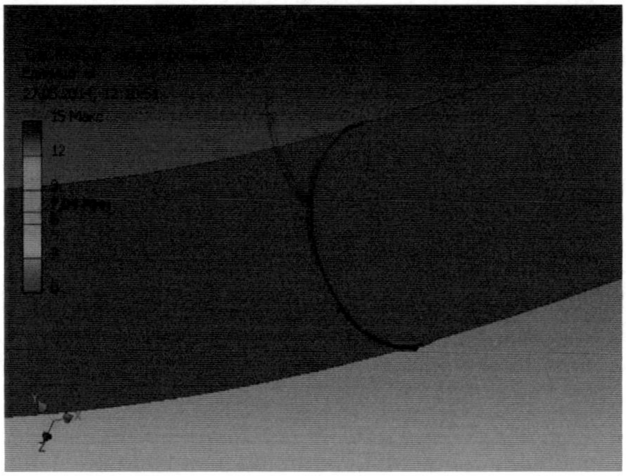

Figure 6. The factor of safety

As a result of this study the following parameters are obtained: von Mises stress (min = 0,00505 MPa; max = 35,51 MPa); safety factor (min = 7,04; max = 15); basic deformation (min=-0,00000093; max = 0,000131). These values were compared with those presented in SNiP 36.13330.2012 'Main pipelines' and no deviations were found.

4. Conclusions

The use of Autodesk Inventor software for the stress and strain state analysis of the pipeline portion in protective steel casing showed that the universal CAD system allows solving a variety of difficult problems.

For example, computer simulation of welded joints allowed detecting the effect of non-metal inclusions on stresses in welded joints of the pipeline. Stresses on the interface between those inclusions and the steel matrix increased due to the physical properties of the both. The linear portion of the pipeline with circular joints was characterized by a non-uniform stress and strain state. Maximum stress values were observed on the interface between the base metal and weld root.

The simulation of the stress and strain state showed that the finite element model of the pipeline portion with the defective metal wall is adequate to the NDT results. The pipeline displacement enabled local stresses. Thus, the increase of the equivalent stresses nearby the maximum displacement depth was 1,4%. The maximum values of membrane, shear, and equivalent stresses do not exceed the theoretical resistance of 35,51 MPa. Thus, in observing the theoretical pressure inside the pipe having the wall thickness corresponding to the actual strength, the displacement type defect has no a significant effect on strength and reliability of the pipe portion.

5. References

[1] Aleshin V V Kobyakov V A Seleznev V B Strength analysis of industrial pipelines in ANSYS El J *SAPR i grafika* 7(2007) pp 57-61
[2] Tarasenko A A Chepur P V Kuzovnikov E V Tarasenko D A Stress-strain analysis of pipe header with defect J Fund Res 9 (2014) pp 1471-1476
[3] Fedoseeva E M Ol'shanskaya T V Ignatov M N Shestakov A P Modeling of non-stationary processes in pipeline weld El J Oil Gas Bus 5(2011) pp 376-381
[4] Ivanitskaya E V The role of pipeline transport in regional development *Trubopr Trans : teor i prak* 2 (2009) pp 4 - 5
[5] Gafarov N A Mitrofanov A V Goncharov A A Tret'yak A Ya Kichenko B V The analysis of the equipment and pipeline damages in extracting processing and transportation in Orenburg oil and gas plant Ser Diag oborud i trubopr Gazprom Moscow 2000
[6] Semin E E Tarasenko A A The use of software packages for technical diagnostics and repair of vertical steel containers *Trubopr Trans : teor i prak* Moscow 2 (2006) pp 5-8
[7] Sil'nitskii P F Tarasenko M A Tarasenko A A Design of foundation ring of reservoir with defects *Izv Vuz Neft' i gaz* 5(2011) pp 76–78
[8] Burkov P V Burkova S P Timofeev V Yu Ashcheulova A A Klyus O V The analysis of stress-strain state of pipeline under permafrost conditions *Vest KuzSTU* 6 (2013) pp 77–79
[9] Burkov P V Chernyavsky D Y Burkova S P Konan A Simulation of pipeline in the area of the underwater crossing *IOP Conference Series: E and Env Sc* 21 (2014) pp 1-5
[10] Burkov P V Kalmykova K G Burkova S P Do T T Research of stress-deformed state of main gas-pipeline section in loose soil settlement *IOP Conference Series: E and Env Sc* 21(2014) pp 5-7

Ultrasonic plastic deformation of steels

K Zaitsev[1], **V Klimenov**[2,3], **M Loshchilova**[1], **C Polovnikov**[1]

[1] Yurga Institute of Technology, TPU affiliate, 26, Leningradskaya Str., 652050, Yurga, Russia
[2] National Research Tomsk Polytechnic University, 30, Lenin Ave., 634050, Tomsk, Russia
[3] Tomsk State University of Architecture and Building, 2, Solyanaya Sq., 634003, Tomsk, Russia

E-mail: zaitsev81@rambler.ru

Abstract. In the work we demonstrate the possibility of forming thin surface and near-surface layers with submicrocrystalline structure under ultrasonic plastic deformation of machined steels. Formation of fine gradient textures up to nanocrystalline allows significant changing physical and mechanical properties of machined steels. Ensures improvement of microstructure, produces internal compressive strains which improves the cyclic strength of machine parts.

1. Introduction

In recent times attention of Russian and foreign scientists is attracted to the problem of structural materials modification due to external high-energy impacts [1]. A characteristic feature of such technologies is insignificant time of impact upon the processed surface and high energy density (dozens and hundreds of J/cm2) of input energy. As a result the processed surface layer obtains modified and gradient structure.

One of the methods of high energy impact upon metal surface is ultrasonic metal deformation (ultrasonic treatment) when the surface of the part after machining is impacted by a tool (a ball) oscillating with ultrasonic frequency [1-3, 7-11].

The oscillating system of the ultrasonic device for strengthening metallic parts is additionally loaded with static force determining the hold-down pressure. For the purposes of the whole surface treatment motion is transmitted to the part or to the tool. The parts to undergo ultrasonic treatment can be of varying forms: round, flat and profiled [4-6, 10, 11].

For the practical application of the ultrasonic treatment method we need ultrasonic equipment – special ultrasonic unit and ultrasonic low power generator (0.2-0.4 kW), tool. Treatment of outside and inside surfaces can be completed on metal-working (turning, milling) machines of normal accuracy.

The basic parameters of ultrasonic strengthening are: value of static pressure of the tool upon the machined part, speed of the main oscillating motion, longitudinal and traverse feed, number of passes, size and form of the tool. Oscillating motion of the tool produces internal stresses in the surface layer of the machined part [2, 3, 9, 10].

Assisting motion (feed motion) and the number of passes characterize the number of impacts per unit of machined area in a unit of time. The size and the form of the ultrasonic tool influence the impact force and, thus, the pressure the machined part surface undergoes [11].

Ultrasonic treatment is characterized by a number of peculiarities which distinguish this method from other types of surface deformation. The first characteristic is shock loading which periodically repeats and involves comparatively small area. The second peculiarity is microhardness increase by 1.5 and more times and ensuring its favorable distribution over the depth of the cold-worked layer, reducing the roughness of the surface and producing residual compressing stress. The third peculiarity of ultrasonic treatment is producing quickly alternating compressive and shear deformations.

The physical and mechanical condition of the surface layer (roughness, microhardness, internal residual stresses) is strongly influenced by specific for this process elements: the value of the static pressure of ball hold-down towards the part, shift amplitude of the part, oscillation frequency and other. The static force ensures contact between the ball and the machined part. Shift amplitude and static pressure produce the main influence upon the surface roughness, the degree and the depth of cold-working, value of residual stresses in the surface layer of metal and its other physical and mechanical properties. Choice of shift amplitude and static hold-down pressure of the ball determine the productivity of machining, feed and rotation velocity of the part, number of passes. The value of the static force must be the larger the less ductile part material is, the higher original roughness is, the larger are diameters of the strengthened part and ball, the larger are feed and rotation speed.

In the given work we study the influence of ultrasonic plastic deformation upon structure and properties of surface layers of low-carbon, medium-carbon and high-carbon steels of ferrite, ferrite-pearlite and pearlite classes. The possibility of formation submicrocrystalline up to nanocrystalline structure of surface layer of machined metals [12-14], improving cyclical strength, wear resistance as well as reliability and life time of various machine parts.

We selected the original roughness of the surface as the technological factor upon which the efficiency of submicrocrystalline and nanocrystalline structure formation may depend when applying ultrasonic plastic deformation. As the planishing effect is localized in the thin surface layer the original roughness of the surface will have significant influence upon the degree of deformation.

2. Materials and methods

The ultrasonic impact upon the surface of metal workpieces was caused with the use of technological complex [15]. The complex is of small size, it is characterized with reliability and stability of work. Machining was completed with the power output of 200-300 W and with indenter amplitude of oscillation $\xi_{ИНД}$=20 μm. The diagram of indenter impact upon the surface of the turning part is presented in Figure 1. The static force F_N, ensuring the physical contact between the indenter and the machined surface, and the shift amplitude (under the given frequency of oscillations $f=2\cdot10^4$ c^{-1}) determine the value of stresses, the number of singular acts of deformation and their speed.

Machining of metal samples was completed in the following way: preliminary machining of the workpiece on the turning machine [16, 17] with further ultrasonic plastic deformation. Sample 2 (Figure 2, position 1) was mounted into three-jaw universal 1 of the turning machine and tightened with back center 3. We completed turning of journals of the samples under the following conditions: longitudinal saddle feed – from 0.04 to 0.28 mm/rev; cutting depth – from 0.5 to 1 mm; the angle in the plane of the turning tool φ from 40° to 85°. Machining formed the surface with various microrelief and roughness Rz 10-20 μm and keystone form rims of roughness profile with base width of 25-35 μm and angle of lateral side inclination of 5^0-10^0 (Figure 3 a).

Then, instead of cutter 4 we insert ultrasonic impact driver 5 into the toolholder (Figure 2, position 2). The surface of workpiece 2 is machined with hard-alloy indenter oscillating at ultrasonic frequency. In the zone of local contact between the indenter and the machined surface the deformation region appears which travels with the indenter. That kind of machining results in formation of a surface with peak to trough height Rz 3.5-5.5 μm (Figure 3 b). High grade finish is achieved due to localization of finishing action of indenter in the thin surface layer, namely on the peaks remaining after preliminary turning machining.

Further in the work we studied the microstructure of the samples in the area of formation gradient structures with the application of optical and transmission electron microscopy. Combined application of the given methods of research allowed completing the most comprehensive analysis of structure which is rather difficult to do with application of only one of these methods. The detailed study was completed in work [18].

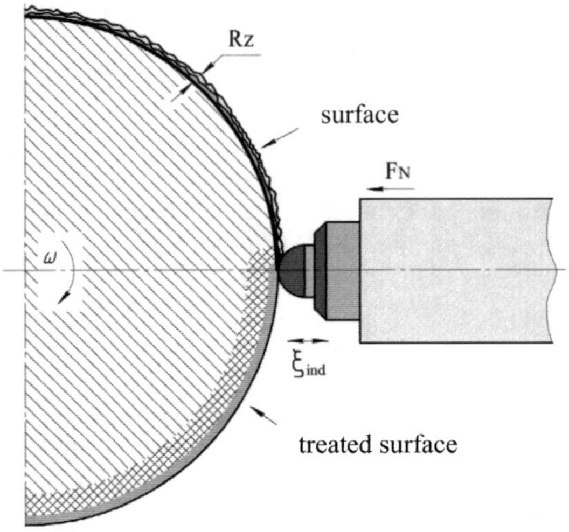

Figure 1. Diagram of ultrasonic finishing: ξ_{ind} – amplitude of indenter oscillations; F_N – static force; ω – angular rate.

The structure study of metallographic sections prepared according to the standard techniques [19, 20] was carried out with microscope "MIM-10". To reveal the grain structure we completed chemical etching according to [21, 22]. Thin steel foils were studied with device "EM-125" under accelerating voltage 125kV. The operating magnification in the microscop column was 8000 – 80000 power.

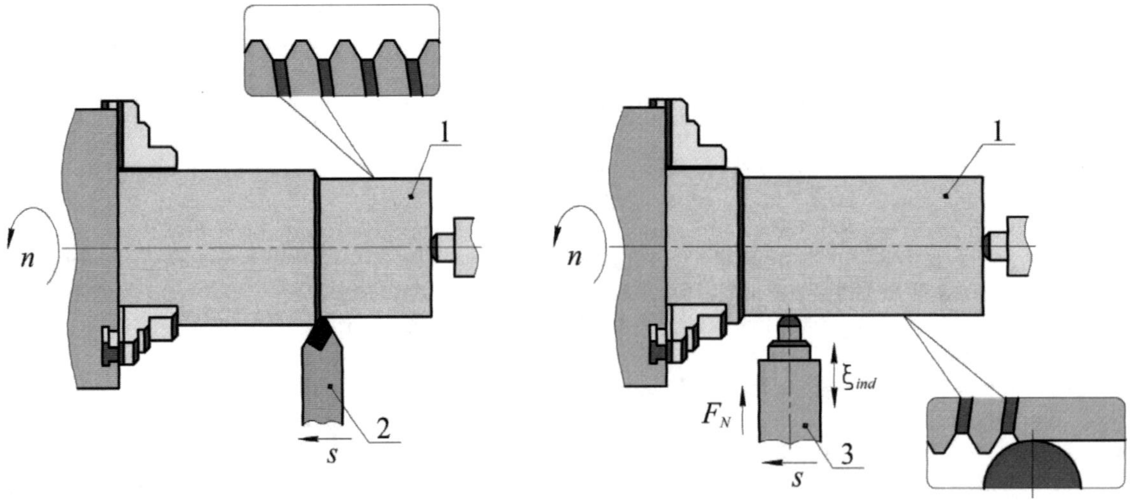

Figure 2. The diagram of finishing with ultrasonic tool: 1 – workpiece; 2 – cutter; 3 – ultrasonic tool

To measure surface roughness and to study surface morphology we applied optical 3D-profilometer " Micro mesure 3D station". Measuring was completed by non-contact method with a scanning laser beam. Line-by-line scanning allows obtaining high-accuracy 3D-image of the measured region of the surface.

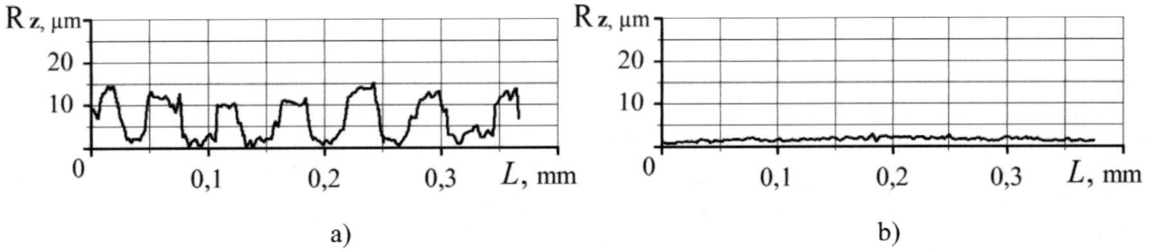

Figure 3. Surface profilogram after: a – preliminary machining; b – after ultrasonic finishing

The results were processed with the application of software "Mountains Map" («SARL Digital Surf», France). Surface roughness was estimated according to Rz and Ra (GOST 2789-73). Microhardness estimation according to GOST 9450-76 was completed with application of "Nano Hardness Tester" and "PMT-3" with indenter load of 0.49÷0.98 N. Microhardness was measured according to the method of reconstructed image, by pressing a diamond quadrihedral pyramid with the angle of 136 between opposite planes.

3. Results and discussion

As a result of ultrasonic plastic deformation of metal samples we observe cardinal changes in the structure of surface and subsurface layers and significant increase of microhardness. In the metallographic image of the structure (Figure 4 a), in the surface layer with the thickness of 2-3 µm we observe a nonerodible layer, and at the depth of 3-12 µm – non-equiaxial grains protruded in the direction of machining. Then there is smooth transition to the original structure of steel.

Figure 4. Metallographic (a) and electron-microscopic images of steel 60 after ultrasonic finishing at the depth of 3 µm (b), 10 µm (c), 450 µm (d)

For more detailed study of the microstructure of samples (for steel 60) after the ultrasonic plastic deformation electron-microscopic investigations were completed. Their results are presented in Figure 4. In the original state (in the core of the sample) steel 60 mainly has pearlite structure. Pearlite is lamellar, ferrite grains are situated between the pearlite colonies and gravitate toward the original austenite grain junction line. Grains of free microcrystalline ferrite can also be found in pearlite colonies (Figure 4 d).

After turning machining and further ultrasonic plastic deformation, in the 2-3 μm thick surface layer of the sample we observe nanocrustalline structure consisting of a mixture of α-phase crystalline particles chaotically situated cementite particles (Figure 4 b). Regions of free ferrite characteristic of the original state structure are practically absent. At the depth of 3-10 μm the grain, although deformed, structure of the original material is retained and substructural transformations in lamellar pearlite are observed, for example, fragmentation of ferrite lamels and cementite lamels (Fig. 4 d). The structure of free ferrite regions becomes subgrain. It should be noted that in the structure of ferrite subgrains, in contrast to original grains, carbide phase is found indicating higher diffusion mobility of carbon atoms under ultrasonic plastic deformation and their mass transfer into the regions of large enough ferrite grains.

The results of study of microhardness values distribution in the sample section are presented in Figure 5. For all studied steels maximal microhardness values after the ultrasonic plastic deformation are found on the surface. The microhardness value uniformly decreases depthwise up to the microhardness level of original alloy material. Uniform decrease of microhardness (the studied depth was up to 400 μm) is determined by the changes of macro-, micro- and submicrostructure of steel as well as by behavior of residual variation of compressive stress under ultrasonic plastic deformation. Maximum absolute microhardness values on the surface of the studied samples and the depth of the hardened layer depend upon carbon concentration in steel and increase with the growth of carbon content.

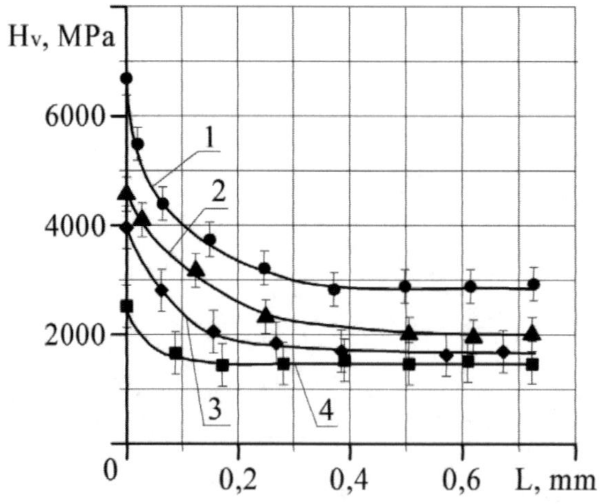

Figure 5. Microhardness depthwise distribution in steel samples after ultrasonic finishing: 1 – steel 60; 2 – steel 45; 3 – steel 20; 4 – steel 3

Figure 6 presents dependence of surface layer microhardness after ultrasonic plastic deformation upon the value of original roughness after preliminary turning. Reduction of original roughness contributes to reduction of microhardness values up to the values which are characteristic of original metal. It indicates the low degree of deformation of roughness peaks obtained after preliminary turning machining. Growth of original roughness up to certain level (Rz 10-25 μm) contributes to increase of microhardness values in the surface layers after the ultrasonic plastic deformation due to the high degree of roughness peaks deformation. At the same time further increase of roughness leads to decrease in surface microhardness due to intensive accumulation of macrodefects of the surface layer in the form of incompletely deformed roughness peaks after the ultrasonic plastic deformation With the values of surface roughness less than Rz 4-8 μm deformation is less intensive due to small size of roughness peaks.

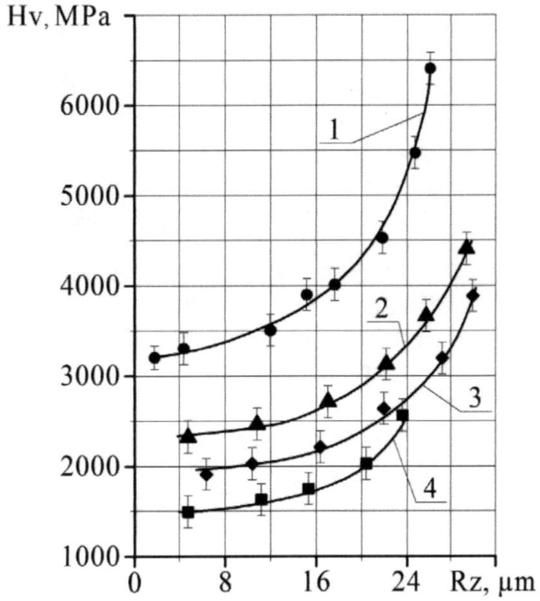

Figure 6. Dependence of surface layer microhardness after ultrasonic treatment upon the original roughness value obtained after preliminary turning:
1 – steel 60; 2 – steel 45; 3 – steel 20; 4 – steel 3

Forming on the surface a plastically deformed modified layer with a large concentration of dislocations and a diffused interface with the basic material distinctly slows down formation of dangerous stress concentrators and fatigue cracks nucleation blocking development of fatigue processes on the surface. At the same time abrasive-adhesive wear is also reduced due to surface hardness increase. On the finished surface a surface layer with nanocrystalline structure is formed and the depth of the modified layer with elastically plastic stress fields and net dislocation substructure increases up to 400 μm for steel 60[23, 24].

Conclusion

With consideration to the fact that the problem of producing nanocrystalline structures with corresponding improvement of physical and mechanical properties on massive parts is not yet solved, the obtained results are rather important. They give a possibility of producing fine nanocrystalline structure on the parts of any sizes and any geometry (leading spindles, stamping tools, piston plungers, etc.) which significantly increases hardness, wear resistance, cyclic strength and operational life of various parts and machine units.

References

[1] Odintsov L G Strengthening and finishing of machine parts by surface plastic deformation: Reference book – M : Mashinostroyeniye 1987 – 328 p il
[2] Abramov O V Khorbenko I G Shvegla Sh Ultrasonic finishing of materials / Under the editorship of O V Abramov – M : Mashinostroyeniye 1984 – 280 p il

IOP Conf. Series: Materials Science and Engineering **91** (2015) 012056 doi:10.1088/1757-899X/91/1/012056

[3] Severdenko V P Klubovich V V Stepanenko A V Metal forming with application of ultrasound minsk: Nauka I tehnika 1973 p 288
[4] Abramov O V Gurevich L B Influence of ultrasound upon the structure and properties of pure metals // FHOM – 1972 – №3 – P 18-20
[5] Abramov O A Dobatkin V I Kazantsev V F and others Influence of high power ultrasound upon the inter-phase surface of metals M : Nauka 1986 – 227 p
[6] Abramov O V Khorbenko I G Shvegla Sh Ultrasonic finishing of materials / Under the editorship of O V Abramov – M : Mashinostroyeniye 1984 – 280 p il
[7] Klimenov V A Panin V Ye Bezborodov V P and others Substructure and phase transformations under ultrasonic machining of martensitic steel // FHOM – 1993 – № 6 – P 77-83
[8] Klimenov V A Ivanov Yu F Perevalova O B and others Structure phase composition and mechanisms of strengthening austenite steel treated by ultrasound // FHOM -2001 – P 90-97
[9] Klimenov V A Nekhoroshkov O N Uvarkin P V Kovalevskaya Zh G Ivanov Yu F Structure phase composition and properties of steel 60 after ultrasonic finishing // Physical mesomechanics – 2006 – V 9 – Special issue – P 173-176
[10] Markov A I Ultrasonic treatment of materials – M : Mashinostroyeniye 1980 – 237 p il
[11] Markov A I Application of ultrasound for machining and surface strengthening of hard-to-machine materials // Application of ultrasound in production Under the editorship of Makarov A I M : Mashinostroyeniye 1975 – P 157-180
[12] Lotkov A I Baturin A A Grishkov V N Kuznetsov P V Klimenov V A Panin V Ye Defects of structure and mesorelief of the surface of titanium nickelide after intense plastic deformation according to ultrasonic method Physical mesomechanics – 2005 – V 8 – Special issue – P 109-112
[13] Klimenov V A Nekhoroshkov O N Uvarkin P V Ivanov Yu F Ultrasonic finishing as the method of producing nanocrystalline structures // Physicochemistry of ultradisperse (nano-) systems Works of VI All-Russian (international) conference M : MEPhI – 2002 – P 168
[14] Klimenov V A Nekhoroshkov O N Uvarkin P V Ivanov Yu F Formation of nanocrystalline structure on the surface of railway power head wheelset tyres with application of ultrasonic machining // Experience of innovation enterprises in commercialization of nanotechnologies Works of All-Russian (international) conference " Physicochemistry of ultradisperse (nano-) systems" August 19-23 2002 TPU publishers – 2002 – P 57-61
[15] Ultrasonic technological package UZTK-02 Operations manual
[16] V A Petrova A A Bakanov and A V Walter "Crack Resistance of Weld Seals Hardening Methods and Composite Tools Quality " Applied Mechanics and Materials vol 682 pp 431–437 2014
[17] A A Lasukov P A Chazov A V Barsuk Investigation on the Elemental Chip Formation Process in Hard-to-Machine Material Cutting Applied Mechanics and Materials Vol 682 (2014) pp 504-509
[18] Alfyorova E A Lychagin D V Characterization of deformation pattern structure elements generated in uniaxial compression of nickel single crystals // Applied Mechanics and Materials – 2013 – v 379 – pp 66-70
[19] Tushinsky L I Plokhov A P Study of structure and mechanical properties of coating – Novosibirsk: Nauka 1986 – 200p
[20] Tushinsky L I Plokhov A P Tokarev A O Sindeyev V I Methods of material study: Structure properties and processes of applying inorganic coatings – M : Mir 2004 – 384 p : il
[21] Kovalenko V S Metallographic reactives Reference book – M : Metallurgiya 1981 – 120 p
[22] Baranova L V Demina E L Metallographic etching of metals and alloys – M : Metallurgiya 1983 – 256p
[23] Klimenov V A Zaitsev K V Borozna V Y Formation of metal –polymer composite on titanium alloy surfaces // Applied Mechanics and Materials 2014 – v 682 pp 453-458
[24] Kovalevskaya Z G Klimenov V A Zaitsev K V Interfacial adhesion between thermal spray coating and substrate achieved by ultrasonic finishing // Applied Mechanics and Materials – 2014 v 682 pp 459-463

IOP Conf. Series: Materials Science and Engineering **91** (2015) 012057 doi:10.1088/1757-899X/91/1/012057

Research of surface activating influence on formation of adhesion between gas-thermal coating and steel substrate

Z Kovalevskaya[1,2], **V Klimenov**[1,3], **K Zaitsev**[4]

[1] National Research Tomsk Polytechnic University, 30, Lenin Ave., 634050, Tomsk, Russia
[2] Institute of Strength Physics and Material Science SB of RAS, 2/4, Akademichesky Ave., 634021, Tomsk, Russia
[3] Tomsk State University of Architecture and Building, 2, Solyanaya Sq., 634003, Tomsk, Russia
[4] Yurga Institute of Technology, TPU affiliate, 26, Leningradskaya Str., 652050, Yurga, Russia

E-mail: kovalevskaya@ispms.tsc.ru, nauka@tsuab.ru, zaitsev81@rambler.ru

Abstract. Estimation of influence of physical and thermal activating on adhesion between steel substrates and thermal coatings has been performed. The substrates with surfaces obtained by and ultrasonic surface plastic deformation were used. To evaluate physical activating, pre-heating of the substrates to 600°C was performed. To evaluate the effect of thermal activating, the substrate surfaces after interfacial detachment were examined. Bonded areas on the substrate surfaces were measured by means of optical profilometry. The experiments have shown that surface physical activating is the main factor in formation of the adhesive bond between the coating and the substrate processed with the proposed methods.

1. Introduction

Deposition of particles on the substrate is based on impact force. Heated particles move to the substrate at a high velocity, so a high pressure occurs in the moment of particle impact. The pressure has two components: forward pressure and short pulse pressure produced by a hydraulic impact. A large magnitude of the pulse pressure contributes to cleaning the substrate surface in the zone of the impact and produces a physical contact between the particle and the substrate.

Chemical interaction between the particles and the substrate occurs under influence of forward pressure, which acts during the entire particle solidification and deformation. At the moment of collision, the particle undergoes deformation and its kinetic energy is converted into heat and deformation of the substrate.

Physical and chemical interactions between materials of the particles and the substrate occur as the result of heat transfer and plastic deformation of the surface micro relief at the moment of particle impact. The physical-chemical processes form adhesion i.e. interface bonding of the coating and the substrate [1].

Completeness of physical and chemical processes, proceeding in the contact zone, is defined by heat and kinetic energies of deposited particles and physical properties of contact materials [2]. The first two indicators are set by thermal spraying [1]. The third indicator is defined by the nature of contact materials and condition of the substrate surface. The simplest technique of adhesion control is the change of topography and stress condition of the deposited surface by its preconditioning [3-5].

The preconditioning increases surface atom activity, which enhances their contact with atoms of the deposited material. There are a number of methods for surface activating: chemical activating - by clearing of absorbent substances from the surface; physical activating - by mechanical processing of the structure of the surface layer; thermal activating – by prior or concomitant heating of the substrate surface; that facilitates overcoming of the activation barrier [1, 3].

At physical activating, adhesion strength is defined by the actual contact area of the coating and the substrate, uniformity of surface topography and the degree of surface layer modification. At thermal activating, substrate heating is required up to a temperature ensuring chemical interaction between the coating material and the substrate. Substrate preheating increases the area of physical contact by means of better spreading or deformation of the particles on the substrate surface. Wherein, the adhesion in the area of physical contact increases due to increased local bonded areas. Activating is possible with increasing temperature, if there is no rapid oxidation. Oxide layers on low-heated metal surfaces do not limit contacts of the deposited metal particles. If there is a strong oxidation of the substrate, the layer on its surface may undergo interfacial detachment [2, 3].

The objective of this work was to evaluate influence of physical and thermal activating on adhesion experimentally. For this purpose preconditioning of substrate surfaces by rough cutting machining and ultrasonic surface plastic deformation was performed. To evaluate the effect of thermal activating, heating of the substrates to 600°C was performed before spraying.

2. Materials and methods

For the research, flat samples of steel 45 were prepared. The sample surfaces were processed in two ways: rough cutting and ultrasonic finishing (USF). USF as a method of substrate processing is described in [6-8]. The preconditioning facilitated chemical and physical activating of the sample surfaces by removing oxide layer and producing hardening. The areas of effective contact between the substrate and the deposited particles vary due to various processes. Deposition of powdered nickel-based alloy was performed with a thermal spraying device using an argon-nitrogen plasma jet [9]. Deposition was performed in layers to form a coating with thickness of 300 microns. To evaluate the effect of thermal activating on adhesive bond formation, depositions were performed at a room temperature and at 600°C.

Using MICRO MEASURE 3D station optical profilometer, surface conditions were examined before coating deposition and after its detachment. The device allowed examining three-dimensional image of the surface conditions and describing quantitatively content and morphology of the bonded areas [6].

3. Results and discussion

As it is known, the actual bonding between the particle and the substrate does not occur in the entire area of chemical contact, but in local bonded areas. The bonded areas are sites on the contact surface, which form stable metal bonds between the material particles and the substrate material. Local interaction between contacting materials is due to localization of chemical interaction in areas with the highest reserve energy. Reserve energy occurs due to presence of structural defects. Active centers on the substrate surface may be impurity atoms, vacancies, dislocation steps [1, 2]. At detachment, bonds between the coating and the substrate exceed cohesive forces in bonded areas. Failures occurred in the coating volume itself, and some fragments of material remained bonded to the substrate surface (Fig. 1).

Thus, the bonded areas form the actual content of contact between the coating and the substrate. Expansion of the bonded areas throughout contact surface demonstrates gradation of physical and chemical process development at the interface "coating - substrate". As shown in [4], a stable bond of particles with the substrate occurs if bonded areas occupy 40 - 70% of the contact surface.

Let us examine how different methods of substrate surface precondition influence formation of bonds between the coatings and the substrate. To improve the adhesion it is advisable to apply coating with a higher surface roughness. A rough surface has better thermo-physical conditions for activating

IOP Conf. Series: Materials Science and Engineering **91** (2015) 012057 doi:10.1088/1757-899X/91/1/012057

behavior, which increases the adhesion of deposited particles. On tops of projections temperature will be higher than in grooves, and period of heat action on the micro-projections will be longer as well [3].

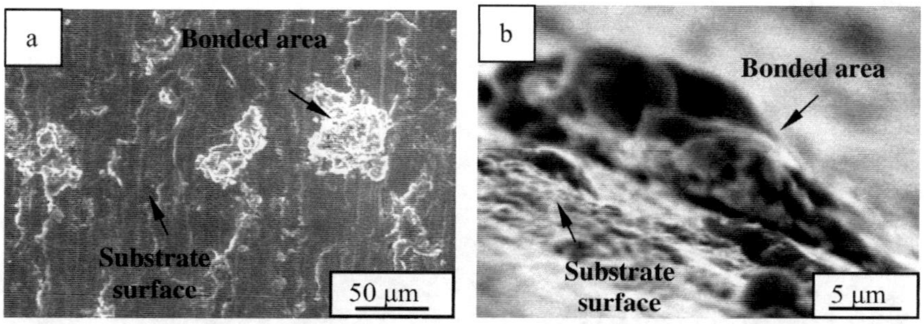

Figure 1. SEM image of bonded areas on surface of steel substrate after coating detachment

To perform the rough surface, rough machining was first chosen for preconditioning of the substrate surface. Exterior surface is shown in Figure 2 a.

The substrate surface roughness is formed by cutting tool advance. When advancing, the cutting edge removes chips spirally layer by layer and forms a microprofile. Cross section of the microprofile is specified by the cutting tool geometry. To obtain a ragged surface, the cut tip is displaced below the part axis; this displacement increases cutter vibration, thereby the projections on the substrate surface are crushed. As shown in Figure 2b, surface topography is formed of circulated lines of surface projections. The projections have a height of about 10 micron and a pitch of 200 micron. Tops of projections are partially smoothed, partially cut. Grooves have sharp forms. Line anisotropy of surface topography is strong.

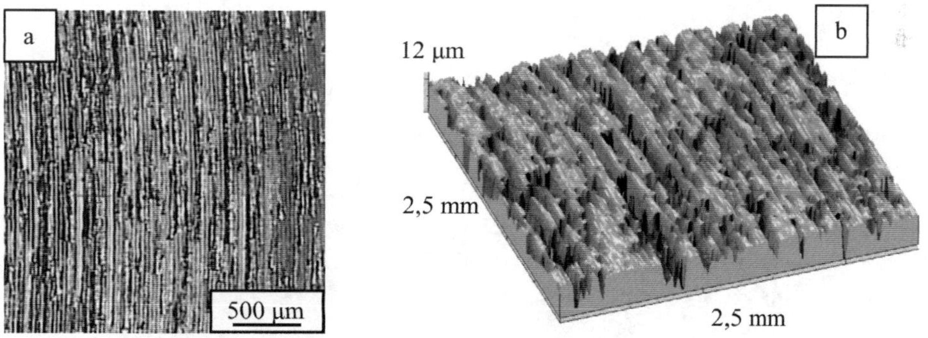

Figure 2. Image of substrate surface after rough cutting

It is believed, that surface microrelief in the form of closely spaced wells with profile radii greater than radii of sprayed particles is the most appropriate for formation of stable adhesion. Such a surface is formed by ultrasonic finishing (USF) [6].

Inclinations of the sides of surface projections are important for deformation and spreading of the sprayed particles. The cuts between the projections should be wide, with sloping lateral surfaces; which enable the particle to be deformed to a larger size. Projections contribute to improvement of thermal conditions for contact between the spray particles and the substrate [3]. Figure 3 shows the USF-processed surface

Surface topography after USF is a trace, shaped by two movements of the tool. The first movement is superposition of grooves left on the workpiece with the deforming tool (Fig. 3a). The second movement is the plastic flow of metal from each individual impact of the tool (Fig. 3b). A tool indenter has a spherical shape. As a result, the surface is formed, which comprises cuts and projections flowing into each other, with height up to 1 micron and pitch of roughness in transverse direction - approximately 100 microns, longitudinally - about 10 microns. Line anisotropy is flattened by plastic metal flow caused by individual tool strokes [6]. USF is also characterized by surface deformation, which gives access of structural defects such as dislocations and grain boundaries to the surface [7].

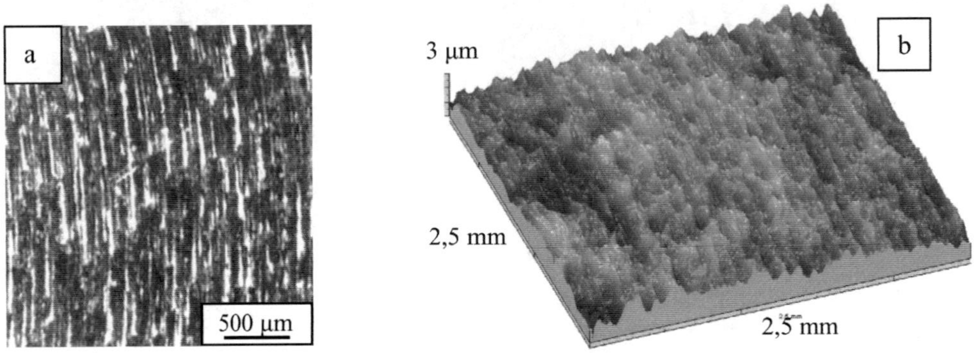

Figure 3. Image of substrate surface after USF

Thus, before deposition, substrates with different surface conditions were obtained. They differed from each other in roughness, surface topography, the effective area of contact between the sprayed particles and the substrate. Then surface conditions after interfacial detachment of the coating were analyzed.

The state of the substrate surface processed by rough machining after detachment of the coating was examined.

Figure 4a shows the surface after detachment of the coating deposited to the substrate at room temperature.

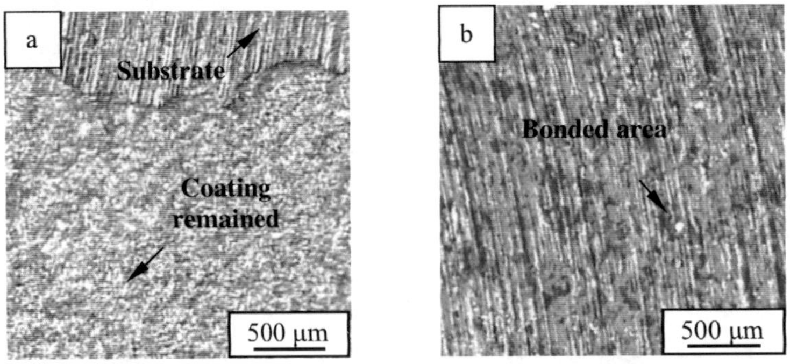

Figure 4. Surface processed by rough machining after detachment of coating deposited to substrate: a) at room temperature, b) preheated to 600°C

When detaching of the coating, destruction occurred at the interface fractionally. In this case adhesive and cohesive forces correlated in the coating. In the substrate areas, where detachment occurred at the interface, some bonded areas can be observed. Their content is about 4%. It can be noted, that the bonded areas consist of deposited particles, attached to the grooves of the substrate,

IOP Conf. Series: Materials Science and Engineering **91** (2015) 012057 doi:10.1088/1757-899X/91/1/012057

rather than to the projections. Consequently, contact between the particles and the substrate occurs due to sticking of the sprayed material into the grooves of the substrate surface. Flat surface projections do not form adhesion bonds with the deposited particles because of unfavorable topography of the roughness peaks. It is known that the surface with roughness peaks, isolated from each other, is characterized by maximum temperature gradient and cooling rate of the sprayed material [3]. Then, temperature increase on the substrate peaks leads to formation of chemical bonds and bonded areas.

However, despite absence of physical and chemical bonds at the "coating - substrate" interface, the coating has high adhesion. This is conditioned by mechanical engagement between the sprayed material and the effective substrate surface. One can not exclude formation of residual stresses favorable for interface adhesion, which arise due to heterogeneity of coating and substrate properties and temperature gradient [1]. Because of residual stresses, the spraying material is attached into the grooves of the substrate surface.

Substrate preheating to 600°C changes conditions of adhesion bond formation significantly. When heating, a continuous surface oxide layer occurs on the surface which can be classified as a scale diffusion layer. Such oxide layers on iron and carbon steels are characterized by low heat resistance and tendency to flaking. Therefore, when sprayed particles impact the substrate surface, local clearance of iron oxides and its activation can occur [2].

The coating deposited on the pre-heated rough cut surface is detached *completely* over the interface (Fig. 4b). After coating detachment the bonded area content is only about 1%. Destruction of base oxides occurs only on certain peaks of the surface microrelief, and single bonded areas are formed there. Then, low surface activating is conditioned by complex anisotropic topography of the rough cuts and high surface roughness. Process of oxide layer destruction under hydraulic impact of the liquid drop is hindered. Mechanical engagement of the deposited material with the substrate also does not occur because of change of surface condition at oxidation. The surface loosens, its roughness decreases. Furthermore, the heating partly relieves residual stresses which reduce attaching of the deposited particles into the grooves of the substrate surface. As a result of all the factors, surface activating does not occur during spraying, and coating does not form adhesion bonds with the substrate.

The coating deposited on the pre-USF surface is detached *partially* over the interface (Fig. 5a). The resulting coatings, as well as the coating sprayed onto the rough cut substrate, have a high adhesion to the substrate. However, mechanisms of adhesive bond formation differ. This is proved by the topography of interfacial detachment "coating - substrate" (Fig. 5 a). On the substrate surface a large number of bonded areas are observable. Sizes of the bonded areas vary from 15 to 40 microns, which are smaller than those of the sprayed particles. Distances between bonded areas coincide with the distances between the surface projections i.e. about 100 microns. Content of bonded areas formed from separate deposited particles is about 10%.

In some surface regions, sprayed particle agglomerates are observable, which are held by complexes of closely spaced bonded areas. The total content of both the bonded areas and the agglomerates is about 40%.

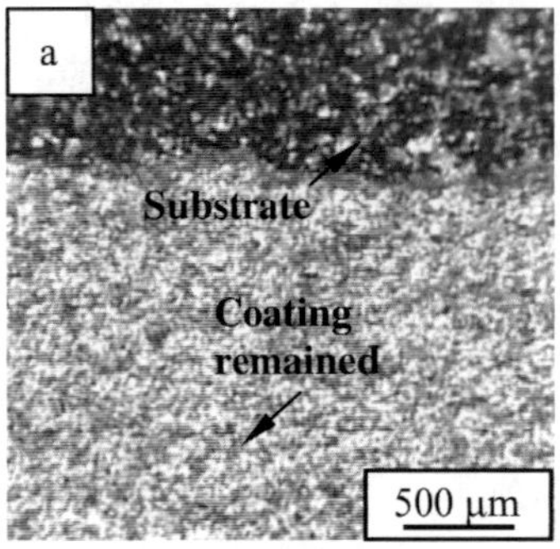

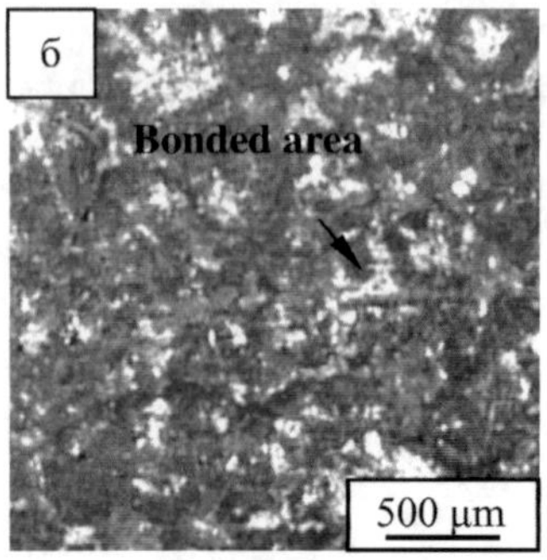

Figure 5 Image of pre-USF substrate surface after interfacial detachment of coating deposited: a) at room temperature; b) preheated to 600°C

In the case under study a high adhesion between the coating and the substrate is ensured by several factors, including formation of favorable substrate topography. Surface microrelief in the form of closely spaced wells with profile radii commensurable with radii of sprayed particles is the most appropriate for formation of stable adhesion. Inclinations of the sides of surface projections are important for deformation and spreading of the sprayed particles. Grooves between projections should be wide, with sloping lateral surfaces, which enable particles to be deformed to larger sizes. Projections should be separated from each other. Such a surface is formed by ultrasonic finishing (USF). Roughness peaks, during depositing of molten particles, are heated and actually welded with crystallizing particles to form bonded areas. This can be seen in the photograph of interfacial detachment of the coating (Fig. 5a). Arrangement of the bonded areas corresponds to the distance between roughness peaks.

One more factor ensures intensive chemical interaction between spray particles and the pre-USF substrate material. A large amount of surface crystal structure defects ensure formation of stable chemical bonds between the sprayed material and the substrate material. Evidence of chemical activating mechanism is proved by "coating-substrate" interface uncontrasty.

As it is described in our previous work, metallographic analysis of coating cross-sections does not reveal the interface in a non-etched micro-section [10]. After chemical etching a stable interface of materials and single micro-pores become observable. Coating detachment occurs by growth of cracking in the coating solid body, just above the "coating - substrate" interface. A thin layer of sprayed material remains on the substrate surface; which is commensurate with the height of two or three malformed particles [10]. Such areas of stable particles have the appearance of agglomerates on the detachment surface.

Preheating of the pre-USF substrate to 600°C increases a contact temperature and reduces an energy barrier of chemical interaction of materials. However, an oxide layer in the contact area interferes the depositing; so it should be removed. As it is proved by the above experiment, when using the rough cut surface, coating failed. Using pre-USF substrate, the task was completed.

Figure 5b shows "coating - substrate" interfacial detachment. On the coating surface the bonded areas are observable. Their size varies from 10 to 100 microns. Some bonded areas take the form of some deposited particles. Most of the bonded areas are represented as agglomerates of the deposited particles. A total content of the bonded areas is about 45%; therefore about half of the surface is activated despite of the oxide layer.

Surface topography obtained by USF ensures activating. Destruction of the oxide layer occurs on the peaks of micro-roughness at collision with sprayed particles. Impact and deformation of liquid particles perform local clearance of the substrate surface from iron oxides and surfece activating. Time of particle action onto the substrate is sufficient for forming of chemical bonds between the contacting materials [11-13].

IOP Conf. Series: Materials Science and Engineering **91** (2015) 012057 doi:10.1088/1757-899X/91/1/012057

Increased contact temperature also provides relaxation of residual stresses. Then, during formation of the chemical bond between the sprayed material and the substrate material, maximum of inter-atomic bonds are kept as a result of activated diffusion processes, which flow rates are set by on-process temperature.

Conclusion

When applying plasma coatings to the substrate preconditioned by rough cutting, the main mechanism of adhesive bond formation is sticking of spray particles in the grooves of the surface. If the substrate is preheated, the oxide layer interferes with sticking of the spraying material to the substrate surface. Adhesion does not occur between the substrate heated to 600°C and the coating.

Surface physical activating followed by chemical activating are basic mechanisms of formation of adhesion between coatings and pre-USF substrates primarily at the roughness peaks. The coating, sprayed on the substrate at a room temperature, has an enforced adhesive bond. This occurs because the additional energy of 600°C substrate heating partially compensates for the loss of energy required for passing the potential barrier caused by the oxide layer interference.

For the above-described methods of substrate preconditioning, physical activating of the surface is the main factor that ensures formation of the adhesive bond at the interface "coating - substrate". Thermal activating of the substrate surface is relevant if, during spraying, system reserve energy exceeds the potential barrier, which is to be passed for oxide layer removal.

References

[1] Pawlowski L Science and Engineering of Thermal Spray Coatings Wiley-Blackwell 2008 656

[2] Matejicek J Vilemova M Musalek R Sachr P Hornik J Influence of interface characteristics on adhesion/cohesion of plasma sprayed tungsten coatings Coatings 3 (2013) 108-125

[3] Paredes R S C Amico S C D'Oliveira A S C M Effect of roughness cut and pre-heating of substrate on morphology of aluminum coatings deposited by thermal spraying Surface and Coatings Technology 9 (2006) 3049-3055

[4] Klimenov V A Kovalevskaya Z G Ul'yanitskii V Y Zaitsev K V Borozna V Y Effect of ultrasonic treatment of substrate on formation of coating in detonation spraying Welding International 24 (2010) 803-807

[5] Klimenov V A Kovalevskaya Zh G Zaitzev K V Tolmachov A I Adhesion obtained by high velocity oxygen fuel thermal spray process Bulletin of Tomsk Polytechnic University 310 (2007) 52-55

[6] Kovalevskaya Zh G Uvarkin P V Tolmachev A I Some features of formation of surface microrelief of steel under ultrasonic finishing Russian Journal of Nondestructive Testing 48 (2012) 153-158

[7] Kovalevskaya Zh G Ivanov Yu F Perevalova O B Klimenov V A Uvarkin P V Study of microstructure of surface layers of low carbon steel after rough cutting and ultrasonic finishing The Physics of Metals and Metallography 114 (2013) 41–53

[8] Lotkov A I Baturin A A Grishkov V N Kovalevskaya Zh G Kuznetsov P V Effect of ultrasonic plastic treatment on surface structure and phase state of nickel titanium Technical Physics Letters 31 (2005) 912-914

[9] Klimenov V A Ivanov Yu F Perevalova O B Senchilo Z G Structural and phase transformation in Ni- and Fe- based plasma coating under high energy Materials and manufacturing 12 (1997) 849–861

[10] Kovalevskaya Z G Klimenov V A Zaytsev K V Interfacial Adhesion Between Thermal Spray Coating and Substrate Achieved by Ultrasonic Finishing Applied Mechanics and Materials 682 (2014) 459-463

[11] Ovcharenko V E Psakhie S G Ivanov Yu F Ivanov K V Boyangin E N Bulk nanostructuring intermetallic composite material AIP Conference Proceedings 1623 (2014) pp 459-462

[12] Ovcharenko V E Mohovikov A A Yu B H Zhao Y H Structural state scale-dependent physical characteristics and endurance of cermet composite for cutting metal Applied Mechanics and Materials 682 (2014) pp 405-409

[13] Ovcharenko V E Mokhovikov A A Ignat'ev A S Influence of surface nanostructure on the life of cermet in metal cutting Steel in Translation 43 (6) 2013 pp 348-350

IOP Conf. Series: Materials Science and Engineering **91** (2015) 012058 doi:10.1088/1757-899X/91/1/012058

New design of cutters for coal mining machines

S Prokopenko [1,a], A Sushko[1b], I Kurzina [2,c]

[1]652055, Russia, Kemerovo Region, Yurga, Leningradskaya St., 26
Yurginsky Technological Institute (Branch), Tomsk Polytechnic University;
[2]634050, Russia, Tomsk, Lenin Ave. 36, Tomsk State University;

e-mail: sibgp@mail.rua; sushko.a.v@mail.rub; Kurzina99@mail.ruc

Annotation. The author designs an up-to-date rock cutter for mining machines and presents a matrix to determine its service life depending on the strength of the rock being broken. The dependence of the cutter service life on the strength of the rock being broken is proved. Advanced designs of rock cutters by the leading companies are described. The author develops an original design of a rock cutter which provides greater cutting surface, increased strength and prolonged service life. Pilot samples of the cutters ready to the test run are presented.

Introduction

Rocks breaking in mines during the mining and clearing activities is conducted by special coal mining machines; their working bodies are equipped with cutters. The cutters are placed on the coal mining machines in the amount of 38-96 pieces, depending on the design of the executive body (Fig. 1).

Figure 1. Working element of coal mining machine with cutting tools

In 2011 134 coal mining machines operated in Russian mines, 108 of them being produced in Ukraine, Poland, Germany and the USA [1]. Coal mining machine cutting tools play a key role in

IOP Conf. Series: Materials Science and Engineering **91** (2015) 012058 doi:10.1088/1757-899X/91/1/012058

mining technology. If a rock cutter is out of order the whole complex of mine machines has to stop work as far as coal cutting is impossible. An average coal mine in Kuzbass spends 15,000 – 18,000 rock cutters per year which increases the cost of mining. In this view improving strength and prolonging service life of rock cutters are important.

Research Results

Currently coal mining machines are equipped with swivel tangential cutter which is designed as a cylinder swing holders made of high strength alloy steel and hard-tipped with hard-metal nose (Fig. 2).

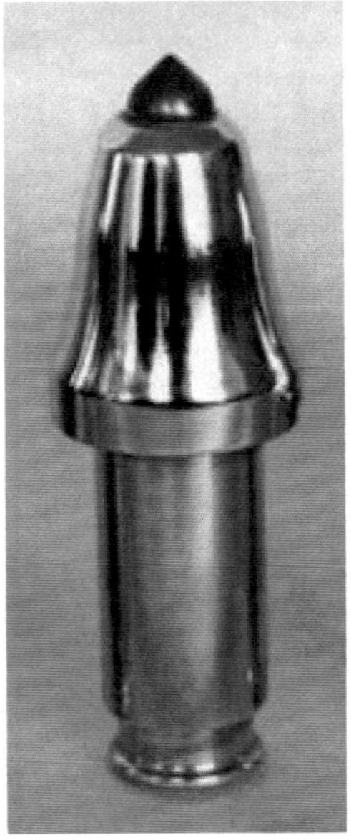

Figure 2. Swivel tangential cutter of coal mining machine

The nose is made of strong wear-resistant tungsten-rhenium alloy and is joined to the holder by means of brazing. The efficiency of rock braking depends on the depth and the speed of penetration into the rock, strength and abrasiveness of the rock, cutter wear resistance, a rock breaking technique, mining machine driver's skills and some other factors.

Mechanical characteristics such as rock strength and cutter strength are crucial for effective coal breaking. Both the rock and the cutter can have high, average and low strength. Various combinations of these properties define different rate of reciprocal destruction of the rock and the cutter. Rock cutter wear rate is stipulated by different variations in strength ratio. (Fig.3).

Low strength cutters have limited applications and can be used only with coal and soft rock. Using such cutters with average and high strength rocks leads to quick wear out and inefficient performance. Soft and average strength rocks can be effectively cut by an average strength cutter. When using high strength cutters with strong rocks, their reciprocal impact matches. If a rock becomes softer wear rate of a high strength cutter decreases, which provides wide application field of such cutters.

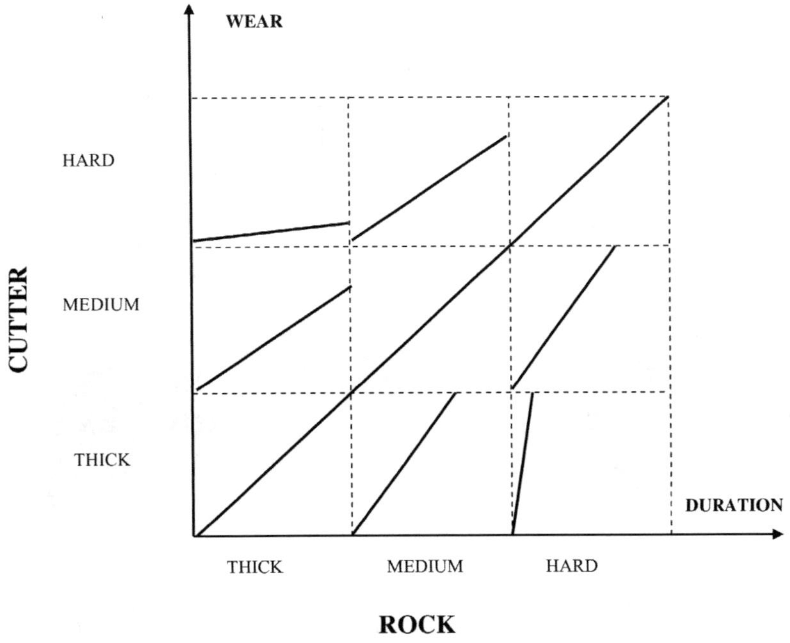

Figure 3. Rock cutter wear rate matrix depending on the strength of destroyed rock

Strength of the rock being destroyed influences rock cutter durability as well as coal mining machine service life. The dependence in Fig. 4 build up according to data [2] shows that as rock strength rises from 30 to 80 MPa, coal mining machine service life is shortened and it is 80 % worn out and needs general overhaul after breaking 55,000-60,000 m^3 in average instead of 10,000-12,000 $м^3$.

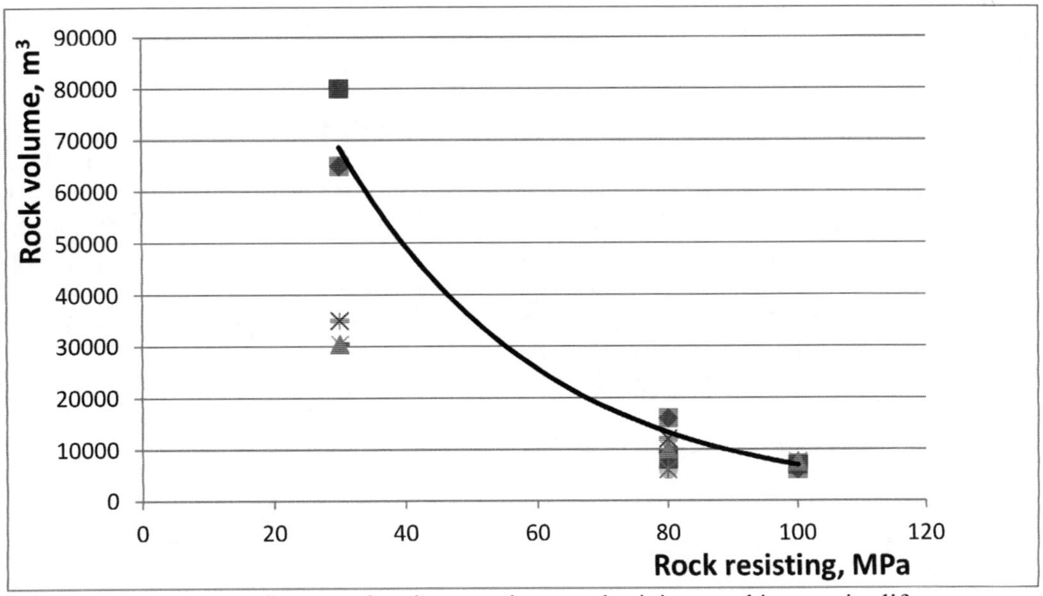

Figure 4. Influence of rock strength on coal mining machine service life

To reduce the negative influence of rocks on mining machines designers increase power of coal mining machine drives, strengthen assembly units and joint points, change operating device design. Improving rock cutting tools is an important measure to increase efficiency of rock cutting. Leading companies all over the world are concerned in designing new and better cutting tools.

Among the latest innovation by Sandvik company there are ribbed cutter heads to provide easy rotation and uniform wearing [3]. The cutter heads has four ribs longwise (Fig. 5a).

a

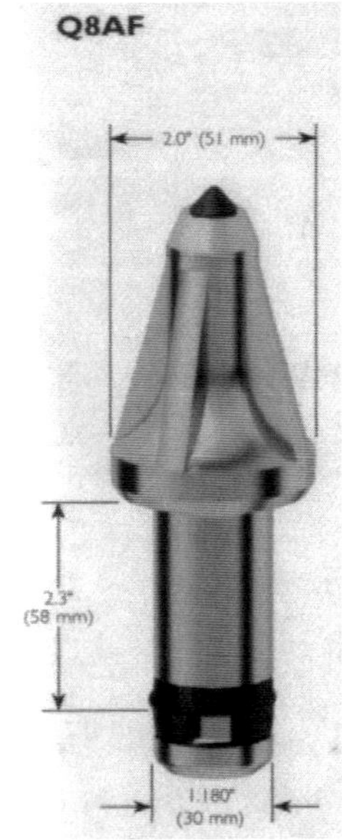

b

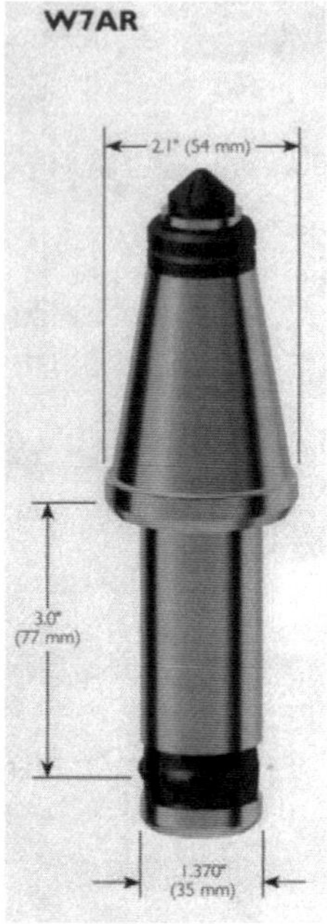

Figure 5. Sandvik cutters with ribbed heads (a) and protecting ring (b)

Kennametal company adopted cutter heads of similar shape. Cutters U83K 3.1 48,UR118 3.5 12.5NB, AM512FG have longwise ribs on their heads.

Another promising design by Sandvik company is equipped with TriSpecTM carbide ring to protect the area where the nose is fastened to the head body (Fig. 5b) [3]. The ring entrains about the front part of the cutter head where the nose is brazed-in and prevents the steel body from early wearing, thus prolonging service life of the carbide insert. Kennametal company offers a similar design of rings U119HF 80 16 and AM521HF 74 with protecting ring on the head nose.

Some designers suggest using a hard steel insert (for example X12MΦ) as a cutting element in order to uniform the rate of wearing of cutter body and insert. Work tests of such cutters showed their efficiency and uniform self-sharpening [4]. There are a number of researches aimed at improving design and increasing strength of cutters in various ways [5-8].

In mines of Kuzbass region some tests have been carried out to study the degree and the pattern of cutters wearing. The tests showed that the upper tapered section of the cutter head undergoes intensive wear. Fig. 6 shows a cutter which has lived its lifespan on KP-21 coal mining machine.

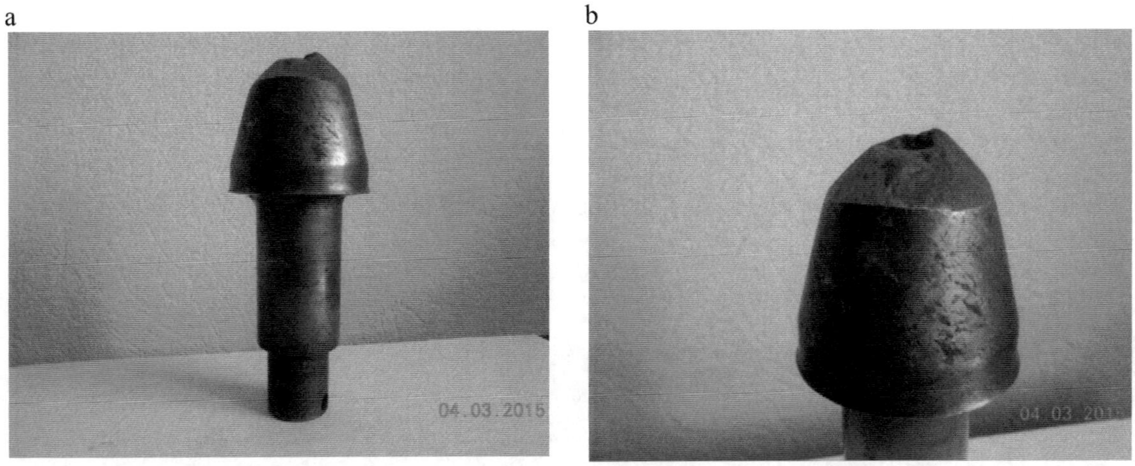

Figure 6. Worn out coal mining machine cutter (a) and its head (b)

We can see that the tapered area of the cutter head is uniformly worn out by 36 %. The rest of the head is not changed except for surface wear as deep as about 1mm.Wear rate (reducing head length) can be estimated with the help of graphs in Fig. 7, where curve 1 shows reducing head length from initial 70mm, and curve 2 shows an aggregate wearing of the head in per cent on a cumulative total.

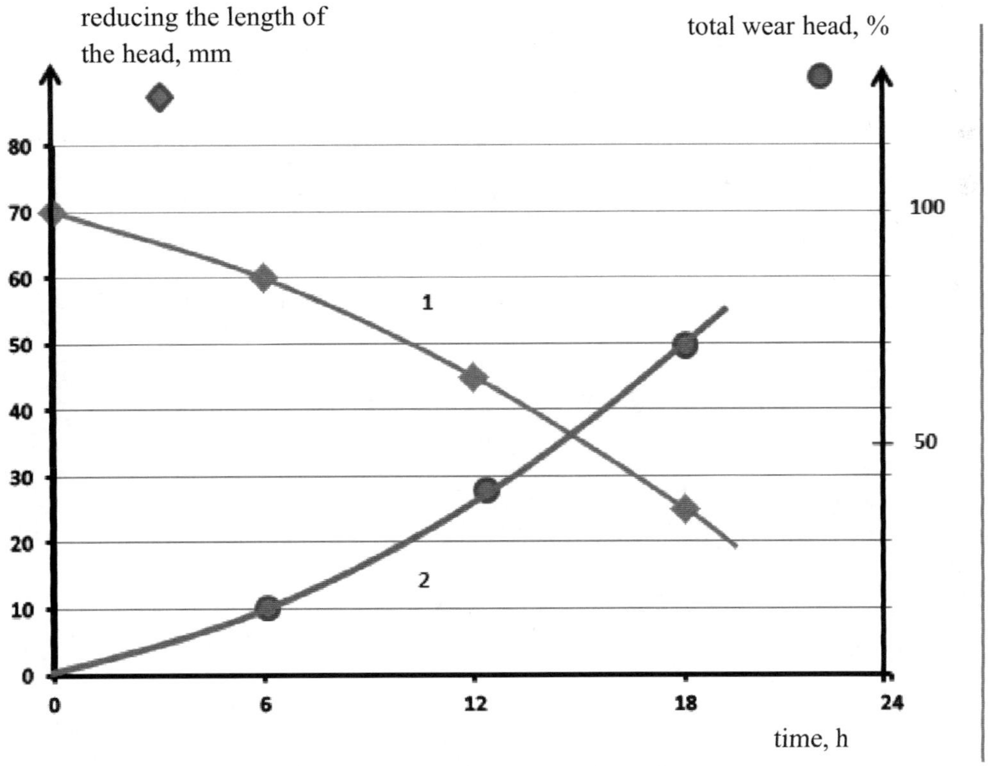

Figure 7. Rate of cutter head wearing

It has been found out that among other factors accelerated wearing of the tapered part of the cutter head is caused by irregular hardness of the cutter material, which decreases from 55 to 42 HRD on the surface and from 50 to 35 HRC in the center in the direction from the wide part of the head to the tapered one [9].

Based on the study of the rate and the mode of wearing of coal machine cutter we have developed a new design of the cutter which has high wear resistance and efficiency. We suggest reinforcing the head with extra reinforcing elements alongside with providing it with a tampered carbide nose. We recommend using carbide blades brazed in notches along the cutter axis. The length of the blades is 30-35 mm and their thickness is 5 mm. Such design provides a larger cutting surface, higher strength and greater durability. There can be from 2 to 4 brazed-in blades. Fig. 8 shows a photo of new cutters.

Figure 8. Coal machine cutters with heads reinforced by carbide blades

The left cutter has 3 cutting blades, the right one is equipped with two blades. These cutters are innovative and we are currently arranging test runs in the mines of Kuzbass.

Carbide cutting blades improve wear resistance of cutter surface. Table 1 shows calculations of the reinforced area depending on the number of cutting blades per a head (table 1).

Table 1. Increasing reinforced area of cutter head

Reinforced area	Increasing reinforced area with different number of blades, %			
	0	2	3	4
Cutter head $S=8356\ mm^2$	0	3,6	5,4	7,2
Working area of the head $S=1885\ mm^2$	0	16	24	32

When 2 - 4 blades are inserted, the area protected by hard alloy increases from 4% to 7%. As for working part of the head which undergoes the greatest wearing two inserted blades increase the hardened area by 16%, with three blades this area grows by 24%, and with four blades it grows by

32%. Fig. 9 shows changes in the average hardness of the working part of the head as the number of inserted plates is increased. The graph shows that average hardness can be increased from 55 up to 60 HRC, which significantly raises cutter heads durability.

Figure 9. Changes in average hardness of cutter blade

The cutting blades serve not only to harden the holder but also to fasten the nose in the body. They prevent quick wearing of the body around the nose and its early withdrawal. The chosen length of the blades allows their inserting into the wide part of the head which does not undergo great wearing which provides reliable fastening and long service life of the cutting blades.

Increased service life of the cutters allows breaking great volumes of rock. The cost of a unit of mined coal decrease and coal mining efficiency grows.

Conclusion

Coal and rock cutting in modern mines is carried out by means of coal mining machines equipped with tangential cutters. Efficiency of their performance mainly depends on correlation of strength of a cutting metal and that of a rock.

The design of coal machine cutter reinforced by longwise cutting blades suggested in the article provides it with greater cutting surface, increased strength and longer service life. Reinforced area on the working part of the head is increased to 32%, average strength rises to 60HRC. Test samples of this design are ready for testing in the mines of Kusbass.

Reference

[1] Linnik Yu N Analysis of engineering-and-economical performance of domestic and foreign mechanized coal mining complexes / Mining Journal - 2012 № 8 – P 19-23

[2] Gabov V V Zadkov L A Lykov Yu V Gurimsky A I Shpilko S I Special characteristics of operation of coal mining machines in mines of joint stock company "Vorkutaugol" // Mining equipment and electrical mechanics 2008 - №12 – P 2-6

IOP Conf. Series: Materials Science and Engineering **91** (2015) 012058 doi:10.1088/1757-899X/91/1/012058

[3] Cutting tools for rocks Mining RC 100/ Sandvik product catalogue – 2008 – 52 p
[4] Krestovozdvizhensky G D Some results of watching of shear-loaders in Kuzbass mines //Mining informational analytical bulletin 2009 №6 – P 120-123
[5] Aksenov V V Blashchuk M Y Dubrovsky M V Estimation of torque variation of geohod transmission with hydraulic drive // Applied Mechanics and Materials – 2013 - Vol 379 - p 11-15
[6] Ryabchikov A I Stepanov I B Equipment and methods for hybrid technologies of ion beam and plasma surface materials modification // Surface and Coating Technology – 2009 – V 203 № 17/18 – P 2784–2787
[7] Chinakhov D A Study of thermal cycle and cooling rate of steel 30HGSA single-pass weld joints [Electronic resorces] // Applied Mechanics and Materials – Vols 52-54 – 2011 – p 442-447 - Mode of access: http://www scientific net/AMM 52-54 442
[8] Prokopenko S A Multiple service life extension of mining and road machines' cutters // Applied Mechanics and Materials – 2014 - Vol 682 - p 319-323
[9] Bolobov V I Bobrov V L Talerov M P Bochkov V S the reason of quick wearing of tangential cutters //Notes of Mining Institute 2009 №6 – C 120-123

IOP Conf. Series: Materials Science and Engineering **91** (2015) 012059 doi:10.1088/1757-899X/91/1/012059

Estimation of cutting ability of cubic nitride boron wheels by surface microhardness criteria for pendulum grinding of high-speed steel plates

Ya I Soler [1] and A I Shustov[2]

[1] Associate Professor, Irkutsk national research technical university, Irkutsk, Russia
[2] Postgraduate Student, Irkutsk national research technical university, Irkutsk, Russia

E-mail: solera@istu.irk.ru

Abstract. The grinding wheel LKV50 B107 O V K27 100 KF40 is chosen as preferable according to the statistical interpretation of cases of surface microhardness of high-speed steel plates W6Mo5.

1. Introduction

Grinding is a major process that determines endurance and workability of machine tools, therefore grinding is often a completion phase of an engineering tool production. In mechanical engineering, high-speed steels (HSS) are referred to hard-to-machine materials due to the high content of carbide-forming elements. Therefore, in order to ensure optimal grinding conditions, it is essential to take into account not only a reserve strength coefficient of an abrasive wheel but high heating temperatures as well. In real-time grinding, more than 80% of power transforms into heat. The heat sources are abrasive grains which can be classified into cutting, pressing and non-cutting ones by their different radial orientation. The most projecting grains first cause an elastic deformation of metals, then – a plastic one, and achieving the ultimate strength - chipping. The instantaneous heating of a workpiece surface reaches a temperature of 700-800 °C, in case of intensive grinding - 1200-1500 °C. Heat rate reaches 10000 °C/s, and temperature decreases in a split second as well, as most heat passes to the underlying metal. It causes phase conversions in metal surfaces increasing or decreasing microhardness.

Microhardness of a metal surface is a complex thermal physics and dynamics criterion of cutting characterizing surface hardening and caused by plastic deformation.

Grinding wheels made from traditional abrasives cause such major surface defects as burns (a surface area with a low microhardness and a changed structure). Use of cubic nitride-boron (CBN) abrasive wheels might solve the problem. CBN has a high hardness that exceeds the hardness of traditional abrasives several times. Moreover, CBN has a high thermal resistance: grain oxidation begins only at temperature of 1100-1200 °C. As HSS is high alloyed steel, another advantage of CBN is their chemical resistance that makes it possible to grind avoiding adhesive and diffusion grain abrasive wear [1;6].

For example, in [2;3], the authors having estimated the grinding abilities of W6Mo6 test samples concluded that nitride-boron wheels showed better results than wheels made from fused alumina. Major structural changes in the metal surface were observed when grinding with fused alumina wheels as evidenced by decreasing microhardness. On the other hand, CBN-wheel grinding did not cause

IOP Conf. Series: Materials Science and Engineering **91** (2015) 012059 doi:10.1088/1757-899X/91/1/012059

visible structural changes in the steel surface and microhardness by grinding depth remained unaltered. We have estimated cutting abilities of CBN grinding wheels to provide the highest microhardness and grinding stability.

2. Statistical analysis methods of experimental data

Surface microhardness parameter is estimated statistically using software Statistica 6.1.478.0. Applied statistics makes it possible to analyze the sequences being the sets $e = 1,\ldots k$ where $v = 1,\ldots n$.

$$\{y_{ev}\} \tag{1}$$

There are two statistical methods - parametric and non-parametric ones. Each of these methods has more stability zero-hypothesis «on the home court» [4]. To present and analyze output data, homogeneous distribution parameters [4] have been used. Among them, one can mention position measures (State Standard GOST R ISO 5721-1-2002): averages and medians for parametric and non-parametric methods respectively; dispersion (precision): measures: deviation standards (variances) and ranges - for a parametric statistical method; quartile latitude - for a non-parametric statistical method. Distribution asymmetry (skewness) based on the following equation:

$$As_e = \left[3(\bar{y} - \tilde{y})/SD\right]_e, \tag{2}$$

where A_{Se} is a positive value when $\tilde{y}_e < \bar{y}_e$, otherwise, skewness is negative.

For a parametric statistical method «on the home court» we assumed the conditions when (1) is a normal distribution with homogeneous variances of deviations. In this case, their criteria are robust to an insignificant random variables (RV) distribution violation. However, it cannot be applied to the homoscedasticity of distributions (1). The violation of distribution laws can cause accepting a wrong hypothesis, because maximum permissible deviation values are not stipulated anywhere. In these cases, a non-parametric method which does not depend on the properties of any distribution family is the most suitable.

Kolmogorov-Smirnov (D) and Shapiro-Wilk (W) test statistics can be used to prove Ho for the normal distribution in software. D-test is widely used in technical applications. However, in [6], the authors have revealed the weakness of the criterion. Therefore, in our paper H_o for (1) has been tested according to W-statistics which satisfies the inequality:

$$\alpha > 0.5 \tag{3}$$

There are three tests for homogeneity variance (q=1,…3): 1 – Hartley, Cohran and Bartlett's tests; 2 – Levene's tests; 3 – Brown-Forsythe's tests. In all cases, they satisfies the inequality (α – acceptance reliability for H_o):

$$\alpha > 0.05. \tag{4}$$

Stochastic nature of hypothesizes assumes that decisions f, based on criteria q=1,…3, may differ. Thus, final results on homogeneity (H_o) and heterogeneity (H_1) deviations variances have been accepted for following reasons:

$$H_o \text{ at } f_o \in [2;3]; \tag{5}$$

$$H_1 \text{ at } f_1 \in [2;3]. \tag{6}$$

Predicted averages and medians have been obtained by multiple comparison [4,5] and designated as $\hat{y}_{e\bullet}$ and $m\hat{y}_e$ respectively.

3. Methodology

The experiments have been carried out under the following conditions: plan grinding-machine 3E11V and 3G71M; HSS plates made from steel W6Mo5 (63-64 HRC); grinding scheme – by the side of a wheel without nursing courses; cutting parameters: cutting speed v_w=28 m/s, traverse motion s_{tr} = 7 m/min, cross-feed motion $s_{c\text{-}f}$=4 mm/double pass, cutting depth t=0.01 mm, stock allowance z=0.10 mm. The lubricoolant is a 5 % Akvol–6 emulsion (Technical Specifications TU 0258-024-00148845-98) supplied by watering to the workpiece at a rate of 7-10 l/min. The wheel is being lowered at depth

IOP Conf. Series: Materials Science and Engineering **91** (2015) 012059 doi:10.1088/1757-899X/91/1/012059

t when the bench with HSS plate is being displaced fully to the left relative to the operator. Its movement from left to right is a working stroke. Otherwise, it is a nursing stroke with a final formation of a HSS plate surface micro-relief according to the wind grinding scheme as the spindle rotates clockwise. HSS plates are samples sized D×H = 40×50 mm. They are fixed on the magnetic bench and undergone face grinding. The number of repetitions of experiments is n=30. Microhardness parameter *HV* (State Standard GOST R ISO 6507-1 2007) is measured by instrument PMT-3. Output process data are presented as y_{ev}, where e = 1,…3 is a code of high-porous CBN wheels (HPCBNW): 1 - for CBN50 B107 O V K27 100 KF40 (the basic wheel), 2 - for LKV50 B107 O V K27 100 KF40, 3 - for CBN50 B126 O V K27 100 KF40. The original microhardnes of HSS plate surface is HV_o = 9000 MPa [3].

For a quantitative analysis of process stability, we use:

$$\left(K_{SD}\right)_e = \left(SD\right)_1 / \left(SD\right)_e, e = \overline{2;3}. \quad (7)$$

The value $K_{SD} < 1$ corresponds to the higher process stability for the basic wheel (e=1), the value ($K_{SD} > 1$) – for wheels e = 2,3.

Additionally, we have analyzed the process stability according to the ranges of experimental cases and obtained:

$$\left(K_R\right)_e = R_1 / R_e. \quad (8)$$

4. Result and Discussion

The results in table 1 prove the hypothesis H_o for the homogeneity of variances for all wheels e = 1,…3, and the decision (5) is made for all three tests. However the distribution normality condition is fulfilled only for wheel LKV50 B107 (e = 2). Therefore, a non-parametric method based on Kruskal-Wallis and Dunn's range criteria is the most suitable.

Table 1. Results of tests for homoscedasticity and normality of distributions for a *HV* parameter.

Inequality	CBN50 B107 (e = 1)	LKV50 B107 (e = 2)	CBN50 B126 (e = 3)
(3)	H_1	H_0	H_1
(4)		H_0	

Fig. 1 illustrates case results of (1) based on parametric (figure 1A) and nonparametric (figure 1B) statistics methods for *HV* parameter at grinding with e = 1,…3 wheels where "square" means position measure for $\overline{HV}_e$ (figure 1A) and $\tilde{HV}_e$ (figure 1B). The other marks have different meanings. In figure 1A, "rectangle" means dispersion borders for error standards ($\pm SDE_e$) and "moustache" means deviation standards ($\pm SD_e$) for. In figure 1B, the quartile latitude $\left|y_{0.75} - y_{0.25}\right|_e$ and ranges $R = \left|y_{\max} - y_{\min}\right|_e$ are shown. Median and ranges in figure 1B show a form of a distribution curve (2).

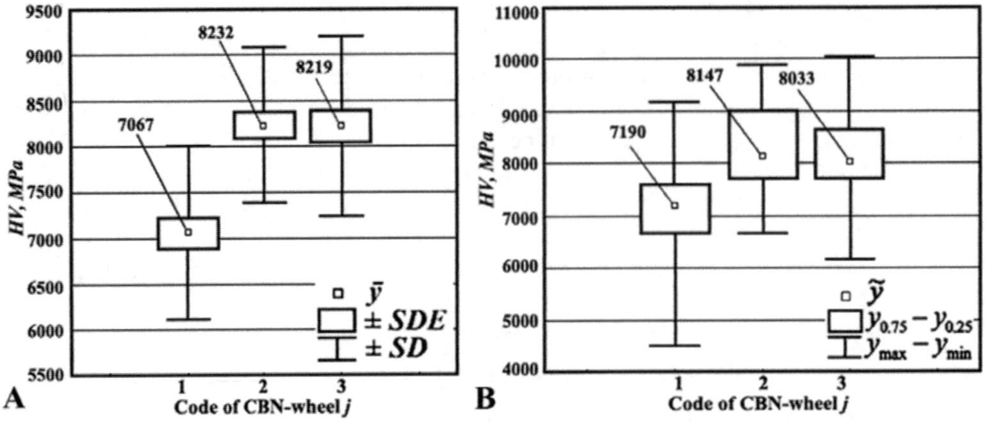

Figure 1. Parametric (A) and nonparametric (B) statistics of wheel characteristic influence to *HV* parameter

First, let us analyze the results for microhardness position measures. As seen from table 2, the wheels e = 2,3 showed the best W6Mo5 steel grinding results with a minimum hardness loss (HV_o = 9000 MPa) for expected medians of *HV* parameter. At the same time, the wheel showed the worst result.

Nevertheless, there is a positive asymmetry of distributions for LKV50 B107 (e = 2) and CBN50 B126 HPCBNW. It has a negative impact on process reliability as it reduces position measures. The asymmetry is negative for CBN50 B107 wheels. However, the worst is when cases (1) are located below the medians. For example, while CBN50 B107 wheel grinding (e = 1), 26 cases from n = 30 turned out to be smaller than an experiment median with the same n = 30, the majority of cases turned out to be bigger than experiment medians - 20 with $\tilde{HV}_2 8147$ and 19 with $\tilde{HV}_3 8033$ for LKV50 B107 HPCBNW respectively (figure 1, table 2).

Wheel grinding stability analysis e = 1,...3 showed that reproducibility ratios (7), (8) coincided (table 2). Wheel LKV50 (e = 2) demonstrated the highest level of stability of deviation standards and ranges. The lowest level of stability of deviations standards was found in *CBN*50 160/125 wheel (e = 3). The lowest level of stability of ranges was found in *CBN*50 100/80 wheel (e = 1).

According to our experiment, grinding wheel LKV50 B107 O V K27 100 KF40 (e = 2) is preferable.

Table 2. Prediction results and wheel influence on position and dispersion measures for *HV* parameter

Grinding wheels	$\hat{y}_{e\bullet}$	$m\hat{y}_e$	SD_e	$As_e > 0$	$(K_{SD})_e$	y_{emin}	y_{emax}	K_{Re}
	[MPa]			(2)	(7)	[MPa]		(8)
CBN50 B107 (e = 1)	7067.34	7190.38	949.39	–	1.00	4515.60	9158.20	1.00
LKV50 B107 (e = 2)	8226.09	8090.31	831.74	+	1.14	6646.62	9902.50	1.42
CBN50 B126 (e = 3)	8226.09	8090.31	985.94	+	0.96	6158.22	10055.00	1.19

5. Conclusion

1. The rationale of grinding high-alloyed tool steels using HPCBNW instead of traditional abrasives has been proved.

IOP Conf. Series: Materials Science and Engineering **91** (2015) 012059 doi:10.1088/1757-899X/91/1/012059

2. It has been proved that the cutting ability of abrasive wheels in assumed technological conditions should be estimated using non-parametric statistical method «On the home court».
3. CBN-grains influence on the process stability has been established. CBN50 wheels have lower reproducibility coefficients than LKV50 wheels.
4. LKV50 B107 O V K27 100 KF40 has shown the best results according to position measures and precision of microhardness when grinding HSS plate W6Mo5.

References

[1] Kremen Z I and Baboshkin A F 2007 *Grinding technology in mechanical engeneering* (St. Petersburg: Politehnika)
[2] Zubarev Yu M and Priemyshev A V 2010 *Theory and practice of increasing grinding efficiency of materials* (St. Petersburg: Lan')
[3] Kremnev L S 2005 Analysis of grinding wheels influence on a structure of the high speed steel W6Mo5 when forming tools *Russian Engineering Research* **9** 29-31
[4] Hollander M and Wolfe D A 1999 *Nonparametric statistical methods, Second Edition* (New-York: Wiley-Interscience)
[5] Sachs L 1976 *Statistical estimation* (Moscow: Statistica)
[6] Soler Ya I and Prokop'eva A V 2009 Information base for regulating the grinding of high-speed cutting plates of variable rigidity by a cubic-boron-nitride tool *Russian Engineering Research* **29** 9 955-956

IOP Conf. Series: Materials Science and Engineering **91** (2015) 012060 doi:10.1088/1757-899X/91/1/012060

Estimating the cutting force when skiving with a radius cutter

A V Filippov [1,2] **and E O Filippova**[1]

[1] Tomsk polytechnic university, Tomsk, Lenina av. 30, Russia
[2] Institute of strenght physics and materials sciencs SB RAS, Tomsk, per. Academicheskii 2/4, Russia

E-mail: andrey.v.filippov@yandex.ru

Abstract. The paper considers the method of determining the components of the cutting force under cutting completed with a radius cutter. The authors provide the design of the experimental study. The authors provide the data on the cutting force change in the process of turning with a radius cutter according to the cutting edge inclination, cutting depth and supply obtained experimentally and calculated analytically. The paper also provides the results of experimental work approximation and theoretical error checking related to experimental data.

1. Introduction

Theoretical estimation of cutting force components is an important problem for the study and analysis of machining process. The complexity of finding the theoretical values is determined by the wide ranges of initial parameters (tool geometry, cutting conditions, properties of the machined and tool material) which have various influence upon the force dependences in terms of character and value.

One of the promising methods of machining is cutting with radius cutters. Such tools are known [1-3] to ensure high productivity of machining and high quality of the surface of machined item. At the same time the problem of tool geometry and setting influence upon the force parameters remains understudied. Radius cutters are traditionally applied as rotary cutting tool with driven cutter or as auto rotary tool which rotation is caused by friction [3].

In his work [3] V.F. Bobrov notes that cutting with auto rotary cup-tip (radius) tools is often accompanied by vibrations due to the bearing clearance which allows rotation of the tool cutting part. That is why it is more preferable to consider the case of machining with non-rotary cutters which have more rigid construction and are less liable to vibrations.

Force dependences are an important parameter of machining as they determine the possibility of achieving the required accuracy of nonrigid axles. Currently the theoretical analysis of machining accuracy based on estimation of elastic displacement of the workpiece influenced by the cutting force is regarded as generally accepted. In the works of V.V. Podporkin [4], A.A. Matalin [5], V.S. Korsakov [6] the thrust force Py is considered as the force which determines this displacement. K.S. Kolev [7], in his turn, to analyze the accuracy of machining suggests considering the bending force of cutting which is equivalent from radial (Py) and tangential (Pz) components. In general the formula for estimating the value of elastic displacement of the workpiece is as follows:

$$y=\frac{P\cdot L^3}{k\cdot E\cdot J}, \qquad (1)$$

[1] Corresponding author: A V Filippov - andrey.v.filippov@yandex.ru.

where k – coefficient taking into consideration the method of workpiece setting,

P – the force effecting the workpiece,

E – elastic modulus of the workpiece material,

J – second moment of area of the workpiece,

L – the distance from the fixed part of the workpiece to the place of bending force application.

During the process of cutting dynamic change of the cutting force occurs due to the nonuniformity of properties of the workpiece material, variations of tolerance and other dynamically changing processes. Despite all these, equation (1), taking only the static displacement of the workpiece allows estimating the possible accuracy of machining under the known value of the cutting force.

Relevance of the given work is in the lack of data on the impact of supply, cutting depth and cutting edge inclination upon the force dependences under turning with radius cutters with the blade which does not overtravel. Besides, the known works do not provide data for the theoretical estimation of the force components.

2. Results and discussions

Figure 1 presents the scheme of oblique turning with radius cutter. The cutting edge of the tool cut off a layer of material which is formed by three intervals: AB – the working area of the cutting edge, AC – belongs to the cut surface, CB – belongs to the machined surface. The working area of the cutting edge is placed below the axis of rotation of the workpiece which makes a considerable difference from the traditional method of machining with single point tools. As a result the cutting conditions and the geometry of the tool undergo a considerable change. The method of estimation of cut off layer cross-section parameters and operating length of the cutting edge is described in work [8].

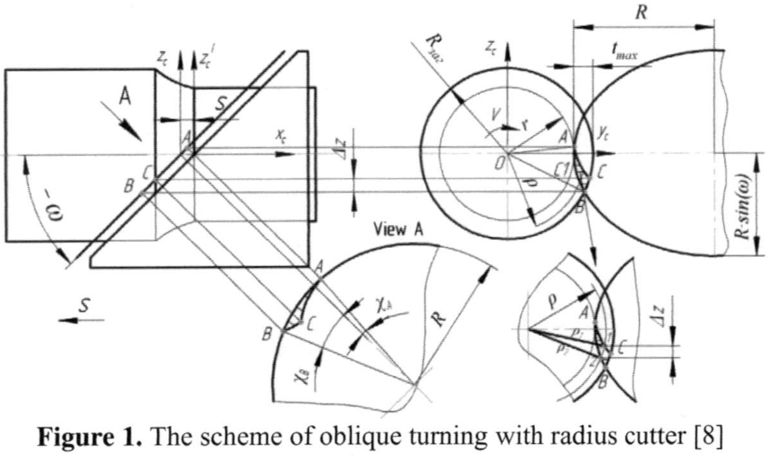

Figure 1. The scheme of oblique turning with radius cutter [8]

In the theory of cutting the components of the cutting force are found through correction factors taking various variables into consideration: work conditions, cutting conditions, workpiece and tool material [4]. Such factors are chosen on the base of experimental results and are attributed to some value of the considered parameter (for example, to the material of the tool cutting part). At the same time these factors cannot be attributed to the whole range of the studied parameters alteration.

In relation to the above-mentioned the given work suggests applying the approach based on summing up the spesific components of the cutting force for estimation of the cutting force components. The essence of the approach can be described as follows: the cross-section of the cut off layer considered as projection on the front surface of the tool is divided into surface elements with an equal interval Δ_i and the thickness value a_i of the cut layer is estimated for the given surface element (see Figure 2). Thickness, according to the common approach is determined normally in the considered point of the cutting edge. For the singled out element of the cutting edge the scheme of free cutting is accepted when the cutting edge is effected only by two components of the cutting force Pz_i

and Py_i. Then it is enough to study the impact of tool geometry and thickness of the cut layer upon the specific force dependences under free cutting than complete the studies for all possible combinations of the impact of considered machining parameters.

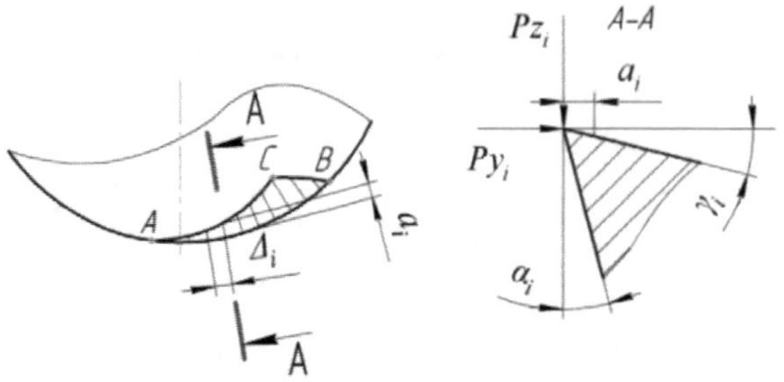

Figure 2. The scheme of the way specific components of the cutting force impact the cutting edge of the radius cutter

According to the chosen approach we completed experimental study of the impact the front and the clearance angles of the tool under various width of the cut layer have upon the specific components of the cutting force Pz and Py. During the experiment the front angle was varied within (-45...35)°, the clearance angle – within (2…40)° and the thickness of the cut layer altered from 0,035mm to 0,115mm. These ranges were obtained by means of analysis of the tool cutting edge geometry parameters and of the cut layer cross-section parameters typical for turning with radius cutters [8-11]. After the experiments the obtained dependences were approximated with division into the positive and the negative ranges of front angle change. Such division into ranges was completed to achieve better accuracy and reliability of results approximation.

Below we present formulas for calculating the specific components of the cutting force Pz_i and Py_i depending upon the thickness of the cut layer and front and clearance angles of the tool.

To estimate the specific force Pz_i:

a) within the positive range γ:

$$Pz_i = k(\gamma)\cdot a\cdot k(\alpha), \qquad (2)$$

where $k(\gamma)$ – coefficient of the front angle impact:

$k(\gamma) = 2315.6 - 18.02\cdot\gamma$;

$k(\alpha)$ – coefficient of the clearance angle impact:

$k(\alpha) = 0.0001\cdot\alpha^2 - 0.0079\cdot\alpha + 1.0512$;

a – thickness of the cut layer.

b) within the negative range γ:

$$Pz_i = k(\gamma)\cdot a\cdot k(\alpha), \qquad (3)$$

where $k(\gamma)$ – coefficient of the front angle impact:

$k(\gamma) = 24.101\cdot\gamma + 2081.8$;

$k(\alpha)$ – coefficient of the clearance angle impact:

$k(\alpha) = 0.0003\cdot\alpha^2 - 0.0153\cdot\alpha + 1.1028$;

a – thickness of the cut layer.

For estimation the specific force Py_i:

a) within the positive range γ:

$$Py_i = (-A\cdot a^2 + B\cdot a)\cdot k(\alpha)\cdot k(\lambda), \qquad (4)$$

where A and B – coefficients of the front angle impact:

$A = 1.6608\cdot\gamma^2 + 0.3345\cdot\gamma + 3759.1$;

$B = 0.093\cdot\gamma^2 - 21.904\cdot\gamma + 1769.2$;

$k(\alpha)$ – coefficient of the clearance angle impact:

IOP Conf. Series: Materials Science and Engineering **91** (2015) 012060 doi:10.1088/1757-899X/91/1/012060

$k(\alpha) = 1.196 - 0.0142 \cdot \alpha$;

k(λ) – coefficient of cutting edge inclination impact:

$k(\lambda) = -0.0005 \cdot \lambda^2 + 0.037 \cdot \lambda + 0.7577$ within the positive range of angle alteration λ – (30…60)°;

$k(\lambda) = 0.0375 \cdot \lambda + 2.9772$ within the negative range of angle alteration угла λ – (-30…-60)°;

a – thickness of the cut layer.

b) within the negative range γ:

$$Py_i = A \cdot a^B \cdot k(\alpha) \cdot k(\lambda). \quad (5)$$

where A and B – coefficients of the front angle impact:

$A = -0.0022 \cdot \gamma^4 - 0.085 \cdot \gamma^3 + 15.05 \cdot \gamma^2 - 103.23 \cdot \gamma + 835$;

$B = 8E - 07 \cdot \gamma^4 - 1E - 04 \cdot \gamma^3 + 0.0033 \cdot \gamma^2 - 0.0109 \cdot \gamma + 0.7222$;

k(α) – coefficient of the clearance angle impact:

$k(\alpha) = 1.126 - 0.0196 \cdot \alpha$;

k(λ) – coefficient of the cutting edge inclination impact:

$k(\lambda) = -0.0005 \cdot \lambda^2 + 0.037 \cdot \lambda + 0.7577$ within the positive range of angle change λ – (30…60)°;

$k(\lambda) = 0.0375 \cdot \lambda + 2.9772$ within the negative range of angle change угла λ – (-30…-60)°;

a – thickness of the cut layer.

To test the obtained calculating formulas (2-5) we completed the experimental study of the impact produced by supply, cutting depth, cutting edge radius and cutting edge inclination upon the cutting force components Pz и Py. Figure 3 shows the skiving radius cutter developed for the experimental study. The suggested design allows changing the cutting edge inclination and applying replaceable indexable inserts.

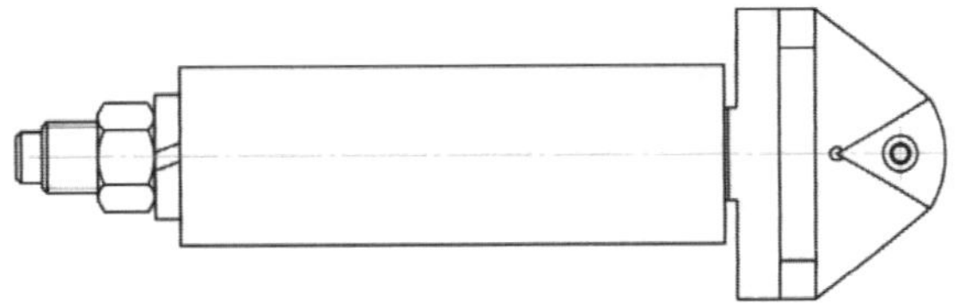

Figure 3 Skiving radius cutter

Figure 4 shows the design of the experimental device. For the experimental study we used axels 38mm in diameter from 45 steel which were previously turned to obtain the necessary diameter. The cutting speed for all the experiments was 100 m/min.

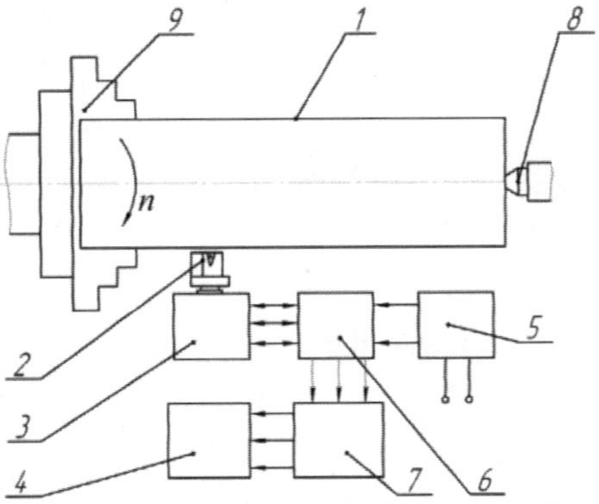

Figure 4 The experimental device for measuring the cutting force components:

1 – workpiece, 2 – radius cutter, 3 – turning dynamometer, 4 – personal computer, 5 – direct-current power supply, 6 – strain-gauge station, 7 – ADC (analog-digital converter) USB 3000, 8 – turning center, 9 – lathe chuck

Figures 5-7 present dependences of the cutting force components P_z and P_y upon the cutting edge inclination, supply and cutting depth estimated (the dashed line) according to the formulas (2-5) with insertion of corresponding coefficients and found experimentally (continuous line).

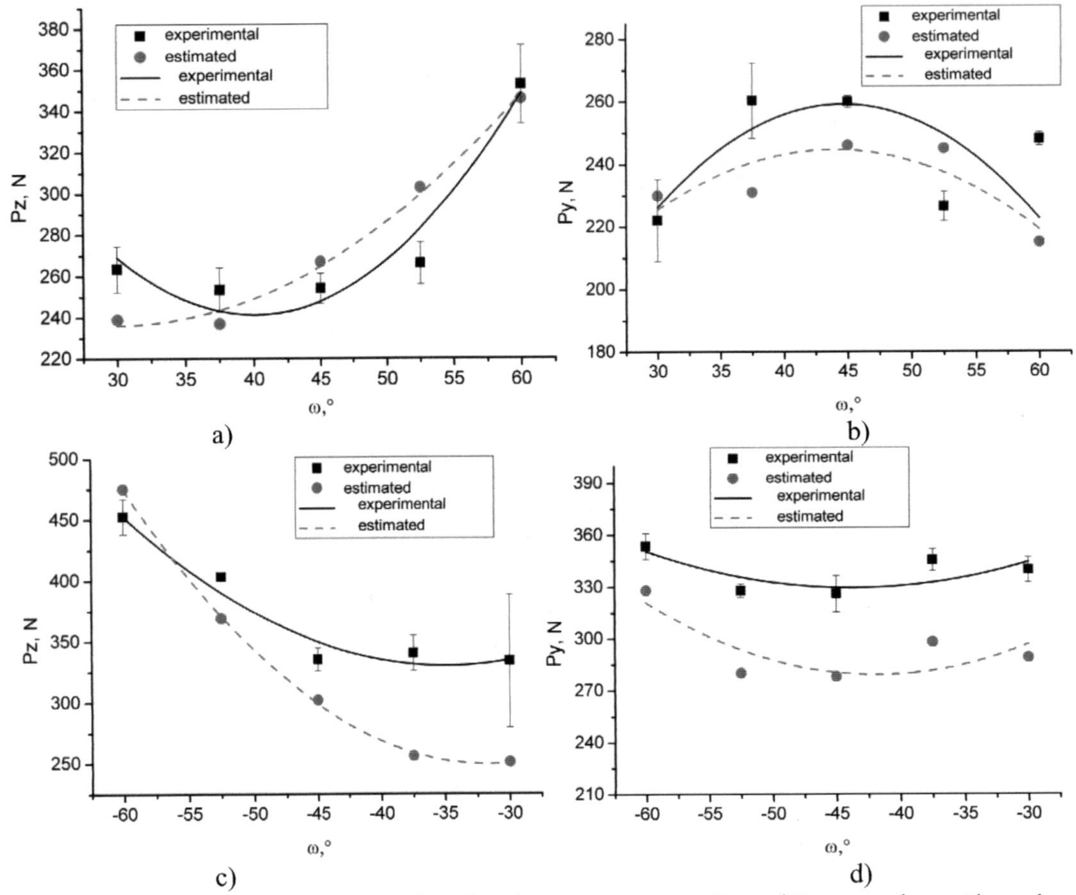

Figure 5. Dependence of changes of cutting force components Pz and Py upon the cutting edge inclination under cutting according to reverse (a, b) and direct (c, d) schemes

As we can see from the given graphic charts (see Figure 5) the cutting force components Pz and Py change differently. Increase of angle ω from 30° to 45° under cutting according to the reverse scheme does not cause significant change of component Pz. Similar situation is observed under cutting according to direct scheme. Further increase of angle ω from 45° to 60° is accompanied by significant growth of component Pz, both for direct and for reverse schemes. This kind of dependence Pz=f(ω) is associated with the peculiarities of changes of cut layer cross-section parameters and parameters of the operating length of the cutting edge [8]. Under the values ω=30°...45° insignificant increase of cut layer thickness and decrease of operating length of the cutting edge occurs.

Changing of component Py due to the front angle change does not demonstrate any unique dependence for designs under consideration. Py within the range of angle ω changes by 20...40N which is not significant under the given parameters of machining. Thus, change of angle ω does not have significant impact upon the value of component Py.

The obtained calculation dependences correspond well to the experimental ones. At the same time we observe some disarrangement of their value within the range of angle ω from -30° to -45° for the direct scheme of cutting. This must result from the conditions of the cut layer deformation [12-14] when under small thickness of the cut layer cutting is completed within the radius of the cutting edge with negative values of the front angle.

Changing the scheme of cutting from the direct to the reverse one contributes to the decrease of the values of cutting force components Pz and Py. It is related to the change of the conditions of cutting the layer of material which are expressed in the free passage of the cutting waste along the front surface of the tool under cutting according to reverse scheme and positive range of front angle changing.

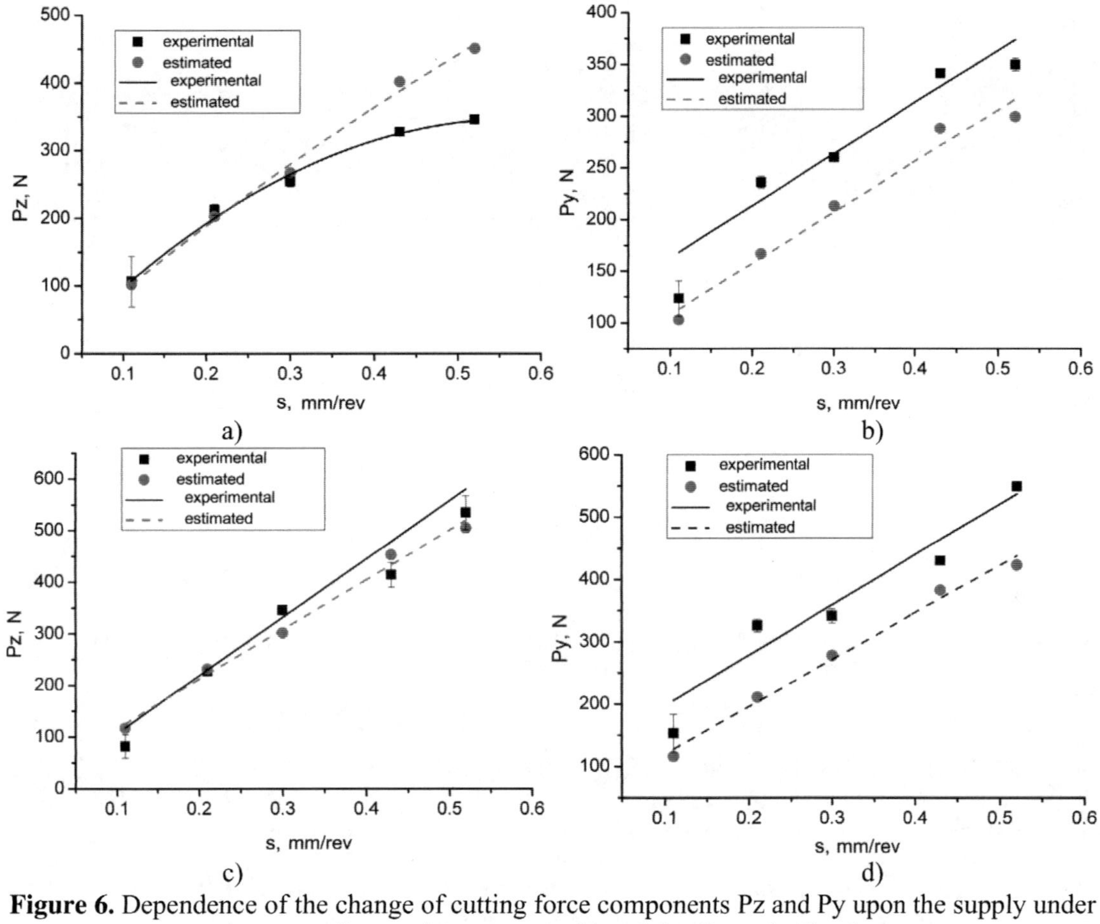

Figure 6. Dependence of the change of cutting force components Pz and Py upon the supply under cutting according to the reverse (a, b) and the direct (c, d) schemes

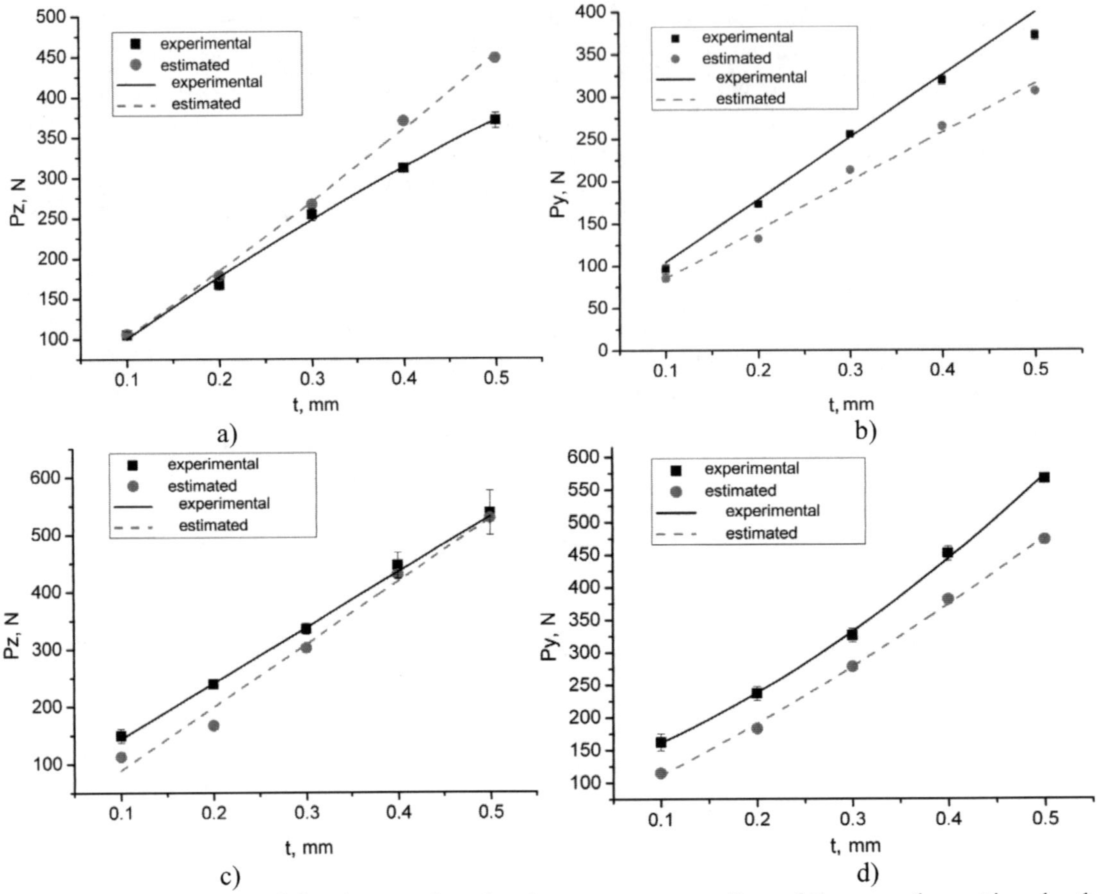

Figure 7. Dependence of the change of cutting force components Pz and Py upon the cutting depth under cutting according to the reverse (a, b) and the direct (c, d) schemes

According to the obtained graphic charts (see Figures 6 and 7) increase of supply and cutting depth leads to almost linear increase of cutting force components Pz and Py. It is a natural phenomenon as with the increase of supply and cutting depth the thickness of the cut layer and operating length of the cutting edge also grow. Changing the scheme of cutting from the direct into the reverse one leads to the decrease of the values of cutting force components while the obtained dependences upon cutting depth and supply remain constant. The calculated dependences Pz=f(s), Pz=f(t), Py=f(s), Py=f(t) obtained according to formulas (2-5) agree well to the experimental results.

3. Conclusion

Comparison of the obtained dependences of cutting force components upon supply, cutting depth and cutting edge inclination allows making an important conclusion. From the point of view of the energy consumption of the process, machining according to the reverse scheme will be more efficient than machining according to the direct one. The given effect is achieved due to the fact that with the five-time increase of the cutting conditions (supply and/or cutting depth) under cutting according to the reverse scheme the value of the cutting force components will be two-time less than when working according to the direct scheme, thus, the machine tool will have to perform less work during the machining process.

To estimate the validity of the method for calculating the cutting forces by the specific components of the cutting force we calculated the amount of average relative error of theoretical values from the experimental ones and obtained the following results:

IOP Conf. Series: Materials Science and Engineering **91** (2015) 012060 doi:10.1088/1757-899X/91/1/012060

1. Average relative error of component Pz calculation for the reverse scheme made 10.3% and for the direct one – 14.3% under the correlation ratio about 0.96.

2. Average relative error of component Py for the reverse scheme made 15% and for the direct one – 18.3% under the correlation ratio about 0.95.

3. The high level of correlation between the theoretical and experimental dependences proves the agreement of the obtained dependences at the qualitative level.

So we can consider the application of the presented method appropriate for estimating the cutting force under cutting with radius cutters.

References

[1] Bobrov V F 1962 *Influence of the inclination angle of the main cutting edge upon the process of metal cutting* (Moscow: Mashgiz)
[2] Grzesik W and Żak K 2011 *Archives of Mat. Sci. and Eng.* **52** 46-53
[3] Bobrov V F 1972 *Metal-cutting with auto rotary cutters* (Moscow: Mashinostroyeniye)
[4] Podporkin V V 1959 *Machining of nonrigid parts* (Moscow: Mashgiz)
[5] Matalin A A 1977 *Accuracy of machining* (Leningrad: Mashinostroyeniye)
[6] Korsakov V S 1961 *Accuracy of machining* (Moscow: Mashgiz)
[7] Kolev K S 1976 *Accuracy of machining and cutting conditions* (Moscow: Mashinostroyeniye)
[8] Filippov A V 2014 *Russ. Eng. Res.* **35** 385-388
[9] Petrushin S I and Filippov A V 2013 *Machining of met. Tech. Equip. Tools.* **2** 8-14
[10] Filippov A V 2013 *Appl. Mech. and Mat.* **379** 139-144
[11] Filippov A V 2014 *Russ. Eng. Res.* **34** 718-721
[12] Filippov A V 2014 Bull. of MSTU N.E. Bauman. Ser. "Mashinostroyeniye" **2** 100-113
[13] Fillipov A V, Proskokov A V and Verbitskaya O Yu 2013 *Nauchnoye obozreniye* **5** 53-56
[14] Filippov A V and Gorbatenko V V 2014 *Appl. Mech. and Mat.* **682** 525-529

IOP Conf. Series: Materials Science and Engineering **91** (2015) 012061 doi:10.1088/1757-899X/91/1/012061

Experimental estimation of chip shrinkage under cup-tip cutting with straight and radius cutters

E O Filippova[1] and A V Filippov [1,2]

[1] Tomsk polytechnic university, Tomsk, Lenin av. 30, Russia
[2] Institute of strenght physics and materials sciencs SB RAS, Tomsk, per. Academicheskii 2/4, Russia

E-mail: andrey.v.filippov@yandex.ru

Abstract. In the given work we present the results of experimental estimation of chip shrinkage under turning with straight and radius cutters for two schemes of cutting: "direct" and "reverse". Ratios of thickening, broadening and longitudinal shrinkage of the chip were established as shrinkage parameters. The given paper presents the graphic charts showing dependences of shrinkage upon cutting depth, supply and cutting edge inclination which indicate the deformations taking place in the cutting area when cutting with cup-tip cutters. The authors show the significant difference in the character and degree of chip shrinkage when working according to various schemes of cutting.

1. Introduction

Monitoring of the chip formation process is a convenient means of monitoring the machining process [1-6] as the chip allows making conclusions about such parameters as stability of chip formation, temperature and deformation in the cutting area. To determine the temperature in the area of tool-workpiece contact the experts estimate the oxidizing colors of the chip and to estimate the deformation they study the chip shrinkage [5, 6]. Such approach allows fast estimation of the processes taking place in the cutting area

It is traditional to estimate the chip shrinkage along longitudinal and transverse directions. In the first case the longitudinal shrinkage is estimated as ratio of chip length to the cutting length and in the second transverse shrinkage (in terms of thickness and width) is estimated as ratio of thickness and width of chip to thickness and width of the cross-section of the cut layer of material.

Cutting with cup-tip cutters is a promising method of mechanical cutting of bodies of rotation. Its characteristics were studied by the authors of works [7-13] where they established that the given method is characterized by high productivity, accuracy and quality of machined surface. But the influence of cutting conditions, tool geometry and form upon the deformations of the cut layer of material is not studied enough. Due to the given fact the aim of this work is experimental estimation of dependences of chip shrinkage when the chip is produced by cutting with straight and radius cup-tip cutters, upon the cutting conditions (supply and depth) and cutting edge inclination.

1 Corresponding author: A V Filippov -andrey.v.filippov@yandex.ru.

IOP Conf. Series: Materials Science and Engineering **91** (2015) 012061 doi:10.1088/1757-899X/91/1/012061

2. Research method

The study was completed for cutting according two schemes (Figure 1): "direct" – the operating area of the cutting edge (lp) is below the axis of the workpiece rotation (see Fig 1a) and "reverse" – lp is above the axis of the workpiece rotation. The material of the workpieces is 45 steel.

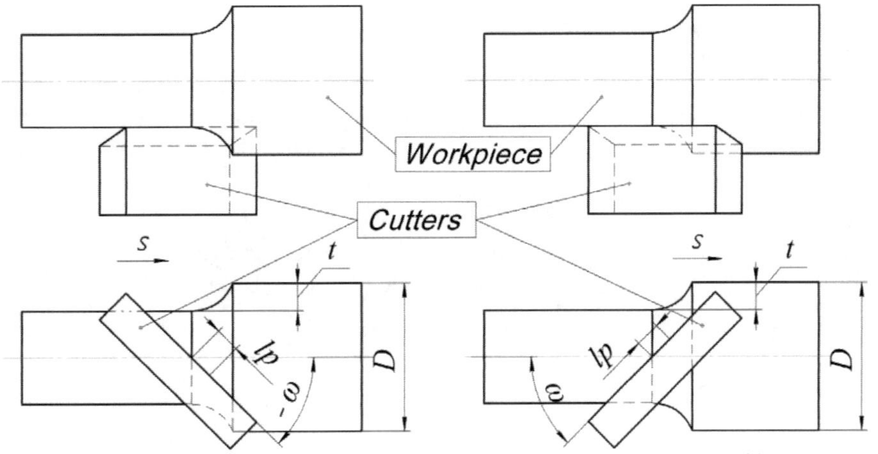

Figure 1. "Direct" (a) and "reverse" (b) schemes of cutting with radius cutters

For the experiments we used cup-tip cutters with straight and radius cutting edges (Figure 2) equipped with carbide triangle blades. The blades were sharpened along top and face surfaces.

Works [14, 15] noted that an important characteristic of cutting with cup-tip cutters is negative clearance angle under a certain combination of cutting edge geometry, parameters of tool setting and cutting conditions with consideration to the chosen scheme of cutting ("direct" or "reverse", see Fig. 1). Basing on this fact and according to the method presented in the work [14] the values of the tool clearance angle were estimated, values which ensure operability of the tool within the studied range of supply, cutting depth and cutting edge inclination for the considered diameter of the workpiece. Basing on the said above we obtained the value of clearance angle $\alpha_и$=7,5° for the "direct" scheme and $\alpha_и$=25° for the "reverse". The value of the face angle for all blades is $\gamma_и$=0°.

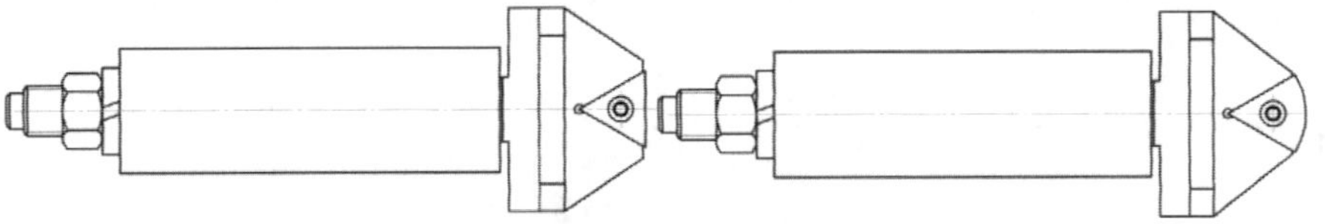

Figure 2. Cup-tip cutters with straight and radius blade

To estimate the longitudinal shrinkage of the chip we applied the gravimetric method [5]:

$$\xi_L = \frac{G}{L_1 \cdot \rho \cdot S}$$

where G – weight of the chip;

L_1 – length of the chip;

ρ – density of the machined material;

S – cross-section of the cut layer of the material.

The weighing machine ADAM HCB 302 with the accuracy of up to 0.01 was used to estimate the weight of the chip and its square s was estimated according to the data provided in works [16, 17].

Transverse chip deformation was found according to the well-known formulas [5]:

1) ratio of the chip thickening –

where a – maximum cross-section thickness of the cut layer; a1 – chip thickness.

2) chip broadening ratio –

where b – maximum cross-section thickness of the cut layer; b1 – width of the chip.

Thickness and width of the chip were estimated by the microphotographs of chip cross-section obtained with the microscope LabotMet-1.

3. Main results and discussions

In the given part we present the dependences of chip shrinkage ratios ξ_L, ξ_a, ξ_b upon the cutting depth, supply and cutting edge inclination obtained when considering "direct" and "reverse" schemes of cutting. Figures 3-5 show the dependences for the longitudinal chip shrinkage.

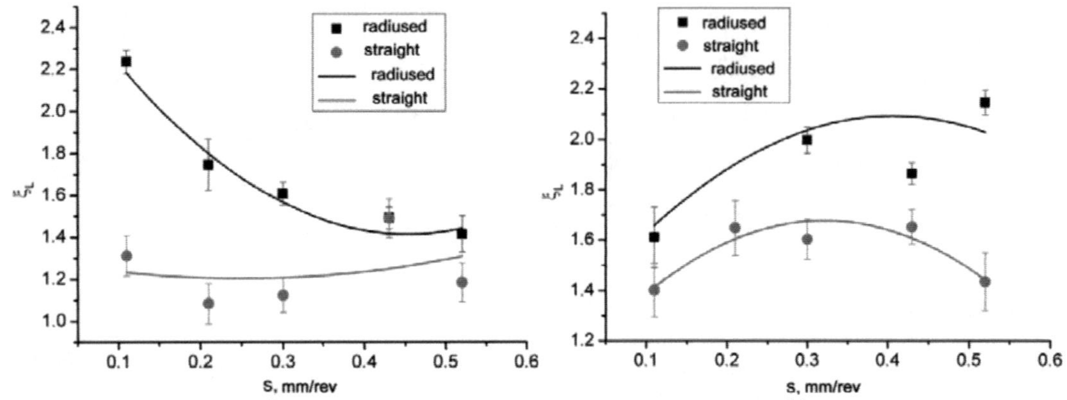

a) b)

Figure 3. Dependence of chip longitudinal shrinkage changing upon the supply for "reverse" (a) and "direct" (b) schemes of cutting: t=0.3mm, D=38mm

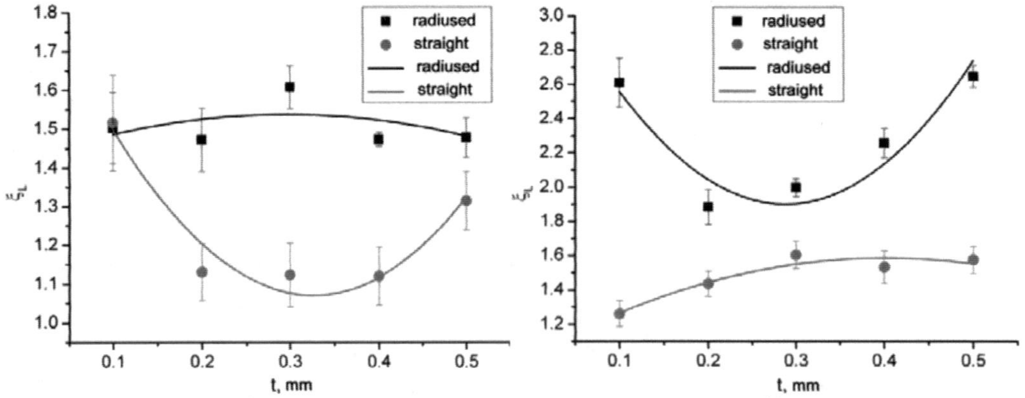

a) b)

Figure 4. Dependence of chip longitudinal shrinkage changing upon the cutting depth for "reverse" (a) and "direct" (b) schemes of cutting: s=0.3mm/rev, D=38mm,

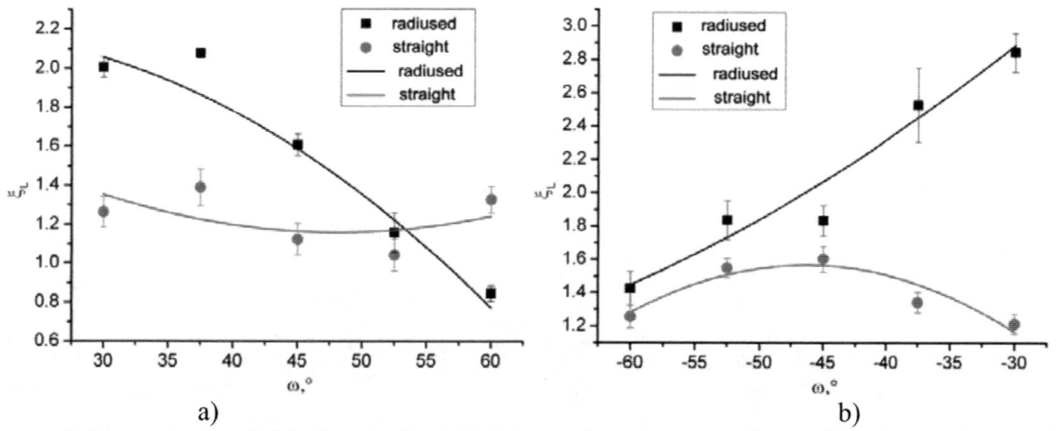

Figure 5. Dependence of chip longitudinal shrinkage changing upon the cutting edge inclination for "reverse" (a) and "direct" (b) schemes of cutting: t=0.3mm, s=0.3mm/rev, D=38mm

The obtained graphic charts allow making a number of conclusions. First, in most considered cases cutting with straight cutters is characterized by less longitudinal shrinkage in comparison to cutting completed with straight cutters. Second, the dependences of the longitudinal shrinkage for the considered schemes of cutting and cutting edge forms have different character. In separate cases we observe extremum, smooth function increase and decrease, but also, for a number of graphic charts we do not observe any obvious influence, for example: $\xi_L=f(\omega)$, $\xi_L=f(t)$ for the radius cutter and "reverse" scheme, . $\xi_L=f(s)$ for the straight cutter and "direct" scheme. Such differences in the character of dependences can occur due to the significant change of the cutting edge geometry and parameters of the cut layer influenced by the considered parameters (t, s, ω) and the chosen scheme of cutting. The last, in its turn, leads to the transformation of the workpiece material deformation and friction in the contact area conditions.

Figures 6-8 show the dependences for the chip thickening ratio ξ_a.

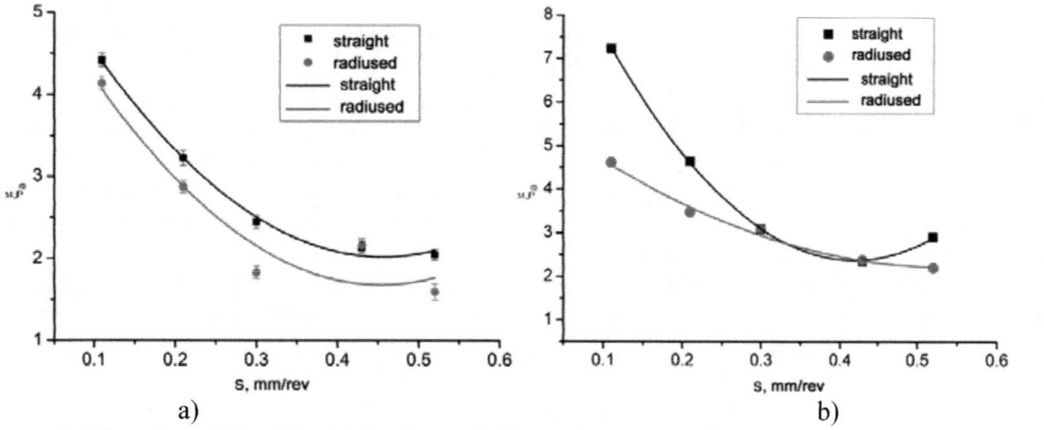

Figure 6. Changing of the chip thickening ratio dependence upon the supply for "reverse" (a) and "direct" (b) schemes of cutting: t=0.3mm, D=38mm

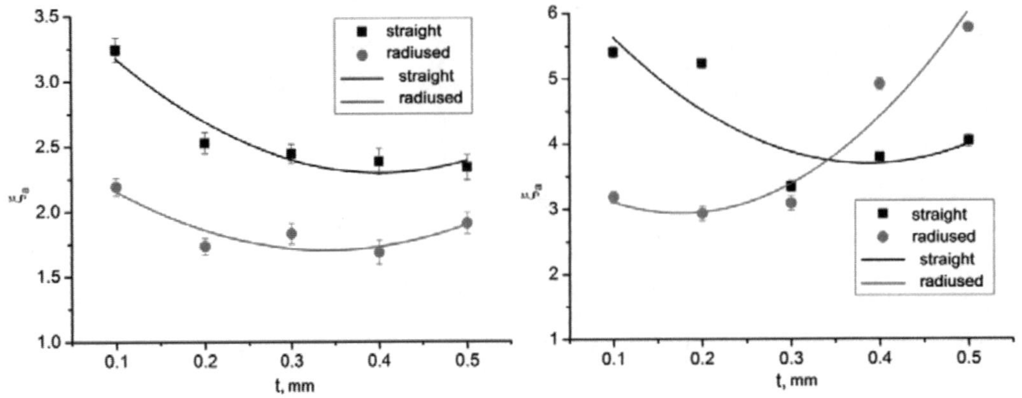

a) b)

Figure 7. Changing of the chip thickening ratio dependence upon the cutting depth for “reverse” (a) and “direct” (b) schemes of cutting: s=0.3mm/rev, D=38mm

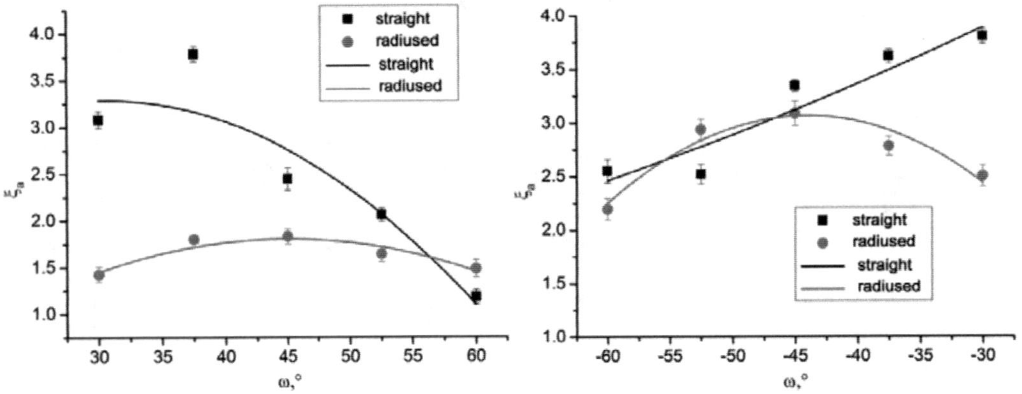

a) b)

Figure 8. Changing of the chip thickening ratio dependence upon the supply for “reverse” (a) and “direct” (b) schemes of cutting: t=0.3mm,s=0.3mm/rev, D=38mm

The data on the influence of the cutting conditions and the cutting edge inclination upon the thickening ratio ξ_a (see Fig. 6-8) prove its more expressed character than the data on longitudinal shrinkage (see Fig. 3-5). For the considered schemes of cutting and tool cutting edges the change of supply produces a similar effect upon ratio ξ_a: with the growth of s up to 0.4 mm/rev transverse shrinkage decreases and goes over the minimum with further increase.

In the case of “reverse” scheme of cutting for the considered tools changing the cutting depth has a similar effect upon ratio ξ_a, but its numerical value for the radius cutting edge is approximately by 1.5 times smaller than that for the straight cutting edge. In the case of “direct” scheme of cutting dependence $\xi_a=f(t)$ for the considered cutting edges is different.

With increase of cutting edge inclination for “direct” and “reverse” schemes ξ_a decreases for the straight cutting edge and for the radius one the dependence has an extremum under $\omega=45^0$.

In general chip thickening when cutting with straight cutters is larger than that when working with radius ones. Supposedly, it is associated with the fact that when cutting with straight cutters under the considered conditions the thickness of the cut layer is significantly smaller than that for radius cutters and its value approaches the corner radius of the tool. As a result of cutting a thin layer of material by the radius part of the cutting edge the material in front of the cutting edge is pressed and the thickness of the chip increases. This process (pressing of material) also contributes to the growth of ξ_a under small values of cutting depth and supply characterized by little thickness of the cut layer.

It is also worth mentioning that chip broadening when cutting according to the "direct" schemes has smaller values and this fact can be associated with the conditions of chip flowing along the front surface of the tool. Due to the fact that under working according to the "reverse" scheme the chip flows towards the previously machined surface the amount of obstructions for the chip flow is less in comparison to the "direct" scheme when the chip flows towards the surface to be machined and the pressing of material in front of the tool is larger.

Figures 9-11 show the dependences for chip broadening ratio ξ_b.

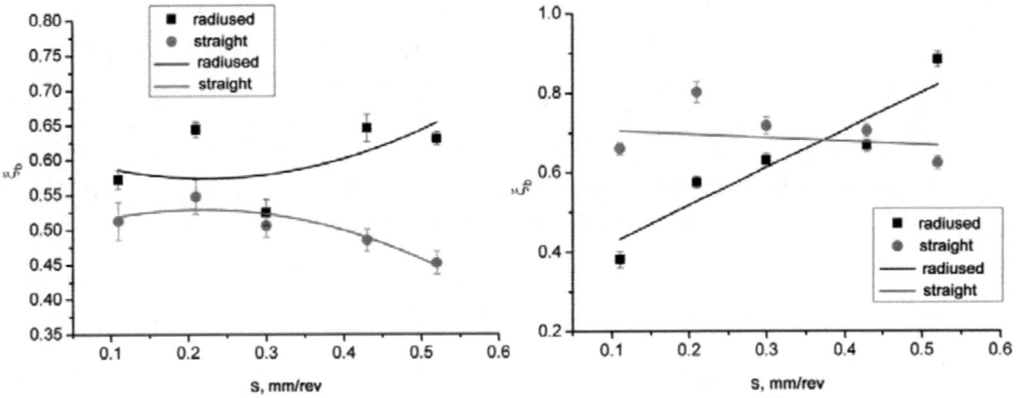

a) b)

Figure 9. Changing of the chip broadening ratio dependence upon the supply for "reverse" (a) and "direct" (b) schemes of cutting: t=0.3mm, D=38mm

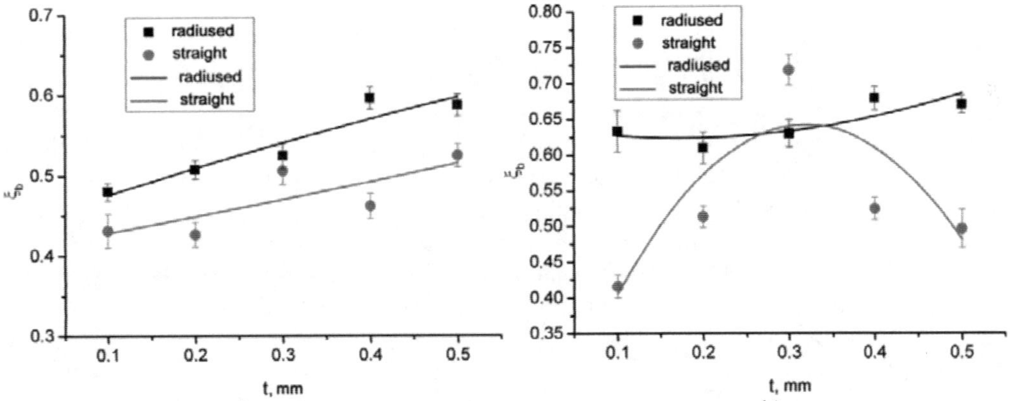

a) b)

Figure 10. Changing of the chip broadening ratio dependence upon the supply for "reverse" (a) and "direct" (b) schemes of cutting: s=0.3mm/rev, D=38mm

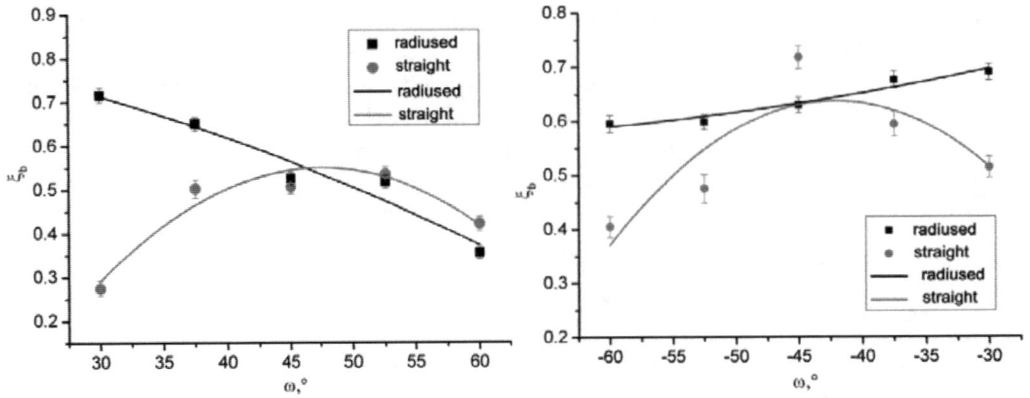

a) b)

Figure 11. Changing of the chip broadening ratio dependence upon the supply for "reverse" (a) and "direct" (b) schemes of cutting: t=0.3mm,s=0.3mm/rev, D=38mm

The obtained data on ξ_b under the considered values of cutting conditions and cutting edge inclination prove that in most cases ξ_b is smaller when cutting with radius cutters than when cutting with straight ones. At the same time in a number of cases there isn't any obvious dependence of ξ_b upon the considered parameters, for example, for the straight cutter $\xi_b=f(t)$ – for the "direct" scheme of cutting and $\xi_b=f(s)$ – for the "reverse" scheme of cutting, for the radius cutter $\xi_b=f(s)$ – for the "direct" scheme. Dependence $\xi_b=f(\omega)$ for the radius cutters has an extremum under ω being about 45^0, and for the straight cutters ξ_b increases almosr linearly with the growth of cutting edge inclination.

4. Conclusion

Basing on the obtained data we can make the following conclusions:

1) An important characteristic of cutting with cup-tip tools is value $\xi_a>1$, $\xi_b<1$, while under traditional cutting with single-point cutters their values are usually $\xi_a<1$, $\xi_b>1$ [4-6]. First, it is conditioned by the thickness of the cut layer which is many times smaller for cup-tip tools than for the single-point ones. Under the little thickness of the cut layer the workpiece material is pressed against the front part of the tool which results in forming chip with larger thickness. Second, the thickness is additionally influenced by kinematic and geometric characteristics of cup-tip cutting which supposes pressing the workpiece material against the inclined front surface of the tool leading to overlying layers lapping the underlying ones. as a result the chip becomes thicker and narrower.

2) Difference in the obtained dependences is associated with the change in the cutting edge geometry and parameters of the cut layer cross-section which will be different in terms of value and character for the straight and radius cutters, besides the geometry parameters demonstrate extra change with transition from the "direct" to the "reverse" scheme of cutting. The previously completed studies show that geometry change (front angle of the tool and cutting edge inclination) have a significant influence upon the conditions of cut material deformation [18, 19].

3) Absence in some cases of obvious dependence of chip shrinkage upon cutting depth, supply, cutting edge inclination results from the change of not only deformation processes and heat generated as a result but also from the size of contact areas which will significantly vary according to the considered parameters. As a result the amount of heat generated in the process of friction will produce additional effect upon the processes in the cutting area.

4) Transverse shrinkage when cutting with radius cutters is smaller than that when working with the straight ones, while the longitudinal is larger. It is conditioned by smaller thickness of the cut layer when cutting with straight cutters. Thus, the process of cutting with straight cutters can be described as the process characterized by pressing of material which leads to strong deformation of the cut layer and changing of the cutting edge geometry.

IOP Conf. Series: Materials Science and Engineering **91** (2015) 012061 doi:10.1088/1757-899X/91/1/012061

References

[1] Armarego E A and Brown R H 1977 *The machining of metals* (Moscow: Mashinostroyeniye)
[2] Merchant M E 1945 *J. of Apl. Phys.* **16** 267-275.
[3] Kronenberg, M. 1954 *Grundzüge der zerspanungslehre: 2nd Ed.* (Berlin: Springer)
[4] Kufarev G L , Okenov K B and Govorukhin V A 1970 *Chip-formation and the quality of the machined surface under restricted cutting* (Frunze: Mektep)
[5] Poletika M F 2001 *Theory of cutting: P.1: Mechanics of the cutting process* (Tomsk: Tomsk polytechnic university)
[6] Rozenberg A M and Eremin A N 1956 *Elements of the theory of metal-cutting process* (Moscow: Mashgiz)
[7] Ryvkin G M and Samoilov B I 1952 *STIN* **4** 20-22.
[8] Podgorkov V V 1974 *STIN* **1** 30-31.
[9] Galoyan G P 1986 *Theoretical basis of the new process of diagonal turning with grounding of the ways of its implementation.* Thesis of the Cand. of techn. sc. (Leninakan: Leninakan polytechnic university)
[10] Minasyan G S 1970 *Bull. of the Academy of Sci. of Armenian SSR.* **V. XXIII. 4** 3-6.
[11] Klimenko S A and Manokhin A S 2009 *Ultra-hard materials* **1** 58-74.
[12] Grzesik W 1991 *Int. J. Mach. Tools Manuf.* **31** 329-344.
[13] Grzesik W and Żak K 2011 *Archives of Mat. Sci. and Eng.* **52** 46-53.
[14] Petrushin S I and Filippov A V 2013 *Machining of metals. Tech. Equip. Tools.* **2** 8-14.
[15] Filippov A V 2013 *Appl. Mech. and Mat.* **379** 139-144.
[16] Filippov A V 2014 *Russ. Eng. Res.* **34** 718-721
[17] Filippov A V 2014 *Russ. Eng. Res.* **35** 385-388
[18] Filippov A V and Gorbatenko V V 2014 *Appl. Mech. and Mat.* **682** 525-529.
[19] Bobrov V F 1962 *Influence of the main cutting edge inclination upon the process of metal cutting* (Moscow: Mashgiz)

Processing of brittle materials in the nanometer range of thickness of layers cut[1]

M A Shavva[1], V V Lapshin[1], S V Grubyy[2]

[1]Russian Research& Development Tooling Institute VNIIINSTRUMENT, 49, Bolshaja Semyonovskaya, Moscow, 107023, Russia
[2]Bauman Moscow State Technical University, 5, Baumanskaya 2-ya, Moscow, 105005, Russia

E-mail: katapu@mail.ru

Abstract. The article deals with the results of experimental research in the ultra-precision diamond machining of brittle materials – potassium dihydrogen phosphate, single-crystal quartz, quartz glass, crystalline glass, sapphire, and germanium. The processing modes at which the nanometer roughness of the processed surface was achieved are specified.

Introduction

The traditional technology to process surfaces of details made of brittle optical materials includes the pregrinding with diamond wheels with the granularity of various granulometric compositions in strict consecutive order from large to small fractions. Then the final chemical and mechanical polishing with classified flours (loose abrasive) is applied. This technology has major deficiencies – rather low production and stability of the process, complexity of automation and control, influence of the polisher's skills on the processing quality.

1. Plasticity of brittle materials

At Russian Research & Development Tooling Institute VNIIINSTRUMENT and Bauman Moscow State Technical University, a complex of technological research and design developments directed at the improvement of performance, accuracy and quality of the processing of brittle optical materials is being performed within the framework of the Agreement for the Federal Target Program. The research is based on the clause relating to the quasi plasticity of brittle materials. The quasi plasticity is the manifestation of plastic properties of a surface layer of brittle materials in some processing modes. When cutting a layer with nanometer thicknesses, the contact interaction of the tool with the surface processed creates the conditions when there is a directed quasi-plastic removal of a surface layer of a material with forming some roughness within a few nanometers and a minimum defective layer.

The first installation specially created to research the processes during diamond grinding in the nanometer range of thickness was designed by Yoshioko [1]. Later the research was continued by T.G.

[1] The article was prepared within the framework of Agreement No. 14.579.21.0042 dated 8/25/2014 (unique identifier: RFMEFI57914X0042) between OAO VNIIINSTRUMENT and the Ministry of Education and Science of the Russian Federation in the subject 'Development of Technology and Equipment for Nanoscale Processing of Optical Materials by Diamond Single-Crystal and Abrasive Tools in the Quasi Plastic Cutting Mode'

IOP Conf. Series: Materials Science and Engineering **91** (2015) 012062 doi:10.1088/1757-899X/91/1/012062

Bifano and T.A. Dow in the eighties of the twentieth century with the PEGASUS machine [1]. As a result of the conducted experiments with a considerable quantity of amorphous glasses, single-crystals and ceramic materials, the authors found the conditions of the brittle-ductile transition when all the materials, irrespective of their hardness and brittleness, transfer from a brittle mode of destruction to the quasi-plastic one during mechanical surface treatment [2].

The diamond machining of brittle optical materials in the quasi-plastic cutting mode can be performed if there are:

- Ultra-precision rigid equipment providing the cutting depth within the limits of 0.5 to 2 μm, blank feed within the limits of 0.1 to 5 μm/rev, cutting kinematics with thickness of the layer cut from a few nanometers to a few tens of nanometers;
- Single-crystal diamond cutters with the corner radius of cutting edges of 30 – 50 nm;
- Diamond wheels on a metal or ceramic holder, the granularity of a diamond-bearing layer of which is 1 – 5 μm.

2. Experimental research

Experimental research was conducted with special ultra-precision test stands for the grinding and milling by a single-cutter diamond head – Fig. 1 and 2. Parameters of the stands are specified in Tables 1 and 2.

The objects of research during the diamond grinding and milling were the blanks of brittle optical materials: single-crystal potassium dihydrogen phosphate (KDP), single-crystal quartz, quartz glass, crystalline glass, single-crystal germanium, and sapphire.

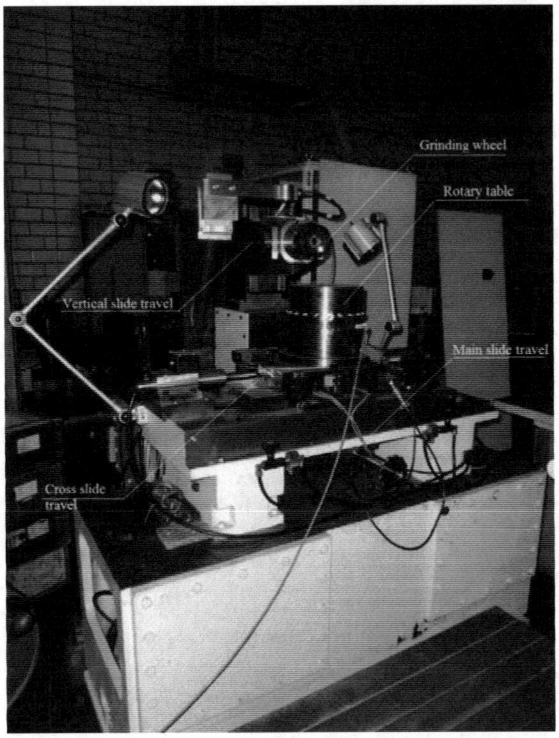

Figure 1. Special ultra-precision grinding stand.

Figure 2. Special ultra-precision milling stand with single-cutter diamond head.

Table 1. Parameters of the ultra-precision grinding stand.

No.	Parameter name	Parameter value
1	Grinding wheel RPM range, min^{-1}	200…6000
2	Rotary table RPM range, min^{-1}	0.01…300
3	Main slide travel, Z axis (penetration slide), mm	100
4	Cross slide travel, X axis, mm	150
5	Vertical slide travel (wheel height adjustment), Y axis, mm	100
6	Rotary table travel (setting motion), B axis, angular degree	20
7	Number of axes, pcs X axis – cross slide; Z axis – main slide (penetration slide); B axis – rotary table; S axis – grinding wheel rotation	4
8	Cutting feed of main slide, mm/min	5…200
9	Movement resolution of cross slide, μm	0.1
10	Movement resolution of main slide, μm	0.1
11	Positioning accuracy of Z and X axes at length of 100 mm, μm	0.5
12	Positioning accuracy of B rotation axis, arcsec	1
13	Power of electric motor of grinding wheel, kW	1

Table 2. Parameters of the ultra-precision stand for diamond milling.

No.	Parameter name	Parameter value
1	Spindle RPM range, min^{-1}	200…6000
2	Main slide travel, Z axis (penetration slide), mm	0.01…300
3	Cross slide travel, X axis, mm	100
4	Number of axes, pcs X axis – cross slide; Z axis – main slide (penetration slide); S axis – grinding wheel rotation	150
5	Cutting feed of main slide, mm/min	100
6	Movement resolution of cross slide, μm	20
7	Movement resolution of main slide, μm	4
8	Positioning accuracy of Z and X axes at length of 100 mm, μm	5…200
9	Power of electric motor of grinding wheel, kW	0.1

To control the processed surface roughness, the NanoFocus® μSurf® system was used. The system is based on the confocal measurement principle with a white-light source. It is an

IOP Conf. Series: Materials Science and Engineering **91** (2015) 012062 doi:10.1088/1757-899X/91/1/012062

autonomous and complete workplace for 3D measurements and is intended for the quality control of surface roughness of up to 1 nm. Triple roughness measurements of the surface were performed; Table 3 includes the processing modes, maximum thickness of the cut layer for the diamond wheel, as well as results of measurement of the processed surface roughness.

Table 3. Diamond grinding modes for brittle optical materials.

	Material processed	Speed of rotation of rotary table, rpm	Blank feed per grinding wheel turnover, µm/rev	Maximum thickness of cut, nm
Potassium dihydrogen phosphate	0.07	33	31	29
Single-crystal quartz	0.02	9.4	8.9	34
Quartz glass	0.05	24	22.6	6
Crystalline glass	0.07	33	31	3
	0.05	24	22.6	3
	0.03	14.1	13.3	3
	0.01	4.7	4.1	3
Sapphire	0.07	33	31	1
	0.05	24	22.6	2
	0.03	14.1	13.3	2

Note: surfaces with roughness less than 10 nm belong to the optical class

3. Results

Fig. 3 shows the processed surface of a sample of potassium dihydrogen phosphate.

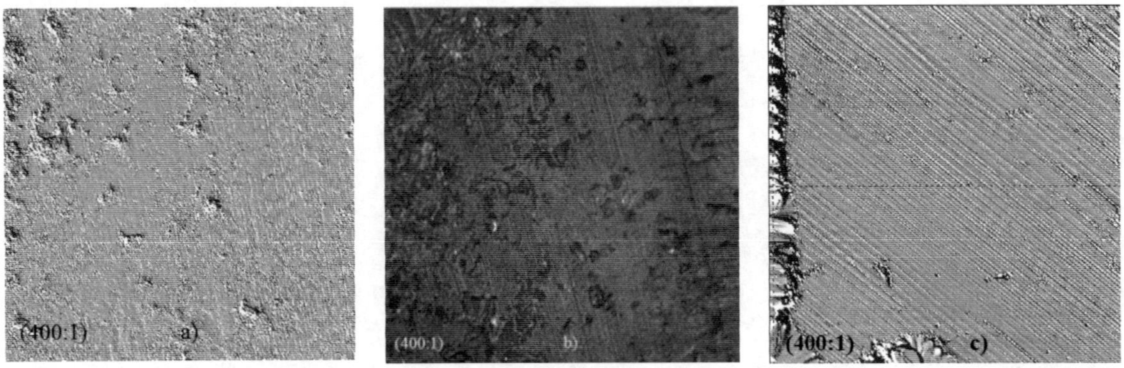

Figure 3. Surfaces of brittle material blanks processed by the method of ultra-precision diamond grinding (400:1 scale): a) potassium dihydrogen phosphate; b) single-crystal quartz; c) quartz glass.

As seen from the figure, during the processing the material was in a borderline state, i.e., in the brittle-ductile transition state. This is evidenced by the simultaneous availability of chips on the processed surface and of feed marks made by the cutting tool. The feed marks are the machining marks on the processed surface in the quasi-plastic mode, which coincide with the motion trajectory of the cutting tool. Since the material in the cutting zone was in a borderline state, the processed surface roughness is rather large – 29 nm. It is possible to assume that the thickness of the cut 31 nm layer does not ensure the full transition of the material into the quasi plasticity state.

Fig. 3-b shows the processed surface of a single-crystal quartz. As in the case with the potassium dihydrogen phosphate processing, the single-crystal quartz also is in the brittle-ductile transition state, therefore the processed surface roughness is 34 nm. Fig. 3-c shows the processed surface of quartz glass. On the processed surface, there are practically no chips and cracks; the feed marks prevail. The surface is processed in the quasi-plastic cutting mode. The processed surface roughness was 6 nm. Similar results were obtained after the diamond grinding of crystalline glass with the 4 μm/rev feed (thickness of cut of 4.1 nm, roughness Ra = 3 nm) – Fig 4-d.

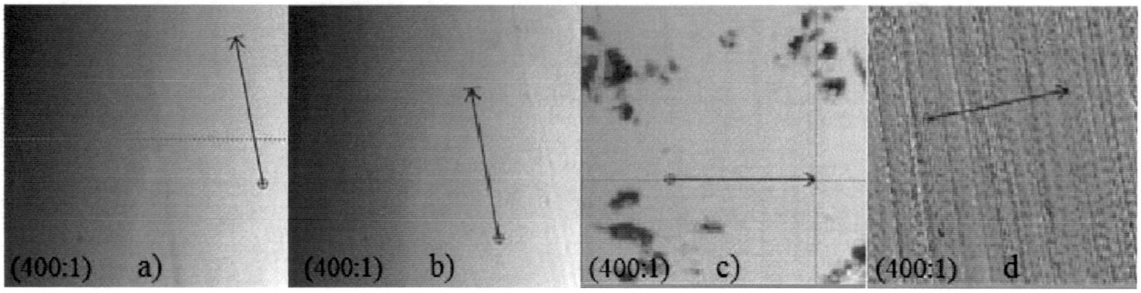

Figure 4. Processed surface of crystalline glass in case of longitudinal feed (250:1 scale): a) – 33 μm/rev; b) – 24 μm/rev; c) – 14.1 μm/rev; d) – 4.7 μm/rev

According to the results of the ultra-precision diamond grinding of various brittle materials, the processed surface roughness was 1 to 34 nm. In case of the grinding in the quasi-plastic cutting mode, the processed surface roughness of quartz glass, crystalline glass, sapphire does not exceed 6 nm, i.e., it belongs to the optical class.

Blanks of single-crystals of potassium dihydrogen phosphate and germanium by a diamond single-crystal tool in the nanometer range of thickness of the layer cut were processed in a special ultra-precision stand with the use of the milling method with a single-cutter head. The parameters of the diamond cutter used in the head, the milling modes and the processed surface roughness are specified in Table 4. The diameter of the single-cutter head is 35 mm. All the experiments were conducted with the spindle rotational speed of 1000 rpm.

Table 4. Diamond milling modes for brittle optical materials

Material processed	Radius of diamond cutter at cutting point, mm	Corner radius of cutting edge, nm	Front rake angle, degree	Back rake angle, degree	Cutting depth, μm	Thickness of cut layer, nm	Number of passes	Ra, nm
Potassium dihydrogen phosphate (KDP)	2					77.4	3	12
	2					31.3	8	16
	2	50	0	-7	1	5.6	3	3
Germanium	3.6					5.6	3	1

Fig. 5 shows photos of surfaces of the brittle materials processed by a single-cutter diamond milling head with various thickness values of the layer cut.

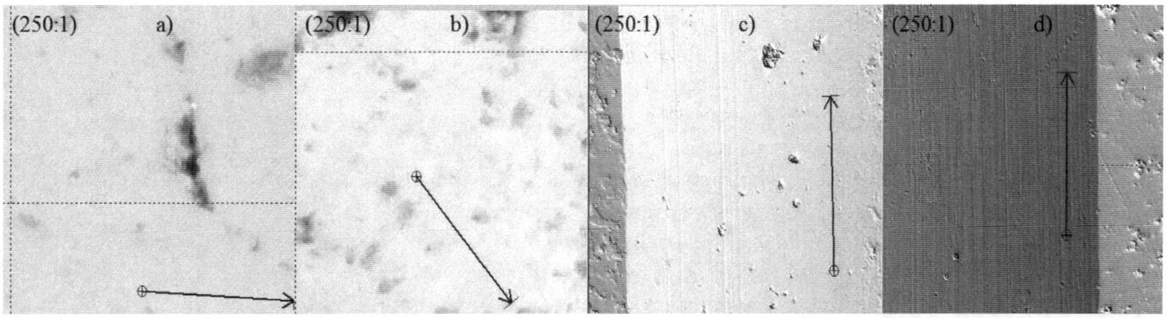

Figure 5. Surfaces of samples after diamond milling with various thickness values of the layer cut (250:1 scale): a) KDP – 77.4 nm; b) KDP – 31.3 nm; c) KDP – 5.6 nm; d) germanium – 5.6 nm

It follows from the photos in Fig. 5 that if the thickness of the layer cut is reduced the feed marks made by the cutting tool are clearly seen on the processed surface, i.e., the surface layer passes into the quasi plasticity state [3]. With reducing the thickness of the layer cut, the processed surface roughness is decreasing. Optical and microelectronic elements of potassium dihydrogen phosphate and germanium can be processed by the diamond single-crystal tool with achieving the optical class roughness [4, 5, 6].

In case of the ultra-precision diamond machining of brittle materials in the nanometer range of thickness of cut, cuttings with a variable nanometer thickness are formed. Appearance of the cuttings is shown in Fig. 6.

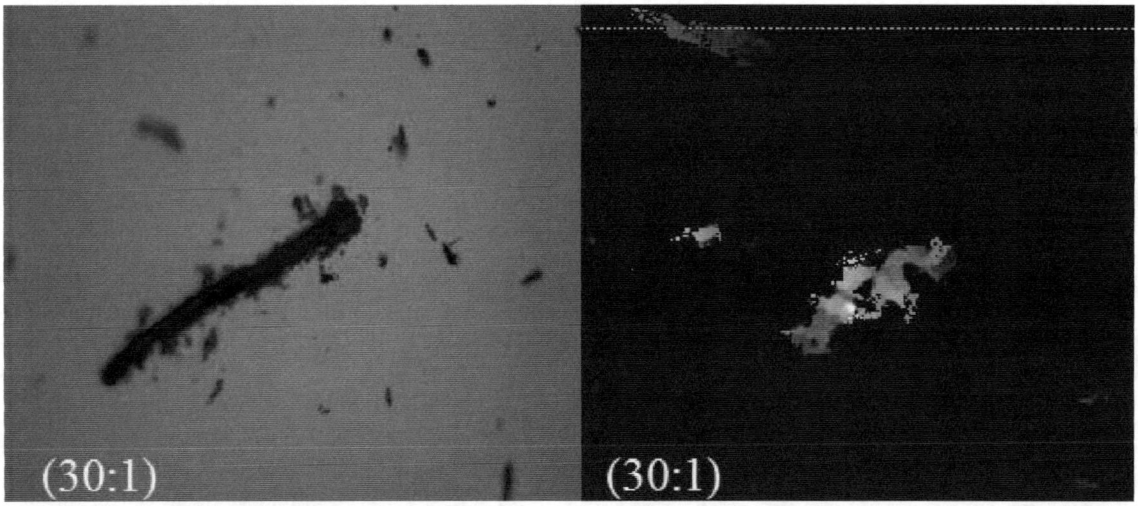

Figure 6. Cuttings after the diamond ultra-precision edge cutting machining of potassium dihydrogen phosphate (30:1 scale)

During the diamond grinding, such cuttings are accumulated in the intergranular space of the diamond-bearing layer, which leads to the glazing of the wheel. It is possible to avoid the glazing of the wheel surface by dressing the wheel. The diamond grinding wheels the granularity of which is less than 5 μm cannot be dressed by a mechanical method for the reason of the impregnation of the diamond-bearing layer with particles of the dressing tool. Since the dressing of the grinding wheel outside the machine tool affects the accuracy of the processed surface and reduces the processing productivity, it is recommended to use continuous electrochemical dressing (ECD dressing) [7].

IOP Conf. Series: Materials Science and Engineering **91** (2015) 012062 doi:10.1088/1757-899X/91/1/012062

According to the results of design studies, an installation was simulated, developed and made to provide the continuous electrochemical dressing of the diamond wheel directly during the ultra-precision diamond grinding of brittle material blanks.

Conclusion.

The conducted experimental research shows that in the cutting modes providing nanometer values of the layer cut, the brittle material surface layer passes into the quasi plasticity state, the surface is formed without chips and cracks, with the roughness of a few nanometers.

To obtain the required quality of the surfaces made of brittle materials: roughness Ra = 1 nm, minimum working error – 10 nm on the area of 100x100 mm, and minimum size of the fissured layer – it is necessary to increase the stiffness and accuracy of the equipment used. Within the framework of the further research, it is planned to create a special ultra-precision stand for the nanoscale processing of optical materials by the diamond single-crystal and abrasive tool in the quasi-plastic cutting mode with air-bearing units having increased stiffness.

References

[1] Bifano T, Dow T, Scattergood R 1991 Ductile-regime grinding: a new technology for machining brittle materials *Transaction of ASME* **113** 184-189.

[2] Nazwa T 2010 Precision surface grinding of silicon carbide with fine grain diamond cup wheel *Engineering e-Transaction* **5** 19-26

[3] Shariff Udin M, Seah K H W, Li X P, Rahman M, Liu K 2004 Effect of crystallographic orientation on wear of diamond tools for nano-scale ductile cutting of silicon *Wear* **257** 751-759

[4] Wang J, Fang F, Zhang X 2015 An experimental study of cutting performance on monocristalline germanium after ion implantation *Precision Engineering* **39** 220-223

[5] Evans J, Rhorer L 2010 Fabrication of optics by diamond turning *Handbook of Optics, 3rd Edition* **37** 843

[6] Chen H, Yifan D, Zheng Z 2011 Effect of crystallographic orientation on cutting forces and surface finish in ductile cutting of KDP crystals *Machining Science and Technology: An International Journal* 231-242

[7] Zhao Q, Guo B 2015 Ultra-precision grinding of optical glasses using mono-layer nickel electroplated coarse-grained diamond wheels. Part 1: ELID assisted precision conditioning of grinding wheels *Precision Engineering* **39** 56-67

Review of modern biometric user authentication and their development prospects

Z V Boriev[1], S S Sokolov[2] and A P Nyrkov[3]

[1] Admiral Makarov State University of Maritime and Inland Shipping, Dvinskaya st. 5/7, St. Petersburg, Russia, 198035
[2] Admiral Makarov State University of Maritime and Inland Shipping, Dvinskaya st. 5/7, St. Petersburg, Russia, 198035
[3] Admiral Makarov State University of Maritime and Inland Shipping, Dvinskaya st. 5/7, St. Petersburg, Russia, 198035

Abstract. This article discusses the possibility of using biometric information technologies in management. Made a brief overview of access control and time attendance. Analyzed biometrics and identification system user. Recommendations on the use of various systems depending on the specific tasks.

Introduction

The use of biometric identification systems these days is becoming more urgent. However, as the reality shows, traditional methods of personal identification and authorization based on the use of passwords or physical media (in the form of pass, passport, driving license, electronic keys and cards), do not meet modern requirements for reliability in the determination of the individual. The password can be forgotten or intercept material carrier - to copy, lose or give to another person. Especially it does not allow the traditional access control systems to ensure an adequate level of reliability. As a result, companies are looking for more efficient ways of ensuring security. One way to solve this problem is to move to biometric identification systems. [1]

1. Information and Control Systems

Effective management requires constant monitoring and planning of all processes in the organization. Planning work is carried out taking into account the fact that it is followed by the monitoring function. A control need timely, accurate information, effective with respect to the cost to develop it. To implement these requirements necessary information management system.

Management Information System (MIS) - a set of hardware and software that allows the company management to obtain timely information necessary for effective decision making.

Daily administration of the enterprise is faced with the decision of questions about payroll, time tracking, providing access to certain resources staff, whose number is constantly changing. By itself, the automation of management processes in the company is not a big problem. These difficulties arise due to the fact that the components of any organization are constantly changing, upgraded or even deleted. Changing environment, changes the structure, process technologies modernized enterprise. If the administration of the organization will be deprived of the opportunity to react promptly to these changes, the consequences for the enterprise can take a serious nature. [2]

IOP Conf. Series: Materials Science and Engineering **91** (2015) 012063 doi:10.1088/1757-899X/91/1/012063

2. Access control.

Access control systems (ACS) - an essential component of information management system of any business or office building. ACS is a very complex hardware-software system that is responsible for the registration and restriction of entry and exit of people and vehicles in a certain area through check points (gates, doors). The main objectives of the ACS is to determine the time of admission to the territory and for the identification of objects that are eligible for admission. In addition, the access control system can perform a range of additional features, which include:

• keeping records of working time;

• Conduct payroll employees (if possible integration with the accounting system);

• maintaining a database of visitors and employees.

In any ACS has a email id (key), which is used to determine the rights of the person who owns them. As a key in Access control can be used: barcode, magnetic or smart (smart cards) Plastic cards; "Electronic tablets» (Touch Memory); Wiegand cards; card remote reading (proximity); PIN-code. Moreover, the identifier may be used as the biometric parameters of the person.

"Electronic tablets» (Touch Memory) chip are arranged in a sturdy metal case. The code information recorded in the memory of the circuit. To identify the need to attach the tablet to the reader. Reading Speed - 0.1. Some models allow you to enter information about the user. The advantages are compactness, high resistance to mechanical damage, corrosion, temperature extremes and low cost (comparable to the cost of magnetic stripe cards).

Card with a bar code is a plate coated with a black stripe (strokes). Code information contained in the variable stroke width and the distance between them. Code of the card is read by the optical reader. On a magnetic card code information is recorded on the magnetic strip. The bar code can simply dirty mud. Magnetic card is easily damaged in daily use.

Perforated card is a plate (plastic or metal). The code information is applied to a perforated card as the holes arranged in a specific order. Code card is read by mechanical or optical readers.

Code information on the Wiegand card contains a certain way on the thin metal wires arranged, glued with special glue. Information from the card is read by electromagnetic reader.

Proximity technologies have many advantages - much greater than with other methods of identification reliability and durability, lack of power (in passive cards). Proximity reader constantly sends a radio signal. Map in contact with the zone of the reader takes it in response to radiation and sends a signal containing recorded on a chip code. The distance between the reader and the card depends on the power of the reader varies from 5 cm to several meters. The lack of mechanical contact during operation allows for an arbitrary form identifiers, the identifier can be implanted into the object identification. The advantages are the complexity of forgery, the possibility of using encryption algorithms.

Smart card ("smart card") is a plastic card has an embedded microcontroller with all its attributes (processor, RAM, nonvolatile memory with a file system tools to input additional coprocessors). The main advantages of smart cards - a large memory capacity and high security information from trying to modify and duplication. Drawback - the high cost. Is the standard equipment of appropriate automated systems can be simply implemented in any arbitrary system.

PIN-code. Carrier code information is the memory of man. User independently from the keyboard code and it gives the signal for the end device. [3]

3. Biometric identification systems. Legislative regulation

In the Russian Federation in the field of protection of human subjects of personal data biometric data given special attention. Article 11 of the Federal Law "On Personal Data" gives the following definition of biometric data - a "data that characterize the physiological and biological characteristics of the person on the basis of which it is possible to establish his identity (biometric personal data) and are used by the operator to identify the subject of personal data can be processed only with the consent in writing of the subject of personal data, except as specified in paragraph 2 of this Article. "

IOP Conf. Series: Materials Science and Engineering **91** (2015) 012063 doi:10.1088/1757-899X/91/1/012063

Part 2 of Article 11 of the Federal Law "On Personal Data" states that the processing of biometric personal data can be carried out without the consent of the subject of personal data in connection with the implementation of international treaties of the Russian Federation on readmission, in connection with the administration of justice and all that is connected with it.

By passing this law provides specific requirements for material media biometric personal data and storage technologies such data is personal data information systems.

Material carrier must provide:

• protection against unauthorized re-entry and additional information after its removal from the information system of personal data;

• access to recorded material carrier on biometric personal data carried out by the operator and persons authorized in accordance with the legislation of the Russian Federation to work with biometric PD;

• the ability to identify personal data information system, which was implemented biometric record of PD, as well as the operator, to carry out such a record;

• preventing unauthorized access to biometric personal data contained in a tangible medium.

The operator of biometric data used is personal data information systems, must:

• maintain records of the number of copies of material carriers;

• Provide appropriate material carrier of a unique identification number that allows the operator to accurately determine, to record biometric personal data on the material carrier.

4. Biometric identification systems. Hardware and software

Biometric identification systems are most effective because they cannot recognize the physical media, and the unique physical characteristics of a human. Access and protection of information, based on such technologies are not only the most reliable but also the most user-friendly to date. All biometric devices have specific requirements for hardware and software. In any system user identification must first be registered. Many biometric systems allow users to do it themselves.

Biometric Access control, allow to organize an efficient, reliable and cost-effective system of access control. The use of biometric technology avoids problems such as loss, re-release and transfer to third parties badges.

Timesheet is no less important issue for the management of the companies. Time tracking system supplies reporting data corresponding to the time of arrival and departure at the workplace staffing, delays in the workplace, processing and defects. The presence of such information allows, in turn, greater flexibility to manage the distribution of wages, bonuses, penalties, increases discipline in the enterprise and, of course, translates into economic benefits. Importantly, the time tracking system constructed on the basis of biometric identification allows, compared to conventional systems, to eliminate instances of intentional abnormality associated with the lack binding - e.g., the magnetic card to a specific person, whereby the card can easily be transferred from one person to another.

Identify several ways of accessing the necessary resources:

1. The verification based on biometrics and unique identifier that identifies a particular person (eg, ID number), ie, This method is based on a combination of identification and authentication techniques.
2. Authentication, which in turn is based only on the biometric measurements. The measured parameters are compared with all entries from the database of registered users, rather than one of them selected on the basis of that ID. [4]

Generally, both methods are used equally. Select the Authentication method is carried out directly on the basis of security requirements, as well as funds allocated for these purposes. But the reliability of access depends not only on the way of authentication. An important factor in this regard is the same biometric selection parameters based on which will be accessed.

5. Biometric parameters

There are a number of biometric parameters, on which the identification and authentication of identity. These include:

IOP Conf. Series: Materials Science and Engineering **91** (2015) 012063 doi:10.1088/1757-899X/91/1/012063

• Iris.
• The retina of the eye.
• Vienna hands.
• Fingerprint.
• Handprint.
• The geometry of the hand.
• The geometry of the face.
• 2-D / 3D face model.
• Signature.
• Voice.
• Gait.
• The shape of the ears.

Every year there are more biometric parameters, which can identify and authenticate identity. But to study them, and find ways of processing algorithms takes a lot of time and money. Therefore, the degree of scrutiny of biometric parameter directly affects the extent of its use.

The most-used options today is the fingerprint. This is due to the fact that fingerprinting got its spread in the late 19th and early 20th century. The cost of fingerprint scanners every day more and more reduced, and the number of additional parameters, such as temperature finger pressing force and the like, increases. We can say that in the near future by the scanner manufacturer reached the limit of opportunities in this area.

The most statistically robust and resistant to forgery access parameters are iris and veins in the arms. But the process of collecting and processing parameters such takes a bit longer than the processing of fingerprint or face shape. Reliability indices are such features as false rejection and false access access. Biometric data for a false denial of access tends to 0%, and a false access does not exceed 1-2%. [7]

Despite all the differences between the biometric parameters, as the basic characteristics of biometric systems was made to use the False Acceptance Rate (FAR) and False Rejection Rate (FRR). Obviously, the smaller these ratios, the stronger or the other option. These factors help to decide whether to use a particular method of identification and authentication, depending on the task.

Table 1 shows the average values for the various biometric systems:

Table 1. Mean values for different biometric systems. [5]

	Fingerprints	Geometry face	Iris
FAR, %	0.001 %	0.1 %	0.00001 %
FRR, %	0.8 %	7 %	0.10 %

6. Structure of biometric systems of access control

In the majority of ACS and the system of working hours are networked distributed systems with the access rights of users who need to build, open to integration with other manufacturers' equipment. At the points of passage set by the controller connected to the control PC or server via USB or LAN Ethernet.

User registration is performed using special software. To register a fingerprint reader applies a control that connects via the USB port of a personal computer. Users can also register independently at the terminals of the system using specific procedures.

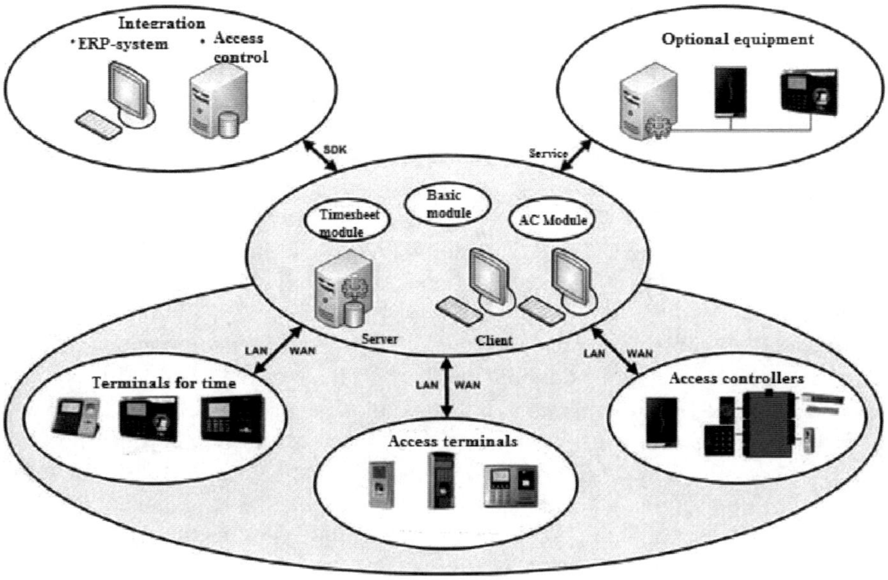

Figure 1. General structure of biometric systems of access control

As seen in figure 1, the ACS has a modular structure. This allows the system to be scalable and more stable. In the event of failure of one of the elements, the system continues to work, of course, in limited functionality mode. Software in most cases consists of two modules "working time control system" and "access control system". Time tracking system works with personnel shifts, schedules, events, or obtained automatically entered in manual mode allows you to build reports. Access control program is responsible for connecting and configuring devices, setting the rights of passage and time zones, synchronization of data between the database and terminals.

7. Comparative evaluation of biometric systems.

Select several empirical characteristics for assessing the quality of the system:

1. Resistance to forgery.
2. Stability of Environmental - characteristic estimating system stability under various environmental conditions, such as changing the lighting or room temperature.
3. Ease of use - shows how difficult it is to use a biometric scanner whether it is possible to identify "on the fly".
4. The speed of the
5. The cost of the system

Below is a comparative evaluation of biometric systems for the ten-point scale.

Table 2. Comparative evaluation of biometric systems

	Resistance to forgery	Resistance to environment	Ease of use	Speed	Cost	Stability characteristic cost Time
Iris	10	9	8	7	10	10
Fingerprinting	6	10	9	10	10	9
2D face	4	6	6	10	10	8
3D face	9	8	10	5	7	10
An arm vein	10	7	9	7	8	7
Retina	10	10	6	3	6	9

It can be concluded that the resistance to forgery or that the biometric parameter is inversely proportional to the speed of its processing by the system.

IOP Conf. Series: Materials Science and Engineering **91** (2015) 012063 doi:10.1088/1757-899X/91/1/012063

Summary

Opting for the use of a parameter when accessing or working hours accounting depends on the system requirements. The most statistically robust and resistant to forgery access systems are the systems of admission iris and veins hands. The least expensive and easiest to use, but having good statistics are of tolerance on one hand. Tolerance on 2D face is convenient and cheap, but it has a limited range of applications due to poor statistics.

In addition to combining the use of biometrics, it is possible to combine the parts of the body from which the parameters are removed. Since Iris can increase the accuracy of the system is almost quadratic, lossless time if complicate the system, making it two eyes. For fingerprint method - by combining several fingers, and recognition of the veins, by combining the two hands, but in this case, the improvement is only possible by increasing the time of human interaction with extractors.

Summarizing the results of the methods that can be said that for medium and large objects as well as objects with maximum safety requirement should be used as an iris biometric access and possibly recognize hand veins. For objects with the number of staff to several hundred people will be optimal access fingerprint. Recognition system for 2D face image is very specific. They may be required in cases where the detection requires the absence of physical contact, but to put the system on the iris control is not possible. For example, if authentication is required for a person without his participation, hidden camera, camera or external detection, but this is only possible with a small number of subjects in the database and a small stream of people shoot camera.

References

[1] Boreysho A A Biometrics and software features of modern ACS *J Security of Business Information* 59 (2005) 40-43
[2] Aleshin A P Maintenance business security Dashkov & Co Moscow 2008
[3] Balle R M Pankanti S Ratha N K Senior A W Guide to Biometrics Springer Sciense+Business Media New York 2004
[4] Information on http://habrahabr ru/post/126144/
[5] Kharitonov A V Overview of biometric identification methods *J Cybernetics and programming* 2 (2013) 12-19
[6] Korobeynikov A G Gatchin Y A Application for Determination of Mail Document Spam Level *World Applied Sciences Journal* 30 (2014) 165-169
[7] Chernyi S Zhilenkov A Sokolov S Titov I Self-contained drilling rig automatic control system efficiency improvement by means of assuring compatibility and integration methods development *Metallurgical and Mining Industry* 2015 No 3 P 66-73

IOP Conf. Series: Materials Science and Engineering **91** (2015) 012064 doi:10.1088/1757-899X/91/1/012064

Using information technologies in dredging

D G Mamunts[1], S S Sokolov[2] and A P Nyrkov[3]

[1] Admiral Makarov State University of Maritime and Inland Shipping, Dvinskaya st. 5/7, St. Petersburg, Russia, 198035
[2] Admiral Makarov State University of Maritime and Inland Shipping, Dvinskaya st. 5/7, St. Petersburg, Russia, 198035
[3] Admiral Makarov State University of Maritime and Inland Shipping, Dvinskaya st. 5/7, St. Petersburg, Russia, 198035

Abstract. The article deals with various aspects of using information technologies in dredging. The possibilities of modern information technology in the framework of improving the efficiency of dredging

Introduction

Dredging is a complex process consisting of multiple steps and is characterized by a large number of parameters. The purpose of dredging on inland waterways is to provide guaranteed navigable depths. For prompt and effective management of dredging at all stages of production it is necessary to take into account a variety of parameters and be able to compare different data.

When the pre-preparation and design stages, and production work, the use of dredging projects in geographic information systems (GIS), Global Navigation Satellite Systems (GNSS), and monitoring systems and automated control of hydraulic digging process, to simplify the task of monitoring and control site to be developed.

1. Use of geographic information systems in dredging.

Using comparing the data of engineering surveys in different years in the software environment of GIS provide analysis of the sediment accumulation of water area and the current mode of channel rearrangement. With this approach to the organization of pre-training is possible to combine and analyze a variety of information (data surveys, satellite imagery, design documentation, etc.) and to evaluate the technical and engineering risks, receiving a full and objective picture of the task site. At the design stage the use of GIS technology solves complex hydraulic problem of choosing the shipping route slots in the implementation of this choice guided by parameters such as: safety and comfort conditions for the passage of vessels through it and minimizing the volume seized of the soil. Also an important aspect of project preparation is not ignored - evaluation of soil characteristics and selection of dredging equipment in accordance with them. Thus, such measures of introduction geographic information systems enable organizations to operate more efficiently, increasing the quality of their work and minimizing costs. Visualization of the project structure using GIS is shown in Fig. 1.

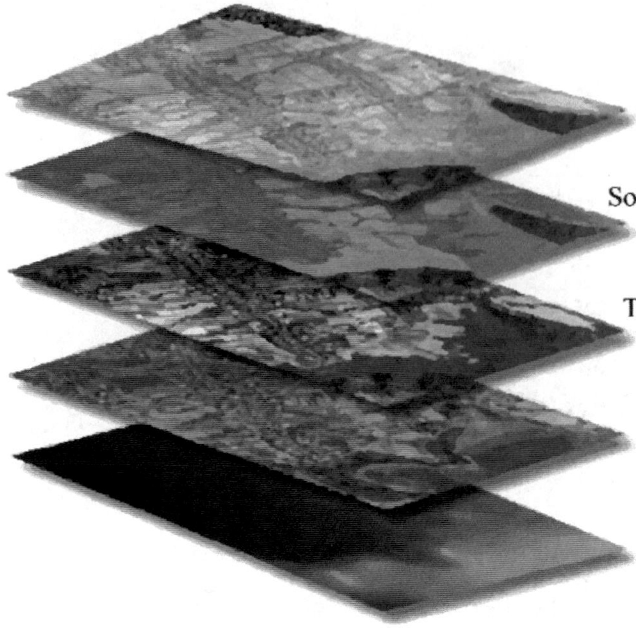

Fig. 1. Visualization of the project structure using GIS

Modern software systems of positioning allow collecting and storing information from a variety of external devices: navigation receivers, depth sounder, magnetometers, heading sensors. It can be directly integrated with GIS (in this case the data exchange is realized in real time) or transfer the data obtained by exporting. Using GIS technology directly in the manufacturing process works, makes it possible to visualize the process of dredging in a timely manner to analyze and evaluate the results of operations and respond quickly to the need to finalize the design decisions.

2. Positioning of dredger.

In the production of dredging dredgers constantly change their location, moving forward along the trench or moving across the slot «swinging» from edge to edge. To make these movements correct and remove dredged soil from the bottom of the reservoir, should be placed where it is needed, avoiding crossing undeveloped plots slots it is necessary to have special orientation and positioning system, allowing to determine where the dredge at this time [4].

To date, technical fleet is using positioning system based on global navigation satellite system (GNSS) such as GLONASS / GPS and differential additions DGLONASS / DGPS. These systems are constantly being improved with the development and optimization of the structure of the satellite segment, segment management and control.

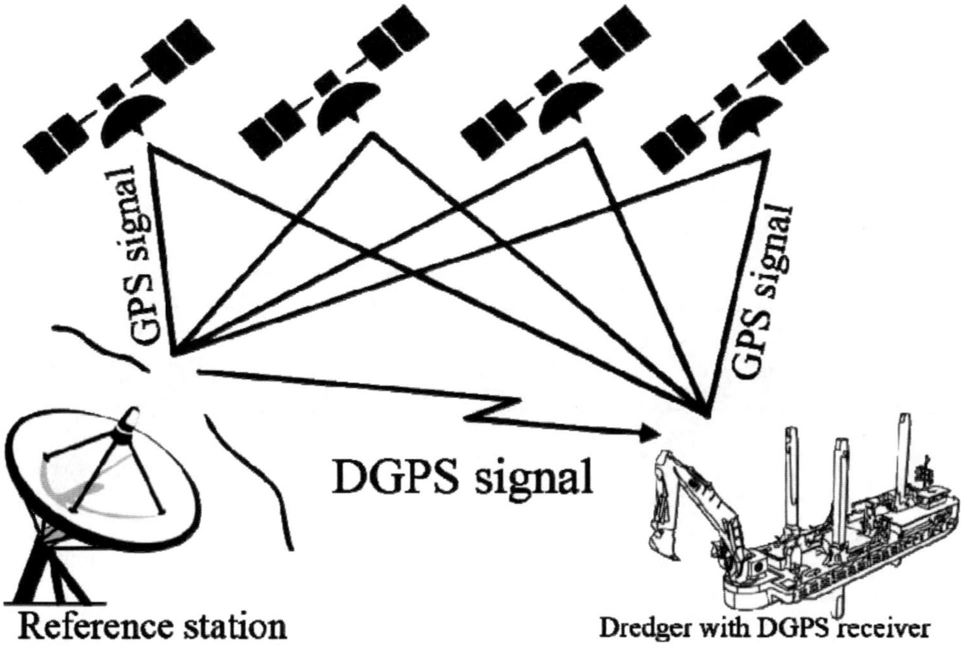

Fig. 2. Schematic diagram of the system DGPS

In positioning systems of dredgers based on GNSS, location of dredger is determined according to the pseudorange measurements between navigation spacecraft and the satellite receiver, established on the dredger. Using differential mode DGLONASS / DGPS positioning systems dredgers identified the need for more accurate positioning than the normal mode, the accuracy of which is tens of meters. Currently, with the operation of the orientation of the dredgers in differential mode DGLONASS / DGPS positioning accuracy is possible to get up to 1 m [3].

The method of differential correction is based on relative constancy of the errors of satellite radio navigation systems in time and space. Differential mode of satellite radio navigation systems presupposes at least two receivers or receiver-navigation spacecrafts. One is located on the receiver-control and correction stations (reference stations), the second is designed to determine the coordinates of the user. Coordinates of correction stations are known, they are defined precisely and geodetically tied to the adopted coordinate system (usually a coordinate system PZ-90 for DGLONASS and the WGS-84 for DGPS) [6]. The scheme of the system is shown in Fig. 1.

There is also an alternative DGLONASS / DGPS positioning technology that allows you to receive a higher positioning accuracy - Real Time Kinematic (RTK). The method uses a differential GPS RTK carrier phase measurements, providing an accuracy of about 1 centimeter in real time.

Measuring carrier phase - the most accurate method of measuring pseudoranges. Carrier phase fluctuations are constant frequency, in contrast to the recorded on the GPS receiver due to the Doppler effect (some "shift" is formed by the transmission signal of the distance from the satellite to the receiver). Thus, the phase of the carrier measured between the satellite and the receiver antenna phase center will consist of an integer number of phase cycles and fractional part. Unfortunately the GPS receiver is unable to distinguish between a carrier cycles. It can measure the fractional part of the phase, and then to monitor its change: the initial phase is uncertain. In order to use the current phase for the pseudorange measurement the unknown number of cycles or ambiguity must be calculated along with the coordinates of the receiver [7]. Positioning technology is shown in Fig. 3.

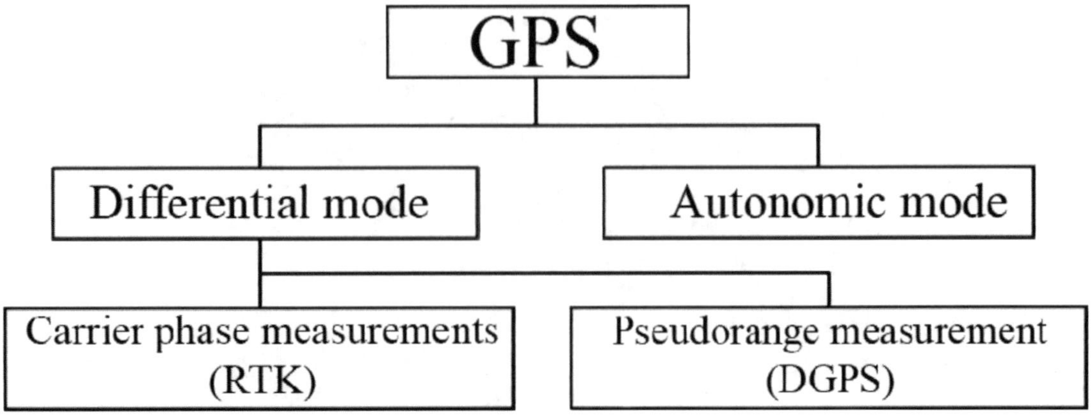

3. Monitoring and automation of hydraulic digging process.

Task of automation process hydraulic digging is to achieve and maintain such relations digging parameters under which the process will be optimal. In this case, there is the task of achieving the best possible performance, consistent with the change in the rate of the working movement of the dredger and the terms of digging and transportation of soil.

Unlike automated positioning systems dredger, which are universal for all types of dredgers, process of automation of hydraulic digging requires a special approach, for example, dredgers have more opportunities to automate the process of control than multi-bucket dredgers.

Automation of the process includes automatic selection of the optimal mode of ladle drive or pump installation and the choice of automatically moving projectile, providing the regime [5].

The first step towards the creation of automated process control system of digging is to build a monitoring system. The objective of the system is the automatic collection of information about the processes in the dredger.

This information includes performance characteristics such as structure and rate of slurry, the level of penetration of device, and others.

The purpose of monitoring these data is to provide accurate data to the operator of the dredger, which will allow him to adjust the process quickly and to achieve maximum performance.

Sensors monitor the operating parameters. Sensors convert physical parameters into electrical signals and low-level transmitter converts electrical signals into the standard low-level electrical signals. These signals may be connected directly to the digital display, which will render information about a particular process.

In the next step, PLC (Programmable Logic Controller) implements the amount of information from various sensors, their comparison and building relationships. PLC is the heart of the monitoring system or control. This device is composed of input and output modules under the control of the microprocessor control program and, if necessary, communication modules to enable it to send and receive data from other devices such as computers and graphic displays. For example, the controller may display the instantaneous density flow in real time.

The program to manage the process of dredging will use the PLC, which will monitor all incoming data and automatically control the on-board devices; it is a transition from a simple display of information about the process to direct control of the process.

Summary

The use of modern geo-information technologies and the creation of complex hardware and software systems will maximize the effectiveness of the design and manufacture of dredging and optimize processes. Thus, the task of overall coordination of the project and its control will be simplified, and at the same time is reached improving the overall efficiency of dredging. In the foreseeable future, we

IOP Conf. Series: Materials Science and Engineering **91** (2015) 012064 doi:10.1088/1757-899X/91/1/012064

should expect the evolution of automated systems into intelligent. Intelligent systems will be able to educate themselves directly in the production process works in parallel with an analysis and processing of statistical data.

References

[1] Chernyi S Zhilenkov A Sokolov S Titov I Self-contained drilling rig automatic control system efficiency improvement by means of assuring compatibility and integration methods developmen Chernyi S Zhilenkov A Sokolov S Titov I *Metallurgical and Mining Industry* Volume 7 Issue 3 2015 p 66-73

[2] Chernyi S The implementation of technology of multi-user client-server applications for systems of decision making support *Metallurgical and Mining Industry* 2015 No 3 P 60-65

[3] Chernyi S Zhilenkov A Analysis of complex structures of marine systems with attraction methods of neural systems *Metallurgical and Mining Industry* 2015 No 1 P 37–44

[4] Zhilenkov A Chernyi S Investigation performance of marine equipment with specialized information technology *Procedia Engineering* Vol 100 2015 p p 1247–1252

[5] Nyrkov A P Belousov A S Sokolov S S Algorithmic Support of Optimization of Multicast Data Transmission in Networks with Dynamic Routing / Nyrkov A P Belousov A S Sokolov S S *Modern Applied Science*; Vol 9 No 5; 2015 p 162-176

[6] Sokolov S S Mamunts D G Automation of dredging process: domestic experience and development prospects / S S Sokolov D G Mamunts *"River transport (XXI century) "* Moscow IS S ue №2 (2014) IS S N: 1729-4258 pp 47-49

[7] Rudykh SV Orientation system dredgers based on global navigation satellite systems /S V Rudykh *Vestnik Gosudarstvennogo Universiteta Morskogo I Rechnogo Flota Im Admirala S O Makarova* 2013 - 186 p (Vol 2)

[8] Gatchin U A Korobeynikov A G Fejer Application for Determination of Mail Dokument Spam Level *World Applied Sciens Journal 30(Magament Economics Technology & Tourism)* – 2014 - P 165-169 – ISSN 1818-4952

[9] Gladkov GL Rebkovets AV Development of navigation equipment inland waterways / smooth GL Rebkovets AV *Vestnik Gosudarstvennogo Universiteta Morskogo I Rechnogo Flota Im Admirala S O Makarova* SPb : IS S ue №1 (2012) IS S N: 2309-5180 pp 16-27

[10] Grishanin KV Degtyarev V V Seleznev V M Waterways - M "Transport" 1986 - 400 p

[11] A typical process flow chart (TTC) dredging

[12] Prikhodko VM Electrical and automation of technical fleet: Proc allowance - SPb : SPGUVK 2009 - 77 p

[13] Sakun AI "Navigation Positioning System: Past Present and Prospects" *Management Rosreestra Kamchatka Territory* 29 10 2013 [http://to41 rosreestr ru/news/media/4589965/]

Microsoft Business Solutions-Axapta as a basis for automated monitoring of high technology products competitiveness

G O Tashchiyan, A V Sushko, S V Grichin

Yurga Institute of Technology (branch) of National Research Tomsk Polytechnic University Leningradskaya 26, Yurga, 652055

E-mail: gtashiyan@mail.ru

Abstract. One of the conditions of normal performance of the Russian economy is the problem of high technology products competitiveness. Different tools of these products estimation are used nowadays, one of them is automated monitoring of the high technology products in mechanical engineering. This system is developed on the basis of "Innovator" software integrated in Microsoft Business Solutions-Axapta.

1. Introduction

During the last 22 years coal mining industry, mechanical engineering and other industries have undergone a number of deep reforms and have changed drastically. Both qualitative and quantitative indexes of these industries have changed, the role of these industries in the system of international economic relations have become more significant [1].

The aim to modernize the economy of Russia based on innovations was declared about six years ago, but the achieved results turned out to be far from those expected by the authors of the concept and the society [2].

The problem is that now the domestic producers do not treat the issue of competitiveness seriously. However, there is a tendency that without the competitive high technology products modern companies cannot retain their market niche.

The major part of the issue of the high technology products competitiveness consists in the problems of a technical character. For example, in paper [3] technical parameters are used to increase the durability of cutters of mining and road-making machines. However, the issues of competitiveness should be considered not only from the technical point of view, but from ecological and social points of view as well.

That is why the problem of reliable estimate of high technology products competitiveness, creating an automated decision-making system on producing and consuming high technology products is of great urgency.

2. Integration of automated monitoring of high technology products competitiveness in Microsoft Business Solutions-Axapta

There are various tools to estimate competitiveness, and one of them is automated monitoring of high technology products competitiveness in mechanical engineering (AMCCE) [4].

The integrated system of company management of *ERPII* class *Microsoft Business Solutions-Axapta* is a scalable system for small and medium-size businesses, corporations, holding companies that require a uniform solution.

The AMCCE system has been developed based on functional module «Questionnaire» of *Microsoft Business Solutions-Axapta* in combination with "Innovator" software (Fig.1).

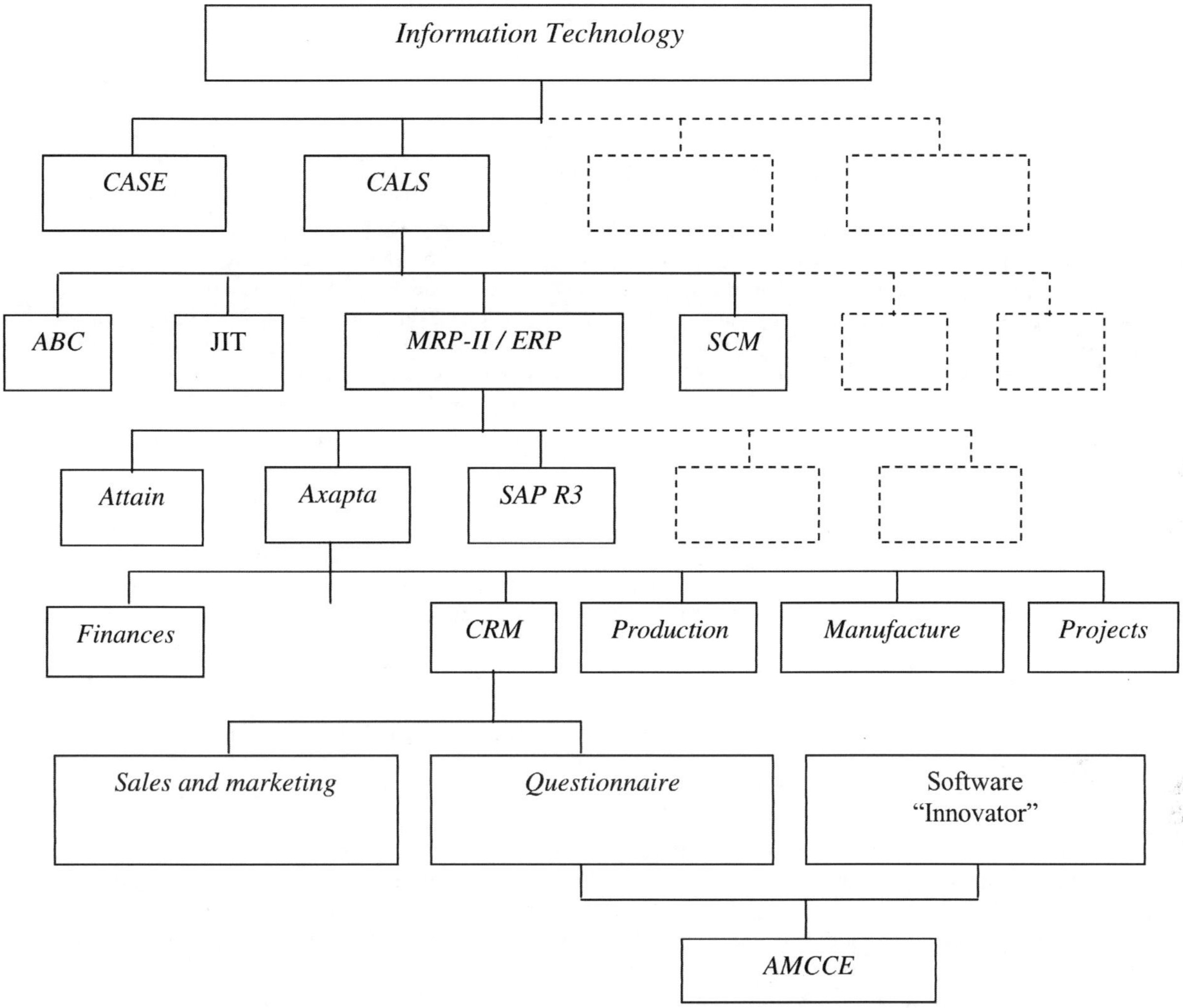

Figure 1. Integration of automated monitoring of high technology products competitiveness in Microsoft Business Solutions-Axapta

The functional modules of the system are Production, Finance, Fixed Assets, Wage, Knowledge Management *(KM*), Human Resources (*HR*), Customer Relationship Management (*CRM*), Logistics, Warehouse Management, Business Analysis, Product Configurator, Projects, Corporate Portals.

The AMCCE system is integrated with *Microsoft Business Solutions-Axapta,* namely its module *CRM,* which allows storing all the information about clients, partners and other contractors centrally. Employees form different departments can input, view and use this information for their purposes. As a result, the coordination of all the divisions of a company can improve, and in the first place of sales and marketing departments. Due to close integration with other modules of *Axapta,* which automate work with customers and clients, the CRM module supports business processes of companies of any type.

Thus, one can build a module in accordance with ones' own specific business processes; there is no need to modify them depending on the system's capacities.

The CRM module allows creating applications for current and prospective customers, and make simulations based on offers. The offers with a selling probability higher than specified can be taken into account in combined planning.

Microsoft Business Solutions-Axapta helps the heads of sales departments manage the operation of certain sales managers and departments in general in a more efficient way, achieving the set goals and increasing the efficiency of operation. The system automates carrying out marketing campaigns, calling and collection of information about the current and prospective clients, suppliers and partners.

Management of sales and marketing in *Microsoft Business Solutions-Axapta* is closely integrated with Customer Relationship Management (CRM). Such integration improves interaction of marketing and sales departments and, as a result, overall performance of a company as a whole. Because all the business information is well structured and stored in one place, the staff of the sales and marketing departments have an opportunity to analyze and estimate various data and to find new opportunities for sales easily. The marketing functions make it possible to define, what information is necessary for clients and the market, to plan, budget and hold marketing events, and to analyze their results.

When organizing marketing campaigns, *Axapta* can segment the target audience. It helps to carry out campaigns more successfully and to analyze their results in more detail, so it is easy to estimate the efficiency of campaigns, comparing their costs to the gained income. Marketing campaigns are presented in treelike structure; the symbol of each campaign is defined by its status. Thus, looking at the "tree", one can immediately define the status of the marketing campaign of a company. In order to regularize the activity of a company, one employee is appointed responsible for a campaign, and others are given individual tasks within its framework. After the campaign is finished, it is possible to produce reports on the results of the campaign, on the resources spent, and to determine the return of investments.

There is a capability of integration with other modules of *Microsoft Business Solutions-Axapta*, for example, with "Projects" module.

Let us assume the management of a company is planning to carry out an extensive marketing campaign, offering clients a new training course. The campaign will cost the company a lot. Ten employees will be working on advertising for two weeks, do direct marketing and make the list of calling.

When carrying out campaigns and questioning on the Internet, answers of respondents get directly to the *Axapta* database. In case answers are received too late or there is not enough time for processing of all the received answers, the marketing campaign may not bring the expected results. In order to avoid such situations, the module contains information on how many people took part in the poll, who exactly answered the questions, and what answers were received.

Integration of marketing, sales and service departments is extremely important for effective work of a company. Information obtained by one of the departments has to be available to other departments. It can be obtained by maintaining "Encyclopedias of marketing". Having access to a larger volume of resources, these departments have more opportunities for effective work and development of business.

The "Questionnaire" module of the *Axapta* system is a powerful tool for creation of questionnaires, carrying out polls and the analysis of the obtained results.

Its main capacities are

- Easiness of creating questionnaires and carrying out polls
- Questioning in the internal corporate *Intranet* network and on the *Web* sites
- Convenient means of data analysis and presentation.
- Integration into "CRM", "Human resource management" modules and corporate portal.
- Integration with *Microsoft Excel* and *Access.*

The "Questionnaire" module of *Microsoft Business Solutions-Axapta* can be used for the solution of the following tasks: identification of clients' loyalty, estimation of competitiveness of high technology products, certification of employees, etc.

In our case, any employee of a company can organize and carry out an expert poll about competitiveness of high technology products by means of the "Questionnaire" module of *Axapta* system. It is possible to create target audience, i.e. to define subjects of dialogue, to create the questionnaire and carry out a poll in a few minutes. No special knowledge or skills are required for this purpose. Integration with the Internet allows carrying out questioning, both within a corporate network, and on the open *Web* sites. When organizing a dialogue of the automated monitoring of competitiveness of the high technology products, a group of independent experts is gathered [4].

When carrying out mass polls. the most important stage of work is the analysis of the obtained results. The "Questionnaire" module contains all the necessary tools for effective calculation, representation and the analysis of results, including summary tables and diagrams.

Due to integration into the "Human resource management" module, it is possible to obtain results of questioning for certain groups of employees. Groups can be created based on tender or age, period of working in the company, wage level and so forth. When questioning third-party respondents (partners, clients, etc.), the obtained results can be synchronized with the "Customer Relationship Management" module.

The "Questionnaire" module can also be used for analyzing management efficiency. In this case, results of questioning are processed in the "Human resource management" module. Along with the analysis of data directly in the Questionnaires module, it is possible to use *OLAP* tools (*On-Line Analytical Processing*).

All the results of polls are stored in the system; therefore, when carrying out another poll, it is always possible to compare the results and to analyze their dynamics of change.

The "Questionnaire" module supports all business functions presented to *Axapta*. It is closely integrated with *CRM* and "Human resource management" modules, and with a corporate portal. It allows using the module for communication with partners, suppliers and clients. The "Questionnaire" module is a part of the integrated system, which makes it easier to compose target audiences for polls, because all the necessary data, including lists of clients, suppliers, employees and information on them is already in the system. In combination with the "Innovator" software, an automated monitoring of competitiveness of high technology products is made.

A wide use of personal computers connected by the local computer network on workplaces in economical, marketing and designing divisions of JSC PA *Yurmash* gives the opportunity to organize on-line individualized information processing. Information capacities are personalized, but the centralized databases supplied with internal and external electronic communication remain. The decentralized data collection and processing on a workplace of the employee allows increasing completeness, accuracy and relevance of the prepared documents, increasing the speed of their preparation. The greatest effect from creation of the system-wide databanks is reached when they are created in the divisions, responsible for carrying out the analysis and assessment of information. In this case, there is a natural accumulation and processing of the data, coming in accordance with the tasks and functions of divisions to the places responsible for assessment and decision-making; completeness and relevance of databanks are provided without breaking the established traditional interrelations.

Sophisticated automation of information processes in quality and competitiveness control systems imply functional and information integration of all of the automated systems operating in production and other organizations. The most labor-consuming operation in creating a system of high technology products competitiveness monitoring is designing software. Labor input is defined by a great variety of data used in a control system, by local nature of automation of individual operating subsystems and management functions.

The information aspect of creating the system of automated monitoring of high technology products competitiveness at JSC PA *Yurmash* consists in solving the following tasks:

- determining the list of data to be automated;
- establishing formal relation between the data in order to build an optimal structure of data base;
- structuring informational messages in accordance with their content and purpose in the management system;
- organization of technological process of acquisition, processing and output of data, providing its reliability and timeliness.

Information processing in the system of automated monitoring of high technology products competitiveness JSC PA *Yurmash* is carried out on a workplace in the course of a dialogue of a person with the PC. There are three main modes of expert dialogue procedures: linear, alternative and cyclic ones. The dialogue processes with preliminary developed well-defined strategy without iterative operations are solved in the linear mode. The typical application field is collecting data and data input. The tasks of a dialogue, which include decision-making situations connected with the dialog with excluding each other stages, require an alternative mode. The cyclic dialogue mode is used for solution of optimizing tasks and choice of alternatives. These main options are implemented in the system of automated monitoring of high technology products competitiveness at JSC PA *Yurmash* in various mixed forms. The form of implementation of the dialogue modes is the so-called "menu technology", allowing experts without thorough training to work with the PC. In dialogue points, the expert gets access to all necessary information by means of corresponding actions

The main advantages of the monitoring system of the high technology products competitiveness at JSC PA *Yurmash* are:

- redistribution of data traffic between the human and the computer with reduction of the load on the human on storing, analysis and information processing;
- the opportunity to use personal experience and knowledge of narrow specialists;
- ease of learning to use, experimenting with the system, psychological comfort of a user in communicating with the computer;
- high adaptability of the monitoring system of the high technology products competitiveness to new expert knowledge in a problem and the level of users training;
- text, table and graphical representation of the data obtained.

These advantages provide the monitoring system of the high technology products competitiveness with the opportunity to be used to determine the level of competitiveness of high technology products based on the following indexes: "the significance of an engineering solution" [5], "the significance of an economic event" [6], "the significance of the ecological solution" [7], and "the significance of social effect" [8].The high efficiency of the monitoring system of the high technology products competitiveness at JSC *Yurmash* is achieved due to the application of up-to-date software and hardware.

The main software for the monitoring system of the high technology products competitiveness design and operation are *Microsoft Excel*, *Microsoft Access* and the *Questioning* module of *Axapta system.* One of the essential advantages of the monitoring system of the high technology products competitiveness is the capacity to work with data of other sources, including DBMS for personal computers (*dBASE, Paradox, FoxPro,* etc.). Thus, there is an opportunity to develop these forms and reports based on specific data, connecting them by means of simple macros or *Visual Basic* instructions. This software allows connecting new blocks and subprograms, which provides an increase in functionality of the system.

The hardware of the monitoring system of the high technology products competitiveness at JSC PA *Yurmash* includes: PCs, peripheral equipment; network equipment; transmission medium media and data security; communication facilities. The main divisions where the monitoring system is introduced include the marketing management department, the engineering center, and the economic planning department of JSC PA *Yurmash.*

3. Conclusion

Thus, the automated workplaces equipped with the monitoring system of the high technology products competitiveness are personal computers and a complex of peripheral equipment connected with the central computer by means of network devices. They provide collecting, accumulation and transmission data. Depending on the tasks to be solved, specific conditions and requirements, various configurations of the automated workplaces can be created.

At present, the algorithms of competitiveness determination and the volume of investment of a company have been tested based on the monitoring system of the high technology products competitiveness volume of investment at JSC PA Yurmash in terms of group index of competitiveness Ksn [9]. Parameters series of standard size pit barring produced by the company have been simulated.

References

[1] Solovenko I S Trifonov V A Nagornov V I Russian Coal Industry Amid Global Financial Crisis in 1998 and 2008 // Applied Mechanics and Materials - 2014 - Vol 682 - P 586–590
[2] Prokopenko S A Ludzish V S Problems of innovation course of development of Russian mining enterprises // Gornyi Zhurnal - 2014 - №1 – pp 47-49
[3] S A Prokopenko Multiple service life extension of mining and road machines' cutters // Applied Mechanics and Materials - 2014 - Vol 682 - p 319-323
[4] Tashchiyan G O Organizing automated Ted dialogue on the competitiveness of science-intensive products // Applied Mechanics and Materials - 2013 - Vol 379 - p 240-243
[5] Osipov Yu M The "signnificance of engineering solution" index of simulation model of the automated control system and products competitiveness // Automation and modern technologies - M 1994 № 3 P 33-35
[6] Bykov S N Osipov Yu M The "significance of economic event" index of the automated control system of the products competitiveness // Automation and modern technologies - M 1998 № 4 P 36-38
[7] Bykov S N Voronkova I V The 'significance of ecological solution" index // The 5 th regional scientific and practical conference "Modern engineering and technologies" proceedings Tomsk TPU 1999 - P 90
[8] Bykov S N Tashchiyan G O Osipov Yu M A system of social indexes of products competitiveness // Automation and modern technologies - M 2003 №5 P 40-42
[9] Korikov A M Tashchiyan G O System analysis of the problem of competitive products // Natural and intellectual resourses of Siberia The 9th international scientific and practical conference Ulan-Ude 23-24 Sept 2003 – Tomsk TPU 2003 P 312-315

Forecasting risk of bankruptcy for machine-building plants[1]

E Telipenko[1], A Zakharova[2], Svetlana Sopova[3]
[1] Cand.Tech.Sci., associate professor of information systems NR TPU, Tomsk, Russia
[2] Cand.Tech.Sci., associate professor of information systems NR TPU, Tomsk, Russia
[3] Senior lecturer in humanities education and foreign languages NR TPU, Tomsk, Russia

E-mail: kochetkovaev@mail.ru

Abstract. The paper presents an overview of well-known bankruptcy risk forecasting models, elaborated as by Russian so by foreign authors, on the basis of the data about financial and business activities of the biggest machine-building Russian plants. The authors substantiate and confirm appropriateness of a fuzzy set model to the problem of bankruptcy risk forecasting. This model is worked out on the basis of 10 most important factors, which have the greatest influence on sales proceeds as the main financial source for a production plant.

1. Introduction

One of the most important market system institutes, which allow attracting bailout money and encouraging redeployment of production resources from inefficient economy sectors to efficient ones, is corporate bankruptcy. In any developed economy of market type bankruptcy is used not only for indicated purposes but also on behalf of agents-debtees and owners of a problem enterprise. It is a comparatively young institute in Russia, which had first legal confirmation in RF Law, dated 19 November, 1992, N 3929-1 "About Business Failure (Bankruptcy)". During this time it showed some ambiguity from the point of view of its fitness for the "right" purpose. In Russia corporate bankruptcy very often had a raider character and was used as an instrument for business take-over.

In this context tested, mathematically sound ways for bankruptcy risk forecasting have extrinsic value, because their elaboration and application is aimed at self-insurance system development of an enterprise for eventual bankruptcy.

2. Main results

2.1 Selection of the most important factors, which have an influence on sales proceeds as the main financial source for an enterprise, for bankruptcy risk forecasting by means of a principal components method

The method of principal components is meant for data structuring through the medium of reduction of many test variables to fewer number of variables (components or factors), which would explain most of studied data value variability.

[1] *The article is prepared with financial backing of Russian Foundation for Basic Research (project № 14-01-31208).*

For the analysis the authors selected 33 components (including sales proceeds), which characterize more precisely all aspects of enterprise financial and business activities. Selected indicator values were calculated for 33 machine-building plants on the basis of financial statements (Bookkeeping balance sheet (form №1), Supplement to the Bookkeeping balance sheet (form №5), Profit and loss account (form №2)) by 1st quarter 2010.

Then according to the main steps of the principal components method everything was calculated. In the result 5 basic components were worked out. They explain 73,607 % of all variables dispersion. It is necessary to notice that indicator selection was done with a glance to their influence on sales proceeds as the main financial source for a production plant.

After analyzing component eigen-vectors with the help of information coefficient the authors clarified which varieties contributed the most to their formation: x_5 – working capital ratio; x_6 – acid test ratio; x_8 – absolute liquidity ratio; x_9 – liquid capital ratio in assets; x_{14} – concentration ratio of capital loans; x_{16} – long term fund raising ratio; x_{17} – borrowed and own funds ratio; x_{21} – turn-round of float; x_{24} – turn-round of owned capital; x_{25} – product profitability.

Other variables can be excluded.

Therefore application of the principal components method allows diminishing feature space and moving from 33 factors, influencing sales proceeds as the main financial source for an enterprise, to 10 the most important, which explain most of studied data value variability [1].

2.2 Application of machinery of a possibility theory for bankruptcy risk forecasting of an enterprise

Bankruptcy risk assessment with the help of fuzzy sets includes 9 basic steps. You may see a simplified method description in Figure 1.

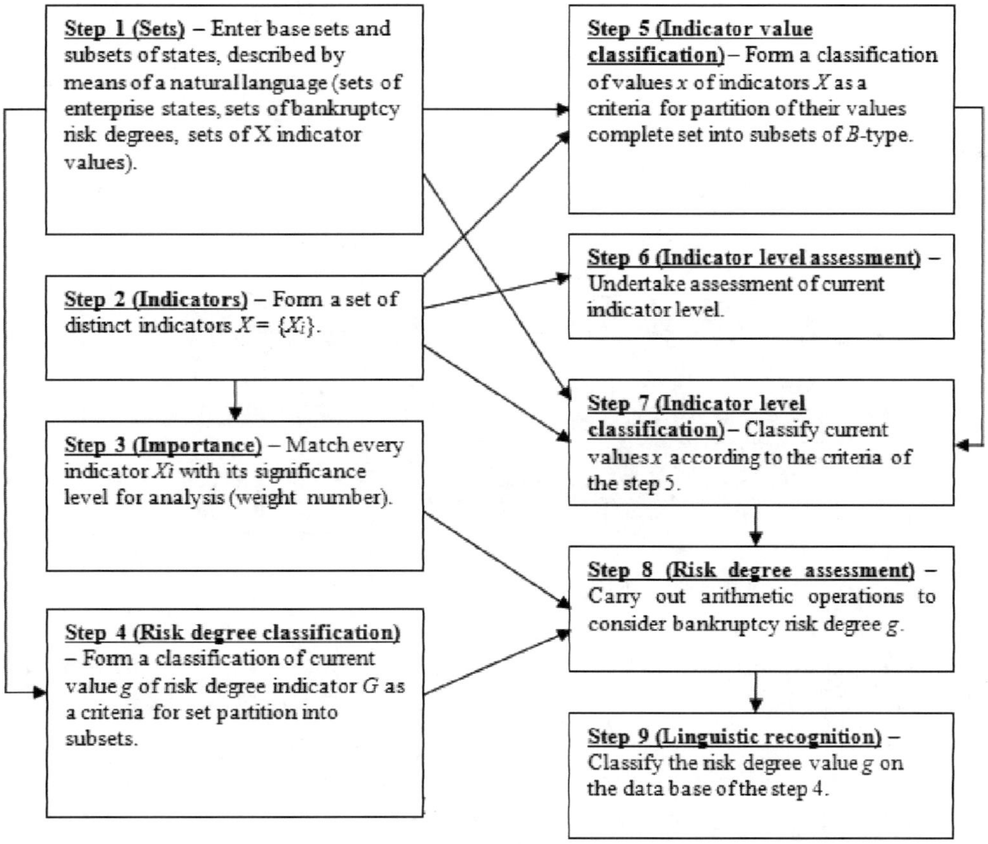

Figure 1. Simplified description of a fuzzy logical method.

Application of machinery of a possibility theory to the problem of bankruptcy risk forecasting allows:

- considering quantitative and qualitative indicators, which could be as standardized so non-standardized;
- considering an acceptable risk level for an enterprise and its property;
- recognizing in time possible beginning of bankruptcy risk.

2.3 Forecasting models for basic indicator evaluations of enterprise financial and business activities
To manage effectively bankruptcy risk of an enterprise it is necessary to be oriented not only onto current enterprise state assessment, but also onto its time history.

To calculate forecasting indicator values it is supposed to use a time-series forecasting method on the basis of polynomial growth curves. For 10 statistically based factors they form data selections over some periods, and then forecasting models are developed on this basis. Obtained models are needed to forecast indicator values for some periods of time, and also for time lag calculation before they take their critical values.

For example, there is a forecasting model for the factor - working capital ratio for open joint-stock company "Ishimbaiskiy Machine-building Plant" with the data from the table 2.

Table 2. Values of working capital ratio over 2007-2011. Open joint-stock company "Ishimbaiskiy Machine-building Plant"

Year	1 quarter	2 quarter	3 quarter	4 quarter
2007	2,65	2,2	2,17	1,7
2008	1,6	2,6	1,95	1,9
2009	1,9	2,2	2,3	2,7
2010	2,2	2,4	2,2	1,9
2011	1,4	1,1	0,96	0,84

Construct x_1-t diagram (fig. 3). It is evident from the diagram in fig. 2 the coefficient has a downward trend. Polynomial regression is the best in this case to characterize the trend.

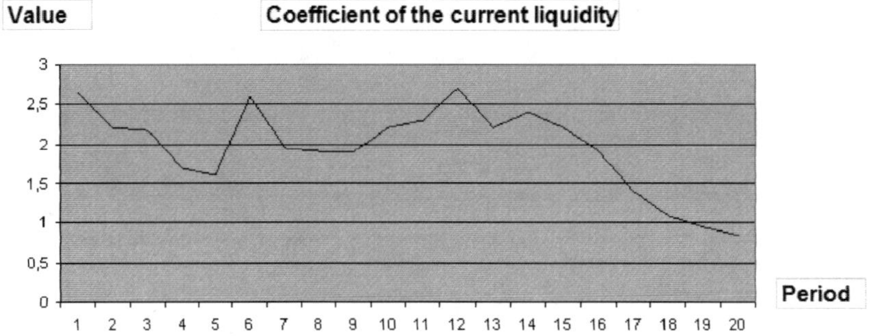

Figure 2. Changes trend of working capital ratio.

Then according to finite-difference method (Tintner method) a polynomial degree is defined, and with the help of least-square method one finds equation coefficients:

$\hat{y}_t = 1,92 + 0,107 \cdot t - 0,008 \cdot t^2$, where *"t"* is time in quarters.

It is possible to assess quality of the model by means of a relative approximation error, which is 14,96% in this case. An average approximation error is in interval 10-20%, consequently the model fidelity is quite good.

In a similar way one can form forecasting models for the rest factors.

On the basis of deduced equations it is possible to define forecasting factor values by 1 quarter 2012 (tab. 4).

2.4 Application of a model on the basis of fuzzy logic for enterprise bankruptcy risk forecasting

The authors calculated bankruptcy risk level of 27 enterprises on the basis of financial accounting data over the first quarter 2010-2012 with the help of six models: Altman's, Taffler's, Fulmer's, Springate's, Davydova-Belikov's, Telipenko's fuzzy set model (using 10 factors) [2-5]. A part of the calculations for the most prominent selected factors is presented in tab. 3.

At the result of analysis the authors drew a conclusion:

1) All reviewed models define precisely bankruptcy risk level of financially troubled enterprises[2]. Fig. 4 presents results of the models application which help to assess bankruptcy risk level of the enterprises, against which they instituted bankruptcy proceedings in 2012.

2) Assessments of bankruptcy risk level obtained after Altman's model application are very pessimistic: 20 enterprises out of 27 are bankrupts following the results of 1 quarter 2012, but it is not true. This fact is confirmed by quarterly enterprises reports.

3) Calculations of bankruptcy risk level by Fulmer's model are binary[3], as the result risk level is underestimated.

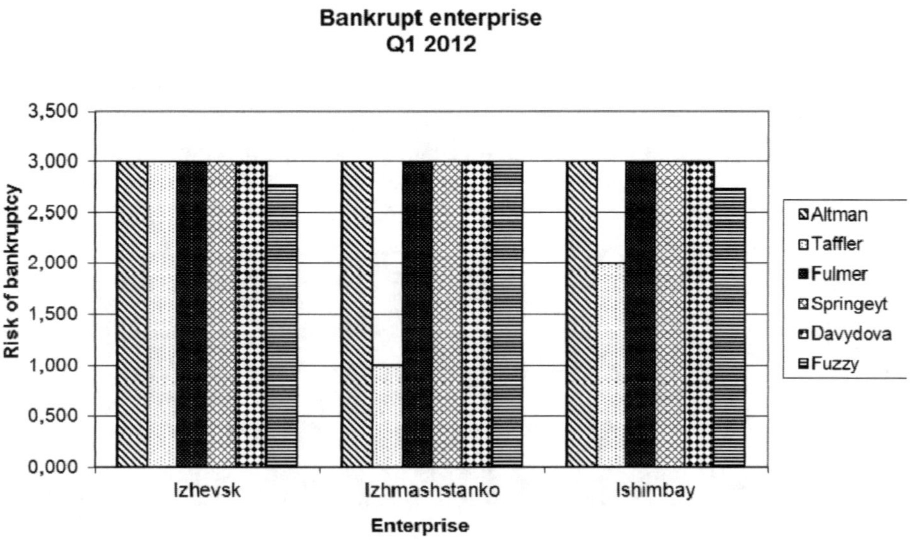

Figure 3. Results of the models application which help to assess bankruptcy risk level of the enterprises over 1 quarter 2012.

4) Fuzzy model gave appropriate assessments of bankruptcy risk level in respect of all analyzed enterprises. It is confirmed by quarterly enterprises reports.

For example, the method of fuzzy logic helped to correctly assess enterprise state (open joint-stock company "Ishimbaiskiy Machine-building Plant", Russia, Ishinbay city) not only in the time of crisis, but long before it (fig. 4). Against the plant they instituted bankruptcy proceedings in 1 quarter 2012.

You can see gradual degradation of the situation on the enterprise (fig. 4), the fuzzy model fairly presented it.

---

[2] Against these enterprises they instituted bankruptcy proceedings in different times in 2011-2012.

[3] Assessment of risk level has only two linguistic interpretations: either low or high without intermediate result.

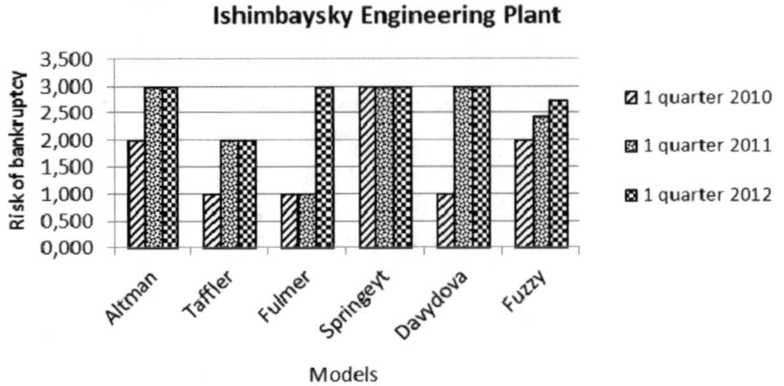

Figure 4. Bankruptcy risk assessment results of Open joint-stock company "Ishimbaiskiy Machine-building Plant".

Advantages of a fuzzy model are evident even when analyzing a successful enterprise (fig. 5).

Four models out of six assessed bankruptcy risk level of closed joint-stock company "Sibkabel" (Russia, Tomsk city) as low. But it is known that since 2000 financial condition of the enterprise has changed for the worse because of downswing of production output and serious contraction of market channels. Situation gradually improved when closed joint-stock company "Sibkabel" merged into Ural mining and metallurgical company (Russia, Verkhnaya Pyshma city). It was adequately shown by means of a fuzzy model of risk assessment and it is clearly seen in a block diagram.

Therefore most preferable model for bankruptcy risk forecasting of an enterprise is a fuzzy set model, because it is most sensitive to enterprise state changes and it gives a reliable assessment not only in the time of crisis but long before it.

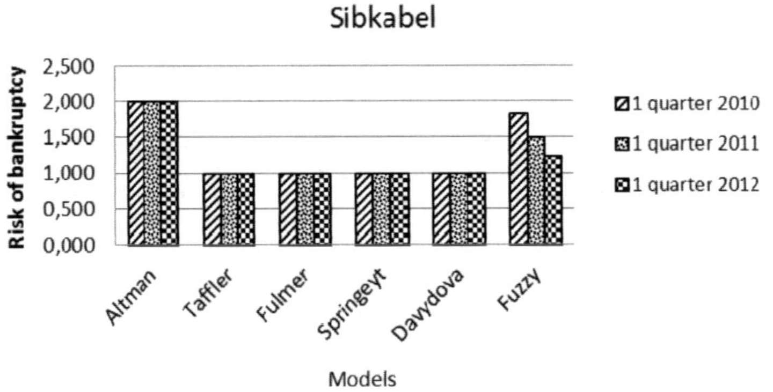

Figure 5. Bankruptcy risk assessment results of closed joint-stock company "Sibkabel".

2.5 Verification of forecasting model application on the basis of polynomial growth curves for calculation of indicator values in the future and with their help bankruptcy risk forecasting

For verification the authors made calculations for four enterprises, two of which were bankrupts.

10 indicators of enterprise financial and business activity over 2010-2011 were calculated to conduct analysis. They had been selected on the first step by means of the principal components method. On the base of obtained data forecasting models were formed and forecasting indicator values were calculated for 1 quarter, 2012.

Then on the basis of real and forecasting indicator values over 1 quarter, 2012, enterprises bankruptcy risk level was assessed and obtained results were compared (tab. 4).

Analysis of the table 4 shows:

1) assessment results of bankruptcy risk level with the help of a matrix method based on forecasting indicator values of enterprise financial and business activity almost coincide with assessment results based on real values; and when one classifies them they are in the same interval and have relatively equal assessment reliance;

2) assessment results of bankruptcy risk level confirm real situation of analyzed enterprises at 1 quarter 2012, i.e.: Commercial Court decision of the Udmurtian Republic as of 02.20.2012 was to declare OJSC "Izhmashctanko" (Russia, Izhevsk city) a bankrupt; Commercial Court decision of the Udmurtian Republic as of 04.06.2012 was to declare OJSC "Izhevsk machine-building plant" (Russia, Izhevsk city) a bankrupt; OJSC "Machine-building plant" (Russia, Electrostal city) and CJSC "Sibkabel" (Russia, Tomsk city) function today.

3. Conclusion

Taking into consideration everything what was said it is necessary to point out some advantages of a fuzzy set model for enterprise bankruptcy risk forecasting:

1. Application of 10 factors selected by means of the principal components method for analysis and bankruptcy risk level forecasting allows obtaining adequate results for machine-building enterprises, and it is confirmed by real situation in analyzed enterprises.
2. Application of forecasting models for indicator values calculation at a particular moment in the future allows fairly forecasting of situation development and identifying forecasted bankruptcy risk level.
3. Application of a matrix (fuzzy set) method at the step of assessment and forecasting of enterprise bankruptcy risk allows not only classifying risk level values as low or high but also considering their percentage. It helps to more fully comprehend the situation and come to the right management decision. Matrix method is also graphical, that is why it allows following dynamics of indicator values changes.
4. Regular application of a forecasting model for enterprise state monitoring will help to come to management decisions well-timed, but not post factum. It will help to reduce or escape bankruptcy risk.
5. Presented in the paper fuzzy set model has software support now, it is a module part of "Information system of bankruptcy risk management for an enterprise". This module can be used by owners, creditors, investors, and others for enterprise bankruptcy risk forecasting. More information about "Information system of bankruptcy risk management for an enterprise" is in [6-8].

References

[1] Telipenko E Zakharova A Bankruptcy risk management of a machine builder // Applied Mechanics and Materials 682 (2014) 617-622

[2] Altman E I Haldeman R G Narayanan P ZETATM analysis A new model to identify bankruptcy risk of corporations Journal of Banking and Finance Volume 1 Issue 1 June (1977) 29-54

[3] Toffler R Tishaw H Going going gone ñ four factors which predict // Accountancy March (1977) 50-54

[4] FULMER J G JR MOON J E GAVIN T A and ERWIN M J A bankruptcy classification model for small firms Journal of Commercial Bank Lending - 66 (1984) 25–37

[5] Springate G L V Predicting the Possibility of Failure in a Canadian Firm Unpublished M B A Research Project Simon Fraser University January 1978

[6] Zakharova A Telipenko E Information system of bankruptcy risk management of an enterprise Proceedings - 2012 7th International Forum on Strategic Technology IFOST 2012 – 2012

[7] Zakharova A A Fuzzy swot analysis for selection of bankruptcy risk factors Applied Mechanics and Materials Volume 379 (2013) 207-213

[8] Grigoreva A Information system of innovative products competitiveness determining Proceedings - 2012 7th International Forum on Strategic Technology IFOST 2012 – 2012

Table 3. Model approbation and comparison results for bankruptcy risk level determination.

Enterprise	Period of time	Altman's model		Taffler's model		Fulmer's model		Springate's model		Davydova-Belikov's model		Fuzzy logic model	
		Z	Linguistic interpretation	Z	Linguistic interpretation	H	Linguistic interpretation	Z	Linguistic interpretation	Z	Linguistic interpretation	G	Linguistic interpretation
OJSC "Izhevsk machine-building plant"	1qtr 2010	2.231	uncertain	0.093	high	-3.574	bankrupt	-0.040	high	-1.546	maximum (90-100%)	0.524	100% acceptable
	1 qtr 2011	0.801	bankrupt	0.100	high	-5.956	bankrupt	0.180	high	-1.257	maximum (90-100%)	0.687	56.5% acceptable; 43.5% high
	1qtr 2012	0.179	bankrupt	0.136	high	-10.391	bankrupt	0.186	high	-3.368	maximum (90-100%)	0.754	23% acceptable; 77% high
CJSC "Sibkabel"	1 qtr 2010	1.778	uncertain	0.496	low	3.209	low	1.444	low	4.041	insignificant (до 10%)	0.367	16.5% low; 83.5% acceptable
	1 qtr 2011	1.694	uncertain	0.534	low	3.453	low	1.224	low	2.043	insignificant (до 10%)	0.299	50.5% low; 49.5% acceptable
	1 qtr 2012	2.031	uncertain	0.590	low	4.337	low	1.151	low	2.357	insignificant (до 10%)	0.246	77% low; 23% acceptable
OJSC "Machine-building plant Zio-Podolsk"	1 qtr 2010	0.392	bankrupt	0.289	average	0.812	low	0.789	high	1.689	insignificant (до 10%)	0.699	50.5% acceptable; 49.5% high
	1 qtr 2011	0.319	bankrupt	0.263	average	1.688	low	0.881	low	1.802	insignificant (до 10%)	0.616	80% acceptable; 20% high
	1qtr 2012	0.264	bankrupt	0.237	average	1.120	low	0.846	high	1.768	insignificant (до 10%)	0.746	27% acceptable; 73% high
OJSC "Machine-building Plant", Electrostal city	1 qtr 2010	2.774	uncertain	0.819	low	5.713	low	0.833	high	3.577	insignificant (до 10%)	0.238	81% low; 19% acceptable
	1 qtr 2011	1.974	uncertain	0.362	low	5.513	low	0.502	high	1.948	insignificant (до 10%)	0.395	100% acceptable
	1qtr 2012	2.275	uncertain	0.343	low	5.369	low	0.399	high	2.729	insignificant (до 10%)	0.353	23.5% low; 76.5% acceptable
OJSC "IZHMASH-Stanko"	1 qtr 2010	-0.661	bankrupt	0.216	average	-3.547	bankrupt	0.101	high	-4.506	maximum (90-100%)	0.837	100% high
	1 qtr 2011	-0.456	bankrupt	0.297	average	-2.923	bankrupt	0.654	high	-6.845	maximum (90-100%)	0.891	100% high
	1qtr 2012	-1.617	bankrupt	0.349	low	-6.494	bankrupt	0.167	high	-9.117	maximum (90-100%)	0.819	100% high
OJSC "Ishimbaiskiy Machine-building Plant"	1 qtr 2010	1.937	uncertain	0.386	low	1.448	low	0.496	high	2.937	insignificant (до 10%)	0.469	100% acceptable
	1 qtr 2011	0.826	bankrupt	0.286	average	0.851	low	0.581	high	-1.918	maximum (90-100%)	0.684	58% acceptable; 42% hihg
	1qtr 2012	0.280	bankrupt	0.256	average	-0.025	bankrupt	0.513	high	-3.882	maximum (90-100%)	0.747	26.5% acceptable; 73.5% high

IOP Conf. Series: Materials Science and Engineering **91** (2015) 012066 doi:10.1088/1757-899X/91/1/012066

Table 4. Assessment results of enterprise bankruptcy risk level on the basis of real and forecasted indicator values over 1 quarter, 2012.

Indicator name	OJSC "Izhmashstanko", Izhevsk city		OJSC "Izhevsk machine-building plant", Izhevsk city		OJSC "Machine-building plant", Elektrostal city		CJSC "Sibkabel", Tomsk city	
	fact	forecast	fact	forecast	fact	forecast	fact	forecast
x_1 – working capital ratio	0.123	0.126	0.387	0.419	3.000	3.15	2.98	2.83
x_2 – acid test ratio	0.21	0.24	0.0059	0.0055	1.293	2.44	1.41	1.9
x_3 – absolute liquidity ratio	0.0003	0.0002	0.00016	0.00023	0.394	0.444	0.00223	0.00198
x_4 – liquid capital ratio in assets	0.188	0.201	0.089	0.129	0.503	0.492	0.771	0.757
x_5 – concentration ratio of capital loans	1.783	1.813	0.481	0.605	0.170	0.100	0.251	0.264
x_6 – long term fund raising ratio	0.063	0.049	0.0184	0.0203	0.026	0.019	0.0199	0.0201
x_7 – borrowed and own funds ratio	2.276	2.571	0.925	0.981	0.205	0.199	0.334	0.381
x_8 – turn-round of float	0.363	0.514	0.032	0.065	0.476	0.392	1.726	1.958
x_9 – turn-round of owned capital	0.219	0.322	0.006	0.012	0.059	0.057	0.621	0.786
x_{10} – product profitability, %	0.563	0.478	0.03	0.01	32.9	27.2	6.25	6.13
Generalized estimator of a risk level	**0.819**	**0.843**	**0.754**	**0.765**	**0.403**	**0.301**	**0.246**	**0.237**
Linguistic interpretation	**High 100%**	**High 100%**	**Acceptable 23% High 77%**	**Acceptable 17,5% High 82,5%**	**Low 23,5% Acceptable 76,5%**	**Low 29,5% Acceptable 70,5%**	**Low 77% Acceptable 23%**	**Low 81,5% Acceptable 18,5%**

IOP Conf. Series: Materials Science and Engineering **91** (2015) 012067 doi:10.1088/1757-899X/91/1/012067

Software selection based on analysis and forecasting methods, practised in 1C

A N Vazhdaev[1, a], **T YChernysheva**[1, b], **E I Lisacheva**[1, c]

[1, 2, 3] 26, Leningradskaya street, Yurga, Kemerovo region, 652055, Russia
Yurga Technological Institute (affiliated) National Research Tomsk Polytechnic University

e-mail: [a]wazdaev@ngs.ru, [b]chernyshevat@tpu.ru, [c]djlena.91@mail.ru

Abstract. The research focuses on the problem of a "1C: Enterprise 8" platform inboard mechanisms for data analysis and forecasting. It is important to evaluate and select proper software to develop effective strategies for customer relationship management in terms of sales, as well as implementation and further maintenance of software. Research data allows creating new forecast models to schedule further software distribution.

1. Introduction

One of the current requirements imposed to organizations is the utilization of information systems and technologies. Many organizations place confidence in businesses that specialize in software supply and maintenance to select proper software. In fact, the process of evaluating and selecting proper software is a complex and multistep process. Mathematical evaluation is performed by both statistic and expert methods [1]. When organizations use inboard mechanisms for data analysis and forecasting while working in authors information system of product programs evaluation and selection, they can develop an effective strategy for customer relationship management in terms of sales [2], as well as implementation and further maintenance of software [3].

2. "1C: Enterprise 8" platform inboard mechanisms

Mechanism for data analysis and forecasting is one of the "1C: Enterprise 8" platform options that help to develop commercial and analytical reports.
This mechanism allows [4]:
- to find consistent patterns of the information database input data;
- to control evaluation parameters by means of software-base method and interactively;
- to implement of programmic access to the evaluation results;
- to convert evaluated data into tabular style automatically;
- to create forecast models that automatically forecast subsequent events or some new objects values.

Initial data for evaluation can be obtained from information databases as well as from external sources. Recommendations, developed due to evaluation, make further operations more efficient.

Evaluated data is a sample of data behavior which can be displayed in a final document, or saved for future utilization. Further utilization of evaluated data allows creating a trend model to forecast new datum behavior in accordance with the existing model [5].

IOP Conf. Series: Materials Science and Engineering **91** (2015) 012067 doi:10.1088/1757-899X/91/1/012067

3. Different types of platform "1C: Enterprise 8"data analysis

Platform "1C: Enterprise 8" mechanism of data analysis and forecasting includes several types of data evaluation [4]:

1) General Statistics. This mechanism serves to gather information about data selected to be analyzed. This type of evaluation is intended for preliminary investigation of evaluated data source.

Evaluation shows a number of characteristics of discrete and continuous fields. When the report is converted into tabular style, pie charts are filled to display fields' structure. Figure 1 shows the use of this type of data evaluation in the information system developed by the authors of the paper.

Datum in a source can be of continuous or discrete types. Numbers and dates are continuous, while other types refer to discrete one. Each type contains its own specific information [4].

Discrete datum:

- Number of values. Number of values in a data source field;
- Number of unique values (with the exception of duplicate values);
- Mode (the most frequent value in a data source). If there is more than one value with equal frequency, the first one is to be taken as a mode;
- Frequency (depends on the number of times the value occurs in a data selection list);
- Relative frequency (is the ratio of a number of volume occurrence to the total number of values);
- Cumulative frequency (is a sum of values frequency and frequency amount of previous values in a data selection list);
- Cumulative relative frequency (is a sum of a cumulative frequency of a value and a sum of relative frequencies of previous values).

Continuous data:

- Number of values;
- Minimal value;
- Maximal value;
- Average;
- Swipe. Difference between maximal and minimal values;
- Standard deviation (meansquare deviation);
- Median value (value that lies in the middle of a data selection list).

It should be noted that if several fields of different types are analyzed simultaneously, evaluation is to be done independently so to avoid intercorrelation.

2) Search for associations. This type of evaluation finds groups of objects that often go together or property value, and searches for association rules [4]. Search for associations can be used, for example, to determine items and services frequently purchased together. Figure 2 demonstrates search for associations in an information system. Evaluation helps to obtain:

- data information (number of objects and elements, average number of elements in an object, number of groups and association rules found);
- found groups elements. Group items are specified as well as an amount of cases and percentage of cases in which the group presents;
- found association rules. Indicates initial composition of elements, consequence (elements composition) and percentage of cases, credibility and relevance rules.

IOP Conf. Series: Materials Science and Engineering **91** (2015) 012067 doi:10.1088/1757-899X/91/1/012067

Data analysis and forecasting

Data analysis | Prediction

General Statistics | Search Associations | Search Strings

The Result:

General statistics

Information about the data

Number of objects: 13

Continuous fields

Field	Values	Minimum	Maximum	Average	Sweep	STD. off.	Median
The number of	13	1	5	1,6154	4	1,3253	1

Discrete field

Nomenclature

Number of values: 13
The number of unique values: 5
Fashion: Salary and personnel management 8

Frequency table

The value of	Frequency	Relative frequency	Cumulative frequency	Cumulative relative frequency
Salary and personnel management 8	4	30,77	4	30,77
Trade management 10.3.	3	23,08	7	53,85
Accounting 8 Prof	3	23,08	10	76,92
Retail 2.0	2	15,38	12	92,31
8 Basic accounting	1	7,69	13	100,00

The counterparty customer

Number of values: 13
The number of unique values: 5
Fashion: IP Kurochkin P.s.

Frequency table

The value of	Frequency	Relative frequency	Cumulative frequency	Cumulative relative frequency
IP Kurochkin P.s.	8	61,54	8	61,54
Momentum LLC	2	15,38	10	76,92
IB Petrov Sergey Mikhailovich	1	7,69	11	84,62
Siberia LLC	1	7,69	12	92,31
IP Pavlov A. M	1	7,69	13	100,00

Industry

Number of values: 13
The number of unique values: 3
Fashion: Retail, warehouse, logistics, transport

Frequency table

The value of	Frequency	Relative frequency	Cumulative frequency	Cumulative relative frequency
Retail, warehouse, logistics, transport	8	61,54	8	61,54
Professional services	4	30,77	12	92,31
Production, ENERGY	1	7,69	13	100,00

Figure 1. General Statistics type of evaluation.

3) Search for sequences. This type of evaluation helps to find successive chains of events in a da-ta source [4]. This type of evaluation allows to search through the hierarchy, which allows monitoring not only a specific sequence of events, but also a sequence of parent groups. Figure 3 shows the use of search sequences in an information system.

The main result of this type of evaluation is templates sequences. These templates contain following information [8]:

- composition of template sequence;
- number of cases with this sequence;
- maximum intervals between events (two events - interval a);
- minimum intervals between events (two events - interval a);
- percentage of cases in which the sequence was executed;
- average intervals between events (two events - interval a).

IOP Conf. Series: Materials Science and Engineering **91** (2015) 012067 doi:10.1088/1757-899X/91/1/012067

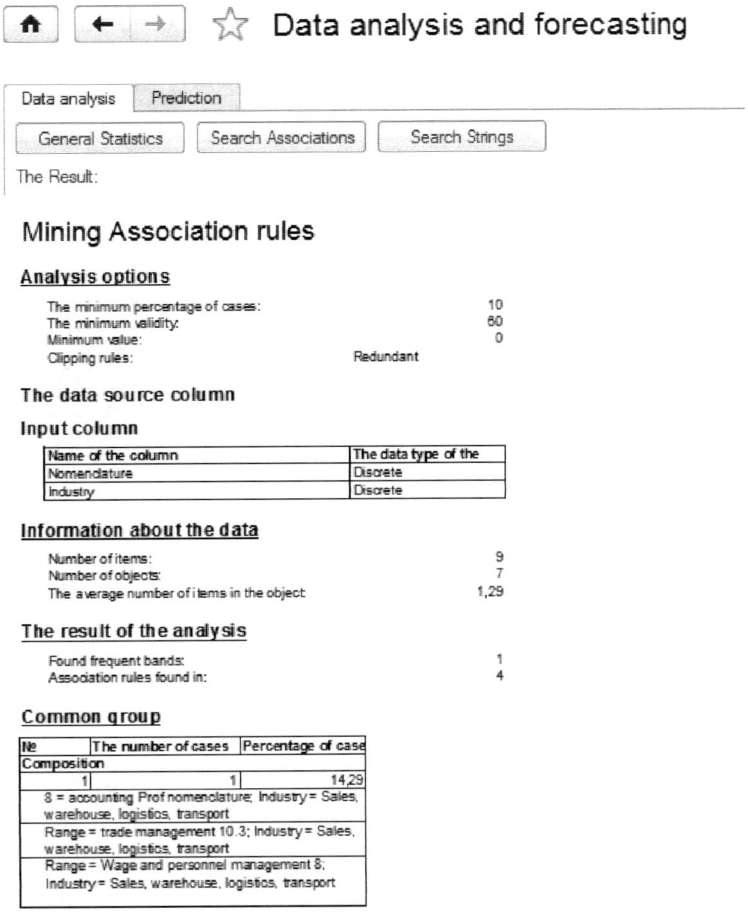

Figure 2. Search for associations type of evaluation.

4) Cluster analysis. Clustering helps to select a number of relatively homogeneous groups, which are segments or clusters, among objects of the same nature [5]. The analysis is based on calculating distance between groups, which are clusters. It is possible to calculate distance with different methods (based on different metrics). Platform "1C: Enterprise 8" supports following metrics:

• Euclidean metric. Distance between two objects is calculated by the equation:

$$S = \sqrt{\sum (W_i * (X_i - Y_i)^2)} \quad , \tag{1}$$

where

X_i, Y_i –two objects attribute values (distance between these objects is calculated);

W_i – weight number of an attribute;

i – attribute number from 1 to n;

n – number of attributes.

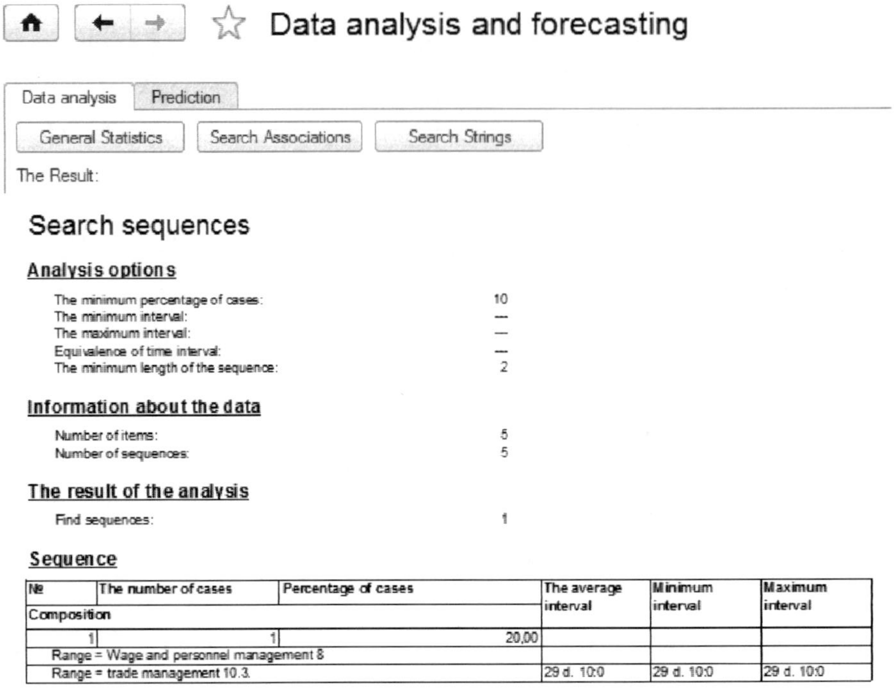

Figure 3. Presentation of a Search for sequences evaluation method results.

• The Euclidean metric squared. Distance between two objects is calculated by equations:

$$S = \sum (W_i * (X_i - Y_i)^2) \tag{2}$$

where

X_i, Y_i – two objects attribute values (distance between these objects is calculated);

W_i – weight number of an attribute;

i – attribute number from 1 to n;

n – number of attributes.

• City metric. Distance between two objects is calculated by equations:

$$S = \sum (W_i * |X_i - Y_i|) \tag{3}$$

where

X_i, Y_i – two objects attribute values (distance between these objects is calculated);

W_i – weight number of an attribute;

i – attribute number from 1 to n;

n – number of attributes.

• Dominance metric. This distance metric is calculated between two objects according to equations:

$$S = MAX(W_i * |X_i - Y_i|) \tag{4}$$

where

IOP Conf. Series: Materials Science and Engineering **91** (2015) 012067 doi:10.1088/1757-899X/91/1/012067

X_i, Y_i – two objects attribute values (distance between these objects is calculated);

W_i – weight number of an attribute;

i – attribute number from 1 to n;

n – number of attributes.

When distance between objects is calculated, several algorithms that allow ranging objects into clusters can be opted.Clusterization method is chosen on the base of principals through which an object refers to this or that group and algorithms of clusterization [6]. Any clusterization algorithm aims at minimizing variability within clusters, and at maximizing variability between clusters. Following clusterization methods are supported:

• Short-range communication – object: we tap to a group which is at a minimal distance to the nearest object;

• Long distance – object: we tap to a group which is at a minimal distance to the furtherest object;

• Middle-end: we opt for objects which are at the top of a data selection list. They are considered as cluster centers. Then we opt for another object in accordance with a distance to the cluster centers. Cluster center, to which an object was added, is to be recalculated. The procedure is repeated until the complete enumeration of all objects. Then we start a new object selection (starting with the first one). The procedure is repeated as long as cluster centers vary;

• Center of gravity: we tap objects to a group with a minimal distance from the center of gravity.

Currently, the authors of cluster analysis implemented in its information system [7].

5) Decision tree. This method of evaluation allows developing a hierarchical structure of classification rules in a form of a tree. To make a decision tree, it is necessary to choose a target attribute, according to which it will be possible to make a classifier, and a number of input attributes that will be used to develop rules.

The research product is a decision tree each joint of which contains a condition or classification error that reveal conditions at which developed rules contradict to reality. To classify a new object, it is necessary to answer questions in each joint and pass all the chain from the root up to the leaves of the tree, going to daughter joints if the answer is positive or to neighbor joints if the answer is negative. An evaluation criteria set helps to regulate accuracy of the tree [4,5].

It is possible to make the tree less complicated. Some rules allow converting tree joints into leaves, excluding excessive branching [9].

During the process of converting a joint into a leaf it is necessary to take into account:

$$O = \sqrt{\frac{\left(b+\frac{k}{2}\right) * \left(s-\left(b+\frac{k}{2}\right)\right)}{s}} \tag{5}$$

where

a - number of errors in a joint, b - number of errors in daughter joints, k - number of leaves in a joint, s - number of cases.

We decide to convert a joint into a leaf if following conditions are fulfilled:

$$a + 0,5 \prec= b + \frac{k}{2} + O \tag{6}$$

4. Results and Discussion

The task of forecasting on a "1C: Enterprise 8" platform is generally divided into two basic procedures [10]:

1. Training the model based on a definite data selection list.

2. Utilization of a trained model in the process of actual data forecast.

Results of a S*earch for Associations* analysis help to develop a model of association forecast automatically. It means that a buying organization, based on other cases analogy, will more likely purchase a recommended item alongside with those software products it bought before. Sequence forecast model is developed through the evaluation of a Search for associations analysis.

5. Conclusion

Datum obtained from a database and external sources with their further evaluation by means of different types of analysis available on a "1C: Enterprise 8" platform allows developing specific recommendation for further actions. Evaluation helps to develop a trend model, which allows businesses to forecast new data behavior in accordance with a model which is already in operation.

Using the described 1C: "Enterprise 8" platform mechanisms of analysis and forecasting data during the process of choosing proper software for franchisee firms' clients, allows businesses to find patterns in sale programs and develop forecast models able to plan further sales of software automatically.

References

[1] Vollmann T, Berry W, Whybark D C, Jacobs F R 2004 *Manufacturing planning and control systems for supply chain management: The Definitive Guide for Professionals (5th ed.)* (McGraw-Hill)

[2] Saloner G, Shepard A, Podolny J 2001 *Strategic Managemen* (Wiley)

[3] Toporkov V V, Yemelyanov D M 2014 *Economic model of scheduling and fair resource sharing in distributed computations* Programming and computer software (USA: Pleiades Publishing) pp 35-42

[4] Kataev M Y, Emeljyanenko V A, Ponomareva M G 2014 *The model of strategic planning at the enterprise with process management approach* Applied Mechanics and Materials (Switzerland: Elsevier) pp 572-578

[5] Kataev M Y, Bulysheva L A, Emeljyanenko A A, Emeljyanenko V A 2013 *Enterprise systems in Russia: 1992–2012* Enterprise Information Systems (Great Britain: Elsevier) pp 169-186

[6] Telipenko E V, Zakharova A A 2014 *Bankruptcy risk management of a machine builder* Applied Mechanics and Materials (Switzerland: Elsevier) pp 617-622

[7] Chernysheva T Y, Gnedash E V, Zorina T Y, Lenskaya N V 2014 *Information systems project risk assessment: expert approach* Applied Mechanics and Materials (Switzerland: Elsevier) pp 539-543

[8] Chernysheva T Y 2013 *Preliminary risk assessment in IT projects* Applied Mechanics and Materials (Switzerland: Elsevier) pp 220-223

[9] Chernysheva T Y, Zhukov A G 2013 *Hierarchical model evaluation and selection of software prod-ucts for the enterprise* Polzunovsky Gazette (Krasnoyarsk: AltSTU) pp 25-28

[10] Lisacheva E I, Vazhdaev A N 2013 *Decision Support System franchisee organization for the selection of software for buyers* Polzunovsky Gazette (Krasnoyarsk: AltSTU) pp 224-228

IOP Conf. Series: Materials Science and Engineering **91** (2015) 012068 doi:10.1088/1757-899X/91/1/012068

Strategic Concept of Competition Model in Knowledge-Based Logistics in Machinebuilding

O V Medvedeva
Russia, Kemerovo region, Yurga, Leningradskaya str.26,
Yurga Institute of Technology of National Research Tomsk Polytechnic University

E-mail: AlisaSten@mail.ru

Abstract. A competitive labor market needs serious changing. Machinebuilding is one of the main problem domains. The current direction to promote human capital competition demands for modernization. Therefore, it is necessary to develop a strategy for social and economic promotion of competition in conditions of knowledge-based economy, in particularly, in machinebuilding. The necessity is demonstrated, as well as basic difficulties faced this strategy for machinebuilding.

Introduction
Current employer-employee relations demand for fundamental modernization. Machinebuilding is the most problem domain, where the symbiosis of contemporary model of industry development and its traditional form can be seen. Labor force of machinebuilding doesn't meet the challenges of current labor market. For correct and timely promotion of labor force competition in machinebuilding we need to redefine completely the current market of human capital competition in this area. A strategy of social and economic competition model in conditions of knowledge-based economy can be applied in order to promote labor force competition in machinebuilding.

Main body
The strategy of social and economic competition model in conditions of knowledge-based economy is to include various directions of economy development and has to contribute to stabilization of country positions at the world market. A number of serious transformations are to be made for this purpose (Figure 1).

The data in Table 1 make it evident, that the government has been increasing expenses on social policy of the country; therefore, revenues of this direction are to be controlled and raised.

The number of small and medium-sized businesses in the country is to be controlled over, because they, competing with similar companies, influence on competition at the domestic market. According to the information provided by Federal State Statistic Agency the number of small and medium-sized businesses is reducing. Therefore, it is necessary to develop this sphere, as well as machinebuilding, because this industry will have strong positions both in small and medium-sized business and influence on big business.

Similarly to the shortening number of businesses in the country enterprises suffer from demanning. Employers and employees have to pay much attention to research activities. It is of primary importance for machinebuilding because scientific and technical basis will provide both training of skilled personal, and designs needed by machinebuilding enterprises on the whole. Moreover, research will promote economy of the country on the world stage. It's also important that research is very valuable for Russia.

IOP Conf. Series: Materials Science and Engineering **91** (2015) 012068 doi:10.1088/1757-899X/91/1/012068

Our country sells raw materials only, and research will help Russia to be in the top countries of the world even in terms of raw materials depletion.

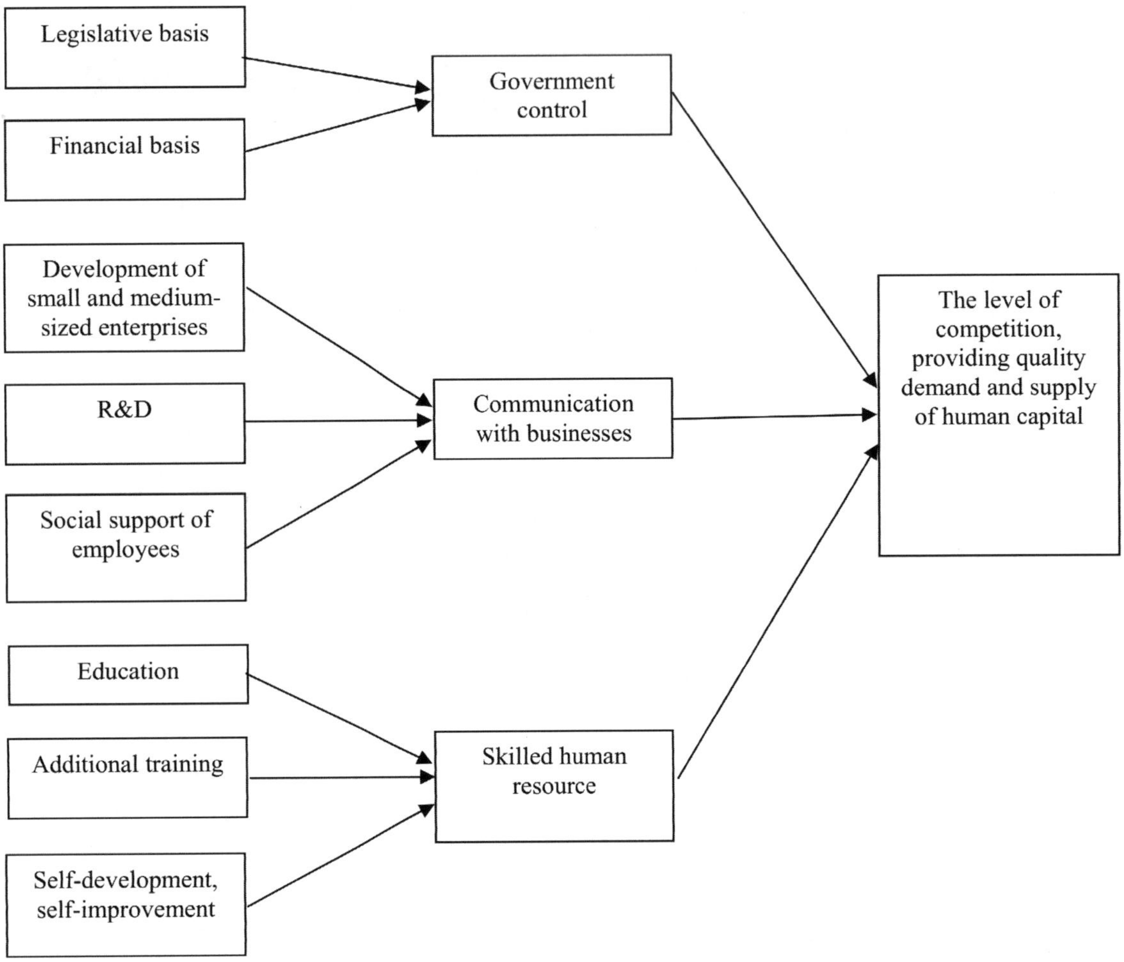

Figure 1. Development of competition level in conditions of knowledge-based economy

Table 1. Main indexes of current competition level in conditions of knowledge-based economy

Main indexes	2009	2010	2011	2012	2013	2014
Social policy expenses, milliard rubl.	310.905	344.944-	3128.5	3859.8	3839.1	1927.1
Small and medium-sized businesses, unit	69274	66574	63803	60260	58245	58959
R&D expenses, milliard rubl.	485.09	523.29	608.22	696.85	674.23	447.06

Main indexes	2009	2010	2011	2012	2013	2014
Education expenses, milliard rubl.	387.9	386.4	552.4	603.5	558.9	499.5
Social support of employees expenses, million rubl.	930.9	3370.6	3059.6	7824.0	8440.0	9092.0

After expenses on R&D rose they have been gradually reducing at the moment. This declining has a negative character, because the government spends less on research activities. Having considered the index of population scholarship a negative tendency has been also revealed over the last years.

First, the decline of University graduates with higher technical education is to be noted: 2010 – 1177.8 thousand people, 2011 – 1157.3 thousand people, and 2012 – 1125.3 thousand people [1]. Second, over the last years the expenses on education have been shortened.

One more important factor of competition development of knowledge-based economy is social support of employees, including those of machinebuilding industry. It is social support that will be an influential factor for a skilled employee of any branch; a person besides a salary, paid for his/her work wants to obtain additional compensation in form of social benefits for him/her and their family.

The government allocates funds annually aimed at social support of population. Expenses on social support of employees have been raised recently.

The general dynamics of current competition level in conditions of knowledge-based economy is shown in Figure 2.

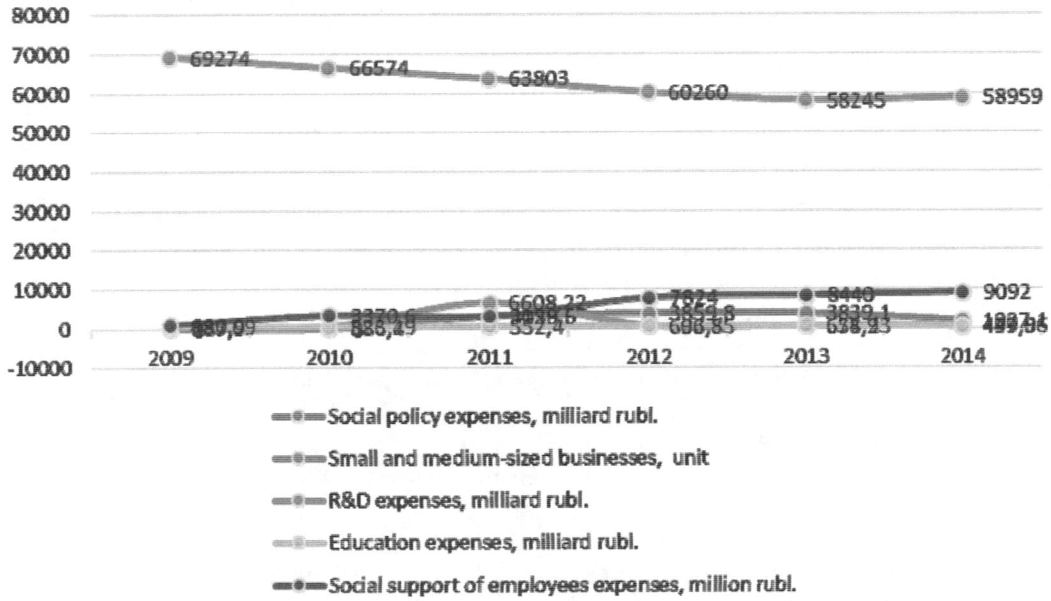

Figure 2. The dynamics of main factors of current competition level in conditions of knowledge-based economy from 2009 to 2014

All aforementioned statistics are relevant for forming competition level of human capital in the country nowadays; however, a social and economic model of competition is to be developed in conditions

IOP Conf. Series: Materials Science and Engineering **91** (2015) 012068 doi:10.1088/1757-899X/91/1/012068

of knowledge-based economy. The government has this model at present; it is subjected to constant transformations, but doesn't meet the requirements of human capital market.

So an optimal model is necessitated for social and economic control over competition in terms of knowledge-based economy. A number of revolutionary changes are to be made in this area of country's economy. Legislative and financial systems of the country are to be changed to form an optimal level of governmental control.

Legislative basis can be changed as follows to form governmental support:

1) The government considers and creates new terms and laws in the sphere of human resources development;
2) The government has to influence on financial basis to support competition.

If the mentioned above conditions are met the government will have a definite level of control over the level of competition, it will guarantee a correct regulation of human resources demand and supply, these recourses form the main capital of the country needed for economy development.

Then, communication with businesses is to be considered, that is relations to the skilled personal.

This stage involves possibility of creating additional interest of workers in the development of their enterprise. The government must not develop this model by budgetary means to encourage labor force, it has to motivate and demonstrate the necessity to implement this model for business development in the country.

The government has to pay much attention to deterioration; businesses must rely on the government and its support. We have already mentioned that the government has to appreciate the necessity and value of this support in spite of determining a common threshold for all businesses.

The governmental support aims at development of social economic and judicial conditions for increase in economic growing power of small and medium-sized businesses. In addition, we can say that governmental support is to include encouragement of businesses to train skilled personal through seminars and workshops, as well as through additional training at plants and in branches, which employees are interested in.

In any country businesses influence on its economy, and the level of the development of the country depends on that of businesses. This fact is of a primary importance for development of human capital competition, because the majority of employees work for these businesses.

To train new skilled personal the government has to take some measures for reforming the system of education in the country, adopting useful experience of other countries.

Moreover, the government has to encourage businesses for R&D. An employee stars suggesting his conditions for further work of the business provided that R&D is paid attention to at this enterprise. An employee will also get benefits of other businesses, which were supported by the government, so in this case the government has an indirect influence. The separation of an employee from the business can provide rather more possibilities. For instance, a separate sub-division, an employee is more independent, therefore, more valuable.

Considering social support of human resources in terms of their links with businesses we have to emphasize:

– Employers have had quite new attitude to motivation of employees to their work, it has been determined that personal is the main factor effecting on business success. Material encouragement of employee along with moral incentives is very important at present;
– The government has to promote the development of human potential at the enterprise.

Studying human resources in this context it is necessary to take into consideration abilities of a person (as an employee) for a particular business, for particular sphere.

Finally quality of human capital and its constituents are to be considered which influence on competition development. We start this consideration with the system of education of the country. The system shown in Figure 3 is applied at present.

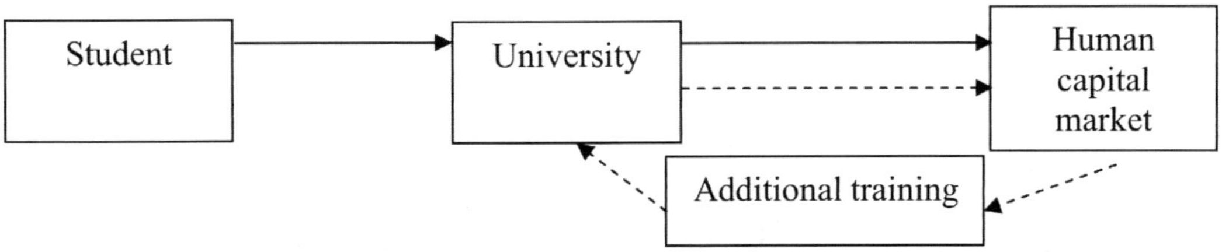

Figure 3. A wide-spread model of education

In this model a special attention isn't paid to the number of factors influencing on the educational process, although they are very important for the development of the system of education in the country.

We use additional factors in the model of the educational system of the country (Figure 4).

First, we have to keep in mind, who needs training: a student with no experience or a specialist with higher education, obtaining additional training. We also have to respect their wishes.

Second, a staff of the institution of higher education is to be focused on, as well as forms of student support. It is included in the process of education obtaining. A student can't be interested in obtaining of education anymore, he/she needs motivating.

Third, we have a specialist having particular knowledge, skills and practices, he/she can either have work experience or not.

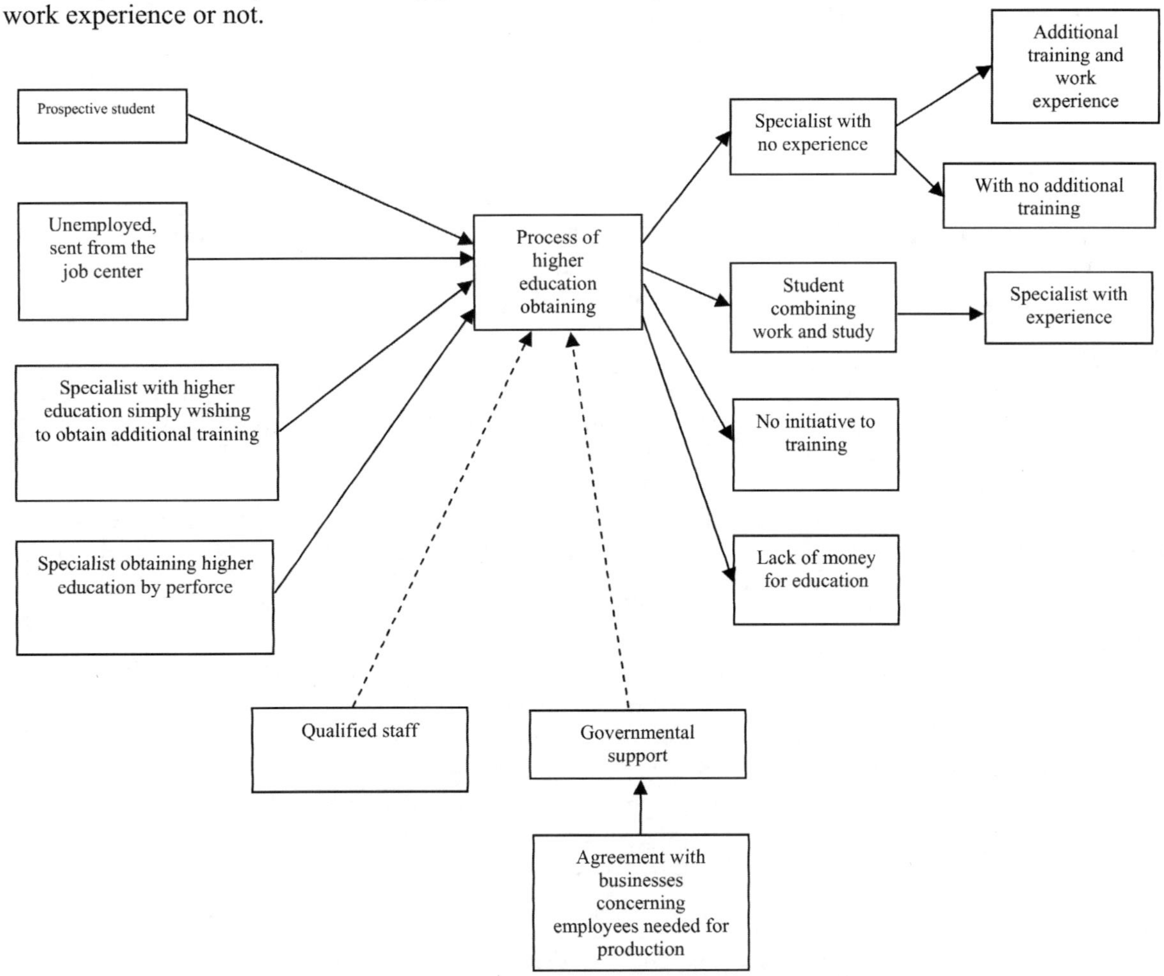

Figure 4. Model of education obtaining in details

IOP Conf. Series: Materials Science and Engineering **91** (2015) 012068 doi:10.1088/1757-899X/91/1/012068

Government support is a main component of this model; it is to focus on accessibility of higher education for everybody who wants to get it. However, we should give attention to the work of the government with businesses aimed at obtaining timely information on specialists needed for development of businesses.

It's also to say that a specialist with higher education but with no wok experience isn't as required as a specialist with the same education and work experience. An exception is an age of the specialist. The older the specialist is, the harder is to save a place of work, because a business will get a new specialist who masters the specifics of work quicker and promotes the development of the business through providing it with innovations.

Additional training of a specialist is other factor of human capital.

The government and businesses have to encourage a specialist for obtaining additional training. We have to realize that a high level of knowledge and work experience an employee has is relevant for the development of our country's economy.

If a necessary level of additional training and work experience is provided, trained personal will influence on other employees. At this point competition of knowledge-based economy goes beyond the businesses.

The government has to promote development of research institutions, where research is carried out, new technologies are developed; it will be useful for industry and improve the level of specialist training in a particular branch.

In addition, the government has to give an employee wider opportunities and let him/her compete at the labor market. An employee can also be sent to more successful regions of the country.

However, self-development and self-improvement are the most important factors. As an employee wants to get new and productive knowledge he can obtain the sufficient qualifications in his sphere (self-development, improvement of own competitive ability in order to minimize risks of being unemployed or for making a carrier).

Self-development and self-improvement are the most complex factors of qualified human capital formation, an employee has to admit it and focus on self-development in his area. The businesses are to promote specialists and give him a possibility for self-development. The government, in its turn, is to provide conditions, opportunities for human resources development, obtaining self-education, carrying our research activities.

In conclusion, we can say that complete and comprehensive modernization of human resources market by the government doesn't cause changes at this market now, but in future. Difficulties resulted from modernization of the market of human resources arise in machinebuilding, because machinebuilding tends to use a traditional model of raw materials utilization and doesn't meet quickly the economic challenges of the country.

Nevertheless, competition is necessary for all market agents. A person investing in his/her self-development gets an advantage at this market and a business improves its competitive ability employing well-qualified specialists in their branches.

Conclusions

Quality competition at human resources market is a complex multi-component mechanism, the efficiency of its work depends on the governmental support. In machinebuilding industry improving of competitive ability is associated with scientific and technical progress, the importance of human resources isn't taken into consideration of general strategy of business development though.

This paper demonstrates the strategy of social and economic competition model of knowledge-based economy; its implementation will have a positive effect on the situation at human recourses market. Therefore, positions of machinebuilding enterprises that apply it will be strong.

References

[1] Volume and structure of revenues [Electronic resource]. – Accessed at: http://www.gks.ru/bgd/regl/b13_13/IssWWW.exe/Stg/d1/06-06.htm

[2] Shvedova M.F. Structure of governmental support of small business: theoretical aspect of system approach: Fundamental research. 2012. № 9-1. pp. 135-139

[3] Barbarskaya M.N., Rybalkina S.M. Development of the system for human potential management in an organization // Human resources: issues of innovative development and utilization: collected works / edited by I.P. Povarich. – Issue 2; Kemerovo State University. – Kemerovo, 2013. – pp. 25-31

[4] Various statistical information on the budget of Russia in 2013 [Electronic resources] – Accessed at: http://inance.ru/2014/02/budzet2013/

[5] Structure and dynamics of outcomes [Electronic resource] – Accessed at: http://info.minfin.ru/fbrash.php

[6] Institutional transformations in economy. Big and medium-sized businesses and enterprises [Electronic resource] – Accessed at: http://www.gks.ru/dbscripts/cbsd/dbinet.cgi

[7] Institutional transformations in economy [Electronic resource] – Accessed at: http://www.gks.ru/wps/wcm/connect/rosstat_main/rosstat/ru/statistics/enterprise/reform/#

[8] Specialists trained by state and municipal educational institutions in specialties and training programs in 2005 – 2012 [Electronic resource] – Accessed at: http://www.gks.ru/bgd/regl/b13_13/IssWWW.exe/Stg/d1/07-50.htm

[9] Education in Russia [Electronic resource] – Accessed at: https://ru.wikipedia.org/wiki/%D0%9E%D0%B1%D1%80%D0%B0%D0%B7%D0%BE%D0%B2%D0%B0%D0%BD%D0%B8%D0%B5_%D0%B2_%D0%A0%D0%BE%D1%81%D1%81%D0%B8%D0%B8

IOP Conf. Series: Materials Science and Engineering **91** (2015) 012069 doi:10.1088/1757-899X/91/1/012069

Methods for control over learning individual trajectory

A A Mitsel [1], N V Cherniaeva [2]

[1] Professor, Doctor of Engineering Sciences,
Yurga Institute of Technology Tomsk Polytechnic University
[2] Postgraduate student, Tomsk Polytechnic University

e-mail: maa@asu.tusur.ru, nina.turalina@yandex.ru

Abstract. The article discusses models, methods and algorithms of determining student's optimal individual educational trajectory. A new method of controlling the learning trajectory has been developed as a dynamic model of learning trajectory control, which uses score assessment to construct a sequence of studied subjects.

1. Introduction

In connection with the transition of the educational system to a competence-oriented approach, the problem of learning outcomes assessment and creating an individual learning trajectory of a student has become relevant. Its solution requires the application of modern information technologies.

The third generation of Federal state educational standards of higher professional education (FSES HPE) defines the requirements for the results of mastering the basic educational programs (BEP). According to FSES HPE up to 50% of subjects have a variable character, i.e. depend on the choice of a student. It significantly influences on the results of developing various competencies.

The object of research is the process of formation of individual student's learning trajectory and the subject is dynamic model of formation of individual student's learning trajectory based on linear dynamic model of portfolio management of securities under constraints.

The problem of the research is development of a model (algorithm) to form individual learning student's trajectory. The aim of research work is to develop dynamic models of formation of individual student's learning trajectory.

Scientific novelty consist in adaptation of the linear dynamic model of portfolio management of securities under constraints to our domain to build individual student's learning trajectory.

The article discusses models, methods and algorithms of determining student's optimal individual educational trajectory. A new method of controlling the learning trajectory has been developed as a dynamic model of learning trajectory control, which uses score assessment to construct a sequence of studied subjects.

---

[1] A.A. Mitsel, Russia, Kemerovo region, Yurga, Leningradskaya street 26

[2] N.V. Cherniaeva, Russia, Kemerovo region, Yurga, Kirova street 27B, apt. 18

IOP Conf. Series: Materials Science and Engineering **91** (2015) 012069 doi:10.1088/1757-899X/91/1/012069

2. Approaches to solve the problem of formation and selection of student's individual learning trajectory

DSSs are often used when building individual learning path, because this task can be considered as a discrete multi-criteria problem, creating a significant burden on the decision maker (DMs). In the paper [1] three ways of representing data in the DSS were analyzed, including tables, heatmaps and parallel coordinate plots. The data in a DSS are represented both by objective and subjective indicators. In heatmaps each column represents the criterion, and each row represents an alternative. The cell color represents the significance of a criterion for a concrete solution. The criterion with low significance is indicated red, the medium one is indicated yellow, and the high one is indicated green. When using parallel coordinate plots alternatives are depicted as a line connecting points on the corresponding axes. All feasible solutions are superimposed on each other. Therefore, multiple criteria can be displayed, and reducing the load on the DMs.

In the experiment, the authors found out that the sequential structure of the spatial information representation facilitates decision-making provided that the amounts of data are large.

The work by G. Setlak [2] is of a significant interest for our research. It considered the Neuro-Fuzzy Method for Knowledge Modeling. The model is based on neural networks and artificial intelligence, which can be used to build fuzzy inference systems, aimed at forming an individual learning path of a student.

The fuzzy neural network, used for classification in this paper, has the following features:

- each neuron represents one fuzzy IF-THEN rule,
- the number of neurons equals to the number of rules in the base of rules,
- weights of the neurons have an interpretation concerning parameters of the membership functions of the corresponding neuro-fuzzy system.

In [3], the author compares various algorithms of curriculum formation: KBS, LS-Plan and IWT. For comparison the following criteria and metrics were used: Overall Effort metrics, Overall Acquired Knowledge metrics and Overall effort metrics. The curriculum was presented as an algorithm or graphs. It was stated that the LS-Plan has the longest distance of educational trajectories and the greatest number of errors, whereas the algorithm IWT generates the shortest path.

We can classify incoming data using the method of hierarchies analysis. An example of this classification is presented in [4] by J. Sivic, B. Russell et al.

Another way to classify data are Data Mining Techniques, successfully applied in [5] for the Student Recruitment Classification.

The following techniques are applied to profiling and classifying the student database:

- K-Means Clustering;
- J-48 Algorithm;
- LMT (Logistic Model Trees) Algorithm;
- Bayesian Network Algorithm.

The Bayes algorithm is often used for processing and analyzing the data. In [6] this method is used to classify text information, reducing uncertainty and ambiguity of natural language.

The proposed ensemble method is based on Bayesian Model Averaging, where both uncertainty and reliability of each single model are taken into account. The authors resolved the issue of classifier selection by proposing a "greedy" approach that evaluates the contribution of each model to the ensemble. Experimental results on "gold standard" datasets show that the proposed approach outperforms both traditional classification and ensemble methods.

The analysis of scientific works relevant for our research problem provided the foundation for developing our own dynamic limitation model to form student's individual learning trajectory.

3. The dynamic model for control over student's learning individual trajectory

Let us denote the number of disciplines by N_t, $t = 1,...,T$, student has to study per semester t . Here, T is the period of study (number of semesters).

IOP Conf. Series: Materials Science and Engineering **91** (2015) 012069 doi:10.1088/1757-899X/91/1/012069

As the results of the studying disciplines a student acquires many competencies. The structure of the model of graduate competence will be presented in the form of three levels, as written in FSES -3:

- The first level — competence;
- The second level — general cultural and professional competences;
- The third level — private competence.

The competence of the student can be evaluated on the basis of multiple estimates obtained by the student in the process of learning $N=\sum_{t=1}^{T}N_t$ disciplines of the chosen specialty.

We denote the valuation on subjects as $V_j(t),\ j=1,...,N_t$, where N_t – the number of courses to be studied in the semester t in according to the curriculum.

Variables V_j can be estimated in points, for example on a 100-point scale. Here traditional assessment offered in the examination sheet, is determined by each University according to it own scale.

Integral score of student $V(t)$ at the time t equal $V(t)=\sum_{j=1}^{N_t}w_jV_j(t),\ t=1,...,T$, where w_j – the importance weight of discipline.

The dynamics of the student's academic performance in discrete time will be described by the equation:

$$V_j(t+1)=\left(1+\mu_j(t)+\eta_j(t)\right)\left(V_j(t)+u_j(t)\right),\ j=1,...,N_t. \quad (1)$$

Here $\mu_j(t)$ is the average complexity of learning discipline j; $\eta_j(t)$ – the random component (deviation) of the learning discipline j complexity with parameters: $M\left(\eta_i(t)\right)=0,\ M\left(\eta_i(t)\eta_k(t)\right)=\Sigma_{ik}(t),\ i,k=1,...,n$, where $\Sigma_{ik}(t)$ – the covariance matrix of the development disciplines complexity. Values $\mu_j(t)$ are determined on the basis of semester evaluation's historical data. $u_j(t)$ – Scores in the discipline obtained during the semester ($u_j(t)>0$), or penalty (minus) scores ($u_j(t)<0$).

The complexity $\rho_{jk}(t)$ of the learning discipline j by the student k in this semester of the year t we define as $\rho_{jk}(t)=\dfrac{1}{V_{jk}(t)}$, where $V_{jk}(t)$ is the final score in the discipline in the semester t. Then, the average value μ_j and the covariance matrix Σ_{ij} are calculated by the formulas $\mu_j=\dfrac{1}{T_h\cdot N_g(t)}\sum_{k=1}^{N_g(t)}\sum_{t=1}^{T_h}\dfrac{1}{V_{kj}(t)}$, $\Sigma_{ij}=\dfrac{1}{T_h\cdot N_g(t)-1}\sum_{k=1}^{N_g(t)}\sum_{t=1}^{T_h}\left(\rho_{ik}(t)-\mu_i\right)\left(\rho_{jk}(t)-\mu_j\right)$.

Here, T_h – historical horizon (quantity of years); t – sequence number of year; $N_g(t)$ – the number of students in the group in the year t.

Let us introduce "reference" integrated scoring and write the equation of the reference student, as follows:

$$V^0(t+1)=[1+\mu_0(t)]V^0(t), \quad (2)$$

Where, $\mu_0(t)$ - the given complexity of the reference student (set by an expert).

The initial condition $V^0(0)=V(0)=0$, i.e., in the initial moment of time the score of the reference student, as well as the scoring actual student equal to zero.

The task of managing the student's learning trajectory is to select disciplines and tasks on the basis of the outcome's estimates of the curriculum so that the generated learning trajectory is to follow the reference one on the horizon management T, where T – the period of time for which the student develops a program of the specialty.

Let us introduce the vector $y(t)=\left(V_1,...,V_N\right)^T$ and vector $z(t)=\left(y(t),\ V^0(t)\right)^T$. Then, equations (1), (2) can be rewritten in the form

$$z(t+1)=A(t)z(t)+B(t)u(t), \tag{3}$$

Where, $A(t)=\overline{A}(t)+\widetilde{A}(t)$; $\overline{A}(t),\ \widetilde{A}(t)$ – diagonal matrix of dimension $(N+1)\times(N+1)$.

Matrix $B(t)$ with dimension $(N+1)\times N$ has the structure:

$$B(t)=\begin{pmatrix} A_{11}(t) & 0 & \dots & 0 \\ 0 & A_{22}(t) & \dots & 0 \\ \dots & \dots & \dots & \dots \\ 0 & 0 & \dots & A_{NN}(t) \\ 0 & 0 & \dots & 0 \end{pmatrix}.$$

As the objective function a linear functional is chosen

$$J=M\left\{\sum_{t=1}^{T-1}\left(V(t)-V^0(t)\right)+\sum_{t=0}^{T-1}b^T(t)\cdot u(t)+\left(V(T)-V^0(T)\right)\right\}\to\min_{u(t)},$$

Where, $b(t)=\left(\mu_1(t)d_{1t},...,\mu_N d_{Nt}(t)\right)^T$.

Using $z(t)$, let's rewrite $\left(V(t)-V^0(t)\right)$ in the form $\left(V(t)-V^0(t)\right)=Cz(t)$, where $C=\left(1,1,...,1,-1\right)\in R^{N+1}$. The quality criterion J takes the form

$$J=M\left\{\sum_{t=1}^{T-1}Cz(t)+\sum_{t=0}^{T-1}b^T(t)\cdot u(t)+Cz(T)\right\}\to\min_{u(t)}. \tag{4}$$

So, we have the optimal control task, in which the equation of state is described by a multistep process (3), and the functional of quality by expression (4). The control vector is given by $u(t)$. The problem is solved with the constraint $V(t)\geq V^0(t)$ or $C\cdot z(t)\geq 0$.

4. The restrictions of the task

The restriction related to the prohibition of penalty (minus) points, looks $u(t)\geq 0,\ t=0,...,T-1$.

Let us introduce restrictions on scoring disciplines

$$c_j^{\min}\leq y_j(t)+u_j(t)\leq c_j^{\max},\ j=1,...,N. \tag{5}$$

Here $c_j^{\min},\ c_j^{\max}$ - the minimum and maximum number of points that a student can get to receive a grade in the gradebook ($c_j^{\min}=55,\ c_j^{\max}=100$).

In terms of $z(t)$ restrictions (5) take the form of a $c^{\min}\leq z(t)+Y(t)u(t)\leq c^{\max},\ t=0,...,T-1$, where Y – matrix with dimension $(N+1)\times N$ with elements $Y=\begin{pmatrix}1 & 0 & \dots & 0\\ 0 & 1 & \dots & 0\\ \dots & \dots & \dots & \dots\\ 0 & 0 & \dots & 1\\ 0 & 0 & \dots & 0\end{pmatrix}$.

The restriction of the semester discipline complexity can be viewed as

$$\frac{M_{\min}^{g}(t)}{N_t}\le\frac{1}{100N_t}\sum_{j=1}^{N}k_j d_{jt} y_j(t)\le\frac{M_{\max}^{g}(t)}{N_t},\ t=1,\dots,T, \tag{6}$$

Where, k_j – the number of credits in the discipline j;

$N_t=\sum_{j=1}^{N}d_{jt}$ – the number of disciplines studied in semester t.

$M_{\min}^{g}(t),\ M_{\max}^{g}(t)$ – the minimum and maximum values of credits on subjects defined by the curriculum. For example, for a bachelor degree in technical specialties humanities, social and economic cycles are allocated ($g=1$), mathematical and natural-science cycle ($g=2$) and professional cycle ($g=3$). In each of these cycles, as well as in FSES -3 the minimum and maximum number of credits are determined.

Let us introduce the vector $D(t)=\frac{1}{100N_t}\left(d_{1,t}k_1,d_{2,t}k_2,\dots,d_{N,t}k_N;0\right)$.

Then the restriction (6) takes the form $\frac{M_{\min}^{g}(t)}{N_t}\le D(t)z(t)\le\frac{M_{\max}^{g}(t)}{N_t},\ t=1,\dots,T$.

For accounting courses-prerequisites we will enter the coefficients r_{ij}. These coefficients are called the coefficients of the narrowness of interdisciplinary communication. The coefficients r_{ij} will be equal to 0 or 1. The coefficients $r_{ij}=1$, if for studying of discipline j you must learn discipline with number i and $r_{ij}=0$ — vice versa. For accounting disciplines-co requisites we introduce a coefficient f_{ij}, which is equal to 1 if discipline i will be used further to study discipline j, and $f_{ij}=0$ — vice versa. Using these coefficients it is determined by the order of the studies, which specifies the individual educational plan. If $r_{ij}=0$ and $f_{ij}=0$, then there may be a parallel study of subjects.

Restrictions related to the procedure of study subjects in each semester t, we write in the form

$$\begin{aligned}&N_t c_k^{\min}\le\sum_{i=1}^{N}r_{ik}d_{it}y_i(t)\le c_k^{\max}N_t,\ k=1,\dots,N_t;\\ &N_t c_i^{\min}\le\sum_{k=1}^{N}f_{ik}d_{kt}y_k(t)\le c_i^{\max}N_t,\ i=1,\dots,N_t.\end{aligned} \tag{7}$$

Here $c_k^{\min}, c_k^{\max}$ – the minimum number of scores allowed in the discipline. If you restrict your studies during the entire period of study T is not attributed to a specific semester, then instead of (7) we will have

IOP Conf. Series: Materials Science and Engineering **91** (2015) 012069 doi:10.1088/1757-899X/91/1/012069

$$Nc_k^{\min} \le \sum_{i=1}^{N}\sum_{t=1}^{T} r_{ik} d_{it} y_i(t) \le c_k^{\max} N,\ k = 1,...,N;$$
$$Nc_i^{\min} \le \sum_{k=1}^{N}\sum_{t=1}^{T} f_{ik} d_{kt} y_k(t) \le c_i^{\max} N,\ i = 1,...,N. \quad (8)$$

We introduce the vectors: $R^k(t) = \left(r_{1,k}d_{1,t}, r_{2,k}d_{2,t},..., r_{N,k}d_{N,t}, 0\right)$, $F^i(t) = \left(f_{i,1}d_{1,t}, f_{i,2}d_{2,t},..., f_{i,N}d_{N,t}, 0\right)$ and we can rewrite the constraints (7) in terms of $z(t)$. Get: $N_t c_k^{\min} \le R^k(t)z(t) \le c_k^{\max} N_t,\ k = 1,...,N$, $N_t c_i^{\min} \le F^i(t)z(t) \le c_i^{\max} N_t,\ i = 1,...,N$.

Restrictions (8) take the form of a: $Nc_k^{\min} \le \sum_{t=1}^{T} R^k(t)z(t) \le c_k^{\max} N,\ k = 1,...,N$, $Nc_i^{\min} \le \sum_{t=1}^{T} F^i(t)z(t) \le c_i^{\max} N_t,\ i = 1,...,N.$

For solving the problem of tracking, we must specify the initial state of the system $z(0) = \begin{pmatrix} y(0) \\ V^0(0) \end{pmatrix}$. The initial scores of real and reference student are considered equal to zero $V(0) = V^0(0) = 0$.

So, finally we will state the problem for control of an individual training trajectory:

$$J = M\left\{\sum_{t=1}^{T-1} Cz(t) + \sum_{t=0}^{T-1} b^T(t)\cdot u(t) + Cz(T)\right\} \to \min_{u(t)}, \quad (9)$$

$$z(t+1) = A(t)z(t) + B(t)u(t), \quad (10)$$

$$\begin{aligned}
&C\cdot z(t) \ge 0,\\
&c^{\min} \le z(t) + Yu(t) \le c^{\max},\ t = 0,...,T-1;\\
&N_t c_k^{\min} \le R^k(t)z(t) \le c_k^{\max} N_t,\ k = 1,...,N;\\
&N_t c_i^{\min} \le F^i(t)z(t) \le c_i^{\max} N_t,\ i = 1,...,N;\\
&\frac{M_{\min}^g(t)}{N_t} \le D(t)z(t) \le \frac{M_{\max}^g(t)}{N_t},\ t = 1,...,T;\\
&u(t) \ge 0,\ t = 0,1,...,T-1
\end{aligned} \quad (11)$$

5. Solution of the problem

Task (9) - (11) is a linear dynamic programming task. It can be solved by Bellman method or converted to an equivalent model of linear programming, for which a reliable solution methods are developed. To solve large-scale problems, you can use the control method with the predictable model.

6. Conclusion

Today, many of the issues in the domain have been solved, but still there are contradictions in the application of the proposed models in practice, i.e. how formed individual learning trajectory meets the interests and need of a student as well as the requirements of the Federal State education standards.

The problem of forming student's learning trajectory is analyzed in general and the choice of an individual direction was studied in details. Various methods, models and algorithms of the student's individual learning trajectory formation were described.

A new model of management learning trajectory based on dynamic models for tracking the reference trajectory was developed.

References

[1] Gettinger J Kiesling E Stummer C Vetschera R A comparison of representations for discrete multi-criteria decision problems *Decision Support Systems* 54 (2013) pp 976–985
[2] Setlak J G The fuzzy-neuro classifier for decision support *Information Theories & Applications* 15 (2008) pp 21-26
[3] Limongelli C Sciarrone F Temperini M Vaste G Comparing Curriculum Sequencing Algorithms for Intelligent Adaptive (e)-Learning *Conference ICL2010* September 15 -17 2010 Hasselt Belgium pp 1013 – 1020
[4] Sivic J Russell B Zisserman A Freeman W and Efros A Unsupervised discovery of visual object class hierarchies *IEEE Conference on Computer Vision and Pattern Recognition* pp 1–8 2008
[5] Kumar A Singhal S Dhankher P Gulia A Student Recruitment Classification using *Data Mining Techniques IJCSMS (International Journal of Computer Science &Management Studies)* Vol 14 Issue 08 (August 2014) pp 24-31
[6] Fersini E Messina E Pozzi F A Sentiment analysis: Bayesian Ensemble Learning *Decision Support Systems* 68 (2014) pp 26–38

Formalization model of expert knowledge about a technical index level of engineering products

A A Zakharova[1], V V Ostanin[1]

[1] Yurga Institute of Technology (affiliate) of Tomsk Polytechnic University, Leningradskaya Str., 26, Yurga, the Kemerovo region, Russia

E-mail: aaz@tpu.ru

Abstract. The authors set a timely problem that concerns development of decision making models, which allow formalizing expert subjective ideas about technical index level of engineering products. The authors proposed a formalization model of expert knowledge about technical index level of engineering products on the basis of fuzzy sets. The model has a method of membership-function construction for linguistic variable terms on the basis of exponential functions.

1. Introduction

The first paragraph after a heading is not indented (Bodytext style). Manufacture of innovative products is connected with many science-based engineering solutions, some of which could not be provided with a pure mathematical tool. That is why some of technological, constructive, organizational, and other decisions when people create and produce engineering products need experts in every branch. To support a decision making process it is necessary to form a method and model complex for decision support, which can process expert estimations and knowledge [1].

One of the problems, which experts have to solve when estimating a level and marketability of complex science-based units, is estimation of some technical indexes, quality merits and so on. During the process experts have to use qualitative (word) evaluations, for example, "high lifting capacity", "weakness", etc. [2]. That is why a timely problem is development of methods and models for decision making, which allow formalizing qualitative and quantitative evaluations, formalizing subjective expert ideas about any engineering product characteristic. It is proposed to use methods of a fuzzy set theory, which helps model smooth change of an object, and also unknown functional relationships presented as qualitative connections.

2. Formalization model of expert knowledge about a technical index level of engineering products

When scientists describe a decision making process in its hard formalized stages they take into account the following:

- a decision making process is characterized by some input parameters and one output parameter;
- some information given by experts about decision making strategies in standard situations is described by a set of conditional statements in terms of fuzzy and linguistic variables that connect input and output variables [3].

IOP Conf. Series: Materials Science and Engineering **91** (2015) 012070 doi:10.1088/1757-899X/91/1/012070

For example, if an expert arranges statements about a value of an output variable depending on values of two input variables, the system of standard fuzzy statements can be as follows (1):

$$\tilde{L}^{(1)} = \begin{cases} \tilde{L}_1^{(1)} :< \text{IF } E_{11} \text{ OR } E_{12} \text{ OR } E_{13} \text{ THEN } \beta_V \text{ is } a_{V_1} >; \\ \tilde{L}_2^{(1)} :< \text{IF } E_{21} \text{ OR } E_{22} \text{ OR } E_{23} \text{ THEN } \beta_V \text{ is } a_{V_2} >; \\ \tilde{L}_3^{(1)} :< \text{IF } E_{31} \text{ OR } E_{22} \text{ OR } E_{33} \text{ THEN } \beta_V \text{ is } a_{V_3} >. \end{cases} \quad (1)$$

Statements E_{ji} are statements of the type: $< \beta_X$ is a_{Xi} AND β_Y is $a_{Yk} >$.

The statement E_{ji} represents itself an input fuzzy situation i, which can be if the linguistic variable β_V has a value of a_{V_j}.

For example, $E_{11} :< \beta_X$ is a_{X_1} AND β_Y is $a_{Y_1} >$.

X, Y are the range of input parameters (technical data of products) that tangibly influences the output parameter V.

$< \beta_X, T_X, X >, < \beta_Y, T_Y, Y >, < \beta_V, T_V, V >$ are linguistic variables that are defined on the range *X,Y* and *V*.

A fuzzy variable is a variable of the type:

$< \alpha, X, C_\alpha >$,

where α is a name for a fuzzy variable;

$X = \{x\}$ is a range of definition;

$C_\alpha = \{\mu_\alpha(x)/x\}$ is a fuzzy set on *X* that describes restriction to possible values of the fuzzy variable α (its semantics). [4].

A linguistic variable is a variable of the type: $< \beta, T, X >$ where:

β is a name of a linguistic variable;

T is its range (term-set) that represent fuzzy variable names;

X is a range of definition for fuzzy variables [5].

Linguistic variables play important role in fuzzy model designing. With their help it is possible to formalize qualitative information about a decision making object, which is verbalized by specialists-experts [6].

Therefore the most important stage in a decision making process on the basis of fuzzy methods is the selection of methods for membership function construction of linguistic variables. Some indexes of engineering products should be presented as linguistic variables. It provides comparison of qualitative and quantitative expert evaluations. Depending on a factor type one applies different methods for membership function construction: for example, on the basis of pair-wise comparison, statistical data processing, standard functions, and so on [7, 8, 9]. This article considers the method based on standard functions.

During the process of expert estimation formalizing about the level of some technical indexes, in the majority of cases it is not necessary to have a model of high accuracy, rough description is enough. To construct membership functions of such notions it is possible to use direct methods based on direct expert assigning of a membership or function degree, which allows calculating a value.

To describe a membership function it is proposed to use a normal distribution frequency function of a continuous random quantity (Gaussian curve):

$$y = \frac{1}{\sigma\sqrt{2\pi}} e^{-(x-a)^2/2\sigma^2}, \quad (2)$$

where a is an expected value of a random variable;

IOP Conf. Series: Materials Science and Engineering **91** (2015) 012070 doi:10.1088/1757-899X/91/1/012070

σ is a mean-square deviation.

Membership function μ_x takes on values in the interval [0, 1], but conditions for terms description of linguistic variables suppose to use only normal fuzzy sets that describe basic values of linguistic variables. That is why it is necessary to normalize the frequency function of normal distribution. To do this it is necessary to divide it by a maximum value. The frequency function of normal distribution has its maximum value when *x=a*, here.

$$y = \frac{1}{\sigma\sqrt{2\pi}} \tag{3}$$

Consequently:

$$\mu_x = \exp\left[\frac{-(x-a)^2}{2\sigma^2}\right] \tag{3}$$

The parameter a is such x value as completely corresponds with a term described notion according to an expert point of view.

The parameter σ (i.e. the multiplier factor is $\alpha = -\frac{1}{2\sigma_2}$) characterizes largeness of a definition range for a membership function or a fuzzy degree μ_x.

Exponential functions are wildly used for membership function description. Let us consider a possibility to specify function parameters (3) by an expert way.

The parameter *a* is a dominant element of a fuzzy set, the membership function is $\mu_x = 1$. An expert should select from a definition range of a linguistic variable the value that is "perfect" for describing a necessary notion. The parameter *a* is specified by an expert for every basis value of a linguistic variable.

Every value from a definition range of a linguistic variable should go with at least one notion (basis value of a linguistic variable), that is why membership functions of fuzzy variables, which describe neighbor basis values of a linguistic variable, are supposed to cross over. So the expert can set such a value *x*, when membership functions of neighbor term sets have the same values. It means that the expert set the value with which according to his opinion it is difficult to say what a neighbor value of a linguistic variable it refers to. The expert can also identify a membership degree of the value x to fuzzy sets of a neighbor term sets.

Now it is necessary to specify the parameter σ, that cannot be directly set by the expert, because a person can hardly imagine measure of characteristic dispersion related to its average value. The parameter σ can be expressed in terms of the formula (3):

$$2\sigma^2 = \frac{(x-a)^2}{-\ln \mu_x}; \tag{4}$$

where *x* is the value $x \in X$, when membership functions of neighbor term sets have the same values;

μ_x is a membership degree of the value $x \in X$ to fuzzy sets of a neighbor term sets.

So, all function parameters (3) can be specified either by an expert or on the basis of his information. For terms construction of a linguistic variable the expert should set n values a_i, (n–1) of the values x_{k_j} and μ_{k_j}.

As membership functions have to have a final range of definition and exponential functions are infinite, we take up the range of α-level when α=0.05 as a range of definition for fuzzy variables, which define basis values of a linguistic variable.

IOP Conf. Series: Materials Science and Engineering **91** (2015) 012070 doi:10.1088/1757-899X/91/1/012070

Membership functions of linguistic variable terms are set by the following functions (5):

$$\mu_{x_1} = \begin{cases} 1 \text{ when } x \le a_1 \\ \exp\left[\dfrac{-(x-a_1)^2}{2\sigma_{11}^2}\right] \text{when } x > a_1; \\ 0 \text{ when } x \ge x_{11} \end{cases}$$

$$\mu_{x_2} = \begin{cases} \exp\left[\dfrac{-(x-a_2)^2}{2\sigma_{21}^2}\right] \text{when } x \le a_2 \\ \exp\left[\dfrac{-(x-a_2)^2}{2\sigma_{22}^2}\right] \text{when } x > a_2; \\ 0 \text{ when } x_{21} \ge x \ge x_{22}; \end{cases}$$

$$\mu_{x_n} = \begin{cases} 0 \text{ when } x \le x_{nn-1}; \\ \exp\left[\dfrac{-(x-a_n)^2}{2\sigma_{nn-1`}^2}\right] \text{when } x \le a_n \;; \\ 1 \text{ when } x > a_n \end{cases} \quad (5)$$

where: μ_{x_i} is a membership function of *i*-term of a linguistic variable;

$i = \overline{1,n}$ is a number of a linguistic variable terms, term numeration is from left to right;

a_i is a dominant element of i-term fuzzy set.

$$2\sigma_{ij}^2 = \frac{(x_{k_j} - a_i)^2}{-\ln \mu_{k_j}}; \quad (6)$$

where x_{k_j} is a value $x \in X$, when membership functions of neighbor terms have the same values;

$j = \overline{1,n-1}$ is a number of the value x_{k_j}, numeration is from left to right;

μ_{k_j} is a membership degree of the value x_{k_j} to fuzzy sets of neighbor terms (separation value);

x_{nn-1} is x values that define a definition range of fuzzy variables.

This method simplifies the procedure of membership function construction, and function memorization by the computer, when it is provided meeting requirements for membership functions of linguistic variable terms [3]. The expert can also change membership functions, and a definition range of a linguistic variable. One more advantage is that a membership function is defined on a continuous carrier; it allows calculating its value at any variable value.

3. Examples of expert knowledge formalization about technological index level of engineering products on the basis of a fuzzy model

We construct a term sets for a linguistic variable β_Y – item technical complexity (unit of measurement is a ball) with a range of definition Y = [0;100] and basis range T_Y = {low, medium, high} = $\{a_{Y_1}, a_{Y_2}, a_{Y_3}\}$.

Expert estimations of parameters needed for membership function construction of β_Y variable are presented in table 1.

Table 1. Expert estimations of parameters for a linguistic variable β_Y – technical complexity of an item

Basis values	Dominant value y of the fuzzy set that describes a term , a_i	Breakpoint values of neighbor terms y_{k_j}	Membership degree of neighbor values (separation value), μ_{k_j}
a_{Y_1} - low	0		
a_{Y_2} - medium	50	$y_{k_1} = 25$	$\mu_{k_1} = 0.5$
		$y_{k_2} = 75$	$\mu_{k_2} = 0.5$
a_{Y_3} - high	100		

We process the expert data.

1. We calculate the value $2\sigma_{ij}^2 = \dfrac{(y_{k_j} - a_i)^2}{-\ln \mu_{k_j}}$ according to the formula (6).

$$2\sigma_{11}^2 = \frac{(y_{k_1} - a_1)^2}{-\ln \mu_{k_1}} = \frac{(25-0)^2}{-\ln 0.5} = 901.75;\ \ 2\sigma_{21}^2 = \frac{(y_{k_1} - a_2)^2}{-\ln \mu_{k_1}} = \frac{(25-50)^2}{-\ln 0.5} = 901.75;$$

$$2\sigma_{22}^2 = \frac{(y_{k_2} - a_2)^2}{-\ln \mu_{k_2}} = \frac{(75-50)^2}{-\ln 0.5} = 901.75;\ \ 2\sigma_{32}^2 = \frac{(y_{k_2} - a_3)^2}{-\ln \mu_{k_2}} = \frac{(75-100)^2}{-\ln 0.5} = 901.75.$$

2. We calculate the values y_{ik}, when $\mu_{a_{yi}} = 0.05$ according to the formula:

$$y_{11} = a_1 + \left|\sqrt{-2\sigma_{11}^2 \ln 0.05}\right|;\ y_{21} = a_2 - \left|\sqrt{-2\sigma_{21}^2 \ln 0.05}\right|; \tag{7}$$

$$y_{22} = a_2 + \left|\sqrt{-2\sigma_{22}^2 \ln 0.05}\right|;\ y_{32} = a_3 - \left|\sqrt{-2\sigma_{32}^2 \ln 0.05}\right|.$$

y_{11}=51.98; y_{21}= -1.98; y_{22}=101.98; y_{32}=48.02.

As y_{21}= -1.98; y_{22} =101.98 leave Y range of definition, so we take up y_{21}= 0; y_{22} =100.

3. We define membership functions according to the formulae (5) taking into account the limits (7):

$$\mu_{a_{y_1}} = \begin{cases} 1 \text{ when } y \le 0; \\ \exp[-(y)^2/901.75] \text{ when } 0 < y < 51.98; \\ 0 \text{ when y} \ge 51.98; \end{cases}$$

$$\mu_{a_{y_2}} = \begin{cases} 0 \text{ when } 0 \geq y \geq 100; \\ \exp[-(y-50)^2/901.75] \text{ when } 0 < y < 100; \end{cases} ; \quad (8)$$

$$\mu_{a_{y_3}} = \begin{cases} 0 \text{ when } \; y \leq 48.02; \\ \exp[-(y-100)^2/901.75] \text{ when } 48.02 < y < 100; \\ 1 \text{ when } y \geq 100. \end{cases}$$

Figure 1 presents graphs of linguistic variable terms "Item technical complexity".

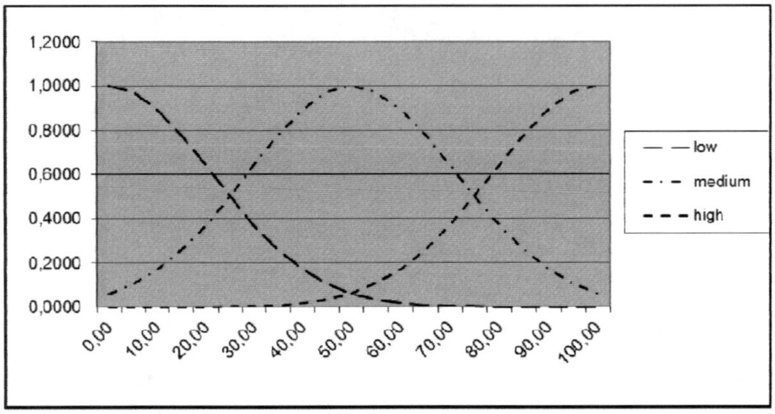

Figure 1. Graphs of linguistic variable terms "Item technical complexity"

In this example dominant values of linguistic variable terms are equally arranged in a defined definition range.

This method allows obtaining membership functions even when term intervals are unequally arranged.

For example, when formalizing expert knowledge for the linguistic variable β_Y – loading capacity (tons) with the range $Y = [0;50]$ and the basis range T_Y = {low, medium, high} = $\{a_{Y_1}, a_{Y_2}, a_{Y_3}\}$, we got expert estimations of parameters, which are presented in Table 2.

Table 2. Expert estimations of parameters for the linguistic variable β_Y – loading capacity

Basis values	Dominant value y of the fuzzy set that describes a term , a_i	Breakpoint values of neighbor terms y_{k_j}	Membership degree of neighbor values (separation value), μ_{k_j}
a_{Y_1} - low	0		
		$y_{k_1} = 10$	$\mu_{k_1} = 0.5$
a_{Y_2} - medium	20		
		$y_{k_2} = 35$	$\mu_{k_2} = 0.5$
a_{Y_3} - high	50		

We process the expert data.

1. We calculate the values $2\sigma_{ij}^2 = \dfrac{(y_{k_j} - a_i)^2}{-\ln \mu_{k_j}}$.

IOP Conf. Series: Materials Science and Engineering **91** (2015) 012070 doi:10.1088/1757-899X/91/1/012070

$$2\sigma_{11}^{\;2}=\frac{(y_{k_1}-a_1)^2}{-\ln\mu_{k_1}}=\frac{(10-0)}{-\ln 0.5}=144.16;\quad 2\sigma_{21}^{\;2}=\frac{(y_{k_1}-a_2)^2}{-\ln\mu_{k_1}}=\frac{(10-20)^2}{-\ln 0.5}=144.16;$$

$$2\sigma_{22}^{\;2}=\frac{(y_{k_2}-a_2)^2}{-\ln\mu_{k_2}}=\frac{(35-20)^2}{-\ln 0.5}=324.1;\quad 2\sigma_{32}^{\;2}=\frac{(y_{k_2}-a_3)^2}{-\ln\mu_{k_2}}=\frac{(35-50)^2}{-\ln 0.5}=324.1.$$

2. We calculate the values y_{ik}, when $\mu_{a_{yi}}=0.05$ according to the formula (7):

y_{11}=20.78; y_{21}= -0.78; y_{22}=51.16; y_{32}=18.84.

As y_{21}= -0.78; y_{22}=51.16 leave Y range of definition, so we take up y_{21}= 0; y_{22}=50.

3. We define membership functions according to the formulae (5):

$$\mu_{a_{y_1}}=\begin{cases}1 \text{ when } y\le 0;\\ \exp[-(y)^2/144.16]\text{when } 0<y<20.78;\\ 0 \text{ when y} \ge 20.78;\end{cases}$$

$$\mu_{a_{y_2}}=\begin{cases}0 \text{ when } 0\ge \text{y} \ge 50;\\ \exp[-(y-20)^2/144.16]\text{when } 0<y<20;\\ \exp[-(y-20)^2/324.1]\text{when } 20\le y<50;\end{cases}\qquad(9)$$

$$\mu_{a_{y_3}}=\begin{cases}0 \text{ when } y\le 18.84;\\ \exp[-(y-50)^2/324.1]\text{ when } 18.84<y<50;\\ 1 \text{ when y} \ge 50.\end{cases}$$

On the basis of the proposed model it is possible to determine values of input variables.

We can show a fuzzification process of three fuzzy statements: "low loading capacity", "medium loading capacity", and "high loading capacity" for the input linguistic variable "loading capacity" when we need to estimate the index "loading capacity" for products that have the index value 30 (tons) out of the possible value interval [0,50]. According to the formulae (9) we calculate values: $\mu_{a_{y_1}}(30)=0$; $\mu_{a_{y_2}}(30)=0{,}735$; $\mu_{a_{y_3}}(30)=0{,}291$.

Therefore, the most membership degree of the input variable value y=30 corresponds to the second statement: medium loading capacity. Obtained crisp values of the true input value for a linguistic variable terms can be used at the next stage of a decision making process, for example, in models of expert knowledge formalization about dependence of an output product parameter from several input parameters, with application of different logical derivation methods (1).

4. Conclusion

The proposed fuzzy model for formalization of expert knowledge about technical index level of engineering products has an inherent value, because with its help it is possible to model a smooth intensive membership change of definite index values to an estimated level. As the result we have a possibility to transform qualitative expert estimations into quantitative values and vice versa, and make fuzzification of fuzzy statements. Expert confidence in a preferable index value is also taken into account in contrast to the general practice when they set threshold values.

In the model we realized the method of membership function construction of linguistic variable terms on the basis of exponential functions, which allow the expert to easily set membership functions with the help of a few number of parameters. The membership function is determined on a continuous carrier, what allows calculating its value at any variable values.

Linguistic variables obtained on the basis of the proposed fuzzy model help characterize input and output factors in decision making models, which formalize expert knowledge about a decision making strategy in standard situations.

References

[1] Tceplit A, Grigoreva A, Osipov Y 2014 Developing the model for assessing the competitiveness of innovative engineering products *Applied Mechanics and Materials* **682** 623-630

[2] Zakharova A A 2013 Fuzzy swot analysis for selection of bankruptcy risk factors *Applied Mechanics and Materials* **379** 207-213

[3] Malyshev N G, Berstein L S, Bozhenuk A V 1998 Fuzzy model of decision making in CAD system Int. Conf. *Fuzzy Sets in Informatics* (Moscow) p.44

[4] Zadeh L A 1965 Fuzzy sets, *Inf. Control.* **8(3)** 338-353.

[5] Zadeh L A 1978 Fuzzy sets as a basis for a theory of possibility *Fuzzy Sets and Systems* **1** 3-28

[6] Telipenko E V, Zakharova A A 2014 Bankruptcy risk management of a machine builder *Applied Mechanics and Materials* **682** 617-622

[7] Frantti T, Mähönen P 2001 Fuzzy logic-based forecasting model *Engineering Applications of Artificial Intelligence* **14(2)** 189-201.

[8] Ivezić D, Tanasijević M, Ignjatović D 2008 Fuzzy approach to dependability performance evaluation *Quality and Reliability Engineering International* **24(7)** 779-792.

[9] Yadav O P, N. Singh Chinnam R B, Goel P S 2003 A fuzzy logic based approach to reliability improvement estimation during product development *Reliability Engineering and System Safety.* **80(1)** 63-74

IOP Conf. Series: Materials Science and Engineering **91** (2015) 012071 doi:10.1088/1757-899X/91/1/012071

Increase of stability of functioning of a production system of machine-building enterprise

A A Rudenko[1] , D V Antipov[2] and M O Iskoskov[3]

[1] Head of Department of “Management of the organization”, Togliatti State University, Belarusian str., 14, Togliatti, 445020, Russia
[2] Head of Department of “Commerce and production management”, Togliatti State University, Belarusian str., 14, Togliatti, 445020, Russia
[3] Director of the Institute of Finance, Economy and Management, Togliatti State University, Belarusian str., 14, Togliatti, 445020, Russia

e-mail: rudenkoa.a@mail.ru, dmitrya@tltsu.ru, maksim250881@mail.ru

Abstract. In article approaches to an assessment of stability of functioning of a production system of machine-building enterprise are considered, criteria of an assessment of stability of functioning of a production system are defined and mechanisms of ensuring its stability are given

1. Introduction

The high competition in such industries as automotive industry and aircraft industry forces to fight constantly for preferences of the consumer of machine-building production and to increase its competitiveness. With the changes happening in the market, change both approaches to the organization and production management. Current trends assume the increase of flexibility of production expressed in implementation of individual requirements of customers (transition from a mass and mass production to production under the order), increase in productivity of productions for possibility of expansion of a share of the market, and also ensuring stability of functioning of a production system of machine-building production.

The Production System (PS) - set of the interconnected elements of system, the including elements of system of the organization and management and the elements of object of management necessary for realization of purpose – implementation of the production program meeting the requirements for productivity and quality.

Stability of functioning of a production system - ability to provide positive dynamics of changes in the set borders of the target indicators of stability characterizing key production factors.

For ensuring stability of functioning of a production system it is necessary to define factors, an assessment technique, and also to develop mechanisms of ensuring stability of functioning of a production system.

2. Results and discussions

The carried-out analysis of factors, allowed to reveal and classify the key factors influencing stability of functioning of PS. All factors were broken by us on internal and external. The external factors and characteristics expressed through quantitative indices are given in table 1.

IOP Conf. Series: Materials Science and Engineering **91** (2015) 012071 doi:10.1088/1757-899X/91/1/012071

Table 1. Environment factors

Factor of influence of the market	The characteristics expressed through indicators
1. Factors of volumes (quantity)	Volumes of deliveries Sizes of party The nomenclature of the delivered production
2. Cost factors	Production price Cost of storage, possession, service and operation of production
3. Quality factors	Deficiency / loss level from use of low-quality production) Innovation of production (constructional, technological, distributive) Additional characteristics
4. Time factors	for placements of the order Time of implementation of the order Time on modification of the order

On the basis of the revealed external and internal factors, the block diagram in which the directions of ensuring stability of functioning of PS (figure 1) are defined is constructed.

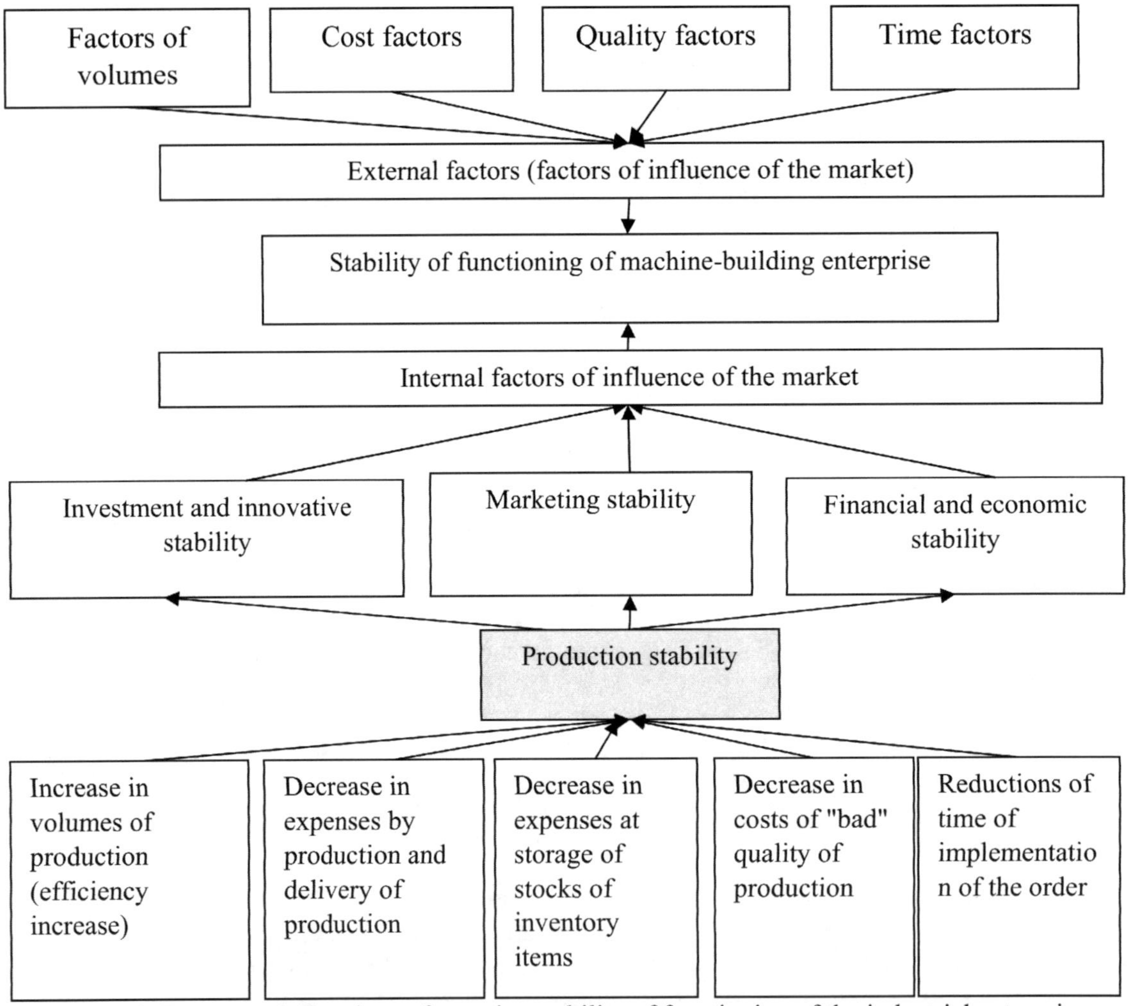

Figure 1. Factors and directions of ensuring stability of functioning of the industrial enterprise

IOP Conf. Series: Materials Science and Engineering **91** (2015) 012071 doi:10.1088/1757-899X/91/1/012071

Stability conditions of functioning of a production stream are:

1. Ensuring flexibility to volumes of deliveries and the sizes of the delivered parties – increase or reduction at the need arising at consumers.
2. Extension of the nomenclature of the made production for satisfaction of requirements of all existing target groups of consumers and convenience of acquisition of production (opportunity to buy production of all standard sizes and the accompanying production in one organization).
3. Minimization of the sizes of the ordered parties, for decrease in expenses during the forming and storage of stocks of machine-building production, thereby increase of turnover in a chain of deliveries.
4. Depreciation of production, including the cost of operation and possession of production.
5. Decrease in time of implementation of the order by production and delivery of machine-building production, including time for placement of the order, if necessary time for modification of the order, time of a production cycle and delivery of production.

It is offered to estimate stability of functioning through the complex indicator consisting of single economic and operational performance (table 2).

Table 2. Complex indicators of stability of functioning

Complex indicator	Single indicator
Capacity of a production system	1. Marginal income
	2. Operational expenses
	3. Work in progress stocks
	4. The volume of the made production
	5. Time of a production cycle
	6. Labor input of production of the order

The complex indicator is on the basis of additive convolution of single indicators [1].

For an assessment of stability we will construct the F(t) function (figure 2). Stability of functioning determines by ability of PS to provide positive dynamics of change of its criterion function in the set borders throughout a long period.

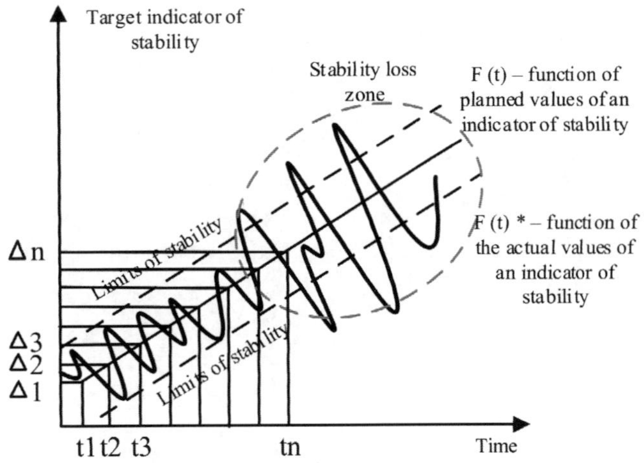

Figure 2. Function of stability of functioning

Stability is defined by two criteria:

1. A stability indicator increment (Δ) on the set time intervals (t).
2. Degree of compliance of the actual value of an indicator of F(t) * to planned F(t).

IOP Conf. Series: Materials Science and Engineering **91** (2015) 012071 doi:10.1088/1757-899X/91/1/012071

Limits of stability determine the acceptable level at which the organization possesses properties of stability and necessary protection against influence of external and internal system factors.

Mathematically the condition of stability is determined by a formula:

$$F(t)=\begin{cases} k1y \mid 0<t<t1 \\ k2y \mid t1<t<t2 \\ k3y \mid t2<t<t3 \\ \ldots \\ kny \mid tn-1<t<tn \end{cases} \tag{1}$$

where k1y, k2y, k3y,… kny – the linear functions defining target values of an indicator of stability on time intervals.

The target indicator of stability of functioning will pay off on a formula:

$$\Delta F(t)=\mid F(t)-F(t)^{*}\mid; \quad \mid \Delta F(t)\mid \leq T \tag{2}$$

where F(t) – the function defining planned values of a target indicator of stability of functioning; F (t) * – the function defining the actual values of an indicator of stability of functioning; ΔF (t) – the size defining stability; T – the admissible value of stability determined by limits of stability.

Target values of indicators and limit of stability are defined on the basis of creation of statistical control cards of averages and amplitude, SPC defined in a technique (statistical managements of processes).

Control of stability is exercised on the basis of logistic function (figure 5.3). Change of parameter of stability in time happens on a logistic curve on which areas of growth, stability and areas of loss of stability are defined.

Logistic function is expressed by Ferkhyudst's equation:

$$F=\frac{A}{1+10^{a+bt}}+C\ , \tag{3}$$

where F-value of function; t – time; C – the lower asymptote (a limit with which function growth begins); a, b – the parameters defining an inclination, a bend and inflection points of the schedule of logistic function.

IOP Conf. Series: Materials Science and Engineering **91** (2015) 012071 doi:10.1088/1757-899X/91/1/012071

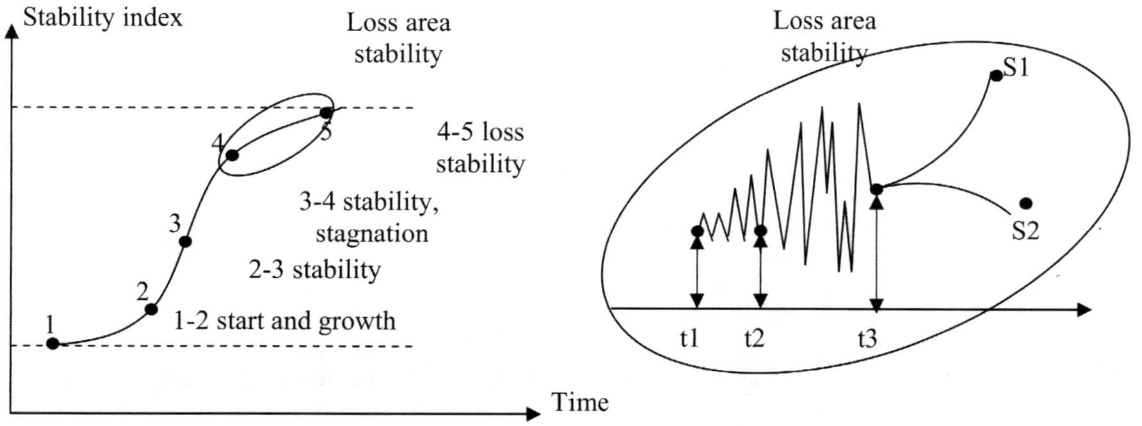

Figure 3. The logistic function defining stability of functioning

The equation of logistic function is expressed in the following logarithmic form:

$$\lg\left(\frac{A}{Y-C}-1\right)=a+bx\text{, or }\lg Z=a+bx \tag{4}$$

"Hit" of parameter in a zone of loss of stability in which there are bifurcation changes transforming its qualitative definiteness, changing a trajectory of its further development leads to an uncontrollable state at which two options of an outcome of events are possible: PS copes with stability loss crisis, increasing the level of organizational development due to introduction of new approaches and technologies to the organization and production management, or PS doesn't overcome stability loss crisis. Thus the enterprise can stop being competitive that will lead to loss of a share or all market or to bankruptcy [2,3,4,5,6,7,8].

Stability of functioning of PS is influenced by external and internal factors – system restrictions. We carried out classification of system restrictions in which three groups are defined: technical, technological and organizational (table 3).

Table 3. Classification of system restrictions.

Group	System restriction
1. Technical	- Capacities
	- Design and technological innovations
2. Technological	- Quality and competitiveness of products
	- Production speed
	- Volumes of reserves of materials, accessories, work in progress and finished goods in warehouses of the organization
	- Reliability of a chain of deliveries
	- Power of a chain of deliveries
	- Cost of a chain of deliveries
3. Organizational	- Speed of design of production
	- Demand for products
	- Efficiency of process of advance of production
	- Financial resources.
	- Competence of the personnel of the organization

For ensuring stability of functioning it is necessary to carry out the coordinating impacts on the system restrictions appearing when functioning PS. We developed the classification of the coordinating impacts on system restrictions given in table 4.

Table 4. The coordinating influences for elimination of system restrictions

System restriction of production	The action directed on elimination of system restriction	Coordination parameter r
Site productivity	- To reduce labor input of performance of operation (change of those. process) - To add capacities on a site (increase in the equipment, the working personnel, operating time) - To change the processing modes on technological operations - - To reduce losses of working hours on a site	- Labor input of technological operation - Quantity of units of equipment - Number of workers - Coefficient of loading of the equipment - Available operating time of a site. - - Site productivity.
Quality of performance of technological operations on a site	- To enter additional control operation - To introduce methods of "the built-in quality" and "protection against mistakes" (PokeYoke) - To automate control operations - - To add additional completion for elimination of discrepancies on quality	- - Labor input on control operations - - Reliability of control operations - - Additional labor input on operations for ensuring quality. - - Award size
Shortcoming or surplus of stocks of NZP	- - To increase / lower reserves on sites - - To introduce system of a pulling - - To reduce the sizes of interoperational parties	- - Volume of stocks of a work in progress - - Size of interoperational party
High labor input of technological operations of a site	- - To change technological processes - - To change the processing modes	- - Processing modes - - Labor input on technological operations
Long time of readjustment of the equipment	- - To introduce methods of fast readjustment. - - To change the sizes of interoperational party	- - Labor input of readjustment - - Size of interoperational party
Losses of working hours	- Change of system of compensation and awarding	- Size of an award of workers
Low administrative competence of the personnel of production	- To train the personnel. - To redistribute responsibility and powers of the personnel. - To develop standards and regulations.	- Integrity of management. - Adaptability of management.
Reliability and power of a chain of deliveries	- To lower a stock rate in knots of a chain of deliveries. - To increase quality and discipline of deliveries.	- Power of knots of a chain of deliveries. - Level of satisfaction of demand. - Reliability of knots of a chain by criteria of quality and terms.

The coordinating influences are provided due to introduction of the modern administrative concepts and technologies applied in mechanical engineering. It is possible to distinguish such concepts from them as Economical production, TQM and JIT. Classification of concepts and methods of ensuring stability of PS (table 5) is developed.

Table 5.Classification of methods of ensuring stability of functioning

Concept	Tools
1. Lean	1.1 SMED (fast readjustment)
	1.2 5 "S" (system of the organization and rationalization of the worker place
	1.3 TPM (general service of the equipment)
	1.4 Poke-Yoke (protection against defects)
	1.5 Visualization
	1.6 Andon (visual system of feedback on production)
2. JIT	2.1 KANBAN (system of regulation of streams of materials and goods in the organization and beyond its limits – with suppliers and customers)
	2.2 ABC-XYZ
	2.3 BBK method
	2.4 Plan of acceptance (selective control)
	2.5 FMEA analysis of a chain of deliveries
3. TQM	3.1 Statistical methods (7 simple methods of management of quality)
	3.2 Standardization
	3.3 QFD
	3.4 Engineering techniques (FMEA, PPAP, MSA, QSA, SPC, APQP)
	3.5 ZERO DEFECT

3. Conclusion

Stability of functioning is important property of a production system of machine-building enterprise. Stability is influenced by external and internal factors, system restrictions. For an assessment of stability it is necessary to define a complex indicator of capacity of PS. By means of the developed mathematical model it is possible to set stability of functioning of PS. At an exit of function of stability for borders of ensuring stability it is necessary to carry out coordination impacts on PS for the account introduction of modern administrative methods.

4. Acknowledgements

This article was prepared as part of the research on the topic goszadaniya number 1120.

References

[1] Bondarenko T 2010 Distinguishing Quantitative and Qualitative Indicators for Specification of Estimation of Dependence between Performance Indicators and Factorial Ones. *Actual Problems of Economics*. **8** 277-82.
[2] Arrow K J 1962 The Economic Implication of Learning by Doing. *The Review of Economic Studies*. **29** 155-123.
[3] Feigenbaum A V 1977 Quality and Productivity *Quality Progress* **7** 18-21.
[4] Forrester J W 1961 *Industrial Dynamics* (Cambridge: MIT Press) p. 95.
[5] Hart, E R 1993 Designing process-based organizations *Plan. Rev.* **5** 39-40.
[6] Ivanov D 2009 Structure dynamics control – based framework for adaptive reconfiguration of collaborative enterprise networks *International Journal of Manufacturing Technology and Management*. **17** 23-41.

IOP Conf. Series: Materials Science and Engineering **91** (2015) 012071 doi:10.1088/1757-899X/91/1/012071

[7] Juran J M 1995 A *History of Managing for Quality. The Evolution, Trends, and Future Directions of Managing for Quality* (ASQC Quality Press) p. 597.
[8] Maskin E 1995 Credit and Efficiency in Centralized and Decentralized *Review of Economic Studies* **62** 541-555.

IOP Conf. Series: Materials Science and Engineering **91** (2015) 012072 doi:10.1088/1757-899X/91/1/012072

Analysis of air pollution from industrial plants by lichen indication on example of small town

K N Orlova [1, a], I R Pietkova [2,a] , and I F Borovikov [3,c]

[1,2]Yurga Technology institute, Tomsk Polytechnic University, Leningradskaya 26, 652055, Yurga, Russia
[3] Bauman Moscow State Technical University, ul. Baumanskay 2-ya, 5, 105005, Moscow, Russia

e-mail: [a]kemsur@rambler.ru, [b]lestaks@rambler.ru

Abstract. According to the research the species of lichens such as ***Parmelia sulcata, Parmeliopsis ambigua, Phiscia stellaris, Xanthoria parietina*** are founded on example of small town**.** Values of clear air index correlated with the average content of sulphur dioxide in the air. These measurement points correspond to residential areas and regions of the objects of food industry. Two zones zero projective coverage are selected. These most polluted zones corresponded to the location of the metallurgical industry and heat electropower station. The roof production and abrasive industry do not show a significant increase in the concentration of sulfur dioxide in the contiguous territory. By method lichen indication on example of small city two zones lichen deserts (sulfur dioxide concentration greater than 0.3 mg/m^3) and one area of critical pollution (sulfur dioxide concentration of 0.1 -0.3 mg/m^3) were founded. The largest area of air pollution allegedly linked to the activities of plants. Thus metallurgical industry and heat electropower station can be called major air pollutants in small towns.

1. Introduction

Anthropogenic sources that cause air pollution, as well as serious violations of the ecological balance in the biosphere are too many [1-5]. However, the most significant of them are two: the industry and transport [6, 7]. Strongest anthropogenic impacts on plant communities exert pollutants in ambient air, such as sulfur dioxide, nitrogen oxides, hydrocarbons, etc. Among them, the most typical is the sulfur dioxide formed by the combustion of sulfur-containing fuel [8] (thermal power plant, boilers, heating stoves populations, as well as transport especially diesel [9-10]).

One of the main objects of global biological monitoring selected lichens.

The goal of the research is an investigation of the state of the atmosphere of the city by method lichen indication.

Objectives:

1. Identify the species composition of lichens found in the city;

2. Qualitative and quantitative analysis of air pollution by projective cover;

3. Make a conclusion about the extent of air pollution in the city.

Lichens are highly peculiar group of spore-bearing plants, consisting of two components - a fungus and single-celled rarely filamentous algae that live together as an integrated organism. When the main breeding and feeding function is due substrate belongs to the fungus, and the function of photosynthesis belongs to the algae. Most Lichen consists of a dense crust are formed by hyphae, which contain the necessary breathing pores (Fig. 1). Cora allows absorbing moisture from the air and protects the lichen from hypothermia or overheating. The hyphae are more porous under the bark. Between them algae cells are located [11]. Normally, photobiont cells are concentrated on the periphery closer to the light forming photosynthetic layer.

According to the anatomy of lichens are distinguished:

- In which algae are more or less uniformly distributed throughout the body;
- In which algae are located just beneath the bark.

Some lichens are involved in symbiosis also with bacteria thereby obtaining nitrogen directly from the air, or are parasitic on mosses and lichens.

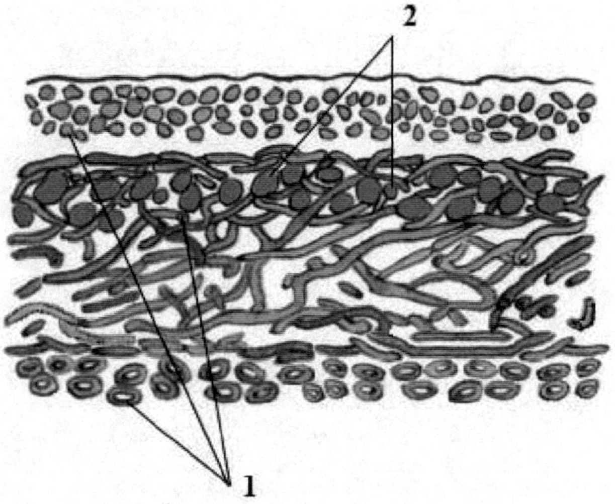

Figure 1. The composition of lichen: 1 - fungal hyphae, 2 - algae cells

The research was conducted on only one type of trees: Poplar (lat. Pópulus). The analysis of scientific evidence revealed the dependence of growth of epiphytic lichens on the different tree species [12]. For example lichens can actively spread on poplars and conifers and less on the birches. Thus, to eliminate the error associated with the tree species composition poplar was selected for investigation.

There are three groups of epiphytic lichens:

- **Crustose**. If thallus fit snugly to the substrate in the granular or like dust form plaque or in the form of scales and crusts of different shapes, such lichens are called crustose. These lichens are similar to the planar crust. It tightly fused with bark, stones, soil. It is difficult to separate the touch, velutinous and moist [5,14].

- **Foliaceous** (leaf like). If lichen thalli have the form of a more or less dissected plates (leaf blades), lichens are called foliaceous. These lichens have the shape of small plates, flakes, adhere to the surface by the thin filament fungus. And these lichens are rather easily separated from it.

- **Fruticose.** Lichens, which are bushy thallus consisting of in varying degrees of vertical branched columns are called fruticose. These lichens are either grow up as small bushes or hang from the tree down like a beard. Not having roots, lichens quite firmly attached to the substrate by special appendages located on the underside of the thallus[3,15].

Study all the lichen species in major cities around the world revealed a number of general laws: the more industrialized cities, more polluted than the air, the less common within its boundaries lichen species, the smaller area covered lichens on tree trunks, the lower the "vitality" of lichens. Found that an increase in air pollution lichens disappear as follows in descending order:

IOP Conf. Series: Materials Science and Engineering **91** (2015) 012072 doi:10.1088/1757-899X/91/1/012072

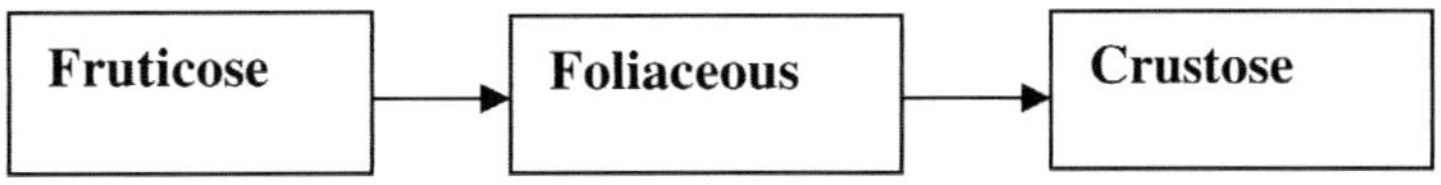

Figure 2. The disappearance of lichens with increasing air pollution [16]

2. Results and discussion

Before the measurements were allocated on a map of the point at which it is necessary to carry out measurements (Fig. 3). Test points were chosen on the location of basic industries and residential areas of the small city (Table 1).

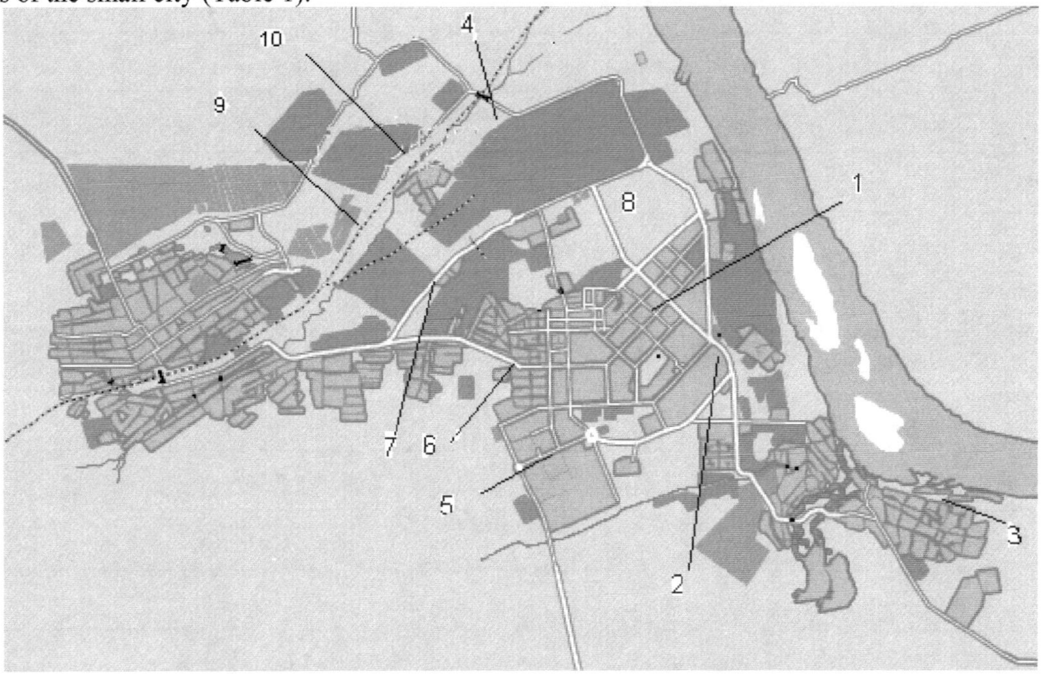

Figure 3. Map of small town with marked measurement points

Table 1 Location of basic industries and residential areas of the small city

Place of measurement (number of points on the -map)	Availability of industrial facilities
1,6	Food processing Industry
4	Heat electro power Station
8	Engineering production
7	Roof production
9	Metallurgical Plant
10	Abrasive production
5,2,3	Residential area

In the city found the following species of lichens:

- ***Parmelia sulcata***

- ***Phiscia stellaris***
- ***Parmeliopsis ambigua***
- ***Xanthoria parietina.***

Groups of epiphytic lichens are recorded in the Table 2.

For measuring the actual number of lichens on trees used mainly two methods - method of estimating projective cover, and methods of linear intersections. We used the method of measurement of projective cover. One of the most common ways to assess the relative abundance of lichens on tree trunks is to define the parameters of the projective cover, ie percentage of area covered with lichens, and areas free of lichens.

Table 2 The species composition of lichens on the example of small towns

Number of measurement points	Species composition
1	Crustose , Foliaceous
2	Crustose , Foliaceous
3	Crustose , Foliaceous
4	No
5	Crustose
6	Crustose
7	Crustose
8	Crustose
9	No
10	Crustose , Foliaceous

Calculation of lichens follows:

First, the number of squares pallets in which lichens occupy more than half an eye on the area of the square (a) is counted, attributing to them conditionally cover of 100%. Then the number of squares in which lichens occupy less than half the area of the square (b) is counted, conventionally ascribing coating of 50% [17]. Data are recorded in the worksheet.

The total projective coverage in the percent (R) is calculated by the formula:

$$R = \frac{(100 \times a + 50 \times b)}{C} \quad (1)$$

a – number of squares pallets in which lichens occupy more than half on the area of the square

b - number of squares in which lichens occupy less than half the area of a square

C - total number of squares in the grid (for example, by using the grid of 10 x 10 cm. The cells with 1 x 1 cm., C = 100).

After determining the projective cover clear air index (CAI) is calculated.

IOP Conf. Series: Materials Science and Engineering **91** (2015) 012072 doi:10.1088/1757-899X/91/1/012072

$$CAI = \sum_{i=1}^{n} \frac{a_{i1} \times c_i}{C_i} \qquad (2)$$

where

n - the number of species described in the test area,

a_i –class type of clear air (1 to 10),

C_i - projective cover of this species in points, C_n - the sum of all coating species.

Clear air index is correlated with the average content of SO_2 in the air (Table 3 and Table 4). Biological principle of quantitative composition of lichen with the amount of sulfur dioxide correlation based on the action of sulfur dioxide on lichens.

Lichens need to be very good at absorbing water and nutrients to grow there. Rainwater contains just enough nutrients to keep them alive. Air pollutants dissolved in rainwater, especially sulfur dioxide, can damage lichens and prevent them from growing.

Many fuels contain small amounts of sulfur compounds. When these fuels are burned sulfur dioxide is released into the air. Sulfur dioxide causes acid rain that can kill plants and aquatic animals [4,13].

Table 3 The species composition of lichens on the example of small towns

Mark	1	2	3	4	5	6	7	8	9	10
Projective coverage, [%]	1-3	3-5	5-10	10-20	20-30	30-40	40-50	50-60	60-80	80-100

For example if there are no lichens present, the air quality is very poor (lichen desert, critical area of contamination), whilst generally only crusty lichens can tolerate poor air quality (zone of strong pollution). In moderate to good air, leafy lichens such as Parmelia caperata can survive and in areas where the air is very clean, rare species lichen may grow.

Table 4 Correlation clear air index containing sulfur dioxide in air [14]

CAI	Concentration SO_2 [mg/m^3]	Conditional zone
1 – 2	Less than 0.01	normal
2 – 5	0.01 - 0.03	small pollution
5 – 7	0.03 - 0.08	medium pollution
7 – 10	0.08 - 0,10	severe contamination
10	0.10 – 0.30	critical pollution
0	More then 0.3	lichen desert

According to the research several zones with high values of projective coverage (over 50%) are identified. And two zones zero projective coverage are selected. That is, on the selected method of study for trees two observation measurement points lichen almost were not observed (Table 5).

It is important to note that the sulphur dioxide levels are increasing. If sulphur dioxide conditions are falling, lichens rarely colonise in exactly the same sequence; lichens are slow growing and may take a year or two to recolonise bark or other substrates following a reduction in air pollution levels, and tiny recolonising specimens can be difficult to spot and identify.

Table 5 Results of quantitative research in the state air pollution on example of small town.

Number point	Projective cover, [%]	CAI
1	52	4
2	56	4
3	67	2
4	0	0
5	19	7
6	22	8
7	15	10
8	16	9
9	0	0
10	23	8

It is need for further studies of air pollution by lichenoindication in the city with the use of combination other techniques and with additional methods to produce lichenoindication maps of the city.

3. Conclusions

According to the research the species of lichens such as ***Parmelia sulcata, Parmeliopsis ambigua, Phiscia stellaris, Xanthoria parietina*** are founded on example of small town**.** The projective cover and clear air index are calculated. Values of clear air index correlated with the average content of sulphur dioxide in the air. According to the research several zones with high values of projective coverage (over 50%) identified. These measurement points correspond to residential areas and regions of the objects of food industry. Two zones zero projective coverage are selected (zone of lichen desert). These most polluted zones corresponded to the location of the metallurgical industry and heat electro power station. The roof production and abrasive industry do not show a significant increase in the concentration of sulfur dioxide in the contiguous territory. Although these factories used chemicals which containing in their composition the sulfur compounds.

As a result, biomonitoring by method lichen indication on example of small city two zones lichen deserts (sulfur dioxide concentration greater than 0.3 mg/m^3) and one area of critical pollution (sulfur dioxide concentration of 0.1 -0.3 mg/m^3) were founded. The largest area of air pollution allegedly linked to the activities of plants. Thus metallurgical industry and heat electro power station can be called major air pollutants in small towns.

References

[1] W. Dathong, N. Thanee, W. Saipunkaew, M. A. Potter, T. Thanee, "Air Pollution Influences Epiphytic Lichen Diversity in the Northeast of Thailand", Advanced Materials Research, Vols 1030-1032, pp. 287-291.

[2] Kuznetsova S. A. , Kozik V. V. , Malchik A. G. Synthesis of transparent conductive coating in2o3:sn films from film forming solutions // Applied Mechanics and Materials. 2014 - Vol. 682. pp. 401-404.

[3] P. Wang, C. C. Wang, L. L. Feng, X. J. Tian, "Investigation of Lead and Cadmium Content of Plant Leaves along a Ring Road in Beijing", Advanced Materials Research, Vol 281, pp. 17-20.

[4] Showman, R. E. Continuing lichen recolonization in the Upper Ohio River valley. // Bryologist. 1997. pp. 478-481.

[5] J. van der Geer, J.A.J. Hanraads, R.A. Lupton. The art of writing a scientific article // J. Sci. Commun. 163, 2000, pp. 51-59.

[6] Loppi, S., Giovannelli, L., Pirintsos, S. A., Putortì, E., Corsini, A. Lichens as bioindicators of recent changes in air quality (Montecatini Terme, Italy). Ecologia Mediterranea 14, 1997, pp. 53-56.

[7] Cislaghi, C., Nimis, P. L. Lichens, air pollution and lung cancer. Nature 14, 1997, pp. 463-464.

[8] S.V. Kakareka, L.A. Kravchuk. Use of Lichens for the Indication of Air Pollution Dynamics in Cities. Urban Ecology, 1998, pp 168-170.

[9] McCune, B.. Lichen communities as indicators of forest health. The Bryologist 103, 2000, pp. 353-356.

[10] Pettersson, R.B.. Effect of forestry on the abundance and diversity of arboreal spiders in the boreal spruce forest. Ecography 19, 1996, pp. 221-228.

[11] Frederickson, R.W.. The zoology of epiphytic lichens: food webs in an algae based system. American Zoologist. 23, 1983, pp. 732-742.

[12] Hodgeman, T.P. and R.T. Bowyer. Winter use of arboreal lichens, Ascomycetes, by white-tailed deer, Odocoileus virginianus, in Maine. Canadian Field Naturalist 99, 1985 pp. 313-316.

[13] J. R. Ikingura, H. Akagi Lichens as a Good Bioindicator of Air Pollution by Mercury in Small-Scale Gold Mining Areas, Tanzania Bulletin of Environmental Contamination and Toxicology May 2002, Volume 68, Issue 5, pp 699-704.

[14] M. Gao, Y. Y. Li, "Energy and Environmental Challenges and Coping Strategies", Applied Mechanics and Materials, Vol 535, pp. 489-494.

[15] Muir, P.S. and B. McCune. Lichens, tree growth, and foliar symptoms of air pollution: are the stories consistent? Journal of Environmental Quality 17, 1988, pp. 361-370.

[16] Van Herk, C.M., A. Aptroot and H.F. van Dobben.. Long-term monitoring in the Netherlands suggests that lichens respond to global warming. Lichenologist 34, 2002, pp. 141-154.

[17] V. N. Zykov, A. I. Kurbatova, N. A. Chernyh, N. N. Savkova, "Environmental Monitoring and Analysis on Air Pollution Issue in China", Advanced Materials Research, Vols 1073-1076, pp. 588-591, Dec. 2014.

IOP Conf. Series: Materials Science and Engineering **91** (2015) 012073 doi:10.1088/1757-899X/91/1/012073

Environmental management on the basis of Complex Regional Indicators Concept: case of the Murmansk region

A Kozlov [1,2], S Gutman [1,3], I Zaychenko [1,4], E Rytova [1,5] and P Nijinskaya [1,6]

[1] St. Petersburg State Polytechnical University, 29, Polytechnicheskaya st.,195251, St. Petersburg, Russian Federation
[2] E-mail: aavk55-spb@yandex.ru (corresponding author)
[3] E-mail: bgsvs@mail.ru
[4] E-mail: cimz.fem.spbpu@mail.ru
[5] E-mail: dlery@list.ru
[6] E-mail: epnizhinskaya@mail.ru

Abstract. The article presents an approach to sustainable environmental development of the Murmansk region of the Russian Federation based on the complex regional indicators as a transformation of a balance scorecard method. The peculiarities of Murmansk region connected with sustainable environmental development are described. The complex regional indicators approach allows to elaborate the general concept of complex regional development taking into consideration economic and non-economic factors with the focus on environmental aspects, accumulated environmental damage in particular. General strategic chart of sustainable environmental development of the Murmansk region worked out on the basis of complex regional indicators concept is composed. The key target indicators of sustainable ecological development of the Murmansk region are presented for the following strategic chart components: regional finance; society and market; industry and entrepreneurship; training, development and innovations. These charts are to be integrated with international environmental monitoring systems.

1. Introduction

According to the presidential decree of the Russian Federation, the Murmansk region is considered part of the Arctic zone [1]. Unlike some other Arctic territories such as the Sakha Republic (Yakutia) and the Yamal-Nenetsk autonomous region, which are relatively "clean" from the ecological viewpoint, large industrial areas including Murmansk and the Murmansk region are facing serious ecological problems. The imperfections of the market economy management system are reflected in the fact that some large industrial enterprises are not interested in solving problems of environmental pollution harmful to the environment. Experts believe that the Russian Arctic regions with the critical level of pollution represent at least 15% of the Russian Arctic zone. The Murmansk region belongs to the industrial region of old formation, where the economy has been based on the exploitation of natural resources for more than 80 years. "The areas with the highest pollution levels are located near the cities of Monchegorsk and Nickel. Environmentally, they are man-made wastelands, with no vegetation, with changed soil structure, and with heavily polluted waters" [2].

The JSC "Apatite" alone stores 30 million tons of waste annually. Over 750 million tons of tailings dumps of floatation enrichment have been accumulated for a 75-year period of the development and exploitation of the Khibiny deposit [3]. The mining enterprises of the Kola peninsula produce up to

200 million tons of solid waste annually. More than 6 billion tons of fractured rock mass have been accumulated in the tailings dumps [4]. The nature of the Arctic is simply unable to process such volumes of accumulated waste in hundreds of years to come. Thus the problem of accumulated environmental damage is evident. Being a key problem of the region, it hinders the steady economic development of the region.

The problem of accumulated environmental damage is addressed in the "Strategy of the Russian Arctic zone development and national security for the period up to 2020" [5]. There we read: "16. In order to protect the environment and to ensure environmental security in the Arctic zone of the Russian Federation, it is necessary to b) eliminate the environmental damage caused by the previous economic, military, and other activities in the Arctic zone of the Russian Federation including the evaluation of the environmental damage and cleaning-up of polluted Arctic seas and territories." However the document suggests neither specific mechanisms nor actions aimed at the solution of the problems of the Murmansk region.

Currently, the development of the Murmansk region is being carried out in accordance with the strategy for the socio-economic development of the region up to 2020, which was launched in 2012 [6]. Despite the fact that the strategic environmental goal of the program implies environmental protection and the improvement of its quality and safety, the strategy does not describe any ecologically-oriented urban development model. The measures taken in this sphere are fragmented and of no systematic character; the results do not demonstrate a complex approach to the elimination of the accumulated environmental damage.

Two other regulatory documents [7,8] clearly define the objects and the methods of evaluation of the accumulated environmental damage, but they do not describe complex approach methods of how to solve the problem of the accumulated environmental damage. There is one more official document considered fundamental for the Arctic zone of the RF development [9]. Regretfully, it does not cover environmental issues at all. Thus, the goal of this study is to develop proposals for the integrated management of sustainable development of the Murmansk region on the basis of Complex Regional Indicators (CRI) concept [10].

2. Existing literature

Existing Russian and foreign literature describes three key issues of importance for the enquiry: the application of balanced score cards concept, sustainable development, and environmental problems of the Arctic regions of the Russian Federation.

The authors of [11-13] propose a balanced scorecard approach for the evaluation of the ecological performance of the enterprises, not regions. The study [14] presents a very interesting but noncomplex approach since the focus is only placed on air environment. The authors of [15] define BSC as a tool for sustainable development but make no connection to the regional development. The ideas presented in [16] and [17] are relevant to the region of our enquiry, but no practical tools for sustainable development are proposed. Authors [18,19,20] provide a suite of indicators that can be applied at different scales as sustainability assessment tool, and as tool to measure the implementation of different policies and programs, but there is not based on complex strategy approach.

The environmental aspects of the Arctic and the Murmansk region development attract attention of both Russian and foreign researchers[21]. Y.I. Sokolov [3] justifies the relevance of the integrated approach development for tackling the problem of environmental safety in the Russian Arctic zone. The author has compiled the data about the cumulative environmental damage. He argues that "...storage, disposal and removal of waste out of the newly development regions of the Russian Arctic zone, as well as their recycling and reduction of the sources of waste should be one of the major factors taken into consideration while planning and executing any kind of activity in the Russian Arctic zone".

Well-known researchers F.D. Larichkin and A.E. Cherepovitsyn with co-authors [22,23] examine issues related to the environmental risks and propose a number of measures to eliminate them. However, the authors only focus on the offshore oil fields in the Arctic.

We cannot but agree with [24-27] that the maintenance of ecological safety is largely dependent on the approach taken; it implies either the measures appropriate for the old tradition (unsteady development) or the strategy and the concept of steady development. We support the position of those who believe that ecological security should be ensured within the framework of sustainable development.

The authors of [28] consider sustainability performance management an emerging term which addresses social, environmental, and economic aspects of corporate management in general and of corporate sustainability management in particular. They propose a term "integrative management" as a complex approach to govern any economic system. However their concept is not applied to the specifics of the Arctic regions.

To sum up, the above mentioned studies have made a significant contribution to the research of the ecological and economic problems both at the regional and the enterprise level. However, the tools have not been sufficiently elaborated for the regional level so that to ensure a complex-oriented approach to the problems associated with environmental management and environmental protection when designing and applying the strategy for the sustainable socio-economic development of the Murmansk region and other Arctic regions of the Russian Federation.

3. Methodology

The methodological basis of this research is the method of integrated regional indicators which are grouped according to a certain classification. The method of analogies as a qualitative research method has been used to justify the transition of the balanced score card system into the complex regional indicators method in the description of the environmental component of the Murmansk region. Methods of analysis and synthesis are also considered as qualitative research methods, in particular, the content analysis and the expert analysis. The methods of synthesis are primarily used in summarizing the results of the content analysis. The proposals and findings of the research are carried out in accordance with the program-oriented management methodology. An important methodological basis of the enquiry is a concept of sustainable development.

4. Findings and proposals

Solving the problem of sustainable ecological development and cumulative environmental damage in the Murmansk region is only possible with the program-targeted approach which involves the development of the Complex of Regional Indicators (CRI). The system of balanced indicators proposed by R. Kaplan and D. Norton has been adopted as the basic methodology for the development of the CRI [29,30].

The CRI model was created with account taken of a region's characteristic properties. As the CRI system is based on the purposive approach, the MBO approach to the Arctic Zone of Russia can be developed drawing on strategic plans specially mapped out for each territory. At the regional level, the CRI exercise is intended to formalize the development strategy of the Russian Arctic, to inform every head of a municipal entity (territory, enterprise) about this strategy and to implement it, to provide monitoring (which is an MBO stage) and feedback at every level (specifically, region, territory, municipal entity, enterprise).

The CRI approach facilitates tying the strategy for the Russian Arctic with a basket of interrelating factors tailor-made for different interconnected levels (those of a territory, municipal entity, enterprise). This system shall balance out the economic, social, financial and other performance results of a region (territories, municipal entities, enterprises) as well as aid in coordinating and pursuing strategies at every level. The CRI of the Russian Arctic is a mechanism that allows for converting the AZRF development strategy into a sequence of activities aimed at attaining the objectives set at every management level. It improves the system of management by bringing the objectives that face the head of a municipal entity (territory, enterprise) into line with those of the region.

According to the overall strategic development map of the Russian Arctic zone presented in [10,31], the problem solving proposals related both to sustainable ecological development and to the improvement of environmental safety involve all the components of the CRI as shown on the diagrams (figures1-5). The key target indicators for monitoring the sustainable ecological development of the Murmansk region are included into schemes (figures 1-5).

5. Conclusions

As a strategic perspective, the sustainable development of the region is based on the achievement of social, economic, and environmental objectives. Thus, the environmental component is viewed as an integral part of the sustainable development strategy. The strategic maps of sustainable development proposed in this study enable the creation of the Complex of Regional Indicators. Its regular monitoring forms the information field for the assessment of the current status and dynamics of the ecological situation in the region. This ensures the monitoring system to control the current state of the pollution level and the use of modern means of land, air, and space-based surveillance. The system is also to be integrated into the existing and emerging international environmental monitoring systems in order to ensure the detection and forecasting of dangerous and extreme natural phenomena in the Murmansk region, including the adverse climate changes, as well as the timely detection and forecasting of natural and man-made disasters.

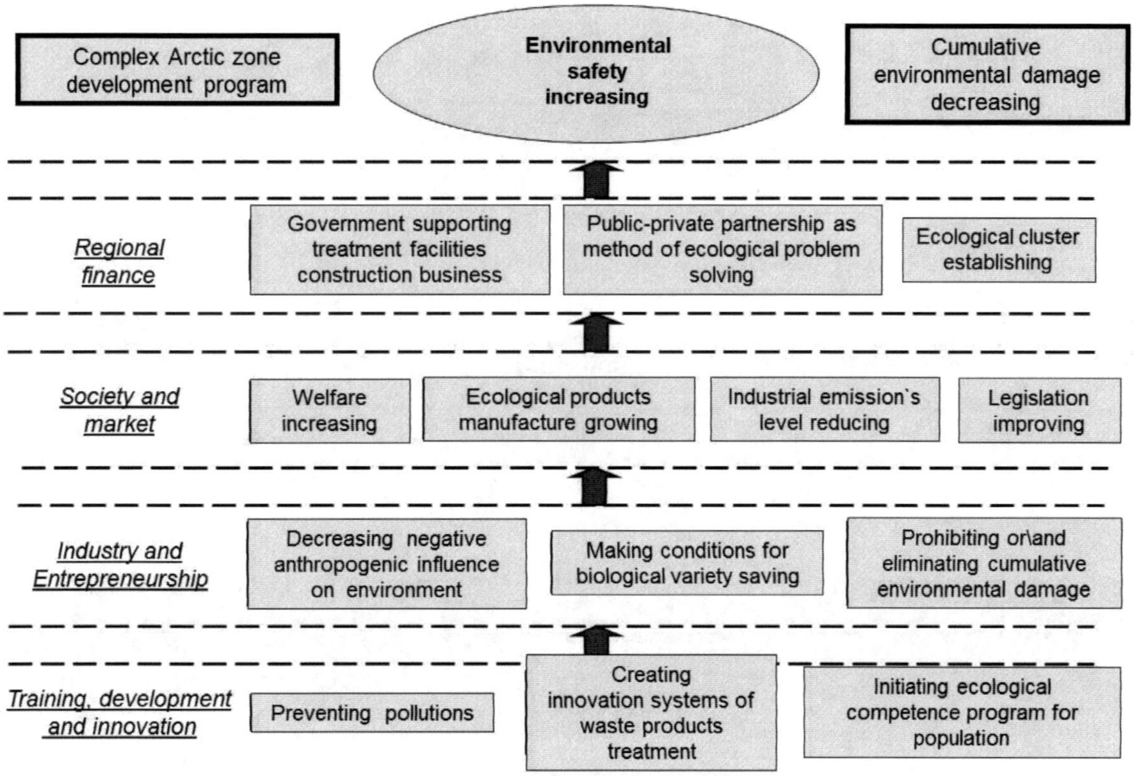

Figure 1. Arctic zone of Russia general strategic map

IOP Conf. Series: Materials Science and Engineering **91** (2015) 012073 doi:10.1088/1757-899X/91/1/012073

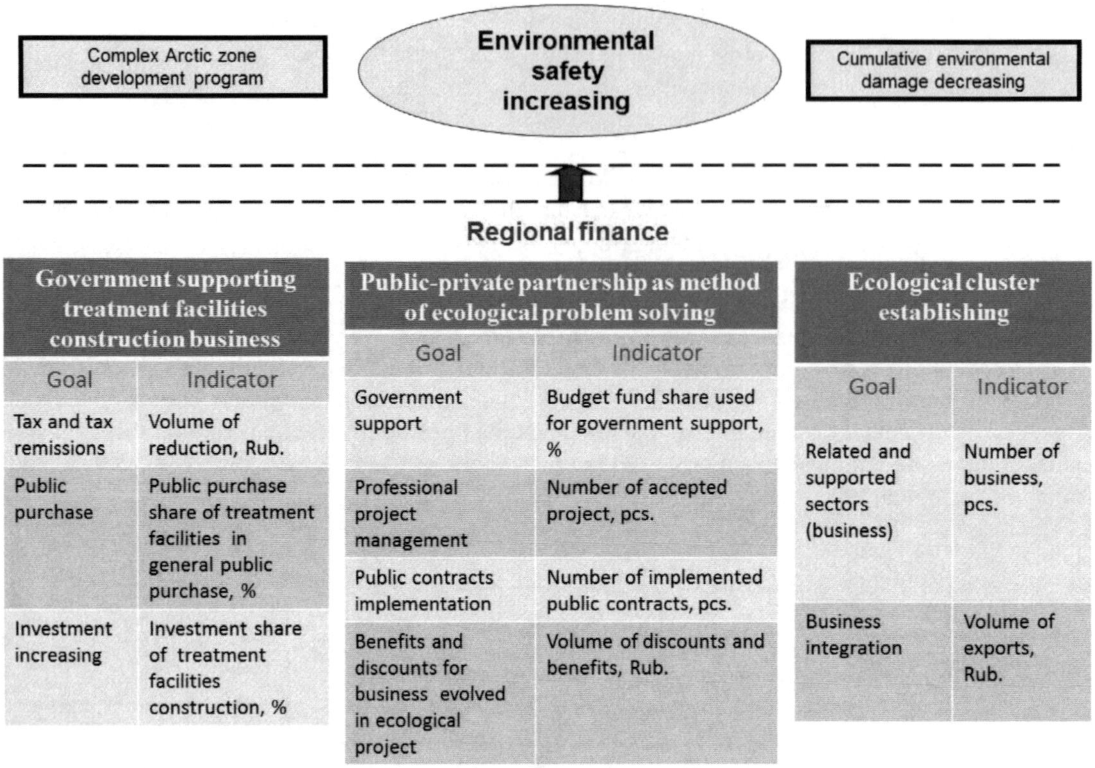

Figure 2. Perspective «Regional finance»

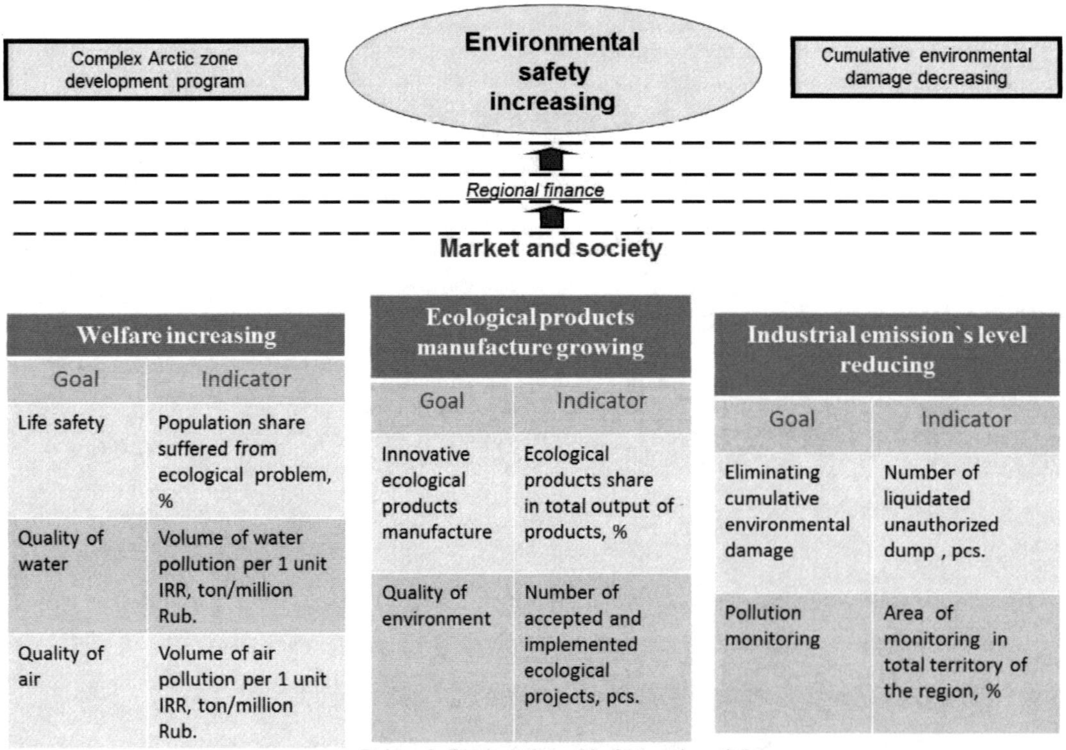

Figure 3. Perspective «Market and society»

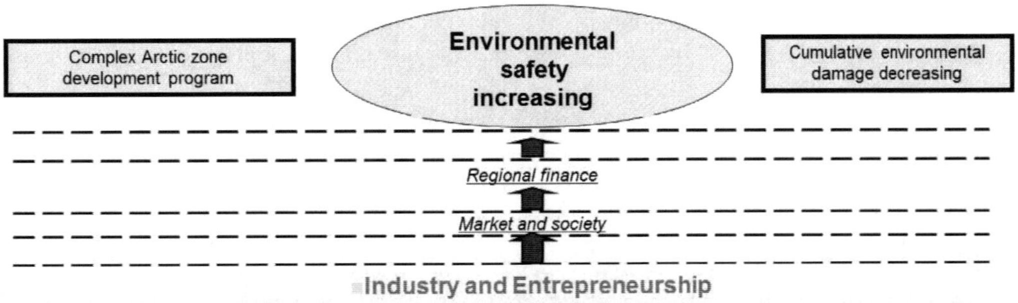

Decreasing negative anthropogenic influence on environment

Goal	Indicator
Increasing responsibility for environment pollution	Violation ecological requirements for implemented projects (esp. for raw materials projects), pcs.
Stimulating new technologies development and implementation	Number of implemented technologies, pcs.
Reducing risks, decreasing emergent consequences	Number of monitoring systems, pcs.

Making conditions for biological variety saving

Goal	Indicator
Extension of special conservation areas	Area of protected territories, hectare
Rehabilitation of flora and fauna	Volume of damage prevented as a result of actions, Rub.

Prohibiting or\and eliminating cumulative environmental damage

Goal	Indicator
Environmental damage assessment	Share of liquidated cumulative environmental damage objects of IV and V danger category, %.
Constructing refuse dump	Availability of implemented construction documents, pcs.

Figure 4. Perspective «Industry and Entrepreneurship»

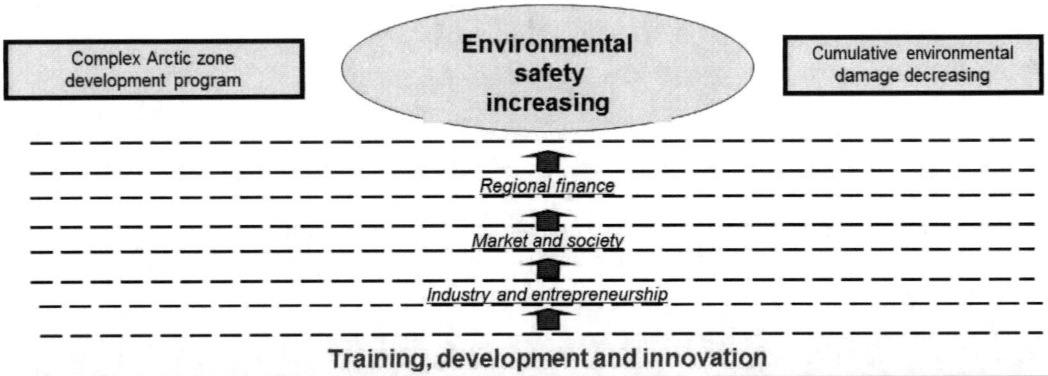

Preventing pollutions

Goal	Indicator
Prohibiting natural objects pollution and rehabilitating pollutant-loaded	Volume of damage prevented as a result of actions, Rub.
Developing preventative arrangements for environment protection	Number of implemented technologies, pcs.

Creating innovation systems of waste products treatment

Goal	Indicator
Updating equipment	Number of new equipment, pcs.
Availability and safety refuse collection system	Number of available container, pcs.
Updating waste recycling system	Availability of appropriated infrastructure objects, pcs.

Initiating ecological competence program for population

Goal	Indicator
Popularizing ecological program	Population share involved in ecological information and education, %
Granting ecological competence program for population	Number of grants, pcs.

Figure 5. Perspective «Training, development and innovation»

This paper is based on research carried out with the financial support of the grant of **the Russian Science Foundation** (project №14-38-00009). **Peter the Great St. Petersburg Polytechnic University.**

References

[1] Ukaz Prezidenta Rossijskoj Federacii ot 2 maja 2014 g. No. 296. [President of Russian Federation Decree] 02.05.2014 **No296**

[2] Diagnosticheskij analiz sostojanija okruzhajushhej sredy Arkticheskoj zony Rossijskoj Federacii (Rasshirennoe rezjume) [Diagnostic analysis Arctic Zone of Russia environment condition (Resume)] 2011 *[Online resource] System request: AdobeAcrobatReader, URL: http://npa-arctic.iwlearn.org/publications/da_res_ru/Content/toc.htm* 11.12.2014

[3] Sokolov J 2013 Arktika: k probleme nakoplennogo jekologicheskogo ushherba [Arctic: problem of cumulative environmental damage] *Arktika: jekologija i jekonomika* **2(10)** pp. 18-27

[4] Informacionnoe agentstvo FlashNord: Zapoljar'e zagrjaznjajut "Kol'skaja GMK", "Apatit", "Kovdorskij GOK" i voennye - vlasti [Information agency FlashNord: Arctic zone is polluted by "Kolskaja GMK", "Apatit", "Kovdorskij GOK" and military authorities] 2014 *[Online resource] System request: AdobeAcrobatReader, URL: http://flashnord.com/news/zapolyare-zagryaznyayut-kolskaya-gmk-apatit-kovdorskiy-gok-i-voennye-vlasti* 10.01.2014

[5] Strategija razvitija Arkticheskoj zony Rossijskoj Federacii i obespechenija nacional'noj bezopasnosti na period do 2020 goda (razrabotana vo ispolnenie Osnov gosudarstvennoj politiki Rossijskoj Federacii v Arktike na period do 2020 goda i dal'nejshuju perspektivu, utverzhdennyh Prezidentom Rossijskoj Federacii 18 sentjabrja 2008 g. No. Pr-1969) [Strategy development and national safety of Arctic Zone of Russian Federation until 2020 year (developed for implementation Basic Government Policy Russian Federation in Arctic zone upon 2020 year and perspective, appointed by President of Russian Federation 18.09.2008, No. Pr-1969] 2008 *[Online resource] System request: AdobeAcrobatReader. URL: http://www.consultant.ru/document/cons_doc_LAW_142561* 25.12.2014).

[6] Prilozhenie k resheniju Soveta deputatov goroda Murmanska «Strategicheskij plan social'no-jekonomicheskogo razvitija goroda Murmanska do 2020 goda» ot 30.05.2012 No 49-660 [Application by decision of Murmansk City Deputy Council «Strategic plan of socio-economic development Murmansk City until 2020 year] 30.05.2012 **No49-660**

[7] Prikaz «Ob utverzhdenii metodicheskih rekomendacij po provedeniju inventarizacii ob`ektov nakoplennogo jekologicheskogo ushherba» ot 25 aprelja 2012 g. No. 193 [Order «Approval methodological guidelines on inventory of cumulative environmental damage»] 25.04.2012 **No193**

[8] Postanovlenie «Ob utverzhdenii perechnja ob`ektov nakoplennogo jekologicheskogo ushherba na territorii Murmanskoj oblasti» ot 29 marta 2013 g. No. 139-PP/5 (v red. postanovlenij Pravitel'stva Murmanskoj oblasti ot 13.12.2013 No 726-PP, ot 26.03.2014 No. 153-PP/5) [Resolution «Approval list of cumulative environmental damage objects in Murmansk region area»] 29.03.2013 **No139-PP/5** (in wording of resolution Murmansk region Government 13.12.2013 **No726-PP**, in wording of resolution Murmansk region Government 26.03.2014 **No153-PP/5**).

[9] Gosudarstvennaja programma Rossijskoj Federacii «Social'no-jekonomicheskoe razvitie Arkticheskoj zony Rossijskoj Federacii na period do 2020 goda» (utverzhdena Postanovleniem Pravitel'stva RF ot 21 aprelja 2014 g. No. 366) [Russian Federation Government Program «Socio-economic development of Arctic Zone of Russian Federation until 2020 year»] 21.04.2014 **No366**

[10] Kozlov A V, Gutman S S and Zajchenko I M 2014 Programma razvitija Arkticheskoj zony Rossijskoj Federacii na osnove kompleksa regional'nyh indikatorov [Program of development of Arctic Zone of Russian Federation on the Basis of Complex Regional

Indicators] *Vestnik Zabajkal'skogo gosudarstvennogo universiteta* **11(114)** pp. 110-120
[11] Dias-Sardinha I, Reijnders L 2005 Evaluating environmental and social performance of large Portuguese companies: a balanced scorecard approach *Business Strategy and the Environment* **14(2)** pp 73-91
[12] Länsiluoto A, Järvenpää M 2010 Greening the balanced scorecard *Business Horizons* **53(4)** pp 385-395
[13] Möller A, Schaltegger S 2005 The Sustainability Balanced Scorecard as a Framework for Eco-efficiency Analysis *Journal of Industrial Ecology* **9(4)** pp 73-83
[14] Wang Q, Yuan X, Zhang J, Mu R, Yang H and Ma C 2013 Key evaluation framework for the impacts of urbanization on air environment–A case study *Ecological Indicators* **24** pp 266-272
[15] Nikolaou I E, Tsalis T A 2013 Development of a sustainable balanced scorecard framework *Ecological Indicators* **34** pp 76-86
[16] Eglington A, Israel R and Vartanov R 1998 Towards sustainable development for the Murmansk region *Ocean and Coastal Management* **41(2-3)** pp 257-271
[17] Vil'chek G Y, Serebryannyy L R and Tishkov A A 1996 A geographic perspective on sustainable development in the Russian Arctic *Polar Geography* **20(4)** pp 249-266
[18] Tasaki T, Kameyama Y, Seiji H, Moriguchi Y and Hideo H 2010 A survey of national sustainable development indicators *International Journal of Sustainable Development* **13 (4)** pp 337-361
[19] Jung W 1997 Sustainable development in industrial countries: Environmental indicators and targets as core elements of national action plans - The German case *Sustainable Development* **5 (3)** pp 139-147
[20] Santoro F, Lescrauwaet A-K, Giraud J-P, Lafitte A and Barbière J 2013 A set of indicators for ICZM *Proceedings of the 10th Global Congress on ICM: Lessons Learned to Address New Challenges* **1** pp 23-34
[21] Revich B A, Kharkova T L, Kvasha E A, Bogoyavlenskii D D, Korovkin A G and Korolev I B 2014 Sociodemographic limitations of the sustainable development of Murmansk oblast *Studies on Russian Economic Development* **25 (2)** p. 201-206
[22] Larichkin F D, Fadeev A M, Cherepovicyn A E and Shishkin A I 2014 Jekologicheskij menedzhment pri osvoenii morskih mestorozhdenij uglevodorodov v Arktike [Ecological management in development of sea hydrocarbons deposits in Arctic] S*ever i rynok: formirovanie jekonomicheskogo porjadka* **1(38)** pp 126-132
[23] Larichkin D F 2008 Kompleksnoe jekologo-sbalansirovannoe ispol'zovanie prirodnyh resursov – osnova razvitija regionov Severa [Complex eco-balanced using of natural resources – basis of North regions development] *Jekonomicheskie i social'nye peremeny: fakty, tendencii, prognoz* **1** pp 58-69
[24] Ozkan U R, Schott S 2013 Sustainable Development and Capabilities for the Polar Region *Social Indicators Research* **114 (3)** pp 1259-1283
[25] Baumgartner R J and Korhonen J 2010 Strategic thinking for sustainable development. *Sustainable Development* **18(2)** pp 71-75
[26] Marcela K, Michaela S and Ondrej S 2011 Dynamic balanced scorecard: Model for sustainable regional Development *WSEAS Transactions on Environment and Development* **7(7)** pp 211-221
[27] Schaltegger S and Wagner M 2006 Integrative management of sustainability performance, measurement and reporting *International Journal of Accounting, Auditing and Performance Evaluation* **3(1)** pp 1-19
[28] Kaenzig J, Friot D, Saadé M, Margni M and Jolliet O 2011 Using life cycle approaches to enhance the value of corporate environmental disclosures *Business Strategy and the Environment* **20(1)** pp 38-54
[29] Kaplan R and Norton D 1996 The balanced scorecard: translating strategy into action (Harvard:

Harvard Business Press) p 322

[30] Kaplan R and Norton D 2004 *Strategy maps: Converting intangible assets into tangible outcomes* (Boston: HBS Press) p 467

[31] Kozlov A V, Gutman S S and Zajchenko I M 2014 Teoreticheskie i metodologicheskie osnovy formirovanija kompleksa regional'nyh indikatorov razvitija Arkticheskoj zony RF [Theoretical and methodological basis of integrated regional indicators of Arctic Zone RF development] *Strategicheskie prioritety razvitija Rossijskoj Arktiki: Sbornik nauchnyh trudov (pod nauch. red. V. V. Ivantera, akademika RAN*) (M.: Izdatel'skij Dom «Nauka») pp 103-112

Engineering of centrifugal dust-collectors based on parallel comparing tests applying computer modelling

Y I Bulygin, D A Koronchik and A A Abuzyarov
Don State Technical University,
Gagarin sq., 1, Rostov-on-Don, Russian Federation

E-mail: bulyur_rostov@mail.ru

Abstract. Currently researchers are giving serious consideration to studying questions, related to issues of atmosphere protection, in particular, studying of new construction of gas-cleaning SPM cyclonic devices effectivity. Engineering new devices is impossible without applying mathematical model methods, computer modeling and making physical models of studying processes due nature tests.

1. Introduction

The main problem in development and improvement of inertial gas filters is absence of the dispersion phase structuring model in aggregates, and as well absence of parallel comparing tests of newly developed dust-collector and certain prevalent aggregates, conducted in equal conditions with the same type of dust. Besides, questions related to exploitation of the dust-collector in term of concrete processing are not in consideration, such that signatures of recommended for the technology aggregate are much worse against signatures, achieved in laboratory conditions. The Cathedra of «Life Safety and Environmental Protection», Don State Technical University, is making R&D on studying aerodynamic characteristics and effectivity of cyclonic equipment with different shapes [1, 2]. Currently the is no uniform requirement for choosing optimal shape of cyclones, however in construction of the most modern cyclones there is a distinct emerging tendency of the cone part evolution. There is a wide range of relations between the cyclone geometrical shape and its effectivity, which is emerging through sophisticated aerodynamic of streams, appearing in this equipment.

One of the proposing technical decisions is previously patented cyclonic equipment, completed as inverse cone without cylindrical part [3]. For solving problems related centrifugal dust-collectors construction improvement it is necessary to develop an experimental installation suitable for making parallel comparing tests of cyclonic equipment with different shapes.

2. Equipment description.

One of the proposing technical decisions is original experimental equipment (application №2014136881 invention RF), allowing parallel and effective exploring of aerodynamic processes and processes of dust precipitation inside centrifugal equipment with various construction (fig.1). The installation is processing in the following way:

Dust-and-air mixture tangentially invades into cyclone through inlet pipe. Due to helically-formed enter, stream crimpling and moving downward spiral line along wallsides of the equipment. As a result of crimple moving, by the action of centrifugal forces dust particles discharging from stream; due

IOP Conf. Series: Materials Science and Engineering **91** (2015) 012074 doi:10.1088/1757-899X/91/1/012074

to existence of electrically conductive layer on transparent material of the cyclone and upper camera of dust-controller it doesn't precipitate on wallsides, but lands in dust-collector.

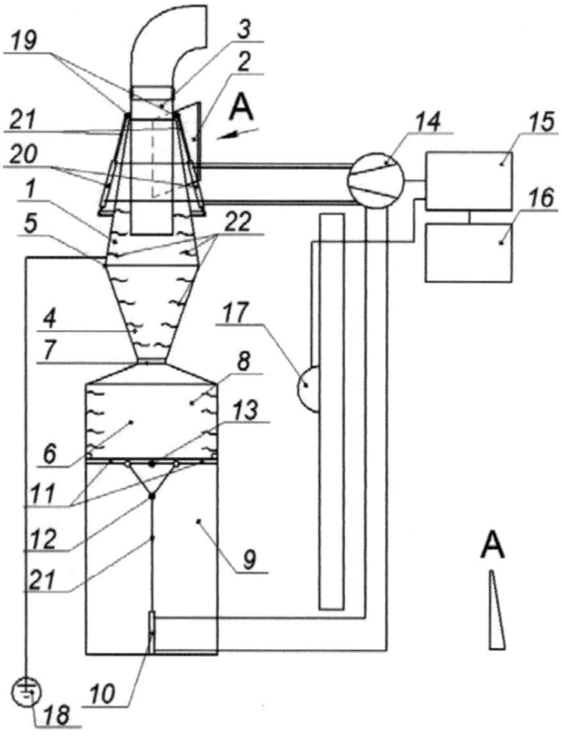

Figure 1. 1-upper cyclonic cone; 2-inlet pipe; 3-exhaust pipe; 4-lower cyclonic cone; 5-cone bases; 6-dust-collector; 7- pipe; 8-upper dust-collector camera; 9-lower dust-collector camera; 10-pneumatic cylinder; 11-semicircular hatches; 12-rotary joint; 13-axis, comparing to which the half-rounded slots are rolling; 14-forcing element; 15-control module; 16-feeding source; 17-video camera; 18-ground path; 19-lugs; 20-pneumatic cylinder's bases; 21-withdrawable pneumatic cylinder's stocks; 22-carbon grains.

Construction of the dust-collector provides carriage of conditions for controlling value in pre-scribed limits. Before entering lower camera of dust-collector, dust, moving rotary, lands on a surface, for example, on baffles in upper camera. Dust grounds onto semicircular hatches, forming uniform layer. Video camera of the equipment fixes achievement of the allowed height of dust layer in upper camera of dust-collector and sends signal to control module, which in its turn gives command to actuate pneumatic cylinder, opening semicircular hatches. Dust disposes into the lower camera of dust-controller and semicircular hatches recover to normal position after some break, allowing the assigned value of duct-collector and effective air cleaning.

The particles discharge into dust-collector, and cleaned gas proceeds into exhaust pipe.

Realization of case and upper camera of the dust-controller of transparent material with application of electrically conductive layer prevents balling of powdery and allows visualizing aerodynamic processes inside the aggregate through using for example carbon grains. By variation of stream-giving speed and immersion depth of exhaust pipe it's possible to explore the dust-landing process and aerodynamic characteristics of stream for detecting its optimal properties.

Experimental installation (fig.2) is realized in the way to allow exploring cyclonic equipment with various shapes. It is attained due quick-disconnecting fixation in area of inlet and in area of convention of lower cyclonic cone to dust-collector bunker. There is a possibility to regulate the heights of connection between inlet pipe and air-duct, conveying dusted gas into cyclonic equipment.

IOP Conf. Series: Materials Science and Engineering **91** (2015) 012074 doi:10.1088/1757-899X/91/1/012074

Figure 2. Experimental installation of exploring dust collecting in cyclonic equipment.

3. Experimental studies are realizing in isothermic terms (at the environmental temperature). Air consumption is in wide range of about 500…2000 m^3/hr, dust concentration is about 3,4…50 g/ m^3. Through experimental action on working units of the installation the next main parameters are measuring: inlet consumption and inlet (exhaust) air speed; cyclone hydraulic resistance; depression in cyclone areas; depression in bunker; blower-made pressure; duration of collecting process; amount of uploaded dust in the feed unit; amount of dust, left in feed unit; amount of collected dust in the cyclone; measuring pressures in the inlet and exhaust of the cyclone.

Some number of tests was realized on exploring cyclone equipment to determinate influence of the cyclone shape and its construction on the dust-landing processes effectivity. Examples [1,2] demonstrate, that cyclones with inverse cone (CIC) have higher dust-collecting effectivity, than cylindrical aggregates. Optimal construction parameters CIC (immersion depth of exhaust pipe and correlation between cyclone working unit and bunker), improving dust-collecting intensity, were estimated.

Established, that dust-collector construction with inverse cone and regulating geometrical parameters allows effective leading of air-stream cleaning process in term of low hydraulic resistance against traditional smooth-wall cyclones. Demonstrated, that using cone cross section and availability to regulate immersion depth of exhaust pipe into the cyclone corpus encourages cyclone hydraulic resistance reduction against smooth-wall cyclones.

4. Computer modeling is a necessary instrument for modern technical objects engineering. Recently many developers and researchers pay special attention to attract commercial packages of application programs for modeling processes, passing in cyclone aggregates. Computer models in programs SolidWorks [4,5] and ANSYS were made for more detailed studying of the processes inside cyclone aggregates. Computational simulation software ANSYS is a universal program system of final element analysis, disposing much more wider modeling possibilities against SolidWorks. Considering specifity and difficulty of solving problems, ANSYS in line with standard k- ε - turbulence model, considering rotating flows inside the aggregate, it is possible to use k- ω - model to describe processes in near-wall areas. The fundamental equations, describing processes inside cyclone equipment, are: equation of continuity (mass conservation), impulse and energy transfer (heat) [5]. Also it is required to set entry and frontier terms. Range of goals, which were already solved due using computer models introduces profiling of sustainable trapezium-shaped construction of inlet pipe CIC *etc.*

IOP Conf. Series: Materials Science and Engineering **91** (2015) 012074 doi:10.1088/1757-899X/91/1/012074

5. Conclusion.

Ways of improving the centrifugal dust-collectors construction are proposed. The experimental installation, allowing to make parallel comparing tests of cyclonic equipment with various shapes is created. Received experimental data about aerodynamic processes in cyclones and dust-collecting processes allows to realize identification of mathematic models and to improve computer models of studying processes.

References

[1] Meskhi B Ch The influence of cyclonic equipment construction specialtys on it's aerodynamic characteristics / B Ch Meskhi Y I Bulygin A N Mikchajlov A V Potyomkina O S Panchenko *Life Safety* – 2010 – № 12 – P 22–28

[2] Potyomkina A V Theoretical and experimental researching of influencing the immersion depth of exhaust pipe on aerodynamic characteristics and effectivity of cyclonic equipment / A V Potyomkina E S Lucash *DSTU Reporter – Rostov-on-Don* 2010 – T 10 - №4(47) – P 526-533

[3] Patent RF № 2506880 from 20 02 2014 Dust-Collector / B Ch Meskhi A N Mikchajlov Y I Bulygin L N Alekseyenko O V Denisov O S Panchenko

[4] Bulygin Y I et al Features of Velocity and Pressure Fields Formation in the Centrifugal Dust-Collector 2014 *Applied Mechanics and Materials* Volume 698 542-545

[5] Meskhi B Ch et al Cyclone with Controlled Parameters and Self-Emptying Bin for Air Dedusting in Machine Building Plants 2014 *Applied Mechanics and Materials* 682 46-52

IOP Conf. Series: Materials Science and Engineering **91** (2015) 012075 doi:10.1088/1757-899X/91/1/012075

Methodology of research for qualitative composition of municipal solid waste to select an optimal method of recycling

M V Kravtsova [1,a], D A Volkov [2,b]

[1] Togliatti State University, Russia, Togliatti, Belorusskaya st. 14, 445020
[2] LLC "Povtor", Russia, Togliatti, Novozavodskaya st, 2A, 445000

E-mail:[a] M.V.Kravtsova@yandex.ru, [b] ekovozr@yandex.ru

Abstract: the article offers research methodology for qualitative composition of municipal solid waste to select an optimal method of recycling. The resource potential of waste directly depends on its composition and determines effectiveness of using various techniques, including separation and separate collection of refuge. The decision on re-equipment of waste-separating enterprise, which decreases the supply of waste to the burial site and provides economy of non-renewable energy sources, is well-grounded, because it allows to diminish an anthropogenic load on environment.

1. Introduction

The situation, which has developed recently in connection with steady annual growth in the amount of waste produced in Russia, brings to the start of irreversible process resulting in degradation of environment and creates real threat to health of population. At present the most widespread method of recycling is burial of waste. According to Eurostat average statistics figures for 27 EU countries, 40 % of MSW (municipal solid waste) is taken to the burial site, for Russia this figure is more than 90 %.

The key point of our research is connected with the necessity of taking immediate decisions for decrease in supply of MSW to the burial sites, taking into account available capacities of the waste-processing enterprises and the situation, when formation of litter is on the stable increase, but waste-processing industry is not sufficiently developed. The purpose of research is to decrease anthropogenic load on the environment at the expense of modernization and re-equipment of waste-processing enterprises.

We offer the methodology of research for qualitative composition of solid waste, that allows to choose an optimal method of waste recycling. Also we offer the technology of complex processing for municipal solid waste and production of alternative fuel.

Theoretical value of this research is in offering of new research methodology for qualitative composition of solid waste to choose an optimal method of waste recycling.

Practical importance of project implementation: solutions of modernizing waste-processing enterprises, offered in this scientific paper, will allow to decrease anthropogenic load on the environment at the expense of decreasing waste supply to waste-processing enterprises and economizing on non-renewable sources of energy.

IOP Conf. Series: Materials Science and Engineering **91** (2015) 012075 doi:10.1088/1757-899X/91/1/012075

2. Theory

The paper offers **methodology of research for qualitative composition** of solid waste to choose an optimal technique of waste recycling; methodology is based on the analysis of several **levels: morphological, fractional, power and chemical** (figure1).

Results of MSW composition study show, that consumer waste possesses high resource potential, since more than 32 % of their weight is taken by food waste, more than 35 % are non-edible secondary material resources (SMR), which can be recycled. More than 60 % of consumer waste volume is taken by energy fractions (cardboard, paper, wood, textiles, polymeric waste (figure 2)), from which alternative renewable fuel can be produced (table.1).

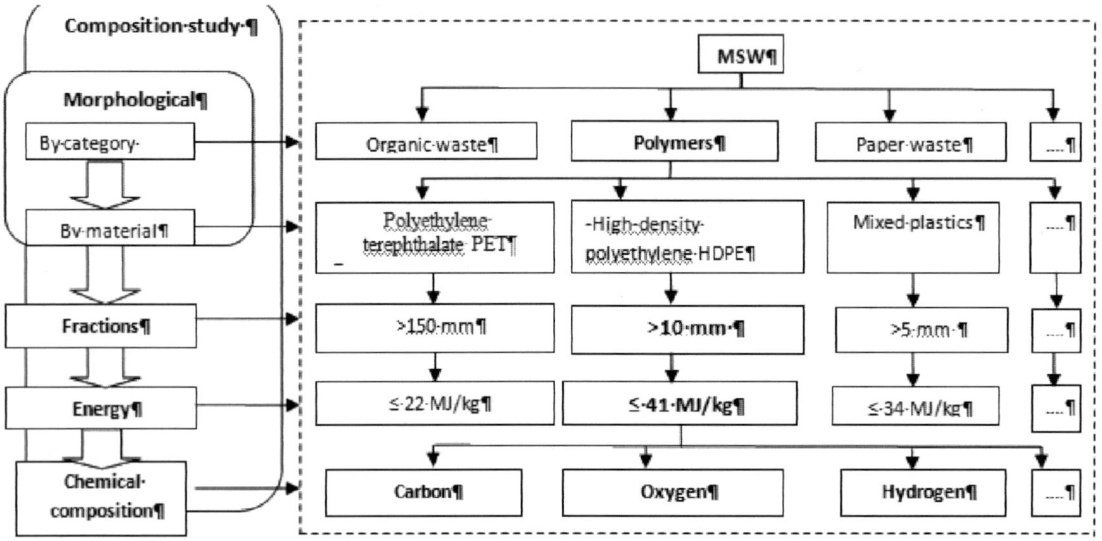

Figure 1. Levels of study for choosing the optimal method of waste recycling

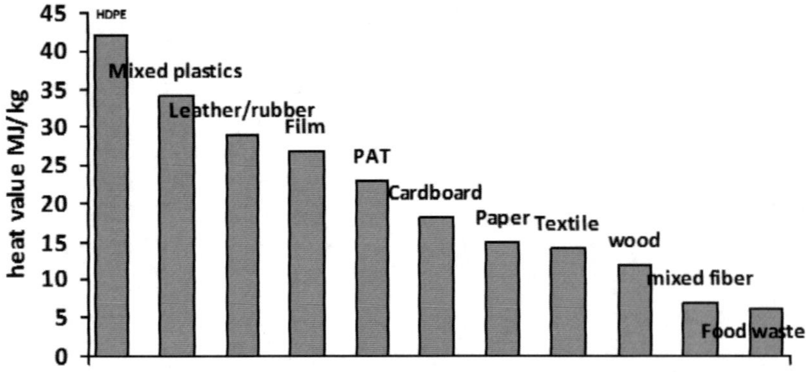

Figure 2. Heat content of MSW components

Table 1. Comparison of MSW energy fraction with other kinds of fuel

Type of fuel	Heat value, MJ/kg	Humidity, %	Ash content, %
MSW Energy fraction	12-16	15-25	10-22
Brown coal	21-32	3-10	5-10
Composite MSW	11-12	30-40	25-35

Comparison of MSW energy fraction with other kinds of fuel shows that SCW possesses less energy potential than coal. But in comparison with composite MSW it possesses a greater energy potential and better ecological characteristics in the process of burning due to smaller ash content. Thus, the major part of MSW can be used for secondary recycling, composting. At the same time it's possible to produce alternative renewable fuel from it.

The analysis of impact, produced by those methods of recycling MSW, which are most widespread in the world, has shown, that the least harmful for the environment are secondary processing, composting and production of solid fuel.

From results of qualitative analysis of MSW one can conclude, that MSW contains plenty of power-intensive components. That, in turn, proves expediency of burning SCW with the purpose of economizing renewable energy sources. As a result of search for patented technologies of burning SCW, there were found modern methods of producing solid fuel from SCW, resulting in production of an alternative kind of fuel for cement furnaces.

The analysis of statuary and regulatory basis in the sphere of specifications and classes of solid fuel, produced from waste, shows, that according to the universally accepted system of classification (table 2) solid fuel, produced from municipal waste, is subdivided into classes depending on values of the following parameters:
- arithmetic average of the lower heat value (Qi);
- arithmetic average of chlorine content (Cl);
- arithmetic average and 80th percentile content of mercury (Hg).

Table 2. Parameters of solid fuel produced from waste according to Russian technical standard GOST P 54236—2010

Parameters	Statistics characteristics	Units of measurement	Class				
			1	2	3	4	5
Lower heat value Q, not less than	arithmetic average	MJ/kg	25	20	15	10	3
Content of Cl, not more than	arithmetic average	%	0.2	0.6	1	1.5	3
Content of Hg, not more than	averaged	mg/MJ	0.02	0.03	0.08	0.15	0.5
	80th percentile	mg/MJ	0.04	0.06	0.16	0.30	1.00

On the basis of Russian technical standard GOST P 54236-2010 one can make a conclusion that for producing fuel from waste, the waste itself should possess high parameters of heat value.

Waste-recycling enterprise LLC "POVTOR " in the municipal district of Tolyatti in Russia gives an example of using methodology of qualitative composition of municipal solid waste. Research of morphological composition of MSW and WAS (waste after separation) allows to come to a conclusion, that after separation in the content of waste there are still plenty of components, possessing high calorific value. It proves expediency of using WAS for production of alternative fuel. Calculations of WAS heat value allow to make a conclusion, that on the basis of heat value parameters, solid fuel, produced as a result of recycling municipal solid waste by LLC "POVTOR", can correspond only to the 5th class of quality - that is the lowest limit of RDF quality (table. 2). For producing fuel of high quality it is necessary to improve technology of fuel preparation, adding the following kinds of equipment to technological process: a pulping engine, roll and air separators, a magnetic separator, an optical sorter, a crusher. The scheme of equipment arrangement is presented in figure 3.

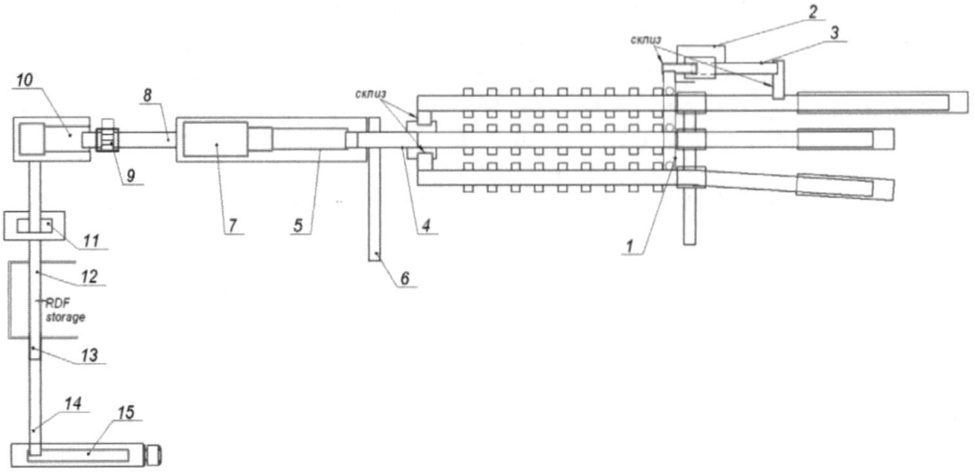

Figure 3. The technological scheme of producing solid fuel from MSW (1-3 - the equipment for gathering and pulping bags with garbage; 5,7,9 and 10 - separators; 11 - crusher; 4, 6, 8, 10, 12-15 - supporting equipment (conveyors))

Table 3. Ecologic and economic efficiency of technological upgrading

Components	**MSW %**	**MSW ton/year**	**SMR (secondary material resources) ton/year**	**Fuel ton/year**	**WAS(waste after separation) ton/year**
Paper for recycling	**5.3%**	**6 625**	**2 800**	**3 213**	**613**
Cardboard	4.0%	5 000	2 800	1 750	450
Composite fiber	1.3%	1 625	0	1 463	163
Plastics	**20.3%**	**25 375**	**14 631**	**10 490**	**254**
PAT	4.3%	5 375	4 838	484	54
HDPE+LDPE	5.0%	6 250	5 625	563	63
Composite plastics	10.5%	13 125	4 069	8 925	131
film	0.5%	625	100	519	6
Organic waste	**27.0%**	**33 750**	**0**	**375**	**33 375**
Wood waste	1.0%	1 250	0	188	1 063
Rubber/leather	1.0%	1 250	0	188	1 063
Food waste	25.0%	31 250	0	0	31 250
Textiles/fabric	**3.0%**	**3 750**	**0**	**0**	**3 750**
Glass	**9.2%**	**11 500**	**7 360**	**0**	**4 140**
Metals	**1.8%**	**2 250**	**1 969**	**0**	**281**
Ferrous metals	1.5%	1 875	1654	0	221
Nonferrous metals - aluminum	0.3%	375	315	0	60
Other waste	**33.4%**	**41 750**	**0**	**0**	**41 750**
Total	**100.0%**	**125 000**	**26 760**	**14 078**	**84 163**

On the basis of undertaken calculations one can make a conclusion, that in case of upgrading the enterprise LLC "POVTOR" with the purpose of producing solid fuel from MSW, this waste-separating enterprise will produce more than 14 000 tons/years of solid fuel. The quantity of the formed tailings will decrease from 110652 to 84 163 tons/year. At the same time the quantity of selected secondary material resources will increase from 14348 up to 26 760 ton/year (table 3).

The scheme of disposing waste after carrying out upgrading of waste-recycling enterprise LLC "POVTOR " is presented in figure 4.

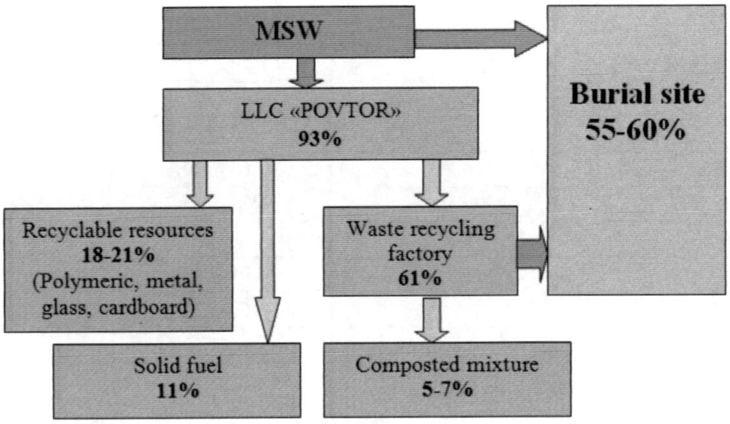

Figure 4. The modernized system of disposing municipal solid waste in the municipal district of Tolyatti city

Calculations of heat value for produced fuel allow to draw a conclusion that on the basis of heat value parameter, according to classification of solid fuel produced from consumer waste in accordance with Russian technical standard GOST P 54236-2010, quality of the solid fuel, produced as a result of recycling MSW on the upgraded conveyer of waste recycling in the enterprise LLC "POVTOR", will raise in 4 points and will correspond to the 1st class of quality - that is the highest limit of RDF quality. Comparison of heat value of the produced fuel with brown coal (figure 5.) evidently proves expediency of using solid fuel, produced from MSW, as an additive to brown coal.

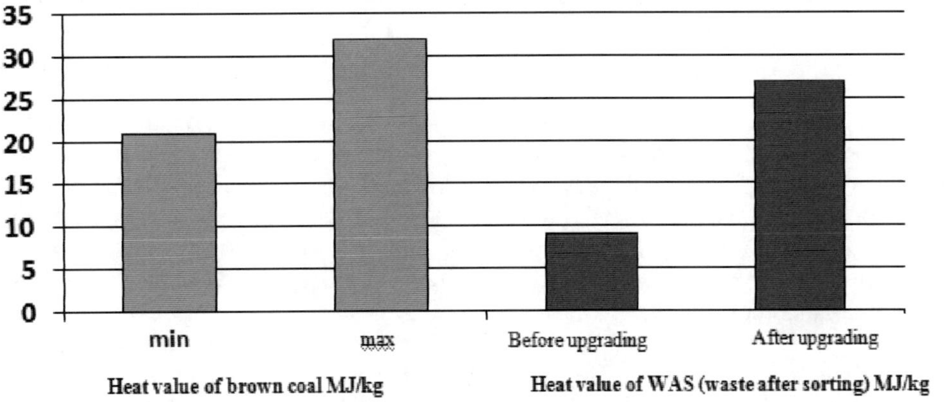

Figure 5. Comparison of heat value of solid fuel made from MSW, produced by the upgraded MSW waste-recycling enterprise LLC "POVTOR", with brown coal

Table 4. Economic substantiation of expenses for upgrading

Parameters, input items	Current technology rub.	Upgraded technology rub.
Production:	**197 195 700**	**293 775 240**
Sorting recyclable resources on conveyer	81 554 900	161 334 440
Payment for recycling MSW	115 640 800	115 640 800
Fuel production	0	16 800 000
Basic wages	26 511 948	80 113 254
Deduction from basic wages	6 627 987	20 028 313
Amortization of equipment	4 822 732	14 242 732
Repair of equipment	2 476 191	4 704 763
Electric energy	1 704 000	5 057 153
Materials for pressing	1 633 377	4 976 506
Load and discharge	3 692 756	8 401 961
Working costs of mechanisms used for loading from the submitting conveyor to the sorting conveyer belt	9 048 310	15 557 296
Transporting MSW to the place of neutralization	57 263 708	43 555 340
Expenses for neutralization and burial of non-processed waste	61 933 950	47 107 572
Overhead charge	17 539 283	21 924 105
Expenses total	193 254 244,20	265 668 995
Profitability	**2%**	**11%**
Profit	**3 941 455.80**	**28 106 245**

Economic calculations allow to make a conclusion, that upgrading waste sorting complex of the enterprise LLC "Povtor" to the status of a wholesale MSW recycling enterprise, producing solid fuel, will be repaid in 3 years and 8 months.

3. Conclusions

During research the following problems are solved and the following **results** are achieved:

1. The research methodology for qualitative composition of MSW, that allows selecting an optimal method of recycling waste in any region, is offered. On the basis of this methodology research of SCW qualitative composition was conducted, it shows, that waste contains 35 % of materials, suitable for secondary recycling; it also shows, that some waste components possess high heat value. Thus, the major part of MSW can be used for secondary processing, composting and production of alternative renewable fuel from it.
2. Analysis of modern methods of waste processing discloses significant backlog of Russia in the sphere of handling waste (Germany has achieved the top position, using three basic methods: secondary processing, composting, safe burning (using MSW as RDF fuel), less than 3 % of municipal waste is sent to burial sites)).
3. On the basis of the undertaken analysis of consumer waste recycling system, it is necessary to note, that the average annual increase of MSW, formed in the city of Tolyatti, is 3 %, while the level of recycling remains unchanged in the course of many years. The existing scheme of processing MSW does not allow recycling more than 15 % of waste weight.
4. During research of technological processes of manufacturing fuel from MSW and capacities of processing MSW, available in the city of Tolyatti, it has been discovered, that there are similarities of

fuel production technologies and technologies of waste sorting complex, used on the enterprise LLC "POVTOR". To follow the purpose of decreasing expenses for the equipment of hi-tech manufacture of solid fuel from MSW, there was made a decision to upgrade waste sorting complex of the enterprise LLC "POVTOR" to a hi-tech complex to manufacture solid fuel from MSW. The developed technology represents multiphase separation of waste with extraction of secondary material resources, suitable for secondary use, further crushing of residual fractions and loading fuel into motor transport.

5. On the basis of the undertaken calculations, the conclusion can be made, that in case of upgrading waste separating complex of the enterprise LLC "POVTOR" to a hi-tech complex, that will manufacture solid fuel from MSW, more than 14 000 tons/years of solid fuel will be produced. The quantity of the formed tailings will decrease from 110 652 to 84 163 tons/years. At the same time the quantity of selected secondary material resources will increase from 14 348 up to 26 760 tons/year.

6. On the basis of the undertaken calculations one can make a conclusion, that according to parameters of heat value, according to classification of solid fuel made from consumer waste in accordance with Russian technical standard GOST P 54236-2010, quality of solid fuel, produced from recycling MSW on the upgraded conveyer of waste recycling enterprise LLC "POVTOR", will increase in 4 points and will correspond to the 1st class of quality - that is the highest parameter of RDF quality.

7. The undertaken calculations of upgrading waste sorting complex of the enterprise LLC "POVTOR" to the state of enterprise, producing solid fuel from MSW, show, that the above described measures are effective for decrease in waste supply and increase in efficiency of a waste-recycling enterprise.

8. The undertaken ecological and economic assessment showed, that the offered upgrading allows to produce more than 14 000 tons/years of solid fuel, to lower quantity of the formed tailings from 110 652 to 84 163 tons/years, to increase quantity of selected secondary material resources from 14 348 up to 26 760 tons/year, to raise profit of the enterprise in more than 24 million rub/year.

References

[1] Vasilyev A V Khamidullova L R Hyukhtina L V Environmental control of toxicity of urban territories using biological monitoring methods *Proc of the international scientific conference (X International forum) Architecture Design Landscape "Le vie dei Mercanti"* 30 May – 4 June 2012 Aversa-Capri Italy (2012) pp 1245-1252

[2] Ezeah C Fazakerley J A Roberts C L Emerging trends in informal sector recycling in developing and transition countries *Waste management* 2013 – Elsevier

[3] Zen I S Noor Z Z Yusuf R O The profiles of household solid waste recyclers and non-recyclers in Kuala Lumpur Malaysia - *Habitat International* 2014 – Elsevier

[4] Bernal M P Alburquerque J A Moral R Composting of animal manures and chemical criteria for compost maturity assessment *A review Bioresource Technology* 100 (22) pp 5444-5453

[5] Moniruzzaman S M Bari Q H Fukuhara T Recycling Practices of Solid Waste in Khulna City Bangladesh *The Journal of Solid Waste Technology and Management* 37 (1) pp 1-15

[6] Franchetti M J The Solid Waste Analysis and Minimization Research Project-A Collaborative Economic Stimulus and Environmental Protection Initiative in Northwest OHIO USA *The Journal of Solid Waste Technology and Management* 35 (2) pp 88-94

[7] Fruergaard T Ekvall T Astrup T Energy use and recovery in waste management and implications for accounting of greenhouse gases and global warming contributions *Waste Management & Research*

[8] Kopyscinski J Schildhauer T J Biollaz S Production of synthetic natural gas (SNG) from coal and dry biomass–*A technology review from* 1950 to 2009 Fuel 89 (8) pp 1763-1783

[9] Boldt E Franchetti M Total Sustainability Assessments for Manufacturing Operations Using the Lean Six Sigma Approach Science *Journal of Environmental Engineering Research* 2013

[10] Franchetti M J Spivak A Concepts Methods and Strategies for Zero-Waste in Manufacturing Green Manufacturing Processes and Systems pp 73-102

IOP Conf. Series: Materials Science and Engineering **91** (2015) 012076 doi:10.1088/1757-899X/91/1/012076

The Influence of Various Factors on the Methane Fermentation Process

M G Kurbanova[1], E A Egushova[2], OG Pozdnjakova[3]

[1] Doctor of technical sciences, professor;
[2] Candidate of technical sciences, associate professor
3 Candidate of technical sciences, associate professor

5, Markovtseva Street, Kemerovo, 650056, Russia

e-mail: thp@ksai.ru

Abstract. The article describes the stages of the methane fermentation process. The phases of methane formation are characterized. The results of the experimental data based on the study of various factors influencing the rate of biogas production and its yield are presented. Such factors as the size of the substrate particles and temperature conditions in the reactor are considered. It is revealed on the basis of experimental data which of the farm animals and poultry excrements are exposed to the most complete fermentation without special preparation. The relationship between fermentation regime, particle size of the feedstock and biogas yield is graphically presented.

Nowadays the problem of waste disposal is very important. In particular the problem of animal wastes and plants residues having a negative impact on the environment should be solved. In different countries this problem is solved in different ways. For example some farmers try to utilize such wastes by spreading them on to their fields as fertilizers. But few farmers pay attention to the fact that the application of fresh manure to land leads to the infection by helminths and other pathogens. The sanitary and epidemiological norms are violated. Furthermore unfermented manure contains a large amount of ammonia and other nitrogen compounds. Getting into the soil, they began to react with carbon dioxide and compounds contained in the soil. As a result, the compounds leading to the formation of nitrous oxide and various substances are formed. Burning the soil, they unfavorably affect it. As these compounds are volatile, a part of them is concentrated in the atmosphere breaking the ozone layer. As a result, the environmentalists have to impose fines and various types of penalties on farmers. Every year these penalties become higher [1-3, 8, 9].

Biotechnology of biogas production and accompanying products is promising from an environmental point of view. Biogas is a product of the microorganisms' metabolism and it is formed by the decomposition of the organic substrate. The decomposition process can be divided into four stages and at each stage different groups of microorganisms are involved in the process [10]. Methane fermentation process has three phases (Fig. 1).

At the first phase the aerobic bacteria convert macromolecular organic substrates (proteins, carbohydrates, fats, cellulose) using enzymes to lower molecular weight compounds such as sugars, amino acids, fatty acids and water. Enzymes isolated by hydrolytic bacteria attach to the outer wall of bacteria and break organic components of the substrate into small soluble components. The polymers are converted into oligomers and monomers. This process known as hydrolysis proceeds slowly and depends on extracellular enzymes such as cellulase, amylase, protease and lipase. This process is influenced by the level of pH from 4.5 to 6.0 as well as the time in the tank.

At the second phase of biomethanogenesis acid-forming bacteria are engaged in splitting. Individual molecules enter into the cells of bacteria where they continue to decompose. Anaerobic bacteria are partially involved in this process using oxygen residues and forming anaerobic conditions required for methane microorganisms. At pH from 6.0 to 7.5 primarily fatty acids, low molecular weight alcohols such as ethanol and gases such as carbon dioxide, hydrogen sulfide and ammonia are produced. This phase is called oxidation. At the same time the level of pH is lowered. The intracellular conversion of simple sugars causes the formation of pyruvic acid that is a key intermediate product of metabolism (carbohydrates, glycerol, amino acids). Ammonia appears as a result of amino acid decomposition by bacteria and hydrogen sulfide appears in the case of sulfur-containing amino acids decomposition. Hydrolysis products of fat are used by many kinds of acid-forming bacteria. During the enzymatic reactions glycerol is converted into phosphoglyceraldehyde that then takes part in the carbohydrate metabolism.

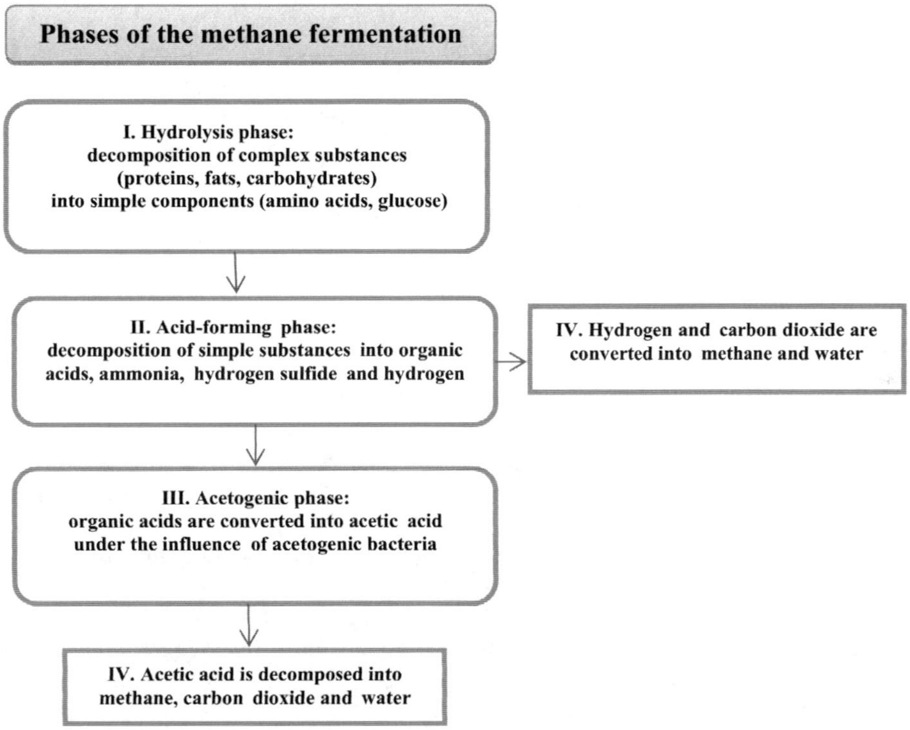

Figure 1 Phases of the methane fermentation process

At the third phase, the acid-forming bacteria produce the starting substrates such as acetic acid, carbon dioxide and hydrogen for the formation of methane from organic acids. It should be noted that the bacteria reducing the amount of carbon are very sensitive to temperature.

The stage of alkaline fermentation characterizing this phase is carried out by methane-bacteria. During the fermentation of acetic acid and methanol the methane is synthesized as a result of methyl group recovery.

$$CH_3COOH \rightarrow CH_4 + CO_2 \quad (1)$$

$$4CH_3OH \rightarrow 3CH_4 + CO_2 + 2H_2O \quad (2)$$

IOP Conf. Series: Materials Science and Engineering **91** (2015) 012076 doi:10.1088/1757-899X/91/1/012076

Another mechanism of methane formation is a characteristic of those types of methanogenic bacteria which are not capable of utilizing acetic acid and methanol. These bacteria synthesize methane by carbon dioxide recovery:

$$4AH_2 + CO_2 \rightarrow CH_4 + 2H_2O + A \quad (3)$$

In the process of methane formation more complex substances such as butyric acid, propionic acid, caproic acid are involved. The conversion is performed according to the type of reaction (3), in which instead of molecular hydrogen the listed organic substrates are involved. For example, while using ethanol is oxidized up to CH_3COOH with simultaneous recovery of carbon dioxide on the type of reaction:

$$2CH_3CH_2OH + CO_2 \rightarrow CH_4 + 2CH_3COOH. \quad (4)$$

These reactions take place at the same time but the methane-forming bacteria are more demanding to the conditions of their existence than the acid-forming bacteria. For example, they require an anaerobic environment and a long time for reproduction. The speed and extent of anaerobic fermentation of methane-forming bacteria depend on their metabolic activity.

A great variety of microorganisms according to some sources up to thousands of different types are involved in this complex set of transformations but the most important of them are methane-forming bacteria. Methane-forming bacteria multiply much slower and they are more sensitive to environmental changes than the acid-forming microorganisms. So initially volatile acids accumulate during the fermentation process and the first stage of methane fermentation is called acid stage.

Further, the speed of acids formation and processing are aligned so that further substrate decomposition and gas formation proceed simultaneously. And it should be mentioned that the intensity of gassing depends on the conditions which are created for the methane-forming bacteria living.

Acid-forming and methane-forming bacteria are found throughout in nature, in particular in animal excrements. It is believed that cattle manure has all the necessary microorganisms for its digestion. Consequently, there is no need to use pure cultures of methane-forming bacteria to cause a fermentation process for biogas production. It's enough to provide the following conditions for the existing bacteria in the substrate: maintenance of anaerobic conditions in the reactor; maintenance of temperature; the availability of nutrients for bacteria; choice of the right time of fermentation and timely loading and unloading of raw materials; compliance with the acid-base balance; compliance with ratio of carbon and nitrogen; choice of the right humidity of raw materials; regular mixing process and absence of inhibitors.

These parameters influence the different types of bacteria involved in the three stages of methane formation in different ways.

At the last fourth phase methane, carbon dioxide and water are produced. 90% of the total methane is produced in this phase, 70% is formed from the acetic acid. Thus, the formation of acetic acid (3 phase of splitting) is the determinant of the methane production speed. Methanogenic bacteria are anaerobes. The optimum pH is 7.0 and the amplitude of the oscillation can be from 6.6 to 8.0.

Splitting of organic matter into individual components and conversion into methane can take place only in a humid environment, because microorganisms can convert substances only in a solute form.

There are different ways to intensify the process of methane fermentation. Among them are the using of catalysts of different nature, the choice of optimal temperature conditions, grinding of raw materials [4-7].

The aim of this work was to study the composition of farm animals and poultry excrements as well as the study of the methane fermentation process under the influence of various factors.

To achieve this goal the following problems were solved:

- to investigate the chemical composition of farm animals and poultry excrements in various farms of Kemerovo region;
- to investigate the biogas formation speed in accordance with the degree of raw materials crushing;
- to investigate the methane fermentation process of farm animals and poultry wastes at different temperatures.

Objects of research

At different stages the objects of research were: cattle manure (national standard 26074-84); pig manure (national standard 26074-84); chicken manure (national standard 26074-84); ferrous sulfate II (national standard 6981-94); biogas produced in the pilot plant.

Research Methodology

Experimental studies were carried out in accordance with modern research methodology of complex phenomena by means of conventional, standard and original methods of biochemical, physicochemical, structural and mechanical analysis using the latest achievements in science and technology. Sampling and preparing the samples for the analysis were carried out in accordance with national standard 26712-85. The mass fraction of moisture and solids in manure was determined according to national standard 26713-85. The mass fraction of ash was determined according to national standard 2671-85. The mass fraction of organic matter in manure was determined according to national standard 27980-88 by thermogravimetric method. The pH indicators of the object and the environment were determined in accordance with national standard 27979-88. The mass fraction of total nitrogen in terms of dry matter was determined in accordance with national standard 26715-85. The mass fraction of phosphorus in terms of dry matter was determined in accordance with national standard 26717-85 by photometric method. The mass fraction of potassium in terms of dry matter was determined in accordance with national standard 26718-85 by flame photometric method. The mass fraction of ammonium nitrogen in terms of dry matter was determined in accordance with national standard 26716-85 by the Kjeldahl method.

Results and discussion

In cattle-breeding complexes of the Kemerovo region the keeping of cattle without litter is used more often and the slurry is used to remove excrements. In spite of this, the content of the straw in manure is from 5 to 15% of the total weight. The average yield of excrements per one head of farm animals of different categories is presented in Table 1.

Table 1 - The average daily amount of farm animals' excrements (per head)

Cattle	Excrements yield, kg	Livestock of pigs	Excrements yield, kg
Sires	40	Breeding boars	11.1
Cows	55	Weaned piglets	2.4
Calves from 4 to 6 months for fattening	7.5	Sows:	
Young replacement cattle, months.		barren	8.8
6-12	14	bred	10.0
12-18 and heifers	27	milking with litter	15.3
Store cattle, months.		Fattening pigs and young replacement pigs of live weight, kg	
4-6	14	up to 40	3.5
6-12	26	40-80	5.1
older than 12	15	more than 80	6.6

The most common way to keep pigs is to use straw bedding (in some cases - sawdust). The percentage of straw in manure is usually from 20 to 40%. Poultry is also kept with the help of the bedding materials such as straw, sawdust, rarely sand. They are mixed with manure and make up 50% of manure weight.

Table 2 shows the average data on the chemical composition of livestock manure and poultry droppings contained in various types of Kemerovo region farms.

Table 2 - Average data of excrements of farm animals and poultry selected from the Kemerovo region farms

Parameter	Unit	Test results		
		Cattle manure	Pig manure	Poultry droppings
Moisture content	%	85.5	72.9	38.6
Ash content	%	13.3	8.6	16.0
Mass fraction of organic matter	%	86.7	91.4	84.0
pH	pH unit	7.3	6.3	7.2
Total nitrogen (in terms of dry substance) not less than	%	1.95	2.37	3.15
Mass fraction of phosphorus (in terms of dry substance) not less than	%	0.84	2.83	5.35
Mass fraction of potassium (in terms of dry substance) not less than	%	0.82	0.15	0.94
Mass fraction of ammonia nitrogen (in terms of dry substance) not less than	%	0.78	0.49	0.61

To produce biogas the most important indicator of the suitability of the soil is the presence of organic substance in it. According to research of livestock manure (Table 2) in all kinds of livestock and poultry manure there is enough organic substance to ensure the normal activity of microorganisms involved in the production of biogas. Another important criteria is the pH level. While both hydrolyzing and acid-forming bacteria in the acidic environment with pH level from 4.5 to 6.3 reach the optimum of its activity, bacteria forming acetic acid and methane can live only at neutral or slightly alkaline level with pH from 6.8 to 8. So, the most enabling substrate for biogas yield is livestock and poultry manure, which has a slightly alkaline reaction.

Besides, the humidity of the environment has a great importance for normal metabolism of aceto- and methanogenic microorganisms. Methane-oxidyzing bacteria can live and reproduce when substrates are soluble in water (at least 50% of the composition of water). The most enabling humidity for the development of methanogens is from 85 to 92%. In livestock manure this requirement is maintained (humidity is 85.5%), so this fact significantly reduces the cost of additional moisturing of the substrate.

For further study we have chosen bedding and non-bedding liquid cattle manure that was brought from Kemerovo region farms. Duration of one experiment has lasted for 25 days. For the purity of the experiment additional portions of raw materials were not added to the samples under investigation. Taking into account the harsh continental climate prevailing in the Kuzbass, we selected two modes of fermentation. They are mesophilic (30-35 °C) and psychrophilic (15-20 °C) modes.

The size of the solid particles of substrate is essential. The larger the interaction area for bacteria and the more fibrous substrate the faster it decomposes under the bacteria's action contained in the cattle or poultry manure. As in the selected material for the study there was a bedding material which composed 12-15% by weight of the substrate, it was milled to the size of 20-25, 30-35 and 40-45 mm before introducing the substrate into the fermenter. The size of particles depended on the time of milling. The size of the solid particles was measured by a ruler and caliper after they were washed under running water through the sieve and after drying the dry solid in air.

The impact of particle size on the biogas-release rate from cattle manure at mesophilic and psychrophilic fermentation modes are shown in Figure 2.

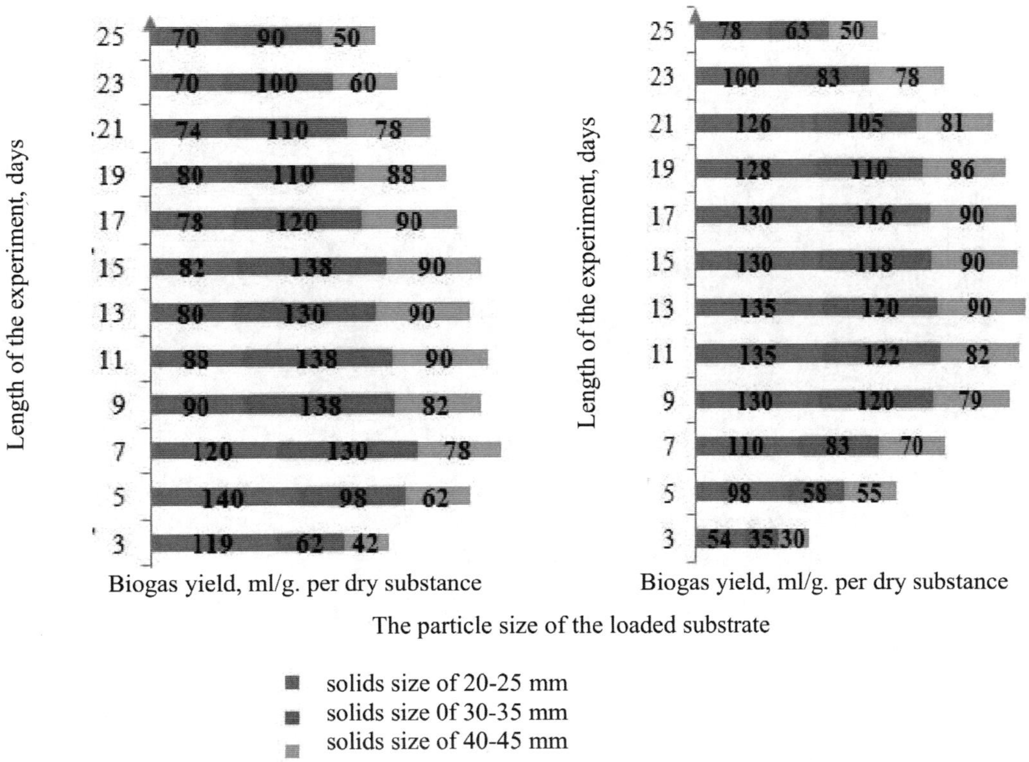

Figure 2. The effect of the particle size of the dry components of manure on the rate of release of biogas

The analysis of the results presented in Figure 2 showed that the first portions of gas at mesophilic fermentation mode is released on the third day, what is more with the particles size of 20-25 mm there was observed maximum biogas yield from 3 to 6 days. With further fermentation the amount of generated biogas was significantly decreased. This can be explained by the fast decomposition of the substrate affected by the bacteria.

The maximum biogas yield was observed at mesophilic mode and with the use of particles sized 30-35 mm. Moreover the period of the gas production increased compared with the first variant for 7-10 days. With the size of particles 40-45 mm the period of active gas production was 19 days but there was minimal amount of released biogas. This is probably due to the fact that solid particles of larger size make redistribution of bacteria and intermixing substrate more complicated and worsen gas yield.

At the psychophilic fermentation mode the beginning of the gas production was noted on the fifth day, and the optimal size of the solid particles was 20-25 mm. The increasing of the solid particles size to 30-35 mm and 40-45 mm reduces the process of the gas outlet to 10-20 ml/l of substrate. The first portions of the gas at psychrophilic fermentation mode and the amount of solid particles of 30-35 mm were observed on the seventh day which is four days longer than it was during the mesophilic fermentation mode. The increasing of the solid particles size to 40-45 mm affected the speed and volume of gas production.

Thus, for optimal psychrophilic fermentation mode the size of the solid particles in the manure was 20-25 mm and for mesophilic mode it composed 30-35 mm.

For further studies we defined the task to compare the amount of produced biogas from excrements of cattle, pigs, and its mixture in a ratio of 1:1 at mesophilic and psychrophilic fermentation modes and the same humidity of raw material at 88%. The process of gas yield from various substrates is shown in Figure 3.

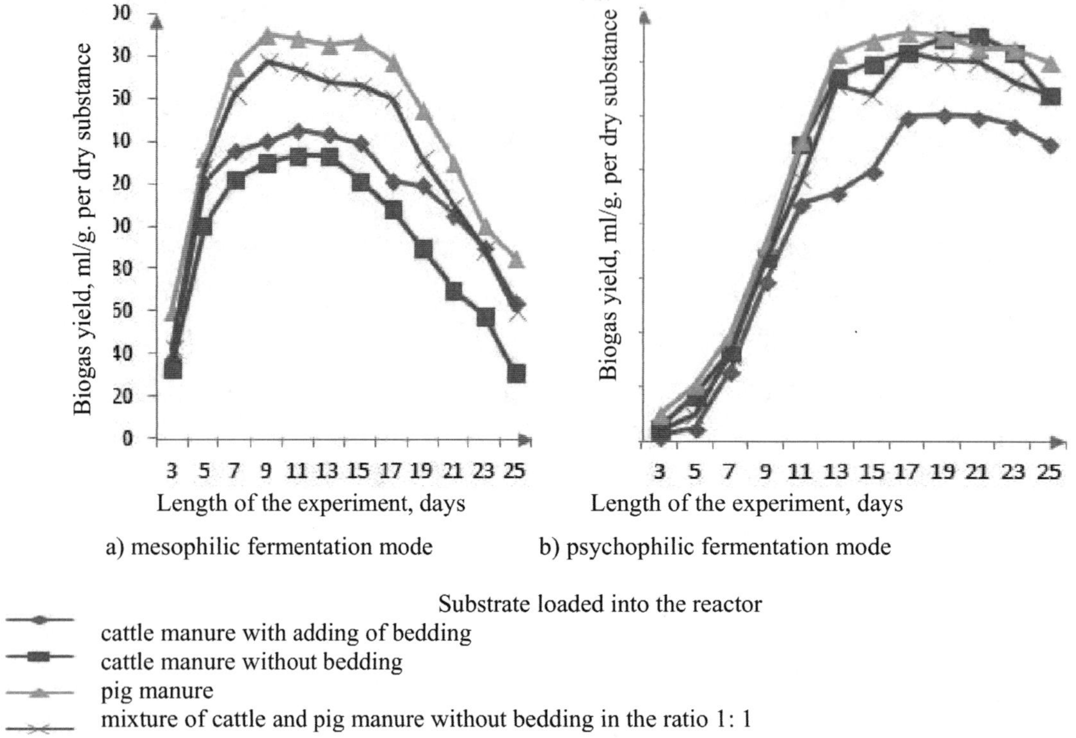

Figure 3. Biogas yield per 1 g of dry organic substances in accordance with the type of substrate

The analysis of the results of this study showed that the gas yeild per unit of time at first increased rapidly, and then, after reaching a maximum, it gradually reduced both at mesophilic and psychrophilic fermentation mode. It should be noted that biogas outlet from cattle manure with adding of bedding was larger by a mean of 10.32%.

On the fifth day already the biogas yield in accordance with the used substrate at mesophilic fermentation was 120 cm^3 for cattle manure with adding of bedding, 100 cm^3 for cattle manure without adding of bedding, 132 cm^3 for pigs manure, 127 cm^3 for a mixture of manure from cattle with adding of bedding and pig manure in the ratio 1:1.

Dynamic equilibrium of substances is determined by the ease of substrate splitting. It should be noted that the inclusion of the litter materials in cattle manure subjected to further processing (grinding to the optimum size) resulted in a greater yield of biogas than the use of pure cattle manure as a substrate. This can be probably explained by the fact that there was a fast decomposition of organic substances by microorganisms in the pure manure at the first stage (hydrolysis) and there was a sharp decrease in pH level from 6.3 to 4.2 due to the activity of acid-forming bacteria that is unfavorable environment for other groups of bacteria.

The optimum pH for the acid-forming and hydrolyzing bacteria is from 4.5 to 6.3 when the bacteria reach their optimum. As for the methane-forming bacteria the optimum environment is neutral or slightly alkaline environment. In addition, the excess concentration of produced substance leads to stunted growth of the bacteria group which produces it. Ideally, a dynamic equilibrium should be established between the splitting phases in concentrations of substances, that is, between the splitting of the nutrients by bacteria and their intake. Thus, the more complicated is the structure of the substrate, the longer splitting of the substrate can be. The maximum yield of biogas from pigs manure was observed that can be explained by the composition of the substrate and the presence of less recalcitrant substances such as lignin in it.

At psychophilic fermentation mode a significant biogas yield was observed with the eleventh day of the experiment and lasted up to 37 days. The rate of the gas formation largely depended on the temperature. The

IOP Conf. Series: Materials Science and Engineering **91** (2015) 012076 doi:10.1088/1757-899X/91/1/012076

higher is the temperature, the faster is the decomposition of the substrate and the intensity of gas formation. Unlike mesophilic fermentation mode under psychrophilic conditions the gas formation was more intense when a liquid cattle manure without bedding and pig manure were used as a substrate. However, the overall yield of gas at psychrophilic mode was lower by 21%. During fermentation, there was an increase in the internal temperature of the substrate by 1.5-3 ° C that indicated the activity of microorganisms.

Conclusions

It can be concluded that the livestock manure is accessible and a good raw material for the biogas production. A number of experiments determining the influence of some factors on the yield of biogas were carried out. It is noted that the maximum yield of biogas was observed at mesophilic mode and particle size of 30-35 mm. At psychophilic fermentation mode the beginning of the gas formation was noted on the fifth day and the optimal size of the solid particles was 20-25 mm. With the increase of the size of the solids to 30-35 mm and 40-45 mm the gas formation process becomes slower and gas yield is reduced by 10-20 ml/l of substrate. A comparison of the amount of generated biogas from cattle and pig excrements and their mixture in a ratio of 1:1 under psychrophilic and mesophilic fermentation conditions and the same moisture of raw material (88%) was done. Biogas yield from cattle manure with adding of bedding was more by 10.32% on the average.

References

[1] Wilson D.C. The problem of waste management in countries with low income per capita // Resource Technologies, 2000. №19. - C. 28 -33.

[2] The nature of landfill gas and his environmental impact / Campbell D.J.V //Wast Manag, 1989.- № 4.- P. 201-208.

[3] [3] Li Han, Ru Ying Li, Min Ji, Enhanced Anaerobic Digestion of Sewage Sludge by Addition of Food Waste, Advanced Materials Research, Environmental Biotechnology and Materials Engineering (2013), 139-143 p.

[4] Zahra Salehi, Francesco Basoli, Igor Luisetto, S. Licoccia, Elisabetta di Bartolomeo, Advances in Science and Technology, 6th Forum on New Materials - Part A, Spin-Coated La0.8Sr0.2Ga0.8Mg0.2O3-δ Electrolyte on Infiltrated Anodes for Biogas Utilization, 7-13 p.

[5] Lidia Pino, Antonio Vita, Cristina Italiano, Concetto Fabiano, Massimo Laganà, Vincenzo Recupero, Biogas Reforming for Hydrogen Production: Performance of Ni/La-Ce-O Catalysts, 19-25 p.

[6] Feng Kong, Xiao Ye Zhang, Jie Hong Cheng, Zai Xin Mao, Yong Qiang Wang, Pan Li Wang, Selected Proceedings of the Eighth International Conference on Waste Management and Technology Performances of Leaching-UBF Anaerobic Digestion for Biogas Production of Kitchen Waste, Advanced Materials Research, p.489-496.

[7] Guang Zhao, Fang Ma, Ting Sun, Shu Ben Li, Kong You, Selected Proceedings of the Eighth International Conference on Waste Management and Technology, Advanced Materials Research, 138-149 p.

[8] The OECD Environmental Data Compendium 2002 / Organisaton for Economic Co-Operaton and Development.- Paris: OECD, 2003.- 27 p.

[9] G. Ajith Weerasundara, A Review of Waste Management in Sri Lanka, p.899.

[10] RusBiogas [electronic resource]. - Mode of access: http://www.rosbiogas.ru/literatura/biogazovie-ustanovki-prakticheskoe-posobie/vozniknovenie-i-etapi-processa-bioagaza.html, free – (10.03.2015).

IOP Conf. Series: Materials Science and Engineering **91** (2015) 012077 doi:10.1088/1757-899X/91/1/012077

Updating of sewage – purification facilities of electroplating enterprises with counterflow ion-exchange filters

V F Torosyan, E S Torosyan, P D Sorokin, A A Telitsyn

Yurga Institute of Technology, Yurga, Russia

E-mail: torosjaneno@mail.ru

Abstract. The paper focuses on work of electroplating sewage-purification facilities of mechanical engineering production; drawbacks caused by specific features of physical and chemical processes of coagulation and technological malfunctions have been revealed. Additional equipment – ion-exchanging filters have been selected on the basis of designed methods, they make it possible for enterprises of mechanical engineering to implement conversion to water rotation systems.

1. Introduction

Materials of different chemical nature are used in mechanical engineering [1], and various physical and chemical processes are applied in surface and deep processing of materials, therefore, enterprises of this branch have both to provide high precision and reliability of produced machines [2], and use raw material reasonably, assure environmental safety of technological and production processes. Electro-plating industry is one of main sewage producers in mechanical engineering. Basic water pollutants in electro-plating industry are ions of heavy metals, non-organic acids and alkalis, cyanides, and surface-active substances. Here, scourage is considered the main source of toxic substances and simultaneously, the main user of water, as well as the source of sewage in electro-plating industry. In most electro-plating industries the volume of sewage is enormous because the method of workpiece washing isn't perfect and requires significant consumption of water (up to 2 m^3 and more per each 1 m^2 of workpieces surface). Nevertheless, none of available methods of electro-plating sewage purification can meet the current standards of industrial sewage; therefore, work optimization of purification plants is thought to be the urgent problem of machinebuilding enterprises.

While workpiece washing, in average 3300 t of zinc, 2400 t of nickel, 2500 t of copper, and dozens of thousands of other metals are taken from operating cell annually. One of the important problems in this industry is also failure of process solutions because extraneous organic and non-organic substances are stored up in electrolytes; and correlation of basic components in electroplating baths gets broken [3].

Sewage of spent solutions makes up 0.2–0.3 % of total sewage volume, but in terms of pollutant concentration its share is about 70 %. Volley of this sewage breaks operating mode of purification plants, and causes irretrievable loss of valuable materials. Unpurified or insufficiently purified sewage and other wastes containing non-ferrous metals, which get into water bodies, bring harm to national economy and environment because used in industry metals get lost, and their impact on environment is increasingly negative [4].

It is worth noting that enterprises of mechanical engineering can completely provide themselves with electrical and thermal energy. They take measures to remove admixtures and impurities of

IOP Conf. Series: Materials Science and Engineering **91** (2015) 012077 doi:10.1088/1757-899X/91/1/012077

industrial sewage, making them meet standards of maximum permissible concentration, and utilize liquid wastes in areas of persistent storage [5].

Chemical composition and concentration of sewage of electroplating industry is of high importance for purification plants of these enterprises, because electrolytes are contaminated by metal impurities in the course of electroplating due to dissolution of workpiece metals. Ions of copper, nickel, ferrum, zinc and other metals are taken by process water into sewage. The increase in concentration of impurity metals has a considerable negative effect on conductivity, hence, electro-chemical purification is damaged (electrical equipment fails).

2. Experimental part

Pursuant to SNiP 2.04.03–85 hexavalent chromium and other metals are removed from sewage as follows:

– concentration of ions of hexavalent chromium Cr+6 only is maximum 100 mg/l;

– total concentration of non-ferrous metals doesn't exceed 100 mg/l, concentration of each of them is maximum 30 mg/l;

– minimum total solid content of sewage is 300 mg/l and concentration of suspended materials is maximum 50 mg/l.

As we can conclude from the information on concentration of ions of heavy metals in chromium-containing sewage of electro-plating production at Yurga Machinebuilding Plant, their qualitative purification from hexavalent chromium can be performed provided that its concentration is maximum 50 mg/l.

A bath for catching valuable or particularly toxic components is placed on the site of electro-plating after the technological bath and before the washing cells in order to reduce contamination of sewage with these components. Used rinsing water from the catching bath is twice diluted process liquor. It can't be subject to purification, because process failure of purification plants is possible, as well as contamination of sewage and quantity of forming slime can increase.

Volleys of concentrated used solutions also disrupt acceptable operation of purification plants. Therefore, neutralization of used solutions at purification plants of enterprises can't be considered as an engineering solution being in line with current level of electro-plating [6].

Electrochemical purification based on the method of electrical coagulation involves electrolytic solution of steel electrodes, which causes formation of ferrous iron ions (Fe^{2+}), as the result bichromate and chromate – ions are reduced:

$Cr_2O_7^{2-} + 14H^+ + 6e^- \rightarrow 2\,Cr^{+3} + 7\,H_2O$,

$CrO_4^{2-} + 4\,H_2O + 3e^- \rightarrow Cr(OH)_3\downarrow + 5OH^-$,

$Fe^{+3} + e^- \rightarrow Fe^{+2}$; $Fe^{2+} + 2OH^- \rightarrow Fe(OH)_2\downarrow$.

As the result of electrochemical reaction hydrogen ions concentration decreases in sewage and hydrogen ion exponent (pH) grows in pretreated waste water [2]. If there are cations of other heavy metals in water, they are reduced by ferrous iron in the course of exchange processes and are deposited alongside with chromium and ferrum hydroxides, some of these metals react with anions OH – transfer to poorly soluble hydroxides or are reduced on the cathode, losing their hydration film and sorbing on its surface.

pH values, providing removal of heavy metals via deposition on cathode, in the course of electrochemical purification are listed in Table 1 [7].

If Cr^{+6} concentration doesn't exceed 20 mg/l, initial value of pH is to be 4.5 – 5. If pH $\leq$ 3 the rate of electrochemical processes of Cr^{+6} reduction increases. If pH $\geq$ 6 specific consumption of electric power increases as conductivity of solution reduces and overvoltage grows on cathode because of hydrogen emission.

Table 1. pH of electrochemical purification

Heavy metals	Cr^{6+} concentration, mg/l	pH
Trivalent chromium	–	3.5–4.0
Nickel	–	Minimum 7.0
Copper, zinc, cadmium	–	Minimum 4.5
Hexavalent chromium, copper, zinc	Maximum 20	6–7
	20–50	5–6
	50–100	4–6
Hexavalent chromium, cadmium, nickel	Maximum 50	6–7
	Minimum 50	5–6
Hexavalent chromium	Maximum 50	6–7

In the course of coagulation passivation of anodes takes place because on their surface a hydroxy film forms and decelerates electro dissolution of metal [8]. That is why, every other 30–60 minutes of electro-coagulators work it is necessary to change the polarity of electrodes, i.d. reversal of current on a rectifier. $Cr(OH)_3$ has amphoteric properties; at pH ≥12 it dissolves in redundant alkali and chromites form, i.d. ions of chromium can be transferred to solution again.

$Cr(OH)_3 + 3NaOH \rightarrow Na_3[Cr(OH)_6]$

Hence, while acid chromium containing waste water is neutralized, excess dispensing of alkali is to be avoided. Polyacrylamide can be applied to remove hydroxides of metals and clarify waste water via flocculation.

As one can conclude from the analysis of experimental data on purification of electroplating waste water at Yurga Machinebuilding Plant, concentration of ferrum ions in pretreated water grows because ions Fe^{2+}, which are excessive over the needed quantity to reduce Cr^{+6}, are transferred from the surface into the solution. In this case pretreated waste water can be blown down in the aeration chamber before being transferred into the sedimentation tank in the aeration chamber.

As the result ions Fe^{2+} in ferrum hydroxide (II) are oxidized up to sparingly soluble ferrum hydroxide (III), and their total concentration in water decreases:

$4Fe^{2+} + 8OH^- + O_2 + 2H_2O \rightarrow Fe(OH)_2$,

$4Fe(OH)_2 + O_2 + 2H_2O \rightarrow 4Fe(OH)_3$.

Water can be circulated twice through electrocoagulators to make purification of sewage the most efficiently. However, it is important to note that none of avaliable methods to purify electroplating sewage can provide independently all current standards of industrial waste waters; the latters are as follows:

- purification until the standards of maximum permissible concentration are met, especially those of ions of heavy metals;
- return of 90-95 % water into reverse cycle, low cost price of purification;
- small-size facilities, utilization of valuable components (acids, alkalis, metals).

Impossibility of meeting the environmental standards is worsened by poor financial situation at many enterprises. Therefore, from our point of view, a way to overcome this situation consists in modernization of means and methods of sewage purification and optimization of purification systems.

3. Discussion of results

Here, we think it is necessary to use additional equipment for afterpurification of sewage water. As the total replacement of equipment being at disposal of the enterprises, is hardly possible because of required investments, these new facilities are to be power-consuming, cost effective and meet the standards of productivity and efficiency of industrial sewage purification.

It is worth saying that among available methods of waters and sewage purification ion-exchanging methods are practically the most efficient ones. Therefore, this work focuses on selection of

IOP Conf. Series: Materials Science and Engineering **91** (2015) 012077 doi:10.1088/1757-899X/91/1/012077

equipment, necessary for afterpurification of sewage water of electroplating production at the enterprises of mechanical engineering. The equipment will be selected on the basis of designed methods of calculation and according to requirements for efficiency of purification system being at disposal of Yurga Machinebuilding Plant [9].

The following characteristics of ion-exchanging filters are thought to be basic structural, technological, as well as technical and economical ones: productivity up to 40 m^3/h, filtering fill – cationite KY–2–8, anionite AB–17–8, which provide practically complete removal of ions of heavy metals in industrial tests. Necessary reconstruction predetermines also the replacement of forward feeding of water to be purified by the counterflow one, from our point of view, it is reasonable and economical, technologically and environmentally efficient. Here we take into consideration the advantages of counterflow feeding technology:

- low layers of ionite (the last ones down pretreated water) are subject to deep regeneration with excessive reagent;
- the second regeneration of ion-exchange resin with untreated exchanging capacity is not required (material isn't used evenly), consumption of the reagent can be reduced.

Conversion of forward feeding ion-exchange filters to the counterflow one makes it possible:

- to increase efficiency of filters by means of available facilities;
- to reduce annual filling of material because of advanced technology;
- to reduce quantity of salt, acid and alkali needed for regeneration (8–10 %);
- to decrease chloride, acid and alkali discharge;
- to cut down water consumption for needs of filters (up to 50 %) [10].

Methods to calculate basic technological characteristics of cationite filter type XB–042–1

Number of flows (Q), transferred to filters – 40 m^3/h. Total concentration of cations – 0.54g·equ/m^3.

Cationite filter XB–042–1 (Table 2) is used for ion-exchanging purification.

Table 2. Technical parameters of cationite filter type XB–042–1

Parameter	Value
Diameter of filter	1000 mm
Area of filtering	0.8 m^3
Volume of filtering fill	1.6 m^3
Filtering fill	Cationite KY–2-8 (Table 3)

Table 3. Chemical characteristics of cationite KY–2-8

Characteristic	Standard	
	Supreme grade	First grade
Granulometric composition: dimensions of grains, mm	0.315–1.25	0.315–1.25
Inclusion volume of power fraction, %, minimum	96	95
Efficient dimensions of grains, mm	0.40–0.55	0.35–0.55
Coefficient of homogeneity, maximum	1.7	1.8
Mass concentration of moisture, %	43–53	43–53
Full static exchanging capacity, mmole/cm^3 (mg equ/cm^3), minimum	1.9	1.8
Specific volume, cm^3/g in H-form, maximum	2.8	2.8
Osmotic stability, %, minimum	94.5	90
Apparent density of commercial cationite, g/dm^3	750–800	750–800

The running facilities of industrial sewage purification are to be equipped with 3 cationite filters – 2 operating ones and 1 in reserve.

IOP Conf. Series: Materials Science and Engineering **91** (2015) 012077 doi:10.1088/1757-899X/91/1/012077

Filtering rate is equal to:

V_f= 40/1.6 = 25m^3/h, permissible – 30 m3/h, where 1.6 – volume of filtering fill.

Working exchanging capacity of cationite is computed pursuant to the formula:

$$E^k_{wc} = \alpha_k E^k_{gen} - K_{ion} q_k \sum C^k_w, \tag{1}$$

where α_k – coefficient of regeneration efficiency, is accepted to be 0.85, E^k_{gen} – full exchanging capacity of cationite, accepted to be 1000 g·equ/m^3, qk – specific water consumption to wash cationite after regeneration, accepted to be 5 m^3 per 1 m^3 of cationite, K_{ion} – coefficient needed to take into consideration the type of ionite, for cationite it is accepted to be 0.5, $\sum C^k_w$ – total concentration of cations in cleaning water (when cationite washing by ionizated water).

Hence: E^k_{wc} = 0.85 · 1000 – 0.5 · 5 · 0.5 4= 840 g·equ/m^3.

Duration of working cycle is determined:

$$t_f = \frac{YE^k_{wc}\, n}{Q \sum C_w^k} \tag{2}$$

where Y – volume of fill filter, is accepted to be 1.6 m^3, n – number of working filters.

t_f = 1.60 · 2 · 840/40 · 0.54 = 12 hours.

Frequency of filter switching on for regeneration is 12/2 = 6 hours.

Reagent consumption for regeneration:

Consumption of 100 % acid for regeneration of one cationite filter is computed according to the formula:

$$T_k = \frac{YE^k_{wc} \cdot B}{1000} \tag{3}$$

where B – specific consumption of acid, it is accepted to be 100 g/g·equ.

T_k = 1.60 · 840 · 100/1000 = 134.4 кг.

Concentration of regenerating solution of acid Cp is 1.2 kg·equ/m^3.

Reclaim is transferred from cationite filter, its volume is:

$$V_{кam} = \frac{YE^k_{wc}}{C_p} \tag{4}$$

V_{kam} = 1.6 · 1/1.2 = 1.3 m^3.

Cationite backwash is carried out by treated water with intensity 3 l/sec·m^2.

Water consumption per one filter is 3 · 1.6 = 4.8 l/sec or 17.28 m^3/h for 15 minutes of backwash or 0.25 h.

Water quantity per one filter is 17.28 · 0.25 = 4.32 m^3.

For after regeneration washing water is used in proportion 5 volumes of water per 1 m^3 of resin, then: Q_{OTM} = 1.6 · 5 = 8 m^3.

It takes 60 minutes to wash.

Total consumption of water per one filter is: 4.32 + 8 = 12.32 m^3/24 hours.

Backwash and washing of filters are carried out by treated water from technological water supply system. As the result of calculations basic technological parameters were obtained for cationite filter (Table 4).

Table 4. Technological parameters of cationite filter

Parameters	Units	Quantity
Efficiency	m^3/h	40
Volume of filtering	m^3/h	25
Duration of filtering cycle	h	12
Between – regeneration period	h	6

Parameters	Units	Quantity
Consumption of H_2SO_4 per regeneration period (2 filters)	kg/24 hours	268.8
Concentration of regenerating solution	kg-equ/m^3	1.2
Water consumption A) preparing of regenerating solution B) for backwash C) for washing	 m^3 m^3 m^3	 2.0 4.32 8.0

Methods to calculate basic technological characteristics of anionite filter type XB–040–1

Number of flows, transferred to filter – 40 m^3/h, initial concentration of anions in water is 0.5 g·equ/m^3.
For ion-exchanging purification we used cation filter type XB–040–1 (Tables 5, 6)

Table 5. Technical parameters of anionite filter type XB–040–1

Parameter	Value
Diameter	1000 mm
Filter area	0.8 m^3
Height of filtering layer	2 m
Volume of filtering fill	1.6 m^3
Filtering fill	Anionite AB–17-8

Table 6. Chemical characteristics of anionite AB–17–8

Parameter	Standard	
	Supreme grade	First grade
Granulometric composition: dimensions of grains, mm	0.315–1.25	0.315 – 1.25
Inclusion volume of power fraction, %, minimum	95	93
Efficient dimensions of grains, mm	0.4–0.6	0.6
Coefficient of homogeneity, maximum	1.7	1.8
Mass concentration of moisture, %	35–50	35–50
Full static exchanging capacity, mole/cm^3 (mg equ/cm^3), minimum	1.15	1.00
Specific volume, cm^3/g in H-form, maximum	3.0±0.3	3.0±0.3
Balanced static exchange capacity, mmole/cm^3, minimum	1.00	0.90
Dynamic exchange capacity with full regeneration, mole/m^3, minimum	700	690
Osmotic stability, %, minimum	92.5	85.0
Apparent density of commercial cationite, g/dm^3	750–800	750–800
Working temperature, maximum, °C	100	
Working range, pH	1–14	

Running facilities of industrial sewage purification are to be equipped with 3 anionite filters – 2 working ones and one in reserve.

Filtering rate is equal to:

$V_f = 40/1.6 = 25$ m^3/h whereas the permissible one is 30m^3/h, where 1.6 – volume of filtering fill.

Working exchange capacity of anionite is computed according to the formula:

$$E_{wc}^{an} = \alpha_{an} E_{gen}^{an} - K_{ion} q_{an} \sum C_w^{an} \quad (5)$$

where αk – coefficient of regeneration efficiency, it is accepted to be 0.9, E_{gen}^{an} – full exchange capacity of anionite is accepted to be 500 g·equ/m^3, q_{an} – specific consumption of water for washing anionite after regeneration, it is accepted to be 24 m^3 per 1 m^3 of anionite, K_{ion} – coefficient, which takes into consideration ionite type; it is accepted to be 0.8, $\sum C_w^{an}$ – total concentration of anions in washing water, g·equ/m^3.

Hence: $E_{wc}^{an} = 425$ g·equ/m^3.

Duration of working cycle is determined according to the formula:

$$t_f = \frac{YE_{wc}^{a} n}{Q \sum C_w^{a}} \quad (6)$$

where Y – volume of fill filter – 1.6 m^3, E_{wc}^{a} – working exchange capacity, C_n– total concentration of anions in water, it is accepted to be 0.5 g equ/l.

$t_f = 1.6 \cdot 425/40 \cdot 0.5 = 34$ hours.

Frequency of filter switching on for regeneration: every 17 hours.

Alkali consumption for regeneration of one filter is:

$$P_щ = \frac{YE_{wc}^{a} \cdot B}{1000} \quad (7)$$

$P_щ = 1.6 \cdot 425 \cdot 40/1000 = 28.8$ kg.

Concentration of regenerating solution

$T_p = 1.2$ kg·equ/m^3.

Regenerate is transferred from anionite filter, its volume is:

$$V_{an} = \frac{YE_{wc}^{an}}{T_p} \quad (8)$$

$V_{an} = 1.6 \cdot 0.425/1.2 = 0.9$ m^3.

Backwash of anionite is carried out by technological water with intensity 2 l/sec·m^2.

Water consumption per one filter is: 2 · 0.8 = 1.6 l/sec or 5.76 m^3/h if backwash lasts 20 min or 0.33 hours.

Water quantity per 1 filter is 5.76 · 0.33 = 1.90 m^3.

For after regenerating washing it is necessary to take 15 volumes of water per one volume of loaded resin.

$Q_{OTM} = 1.6 \cdot 15 = 24$ m^3.

Duration of filter washing is 2 hours.

Total consumption of water for washing of 1 filter is: 1.90 + 24 = 25.9 m^3/24 hours.

As the result of calculation we obtained basic technological characteristics of anionite filter (Table 7).

Table 7. Technological parameters of anionite filter

Parameter	Unit	Quantity
Efficiency	m^3/h	40
Volume of filtering	m/h	25
Between-regeneration period	h	17
Consumption of NaOH for filter regeneration	kg/24 hours	28.8

Parameter	Unit	Quantity
Concentration of regenerating solution	kg equ/m^3	2
Water consumption A) for preparing of solutions B) for backwash C) for washing	 m^3 m^3 m^3	 1.0 1.90 24.0

4. Conclusions

1. Available methods of electroplating sewage purification are insufficient to meet the current standards of industrial sewage independently:
- purification until the maximum permissible concentration is obtained, especially for ions of heavy metals;
- return of 90–95 % of water into reverse cycle;
- small-sized dimensions of facilities, utilization of valuable components (acids, alkalis, metals).

2. Ion-exchange filters as additional equipment for after purification of sewage in the system of coagulation make it possible to stop wastes discharge of enterprises and implement the transition to water circulation system.

3. Conversion of available forward feeding ion-exchange filters to the counterflow conditions of operating will enable:
- to increase the efficiency by means of available equipment;
- to reduce annual filling of material due to advanced technology;
- to reduce the consumption of salt, acid, alkali for regeneration (8–10 %);
- to decrease wastes of chlorides, acids, and alkali;
- to shorten water consumption for needs of filters (up to 50 %).

4. Methods of calculation of basic structural, technological, technical and economic characteristics of ion-exchange filters can also be used for their different productivity and various filtering ionites.

5. Modernization of purification plants carried out by counterflow feeding ion-exchange filters makes it possible for the enterprise to reduce parameters of water consumption and solve the problem of raw materials saving.

References

[1] Chinakhov D A 2011 *Applied Mechanics and Materials* vols 52-54 pp 442-447
[2] Solodsky S A, Brunov O G., Zelenkovsky A F 2012 *Welding International* vol 26 **4** pp 314–317.
[3] Aksenov V I, Ladygichev M G, Nichkova I I, Nikulin V A, Klyan S E, Aksenov E V 2005 *Water economy of industrial enterprises* (M.: Teplotechnik)
[4] Kolesnikov V A and Menshutina N V 2005 *Analysis, design of technologies and equipment for sewage* purification (M.: DeLiprint)
[5] Technological process of industrial sewage purification in workshop № 48 of Yurga Machinebuilding Plant
[6] *Industrial ecology* (Mode of access: http://prom-ecologi.ru/?p=61)
[7] *Center of Education* (Mode of access http://edu.dvgups.ru/METDOC/ITS/GIDRA/TEOROSOV/METOD/UP_2/frame/frame_ tema2. htm)
[8] *Industrial water purification* (Mode of access http://www.teh-g.ru/filtr-fsd.html)
[9] Technological process of electroplating sewage purification at Yurga Machinebuilding Plant
[10] Vinogradov S S 1998 *Environmental friendly electro-plating industry*

IOP Conf. Series: Materials Science and Engineering **91** (2015) 012078 doi:10.1088/1757-899X/91/1/012078

The Study of Soil Protection in the System of the Cultivated Lands of Kemerovo Region

M A Yakovchenko, O B Konstantinova, A A Kosolapova
Kemerovo State Agricultural Institute
5, Markovtseva Street, Kemerovo, 650056, Russia

e-mail: mara.2002@mail.ru

Key words: heavy metals, soil protection, ingress sources, technogenesis, phytoremediation, phytostabilization, phytoextraction.

Abstract. The heavy metal content in the surface soils is characterized with their ingress for the given period of time. The sources of heavy metals in the soil are precipitation, seeds, dust, organic and mineral fertilizers, and others. The paper studies the heavy metal content in the soils of the waste dumps of the open-pit coal mines.

1. Introduction

The heavy metals (HM) represent a large group of chemical elements with atomic weight more than 50 c.u. They get into the soil in different ways: as a part of gas-dust emissions, precipitation, irrigation water polluted with industrial wastes, etc. A man can get "his share" of heavy metals not only directly when breathing the air and soil dust, but through the food produced at the polluted agricultural lands as well. The harmful effect of the heavy metals on human beings is that a number of their compounds are characterized with a high toxicity and cancerogenity.

The higher heavy metal concentrations significantly decrease the activity of the enzymes: amylase, dehydrogenase, urease, invertase, catalase, as well as the number of specific agronomically valuable groups of microorganisms. HM inhibit the processes of mineralization and synthesis of different substances in the soil, depress respiration of soil microorganisms, cause bacteriostatic effect, can act as a mutagenic factor. Under excessive HM content in the soil the activity of metabolic processes reduces, morphological transformations in reproductive organs' structure and other changes in soil biota take place. HM can significantly depress biochemical activity and cause variation in the total number of soil microorganisms [1].

The problem of the influence of the open-pit coal mines on the ecology of the nearby territories is being studied for about 50 years. It's known, for example, that the territories situated in the immediate proximity to the open-pit coal mines gradually become unsuitable for life. It happens because of great unfavourable influence of toxic substances contained in coal and its dust. The group of the toxic substances most dangerous to the environment consists of HM which are accumulated in the soil and are certain to be assimilated by the plants growing there [2].

If great amounts of HM get into the soil, its biological, chemical and physical properties change significantly, which leads to soil fertility deterioration. Besides, HM affect the plants directly, desturbing metabolism, reducing their productivity and products quality. There are a lot of studies

IOP Conf. Series: Materials Science and Engineering **91** (2015) 012078 doi:10.1088/1757-899X/91/1/012078

devoted to the pollution influence on the soil properties. Some authors point to the reduction of biological activity of the soils, others note the increase in the number of separate groups of microorganisms, as well as the change in the enzymatic activity of the soils.

Under HM soil pollution the soil properties deteriorate: the soil structure destroys, its density increases, total soil space and water permeability decrease, water-air conditions deteriorate. HM soil pollution reduces humic acids content in the soil humus and increases fulvic acids content [3].

2. The research objective.

Consequently, the problem of soil purification from heavy metals is urgent for the territories of so-called environmentally unfriendly regions which Kemerovo region can be reckoned among. The research objective is to study the soils of the waste dumps of the open-pit coal mines concerning the content of HM labile forms.

3. The research object and methodology.

The studies were conducted on the soils of the territories nearby the open-pit coal mine of the enterprise LLC «Uchastok «Koksovyi» of Kemerovo region.

During the scientific research on the anthropogenic landscapes it's planned to develop a combined technology of reducing HM content in the soil cover of coal producers' waste dumps of Kemerovo region and nearby territories.

The scientists of Kemerovo State Agricultural Institute took soil samples to analyze heavy metal content and to compare the analysis results with MPC, APC. To take soil samples there were chosen 5 areas at the distance of 100 m from each other. 10 combined samples were taken with the help of the agrochemical drill from the depth of 0-10 cm.

While analyzing the special attention was given to the analysis of the content of labile forms of nickel and plumbum in the soil, that is available to the plants, by means of getting different extractions from it: water, extractions by acetated ammonium buffer solution with pH 4.8 or solutions 0.01 M and 1 n. $Ca(NO_3)_2$, 1 M NH_4NO_3, 1 M $CaCl_2$, 0.01 M KNO_3, diethylene triamine pentaacetic acid, etc.

The HM content was determined on the atomic adsorption spectrophotometer (AAS) in the laboratory of the Federal State Centre of Agrochemical service «Kemerovskiy».

4. Results and discussion.

The first stage of heavy metal transformation in the soil is their interaction with soil solution and its components.

Zinc oxide is more stable and less soluble in comparison with plumbum oxide and cadmium oxide. Its solubility in the range of pH 4 – 8 is more than 100 times lower than that of PbO, and almost 10000 times lower than that of CdO. In contrast to zinc oxide, plumbum oxide and cadmium oxide are unstable in water and are transformed into plumbum hydroxide and (or) plumbum carbonate and cadmium carbonate.

Partial pressure of CO_2 in the soil air exceeds one in the atmosphere many times and therefore more stable hydrocarbonates and zinc carbonate and plumbum carbonate prevail in the soil. After the solution of unstable oxides follow such reactions as cation exchange and specific adsorption.

The heavy metal ions are capable of being specifically adsorbed by soils, with relatively stable bonds of coordination type with some surface functional groups being formed. Specific adsorption is more selective than non-specific one and depends both on the properties of sorbing ions and the nature of surface functional groups, therefore heavy metals are actively adsorbed by the soils from the solutions.

Thus, HM getting into the soil during the technogenesis includes the following transformation stages: heavy metal oxides into hydroxides (of carbonates, hydrocarbonates); solution of heavy metal hydroxides (of carbonates, hydrocarbonates) and adsorption of the relevant cations of heavy metals by the solid phases of the soil; formation of heavy metal phosphates and their compounds with organic substances of the soil.

IOP Conf. Series: Materials Science and Engineering **91** (2015) 012078 doi:10.1088/1757-899X/91/1/012078

MPC of a substance acts as a characteristic of its danger to the environment. Polluting substance concentration in the emissions often turns out to be lower than MPC. However, while evaluating the pollution results under given conditions it's necessary to take into consideration the following transformations with account of MPC of transient substances, because the resultant substances can have stronger toxic properties than the initial processes of accumulation and elimination of substances, as well as synergistic effect at their combined presence.

Heavy metals getting into the soil surface are accumulated in the soil and are slowly eliminated while leaching, absorption by plants, erosion and deflation. The first period of half-elimination (that is elimination of the half of the initial concentration) of heavy metals varies greatly: Zn – from 70 to 510 years; Cd – from 13 to 110 years; Cu – from 310 to 1500 years; Pb – from 740 to 5900 years.

HM absorption by the soils depends much on the reaction of the environment (pH). It was revealed that in the acid environment mainly plumbum, zinc, and copper are sorbed; in the alkaline one – cadmium and cobalt.

HM are protoplasmatic poisons the toxicity of which increases with the growth of atomic weight. Iron, for example, forms complex compounds like chelate with common metabolites, herewith desturbing normal metabolism of the organism. Such metals as cadmium, copper, iron (II) interact with cell membranes, changing their penetrability and other properties. High plumbum content in the soil depresses plant growth, causes chlorosis conditioned by abnormality of iron ingress.

Iron, as well as aluminum, belongs to the macroelements of the Earth crust, and such elements as copper, zinc, cobalt, manganese, nickel, plumbum, cadmium, belong to the microelements. Among the microelements there are both typically biogenic (Cu, Zn, Co, Mn) taking part in the most important enzymatic and metabolic processes in living organisms, and typical xenobiotics (Pb, Cd).

HM content (microelements and iron) in the zonal soils is conditioned, in the first place, by their content in the parent material and the tendency of soil formation. Besides, metal content in the soil depends on the amount of organic matter in its texture, the soil solution reactions and it's connected with the migration in the soil crossover and biological cycle of elements.

Copper background content in the soils of Rostov region accounts for 31-38 mg/kg, and maximum values are characteristic of chernozem. Copper background content for chernozem of Siberia accounts for 14 mg/kg [4,5].

In table 1 there is given the total content of some metals in the soils and approxible permissible concentrations (APC) determined for the loam and clay soils (addition to the list of MPC and APC № 6229-91).

Average zinc content in the Earth crust accounts for 200 mg/kg. The soils not polluted with zinc contain it in the concentrations from 10 to 300 mg/kg. Loess-like loams of Western Siberia contain 71.7 mg/kg of zinc. Background concentration of zinc for the chernozem of Siberia accounts for 45 mg/kg.

Average plumbum content in the Earth crust accounts for 16 mg/kg; in the soils - 10 mg/kg. Plumbum distribution in the soil has significant variations with regard to both the types of soils and regions. Background concentration of plumbum for chernozem and chestnut soils of Rostov region accounts for 21 mg/kg and 27 mg/kg respectively, for chernozem of Siberia - 17 mg/kg.

Table 1. Total metal content in the soils (mg/kg of dry matter)

Metal	Average content	Possible range of variations	MPC	Metal	Average content	APC	Possible range of variations
Cadmium	0.06	0.01-0.7	2	Molybdenum	2.0	-	0.2-5
Cobalt	8.0	1.0-40	-	Nickel	40	80	10-100
Chromium	100	5-3000	-	Plumbum	10	130	2-200
Copper	20	2-100	132	Zinc	50	220	10-300
Iron	38000	7000-55000	-	Strontium	300	-	50-1000
Mercury	0.03	0.01-0.3	-	Barium	500	-	100-300
Manganese	850	100-4000	1500				

Average cadmium content in the Earth crust accounts for 5 mg/kg, in the soils - 0.1-0.3 mg/kg. Cadmium content in the soils depends on their type. In the gray forest soils cadmium content accounts for 0.65 mg/kg, in the sod-podzol soils - 0.7-2.31 mg/kg, in the chernozem – 0.7-1.0 mg/kg. In the chernozem of Siberia background content of the total cadmium accounts for 0.6 mg/kg.

Nickel is a rather widespread element in nature. Its average content in the lithosphere accounts for 80 mg/kg, in the soils - from 10 to 100 mg/kg. Nickel background content depends on texture and organic matter of the soils. Total nickel content in the sod-podzol soils of Moscow region accounts for 20-40 mg/kg. In the soils of Krasnodar region average nickel content accounts for 56 mg/kg. For chernozem of Siberia nickel background concentration accounts for 37 mg/kg.

HM total content in 15 stationary points of top soil horizon (0-10 см) of the experimental areas of the waste dump of LLC «Uchastok «Koksovyi» was determined by the method of the atomic adsorption spectrometry (AAS) (table 2).

Copper content can be evaluated as low relatively to the average one for the chernozem of Siberia (14 mg/kg). Copper minimum content is 5.93 mg/kg, which accounts for about 41% of the average one, and maximum content is 13.08 mg/kg (93%). Average content is 8.96 mg/kg, which accounts for 64 % of the average one for the chernozem of Siberia. Zinc content is close to the the average one for the loess-like loams of Western Siberia, which accounts for 71.7 mg/kg, and is more than background content for the chernozem (45 mg/kg). Zinc minimum content accounted for 42.12 mg/kg, and maximum one – 69.17 mg/kg, average one - 54.81 mg/kg (under APC 220 mg/kg).

Total plumbum content is less than average for the chernozem of Siberia (17 mg/kg). Plumbum minimum content accounted for 9.05 mg/kg, maximum one – 15.57 mg/kg, average one - 12.32 mg/kg (under APC 130 mg/kg).

Total cadmium content corresponds to the background content of total cadmium in the chernozem of Siberia and in average accounts for 0.6 mg/kg. Cadmium minimum content accounted for 0.26 mg/kg, maximum – 0.93 mg/kg (under APC 2 mg/kg). Total nickel content is lower than background concentration for the chernozem of Siberia (37 mg/kg). Nickel average content accounted for 22.84 mg/kg, minimum – 14.36 mg/kg, maximum – 29.56 mg/kg (under APC 80 mg/kg).

Total cobalt content in the soils according to table 1 can vary from 1 to 40 mg/kg, with an average value 8 mg/kg. Average content of total cobalt in the soil of the waste dump accounted for 11.54 mg/kg, minimum – 8.93 mg/kg, maximum – 14.79 mg/kg.

IOP Conf. Series: Materials Science and Engineering **91** (2015) 012078 doi:10.1088/1757-899X/91/1/012078

Table 2. The content of the total and labile forms of heavy metals in the soils of the experimental areas of the waste dump of LLC «Uchastok «Koksovyi» of Kemerovo region

point	copper	zinc	plumb um	cadmium	nickel	cobalt	manga-nese	iron	labile nickel	labile plumbum
MPC mg/kg	132	220	130	2	80	-	1500	-	6	4
1	13.08	51.76	11.41	0.58	19.77	10.71	233.87	14032.26	3.74	0.77
2	10.14	46.87	12.64	0.93	25.54	13.40	261.78	18471.13	2.16	1.92
3	9.62	49.32	12.11	0.75	22.62	12.05	148.15	8101.85	4.45	1.79
4	10.40	48.09	12.43	0.84	24.13	12.73	155.5	7770.27	1.67	1.11
5	9.21	51.43	10.6	0.43	15.17	10.3	192.31	7451.92	3.65	2.41
6	7.41	64.4	9.05	0.60	26.8	11.9	170.07	14285.71	4.39	1.44
7	10.78	53.28	10.69	0.53	29.56	10.65	265.11	18965.52	5.47	1.33
8	8.72	42.12	14.77	0.58	28.85	9.23	263.7	15230.77	2.66	1.77
9	8.22	60.07	10.08	0.63	24.65	14.79	262.98	7172.31	3.03	2.32
10	5.93	69.17	10.81	0.57	17.29	11.53	188.65	4669.17	3.97	1.74
11	6.46	66.59	15.57	0.26	19.07	13.09	181.13	7849.06	3.08	1.67
12	8.49	42.74	11.23	0.47	19.23	8.93	161.54	5000.01	2.65	1.68
13	9.26	62.74	14.92	0.74	14.36	12.51	141.09	6821.71	1.22	1.89
14	7.37	46.85	13.35	0.82	28.47	10.04	153.33	7986.11	3.03	1.68
15	9.39	66.85	15.27	0.33	27.14	11.31	183.19	9046.11	4.28	2.14
Averag e value	8.96	54.81	12.32	0.60	22.84	11.54	197.49	10190.26	3.29	1.71

Total manganese content in the soils can vary from 100 to 4000 mg/kg, under average value 850 mg/kg (table 1). Manganese average content in the samples accounted for 197.49 mg/kg, minimum – 141.09 mg/kg, maximum – 265.11 mg/kg (under APC 1500 mg/kg), therefore it can be evaluated as very low.

A possible range of variations of total iron content in the soils is from 7000 to 550000 mg/kg, with an average content 38000 mg/kg. Iron average content in the samples accounted for 8056.90 mg/kg, minimum – 4669.17 mg/kg, maximum – 10965.52 mg/kg.

The evaluation of the level of soil pollution with heavy metals is connected with the availability of their separate forms to the plants. Total content of metals is a characteristic of their total amount. Labile forms of heavy metals are the most dangerous for the plants. The problems of soil pollution rating, including heavy metals, haven't been solved to the full extent. Nowadays State Standard 17.4.1.02.-83 is valid, according to which chemical elements, including heavy metals, are divided into three classes of hazard with respect to toxic effect on the soil. Arsenium, cadmium, mercury, selenium, plumbum, zinc belong to the first class of hazard; cobalt, nickel, copper, chromium, molybdenum, antimony, boron - to the second class and barium, vanadium, wolfram, manganese, strontium – to the third class.

The chemical composition of the plants, as it's known, reflects the soil composition. Therefore, excessive accumulation of HM by plants is conditioned, first of all, by their high concentration in the soils. During their vital activity plants contact only with available forms of HM, the amount of which, in turn, is closely connected with soil buffer capacity. However, the soil capability to combine and inactivate HM has its limits, and when the plants can't cope with the heavy metal ingress, physiological and biochemical mechanisms of the plants themselves, which prevent heavy metal ingress, become very important.

Some interest is shown to the development of soil remediation technology known as phytoremediation which includes phytostabilisation and phytoextraction. Phytoremediation is important for remediation of agricultural lands in the regions with overlapping industrial and agricultural activity where heavy metal content in the soil is very high. Phytoremediation of polluted soils is a relatively «mild» and cheap way in comparison with revegetation.

Phytostabilisation is understood as a technology under which the plants resistant to heavy metals are grown on the polluted soils in order to reduce metal mobility and thereby further environmental pollution through leaching of these pollutants into the ground waters or preventing their distribution by wind and water soil erosion.

Phytoextraction is understood as a technology of selected plants cultivation on the polluted areas in order to extract heavy metals from the soils and accumulate them in the plants with their further processing. The plants extract plumbum from the polluted soils rather actively, and therefore the methods of plumbum phytoextraction, as well as other heavy metals, have, without any doubt, a great significance for soil detoxication.

The study of the soils of the waste dumps of LLC «Uchastok «Koksovyi» was planned to be done for 3 years. If in the revegetation areas the HM content turns out to be higher than APC, then according to table 3 it will be recommended to use plants-remediators in the areas under revegetation [6,7].

Table 3. Reliable correlation links of metal concentrations in the soil and plants (surface soils)

The name of plants	Pb	Cd	Zn	Ni	Cr	Co	Cu
Top soils							
Cankerwort (Taraxacum officinale)	+	+	+	+	+	+	+
Great bur (Arctium lappa)	+	+	+	+	+	+	+
Couch-grass (Agropyron repens)	+	+	+	+	+	+	+
Green ginger (Artemisia vulgaris)	+	+	+	+	+	+	+
Knotgrass (Polygonum aviculare)	+	+	+	-	+	+	+
Nosebleed (Achillea millefolium)	+	+	-	-	+	+	+
Common wormwood (Artemisia absinthium)	+	+	+	+	-	+	-
Blue-sailors (Cichorium intybus)	-	+	+	+	+	+	-
Fire-leaves (Plantago media)	+	+	-	-	+	+	-
Deep soils							
Cankerwort (Taraxacum officinale)	+	-	+	-	-	+	+
Great bur (Arctium lappa)	+	-	-	-	-	+	+
Couch-grass (Agropyron repens)	-	-	+	-	-	+	+
Green ginger (Artemisia vulgaris)	+	-	-	+	-	+	-
Knotgrass	+	-	+	-	+	+	+

The name of plants	Pb	Cd	Zn	Ni	Cr	Co	Cu
(Polygonum aviculare)							
Nosebleed (Achillea millefolium)	+	-	-	-	+	-	-
Common wormwood (Artemisia absinthium)	+	-	-	-	-	+	+
Blue-sailors (Cichorium intybus)	-	-	-	-	-	+	-
Fire-leaves (Plantago media)	-	-	-	-	-	+	-

5. Conclusions

The research hasn't set the soil pollution higher than MPC and APC with respect to the total content of plumbum, cadmium, zinc and manganese in the territory under study. Most of the territory has a level of the content of the total forms of heavy metals in the soil – less than 0.5 of MPC.

The study of heavy metal content in the plant roughage showed the trace amounts, which represents a very small concentration (less than 0.1 of APC).

References

[1] Belyuchenko I S The problems of soil protection in the system of the cultivated land / *Scientific journal of Kuban State Agricultural University* 2014 - № 95 (01) – P 1-32

[2] Blaylock M J Enhanced accumulation of Pb in Indian mustard by soil-applied chelating agents / M J Blaylock D E Salt S Dushenkov et al // *Environ Sci Technol* – 1997 – V 31 - №3 – P 860-865

[3] Ibanez J J Future of soil science / J J Ibanez // *The future of soil science* / Ed A E Hartemink – Wageningen: IUSS 2006 – P 60-62

[4] Maximum permissible concentration (MPC) and approxible permissible concentration (APC) of chemical substances in the soil: hygienic regulations – M : The Federal Hygienic and Epidemiological Center of the Federal Service on Customers' Rights Protection and Human Well-being Surveillance 2006 – P 15

[5] Kolesnikov S I Kazeev K Sh Valkov The influence of heavy metal pollution on ecological-biological properties of typical chernozem / *Siberian Ecological Journal* 2000 - №3 – P 193-202

[6] Tyler G Heavy metal pollution and soil enzymation activity Plant and soil Biology of North-West Caucasus / Tyler G // *Summaries of Latinarican Congress of Soil Science Chilt* 1999 – P 206

[7] Khasanova G R Abramova L M The possibilities of restoration of biodiversity of steppe vegetation: permanent grasses sowing / *Siberian Ecological Journal* 2000 - №6 – P 473-476

IOP Conf. Series: Materials Science and Engineering **91** (2015) 012079 doi:10.1088/1757-899X/91/1/012079

Statistic analysis of acoustic noise in the industrial and living areas

I N Kirillov and V V Bulkin

Murom Institute of Vladimir state University, Orlovskaya, str., 23, Murom, 602264, Russia

E-mail: dedvanya@mail.ru and vvbulkin@mail.ru

Abstract. In this article the algorithm of statistical analysis was considered providing the representation of the results of the measurement figures on the acoustic noise in the industrial zones and adjacent living areas in the form of time in percentage for the course of which the sound level was exceeded at some frequency at different time moments. The explanation of the analysis principle was mentioned. On the basis of the experimental data on the character of the acoustic noise in the area of the city of Murom curves were obtained reflecting the distribution of the noise levels by the action time.

1. Introduction

Among the hazards of the developed machine-building production, not the least is the acoustic noise that is present typically in indoor areas with active process of mechanical production. However, depending on the intensity of work of the enterprise, the vastitude of the territory and the availability of open areas with active construction and traffic, the acoustic noise may be a very influencing factor on open spaces too. If the enterprise is located in urban realm and not equipped with good protection against the propagation of noise into the urban area, this can be a problem for the entire urban agglomeration. Thus, from the problem of worker safety and health within the enterprise, the noise may become an environmental problem of a big urban land.

Of late acoustic noise has become one of the main problems of the modern urbanized space occupying the third place by its importance among all the city pollutants [1]. So the tasks of protecting a man within the frame of industrial or residential zones are becoming more and more important [1,2]. Thereby the demands of creating the systems of the monitoring of acoustic and noise pollution in such areas become more and more important. Due to it the necessity to create the systems of acoustic and noise monitoring systems in such areas increases, which will give an opportunity to create the noise map of the city, to forecast the distribution of noise in the residential areas and to take effective managerial decisions to protect the population from noise [3].

Together with it because of the fact that acoustic noise in the industrial or residential area does not belong to stationary processes there is a problem of the reliability of the formed noise map, the correspondence of the obtained data not to instant but to relatively constant manifestations of acoustic and noise pollution. It is evident that it is impossible to perform the monitoring without any statistical treatment of the results with the purpose of developing prediction estimates with respect to the prediction estimates of this noise distribution in the residential areas. The solution of this task can be performed by using the abilities of the system of control and the information system in the monitoring system [4,5]. Another problem in this direction is the choice and the substantiation of creation the algorithm of the data treatment.

IOP Conf. Series: Materials Science and Engineering **91** (2015) 012079 doi:10.1088/1757-899X/91/1/012079

In this article an algorithm of statistical analysis was considered giving the results of processing the measurement information in the form of time expressed in percentage, during which each sound level was exceeded at this frequency at different time moments.

2. The possibilities of the statistic treatment of the measurement results

In [6] there is a method of determining the reliability and precision of the noise map based on the combination of the analysis and mathematical statistics. Separate results of measuring the city noise $L=x_{ij}$ are given in the form of the matrix of the values $M(x_{ij})$, in the lines of which in the horizontal direction there are the values x_i and i in different space points with the total amount of n, and in the vertical direction, in the columns, there are x_j values at different time moments j with the total number m.

If the random deviations of the measurement x in the space do not depend on the random deviations of this value, then the matrix of the values $M(x_{ij})$ is transformed into the value matrix $M(x_i+x_j)$, where the value x_i depends only on the space measurements and the value x_j depends only on time measurements. Using this method it is possible to express the average value, the dispersion of the space deviations, the dispersion of the time deviations, the dispersion of the deviations in space and in time

The fault of this method with respect to middle-size and big cities is that for the reliability value P=0.95 and the precision value at the level of ΔL=1dB for one street with the length of 2 km it is necessary to perform the measurement m=24 in n=10 space points, i.e. it is necessary to make n•m=240 measurements of the sound level per day. While performing the measurement ten times per month the number of the measurements of the sound levels per year in one street will be 240•10•12=28800, which is rather difficult for middle or small cities because of the high cost of the work.

3. Time analysis of statistic distribution

Let us consider the so-called analyzer of the statistic distribution. In this case it possible to register the relative time part, during which the measured noise level was within the range of each scale step situated, for example, in 1 dB. The results of such measurements show during which part of the full time each of the sound pressure levels was exceeded.

In the calculating algorithm for the statistic distribution of the matrix acoustic noise

$$M\left(L_{ij}\right)=\begin{pmatrix} L_{11} & L_{12} & L_{1j} \\ L_{21} & L_{22} & L_{2j} \\ L_{i1} & L_{i2} & L_{ij} \end{pmatrix} \quad (1)$$

we use an intermediary matrix of the statistic distribution, in the lines of which in the horizontal direction there are distributed levels of sound pressure l with the total amount k with equal intervals l' between the values

$$M\left(P_{ij}\right)=\begin{pmatrix} P_{11} & P_{12} & P_{1j} \\ P_{21} & P_{22} & P_{2j} \\ P_{i1} & P_{i2} & P_{ij} \end{pmatrix}, \quad (2)$$

where P_{ij} is the indicator of the statistic noise distribution expressed in percentage.

The distributed levels of the sound pressure l are chosen as per the required measurement scale with the step of l', the range of each includes the minimum and the maximum values L_{ij} of matrix $M\left(L_{ij}\right)$.

To create an intermediary matrix it is necessary to calculate the values of the variables P_{ij} for the concrete column j using the algorithm:

– If the identities $\begin{cases} l & \leq & L_{ij} \\ P'_{ij} & \leq & n \end{cases}$ are correct, then $P'_{ij} = P_{ij} + 1$;

– If $l \leq L_{ij}$ is incorrect then $l = l + \frac{l_{max}}{k}$, $i = i + \frac{i_{max}}{n}$;

– if $P'_{ij} \leq m$ is incorrect, and $l = L_{ij}$ is correct then $P_{ij} = \frac{P'_{ij}}{n} \cdot 100\%$,

where P'_{ij} shows the number of elements in the line *i* of matrix $M(L_{ij})$ exceeding the current scale value *l*; l_{max} and i_{max} – the maximum matrix values $M(P_{ij})$ and $M(L_{ij})$ respectively, while the equation $l_{max} \leq i_{max}$ .must be observed.

And then using the intermediary matrix $M(P_{ij})$ the matrix of the full statistic distribution $M(l_{Pj})$ is calculated, in the horizontal lines of which there are values expressing the time *P=1...100* expressed in percentage during which each of levels L_{ij} in the column *j* of the matrix $M(L_{ij})$ was exceeded with the total number *τ=100*, in the vertical direction in the columns, the amplitude and frequency responses at different frequencies *j=6...10000 (Hz)* with the total number *m=1000*:

$$M(P_{ij}) = \begin{pmatrix} P_{11} & P_{12} & P_{1j} \\ P_{21} & P_{22} & P_{2j} \\ P_{l1} & P_{l2} & P_{lj} \end{pmatrix} \geq M(L_{Pj}) = \begin{pmatrix} l_{11} & l_{12} & l_{1j} \\ l_{21} & l_{22} & l_{2j} \\ l_{P1} & l_{P2} & l_{Pj} \end{pmatrix}. \quad (3)$$

It is the most convenient method to use distribution with the values *P*=1; 10; 50; 90; 99 (%), where *P*=1 gives an idea about the maximum distribution of the noise level, *P*=10 – is the characteristic high level; *P*=90 –shows the noise level, so the level to which the noise is reduced in case of a temporary quiet period. The difference between the values *P*=10 and *P*=90 shows in what range at that place the noise level fluctuates. At that the noise indicator equalling to 10% is widely used while measuring and predicting the transport noise and in many countries it is the basic characteristic showing the level of the acoustic and noise pollution of the territory. Usually the transport noise fluctuates in a concrete way so the level *P*=10 is a rather satisfactory noise indicator though it represents a statistic picture only partially

So, we shall get a statistic distribution of the noise levels in case of significantly fluctuating noise climate in the form of matrix $M(l_{Pj})$, where *P*=1; 10; 50; 90; 99 (%):

$$M(l_{Pj}) = \begin{pmatrix} l_{11} & l_{12} & l_{1j} \\ l_{101} & l_{102} & l_{10j} \\ l_{501} & l_{502} & l_{50j} \\ l_{901} & l_{902} & l_{90j} \\ l_{991} & l_{992} & l_{99j} \end{pmatrix}. \quad (4)$$

In Fig. 1 there is a graph of the statistic distribution of one of the measurements of the street noise $M(l_{Pj})$, where *j=6...10000 (Hz),* with the total amount *m=1000*.

From the figure it is possible to see that the difference between the values *P*=10 и *P*=90 showing in which range at that place *j* the noise level fluctuates is from 35 to 45 dB. It says that at that place unstable noises prevail with the duration from 2 to 5 sec.

To evaluate the influence of the acoustic noise on the surrounding territory it is necessary to consider the distance, at which the concrete acoustic signal spreads characterizing the exposure area. So for the statistically distributed noise $M(l_{Pj})$ it is necessary to determine the exposure zone. For this purpose taking into account the correcting (meteorological) data it is necessary to determine the attenuation ratio of the clear tone sound in the atmosphere $M(\alpha_j)$.

In its turn the exposure zone of the acoustic noise $M(l_{Pj})$ can be characterized by the distance S from the point of its source to the point, where the level of the acoustic signal is not more than some zero threshold Q, which, for example, can be equal to the threshold of hearing for the average human ear (GOST R ISO 226-2009 «Acoustics — Normal equal-loudness-level contours») or the values in the sanitary norms and rules.

This threshold can be represented as a matrix $M(I_j)$ and calculating the exposure zone S_{Pj} of the acoustic noise in free space. So, there are matrixes $M(\alpha_j)$ and $M(Q_j)$ have the type $M(\alpha_j)=\{\alpha_1 \;\; \alpha_2 \;\vdots\; \alpha_j\}$ and $M(Q_j)=\{Q_1 \;\; Q_2 \;\vdots\; Q_j\}$.

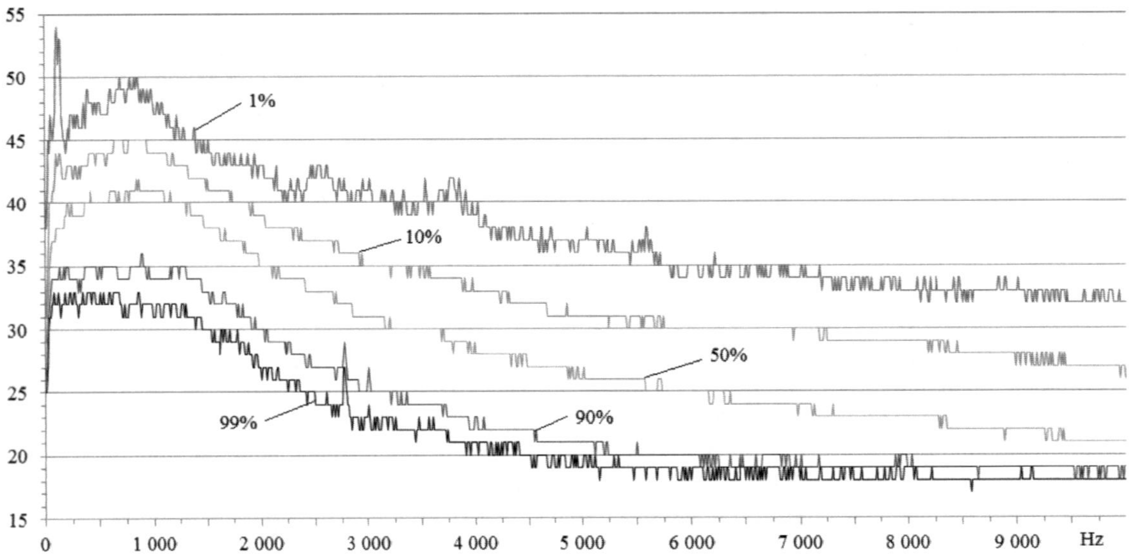

Figure 1. Statistic distribution of acoustic noise $M(l_{Pj})$

The free space S in the form of the distance from the source point to the point, where the identity becomes true $l<Q$ can divided into equal sections S'_{Pj}. So, to determine the distance S_{Pj} it is necessary to use the algorithm:

– if the identity $l_{ij}-\alpha_j \cdot S_{Pj} > Q_j$ is true, then $S_{Pj}=S_{Pj}+S'_{Pj}$;

– if the identity $l_{ij}-\alpha_j \cdot S_{Pj} > Q_j$ is not true then it says that at some distance S_{Pj} from the source of the sound pressure l is not higher than the zero threshold Q.

Finally we get the matrix of the distribution zone $M(S_{Pj})$ of the acoustic noise $M(l_{Pj})$ in free space:

$$M(S_{Pj})=\begin{pmatrix} S_{11} & S_{12} & \vdots & S_{1j} \\ S_{101} & S_{102} & \vdots & S_{10j} \\ S_{501} & S_{502} & \vdots & S_{50j} \\ S_{901} & S_{902} & \vdots & S_{90j} \\ S_{991} & S_{992} & \vdots & S_{99j} \end{pmatrix}. \tag{5}$$

In Fig. 2 there is the graph of the noise carpet $M(S_{Pj})$ of the distributed acoustic noise $M(l_{Pj})$ of one of the measurement results.

In Fig. 2 it is possible that the statistical noise distribution $M\left(l_{Pj}\right)$ where P=99 and P=90 does not represent any threat for the population because of the characteristic peculiarity of the perception of different sounds of different frequencies by the human hearing organs and there is an exposure zone S, close to zero. The zone of the noise distribution with the distribution ratio P=50, P=10 and P=1 is significantly higher. For example, the distance of the maximum noise spreading (P=1) is more than two kilometers and for the noise of a higher level (P=10) it is more than one kilometer.

This graph shows some distance from the source of acoustic noise to the point where it is level does not have any negative influence on a man. In other words at this distance the concrete studied acoustic noise is not perceived any more as a weak acoustic signal in case of complete quietness.

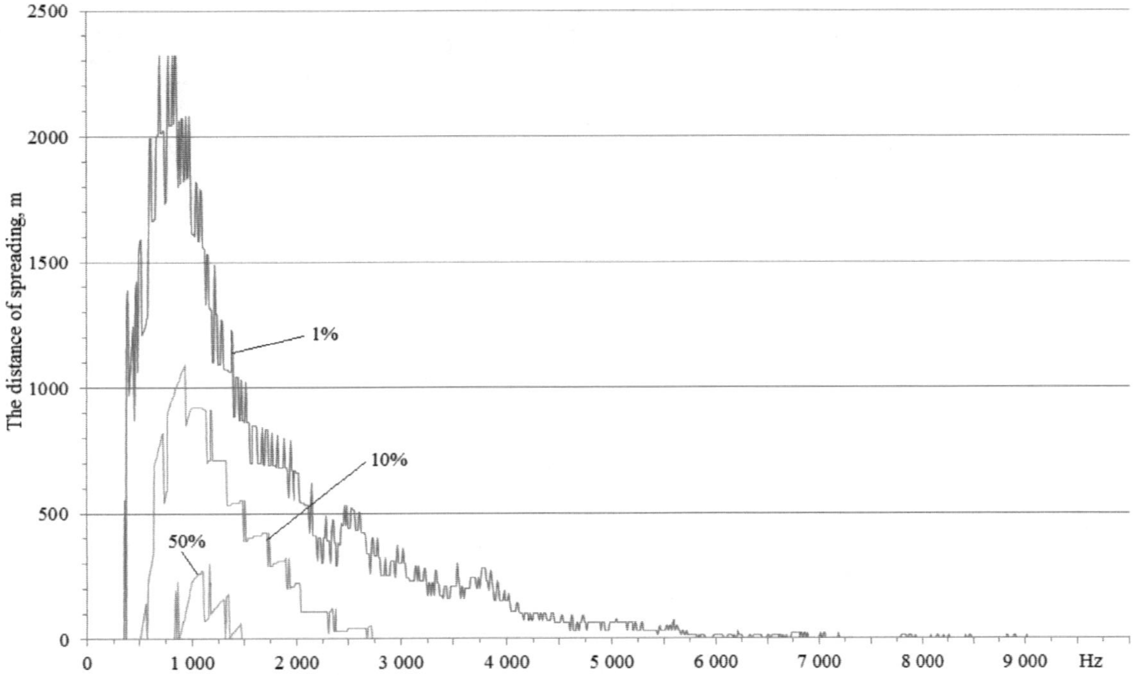

Figure 2. The dependence of the distribution zone of distributed acoustic noise $M\left(l_{Pj}\right)$ from frequency

4. Conclusion

This method gives an opportunity to observe the amplitude and frequency characteristics of the average noise values and the background values (in case of temporary stillness). It gives an opportunity to evaluate in what range in each control zone the noise level fluctuates. The preservation of these data of the amplitude-and – characteristic of the background, average and the maximum ratios gives an opportunity to determine the frequency of their peak values and to find the source in case if it is a point source.

Of particular importance has the method in the analysis of urban areas, which adjoin tightly significant engineering and other manufacturing, highway heavy traffic and residential areas.

The work was supported by the grant of Russian Foundation for basic research №14-08-00186.
The work was supported by the municipality of the city Murom.

References

[1] Solovjev L., Bulkin V. and Sharapov R. (2012) The existence of man in the technosphere, *Engineering industry and life safety*, **1,** 31-39.

[2] Bulkin V., Kalinichenko M., Shtykov E. and Fil'kov D. (2014) To the question of using noise absorbing means in industrial areas, *Bulletin of Tambov State University*, **19**, 1388-1392.
[3] Bulkin V., Solovjev L., Sharapov R., Pervushin R. and Kirillov I. (2014) The problems of creating monitoring systems for acoustic noise pollution in residential areas , *Engineering industry and life safety*, **1,** 48-53.
[4] Bulkin V. and Kirillov I. (2014) Passive-active monitoring system of acoustic contamination of the local urban area, *Radio- and Telecommunication Systems*, **4**, 48-55.
[5] Bulkin V. and Kirillov I. (2015) Environmentally-atmospherical monitoring of industrial machine builder state / *Applied Mechanics and Materials,* **770,** 686-689.
[6] Bogolepov I. and Lapteva N. (2010) Noise map of cities and agglomerations *Engineering and building magazine*, **6**, 5-11.

IOP Conf. Series: Materials Science and Engineering **91** (2015) 012080 doi:10.1088/1757-899X/91/1/012080

The problems of reducing the noise level on open sites of machine building plants and living areas

V.V. Bulkin and M.V. Kalinichenko

Murom Institute of Vladimir state University, Orlovskaya, str., 23, Murom, 602264, Russia

E-mail vvbulkin@mail.ru and marinakali@mail.ru

Abstract. In this article there is an analysis of using noise-absorbing screens on the basis Helmholtz resonator to reduce the level of acoustic noise on open industrial sites of machine-building plants and adjacent living areas. A design realization of this screen was considered, the characteristic of which is a variable depth of the resonator chamber and the change of the clearance of the resonator throat. The evaluation of the efficiency of using the screen was performed using experimental modeling with an acoustic chamber and a screen model. It was shown that in the majority of the cases there is some attenuation of the acoustic signal in the zone in front of the screen. A conclusion was made about the possibility of reducing noise level in industrial and living zone

1. Introduction

The problems of any modern machine building plant are determined by a complicated combination of different processes and tasks. The systematization of those processes was done in the standard GOST R 15288-2005 “Information technology. System engineering. System life cycle processes” (ISO/IEC 15288:2002 “System engineering — System life cycle processes”). The main processes connected with the production consist in the agreement processes, the plant processes, the project processes and technical processes.

Among the hazards of the developed machine-building production, not the least is the acoustic noise that is present typically in indoor areas with active process of mechanical production. However, depending on the intensity of work of the enterprise, the vastitude of the territory and the availability of open areas with active construction and traffic, the acoustic noise may be a very influencing factor on open spaces too. If the enterprise is located in urban realm and not equipped with good protection against the propagation of noise into the urban area, this can be a problem for the entire urban agglomeration. Thus, from the problem of worker safety and health within the enterprise, the noise may become an environmental problem of a big urban land.

The questions of labor protection are included in the group of processes of the plant and they are connected with the media control. The provision of the labor safety provides researching harmful factors, determining the degree of their danger for the personnel of the plants or the inhabitants of the houses in vicinity, measures to reduce the danger level and the implementation of the safety system of the life activity and labor protection.

Until recently special attention was paid to the noise registration and the analysis of its characteristics in the urban areas [1]. Of late the task of reducing noise becomes more and more actual. And because the sources of noise are different and there is no single way of reducing their influence this problem is complicated requiring huge efforts and without simple solutions.

IOP Conf. Series: Materials Science and Engineering **91** (2015) 012080 doi:10.1088/1757-899X/91/1/012080

2. Means and methods of noise absorption

Having analyzed the existing methods of reducing noise and it is possible to single out four main groups [2]:

a. Reducing the noise of the source.

In practice this method is realized using design and technological measures, which give an opportunity to create the mechanism and units with a low noise level.

b. Sound insulation.

It is a complex of measures having the purpose of decreasing noise level penetrating from outside so reducing the noise using adsorbing materials.

c. Building and planning measures.

We mean decreasing the distance between the noise source and the protected object using acoustically nontransparent screening elements (slopes, walls and screening buildings), creating special noise-protecting special green zones, using noise adsorbing covers by means of different methods of planning, rational positioning of industrial and living zones. Besides, to the building and planning measures belong rational building system of the main streets, the maximum possible planting of greenery and separating lines, using the relief of the terrain.

d. Engineering and technical means.

Primarily acoustic screens and resonance absorbers of other types are meant, other noise protecting facilities.

In this article special attention is paid to researching and developing engineering and technical means. The use of such means gives an opportunity to struggle against noise in the existing conditions of the enterprise or city and the existing stationary and moving noise sources.

The most effective engineering and technical means are acoustic (noise protecting screens) and resonant absorbers of different types. Upon the whole noise protecting screens are divided into several types: noise absorbing, noise-reflecting and combined. The choice of the concrete type of the screen is performed using the concrete condition of the considered territory and the characteristics of the noise pollution.

3. Measuring the average noise equivalent in the industrial noise zone in Murom

To evaluate the possibility of using noise protecting screens in territory of industrial zones or city areas situated near industrial sites the research of the noise levels was made in the characteristic zone of the city of Murom. To illustrate the level of acoustic pollution the results of the measurements are given in Fig. 1, which were made during the spring (A, B) and autumn (C, D), periods, in the morning (A, C) and in the evening (B, D). This zone was deliberately selected because the border of the factory territory is near the pedestrian zone of a lively road passing through the city center and connecting several regions of Russia. This road is actively used by heavy trucks including the transportation of the commodities produced by the city plants. The other side of the road is the border of the living zone. Apart from this there is another access point in the zone for heavy trucks performing the import of ready products. So in this zone the industrial and living areas are connected together and there is an in intensive traffic of cars, public transportation and heavy trucks.

In Fig. 1 there are the results of measurements in the moments of the most efficient load on this territory (peak values) and during the most quiet moments (the periods of relative stillness). The measurements were performed using the sound intensity meter ВШВ-003 (VShV-003) in the regime of controlling the average level without using octave analysis. This regime of the measurement is permitted during the total hygienic evaluation of the noise level.

The analysis of the presented results shows that the excess of the level of sanitary norms in the considered zone must be considered to be significant (more than 20 dB). On the other side it is necessary to note the similarity of the acquired results differing not only during different day parts but during different seasons, which underlines the presence of a stable acoustic pollution in the considered area.

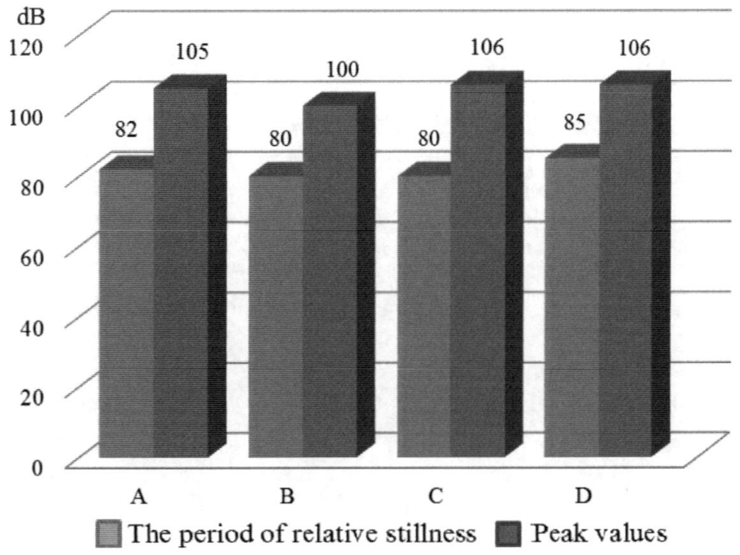

Figure 1. The results of analyzing noise level

4. The possibilities of using noise protection means

On the basis of the obtained information about the level of the acoustic and noise pollution evaluation calculations were performed with respect to the use of noise-protecting screens in the treated area. To provide a required level of the protection the screens must have the height of more than nine meters and they must be installed not only from the side of the plant but from the side of the residential area, which not permissible in city areas.

One of the possible variants of solving partially the task of reducing the noise level is the installation of noise-adsorbing screens representing the combination of usual noise-protecting structure with resonance absorbers of different types.

One of those decisions is provided using the combination of the screen with Helmholtz resonators, which are formed due to the internal volume of the screen. The constructive peculiarity of this screen is the presence of a internal capacity changing by height (on account of changing the profile of the internal cavity depth by height) and the possibility of changing the clearance of the slot-type throats of the resonator. On account of this decision the authors propose the possibility of adjusting the resonators for the required frequencies [3].

It is known that noise absorbers on account of Helmholtz resonator provide some reduction of the acoustic level of the signal in the zone of the resonator position at the frequency corresponding to its own resonance frequency. In case of broadband noise, which usually happens at a plant or when the noise created by vehicles is present this resonator must have polyharmonic characteristics. A practical realization of this noise absorber is possible when resonators adjusted at different frequencies are used. Another variant is the decision proposed by the authors with a variable depth of the internal volume, which provides the acquisition of the combination of different volumes of the internal resonator cavity and getting a combination of the resonant frequencies.

It is evident that the presence of any obstacle in the way of the acoustic signal causes its attenuation. But because of the fact it is known that screens constructed using resonance absorbers are more effective from the point of view of acoustic absorption, it is interesting to evaluate efficiency from the point of view of absorbing the sound in front of the screen.

The experimental check was performed by modeling the signal reflection in the presence of screen model in a small laboratory acoustic chamber using the ВШВ-003 (VShV-003) sound level meter. The measurements were performed within the range of the maximum sensitivity of the human ear so within the interval from 1000 to 4000 Hz. Deviating from the requirements of standard measurement methods, the analysis was performed not using the weight average of octave and part-octave ranges but in all the specified range with the step of 100 Hz. Also some damping material was used inside the case of the screen model to improve the acoustic absorption.

Some results of the experiments are used in Fig.2. The measurements were used for different sizes of the clearances in the slot throats of the resonators, using series by 10 measurements each with a following evaluation of the average statistic value at each frequency. To compare and to evaluate the efficiency of the screen a group of measurements was performed with a control screen representing the same structure but with a closed front wall. In Fig. 2 along the axis of ordinates the values of the noise level reduction are shown in the zone in front of the resonator model calculated in front of the control screen.

The specified information shows that upon the whole there is a real reduction of the noise level in front of the screen. The attenuation value is different at different frequencies and in some cases there is some signal amplification (resonance).

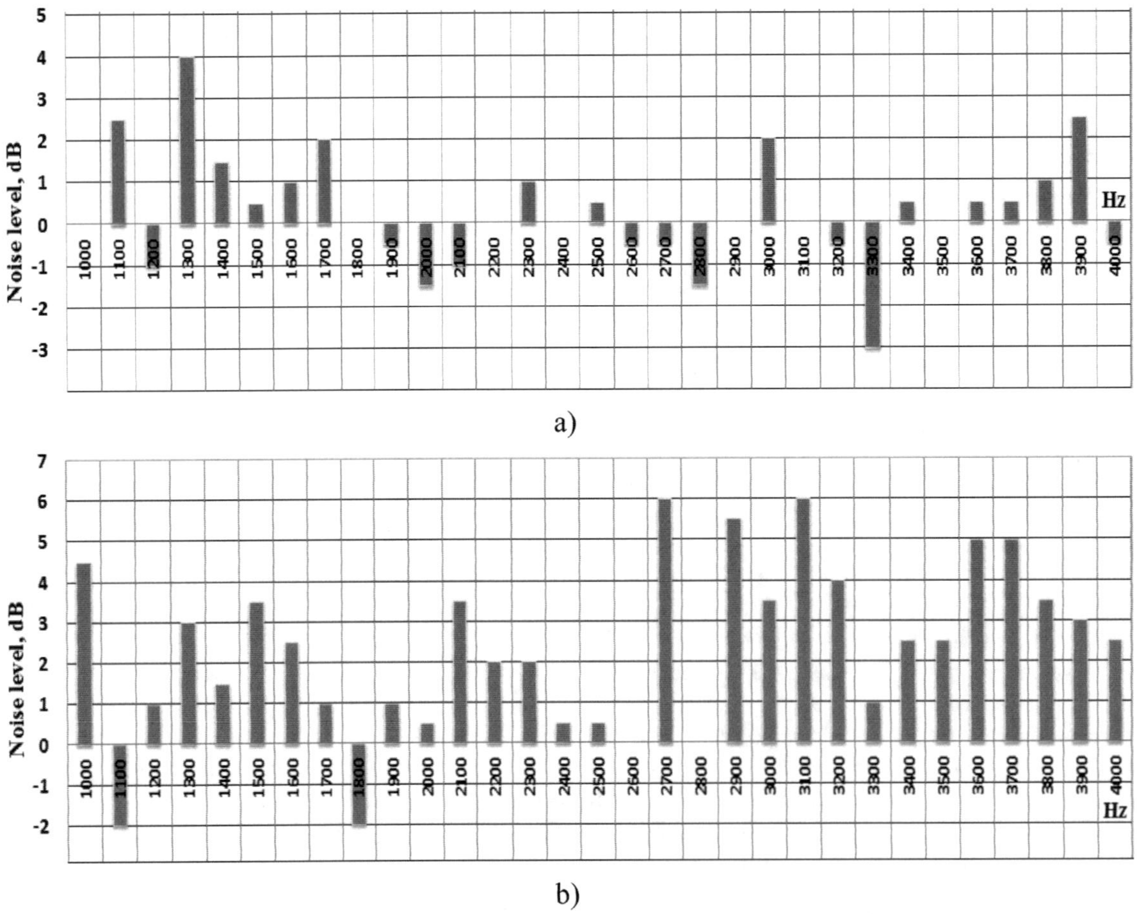

Figure 2. The results of the experiments with a noise protecting screen

5. Summary

The acquired results give an opportunity to make the following conclusions. The method used in this paper cannot provide absolutely precise results. But its use gives an opportunity to evaluate the principal possibility of getting some effect using such structures. In this case it is possible to consider that the statement is correct that in case of using this variant of the noise absorption screen it will be possible to prove a certain reduction of the acoustic noise in the zone in front of the screen which will provide a certain increase of the acoustic and noise pollution of the environment in case of using such structures as the protective facilities of the 1st echelon, for example as the fencing of open industrial sites or near private living houses in vicinity of technological zones, which will provide a certain reduction of the general acoustic and noise pollution of the city environment.

The work was supported by the grant of Russian Foundation for basic research №14-08-00186.
The work was supported by the municipality of the city Murom.

References

[1] Kalinichenko M. (2012) Some aspects of the problem of pollution in urban areas by road (using the example of Murom), *Ecology and industry of Russia*, **12,** 54-59.

[2] Bulkin V., Kalinichenko M., Shtykov E. and Fil'kov D. (2014) To the question of using noise absorbing means in industrial areas, *Bulletin of Tambov State University*, **19**, 1388-1392.

[3] Bulkin V., Kalinichenko M., Shtykov E. and Fil'kov D. (2014) Noise reducing screen, *Patent RF (utility patent) № 139581*, Published: 20.04.2014.

IOP Conf. Series: Materials Science and Engineering **91** (2015) 012081 doi:10.1088/1757-899X/91/1/012081

Investigations of physicochemical properties of bottom-ash materials for use them as secondary raw materials

A G Mal'chik, S V Litovkin, P V Rodionov

652055, Yurga, Leningradskaya st. 26, Yurga Institute of Technology, TPU affiliate, Russia

e-mail: ale-malchik@yandex.ru, protoniy@yandex.ru , rodik-1972@yandex.ru

Abstract. In this paper chemical content and particle size distribution of bottom-ash material are defined; results of differential thermal and X-ray analyzes are given; processes of phase transformations occurring during heating, are examined for possible use of the waste to produce ceramic products. Studies have shown that effective specific activity of radionuclides in the material under examination would have no effect on radiation safety of the finished product.

1. Introduction

On the Russian Federation territory there are more than 350 thermal power plants (TPP). Their proportion of generated energy is 70% from the national overall energy balance [1,2]. About 30% of TP plants burn coal [3]. When burning coal electric energy is generated, and also by-products, namely fly ash and slag, are formed. About 50 million tons per year of the bottom-ash are produced in Russia [1,4]. More than 1.5 billion tons of waste are accumulated at TPP ash disposals; their total area is 2000 km^2 [1,4]. By 2020, the proportion of coal burned at the TPP will increase to 40% [3]; that will cause further formation and accumulation of the bottom-ash wastes.

In Russia only 10% of bottom-ash annual content is recycled. For comparison, Germany utilizes about 100%, India - over 50% [5], Finland, Great Britain - 60%, USA - 25% [3]. In some foreign works it is reported that bottom-ash have been successfully used as the main component of ash bricks [6,7,8] as well as a fertilizer [9].

Taking into account further increase of coal burning and low level of waste recycling, a threat of ash disposal overflows arises, which will cause shutdowns of coal plant [5]. Open ash disposal areas pollute the environment. Dusting of ash has a negative impact on human health, flora and fauna.

In Russia, especially in Siberia, there are certain difficulties in the use of foreign techniques and practical recommendations for ash-and-slag utilization because they exceed actual capabilities of Russian economy. Therefore such recycling techniques are required that can adequately adapt national and international experience as well as have competitive advantages achieved by rational use of natural resources and investments [10].

The object of the examination is bottom-ash and slag materials (BAM) from TPP of Open Joint Stock Company "Yurga Machine Building Plant". The subject of the examination is to determine physical and chemical properties of ASM.

The aim of this work is to examine BAM physical and chemical properties, and its usability as a secondary resource to reduce the anthropogenic impact on the environment.

IOP Conf. Series: Materials Science and Engineering **91** (2015) 012081 doi:10.1088/1757-899X/91/1/012081

2. Experiments

To determine chemical composition of the bottom-ash waste, ash sampling was made. A total of 130 samples were selected. Chemical analysis was performed using X-ray fluorescence spectrometer KevexSpectrace's Quan`X. Analysis of the particle size distribution was performed using Analizette 22 MicroTec Fritsch GmbH (Germany) on the original software. Thermal characteristics were determined using a differential scanning calorimeter DSC 404 F3 Pegasus NETZSCH (Germany). The samples were roasted in a laboratory gradient furnace SP 30/13 (LAC (Czech Republic). The samples were tested for compressive strength with laboratory press PM-20MG4. X-ray diffraction analysis was performed using Rigaku 2500 D-max diffractometer with CuKα-radiation ($\lambda = 1{,}5418$ Å) in the range of $2\theta = 10\text{-}80°$. Identification was performed using catalogs PDF-2. Electron microscopy examinations were performed with JSM-5610 LV JEOL (Japan) scanning electron microscope. RKG-AT1320 gamma radiometer was used to determine the activity of radionuclides).

3. Results and Discussion

Tested ash and slag materials are fine mixtures, mainly gray-coloured. Averaged chemical composition is shown in Table 1. Bottom-ash chemical composition depends on mineral composition of burnt coals, and may differ significantly at different plants. Composition of the materials under investigation is characterized by a low content of calcium oxide, middle - of alumina and high - of iron oxide. Oxides of silicon and aluminum are the main components. This indicates presence of free calcium oxide in bottom-ash composition which, during disposal, reacts with the dissolved carbon dioxide to form calcium carbonate.

Ash samples were also taken from the precipitator. The ash from the precipitator contains twice as much calcium oxide. This indicates presence of free calcium oxide in bottom-ash composition which, during disposal, reacts with the dissolved carbon dioxide to form calcium carbonate. Presence of calcium carbonate is confirmed by XRD data (see Figure 1).

Table 1 Averaged chemical composition of bottom-ash sample.

Element	Ash disposal site,%		Precipitator, %	
	sample 1	sample 2	sample 1	sample 2
SiO_2	55,7	56,25	50,4	50,74
CaO	6,8	6,84	13,96	13,57
Al_2O_3	21,83	21,84	20,52	20,6
MgO	1,95	1,65	1,55	1,67
MnO	0,09	0,1	0,1	0,09
Fe_2O_3	7,44	7,4	8,55	8,53
FeO	6,69	6,66	7,69	7,68
K_2O	3,53	3,44	1,35	1,31
TiO_2	1,11	1,28	0,97	0,81
SO_3	0,72	0,68	0,87	0,82
BaO	0,44	-	0,5	0,59
P_2O_5	0,38	0,4	1,08	1,1

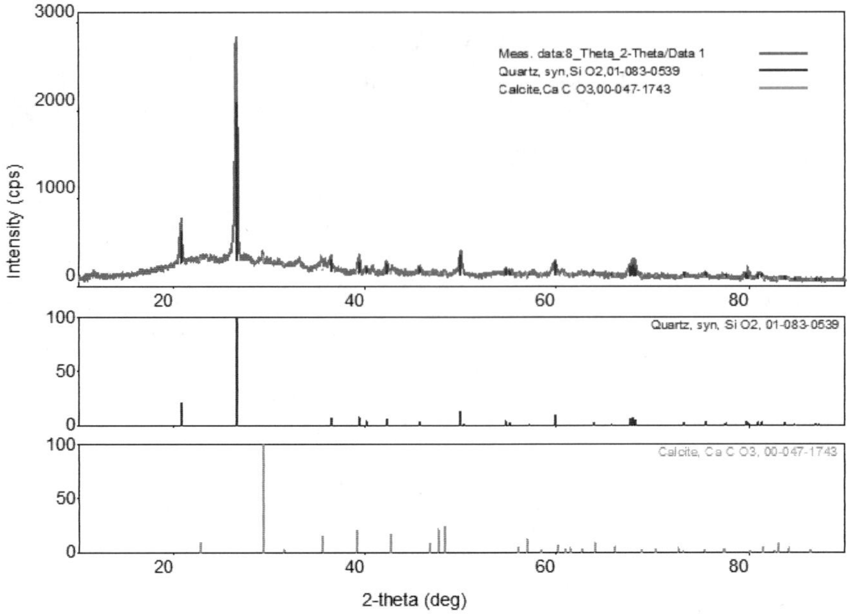

Figure 1. X-ray diffraction analysis of ash sample.

One of the main indicators of raw materials is their particle size distribution. A higher content of micro-dispersed particles increases ductility of the material. Therefore, if a raw material has a high cohesion, it improves the strength characteristics of the finished product; particle size distribution also specifies adsorptive capacity of the material.

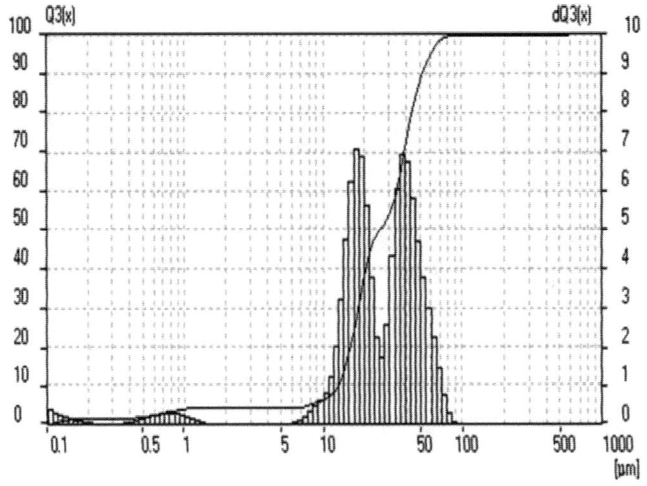

Figure 2. Particle size distribution in bottom-ash.

IOP Conf. Series: Materials Science and Engineering **91** (2015) 012081 doi:10.1088/1757-899X/91/1/012081

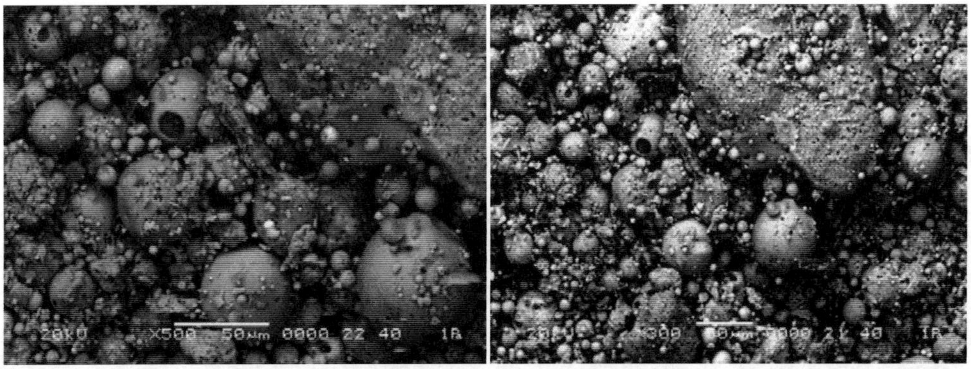

Figure 3. Microphotography of bottom-ash particles.

Results of the particle size distribution are shown in Figure 2. Analysis of particle size distribution shows that 60% of the particles have dimensional characters from 10 to 70 microns in size. Dimensions and morphology of the ash particles are shown in Figure 3.

The picture shows that the bottom-ash particles are beads and compactly shaped aggregates; the particle dimensions vary from 10 to 100 microns. Obviously, the material is very finely dispersed.

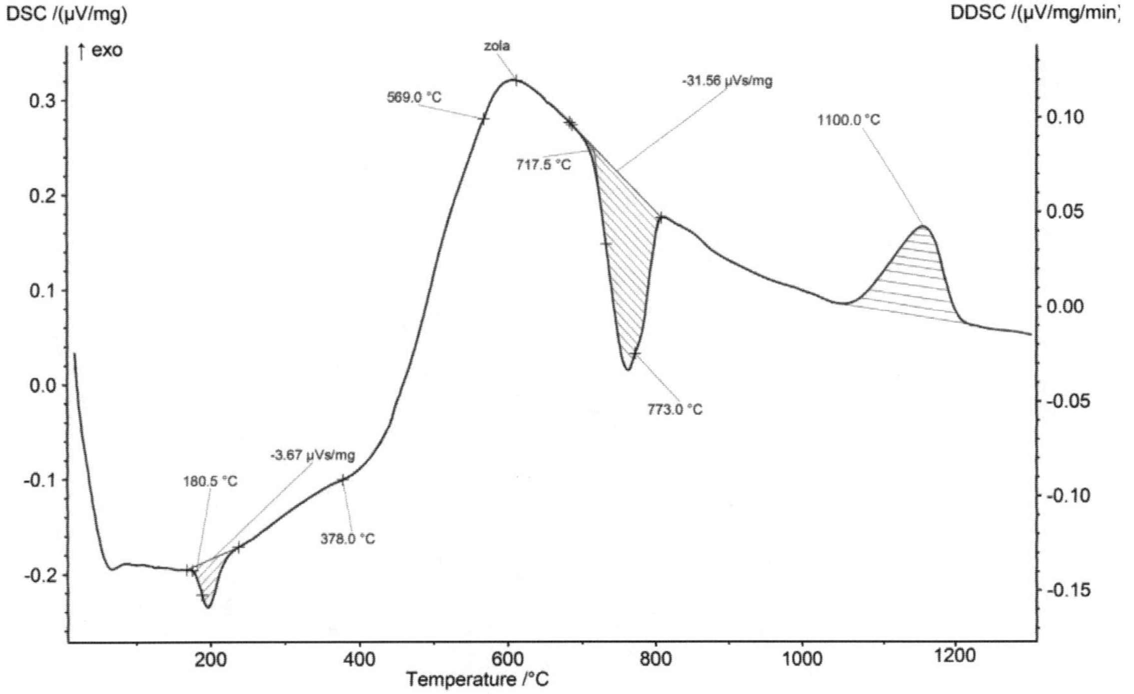

Figure 4. Bottom-ash DTA curve.

Differential thermal analysis (DTA) was performed to examine physical and chemical processes occurring in the ash during heating. On the DTA curve (Figure 4) there is a broad endothermic effect in 180°C area generated by removal of physically-bonded water and dehydration of hydroxides and hydroxide salts. At 180°C the bulk of physically and chemically bonded water is removed; the remainder is removed at a wider temperature range up to 750°C; that proves the presence of stable OH- groups in the BAM. A small exothermic effect at 378°C indicates the start of combustion of organic matter residues in BAM. Exothermic effect at 569ºC confirms presence of quartz; in this temperature range (530-580ºC, as described) quartz polymorphic inversion is observable, which is related to second-order phase inversions.

The next endothermic effect at 773ºC is caused by dissociation of magnesium components of dolomite. Exothermic effect at 1150ºC is caused by formation of mullite from free oxides $3Al_2O_3 + 2 SiO_2 = 3 Al_2O_3•2 SiO_2$. Mullite is the most thermodynamically stable form of the compound wherein some aluminum ions have quaternary coordination, others have sextuple coordination. Analysis of the resulting materials shows, that when ash-and-slag material undergoes heating to1000 – 1200ºC in oxidizing environment, mullite and α-quartz are produced mainly; this is confirmed by X-ray diffraction. For a more detailed examination of physical and chemical processes which occur during bottom-ash roasting, a multi-position heat treatment was performed in temperature range of 950 - 1100°C with steps of 50°C. The resulting sinter density increases with increasing of the processing temperature to 1100°C (Table 2).

Table 2. Influence of sintering temperature on material strength.

Temperature of heat treatment, °C	Strength, MPa
950	2
1000	2
1050	7
1100	15

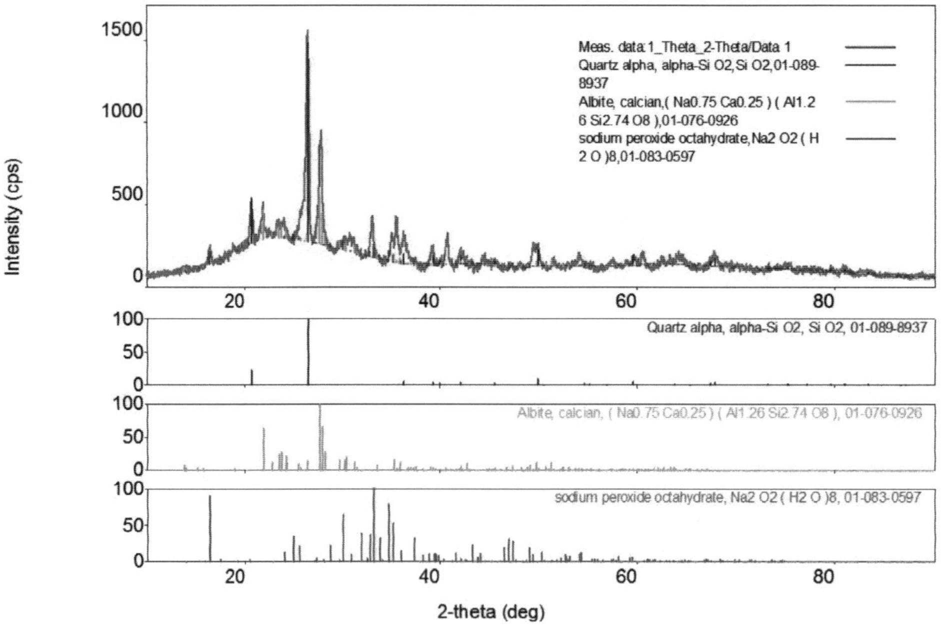

Figure 5. X-ray diffraction analysis of sample roasted at 950ºC.

XRD analysis of ash samples after heat treatment (see Figures 5 and 6) shows that at 950°C albite and sodium peroxide octahydrate are formed, the latter is the most stable crystalline hydrate of sodium peroxide. At 1100°C formation of mullite, which is stronger then albite, is observable. Formation of mullite is also confirmed by the differential thermal analysis described above.

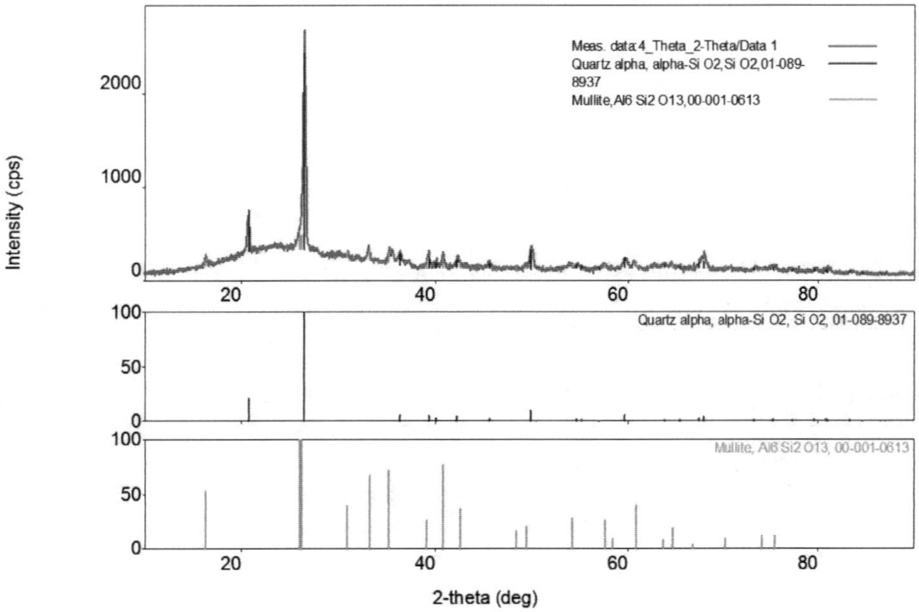

Figure 6. X-ray analysis of sample roasted at 1100°C.

Measured activity of radionuclides showed the following results: ^{40}K – 526 Bq/kg, ^{232}Th – 72 Bq/kg, ^{226}Ra – 37 Bq/kg. Calculation of the specific activity was made according to the formula:

$$A_{эфф} = A_{Ra} + 1.31A_{Th} + 0.085A_K \quad (1)$$

where A_{Ra} ,A_{Th} , A_{K}- Specific activities of radium, thorium, potassium, respectively, Bq/kg.

Calculated ash waste activity is 175 Bq/kg; that does not exceed, requirements of building regulations, according to which bottom-ash can be ranged in the first grade of construction materials, and can be used for all types of construction.

To determine capability of ash and slag further use it is necessary to define basic classification features: indexes of acidity and basicity, silica ratio and quality factor.

To determine acidity index of M_k we calculate ratio of the sum of acidic oxides to the sum of basic ones:

$$M_k = \frac{SiO_2 + Al_2O_3}{Fe_2O_3 + CaO + MgO} \quad (2)$$

Acidity index of samples, taken from the precipitator, is equal to 2.9, samples from the ash disposal - 4.7.

To determine basicity index M_o we calculate ratio of the sum of basic oxides to the sum of acidic ones:

$$M_0 = \frac{CaO + MgO + K_2O + Na_2O}{SiO_2 + Al_2O_3} \quad (3)$$

Basicity index of samples, taken from the precipitator, is equal to 0.2, samples from the ash disposal – 0.16.

Silica ratio M_c is calculated as ratio of silica to the sum of aluminum and iron oxides:

$$M_c = \frac{SiO_2}{Fe_2O_3 + Al_2O_3} \quad (4)$$

Silica ratio of samples, taken from the precipitator, is equal 1.7, samples from the ash disposal – 1.9.

Quality factor K is the ratio of oxides, increasing hydraulic activity, to the oxides, reducing it:

$$K = \frac{CaO+Al_2O_3+MgO}{SiO_2+TiO_2} \quad (5)$$

Quality factor of samples, taken from the precipitator, is equal to 0.7, samples from the ash disposal – 0.54.

Findings show that the bottom-ash wastes are acidic ashes. Acidic ashes have unstable chemical composition, a low content of free calcium oxide and a high content of silica. These ashes do not have cementing properties, but after addition of intensifiers become cementitious.

4. Conclusions

Ash-and-slag wastes were examined for their recycling and usability as a secondary resource to reduce the anthropogenic impact on the environment.

Physical-chemical characteristics of waste behavior during thermal processing have been specified. Chemical analysis has revealed that oxides of silicon (56.25%) and aluminum (21.84%) are the principal components of BAM. Analysis of the particle size distribution has shown that 60% of particles have dimensional characters from 10 to 70 microns. Both X-ray diffraction analysis and differential thermal analysis have shown formation of mullite at temperatures above 1100°C. Effective specific activity of radionuclides in BAM, calculated by isotope activities of radium (^{226}Ra), thorium (^{232}Th) and potassium (^{40}K), is 175 Bq/kg.

Thus, the examinations have demonstrated possibility of BAM utilization as a secondary resource for reducing the anthropogenic impact on the environment.

References

[1] Putilin E 2003 Overview of domestic and foreign experience in use of wastes from TPP coal burning (Moscow: Sousdornii) p 60
[2] Bespalov V 2010 EP technologies for thermal power plants (Tomsk: Publishing house of Tomsk Polytechnic University) p 240
[3] Zyryanov V 2009 Fly ash: man-made raw material (Moscow: IPC "Maska") p 319
[4] Avvakumov E 2009 Mechanical methods of activation in processing of natural and man-made materials (Novosibirsk: Academic Publishing House "Geo") p 155
[5] Lipatov Y 2014 Federal Directory, Volume 27, VI. Energy Efficiency and energy development, government regulation of use of coal-fired power plants waste (Moscow) p 285
[6] Tabin Rushad S 2011 Experimental Studies on Lime-Soil-Fly Ash Bricks (International journal of civil and structural engineering, volume 1, № 4)
[7] Obada Kayali 2005 High Performance Bricks from Fly Ash. (USA, Kentucky, Lexington World of Coal Ash, April 11-15)
[8] Henry Liu 2009 Environment Properties of Fly Ash Bricks, (USA in Lexington World of Coal Ash Conference – May 4-7)
[9] Kevan D. Sharp. 1993 Fly ash for soil improvement: proceedings of sessions sponsored by Committees on Soil Improvement and Geosynthetics of the Geotechnical Engineering Division of the American Society of Civil Engineers in conjunction with the ASCE Convention in Dallas (Texas, October 24-28)
[10] Kozhemyako S 2009 Strategy of reuse of bottom-ash and slag wastes from TPP-members of "Siberian Energy Association." (Novosibirs)

Carbon Nanotubes Influence the Enzyme Activity of Biogeochemical Cycles of Carbon, Nitrogen, Phosphorus and the Pathogenesis of Plants in Annual Agroecosystems

O B Vaishlya[1], N N Osipov[2] and N V Guseva[3]
[1, 2] National Research Tomsk State University, Lenin Av., 36 Tomsk, 634050, Russia
[1] S.Seifullin Kazakh Agrotechnical University, Pobedy Av. 62 Astana, Republic of Kazakhstan
[3]National Research Tomsk Polytechnic University, Lenin Av. 30, Tomsk, 634050, Russia

E-mail: [1]plantaplus@list.ru, [2]nicolay-00@yandex.ru, [3]guseva24@yandex.ru

Abstract
We conducted pre-sowing seed treatment of spring wheat carbon nanotubes modified with thionyl chloride, ethylene diamine, azobenzole, and dodecylamine. CNTs did not disrupt the structure of the crop, but the activity of extracellular enzymes in the rhizosphere of plants in the flowering stage changed: laccase works more poorly in the variant of the CNTs with the amino groups exochitinase and phosphatase activity increased in the case of chlorinated CNTs, OH and COOH groups on the surface of the nanotubes twice accelerate work β-glucosidase. The changes observed in the biogeochemical cycles in the rhizosphere are a possible cause of the effect of nanotubes on the development of epidemic diseases of wheat.

1. Introduction

The expansion of the application sphere of nanomaterial led the humanity to deeper understanding that such materials are unique compositions with unpredictable physical and chemical properties. Contemporary nanotechnologized society absolutely needs the study of both the properties of these materials and the mechanisms of their behavior in natural environments [1-2]. In the course of interaction with bio-objects, carbon nanotubes—like toxicants—behave in an unconventional way, which complicates both the normalization of the content of nanomaterial in living environments and the assessment of their potential harmfulness [3-4]. For instance, the document of year 2010, GN 1.2.2633-10 "Hygienic standards of content of nanomaterial with high priority in environment", covers the standards only for one type of single-walled carbon nanotubes (CNTs) and only the value of Tentative Safe Exposure Levels (TSELs) in the air of work area that equals to 0.01 fiber per 1 cm^3, while this fiber has the length of more than 5 μm. Toxic properties of carbon nanotubes depend on the parameters of the material itself, on the length of nanotubes (the more the length, the more the toxic effect), on their capacity for aggregation and dispersion, on the presence of various metal particles on the surface of nanotubes, on the release by a cell of proteins that wrap nanotubes.

The least studied biological object is soil, which represents the most complex methodological subject to study. The analysis of biological activity of soil allows determining the character and degree of its alteration under any kind of anthropogenic exposure on topsoil. In the recent years, soil science has experienced the dissemination of the concept of soil matrix. According to the existing concept, soil matrix represents the surface of solid soil particles. Near this surface, layers of adsorbed particles form in a specific way. The particles are organic and mineral substances, microorganisms, gases, ions and molecules. The matrix includes three subsystems: mineral, organic and organic-mineral. The most important property of active centers of the soil matrix is their catalytic activity occurring in soil processes. In this case, we consider the enzymes immobilized on mineral, organic and organic-mineral components of soil particles. These particles also adsorb nanomaterial present in soil. In 2009, the research has demonstrated that COOH-modified single-layer CNTs have low sensitivity to the ionic strength of a solution (as compared to fullerenes and small nanoparticles), since they are strongly held by a soil matrix. The enzyme activity was preferably chosen as a diagnostic indicator due to low experimental error rate, simplicity of evaluation, high sensitivity to external exposures and proven role of enzyme activity indicators in assessing the effect of pollution induced by various techno-genic products.

The analysis of the biological aspects of safe application of carbon nanotubes (CNTs) for cultivating strategic grain crops is of particular importance [5]. The work has the aim of studying the influence of various modifications of CNTs on the pathogenesis and activity of extracellular soil enzymes catalyzing the biogeochemical cycles of carbon, nitrogen and phosphorus in the rhizosphere of spring wheat.

2. Methods

Carbon nanomaterial was produced at scientific and educational innovative center "Nanotechnologies and nanomaterials" of Vladimir State University named after Alexander and Nikolay Stoletovs. The synthesis involved the pyrolysis of propane-butane mix on copper-nickel catalyst at the temperature of 600 °C. According to electron microscopy results, the diameter of CNTs varied from to 20 to 40 nm; the length was about 100 nm (figure 1-2).

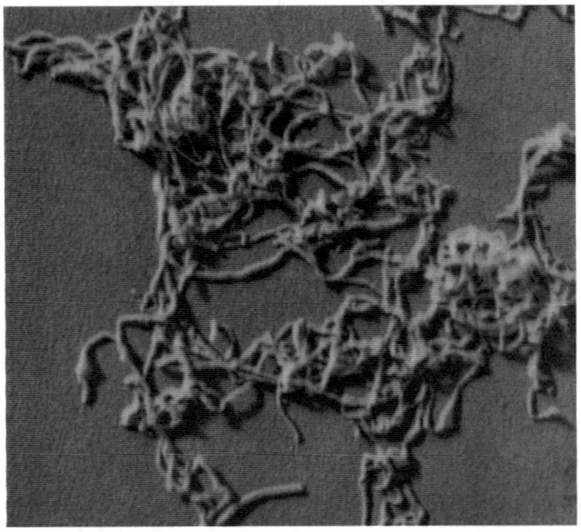

Figure 1. SEM image of CNTs on carbon substrate (70000x magnification)

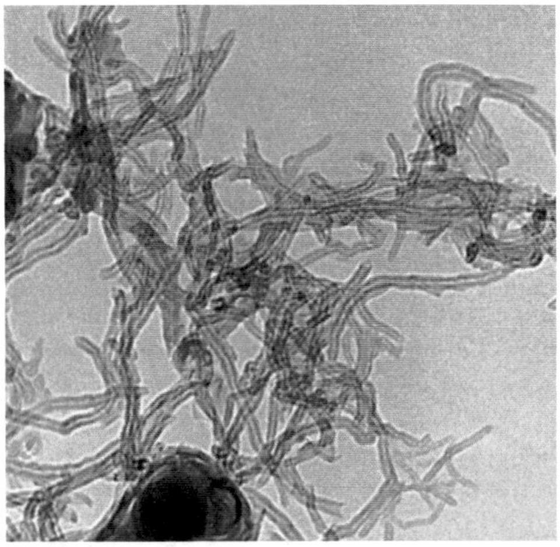

Figure 2. TEM image of CNTs (150000x magnification)

The specific area of material's surface was 232 m^2/g; the poured density of dry material was 0.03 g/cm^3. The content of carbon nanotubes, the content of amorphous carbon and their percentage ration were determined by X-Ray diffractometry (figure 3). According to the analysis of obtained diffractogram and the interpretation of carbon peaks using PDF-2 Release 2011 database, the content of CNTs in the obtained material was 77% (figure 4). The oxidation of CNTs was performed by two methods: a conventional method with the use of the mix of mineral acids and a mechanochemical method. Carboxylation of CNTs was performed using a mix of concentrated nitric and sulfuric acid

IOP Conf. Series: Materials Science and Engineering **91** (2015) 012082 doi:10.1088/1757-899X/91/1/012082

(1:1). A subsample of CNTs was placed into a flask that was consequently filled with a mix of mineral acids. The proportion of masses was 1:5. The mix was intensively stirred at the temperature of 70 °C for 24 hours. Then, the material was filtered off, rinsed with water and ethanol and air dried.

Fabrication of CNTs modified by hydroxyl group was carried out using a mechanic-chemical method. The method involved the milling of equal-weight subsamples of CNTs and alkali (KOH) in a ball mill for 60 minutes. Consequently, the produced material was washed to remove alkali and air dried. In the process of functionalization, CNTs were cleaned from amorphous carbon. The purity degree was determined by interpreting the diffractogram of modified CNTs with the use of PDF-2 Release 2011 databases (figure 3-4). After adding carboxyl and hydroxyl groups to the surface of CNTs, the fraction of amorphous carbon was less than 5 wt%.

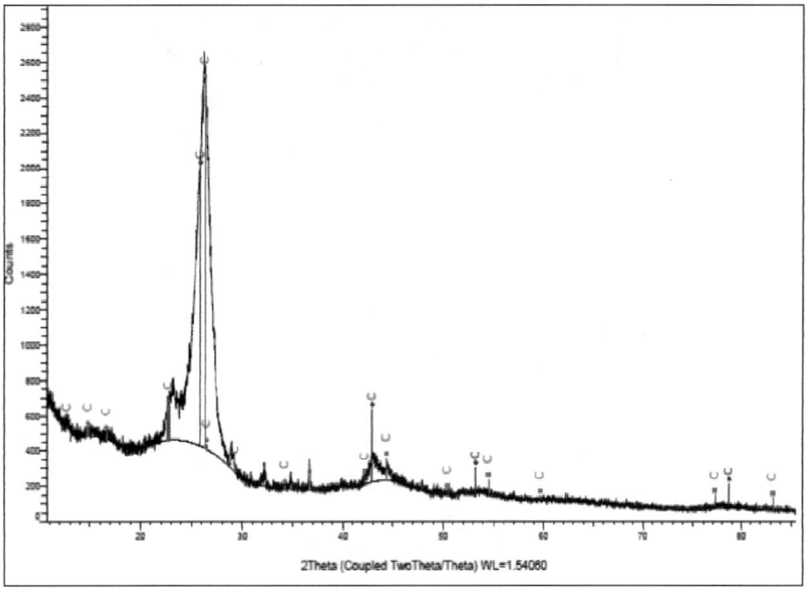

Figure 3. Diffractogram of CNTs after production

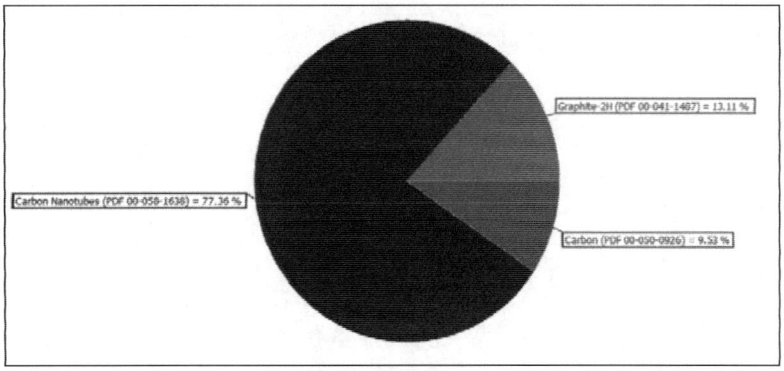

Figure 4. Diagram of carbon state distribution in the material containing CNTs after pyrolysis

The fabricated material was studied by a complex of physicochemical methods (IR and Raman scattering spectrometry, XRD analysis and thermo-gravimetric analysis).

IR spectra were recorded by PerkinElmer Spectrum 100 FT-IR spectrometer (PerkinElmer, USA, 2006). IR spectra of oxidized CNTs demonstrate the peaks that correspond to oxygen-containing functional groups and various vibrations of aromatic ring double bond (figure 5 and 6). The peaks of oxidized CNTs are identified in table 1.

Table 1. Peak identification for oxidized CNTs of IR spectra.

Groups	ν [cm^{-1}]	Identification
C–O(H)	1295°cм$^{-1}$, 1400°cм$^{-1}$	Distinctive for phenol
C=C	1456°cм$^{-1}$, 1540°cм$^{-1}$	Various vibrations of aromatic ring double bond
C=O	1567°cм$^{-1}$, 1636°cм$^{-1}$, 1701°cм$^{-1}$	Corresponds to various carbonyl compounds
–C(O)–O–C(O)–	1861°cм$^{-1}$	Anhydride group
R(OH)	2324°cм$^{-1}$, 2843°cм$^{-1}$ 3276°cм$^{-1}$, 3746°cм$^{-1}$	Corresponds to the vibrations of hydroxyl group in various compounds

The recording of Raman scattering spectra was performed by confocal Raman and fluorescence microscopy with the use of atomic force microscope (AFM) and by spectroscopy with the use of NTEGRA Spectra Probe NanoLaboratory.

The spectra of Raman scattering of CNTs (figure6) modified by hydroxyl and carboxyl groups demonstrate the increased intensity of D and G modes in the regions of 1362 cm^{-1} and 1584 cm^{-1}. This evidences the violation of graphene hexagonal symmetry induced by the occurrence of covalent bond on the sidewall of the external layer of carbon nanotubes.

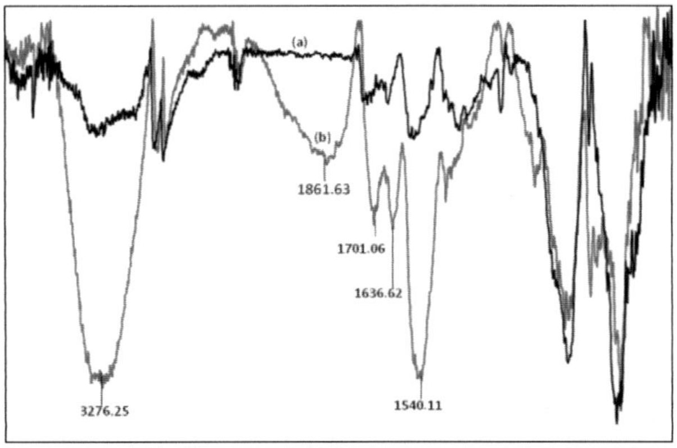

Figure 5. IR spectra. a) pure CNTs and b) CNTs oxidized by mineral acids.

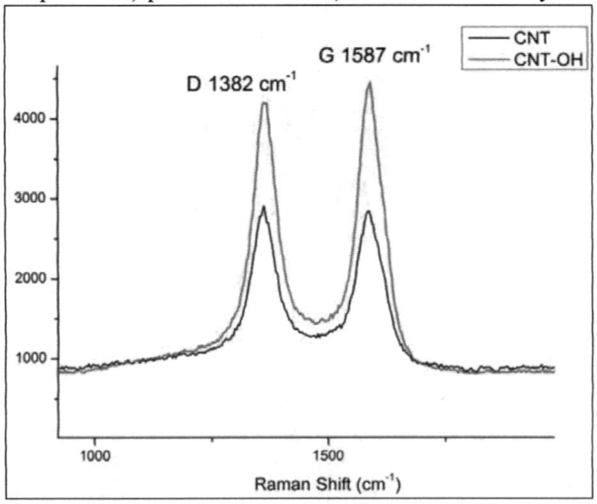

Figure 6. Raman scattering spectrum of CNTs after mechano-chemical treatment

IOP Conf. Series: Materials Science and Engineering **91** (2015) 012082 doi:10.1088/1757-899X/91/1/012082

XRD analysis was carried out with the use of D8 ADVANCE X-ray powder diffractometer (Bruker AXS, Germany) and X-ray tubes with copper anode. The parameters of the systems were as follows: the power was 1.6 kW, Bragg Brentano geometry, 2theta scanning angles from about 15° to 90°, the step of 0.02, the exposure of 0.3 seconds per point, PDF-2 Release 2011 database.

According to the analysis of the obtained diffractogram and the identification of carbon peaks through PDF-2 Release 2011 database, the content of CNTs in the produced material is 77% (figure 3 and 4). After oxidation, the fabricated sample, according to the identification of the diffractogram of carbon peaks, contains 95% of CNTs (figure 7 and 8).

We used the following variations and designations: reference sample without CNTs and carbon, M-200 stands for activated carbon, S-0 stands for initial CNTs, S-1 stands for carboxylated CNTs, S-2 stands for carboxylated nanotubes after chlorination (chlorinating agent was thionyl chloride), S-3 stands for CNTs modified by ethylenediamine (OH-group), S-4 stands for CNTs modified by azobenzene, S-5 stands for CNTs modified by dodecylamine.

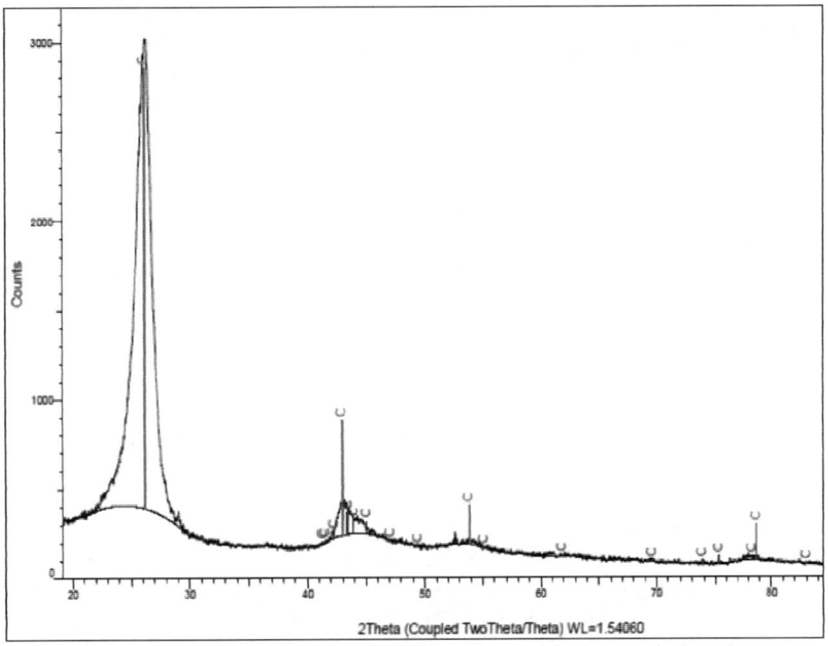

Figure 7. Diffractogram of CNTs after oxidation and cleaning

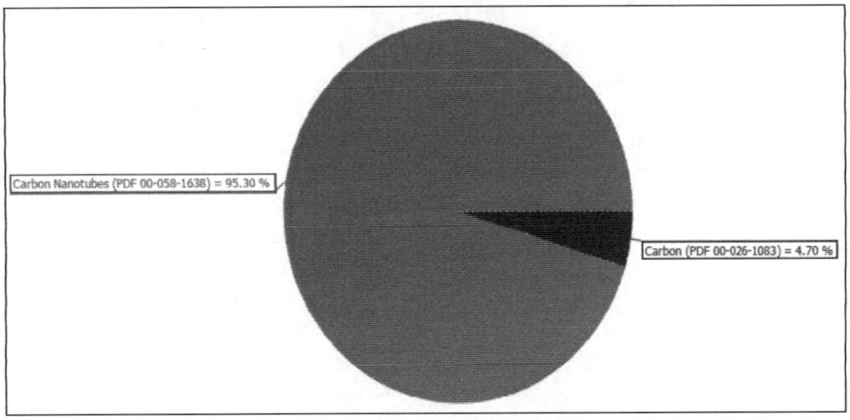

Figure 8. Diagram of carbon state distribution in the material containing CNTs after oxidation and cleaning

IOP Conf. Series: Materials Science and Engineering **91** (2015) 012082 doi:10.1088/1757-899X/91/1/012082

Studied enzymes contribute to mobilize nutrients sequestered in soil organic matter trough the lysis of macromolecules such as cellulose, hemicelluloses, lignin, chitin, proteins, polyphenols, etc.1–2 g of fresh rhizosphere soil and 60 ml acetate buffer (pH 5.0) were blended on the highest speed for 2 minutes. Then we poured the homogenate into two conical retort, closed and put them in a shaker at the regime 180 rpm for 2 or 3 hours at room temperature. After this we removed the retorts and put them in the refrigerator (+5 by Celsius) for overnight. In the morning we put the retorts on a shaker again for 15-20 minutes or more. For enzyme analysis we diluted the suspension by 10 times while stirring, centrifuged at 2000 rpm for 2 minute; supernatant was kept in Eppendorf-tubes in the refrigerator (+5 by Celsius). The activity of laccase was determined by colorimetric method. The activity of ß-glucosidase, cellobiohydrolase, acid phosphatase, exochitinase, xylosidase, glucuronidase and leucine-aminopeptidase were determined by a fluorimetric method, which is a novel approach to analyzing soil enzymes [6]. The method is based on the fact that the substrates being specific for each enzyme, are connected with fluorochrome, methylumbelliferone or aminomethyl coumarin. Fluorochrome, being excited by a light with the wavelength of 355 nm, re-emits the light with the wavelength of 460 nm. Thus, the measurement of the number of re-emitted photons allows evaluating the amount of substrate reacted with an enzyme.

3. Experimental

The field experiment was carried out at the same region of West Siberia as in the previous research and we obtain the same results that no modifications of CNTs damaged the production process of spring [7]. Crop structure was analyzed according to the standard procedure. In terms of agro-climatic zoning, the experimental area belongs to moderately cold and moderately moist climatic region of a forest-steppe zone. The sum of air mean daily temperatures over the period with the temperature of air above 10 °C is 1700–1800; the amount of precipitation over the period is 225–250 mm, while the annual precipitation is 400–450 mm. This area is characterized by late-spring and early-autumn frost and drought conditions during early spring-summer period. The zone exhibits podzolized grey forest loamy soil. Each kilogram of topsoil contains 5.3 % of humus, 100–164 mg of mobile phosphoric acid, 34-43 mg of available potassium and 0.95 mg of mobile nitrogen.

The study involved the seeding of the first-grade Novosibirskaya-29 spring wheat. The seeding rate amounted to 200 kilograms of viable seeds per a hectare. The soil was subjected by fall tillage, harrowing by BDT-7 disk harrow and pre-seeding cultivation by KPS-4 ripper. The area of plot was 1.2 m^2; the accounting area was 1.0 m^2. The location of analytical replication of variants was random. The experiment had three replications. The soil was not fertilized; weeding and crop tending were manual. Potatoes was the previous culture, pesticides were not used. The selection of samples for phytopathological expert examination was carried out as per GOST 12044-93.

In 2009, it was shown that COOH-modified single-walled CNTs were weakly sensitive, compared to fullerenes and small nanoparticles in the ionic strength of the solution as effectively retained by the soil matrix [5, 8]. In this case, a number of experiments proved the role of CNTs as regulators of growth [8-9]. Currently, living organisms and enzyme activity are used as important diagnostic indicators assessing the impact of soil contamination by various products' of techno-genesis [10-12]. Previously, it was shown that the CNTs affect plant phenotype and microbiota, maximal development of which are determined at the flowering stage [5, 9].

The examination involved the study of the activity of hydrolase and oxidase taking part in biogeochemical cycles of carbon, nitrogen and phosphorous. Enzyme activity depends on different plant-microbe interactions. For example, in the rhizosphere of wheat plants infected by Fusarium, the possibility of *in vitro* depolymerization and polymerization of humic substances by fungus laccase was shown. The possibility of synthesizing humic substances in soils by oxidative condensation of low-molecular predecessors on the surface of loamy minerals under the action of immobilized laccase is well known fact. We studied the activity of key enzymes of the cycles include the degradation of plant cell wall – cellulose, hemicellulose, phenol compounds, and pectin; the breakdown of organic nitrogen

– proteins, peptides, and chitin; and of the degradation of organic phosphorous – nucleotides, phospholipids (table. 2).

Table 2. General potential catalytic activity of soil enzymes in the rhizosphere of wheat plants at blossom stage [pmol of substrate per min per mg of dry soil]

Enzyme name	Control	M – 200	S – 0	S – 1	S – 2	S – 3	S – 4	S – 5
Leucine-aminopeptidase	2.25±0.19	*	1.27±0.04	2.08±0.01	4.39±0.24	4.03±0.12	1.34±0.02	1.66±0.02
Acid phosphatase	5.8±0.3	5.2±0.3	8.0±0.6	5.7±0.1	15.8±1.3	10.7±0.2	5.1±0.1	5.2±0.3
Xylosidase	20.3±1.4	*	23.5±1.8	36,9±1.8	*	38,7±2.3	17.1±0.8	*
β-Glucosidase	11.3±0.2	10.9±0.3	10.0±0.4	27.3±0.6	12.7±1.3	20.3±0.5	10.5±0.3	18.7±0.2
Cellobio-hydrolase	195±12	148±8	159±12	135±11	358±29	193±17	148±13	239±12
Exochitinase	18.1±0.9	9.5±0.6	14.5±0.5	5.9±0.7	46.1±2.7	27.0±0.3	16.2±0.9	9.9±0.5
Glucuronidase	89.3±9.7	71.4±3.9	75.2±6.5	39.2±3.0	32.8±4.6	97.6±2.2	92.1±5.2	87.7±3.4
Laccase	77.1±8.3	79.1±5.9	42.9±1.3	62.7±3.8	88.0±5.2	92.1±5.7	45.7 ± 1.4	52.9 ± 3.1

* Sample/reference sample < 1.2, i.e. the results are non-informative, since they are too close to the values of non-transformed substrate fluorescence.

It is worth mentioning that OH modification of CNTs leads to increased activity of all studied enzymes of nitrogen, carbon and phosphorus cycles, except those of cellobiohydrolase activity, which does not change. Perhaps this may be due to an increase in the intensity of the D and G modes in the 1362 cm^{-1} and 1584 cm^{-1} in the Raman spectra of CNTs-OH, which constitutes a violation of the hexagonal symmetry of graphene, caused by the appearance of a covalent bond in the side wall of the tubes. Laccase works poorly in the variants of the CNTs with the amino groups, and exochitinase and phosphatase activity increased in the case of chlorination of CNTs. OH- and COOH-groups on the surface of the nanotubes twice accelerate the work of β-glucosidase. There is no unequivocal opinion on this point: in soil both single and multi-walled CNTs with azo-group increased the enzyme activity in the soil [12] and single-layer inhibited the activity of alkaline phosphatase and invertase for 14 days [13]. This is a possible reason for the differences in the development of fungal diseases as extracellular soil enzymes play a role of elicitors. At the tillering stage symptoms of root rot were significantly decreased in the variants of CNTs with COOH-, OH- and the azo-groups, and all variants of CNTs reduced the incidence of Septoria by 5-9 times. Blight disease was decreased in plants treated with pure CNTs and CNTs with azo-group (table 3).

In the annual agro-ecosystems 75% of fungal and 89% of bacterial pathogens are transferred through seeds, which are the first to occupy ecological niches in the rhizosphere of sprouting seedlings [11]. Phytopathological analysis of seeds showed a two-fold reduction of infection by Fusarium in variants with COOH- and OH- groups; reduction in the incidence of Helminthosporium by three times in almost all cases, and a two-fold decrease in infection of plants by Alternaria in versions with nitrogen, and chloranhydrid groups (table. 4).

It is known that during the growing season the main infection agents of root rot of wheat are spread by airborne droplets using conidia. Germination of conidia is stimulated by exudates of root hairs.

Table 3. Phytopathological expert examination of Nobosibirskaya-29 wheat seeds

CNT modification	Affected by diseases [%]					
	Total	Fusariosis	Helmintho-sporiosis	Alternaria blight	Bacteriosis	Fungi
Reference	43.0	8.0	10.0	25.0	0	0
M – 200	25.0	8.0	5.0	12.0	0	0
S – 0	41.0	15.0	3.0	20.0	1.0	2.0
S – 1	37.0	4.0	6.0	27.0	0	0
S – 2	28.0	11.0	4.0	13.0	0	0
S – 3	46.0	3.0	10.0	33.0	0	0
S – 4	29.0	12.0	3.0	14.0	0	0
S – 5	35.0	15.0	5.0	15.0	0	0

At the same time plants with reduced turgor of cells are more infected, especially on heavy clay soils. Perhaps just the ability of CNTs to work in a cell plasma membrane as a water pump [5, 9] is one of the reasons why the infection of root rots in wheat seedlings treated by CNTs is decreasing.

Table 4. Analysis of the development and dissemination of diseases affecting Novosibirskaya-29 wheat plants at tillering stage*.

CNT modification	Root rot [points]			Septoria spot [points]			Helmintho-sporiosis [points]			Powdery mildew [points]			Brown rust [%]		
	P, %	R	I	P [%]	R	I	P [%]	R	I	P [%]	R	I	P [%]	R	I
Reference	22.70	0.27	1.20	27.20	0.27	1.00	13.60	0.18	1.33	18.18	0.18	1.00	50.00	0.59	1.18
M – 200	18.20	0.22	1.25	4.50	0.04	1.00	13.60	0.13	1.00	18.20	0.22	0.71	36.30	0.80	2.25
S – 0	22.50	0.48	2.14	6.45	0.06	1.00	9.67	0.09	1.00	22.58	0.22	1.00	61.20	0.60	1.00
S – 1	13.50	0.21	1.60	2.70	0.05	2.00	18.90	0.18	1.00	37.80	0.37	1.00	64.80	1.32	2.04
S – 2	25.00	0.50	2.00	3.57	0.04	1.00	14.28	0.14	1.00	17.86	0.25	1.40	64.28	28.90	45.00
S – 3	12.50	0.12	1.00	0.00	0.00	0.00	29.10	0.29	1.00	25.00	0.29	1.16	62.50	0.80	1.40
S – 4	13.50	0.24	1.80	2.70	0.02	1.00	10.80	0.10	1.00	13.50	0.16	1.20	48.60	0.78	1.60
S – 5	36.36	0.86	2.30	4.50	0.04	1.00	18.10	0.18	1.00	9.00	0.09	1.00	77.20	27.90	36.10

*P is the disease extension, the ratio between the number of diseased plants and total number of plants in a testing sample (percent). R is the development of a disease reflecting the averaged degree of affection of the plot or the whole field (points). I is the intensity or degree of plant affection (points).

Thus, the CMTs functionalized by different groups have a significant influence on rhizosphere microbiota of spring wheat, changing the activity of extracellular enzymes of the main biogeochemical cycles and phyto-sanitary state of the soil.

IOP Conf. Series: Materials Science and Engineering **91** (2015) 012082 doi:10.1088/1757-899X/91/1/012082

References

[1] Herrero-Latorre C, Álvarez-Méndez J, Barciela-García J, García-Martín S and Peña-Crecente R M 2015 Characterization of carbon nanotubes and analytical methods for their determination in environmental and biological samples. *Anal. Chim. Acta.* **853** pp 77-94.

[2] Sarma S J, Bhattacharya I, Brar S K, Tyagi R D and Surampalli R 2014 Carbon Nanotube-Bioaccumulation and Recent Advances in Environmental Monitoring. *Crit. Rev. Env. Sci. Tec.* **45** 9 pp 905-938.

[3] Maiti U N, Lee W J, Lee J M, Oh Y, Kim J Y, Kim J E, Shim J, Han T H and Kim S O 2014 25th Anniversary Article: Chemically Modified/Doped Carbon Nanotubes & Graphene for Optimized Nanostructures & Nanodevices. Adv. Mater. 26: 1 pp 40-67.

[4] Saeb M R, Najafi F, Bakhshandeh E, Khonakdar H Al, Mostafaiyan M, Simon F, Scheffler C and Mäder E 2015 An effective approach to functionalization of carbon nanotubes. *Chem. Eng. J.* **259** pp 117-125.

[5] Thul S T, Sarangi B K and Pandey R A 2013 Nanotechnology in Agroecosystem *Dovepress - a SciTechno.* **18** pp 1-7.

[6] Pritsch K, Courty P E,Churin J L, Cloutier-Hurteau B, Damon C, Duchemin M, Egli S, Ernst J, Fraissinet-Tachet L, Kuhar F, Legname E, Marmeisse R, Müller A, Nikolova P. et al. 2011 Optimized assay and storage conditions for enzyme activity profiling of ectomycorrhizae. *Mycorrhiza.* **23** pp 245-255.

[7] Vaishlya O B, Osipov N N and Belenko AA 2014 Use of Modified Carbon Nanotubes in Agricultural Technologies. *Applied Mechanics and Materials.* **682** pp 346-350.

[8] Yakobson B I and Couchman L S 2012 Carbon Nanotubes: Supramolecular Mechanics. *American Chemical Society Journal* **243** 3 pp 212-222.

[9] Khodakovskaya M V, Kim B S, Dervishi E, Mustafa T and Cernigla C E 2013 Carbon nanotubes as plant growth regulators. *Annual Review of Plant Biology.* **270** 6 pp 43-67.

[10] Newton A C, Fitt B D L, Atkins S D, Walters D R and Daniell T J 2010 Pathogenesis, parasitism and mutualism in the trophic space of microbe–plant interactions. *Trends in microbiology.* **18** 8 pp 365–373

[11] McDonald B A and Stukenbrock E 2011 The origins of plant pathogens in agro-ecosystems. *Annual Review of Phytopathology.* **80** 7 pp 54-67.

[12] Pastorin G and Cang-Rong J T 2014 The influence of carbon nanotubes on enzyme activity and structure. *Beilstein Journal of Nanotechnology* **112** 8 pp 77-82.

[13] He F, Wang H, Chen Q, Yang B, Gao Y and Wang L 2015 Short-Term Response of Soil Enzyme Activity and Soil Respiration to Repeated Carbon Nanotubes Exposure. *Soil and Sediment Contamination: An International Journal.* **24** 3 pp 250-261.

Comparison of mathematical methods of geochemical data processing

A Marfin[1], D V Lychagin [1,2], A Shapovalov[1] and E A Alfyorova[2]

[1] National Research Tomsk State University 634050, Russia, Tomsk, pr. Lenina, 36
[2] National Research Tomsk Polytechnic University 634050, Russia, Tomsk, pr. Lenina, 30

e-mail: dvl-tomsk@mail.ru, marfin1309@gmail.com, shpv@phys.tsu.ru, katerina525@mail.ru

Abstract. We have analyzed the data on concentration of geochemical elements within the 3-DV regional geochemical field. We have determined the fractal dimension index of spider diagrams and found the fractal nature of the distribution of chemical elements along the profile. We have compared these data with the results of the correlation analysis and hierarchical clustering methods. The agreement between the grouping of the elements according to the values of fractal dimension and the results of statistical analysis method was considered satisfactory.

1. Introduction

Geochemical work holds an important position in searching and forecasting activities. Due to the increased accuracy of chemical analysis and its widespread implementation in geological practice, it is necessary to improve the methods of statistical analysis and interpretation of the incoming data. In most cases the initial position is considered the view of the thermodynamic equilibrium of geological object or environment. But this approach is not suitable for systems formed as the result of self-organization. Therefore, the usage of the processing techniques based on the ideas of fractals and self-similarity has increased so far.

In the article by Mr. Li Wan et al., the element distribution dynamics was analyzed by examining its fluctuations [1]. Their findings state the relationship between the scale level and the concentration value, which has a fractal nature. An interesting result was obtained with the evaluation of Ag-Cu-Pb-Zn ores [2]. They determined the element distribution patterns that can be well described by the Weibull distribution with fractal characteristics.

Investigation of fractal dimension allowed Mr. Yan Shi Xie and colleagues determine the self-similar nature of Au-Pb-Zn ore areas (Hunan, China) [3]. Knowledge of elements' behavior allowed the authors to assess the perspective areas and identify anomalies more accurately.

Thus, the increasing attention of researchers to this method of analysis is induced by the fact that it is possible to detect relationships between various elements of geological environment that are generally could not be distinguished with the use of traditional methods [4, 5, 6]. While the application of the ideas of self-similarity in geomorphology, geophysics or sedimentology is a firmly established practice, in geochemical research sphere it is not so widely spread. One of the possible reasons, according to the authors, is the lack of interpretative geological database that could connect the fractal dimension indexes and the specific environmental conditions. Undoubtedly, the regional works should form the basis of such framework as they provide an opportunity to highlight major patterns of concentration and dispersion of elements. Anyways, the results of different methods should correspond with each other and a real geological environment.

In this context, the aim of the work is fractal and statistical analysis of experimental geochemical data in order to establish the relationship between the methods' parameters and the possibility of a combined application.

2. Results and Discussion

We have analyzed the distribution of elements in the regional geochemical profile 3-DV. Profile band crosses south-eastern part of Central Siberian plate. Minerogenic structure of territory is determined by its geological tectonic position. Geochemical specialization of the crust blocks is related to their evolution and formation history, as well as the set of structural rock complexes (including the ore-bearing ones). There is a clear correspondence between geostructures of certain hierarchical level and the fields of geochemical specialization of the area.

Initial data consists of 762 analyses of geochemical elements. They identify some transit elements (Ti, Mn, Co, Ni, Cu), alkaline-earth metals (Ba, Be), high field strength elements (Pb, Zr), Li and Ga. To calculate the fractal dimension the 2-d images of spider diagrams, which show the dependence of normalized content of chemical elements in the rock on a point on the geochemical profile, i.e. the distance, were used.

By the term "fractal dimension" we define the degree of preservation of self-similarity under the changes of the scale. The diagrams were covered with a grid with the cell size equal to δ. After decreasing δ two times we counted the amount of cells that cover the spider diagram, a similar problem was solved in the example about the coastline [7].

Statistical grouping was performed by comparing the data obtained by hierarchical clustering, calculating the value of D and distribution laws.

All elements excepting lithium were combined in pairs by hierarchical clustering using complete-linkage method. Following pairs were formed: Cu-Ba, Ni-Be, Ga-Pb and Mn-Co, Zr-Ti. Element grouping plan is shown in Figure 1.

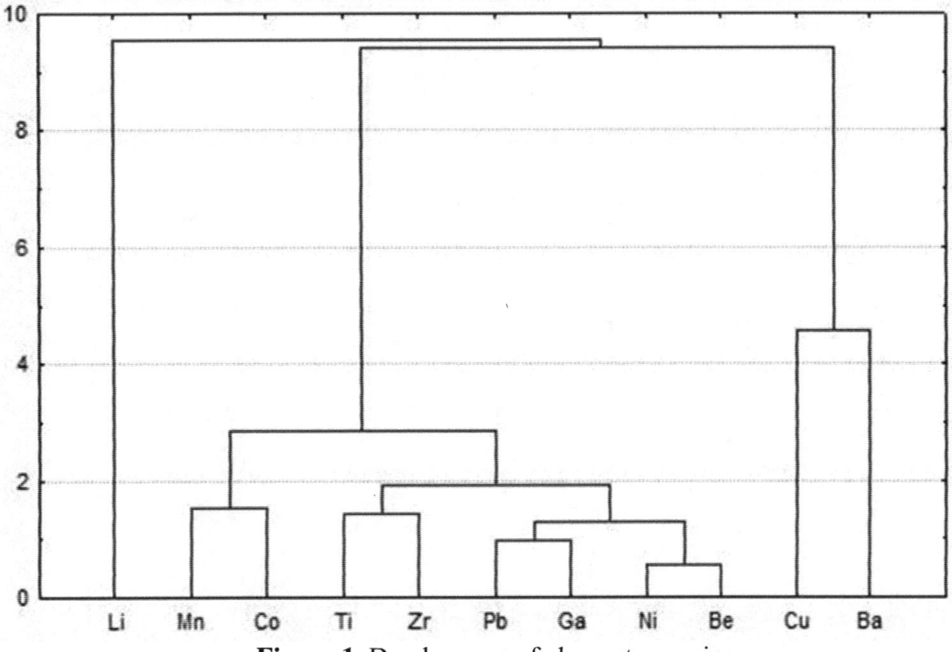

Figure 1. Dendrogram of element grouping

Correlation between the elements was studied; the results are shown in Table 1 and Figure 2.

Table 1. Correlation matrix.

	Ba	Be	Co	Cu	Ga	Li	Mn	Ni	Pb	Zr	Ti
Ba	1	0.13	-0.21	0.76	0.78	-0.14	0.13	-0.07	0.56	0	-0.03
Be	0.13	1	-0.5	0.17	0.28	-0.17	-0.3	-0.7	-0.3	0.26	0.39
Co	-0.21	-0.5	1	-0.03	-0.23	0.3	0.56	0.86	0.29	0.25	0.16
Cu	0.76	0.17	-0.03	1	0.77	0.15	0.25	0.05	0.51	0.09	0.12
Ga	0.78	0.28	-0.23	0.77	1	0.02	0.06	-0.18	0.55	0	0.24
Li	-0.14	-0.17	0.3	0.15	0.02	1	0.13	0.33	0.01	0.01	0.13
Mn	0.13	-0.3	0.56	0.25	0.06	0.13	1	0.61	0.59	0.36	-0.25
Ni	-0.07	-0.7	0.86	0.05	-0.18	0.33	0.61	1	0.43	0.21	-0.1
Pb	0.56	-0.3	0.29	0.51	0.55	0.01	0.59	0.43	1	0.14	-0.1
Zr	0	0.26	0.25	0.09	0	0.01	0.36	0.21	0.14	1	0.27
Ti	-0.03	0.39	0.16	0.12	0.24	0.13	-0.25	-0.1	-0.1	0.27	1

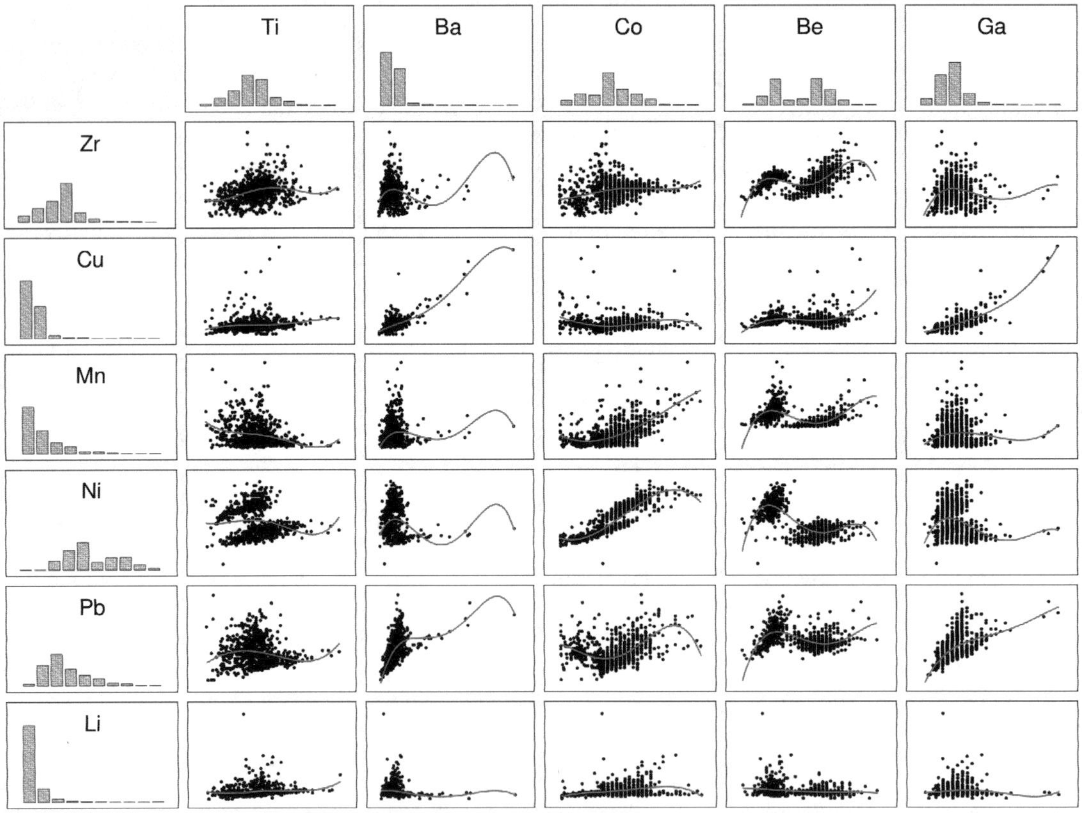

Figure 2. Graphic display of some correlation pairs.

It is possible to observe the presence of high values of the correlation coefficient in some pairs, e.g. Ba-Ga, Ba-Cu, Co-Ni, Mn-Ni and others. It is worth nothing that not all of the values are important.

For instance, the Ga-Pb pair provides seemingly moderate correlation coefficient, but it is clearly visible on Figure 2 that this dependence is poorly modeled by polynomial approximation. There is no connection between the geochemical classification of elements and their distribution laws. Also, we must keep in mind that the correlation itself does not state that one variable affects the other or that they are casually linked. We should also be aware of the possible effect of the third factor.

Therefore, the analysis of the correlation coefficient does not allow us to distinguish groups of elements and reconcile the data with the results of cluster analysis. On this basis, we proposed a method for calculating the cell fractal dimension of the distribution diagrams (spider diagrams), one of which is shown in Figure 3. Grouping in pairs is made considering the values of the fractal dimension of spider diagrams (Table 2).

According to the results of calculation of the fractal dimension (Table. 2) Li diagram has the lowest index. In comparison with the fractal dimension of the other charts, the value does not belong to any group; low correlation coefficient with other elements and these dendrograms can serve as an indirect confirmation of the isolated behavior of the element (Figure 1). At the same time, the difference of fractal dimension values in Cu-Ba, Ni-Be, Ga-Pb and Mn-Co plots is of the order of 10^{-3}.

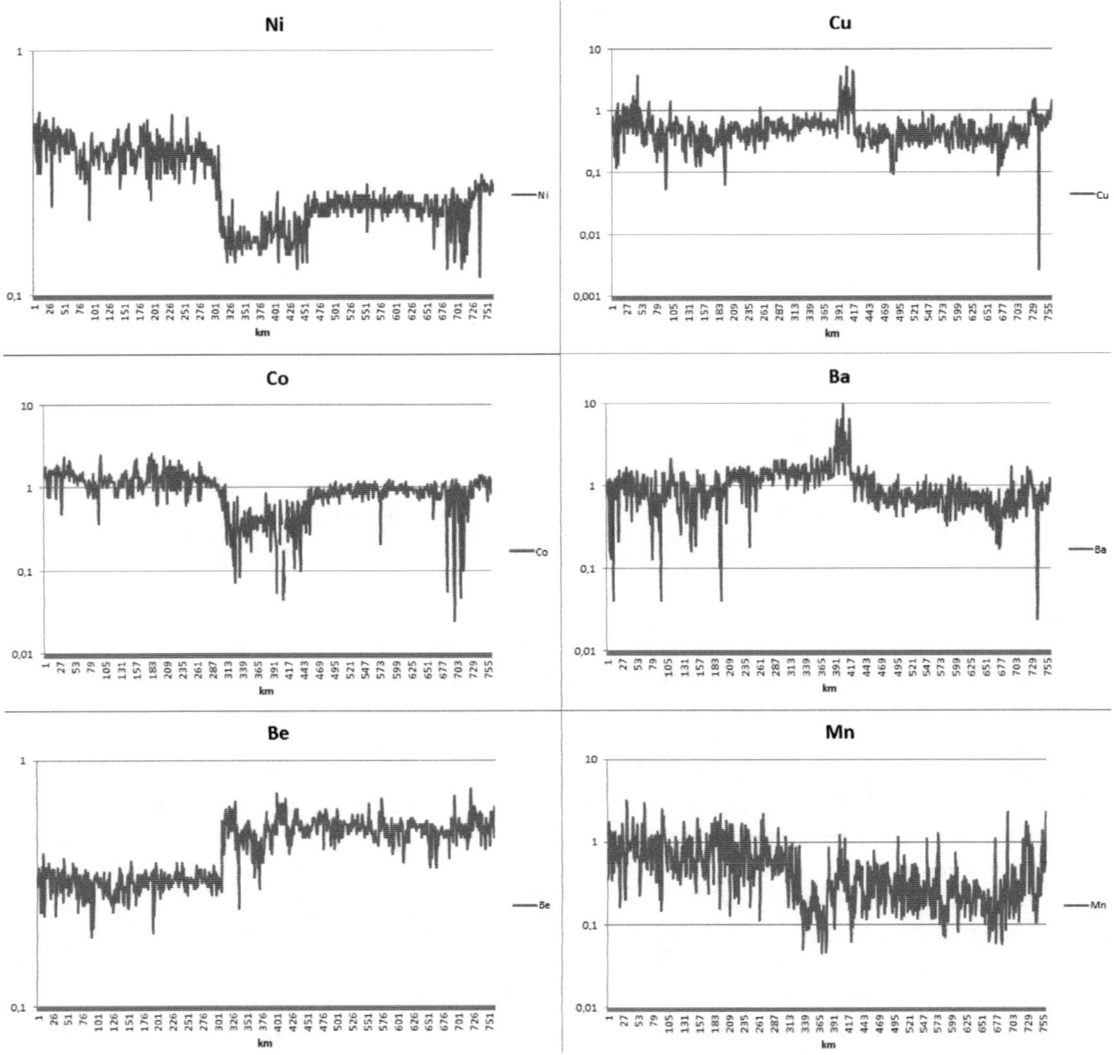

Figure 3. Spider diagrams of Ni, Cu, Co, Ba, Be, Mn

Zr-Ti pair provides the greatest value of ΔD and an absence of correlation which is followed by a low determination coefficient. Thus, the communication within the group is implicit, if existent at all, or may be a subject to more complex patterns.

The Cu-Ba group is characterized by a low ΔD rate, log-normal distribution, high determination value and is well modeled by the polynomial approximation. It is interesting that these elements have the lowest value of the fractal dimension of spider diagrams after lithium.

Ni-Be, Pb-Ga and Ti-Zr groups can be described as moderately related, with a normal distribution law, minor correlation, and, as a result, average determination rate. The curious fact is that in this group, Ti-Zr pair stands in the dendrogram in a separate subgroup, which is also evident through minimal determination and maximal ΔD values.

The last group, Mn-Co, is characterized with average determination coefficient, rather small ΔD, and different distribution laws for Mn and Co.

Table 2. The results of fractal and statistics analysis.

Chemical element	Fractal dimension, D	ΔD	Distribution law	Determination coefficient, R^2
Ba	1.49897	0.00718	Log-normal	0.58
Cu	1.49179		Log-normal	
Ni	1.50768	0.00068	Normal	0.49
Be	1.50700		Normal	
Mn	1.50878	0.00576	Log-normal	0.31
Co	1.51454		Normal	
Ti	1.50230	0.01884	Normal	0.07
Zr	1.52114		Normal	
Pb	1.52683	0.00202	Normal	0.30
Ga	1.52885		Normal	
Li	1.47607		Log-normal	-

3. Conclusion

This work exposes that within the regional geochemical profile analyzed elements demonstrate correlative behavior. Rate of correlation between the specific elements is different. To determine its characteristics a set of different statistic tools was used, such as the correlational and hierarchical analysis and geometrical approach to the study of the spider diagrams. It was possible to trace the relationship between the index of the fractal dimension and the results of the dendrograms analysis and the comparison with the standard statistical indexes. Basing on the properties of elements groups discussed above, it is possible to presume the direct relationship between the value of D and the coefficient of determination.

Therefore, calculation of the fractal dimension of spider diagrams can be an effective tool in conjunction with standard statistics methods used to identify the nature of the relationship between the chemical elements. It is also worth nothing that very often, along the fractal, come the multifractal characteristics of the geological environment [8]. These facts demand the applying the synergetic view on geology and geochemistry. That means studying processes in non-equilibrium systems from the position of self-organization [9]. Some of these processes can be revealed and examined by comparing the data obtained by different statistical methods.

Acknowledgements

The author thanks Siberian Research Institute of Geology, Geophysics and Mineral Resources for providing the information on the 3-DV profile.

This study (research grant No 8.1.76.2015) was supported by The Tomsk State University Academic D.I. Mendeleev Fund Program in 2015.

References

[1] Wan L, Chen P, Gong Z 2013 *Applied Mechanics and Materials* **249-250** 26
[2] Yanshi X, Jianwen Y, Kaixuan T, Liang C, Yang H and Yonghui L 2015 *Advanced Materials Research* **1092-1093** 1398
[3] Wan L, Chen P and Hu X 2012 *Applied Mechanics and Materials* **204-208** 4851
[4] Panahi A, Cheng Q 2004 *Mathematical Geology* **36 7** 827
[5] Cheng Q 1999 *Computers & Geosciences* **25** 949
[6] Xie S, Cheng Q, Xing X, Bao Z and Chen Z 2010 *Geoderma* **160** 36
[7] Feder J 1989 *Fractals, Plenum Press* 283
[8] Xie S, Bao Z 2004 *Mathematical Geology* **36 7** 847
[9] Haken H 1977 *Synergetics* (Berlin: Springer-Verlag)

IOP Conf. Series: Materials Science and Engineering **91** (2015) 012084 doi:10.1088/1757-899X/91/1/012084

The distribution of stresses and strains in the mating elements disk tools working bodies of roadheaders

A A Khoreshok[1,a], **L E Mametyev**[2,b], **A Yu Borisov**[3,c], **A V Vorobyev**[4,d]
[1,2]Prof. dr.-ing., [3]head teacher, [4]candidate of science (engineering), associate professor
[1,2,3]Department of Mining Machines and Complexes/Mining Institute,
T.F. Gorbachev KuzSTU, Russia, 650000, Kemerovo, Vesennyaya street 28,
[1,4]Department of Mining Equipment, Yurga Institute of Technology, TPU affiliate
Russia, 652055, Yurga, Leningradskaya street 26
E-mail: [a] haa.omit@kuzstu.ru, [b,c] bau.asp@rambler.ru, [d] vorob@tpu.ru

Abstract. Presents the results of modeling the stress-strain state in the mating structural elements of the attachment disk tools various design on triangular and tetrahedral prisms working bodies of roadheaders selective action in the destruction of coalface of heterogeneous structure.

1. Introduction

In the world practice of conducting underground mining is a widespread tool, disc and roller tools to equip working bodies roadheaders, shield tunneling complexes and aggregates [1–13]. The Department Mining machines and complexes, T.F. Gorbachev KuzSTU together with the Department of Mining Equipment of YUTI (affiliate) TPU conducted research aimed at developing constructive modules of the attachment disk of the instrument, is in charge of maintaining the extension of the scope of working bodies of roadheaders selective action on the destruction of structurally heterogeneous coal and rock face. Evaluation of the effectiveness of the developed technical solutions is carried out by results of modeling the stress-strain state of the use of the finite element method.

2. The distribution of equivalent stresses in conjugate constructive elements of the attachment of the disk tool to a polygonal prisms

When modeling the stress state of structures of triangular prisms with the attachment of the three options (Figure 1, a, b, c) and tetrahedral prisms (Figure 1, d, e) used four design disk tool diameter D = 160 mm (three biconical with angles of taper: 1 – $\varphi = \varphi_1+\varphi_2 = 5°+25° = 30°$; 2 – $10°+20° = 30°$; 3 – $15°+15° = 30°$ and one tapered 4 – $\varphi = 0°+30°$). It should be noted that the angle of taper φ_1 biconical disc facing tool to the surface of the exposed face of the tunnel excavation.

In table 1 and Figure 2 presents the dependence of the equivalent stress distribution σekv from conjugate diameters D of the structural elements of the attachment of the disk tool to the triangular prisms for the predicted coalface fracture of rocks with $\sigma comp$ = 70 MPa. At the same characteristic cross sections pass through the edge of the disc and following cross mating elements: - for the first variant in Figure 1, a (1 – disc, 2 – axle-pivot), for the second and third variants in Figure 1, b (1 – disc, 2 – pivot, 3 – axle with thrust collar).

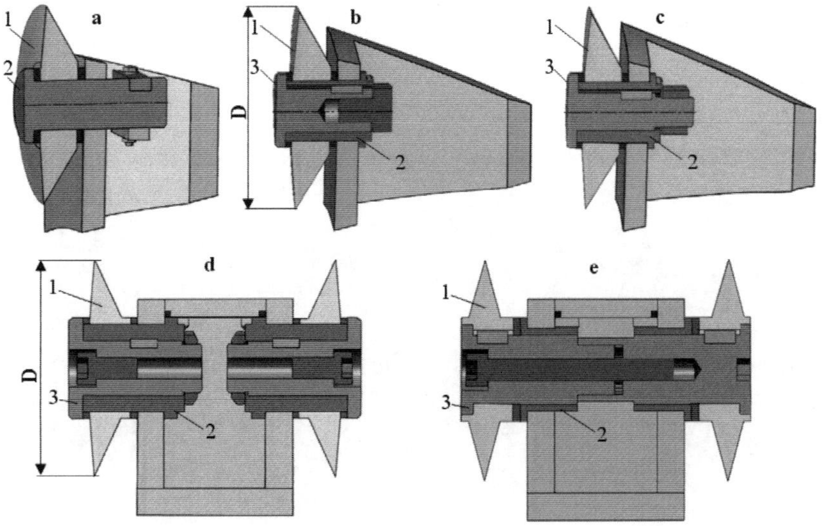

Figure 1. The design of the mating elements of the attachment of the disk tool on the polygonal prisms: a – with strap-lock, b – with screw; c – with nut; d – separated nodes mount disks; e – tandem nodes mount disks

Table 1. The distribution of equivalent stresses in the destruction of the face disk tool on the triangular prism crowns roadheaders

Options nodes attachment	The angles of taperdiscs $\varphi = \varphi_1 + \varphi_2$,°		Polynomial according	Validity coefficients of approximation R^2
I	1	5°+25°	σ_{ekv} = -4E-10D^6+2E-07D^5-4E-05D^4+ +0.0036D^3-0.13D^2+2.1824D-3.364	0.9116
	2	10°+20°	σ_{ekv} = -2E-10D^6+1E-07D^5-3E-05D^4+ +0.0028D^3-0.1138D^2+1.8771D-2.6505	0.9
	3	15°+15°	σ_{ekv} = -4E-10D^6+2E-07D^5-5E-05D^4+ +0.0044D^3-0.1686D^2+2.5206D-4.4542	0.875
	4	0°+30°	σ_{ekv} = 2E-10D^6-1E-07D^5+2E-05D^4- -0.0011D^3+0.0297D^2+0.0692D+0.8159	0.8987
II	1	5°+25°	$\sigma_{экв}$ = -1E-10D^6+8E-08D^5-2E-05D^4+ +0.002D^3-0.0932D^2+1.8159D-3.217	0.9143
	2	10°+20°	σ_{ekv} = 2E-10D^6-2E-08D^5-4E-06D^4+ +0.0011D^3-0.0686D^2+1.5676D-2.8298	0.8901
	3	15°+15°	σ_{ekv} = 5E-10D^6-2E-07D^5+2E-05D^4- -0.0008D^3-0.0002D^2+0.6508D-1.3489	0.8752
	4	0°+30°	σ_{ekv} = 6E-10D^6-3E-07D^5+4E-05D^4- -0.0031D^3+0.0984D^2-0.7694D+2.4977	0.9467
III	1	5°+25°	σ_{ekv} = -1E-10D^6+6E-08D^5-1E-05D^4+ +0.0014D^3-0.0648D^2+1.5159D-3.3668	0.9004
	2	10°+20°	σ_{ekv} = -3E-11D^6+6E-08D^5-2E-05D^4+ +0.0022D^3-0.1041D^2+2.0006D-4.0917	0.9003
	3	15°+15°	σ_{ekv} = 4E-10D^6-1E-07D^5+1E-05 D^4- -4E-05D^3-0.0253D^2+0.921D-1.7204	0.9011
	4	0°+30°	σ_{ekv} = -3E-10D^6+1E-07D^5-2E-05D^4+ +0.0012D^3-0.0291D^2+0.4971D+0.2927	0.8799

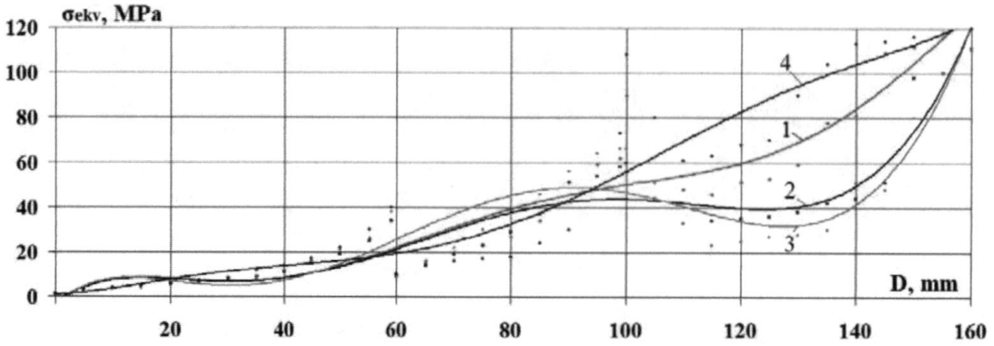

Figure 2. Dependence of the distribution of equivalent stresses σ_{ekv} of the diameter D of the mating structural elements in the cross section passing through the flange of the disk tool to a first variant of the attachment point to the triangular prism (Figure 1, a): 1, 2, 3, 4 – the angles of taper disks $\varphi = \varphi_1 + \varphi_2$ (table 1)

In table 2 and Figure 3, 4 shows the picture and dependence on the distribution of equivalent stresses σ_{ekv} from conjugate diameters D of the structural elements of the attachment with separate disk tools to a tetrahedral prism (Figure 1, d) for the predicted destruction of the face: coal (1 – σ_{comp} = 12.4 MPa), breed (2 – σ_{comp} = 51 MPa; 3 – σ_{comp} = 60.6 MPa; 4 – σ_{comp} = 78.9 MPa). Thus, the mating structural elements in the characteristic cross-section are (fig. 1, d): 1 – disc, 2 – pivot, 3 – axle with thrust collar.

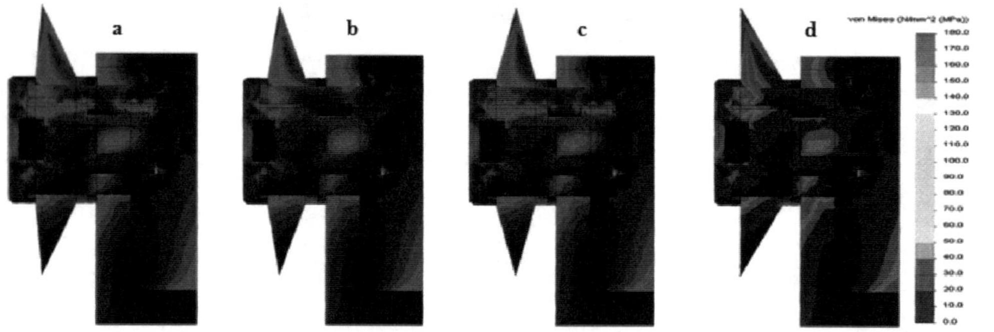

Figure 3. The distribution of equivalent stresses σ_{ekv} on the von Mises criterion in the attachment to the tetrahedral prism with the destruction of the rock mass σ_{comp} = 60.6 MPa separate disk tools with angles of taper: a – φ = 5°+25°, b – φ = 10°+20°; c – φ = 15°+15°; d – φ = 0°+30°

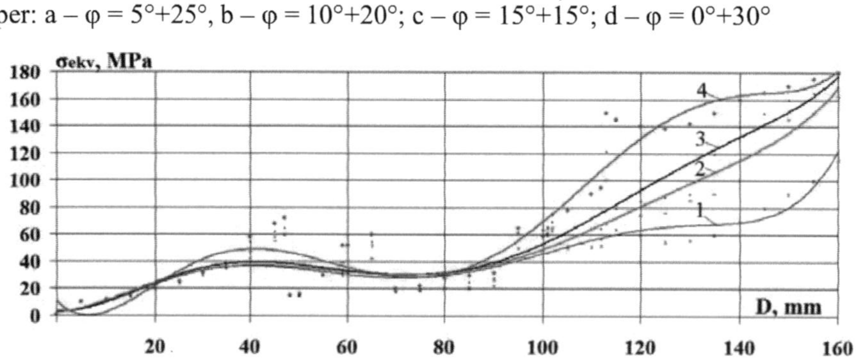

Figure 4. Dependence of the distribution of equivalent stresses σ_{ekv} of the diameter D of the mating structural elements in the cross section passing through the flange disk tool φ = 5°+25° attachment point to tetrahedral prism (table 2): 1 – σ_{comp} = 12.4 MPa; 2 – σ_{comp} = 51 MPa; 3 – σ_{comp} = 60.6 MPa; 4 – σ_{comp} = 78.9 MPa

Table 2. The distribution of equivalent stresses in the destruction of face the separate disc tool on the tetrahedral prisms roadheaders

Corners sharpening two discs $\varphi = \varphi_1+\varphi_2$, °	Face σ_{comp}, MPa		Polynomial according	Validity coefficients of approximation R^2
5°+25°	1	12.4	σ_{ekv} = 9E-10D^6-4E-07D^5+7E-05 D^4--0.005D^3+0.1473D^2-0.4312D+3.5697	0.8307
	2	51	σ_{ekv} = 5E-10D^6-3E-07D^5+5E-05D^4--0.0034D^3+0.0955D^2+0.236D+2.227	0.9124
	3	60.6	σ_{ekv} = 6E-10D^6-3E-07D^5+5E-05D^4--0.0041D^3+0.1219D^2-0.0698D+2.6707	0.9093
	4	78.9	σ_{ekv} = 2E-09D^6-8E-07D^5+0.0001D^4--0.0115D^3+0.3972D^2-3.802D+11.028	0.9305
10°+20°	1	12.4	σ_{ekv} = 9E-10D^6-4E-07D^5+7E-05D^4--0.0051D^3+0.1618D^2-0.792D+4.5062	0.8155
	2	51	σ_{ekv} = 4E-10D^6-1E-07D^5+2E-05D^4--0.0011D^3-0.0011D^2+1.5566D-0.5273	0.8962
	3	60.6	σ_{ekv} = 7E-10D^6-3E-07D^5+5E-05D^4--0.0033D^3+0.0822D^2+0.4756D+1.7038	0.8832
	4	78.9	σ_{ekv} = 5E-10D^6-2E-07D^5+4E-05D^4--0.003D^3+0.0857D^2+0.3425D+1.9343	0.8462
15°+15°	1	12.4	σ_{ekv} = 1E-09D^6-5E-07D^5+9E-05D^4--0.0071D^3+0.2323D^2-1.9484D+4.5644	0.704
	2	51	σ_{ekv} = 7E-10D^6-3E-07D^5+4E-05D^4--0.0025D^3+0.045D^2+0.8656D-0.1547	0.8645
	3	60.6	σ_{ekv} = 6E-10D^6-2E-07D^5+3E-05D^4--0.0017D^3+0.0208D^2+1.1856D-0.8734	0.8712
	4	78.9	σ_{ekv} = 8E-10D^6-3E-07D^5+5E-05D^4--0.0031D^3+0.0604D^2+0.9739D-0.1622	0.763
0°+30°	1	12.4	σ_{ekv} = 8E-10D^6-4E-07D^5+8E-05D^4--0.0059D^3+0.1813D^2-0.9435D+1.7707	0.8275
	2	51	σ_{ekv} = 1E-09D^6-7E-07D^5+0.0001D^4--0.0099D^3+0.3281D^2-2.8843D+6.5965	0.8627
	3	60.6	σ_{ekv} = 2E-09D^6-8E-07D^5+0.0001D^4--0.0108D^3+0.3579D^2-3.1909D+6.9559	0.8572
	4	78.9	σ_{ekv} = 2E-09D^6-9E-07D^5+0.0002D^4--0.0125D^3+0.4166D^2-3.8895D+7.5362	0.8501

In Figure 5 and table 3 presents the dependence of the distribution of equivalent stress σ_{ekv} on the von Mises criterion in the mating structural components of the mount disk tool on tetrahedral prism with the destruction of the mountain massif σcomp. Features of the mountain massif: coal (σ_{comp} = 12.4 MPa; 13.5 MPa; 14.8 MPa) and breed (σ_{comp} = 51 MPa; 60.6 MPa; 78.9 MPa). The analysis of dependencies in the cross section passing through the cutting edge of each of the four disks of diameter D = 160 mm according to the angles of taper: (biconical: φ = φ1+φ2= 5°+25° = 30°; 10°+20° = 30°; 15°+15° = 30° and cone φ = 0°+30°).

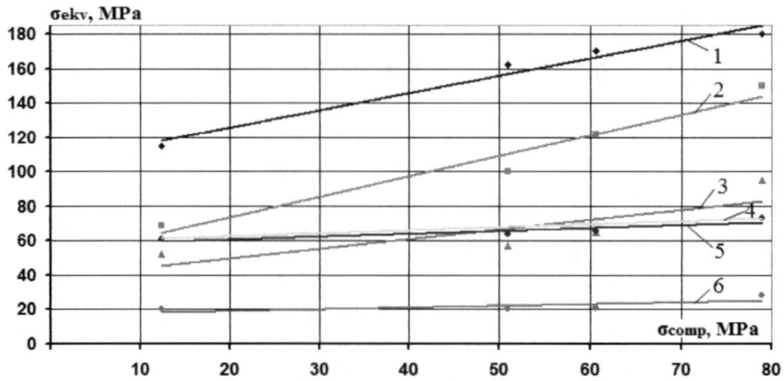

Figure 5. Dependence equivalent stress σ_{ekv} from the tensile strength of the destroyed mountain massif on compression σ_{comp} in diametrical section along the wedge flange of the disk (φ= 5°+25° = 30°) for mating structural elements 1–6 (table 3) of the attachment point to tetrahedral prism (Figure 1, d)

Table 3. The distribution of equivalent stresses σ_{ekv} in conjugate constructive elements of the attachment disk tools to the tetrahedral prisms

Corners sharpening two discs $\varphi = \varphi_1+\varphi_2$,°		The surface simulation of mating structural elements	Dependence	Validity coefficients of approximation R^2
5°+25°	1	along the edge of the disk	$\sigma_{ekv} = 1.0115\ \sigma_{comp} + 105.44$	0.9711
	2	at the hub of the disc	$\sigma_{ekv} = 1.1923\ \sigma_{comp} + 49.771$	0.9519
	3	on the outer surface of the pivot	$\sigma_{ekv} = 0.5683\ \sigma_{comp} + 38.424$	0.685
	4	on the inner surface of the pivot	$\sigma_{ekv} = 0.1728\ \sigma_{comp} + 59.486$	0.8121
	5	on the outer surface of the axis	$\sigma_{ekv} = 0.1645\ \sigma_{comp} + 57.655$	0.819
	6	in the center of the axis	$\sigma_{ekv} = 0.0997\ \sigma_{comp} + 17.194$	0.5241
10°+20°	1	along the edge of the disk	$\sigma_{ekv} = 1.153\ \sigma_{comp} + 94.515$	0.963
	2	at the hub of the disc	$\sigma_{ekv} = 0.7837\ \sigma_{comp} + 59.495$	0.9911
	3	on the outer surface of the pivot	$\sigma_{ekv} = 0.4563\ \sigma_{comp} + 23.353$	0.9969
	4	on the inner surface of the pivot	$\sigma_{ekv} = 0.273\ \sigma_{comp} + 49.9$	0.851
	5	on the outer surface of the axis	$\sigma_{ekv} = 0.921\ \sigma_{comp} + 35.783$	0.5218
	6	in the center of the axis	$\sigma_{ekv} = 0.2206\ \sigma_{comp} + 22.06$	0.7235
15°+15°	1	along the edge of the disk	$\sigma_{ekv} = 0.9396\ \sigma_{comp} + 103.59$	0.9442
	2	at the hub of the disc	$\sigma_{ekv} = 0.5622\ \sigma_{comp} + 53.984$	0.5191
	3	on the outer surface of the pivot	$\sigma_{ekv} = 0.3284\ \sigma_{comp} + 16.344$	0.9024
	4	on the inner surface of the pivot	$\sigma_{ekv} = 0.1354\ \sigma_{comp} + 56.883$	0.4664
	5	on the outer surface of the axis	$\sigma_{ekv} = 0.5446\ \sigma_{comp} + 47.124$	0.4087
	6	in the center of the axis	$\sigma_{ekv} = 0.3868\ \sigma_{comp} + 9.6287$	0.9629
0°+30°	1	along the edge of the disk	$\sigma_{ekv} = 0.9328\ \sigma_{comp} + 116.43$	0.8789
	2	at the hub of the disc	$\sigma_{ekv} = 0.4487\ \sigma_{comp} + 115.99$	0.9884
	3	on the outer surface of the pivot	$\sigma_{ekv} = 0.4338\ \sigma_{comp} + 38.498$	0.8523
	4	on the inner surface of the pivot	$\sigma_{ekv} = 1.2391\ \sigma_{comp} + 38.897$	0.9435
	5	on the outer surface of the axis	$\sigma_{ekv} = 0.4039\ \sigma_{comp} + 98.013$	0.6492
	6	in the center of the axis	$\sigma_{ekv} = 0.0282\ \sigma_{comp} + 9.5708$	0.9368

3. Displacement in conjugate constructive elements of the attachment of the disk tool to a polygonal prisms

For mating structural elements of the attachment of each of the four disk tools to triangular prisms (Figure 1, a, b, c) and the tetrahedral prisms (Figure 1, d) a simulation of displacement (Figure 6, 7). Results of modeling displacements allow us to estimate the stiffness of the mating structural elements of the attachment disk tool of the instrument including clearances, fits and tolerances, linear and diametrical dimensions in the destruction of face tunneling mining. When modeling the displacement eliminates jamming in the work of the structural elements of the mount, which may occur due to elastic deformations of the structure under load.

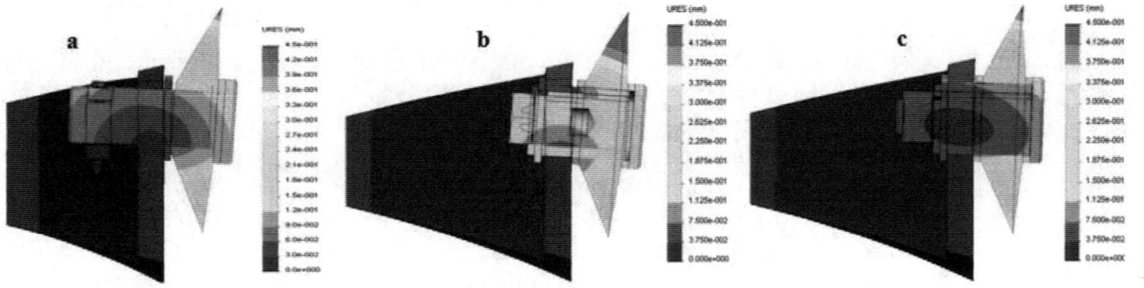

Figure 6. Displacement of structural elements in the attachment biconical disk tool ($\varphi = 5°+25° = 30°$) to the triangular prisms in the destruction of the rock face σ_{comp} = 70 MPa: a – with strap-lock, b – with screw; c – with nut

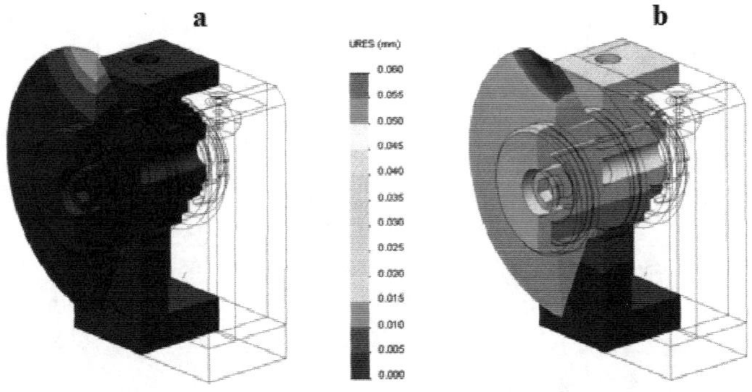

Figure 7. Displacement separate structural elements of the attachment biconical disk tool ($\varphi = 5°+25° = 30°$) to the tetrahedral prisms in the destruction of faces: a – coal σ_{comp} = 12.4 MPa; b – rock σ_{comp} = 60.6 MPa

It should be noted that the modeling of the displacements of two separate disk tools with attachment points on tetrahedral prisms in Figure 7 presents through half the image of paintings of the distribution of displacements in the destruction of coal massif (Figure 7, a) and rock massif (Figure 7, b).

Conclusion

The analysis results presented above on the distribution of equivalent stresses σ_{ekv} from conjugate diameters D of the structural elements of the attachment of the disk tool to triangular and tetrahedral prism showed the presence of zones with maximum values in peripheral downhole portion of discs with different angles of taper and the pair of hub discs with the surfaces of pins or axles on which the disc can rotate freely.

The minimum level of equivalent stress σ_{ekv} and displacements in the destruction of faces (σ_{comp} = 12.4–78.9 MPa) is marked by the installation of a biconical disk tool ($\varphi = 5°+25° = 30°$; $10°+20° =$

30°; 15°+15° = 30°), a maximum value of equivalent stress σ_{ekv} and displacement was observed when using conical disk tool (φ = 0°+30°).

In designs biconical disk tool when you change the angles of taper from asymmetric (φ = 5°+25°; 10°+20°) to symmetric (φ = 15°+15°) recorded a reduction in the estimated level of maximum equivalent stress σ_{ekv} and displacements from a symmetric disk for all cases of loading.

Dependence of the distribution of equivalent stresses σ_{ekv} according to the von Mises criterion on the diameter D of the mating structural elements in the cross section passing through the flange of the disk tool on the polygonal prism describes a polynomial dependency of the sixth order. This dependence of the distribution of equivalent stresses σ_{ekv} according to the von Mises criterion on the edges of the disk tools of the console attachment to the tetrahedral prism indicators are destroying the mountain massif σ_{comp} are described by linear dependence.

Marked decrease in the size of zones of maximum equivalent stress σ_{ekv} and displacements on downhole face of the triangular prism (Figure 1, c) drawn to the face in the third variant of the mount disk tool, compared with the second variant (Figure 1, b), which characterizes the increase in the rigidity of fastening of the nut compared to the screw.

The results of these studies allowed the development of a coupled node of the fastening clip of the double disc tool on the tetrahedral prism according to the patent of the Russian Federation 146845 (Figure 1, e). Here, the mating structural elements in the typical section are: disc, pivot, axle with thrust collar. Distinctive features of this technical solution is that the joint condition of free rotation of the two disks represents the coaxial pins-bushings is achieved by the presence of a single prefabricated structural unit, which is made in the form of rigidly attached to each other two axes with thrust ribs, one of which contains a slotted shank, and the other contains a slotted bushing. Such design implies a reduction process jamming and wear double disc tools, rational redistribution of equivalent stress σekv with the notch of the working body of roadheaders with axial crowns.

Recommended combined scheme of the set disk tool on the casing of the distributing gear between axial crowns of the working body of roadheaders with the placement of conical discs in the central zone, and biconical disk in other zones across the width between the crowns.

Further research aimed at modelling and evaluation of stress-strain state of the mating structural elements of the attachment of the coupled disk tools on tetrahedral prisms (patent of the Russian Federation 146845).

Technical solutions and test results were obtained within the base part of Ministry of Education and Science of Russia state order, project 632 "Investigation of technologies and techniques parameters for selecting and developing innovative designs to improve operating efficiency of multipurpose mining machines in Kuzbass".

Acknowledgements

The authors wish to express their gratitude and appreciation to the first deputy chief mechanic on the mountain-mine equipment in company “SUEK-Kuzbass” Mukhortikov S.G, head of the laboratory of coal geotechnology Institute of Coal, Siberian branch of the Russian Academy of Sciences, doctor Ingeniariae Aksenov V.V., head of Department of Machine tools and cutting tools, professor, doctor Ingeniariae Korotkov A. N., professor of the Department of Mechanical engineering, doctor Ingeniariae Blumenstein V.Yu, head of the Center of Informatization, candidate of science (engineering), associate professor of Department of Mining equipment, Yurga Institute of Technology, TPU affiliate Vorobyov A.V. for active creative cooperation and support.

References

[1] Aksenov V V, Efremenkov A B, Beglyakov V Yu 2013 The influence of relative distance between ledges on the stress-strain state of the rock at a face *Applied mechanics and materials*. vol. 379, pp. 16-19.

[2] Aksenov V V, Khoreshok A A, Beglyakov V Yu 2013 Justification of creation of an external propulsor for multipurpose shield-type heading machine – Geo-walker *Applied mechanics and materials*. vol. 379, pp. 20-23.

[3] Hong K R, Yang R H 2011 The major problems and countermeasures on the shield machine tunneling in the hard rock stratum *Applied mechanics and materials*. vols. 105-107, pp. 1438-1442.

[4] Prokopenko S A 2014 Multiple service life extension of mining and road machines' cutters *Applied mechanics and materials*. vol. 682, pp. 319-323.

[5] Li X H, Du W, Huang Z L, Fu W L 2011 Simulation of disc cutter loads based on ANSYS/LS-DYNA *Applied mechanics and materials*. vol. 127, pp. 385-389.

[6] Zhao B, Zong X M, He B, Zhang L J 2014 Multi variable multi objective optimization for the cutting head of roadheader *Applied mechanics and materials*. vols. 635-637, pp. 358-364.

[7] Zhang Y, Wang X W, Liu H F 2014 Numerical simulation of rock-breaking process by disc cutter in tunnel boring machine *Applied mechanics and materials*. vol. 487, pp. 513-516.

[8] Li F H, Cai Z X, Kang Y L 2011 A theoretical model for estimating the wear of the disc cutter *Applied mechanics and materials*. vols. 90-93, pp. 2232-2236.

[9] Zhou Z L, Li X B, Zhao G Y, Liu Z X, Xu G J 2011 Excavation of high-stressed hard rock with roadheader *Applied mechanics and materials*. vols. 52-54, pp. 905-908.

[10] Wang X D, Shi M Q, Gao S J, Guo Y C 2014 Design of transverse boom-type roadheader remote control system *Applied mechanics and materials*. vols. 701-702, pp. 679-683.

[11] Khoreshok A A, Mametev L E, Borisov A Yu, Vorobev A V 2015 Finite element models of disk tools with attachment points on triangular prisms *Applied mechanics and materials*. vol. 770. pp. 429-433.

[12] Khoreshok A A, Mametev L E, Borisov A Yu, Vorobev A V 2015 Stress state of disk tool attachment points on tetrahedral prisms between axial bits *Applied mechanics and materials*. vol. 770. pp. 434-438.

[13] Khoreshok A, Mametyev L, Bosirov A, Vorobiev A (2014, October) Stress-deformed state knots fastening of a disk tool on the crowns of roadheaders. *Taishan Academic Forum – Project on mine disaster prevention and control*. Atlantis Press, Amsterdam – Paris - Beijing, pp. 177-183.

IOP Conf. Series: Materials Science and Engineering **91** (2015) 012085 doi:10.1088/1757-899X/91/1/012085

Development of math model of geokhod bladed working body interaction with geo-environment

V Yu Sadovets[1,2], V Yu Beglyakov[1], and V V Aksenov[1,3]

[1]Yurga Technological Institute branch of Tomsk Polytechnic University, 26 Leningradskaya, Yurga, Kemerovo region, Russia, 652055.
[2]Kuzbass State Technikal University by T.F. Gorbachev, 28 Vesennyaya, Kemerovo, Russia, 650000.
[3]Coal Institute SB RAS10 Leningradsky, Kemerovo, Russia, 650065.

Abstract. The article describes basic mathematical dependencies for determining of power parameters of geokhods' bladed working bodies both conventional and toothed ones.

Introduction.

Some authors [1,2,3] consider mine working as rigid body motion in solid earth. This approach is the foundation of the geo-winch mining method, geokhods being its basic functional elements.

Geokhods are machines that move in the subsurface using geological environment for advancing. Geokhods are a new class of mining machinery and designed for mine workings of various purpose and spacing.

For geokhod underground workings at shallow depths in unstable rocks a bladed working body (WB) is proposed to use. Lack of design procedure for the bladed working bodies of different designs is a limiting factor of geokhod manufacturing.

Purpose of the work. The aim of the work is to develop a mathematical model of geokhod bladed WB interaction with rock mass.

Special mode of geokhod advancement causes formation of complex surfaces both of the face and of the WB [4]. The face surface has the form of several ledged helical surfaces [5,6,7]. Any point of the blade, located at a distance x (see Figure 1) from the geokhod rotation axis, advances to the face at an angle:

$$\beta_x = arctg\frac{h_B}{2\pi x} \quad (1)$$

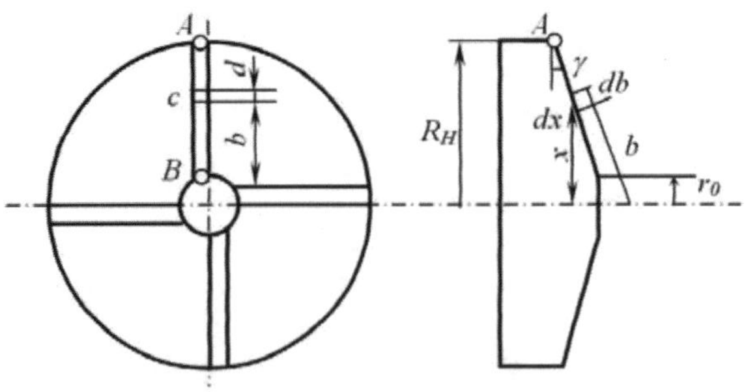

Figure 1. Scheme of the working body blade.

The blade points being closer to the geokhod rotation axis advance to the face at a larger angle than the points being at the blade periphery (Figure 2a). In this way, at helical (screw-like) advancing of the blade to the face, radial blade points form a helicoid (screw-like) surface. Consequently, face surface in the sector between adjacent radial blades takes the form of a helical surface after their passage.

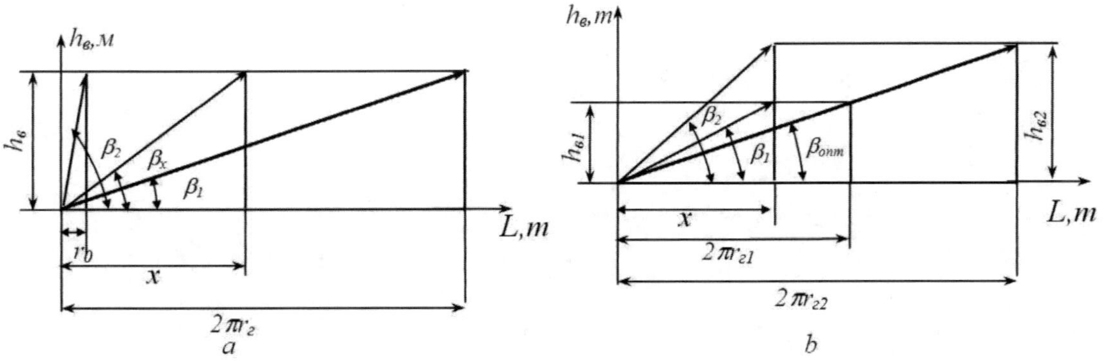

Figure 2 Direction of blade points movement, depending on:
a - location at a radial blade; b - geokhod standard size.

Furthermore, if radius of the geokhod head section is increased at some optimal tilt angle (β_{opt}) of the external helical blade, then any point, being at a distance x from the geokhod center, passes a distance $h_{в2}$ differing $h_{в1}$ per revolution.

Therefore, helicoid geometric parameters, in the form of which a radial blade profile is manufactured, depend on parameters of the external mover ($r_г$ - radius of the geokhod head section, $h_в$ - helical blade step, β - tilt angle of the blade) and are specific for each standard size.

Ledge height h depends on a helical blade step and a number of radial blades mounted on a geokhod working body:

$$h = \frac{h_B}{n}, \tag{2}$$

where n is a number of blades.

Radial blades produce a corresponding amount of mutually conjugate helical areas at the face. A ledge that is formed between two mutually conjugate helical areas is continuously formed and destroyed by the blades as they advance on the face.

The complex helical face is determined only by the nature of geokhod advancing, so helical surface is formed by the work of any radial WB, regardless of its design and tools.

Using fragments of a structural portrait based on integrated approach geokhod bladed WBs are designed [8]. It is noted that the geometric parameters of blades are special for each geokhod radius.

As geokhod advances in the subsurface using geological environment its WB destroys (cuts) both loose and solid rocks. Then a rational number of WB blades must be chosen from the interval [9]:

$$n = \frac{tg\beta \cdot 2\pi r_г}{0{,}1 \div 0{,}2}. \tag{3}$$

Layout of the radial blades referred to the geokhod axis affects cutting forces.When cutting with one WB radial blade, a total blocking cutting force can be represented as a sum of three force components (Figure 3) [10]:

1 Force of earth resistance overcoming by the blade front edge $P_{cв}$, which is proportional to a cross-sectional area of the slot in front of the front blade edge and depends on a cutting angle and rock hardness;

2 Force of earth resistance overcoming in lateral slot widenings which (force) is proportional to these slot parts area and depends on rock hardness but does not depend on cutting angle and cut width;

3 Force of earth resistance overcoming by the blade lateral ribs at the slot bottom, proportional to the cut thickness, which (thickness) depends on rock hardness but does not depend on cutting angle and cut width.

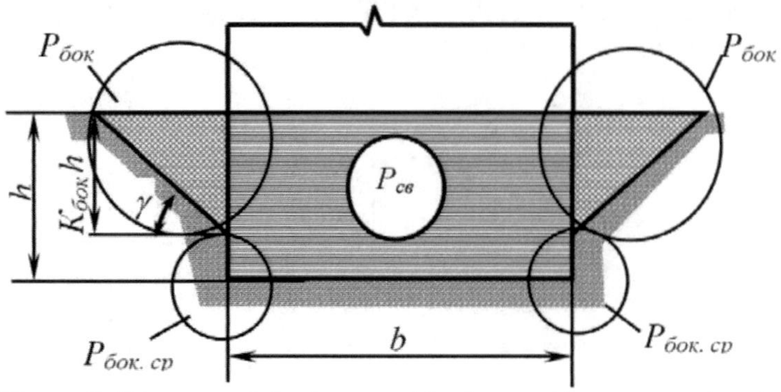

Figure 3. Diagram of the forces at block cut with a conventional blade.

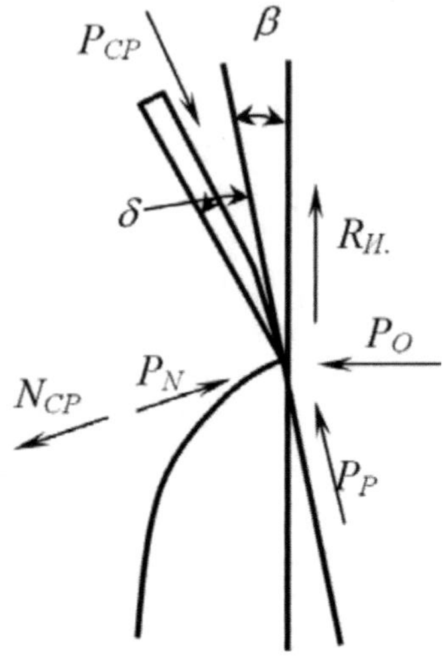

Figure 4. Design model for determination of total force of soil resistance to cutting by blade WB.

The geokhod WB advances to the face in accordance with steps of the helical blade mounted on the outer surface of the head section. Each point of the blade moves along a spiral trajectory. The blade points being closer to the geokhod rotation axis advance to the face at a larger angle than the points being at the blade periphery. So, at helical advancing of bladed WB to the face, radial blade points form a helicoid surface. Consequently, the face takes a helical form.

The blades produce a corresponding amount of mutually conjugate helical areas at the face. A ledge that is formed between two mutually conjugate helical areas is continuously destroyed by the WB blades as they advance to the face.

Total force of blocking cutting by a conventional blade (Figure 4)[11]:

$$P_{ср} = \varphi m_{св} bh + 2m_{бок} h^2 + 2m_{бок.ср} h\,, \tag{4}$$

where $\varphi\ m_{св}bh$ is cutting force component for front resistance overcoming ($P_{св}$ in fig. 3), H;

$2m_{бок}h^2$ - total rock destruction force in lateral slot widenings ($P_{бок}$ in fig. 3), H;

$2m_{бок.ср}h$ - lateral cut forces ($P_{бок.ср}$ in fig. 3), H;

φ - coefficient of cutting angle impact;

$m_{св}$ - specific cutting force for earth resistance overcoming by a front edge at cutting angle 45^0, Pa;

b - blade width, m;

h - cut depth, m;

$m_{бок}$ - coefficient of destruction force in slot lateral parts, Pa;

$m_{бок.ср}$ - coefficient of specific force of cut by a lateral edge, H/m.

Dependencies for determination of cutting forces of bladed WB are developed in [9]. General dependencies are as follows:

$$P_{o.св} = \frac{\varphi m_{св} h_в^2 + h_в n P_{изн}}{2\pi n \cos\gamma} \cdot \left(\ln \frac{tg\left|\frac{\beta_2}{2}\right|}{tg\left|\frac{\beta_1}{2}\right|} \right) - \qquad ; \tag{5}$$

$$- \frac{h_в}{2\pi}\left(\varphi m_{св} \frac{h_в}{n} \frac{ctg(\delta + \varphi_{TP})}{\cos\gamma} - \frac{ctg(\delta_1 + \varphi_{TP})}{\cos\gamma} P_{изн} \right) \frac{\sin\beta_2 - \sin\beta_1}{\sin\beta_1 \sin\beta_2}$$

$$R_{и.о.св} = \frac{\varphi m_{св} h_в^2 + h_в n P_{изн}}{2\pi n \cos\gamma} \cdot \frac{\sin\beta_2 - \sin\beta_1}{\sin\beta_1 \sin\beta_2} + \tag{6}$$

$$+ \frac{h_в}{2\pi}\left(\varphi m_{св} \frac{h_в}{n} \frac{ctg(\delta + \varphi_{TP})}{\cos\gamma} - \frac{ctg(\delta_1 + \varphi_{TP})}{\cos\gamma} P_{изн} \right) \cdot \left(\ln \frac{tg\left|\frac{\beta_2}{2}\right|}{tg\left|\frac{\beta_1}{2}\right|} \right);$$

$$M_{и.о.св} = \frac{\varphi m_{св} h_в^3 + h_в^2 n P_{изн}}{8\pi^2 n \cos\gamma} \left(\ln \left| \frac{tg\frac{\beta_1}{2}}{tg\frac{\beta_2}{2}} \right| + \frac{\cos\beta_2 \sin^2\beta_1 - \cos\beta_1 \sin^2\beta_2}{\sin^2\beta_1 \sin^2\beta_2} \right) + . \tag{7}$$

$$+ \frac{h_в^2}{4\pi^2}\left(\varphi m_{св} \frac{h_в}{n} \frac{ctg(\delta + \varphi_{TP})}{\cos\gamma} - \frac{ctg(\delta_1 + \varphi_{TP})}{\cos\gamma} P_{изн} \right) \cdot \frac{\sin\beta_2 - \sin\beta_1}{\sin\beta_1 \sin\beta_2}$$

Projection of cutting resistance force, independent of cut width, to rotation axis and a plane perpendicular to the rotation axis; and cutting resistance moment of this component:

$$P_{o.бок} = (\sin\beta_1 + \sin\beta_2)\frac{P_{бок}}{2} - (\cos\beta_1 + \cos\beta_2)\frac{N_{бок}}{2}; \tag{8}$$

$$R_{и.о.бок} = (\cos\beta_1 + \cos\beta_2)\frac{P_{бок}}{2} + (\sin\beta_1 + \sin\beta_2)\frac{N_{бок}}{2} \tag{9}$$

$$M_{и.о.бок} = (r_г \cos\beta_1 + r_o \cos\beta_2)\frac{P_{бок}}{2} + (r_г \sin\beta_1 + r_o \sin\beta_2)\frac{N_{бок}}{2}. \tag{10}$$

Insert (5), (6) and (7) in the formulas obtained and after some rearrangements we obtain

$$P_{o.бок} = \frac{h_B}{n}\left(m_{бок} \frac{h_B}{n} + m_{бок.ср} \right)\left[- \frac{\cos(\delta + \varphi_{mp} + \beta_1) + \cos(\delta + \varphi_{mp} + \beta_2)}{\sin(\delta + \varphi_{mp})} \right]; \tag{11}$$

$$R_{и.о.бок} = \frac{h_B}{n}\left(m_{бок} \frac{h_B}{n} + m_{бок.ср} \right)\left[\frac{\sin(\delta + \varphi_{mp} + \beta_1) + \sin(\delta + \varphi_{mp} + \beta_2)}{\sin(\delta + \varphi_{mp})} \right]; \tag{12}$$

$$M_{и.о.бок} = \frac{h_B}{n}\left(m_{бок} \frac{h_B}{n} + m_{бок.ср} \right)\left[\frac{r_г \sin(\delta + \varphi_{mp} + \beta_1) + r_o \sin(\delta + \varphi_{mp} + \beta_2)}{\sin(\delta + \varphi_{mp})} \right] \tag{13}$$

For the bladed WB a total projection of earth cutting resistance force to rotation axis and a plane perpendicular to the rotation axis; and a total cutting resistance moment are equal to:

$$P_o = n\left(P_{o.св} + P_{бок}\right); \tag{14}$$

$$R_{и.о} = n\left(R_{и.о.св} + R_{и.о.бок}\right); \tag{15}$$

$$M_{и.о} = n\left(M_{и.о.св} + M_{и.о.бок}\right). \tag{16}$$

Toothed cutting part of the bladed WB has a complex shape: the teeth are mounted on a radial bar. Therefore, to determine cutting forces, toothed bladed WB will be considered as a complex blade consisting of interacting conventional blades.

Calculation of the complex blade operation is based on conventional blades interaction and values of superposition of rock rapture zones produced by each blade separately.

Mount of the teeth on radial blade reduces rock cutting resistance. As the teeth destroy rock in front of the main cutting edge, that reduces or eliminates cutting by side edges of the main blade of the WB [6].

The value of b (Figure 5) is the width of the tooth, which should be 0,06-0,08m. Wider teeth increase crowd force. Smaller width does not provide sufficient strength to the teeth.

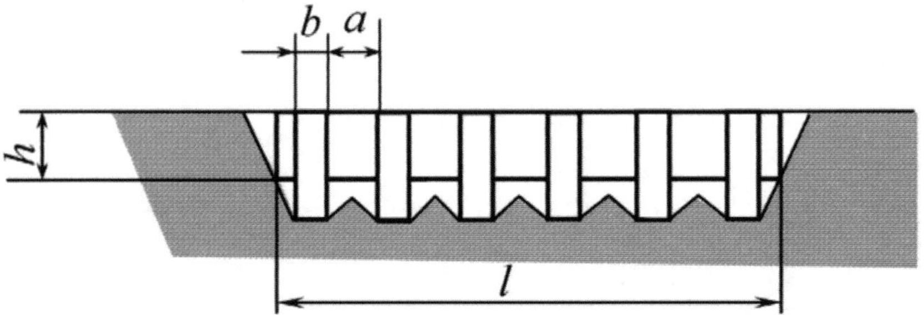

Figure 5. Scheme of teeth arrangement along the blade length.

For the bladed WB a blade length $l = r_г - r_o$ where $r_г$ is a radius of geokhod head section; r_o is the distance from the geokhod rotation axis to the nearest point of the blade. The number of teeth on one blade is:

$$n' = \frac{l}{3{,}5b''}. \tag{17}$$

Cutting force of a sharp complex blade by Y.A. Vetrov is equal [6]

$$P = \sum_{i=1}^{n'} p_{св_i} F_{св_i} + p_{бок} \sum_{i=1}^{n'} F_{бок_i} + p_{бок.ср} \sum_{i=1}^{n'} L_{бок.ср_i}. \tag{18}$$

If teeth are blunted or have a wear pad then the expression (18) takes the form

$$P = \sum_{i=1}^{n'} p_{св_i} F_{св_i} + p_{бок} \sum_{i=1}^{n'} F_{бок_i} + p_{бок.ср} \sum_{i=1}^{n'} L_{бок.ср_i} + \sum_{i=1}^{n'} p_{пл.изн\,i} L_{пл.изн_i}, \tag{19}$$

where $F_{св_i}, F_{бок_i}$ are sub-areas, m^2; $L_{бок.ср_i}, L_{пл.изн_i}$ - cut line lengths, m; $p_{св_i}, p_{бок}, p_{бок.ср}, p_{пл.изн\,i}$ - specified cutting sub-forces.

Normal cutting force of a sharp complex blade by Y.A. Vetrov is equal [6]

$$N = \sum_{i=1}^{n'} p_{св_i} F_{св_i} ctg(\delta_i + \mu) + p_{бок} \sum_{i=1}^{n'} F_{бок_i} ctg(\delta_i + \mu) + p_{бок.ср} \sum_{i=1}^{n'} L_{бок.ср_i} ctg(\delta_i + \mu) \tag{20}$$

Considering the force arising from operation with blunt or having a wear pad teeth, expression (20) takes the form

$$N = \sum_{i=1}^{n'} p_{св_i} F_{св_i} ctg(\delta_i + \mu) + p_{бок} \sum_{i=1}^{n'} F_{бок_i} ctg(\delta_i + \mu) + + p_{бок.ср} \sum_{i=1}^{n'} L_{бок.ср_i} ctg(\delta_i + \mu) + \sum_{i=1}^{n'} p_{пл.изн\,i} L_{пл.изн_i} ctg(\delta_i + \mu) \quad (21)$$

When calculating force dependencies one should consider technical features and specifications of the teethed bladed working body, as well as the calculation scheme of rock cutting resistance total force.

Projection of cutting resistance force to rotation axis and a plane perpendicular to the rotation axis; and cutting resistance moment of this component [12]:

$$P_o^з = \sin\beta_i P - \cos\beta_i N; \quad (22)$$

$$R_{и.о}^з = \cos\beta_i P + \sin\beta_i N; \quad (23)$$

$$M_{и.о}^з = x_i R_{и.о}^з, \quad (24)$$

where x – is a distance from rotation axis to every n' tooth, m.

The final expressions of determining forces for sharp teeth

$$P_o^з = \sum_{i=1}^{n'} p_{св_i} F_{св_i} A_i + p_{бок} \sum_{i=1}^{n'} F_{бок_i} A_i + p_{бок.ср} \sum_{i=1}^{n'} L_{бок.ср_i} A_i; \quad (25)$$

$$R_{и.о}^з = \sum_{i=1}^{n'} p_{св_i} F_{св_i} B_i + p_{бок} \sum_{i=1}^{n'} F_{бок_i} B_i + p_{бок.ср} \sum_{i=1}^{n'} L_{бок.ср_i} B_i; \quad (26)$$

$$M_{и.о}^з = \sum_{i=1}^{n'} p_{св_i} F_{св_i} x_i B_i + p_{бок} \sum_{i=1}^{n'} F_{бок_i} x_i B_i + p_{бок.ср} \sum_{i=1}^{n'} L_{бок.ср_i} x_i B_i, \quad (27)$$

where $A_i = (\sin\beta_i - ctg(\delta_i + \mu)\cos\beta_i)$ and $B_i = (\cos\beta_i + ctg(\delta_i + \mu)\sin\beta_i)$;

For blunt or having a wear pad teeth expressions (25), (26), (27) take the form

$$P_o^з = \sum_{i=1}^{n'} p_{св_i} F_{св_i} A_i + p_{бок} \sum_{i=1}^{n'} F_{бок_i} A_i + p_{бок.ср} \sum_{i=1}^{n'} L_{бок.ср_i} A_i + \sum_{i=1}^{n'} p_{пл.изн\,i} L_{пл.изн_i} A_i; \quad (28)$$

$$R_{и.о}^з = \sum_{i=1}^{n'} p_{св_i} F_{св_i} B_i + p_{бок} \sum_{i=1}^{n'} F_{бок_i} B_i + p_{бок.ср} \sum_{i=1}^{n'} L_{бок.ср_i} B_i + \sum_{i=1}^{n'} p_{пл.изн\,i} L_{пл.изн_i} B_i; \quad (29)$$

$$M_{и.о}^з = \sum_{i=1}^{n'} p_{св_i} F_{св_i} x_i B_i + p_{бок} \sum_{i=1}^{n'} F_{бок_i} x_i B_i + p_{бок.ср} \sum_{i=1}^{n'} L_{бок.ср_i} x_i B_i + \sum_{i=1}^{n'} p_{пл.изн\,i} L_{пл.изн_i} x_i B_i. \quad (30)$$

Determination of force parameters of the blades cutting a peripheral edge helical channel is presented in [9].

For the radial bladed WB with one blade for cutting the helical channel [12,13,14], component cutting forces and resistance moment will be equal:

For conventional WB

$$P_{и.о.общ} = nP_o + \kappa P_o^в$$

$$R_{и.о.общ} = nR_{и.о.} + \kappa R_{и.о}^в$$

$$M_{и.о.общ} = nM_{и.о} + \kappa M_{и.о}^в$$

For teethed WB

$$P_{и.о.общ} = nP_o^з + \kappa P_o^в$$

$$R_{и.о.общ} = nR_{и.о}^з + \kappa R_{и.о}^в$$

$$M_{и.о.общ} = nM_{и.о}^з + \kappa M_{и.о}^в$$

Values of friction force and friction resistance moment are determined depending on rock pressure value [9,10,11,13]:

$$T_{ИО} = \frac{\pi f_{TP} \lambda \gamma_{пор}}{f}\left(1 + 2tg\left(45^0 - \frac{\rho}{2}\right)\right) r_г^3; \quad (31)$$

$$M_{T.ИО} = \frac{2\pi f_{TP} \lambda \gamma_{пор}}{3f}\left(1 + 2tg\left(45^0 - \frac{\rho}{2}\right)\right) r_г^4, \quad (32)$$

where λ - lateral pressure coefficient; $\gamma_{пор}$ – rock specific weight, Hм^3; f - hardness coefficient on the M.Protodjakonov scale; ρ - rock internal friction angle, degrees.

Conclusions

A mathematical model of geokhod bladed WB interaction with geo-environment is obtained on the basis of the developed force parameters calculation method.

This method allows determining:

- Cutting forces on one radial blade both of the conventional and the teethed geokhod working body;

- Cutting forces on the geokhod bladed WB of various designs.

- Helicoid geometric parameters (l - radial blade length, β_i – angle of each i-th point of the blade), in the form of which a radial blade profile is manufactured, depend on parameters of the external mover ($r_г$ - radius of the geokhod head section, $h_в$ - helical blade step, β - tilt angle of the blade) and are specific for each standard size.

References

[1] Gorbunov V F, Aksenov V V, Eller A F, Nagorny V D and Skomorokhov V M 1987 Design and calculation of tunnel complexes

[2] Aksenov V V 2004 Scientific base of geo-winch mining and design of helical-swivel units *Doctor of Technical Sciences thesis.*

[3] Gorbunov V F, Aksenov V V and Sadovets V Yu 2004 Expert assessment of impact of innovative mining machines features on parameter calculation method // *Bulletin KuzGTU* № 6.1 p 41-43.

[4] Aksenov V V, Efremenkov A B, Beglyakov V Yu, Blaschuk M Yu, Timofeev V Yu and Sapozhkova A V 2009 Development of specifications for geokhod Basic Systems *Mining Machinery and Electrical Engineering* №5. p. 3-7.

[5] Aksenov V V, Efremenkov A B, Beglyakov V Yu, Blaschuk M Yu, Timofeev V Y 2009 Geokhod basic specifications *Mining information-analytical bulletin (scientific and technical journal)* Vol. 10. №12. p. 107-118.

[6] Aksenov V V and Sadovets V Yu 2006 Synthesis of geokhod Working bodies technologies *Bulletin of Kuzbass State Technical University* № 6. p. 64.

[7] Aksenov V V and Sadovets V Yu 2006 Synthesis of geokhod bladed Working bodies technologies *Bulletin of Kuzbass State Technical University* № 6. p. 33-37.

[8] Aksenov V V, Sadovets V Y and Beglyakov V Yu 2010 Synthesis of geokhod working bodies design solutions *Mining information-analytical bulletin (scientific and technical journal)* Vol. 3. № 12. pp. 49-54.

[9] Sadovets V Yu 2007 Justification of design and power parameters of geokhod bladed working bodies *PhD thesis*

[10] Vetrov Y A 1985 Calculation of rock cutting and digging forces. Kiev *Publishing House of Kiev University.*

[11] Aksenov V V Efremenkov A B and Beglyakov V Yu 2013 Impact of relative distance between ledges on stress-strain state of rock face *Applied Mechanics and Materials* Vol. 379 pp 16-19.

[12] Prokopenko S A 2014 Multiple service life extension of mining and road machines' cutters *Applied Mechanics and Materials* Vol. 682 pp 319-323.

IOP Conf. Series: Materials Science and Engineering **91** (2015) 012085 doi:10.1088/1757-899X/91/1/012085

[13] Sadovets V Yu, Beglyakov V Yu and Efremenkov A B 2015 Simulation of geokhod movement with blade actuator *Applied Mechanics and Materials* Vol. 762 pp 384–390
[14] Khoreshok A A, Mametyev L E, Borisov A Yu and Vorobiev A V 2014 Stress-deformed state knots fastening of a disk tool on the crowns of roadheaders *Taishan Academic Forum - Project on Mine Disaster Prevention and Control* pp 177-183

IOP Conf. Series: Materials Science and Engineering **91** (2015) 012086 doi:10.1088/1757-899X/91/1/012086

The application of the analytic hierarchy process when choosing layout schemes for a geokhod pumping station

R V Chernukhin, A A Dronov, M Y Blashchuk

Yurga Technological Institute, TPU, 26 Leningradskaya St, Yurga, Kemerovo Region, Russia

e-mail: chernuhin@tpu.ru

Abstract. The article describes one possibility of choosing layout schemes for geokhod systems which is the analytic hierarchy process. There is the essence of the method summarized therein. The usage of the method is considered for the analysis and the choice of layout schemes for a geokhod pumping station. Keywords: geokhod, analytic hierarchy process, pumping station, layout scheme.

1. Introduction

Underground mining processes and construction of underground facilities are demanding and expensive; there is much tension around the tasks of drifting rate increase, performance improvement of tunneling machines, operating safety as well as the reduction of mining costs. The research conducted in Russia as well as abroad around the development of new rock destruction technologies and examples of tunneling machines shows that the improvement of the performance of existing tunneling machine designs is accompanied by their increase in dimensions and weight. Besides, the shaft-sinking and tunneling machines currently used are have quite restricted digging angles.

A prospective and alternative method of mining is the geo-winchester technology. A basic element of the geo-winchester technology is geokhod which is a device that uses the subsurface for underground space drifting [1, 2].

One distinctive feature of a geokhod is an original layout scheme as well as new functional devices and elements which have not been used in the tunneling equipment before. Almost all of the geokhod systems have not only several layout scheme varieties [3, 4, 5], but also manufacturing particularities [6]. The diversity of requirements to geokhod systems and the conditions of its use make the choice of a certain layout scheme quite difficult. The decision-making method concerning the choice of a particular variety can vary from simple intuition to the engagement of highly qualified specialists.

Often, the choice problem of layout arrangements is solved by means of the table procedure. The key idea of the table procedure consists in the presentation of original data in the form of a table, where the columns are requirements to the systems, and the lines present possible variants. Layout schemes are evaluated by an expert on the basis of his/her experience, professional skills, and intuition.

The variants are evaluated according to the following scale:

"+" – the scheme completely matches the criteria;

"+–" – the scheme partially matches the criteria;

"–" – the scheme does not match the criteria.

Finally, the scheme will be chosen that gets the most positive marks.

2. Theoretical basis

In order to maintain the operation of the main systems of tunneling machines, hydraulic power units are effectively used. The system that maintains the hydraulic power unit of a geokhod is a pumping station (PS). It is the absence of scientifically based layout designs of PS and the methods of its parameters that restrains the creation of geokhods of a new technical level. This is why the research is pending concerning the development of layout arrangements and the argumentation of parameters of geokhod PS of new technical level. For the first time, the analytic hierarchy process was used in connection with geokhods when choosing the layout schemes of a pumping station which is described in details in the work [8].

In order to implement the analytic hierarchy process, we need to build a hierarchy structure for the presentation of the selection problem. For the creation of a decision-making problem structure, decomposition and synthesis are used in the AHP. At the top of the hierarchy (figure 1) there is the main goal, further, one level lower, there are the requirements to the systems, and finally, at the lowest level, there are the variants of layout schemes and design solutions of geokhod systems among which the selection and the arranging are conducted.

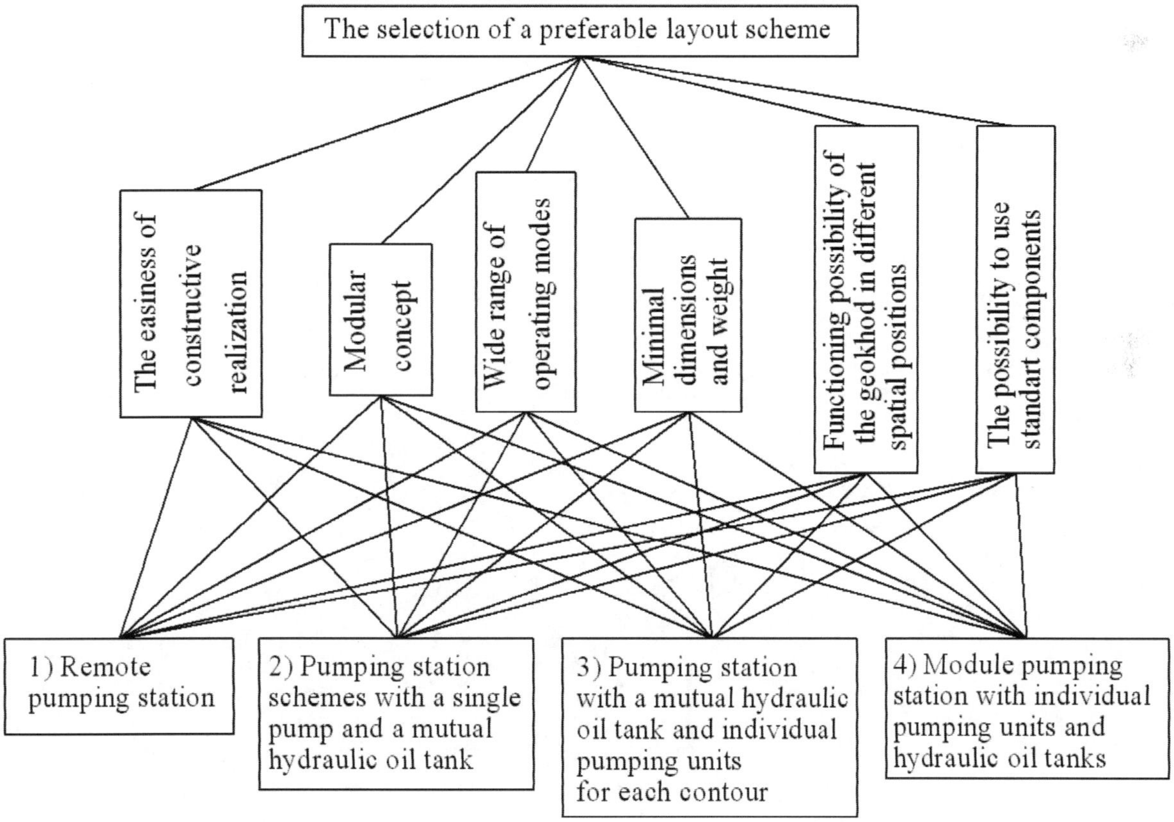

Figure. 1. The hierarchy of the selection problem of appropriate variants of pumping station layout schemes

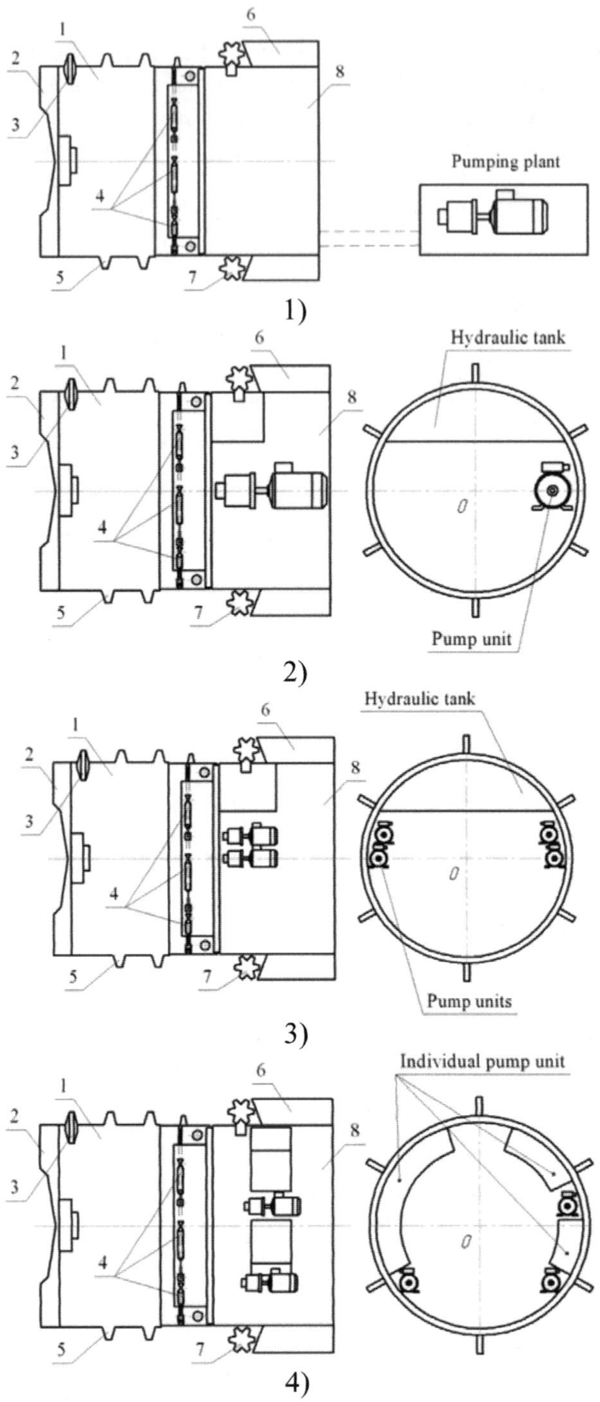

Figure 2. Layout scheme variants (alternative)

1) remote pumping station; 2) pumping station schemes with a single pump and a mutual hydraulic tank; 3) pumping station with a mutual hydraulic tank and individual pump units for each hydraulic contour; 4) module pumping station with individual pump units and hydraulic tanks

1 – head section; 2 – actuating element; 3 – actuating element of the propelling device; 4 – hydraulic cylinder of revolution; 5 – helical blade (propelling device); 6 – counterrotation element; 7 – actuating element of the counterrotation element; 8 – tail section

IOP Conf. Series: Materials Science and Engineering **91** (2015) 012086 doi:10.1088/1757-899X/91/1/012086

Out of all the diversity of layout schemes of pumping stations, we can choose the most applicable ones for certain conditions. As an example, the schemes of horizontal drilling have been picked out (figure. 2).

Within AHP, the hierarchy is the primary way to present the decision framework. Primary function of the hierarchy in AHP is the assessment of higher hierarchy levels on the basis of the cooperation of its different levels.

The next stage is a paired comparison of requirements to the pumping stations, the assessment and creation of the requirements domination matrix. Necessary calculations are performed on the basis of domination matrixes. When comparing, the requirements are evaluated according to the fundamental estimative scale. The types of marks and the principles of evaluation are described in table 1.

Table 1. Fundamental scale of analytic hierarchy process

Degree of preference	Article I. Definition	**Comments**
1	Equal importance.	The two objects equally contribute to the achieving of the goal.
3	Low importance.	The experience and the thinking easily give preference to one of the two objects.
5	High or strong importance.	The experience and the thinking give strong preference to one of the two objects.
7	Very strong and obvious importance.	The preference to one of the two objects is very strong. Its dominance is almost obvious.
9	Absolute importance.	The testimonial to one of the two objects are extremely convincing.
2,4,6,8	Intermediate values between adjoining values of the scale.	Situations demanding compromise decisions.
Reciprocal variables of the above given values	In case we deduce one of the above given values when comparing object **A** with object **B**, the result of the comparison of object **B** with object **A** will be the reciprocal variable.	Given that x is more preferable than y fivefold, i. e. $x=5y$, then $y = x/5$ or $y = (1/5)x$

Table 2 shows the domination matrix for the hierarchy in figure 1. The table lists the results of the comparison of the second level objects with the main goal. There is also the preference evaluated of the given in the line requirement in comparison with the factor in the column.

In case the criterion is not dominating, an inverse value is used and entered into the table.

In order to make the results achieved by means of AHP adequate with the situation, in which the decision is made, it is necessary that required levels of data consistency are achieved in the table of paired comparison. By the consistency of paired comparison matrix the following are meant: numeric (cardinal) consistency and transitive (ordinal) consistency.

Let us consider an example of cardinal inconsistency. May layout scheme A be better than layout scheme B twofold, and layout scheme B be better than layout scheme C threefold; thus, layout scheme A is better than layout scheme C 2x3=6fold. The violation of this equation within the frames of the chosen scale is considered cardinal inconsistency.

IOP Conf. Series: Materials Science and Engineering **91** (2015) 012086 doi:10.1088/1757-899X/91/1/012086

Let us consider an example of transitive inconsistency. May scheme A be more preferable than scheme B, and scheme B be more preferable than scheme C; thus, scheme A is more preferable than scheme C. The violation of the latter in equation is called transitive inconsistency.

For the purpose of determining the consistency, following indicators are identified:

- eigen value of the criteria matrix $\lambda_{\max}$ (EV);
- consistency index CI.

$$CI = \frac{\lambda_{\max} - n}{n-1},$$

where n is the order of matrix;

- consistency relation CR. The consistency relation must not exceed 0,2 (20 %). The consistency relation is determined by means of the comparison of the CI with table values of a random consistency index for mismatched matrixes.

3. Results and Discussion

Paired comparison of the requirements is given in table 2. Within the last column of the table which is called “Priority”, a relative evaluation of the requirements is depicted which is determined by means of ranging.

In the column “Priority” of table 2 a relative estimation of each criterion is depicted.

The most important advantage of the AHP is the consideration of the inequivalence of the requirements to geokhod systems, in particular, to power plants, and thus more specific results. Besides, when changing the number of requirements or exchanging one requirement with another, it is necessary only to compare new pairs and to delete the unnecessary ones from the lines and columns of the matrix of domination. Moreover, the analytic hierarchy process includes an instrument of “quality” evaluation of expert's work.

In order to decrease the complexity of mathematical calculations and to automate the application of the method, it is rational to apply the program system that realizes AHP.

In the following step, paired comparisons of layout schemes had been performed in relation to each criterion. For each criterion, a separate table had been created similar to table 2. Every layout scheme was given an esteem according to the fundamental scale of AHP.

Table 2. Paired comparisons of the criteria of the geokhod pumping stations for horizontal drilling

Criterion	The easiness of constructive realization	Modular concept	Wide range of operating modes	Minimal dimensions and weight	Functioning possibility in different spatial positions	Application of standard components	**Priority**
The easiness of constructive realization	1	1/9	1/5	1/5	9	1	0.0589
Modular concept	9	1	1/9	1/3	9	5	0.1581
Wide range of operating modes	5	9	1	3	9	9	0.4745
Minimal dimensions and weight	5	3	1/3	1	9	3	0.2281
Functioning possibility of the geokhod in different spatial positions	1/9	1/9	1/9	1/9	1	1/9	0.0161
Application of standard components	1	1/5	1/9	1/3	9	1	0.0641
Eigen value of the comparison matrix 7.3229 Consistency index 0.1949 Consistency relation 0.1572							

As a result of the application of AHP for the selection of layout schemes of a geokhod pumping station, it was deduced that the preference should be given to geokhods for horizontal drilling with remote pumping stations which provide power fluid supply in different spatial positions, as well as with integrated modular pumping stations with separate or mutual hydraulic tanks (figure 3).

Variants 3) and 4) cede to other layout arrangements because they do not guarantee the dimensions of the interior space. Besides, the application of one pump unit for the simultaneous feeding of hydraulic cylinders and hydraulic motors is restricted.

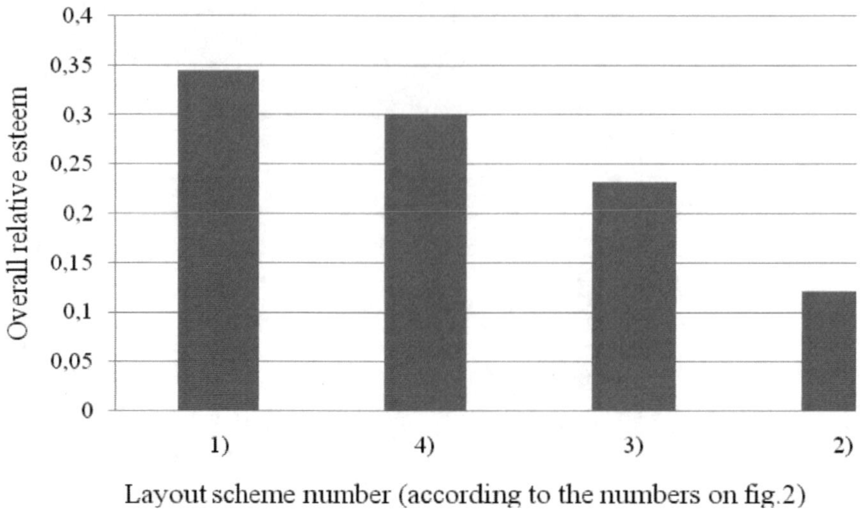

Figure 3. Conclusive result of the AHP based on layout arrangements.

Thus, the application of the AHP for the selection of an optimal decision about layout schemes of geokhod pumping stations includes following steps:

1. The construction of the corresponding hierarchy of a decision-making task.
2. Paired comparison of all requirements to the pumping stations.
3. The elimination of the inconsistency of the domination matrixes of requirements (if necessary).
4. Paired comparison of all layout schemes according to all requirements.
5. The elimination of the inconsistency of domination matrixes of layouts according to the requirements (if necessary).
6. The presentation of a diagram with the variants of layout schemes and the selection of an optimal decision.

It should be noted that, when estimating the preference of one or another layout arrangement, exploitation conditions of the geokhod are an important factor. This means that final results of the AHP about the layout schemes under different conditions (for instance, the usage of the geokhod for slope working, or the application of a small diameter geokhod) can vary.

References

[1] Efremenkov, A.B., 2011. Forming the subterranean space by means of a new tool (geokhod), in: 7th International Forum on Strategic Technology (IFOST – 2012): Proceedings: In 2 Vol. IEEE, p. 348-350.

[2] Aksenov, V.V., Khoreshok, A.A., Beglyakov, V.Y. Justification of Creation of an External Propulsor for Multipurpose Shield-Type Heading Machine – GEOKHOD. Applied Mechanics and Materials (2013) 379, p. 20-23.

[3] Timofeev, V.Y., Aksenov, V.V., Galajamova, I.J. Determination of parameters of roller of wave transmission with intermediate rolling bodies with hollow shaft for geokhod. Applied Mechanics and Materials (2014) 682, p. 246-250.

[4] Aksenov V.V., Blashchuk M. Y., Dubrovsky M. V. Estimation of torque variation of geohod transmission with hydraulic drive. Applied Mechanics and Materials. (2013) 379. p. 11-15

[5] Efremenkov, A.B., Aksenov, V.V., Blashchuk, M.Y.. Force parameters of geohod transmission with hydraulic drive in various movement phases, in: 7th International Forum on Strategic Technology (IFOST - 2012): Proceedings: In 2 Vol. (2012) IEEE, p. 159-163.
[6] Aksenov, V.V., Walter, A.V., Beglyakov, V.Y., 2014. Ensuring the geometric accuracy of shell during assembly of Geohod sections. Obrabotka metallov (Metal working and material science) 4, p.19-28. (2014)
[7] What is the Analytic Hierarchy Process? Mathematical Models for Decision Support NATO ASI Series Volume 48, (1988), p. 109-121
[8] Blashchuk, M.Y., Kazantsev, A.A., Chernukhin, R.V. Capacity Calculation of Hydraulic Motors in Geokhod Systems for Justification of Energy-Power Block Parameters. Applied Mechanics and Materials (2014) 682, p. 418-425.

IOP Conf. Series: Materials Science and Engineering **91** (2015) 012087 doi:10.1088/1757-899X/91/1/012087

Radial deformations of working cylinder of hydraulic Legs depending on their extension

G D Buyalich [1,2,3,a], **K G Buyalich** [1,b], **V V Voyevodin** [1,c]

[1] T.F. Gorbachev Kuzbass State Technical University, Vesennaya Str., 28, Kemerovo, 650000, Russia
[2] Yurga Institute of Technology of Tomsk Polytechnic University, Leningradskaya Str., 26, Yurga, 652055, Russia
[3] Institute of Coal of SB RAS, Leningradskiy Ave., 10, Kemerovo, 650060, Russia

e-mail: [a]gdb@kuzstu.ru, [b]konstantin42@mail.ru, [c]vvoevodin@yandex.ru

Abstract. Current methods of calculation of parameters of hydraulic legs of powered supports are in most cases analytical and do not consider all complex of factors. Finite element model was developed to study this problem and used to analyze the influence of hydraulic legs extension on radial deformations of cylinder of different producers of powered supports at variation of hydraulic fluid pressure. It was revealed that radial deformations of cylinders along the axis of hydraulic legs increase in magnitude in direct proportion to the hydraulic fluid pressure and extension. Research results can be recommended to define optimal geometric parameters of hydraulic legs in respect to the minimal radial deformations of hydraulic cylinder increasing its impermeability and improving the work of cup seals. It is recommended to use the obtained results at power support designing.

Introduction

Hydraulic leg is a hydraulic power cylinder. Its impermeability is determined by the size of the gap between the piston and the cylinder bridged by the seal. Numerical value of the gap includes two components:

1) tolerance for production of piston and cylinder (constant component), which determines minimal Z_{min} and maximal Z_{max} gaps during assemblage;

2) radial deformations of interior surface of cylinder (dR) under the hydraulic liquid pressure (variable component), which depend on the value of this pressure (P), production technique [1–4], support construction [5], hydraulic leg construction [6–8] and their hydraulic extension (l_p).

Methods

To calculate radial deformations finite element method was used as it is one of modern computational approaches allowing to carry out calculations with high accuracy and speed for constructions with complicated configuration using computer aids.

Work Description.

Finite element model was developed for research of the influence of hydraulic leg parameters on their strained state.

Recent researches of radial deformations of the working cylinder [6, 7] show that the curve represented in figure 1 has an expanding section from the side of the head end and a convergent section from the side of the rod end. To pit-point locations of radial deformations along the length of the cylinder during comparative calculations the following key points were used:

point 1 – has maximal radial deformations near the piston (at this point radial deformations are equal to dR_1);

point 2 – is situated in the ares of stable radial deformations in the cylinder main body (dR_2);

point 3 – has negative radial deformations (compression of the cylinder) in the limits of the piston location (dR_3). At this point the contact between the cylinder and the piston and the damage of the working cylinder surface are possible at large strain;

point 4 – is situated in the place of the first seal from the side of the head end. Radial deformations at this point (dR_4) are determined by the sealed gap which influences the work of seal and impermeability of hydraulic leg;

point 5 – is situated in the point of maximal radial deformations near the bottom of hydraulic cylinder (dR_5).

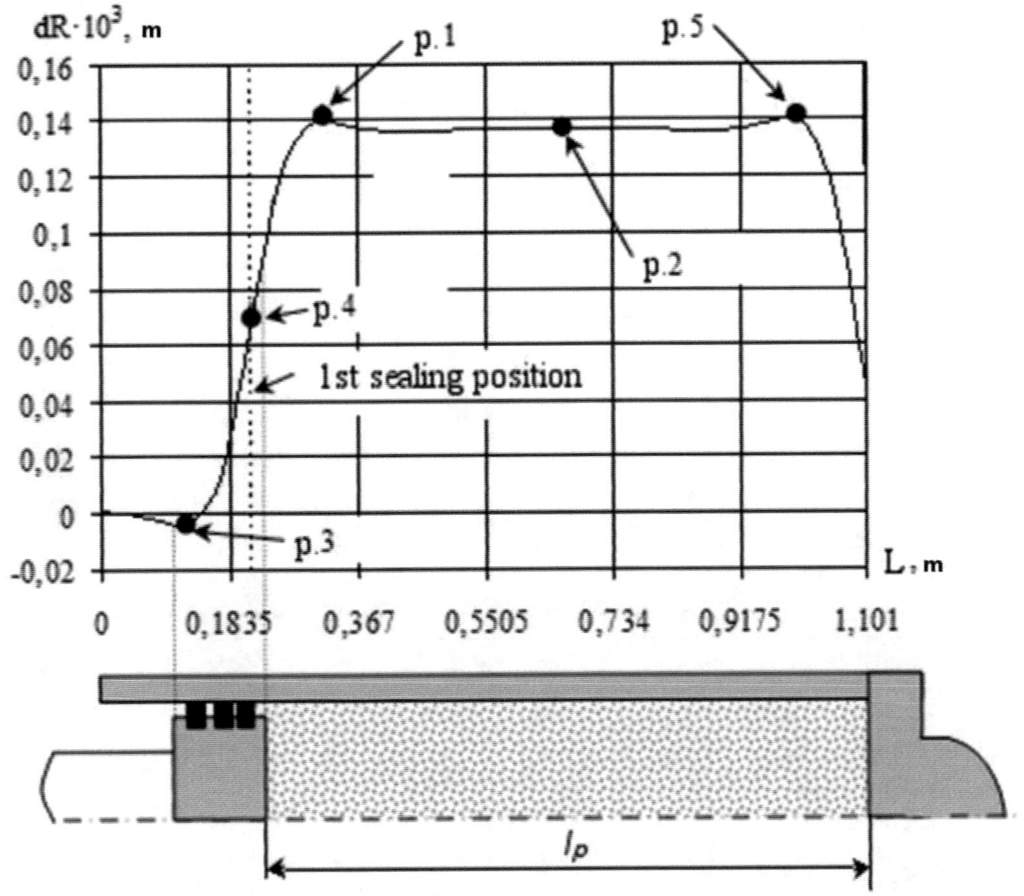

Figure 1. Radial Deformations of the Working Cylinder of Hydraulic Leg (dR) Along the Length of the Working Cylinder (l_c) and Location of Key Points

Radial deformations of working cylinders of hydraulic legs of powered supports at extensions equal to $l_p=(2/3)l_p^{max}$ and $l_p=l_p^{max}$ (fig. 2 and 3) were calculated using the model. It follows from this that extension does not influence the character of deformation curves especially in the piston area. At that, variation of radial deformations in the location of the first seal from the side of the head end (dR_4) owing to the pressure of hydraulic liquid in the head end (P) has linear character at different values of extension (fig.4).

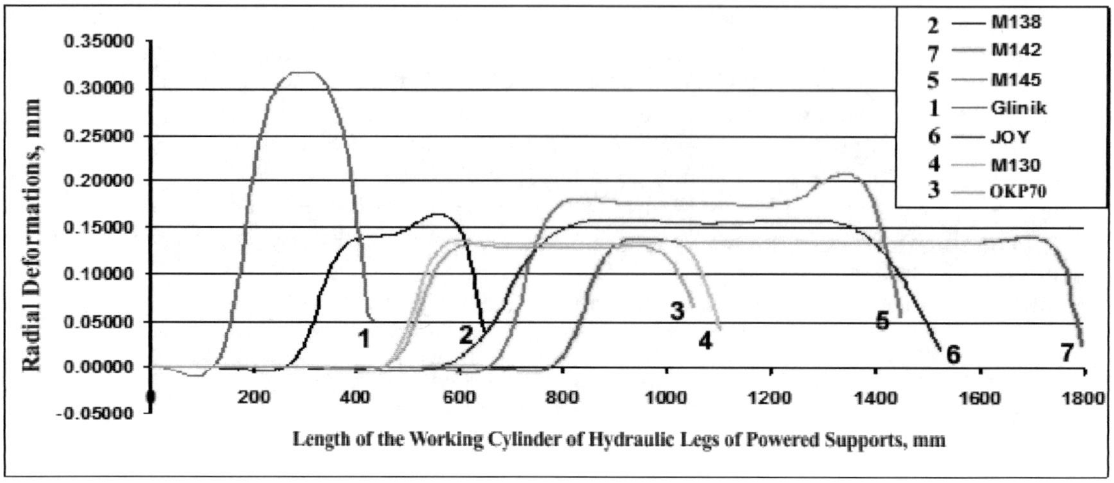

Figure 2. Radial Deformations Along the Length of the Working Cylinder of Hydraulic Legs of Powered Supports at Extension $l_p=(2/3)l_p^{max}$ and $P=50\ MPa$

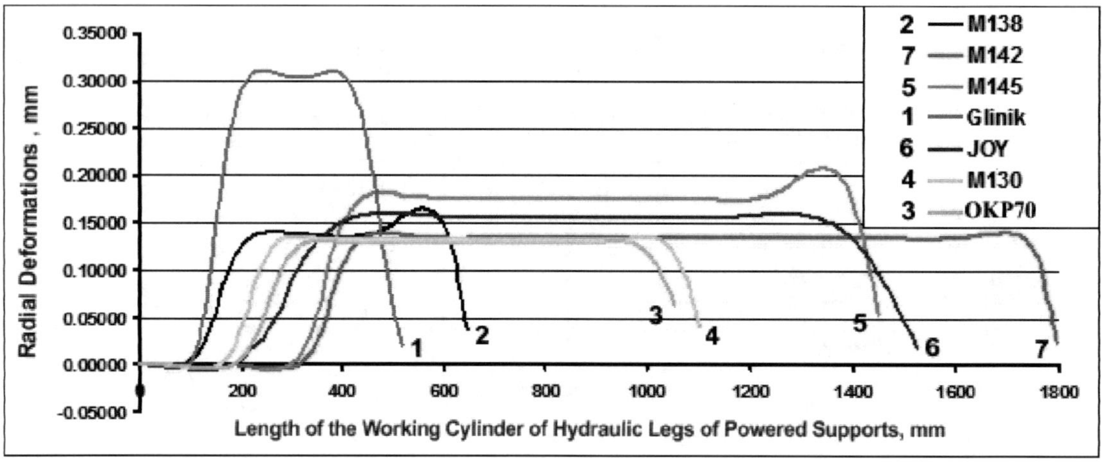

Figure 3. Radial Deformations Along the Length of the Working Cylinder of Hydraulic Legs of Powered Supports at Extension $l_p=l_p^{max}$ and $P=50\ MPa$

Numerical values of radial deformations for hydraulic legs at extension $l_p=(2/3)l_p^{max}$ and $l_p=l_p^{max}$ and pressure P are given in Table 1. Hence it follows that the leg of support M142 has minimal absolute values at extension $l_p=(2/3)l_p^{max}$ and pressure *32MPa*, *50 MPa* and *70 MPa*. At maximal extension l_p^{max} we observe another situation: hydraulic leg Joy has minimal deformations dR_4 at pressure of initial thrust P=32MPa whereas hydraulic leg OKP70 has it at pressure close to nominal (P=50 MPa) and sesquialteral (P=70 MPa). With increase of extension we observe the increase of radial deformations dR_4. It can be explained by the rise of total load on the cylinder wall from hydraulic liquid pressure. As a result of this phenomenon absolute values of constringent radial deformations in the piston area (dR_3) also increase by absolute value.

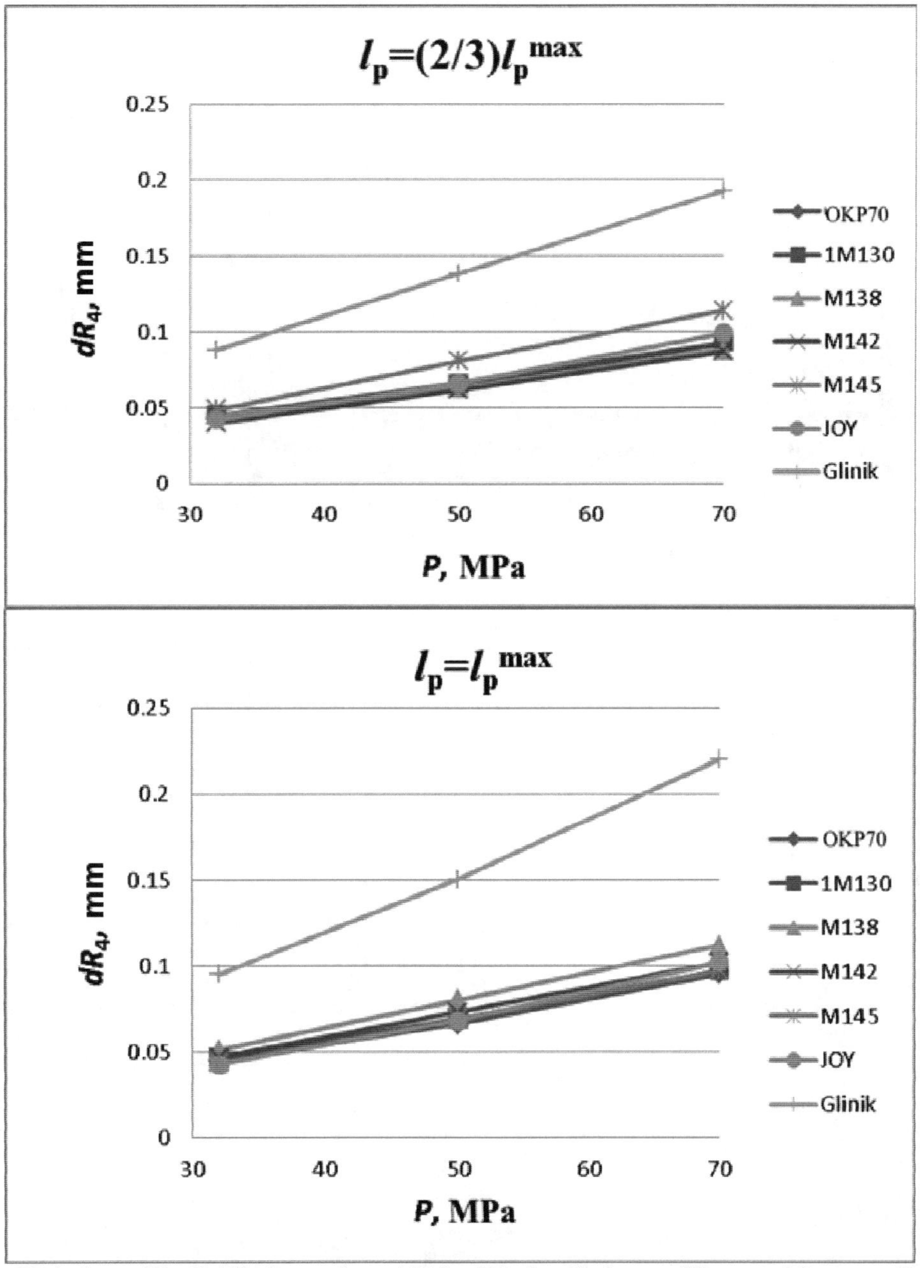

Figure 4. Variation of Radial Deformations (dR_4) of Hydraulic Legs of Powered Supports Due to the Hydraulic Liquid Pressure (P) at Extensions (l_p)

Numerical values of radial deformations of hydraulic leg M138 at the point situated in the limits of piston area (dR_3) are two times less than these of the rest (table 2), as this leg has minimal length of piston (75mm) and maximal value of constringent deformations is beyond its limits. Thus, for example, for this leg deformations in the piston area dR_3= -0,0000214 mm at $P=50$ *MPa* and $l_p=(2/3)l_p^{max}$ whereas minimum value of these deformations is -0,00410 mm. The length of piston of the rest of legs is 90 mm (table 2) and minimum values of radial deformations dR_3 are situated in piston area.

IOP Conf. Series: Materials Science and Engineering **91** (2015) 012087 doi:10.1088/1757-899X/91/1/012087

Table 1.

Numerical Values of Radial Deformations in the Seal Area (at point 4) at Extension l_p and hydraulic liquid pressure P

Powered Support	Radial Deformations (dR_4), mm					
	P=32 MPa		P=50 MPa		P=70 MPa	
	$l_p=(2/3)l_p^{max}$	$l_p=l_p^{max}$	$l_p=(2/3)l_p^{max}$	$l_p=l_p^{max}$	$l_p=(2/3)l_p^{max}$	$l_p=l_p^{max}$
ОКП70	0,042	0,045	0,064	0,066	0,092	0,095
M130	0,045	0,047	0,066	0,069	0,093	0,097
M138	0,045	0,051	0,063	0,080	0,088	0,112
M142	0,040	0,047	0,062	0,073	0,087	0,102
M145	0,049	0,044	0,081	0,069	0,114	0,097
JOY	0,042	0,043	0,066	0,068	0,099	0,102
Glinik	0,088	0,095	0,138	0,150	0,193	0,220

Table 2.

Numerical Values of Parameters and Radial Deformations of Hydraulic Legs of Powered Supports at P=50 MPa

Параметр	OKP70	1M130	M138	M142	M145	Joy	Glinik
Diameter of the cylinder, mm	220	200	220	220	250	350	280
Ehickness of the cylinder wall, mm	26,5	22,5	27,5	27	24,5	65	20
Length of the piston, mm	95	95	75	92	108	296	90
Z_{min}, mm	0,05	0,05	0,05	0,05	0,05	0,062	0,056
Z_{max}, mm	0,28	0,28	0,28	0,28	0,28	0,342	0,316
$l_p=(2/3)l_p^{max}$							
dR_4, mm	0,064	0,066	0,063	0,062	0,081	0,066	0,138
dR_3, mm	- 0,00262	-0,00286	-0,0000214	-0,00233	-0,00510	-0,00440	-0,00643
$l_p=l_p^{max}$							
dR_4, mm	0,066	0,069	0,08	0,073	0,069	0,068	0,15
dR_3, mm	- 0,00385	-0,00402	-0,00224	-0,00242	-0,0053	-0,00548	-0,00961

Variation of radial deformations (dR_3 and dR_4) at extension alteration from $l_p=(2/3)l_p^{max}$ to $l_p=l_p^{max}$ (change of relative extension l_p/l_p^{max} from 2/3 to 1) is shown in Figure 5.

The leg M138 is exposed to maximal variations of constringent radial deformations dR_3 due to small piston length. The change of dR_3 by 40–50 % corresponds to legs OKP70, 1M130 and Glinik. Variations of hydraulic leg Joy make up 25 % and these of legs M142 and M145 – near 4 %.

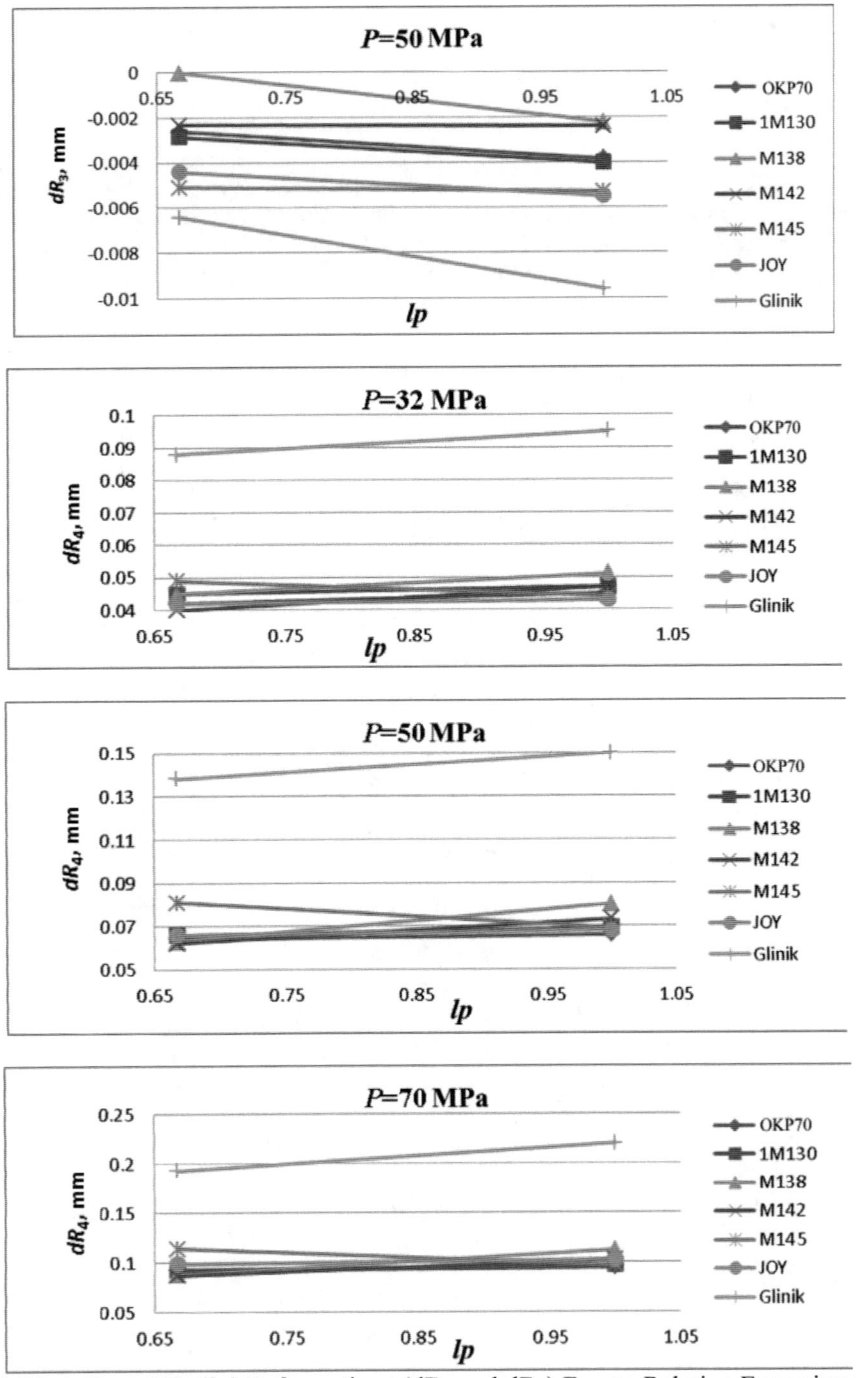

Figure 5. Variation of Radial Deformations (dR_3 and dR_4) Due to Relative Extension (l_p/l_p^{max})

Radial deformations in seal area (dR_4) at $l_p = l_p^{max}$ increase: for hydraulic leg M138 – at initial thrust pressure (32 MPa) by 13 %, approximately, and at high pressure (50 and 70 MPa) – by 27 %, approximately; for hydraulic leg M142- by 17,5 % over all pressure range; for the rest of hydraulic legs over working range of pressure – from 2,5 to 8 %.

For hydraulic leg M145 radial deformations (dR_4) at $l_p = l_p^{max}$ decrease by 10–15 % at different hydraulic liquid pressure.

Conclusions. Realized researches allow to conclude that with increase of hydraulic leg extension the gap between the piston and the cylinder increases in the area of the first seal from the side of the head end ($+dR_4$) and decreases from the side of the rod end in the limits of the piston (dR_3).

References

[1] Chinakhov D A A V Vorobyov A A Davydov and A A Tomchik 2012 Simulation of active shielding gas impact on heat distribution in the weld zone of consumable electrode welding *In the Proceedings of the 7th International Forum on Strategic Technology IFOST2012 Tomsk Polytechnic University* Vol II pp: 136–138

[2] Chinakhov D A 2011 Study of Thermal Cycle and Cooling Rate of Steel 30ХГСА Single-Pass Weld Joints *Applied Mechanics and Materials* Vol 52–54 pp: 442–447

[3] Chinakhov D A 2013 Simulation of Active Shielding Gas Impact on Heat Distribution in the Weld Zone *Applied Mechanics and Materials* Vol 762 pp: 717–721

[4] Kogan B I G D Buyalich and K G Buyalich 2012 Technological support of reliability of cylinders of hydroracks of powered *support Assembling in Mechanical Engineering and Instrument-Making* #10 pp: 29–31

[5] Buyalich G D B A Aleksandrov Yu A Antonov and V V Voyevodin 2000 Increasing the Resistance of Powered Support Brackets *Journal of Mining Science* Vol 36 # 5 pp: 487–492

[6] Burkov P V A V Vorobiev and A V Anuchin 2011 Analysis of Stress Concentrators and Improvement in Designs of Hydraulic Legs Mining *Informational and analytical Bulletin (scientific and technical journal) Separate issue* # 2 pp: 172–183

[7] Klishin V I 1995 Inertial means of protecting hydraulic props from dynamic loads *Journal of Mining Science* Vol 30 # pp: 390–394

[8] Klishin V I T M Tarasik 2001 Stand tests of hygraulic supports with respect to dynamic loads *Journal of Mining Science* Vol 37 # 1 pp: 77–84

Classification of geokhod units and systems based on product cost analysis and estimation for a prototype model production

V V Aksenov[1], A V Walter[2], A A Gordeyev[3] and A V Kosovets[2]

[1] Institute of Coal of the Siberian Branch of the RAS, 10, Leningradskiy Ave., 650065, Kemerovo, Russia
[2] Yurga Institute of Technology, TPU, 26, Leningradskaya Str., 652050, Yurga, Russia
[3] Yurginsky machine engineering plant, 3, Shosseynaya Str., 652050, Yurga, Russia

E-mail: avwalter@rambler.ru

Abstract. The paper considers data systematization on a new shield-type heading machine – geokhod. The target of the paper is to classify geokhod components on the basis of their technical and economical production parameters. A prototype model reveals the structure of a geokhod as an assembly unit and identifies its basic characteristics. The paper overviews the methods of product cost estimations, justifies the application of an operation-based approach for a prototype model, provides the results containing product cost data for various geokhod components and technological processes, and gives the data for a material cost structure. Taking into consideration the product cost analysis, geokhod components are classified according to their technical and economical production parameters. Moreover the paper outlines the ways of a classification application for a geokhod manufacturing techniques improvement.

1. Introduction

A shield-type heading machine – geokhod – is currently being developed to be launched into manufacture [1]. High-volume information concerning the structure of geokhod, its design, technological and economical parameters was obtained in the course of research and development work. Efficient use of the information obtained is reached via its systematization. Information systematization can be based on various principles, that is why relevant objectives of research and development work for a geokhod designing in a short term perspective are worth mentioning.

The main part of work at a current phase is a geokhod prototype model development. Geokhod development and its launching into manufacture is a system of procedures which are aimed at production of competitive products. In any case, this aim is pursued by all researches and developments of new products.

One of the key factors of competitiveness is a product cost. It is known that 70% of costs in a product life cycle is determined by the decisions made at an early stage of a product development [2]. It means that it is necessary to obtain objective data on a significance of an influence of machine components on a cost formation at early stages of development and research work.

As a rule the approaches for constructive solutions in terms of their dependence on a cost formation break in three types [3]: design for manufacturing/assembly, re-design to cost, target costing. Each of the mentioned approaches demands an estimation of future costs in a manufacturing of products at different product life cycles. A prototype model of geokhod can serve as an example for a cost estimation as a complex of geokhod research activities has been conducted and design documents have been prepared.

So for justification of further development of competitive geokhod prototype models it is necessary to classify geokhod components according to prototype model cost estimation and analysis in order to systemize information on a research target and develop a background for improvements of a construction and manufacturing techniques.

A research target is a geokhod prototype model that is going through a test production phase at the moment. Geokhod – a shield-type heading machine that moves in a rock mass using geoenvironment [1]. Geokhod prototype is a large-size machine weighing 19800 kg, consisting of a big amount of components (figure 1). Total amount of component names without standard purchased parts equals 1355. Its diameter is 3200 mm, length (without remote equipment) – 4685 mm. Basic material for geokhod components manufacturing is low-alloy structural steel.

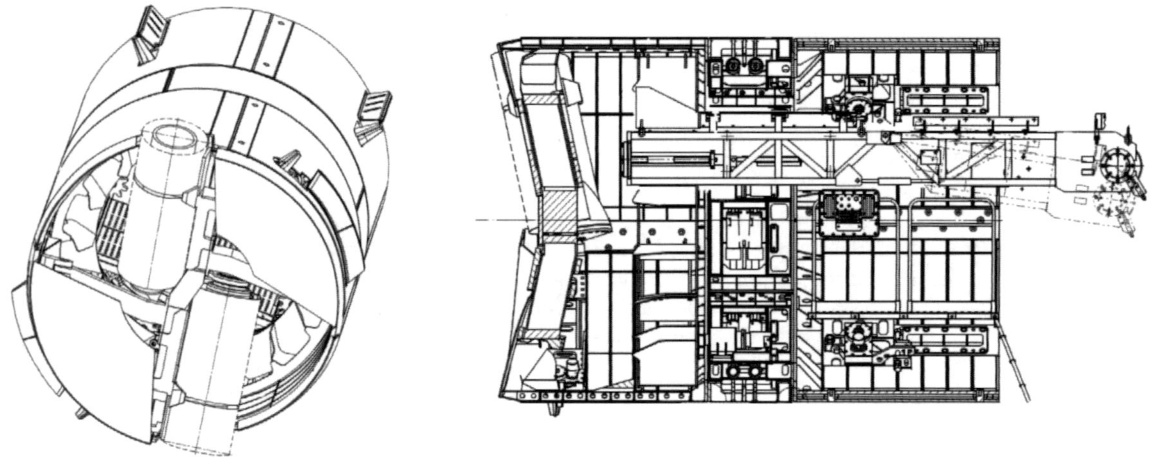

Figure 1. Main view of geokhod prototype model.

Geokhod is an assembly unit and its structure is rather complicated. Structurally the majority of its components are joined into four units: a head unit, a tail unit, an intermediate unit and a transport unit. Geokhod refers to the first level of a product structure tree, the units – to the second level (figure. 2). Lower levels contain assemblies classified according to their functions. Geokhod units consist of ten systems: an operating mechanism, a head section [5], an external propulsor [6], an operating mechanism of an external propulsor, a loading system, a stabilizing section, counterrotation elements, an operating mechanisms of counterrotation elements, an internal body and an external body of an intermediate unit [7]. The systems include assemblies, subassemblies and parts. As a whole the structure of geokhod is complicated, it includes up to eight levels of a product structure tree.

Geokhod construction determines its technological characteristics in many ways:

- the majority of large-size geokhod components are welded assemblies;
- detachable joints are widely used in the construction;
- extended surfaces demand machining when assembling;
- at a pre-assembling phase low hardness of machined components demands deformation control and compensation while machining [8];
- assembling of numerous components is regulated;
- vast majority of welded parts are made of flat steel and pressure-treated in some cases.

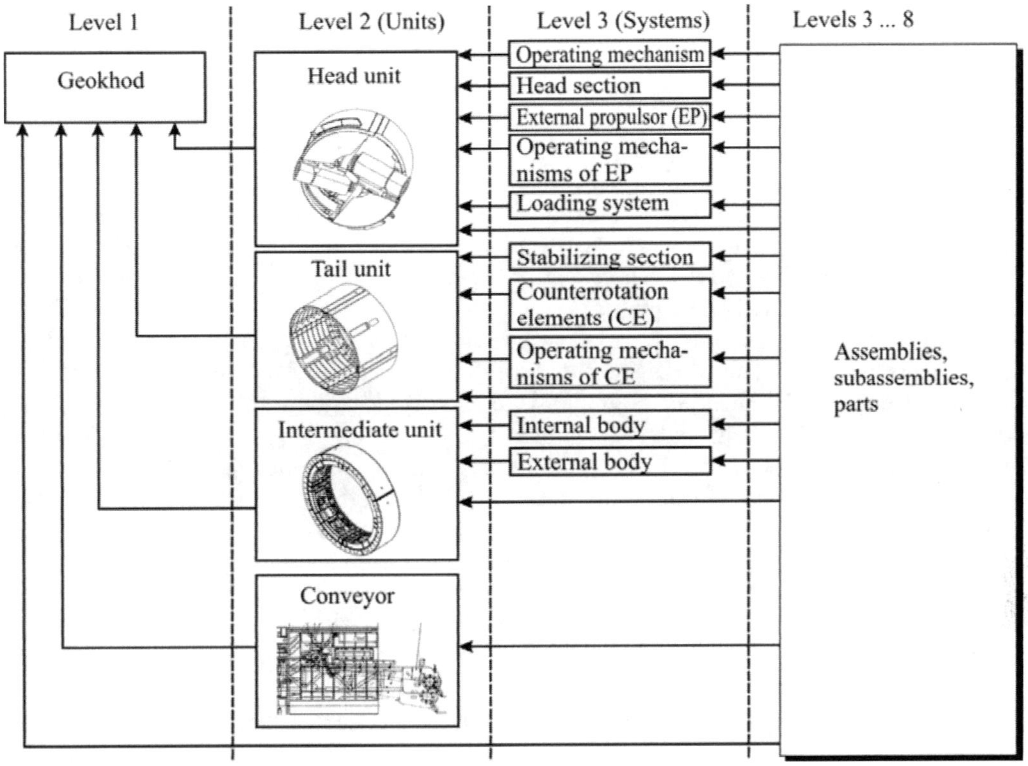

Figure 2. Geokhod structure tree.

That is why assembly techniques, welding, machining and thermal cutting play a significant role in technological processes of construction of geokhod and its components.

2. Product cost estimation technique

There is a great number of various techniques of qualitative and quantitative cost estimation in the process of equipment production. The work [9] overviews and classifies four basic techniques: intuitive technique, analogical technique, parametric technique and analytical technique.

Intuitive and analogical techniques refer to a qualitative product cost estimation which is based on a comparative analysis of a research object and a basic object. A prototype model that is developed and launch into manufacture earlier serves as a basic object. A basic object is similar to a research object in qualitative characteristics. The works [10, 14] demonstrate the application of a qualitative cost estimation when manufacturing a product.

Parametric technique is based on a development of models determining the costs via a function of a number of variables that reveal the characteristics of a prototype model. Model adequacy is estimated using statistical methods of comparing costs with a known cost quantity obtained by an analytical technique or by actual costs in a manufacturing process. Quite often such models are developed on the basis of a regression analysis. The works [4,15,16] give examples of a parametric cost estimation technique application when manufacturing a product.

Characteristics of geokhod prove that the mentioned techniques are not applicable to a prototype model because of the lack of basic objects for carrying out a comparative analysis. Parametric technique is not applicable either due to the lack of cost models for this product type. Development of such models meets the problem of proving their adequacy that demands known cost quantities. That is why at this stage costs are estimated by an analytical technique.

An analytical technique is based on determining of total costs as a sum of fundamental constituents. An analytical technique breaks in five approaches for a product cost estimation [9]: a breakdown approach, tolerance-based cost models, a feature-based cost estimation, an activity-based costing (ABC) system, an operation-based approach.

A break-down approach [17] is based on an estimation of costs as a sum of their constituents at various stages of a product life cycle which break in two categories: relatively well-structured costs (RWSC) and relatively ill-structured costs (RISC). A break-down approach demands the data on cost resources for all stages of a life cycle while a geokhod prototype model production is devoid of the contents of the following stages: delivery, exploitation, reclamation. Also this approach demands a number of parametric values obtained by collecting and processing data of a production process [17,18]. So in a non-existing industry the costs of a geokhod prototype model cannot be estimated without a number of unfounded assumptions.

Tolerance-based cost models are intended to identify relations between the accuracy of product construction units and costs. The works [19,20] present the examples of such models. Tolerance-based cost models are rather efficient at a designing stage as they contribute to establishing the accuracy demands [9]. A thorough result analysis obtained via the accuracy cost models is incorrect as a number of costs influencing factors are not taken into account by these models.

Cost estimation based on construction units involves methods which classify and identify construction units of products, and reveal appropriate cost values. The examples of the methods are found in the works [21–23]. This approach is widely used in the areas of design automation. The work [9] notes that an estimation based on construction units is not objective enough for such complex machines as a geokhod and the vast majority of its components. Moreover an unsolved problem of a systematization of geokhod components hinders the use of this approach when estimating the production costs.

Estimation systems based on actions simulate relations between products and resources. These systems appear to be the tools for making strategic decisions when manufacturing products [24]. As a rule, ABC systems are more accurate in estimating the costs than classical analytical methods [25]. The use of ABC system for a cost estimation of a geokhod prototype model is highly problematic due to heavy costs for an application and maintenance of this system [26]. Cost validity is rather controversial in this case because the manufacture of machines that are similar to a prototype model is still questionable.

Operation-based approach is a traditional technique of PCE [27], it is based on an estimation of time, materials and factory expenses when manufacturing a product [28]. Operation-based approach is a time-taking approach which demands process planning data [9]. In this paper estimation of costs for a geokhod prototype model manufacturing is done with an operation-based approach taking into account that other approaches are inapplicable in a current situation.

The cost estimation is done by the following set of works:

- Studies of design documents:
    - drawing a scheme of a geokhod structure;
    - analysis of an entrance of geokhod components;
    - making a list of purchased parts;
    - data verification of design documents.
- Development of technological processes for geokhod assemblies and components:
    - development of manufacturing routs for each product;
    - designing of operations;
    - choice of techniques.
- Development of material requirements:
    - defining a type and composition of a work piece;
    - functionality of an excess material and cutting charts of sheet materials;
    - estimating costs for basic and auxiliary materials.
- Labour requirements for technological processes.
- Material cost estimation (raw materials, purchased parts, unfinished parts of an own production).

- Cost calculations.

Product cost calculation is done by a recursive equation which is a an extended version of a cost formula mentioned in [28]. The formula calculates a cost value for a of product the level l:

$$C_l = \sum_{k=1}^{m} q_{l+1,k} C_{l+1,k} + C_M + (1 + k_{SI})(1 + k_E) C_S + C_E + C_F \quad (1)$$

m – a number of product names for level $l+1$; $q_{l+1,k}$ – a number of products k for level $l+1$; $C_{l+1,k}$ – costs for a product k for level $l+1$; C_M – material costs for level l; k_{SI} – social costs; k_E – supplementary benefits; C_S – salary for level l; C_E – energy and fuel costs for technological purposes for level l; C_F – factory expenses for level l.

Material costs are calculated by the formula:

$$C_M = C_{SM} + C_{CP} + C_{SF} + C_T - R \quad (2)$$

C_{SM} – raw material costs; C_{CP} – purchased parts; C_{SF} – unfinished parts of an own production; C_T – transport costs; R – waste processing.

Manufacturing activities break in eight types: blank production, fitting, thermal treatment, coating, electrical installations, etc. There is a certain hourly rate, energy costs and factory costs for each activity type. Cost constituents are calculated by the following formulas:

$$C_S = \sum_{i=1}^{n} R_{Oi} t_{\Sigma i}$$
$$C_E = \sum_{i=1}^{n} R_{Ei} t_{\Sigma i} \quad (3)$$
$$C_F = \sum_{i=1}^{n} k_{Fi} R_{Oi} t_{\Sigma i}$$

n – a number of activities; R_{Oi} – an hourly rate for an activity i; $t_{\Sigma i}$ – labour characteristics for an activity i for level l; R_{Ei} – energy costs for an activity i; k_{Fi} – factory expenses for an activity i.

Recursive calculations of a formula (1) going from the value $l = 1$ estimate costs for all geokhod components.

3. Results and discussions

Figure 3 presents a bar chart of total costs for geokhod units and systems. The values in the bar chart are given as a percentage of a total geokhod cost. The cost structure shows that an operating mechanism takes a maximum of total costs. Large-size components – a head section and an intermediate section – take considerable part of costs as well. A geokhod head section consists of numerous systems and mechanisms that is why it turns out to be rather costly.

Costs for a component weight unit serve as estimation for a construction complexity of machine components. Figure 4 shows a percentage of a total geokhod cost for a kilogram of a geokhod unit or system. It is worth mentioning that two groups of systems considerably differ in costs. The first group includes such mechanisms as an operating mechanism, an operating mechanism of an external propulsor, an operating mechanism of counterrotation and a transport unit. These components are characterized by high cost values – (9.00...12.33) $\times 10^{-3}$ [%/kg]. The second group includes body components: a head section, an external propulsor, a stabilizing section, a counterrotation element, an internal body and an external body of an intermediate unit. These components are characterized by relatively low cost values – (2.16...3.11) $\times 10^{-3}$ [%/kg]. The second group may also include a conveyor (cost values – 3.08×10^{-3} [%/kg]) though it is equipped with a motor drive.

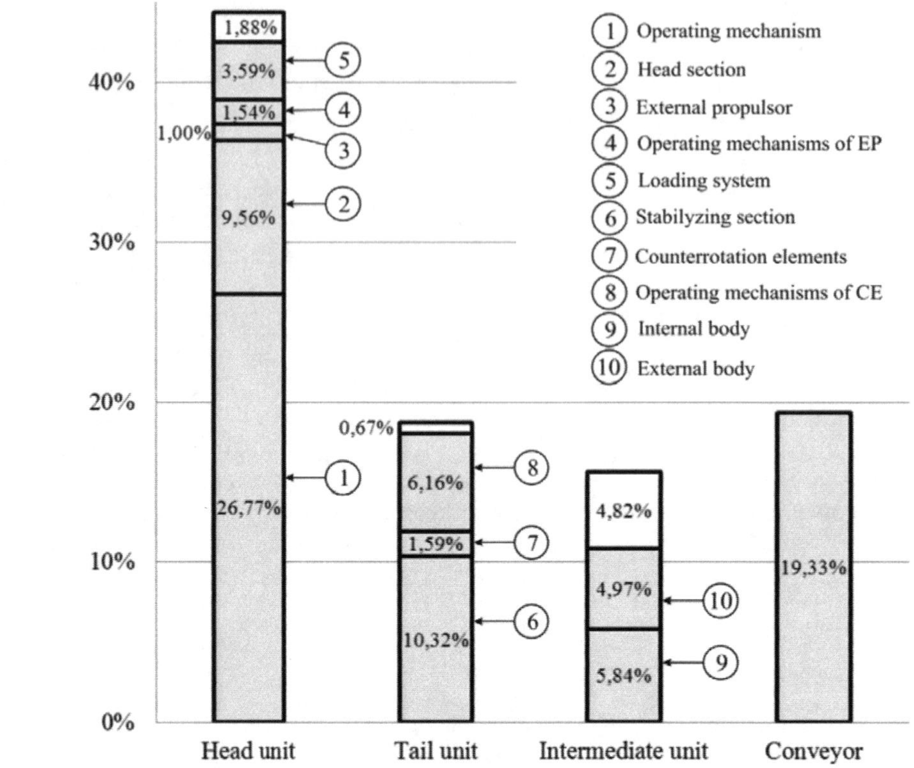

Figure 3. Cost structure for geokhod units and systems.

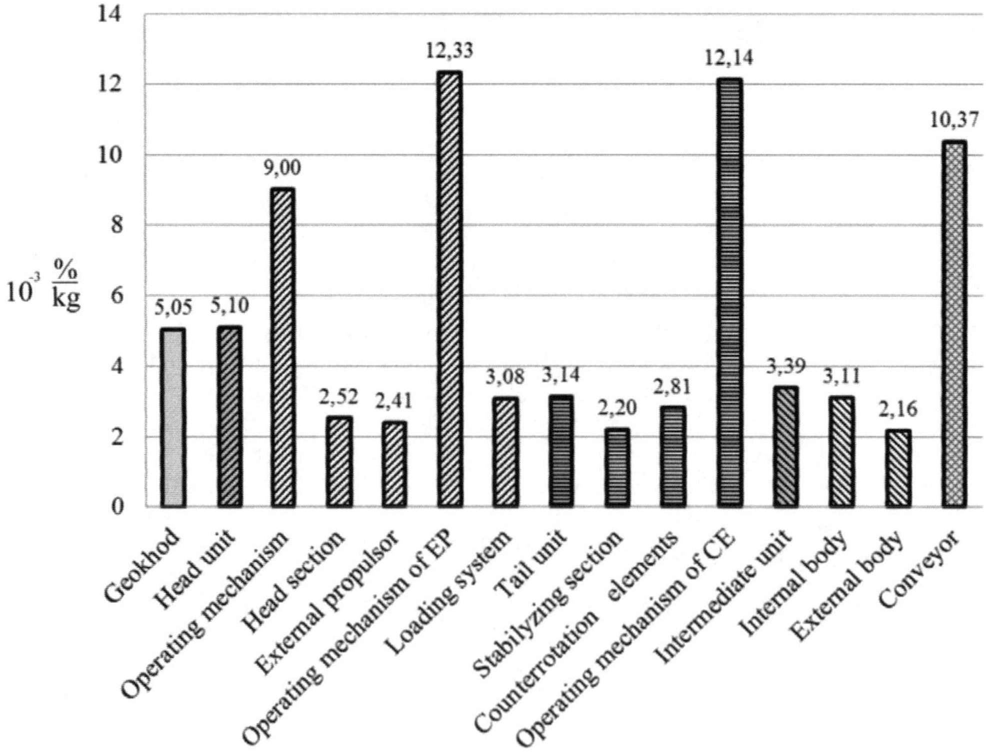

Figure 4. Costs for geokhod units and systems.

Figure 5 presents a bar chart of labour characteristics for technological processes of geokhod units and systems. The bar chart shows that there are no explicit relations between labour intensity characteristics of geokhod components and total product costs. Labour intensity characteristics (for a weight unit of a component) serve as estimation for a work complexity of machine components.

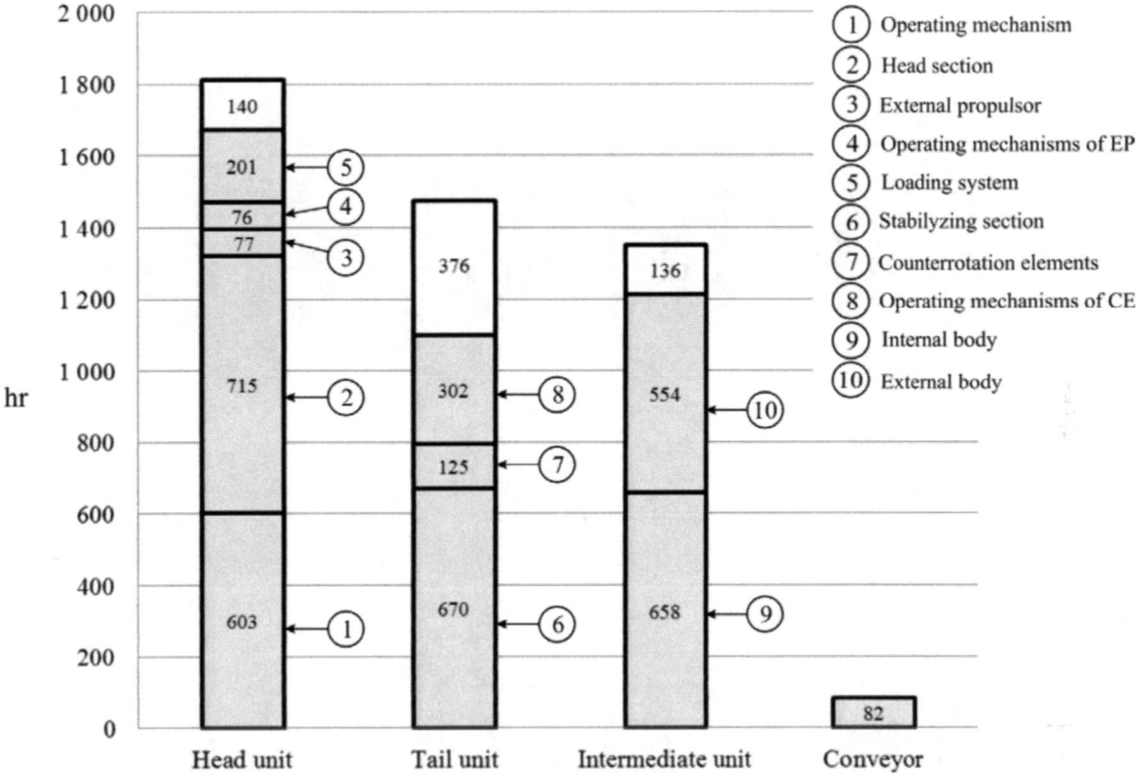

Figure 5. Labour intensity structure for geokhod units and systems.

Figure 6 shows labour intensity values for a kilogram of unit weight or system weight. The bar chart analysis reveals two systems with high values of labour intensity – an operating mechanism of an external propulsor and an operating mechanism of a counterrotation element. At the same time other components – an operating mechanism, a conveyor and a loading system – have moderate values of labour intensity. In the majority of cases the mechanisms of these components are applied in purchased products.

Analyzing the structure of labour intensity for different activity types (figure 7), it is worth saying that technological processes of machining take the biggest values of labour intensity. In average labour intensity of machining takes 57%, components equipped with mechanisms – 62…78% (an operating mechanism, an operating mechanism of an external propulsor, an operating mechanism of a counterrotation element, a transport unit and a conveyor) and intermediate units with high accuracy demands. The mentioned components also demand assembly and welding works (15…26%). Welding of units of an average level of accuracy – a head unit, an external propulsor, a stabilizing section and a counterrotation element – takes 19…29%.

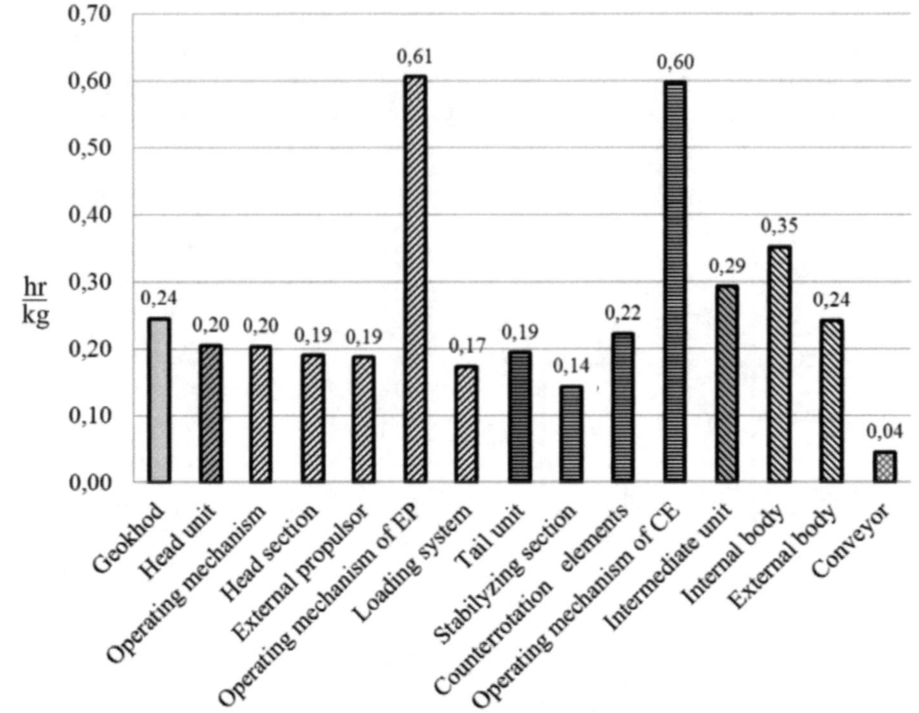

Figure 6. Specific labour intensity for geokhod units and systems.

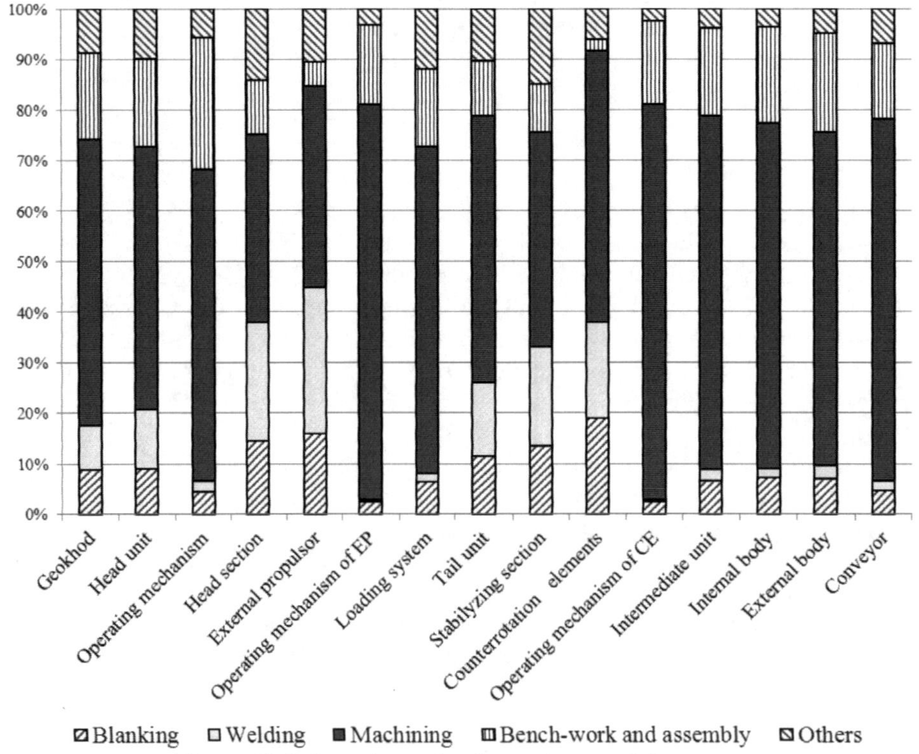

Figure 7. Labour intensity for various activity types

Material costs take an essential part of a production cost (58%). An operating mechanism, a conveyor, an operating mechanism of an external propulsor and a counterrotation element are the most costly (figure 8). Cost structure analysis (figure 9) shows high costs of purchased components mentioned above (85…96%).

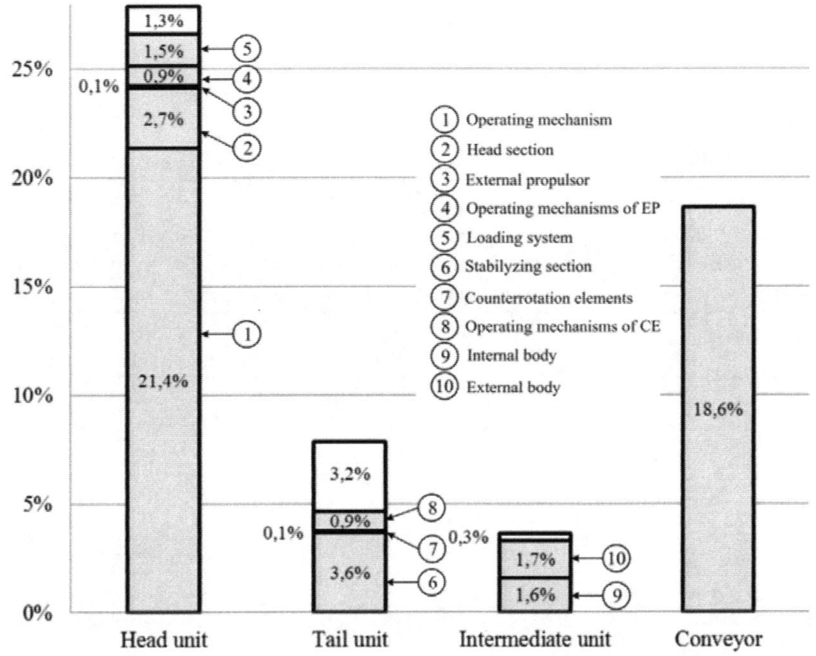

Figure 8. Material costs for geokhod units and systems (a percentage of a total geokhod cost)

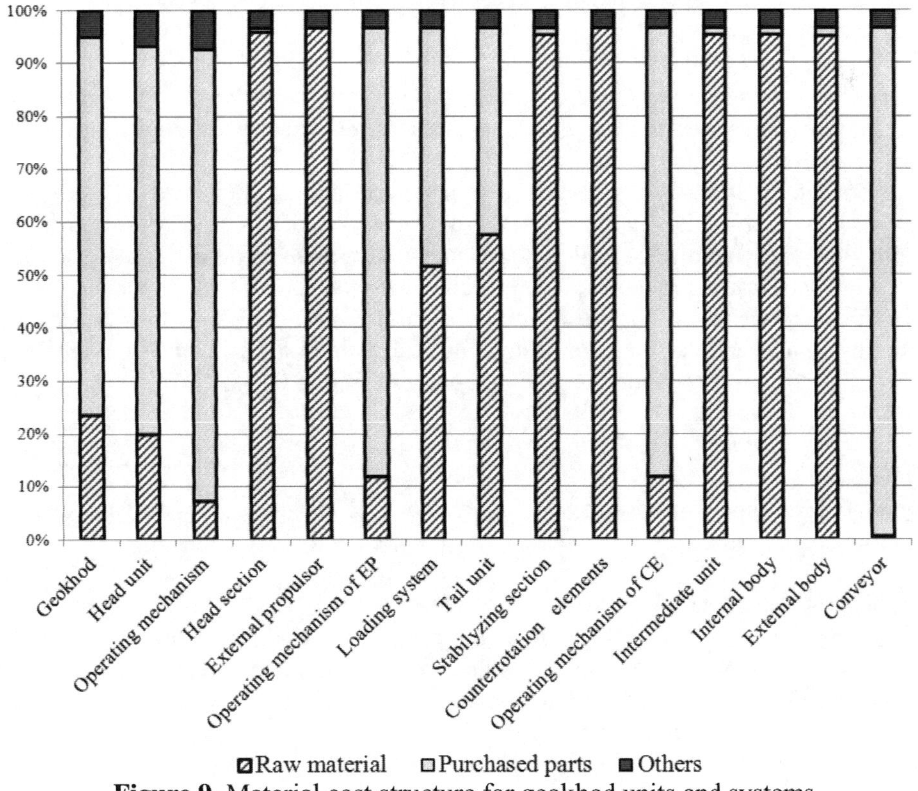

Figure 9. Material cost structure for geokhod units and systems

IOP Conf. Series: Materials Science and Engineering **91** (2015) 012088 doi:10.1088/1757-899X/91/1/012088

Cost structure reveals a number of classification parameters for geokhod components; the categories of parameters characterize efficiency, construction and production complexity of components. Parameters and categories are given in the Table 1.

Table 1. Classification parameters and categories of geokhod components.

Classification parameter	Category	Criteria of a category affiliation	Examples
Construction complexity	Low	$C_{rel} \leq 4\times10^{-3}$ [%/kg]	Counterrotation element
	Average	$C_{rel} = (4\ldots8)\times10^{-3}$ [%/kg]	Head unit
	High	$C_{rel} > 8\times10^{-3}$ [%/kg]	Operating mechanism
Production complexity	Moderate	$T_{rel} \leq 0.4$ hr/kg	Head unit
	High	$T_{rel} > 0.4$ hr/kg	Operating mechanism of external propulsor
Production type	Other units	All the rest	Tail unit
	Welded units	$r_W{}^a > 13\%$ and $r_M{}^b \leq 40\%$	External propulsor
	Accurate units	$r_M > 60\%$ and $r_W \leq 13\%$	Internal body
Purchased components	Low	$r_{CP}{}^c \leq 30\%$	Stabilizing section
	Moderate	$r_{CP} = (30\ldots70)$ %	Loading system
	High	$r_{CP} = (70\ldots100)$ %	Transport section

[a] Labour intensity of machining in technological processes for a considered component.
[b] Welding labour intensity in technological processes for a considered component.
[c] Material costs for purchased products .

Categories within the same classification group are given in the order of increasing product costs. Thus, classification contributes to design and manufacturing techniques improvement in terms of product profitability.

Classification details and various categories are determined by the level of geokhod structure studies. Current classification is worked to the third level (level of systems). Studies of higher levels are connected with the difficulties caused by an exponential growth of product quantity. Classification revision of a detailed structure studies is reasonable for solution of the problems connected with development of design techniques of efficient constructions and methods. Also classification revision can be caused by significant changes in a design and materials of a geokhod (for example, [29]), and an essential increase of new production technologies (for example, [30]).

4. Conclusion

- Geokhod prototype consists of a large number of products varying in constructive and technological parameters, related by complicated hierarchical entrance relations. Representative geokhod structure is created by units and systems – the interrelated components of the second and third design level.
- The product cost assessment for a geokhod prototype including its components is done by an operation-based approach. Obtained values of economic parameters of geokhod and its components serve as basic parameters for a high-quality comparison of various machine designs and manufacturing technologies.

- The structure and the type of product costs of geokhod units and systems show design and technological characteristics of products and form a basis for the formation of quantitative criteria of a product classification.
- The classification of geokhod components is developed on the basis of product cost assessment parameters. The classification can be used for a design and manufacturing techniques improvement of geokhod units and systems.
- Homogeneity of product cost parameters for the same classification group contribute to the development of cost assessment techniques, efficiency of geokhod design and manufacturing techniques based on the developed classification and parameter-based approach or an analogical technique.

References

[1] Efremenkov A B 2011 Forming the subterranean space by means of a new tool (geohod) *6th International Forum on Strategic Technology (IFOST - 2011)* **1** 348–350

[2] Asiedu Y and Gu P 1998 Product life cycle cost analysis: State of the art review *International Journal of Production Research* **36 4** 883–908

[3] Huang G Q 1996 *Design for X Concurrent engineering imperatives* (Netherlands, Dordrecht: Springer)

[4] Cavalieri S, Maccarrone P and Pinto R 2004 Parametric vs. neural network models for the estimation of production costs: A case study in the automotive industry *International Journal of Production Economics* **91 2** 165–177

[5] Aksenov V V, Walter A V and Beglyakov V Y 2014 Ensuring the geometric accuracy of shell during assembly of Geohod sections *Obrabotka metallov (Metal working and material science)* **4 65** 19–28

[6] Aksenov V V, Khoreshok A A and Beglyakov V Y 2013 Justification of creation of an external propulsor for multipurpose shield-type heading machine – GEO-WALKER *Applied Mechanics and Materials* **379** 20–23

[7] Efremenkov A B, Aksenov V V and Blashchuk M Y 2012 Force parameters of geohod transmission with hydraulic drive in various movement phases *7th International Forum on Strategic Technology (IFOST - 2012)* **2** 159–163

[8] Putilova U S, Nekrasov Y I and Lasukov A A 2014 Loading of the Manufacturing Systems Elements in the Process of Unsteady Mode Cutting and the Models of their Arrangement Deviations *Applied Mechanics and Materials* **682** 192–195

[9] Niazi A, Dai J S, Balabani S and Seneviratne L 2006 Product Cost Estimation: Technique Classification and Methodology Review *Journal of Manufacturing Science and Engineering* **128 2** 563–575

[10] Waring C W 1991 Product costing automation: The impact of the learning curve *Computers & Industrial Engineering* **21 1–4** 313–317

[11] Shehab E and Abdalla H Manufacturing cost modelling for concurrent product development *Robotics and Computer-Integrated Manufacturing* **17 4** 341–353

[12] Rehman S and Guenov M D 1998 A methodology for modelling manufacturing costs at conceptual design *Computers & Industrial Engineering* **35 3–4** 623–626

[13] Kingsman B G and de Souza A A 1997 A knowledge-based decision support system for cost estimation and pricing decisions in versatile manufacturing companies *International Journal of Production Economics* **53 2** 119–139

[14] Verlinden B, Duflou J R, Collin P and Cattrysse D 2008 Cost estimation for sheet metal parts using multiple regression and artificial neural networks: A case study International *Journal of Production Economics* **111 2** 484–492

[15] Boothroyd G and Reynolds C 1989 Approximate cost estimates for typical turned parts *Journal of Manufacturing Systems* **8 3** 185–193

[16] Dewhurst P and Boothroyd G 1988 Early cost estimating in product design *Journal of Manufacturing Systems* **7** **3** 183–191
[17] Son Y K 1991 A cost estimation model for advanced manufacturing systems *International Journal of Production Research* **29** **3** 441–452
[18] Son Y K and Park C S 1990 Quantifying opportunity costs associated with adding manufacturing flexibility *International Journal of Production Research* **28** **6** 1183–1194
[19] Yeo S H, Ngoi B K A, L. S. Poh and Hang C 1997 Cost-tolerance relationships for non-traditional machining processes *The International Journal of Advanced Manufacturing Technology* **13** **1** 35–41
[20] Sfantsikopoulos M M, Diplaris S C and Papazoglou P N 1995 Concurrent dimensioning for accuracy and cost *The International Journal of Advanced Manufacturing Technology* **10** **4** 263–268
[21] Feng C-X, Kusiak A and Huang C-C 1996 Cost evaluation in design with form features *Computer-Aided Design* **28** **11** 879–885
[22] Xinsheng X, Shuiliang F and Xinjian G 2006 A model for manufacturing cost estimation based on machining feature *International Technology and Innovation Conference 2006 (ITIC 2006)* 273–278
[23] Zhang Y F, Fuh J Y H, Chan W T 1996 Feature-based cost estimation for packaging products using neural networks *Computers in Industry* **32** **1** 95–113
[24] Özbayrak M, Akgün M and Türker A K 2004 Activity-based cost estimation in a push/pull advanced manufacturing system *International Journal of Production Economics* **87** **1** 49–65
[25] Kee R and Schmidt C 2000 A comparative analysis of utilizing activity-based costing and the theory of constraints for making product-mix decisions *International Journal of Production Economics* **63** **1** 1–17
[26] Chea A 2011 Activity-Based Costing System in the Service Sector: A Strategic Approach for Enhancing Managerial Decision Making and Competitiveness *International Journal of Business and Management* **6** **11** 3–10
[27] Creese R C 1992 *Estimating and costing for the metal manufacturing industries* (New York: M. Dekker)
[28] Jung J-Y 2002 Manufacturing cost estimation for machined parts based on manufacturing features *Journal of Intelligent Manufacturing* **13** **14** 227–238
[29] Alferova E A and Lychagin D V 2013 Characterization of Deformation Pattern Structure Elements Generated in Uniaxial Compression of Nickel Single Crystals *Applied Mechanics and Materials* **379** 66–70
[30] Kovalevskaya Z G, Klimenov V A and Zaitsev K V 2014 Interfacial Adhesion between Thermal Spray Coating and Substrate Achieved by Ultrasonic Finishing *Applied Mechanics and Materials* **682** 459–463

IOP Conf. Series: Materials Science and Engineering **91** (2015) 012089 doi:10.1088/1757-899X/91/1/012089

Spectral analysis of granular material reaction to long-term weak dynamic effect

A A Kazantsev[1], V P Kosykh[2], A F Revuzhenko[2]

[1] Yugra Institute of Technology, TPU affiliate, Yurga, RU
[2] Institute of Mining of Siberian branch of RAS, Novosibirsk, RU

E-mail: kazantsev@tpu.ru, v-kosykh@yandex.ru, revuzhenko@yandex.ru

Abstract. The paper demonstrates the similarity of alternating reaction of rocks to explosions to the response of a granular material to repetitive (up to 330 thousands) light shocks ($3.85 \cdot 10^{-3}$ J, frequency 1.33 Hz). The cause of both reactions is multiple forms of equilibrium of geomaterials. It has been stated that pressure of the sensor in a granular material subject to long-term weak effects doesn't relax to the hydrostatic one, but oscillates about it with the amplitude about its value. The spectral density of this process obeys the power dependence on the frequency; the exponent is typical for black noise.

Introduction

Papers [1, 2] describe the phenomenon of alternating reaction of rock mass to explosion effects. The point of it is that «the process of a natural geophysical material deformation is a heterogeneous one: the deformations localize on the surfaces and in zones with weakening strength, and the structural elements can move independently. Here, the spatial motion of the elements can be alternating» [1]. The change in density of various zones of rock mass is also alternating. For instance, the number of explosions causes a relative rock compaction on particular areas on the rockhole, whereas, the following series of explosions results in relative decompaction of these areas [2].

In generally, the nature of this phenomenon is quite clear. It is caused by multiple forms of equilibrium of geophysical materials. The same edge conditions can be in line with unlimited forms of equilibrium. Therefore, generally speaking, the rock mass has rather good memory for all external effects and all its previous states.

For this reason on can expect that the phenomenon of alternating reaction is also possible in conditions of relatively weak but long-term and systematical effects on rock mass. As the source of these effects one can consider the same explosions made far from the area under consideration, tidal deformations etc.

In [1, 2] explosion effect on the rock mass was considered with the force up to 150 kt (in trinitrotoluene equivalent). Power explosions are unique ones and full-scale research is quite labor intensive even for conventional explosions, and, moreover, not complete. It is interesting to search for models which could be fully investigated in laboratory conditions. It is well known that in most cases the behavior of a rock mass can be modeled with granular materials. This phenomenon can also be used for the issue under consideration in this paper. Actually, mechanical behavior of granular materials is determined by two basic factors: a great quantity of interacting particles and the laws of particles interaction. The major ones are the laws of friction, sliding and rolling of particles. The

IOP Conf. Series: Materials Science and Engineering **91** (2015) 012089 doi:10.1088/1757-899X/91/1/012089

conditions of dry friction are brought to two-side inequality of forces arising at the contacts. As the consequence, one and the same package of particles can have practically unlimited forms of equilibrium. Here, the quantity of the packages of the same particles is also practically unlimited. It implies that original sample of material has a significant capacity of memory responsible for history of its formation. In particular, on the basis of information on the method the sample was formed and preliminary deformed we can obtain an extremely wide range of various stress states of the sample.

Thus, the first condition – probable and unlimited forms of equilibrium – is met here. The second condition concerns the mosaic structure of the material. It is also met in a granular material to a certain extent. In-depth research demonstrates that the particles of granular material are clustered when deforming (actually, in blocks) [3]. Therefore, principal deformations occur on the borders of clusters. Here, the deformation of clusters is relatively small. The clusters for particles with plane faces can be seen perfectly. To illustrate one can make an experiment involving the flattering of a bundle of hexahedral pencils coupled outside by elastic [4].

Here let us discuss the external effects on the material, which can cause the evolution of its internal self-balancing stresses. In this paper we merely consider weak but long-term effects on the material. The influence of rockholes and other cavities on the rock mass is not studied. We confine ourselves to the simplest situation: the sample of granular material is in a cylindrical container. The container is hit slightly (up to 800 thousands for 7 days of continuous experiment). Inside the sample there is a pressure sensor, which records the evolution of stress state. The problem statement is undoubtedly simple and rather trivial. However, its result is quite unexpected and important – in the course of evolution changing stresses could amount to 20 – 150% of their initial values. Let us describe the results. Here, we pay attention to one typical experiment only, which lasted 68 hours ($3.3 \cdot 10^5$ hits were made).

Methods of experiment

The model of experimental facility is depicted in Fig. 1. A 220 mm high cylindrical polystyrene container *1* 100 mm in diameter and with 1.5 mm thick wall was topfull of bank sand *2*; dimensions of particles were 0.5 – 0.63 mm. To measure the vertical pressure two similar $12 \times 25 \times 3$ mm sensors *3* were placed into sand. The first one was placed at the depth of 155 mm, the second one – 115 mm. The container was fixed on the inflexible base. An impacting facility *4* was placed next to the container to effect on the wall. The head of the striker was made of rubber. The distance from the bottom of the container to the impact point was 60 mm.

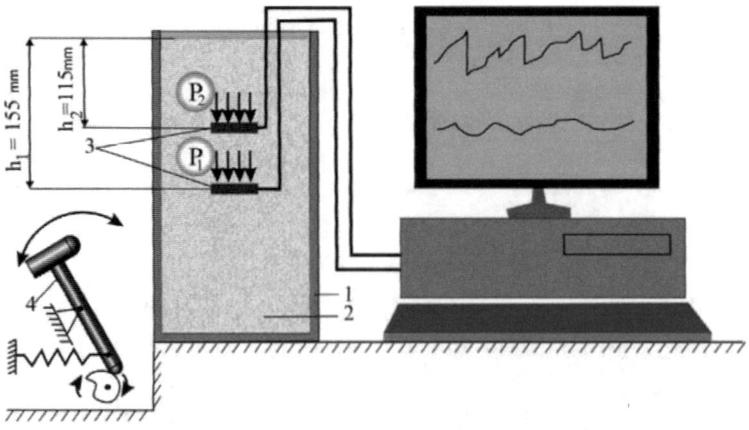

Figure. 1. The model of experimental facility:
1 – cylindrical container; 2 – bank sand; 3 – stress gauges; 4 – impacting facility.

A ready for test experimental facility was placed into the thermostat to exclude the influence of temperature. After the facilities had been warmed up and the indicated values of sensors had been stabilized, an impacting facility was switched on. The energy of impacts E was constant and amounted to $3.85 \cdot 10^{-3}$ J all over the experiment, frequency was - 80 hits a minute. The digitization of signals from the sensors lasted 10 s.

The results of experiment

In the course of experiment the dependencies of pressure on the sensors in time $P_i = P_i(t)$ were recorded on the hard disk of computer, where P_i - the value of the sensor *i* at the moment *t*. According to the obtained data graphs of normalized pressure (Fig. 2) $p_i = P_i / \gamma h_i$, depending on the number of hits *n* were made, where γ, h_i – specific weight and head of the material over the sensor *i*. Horizontal lines depict the pressure γh_i of overlying layers of material.

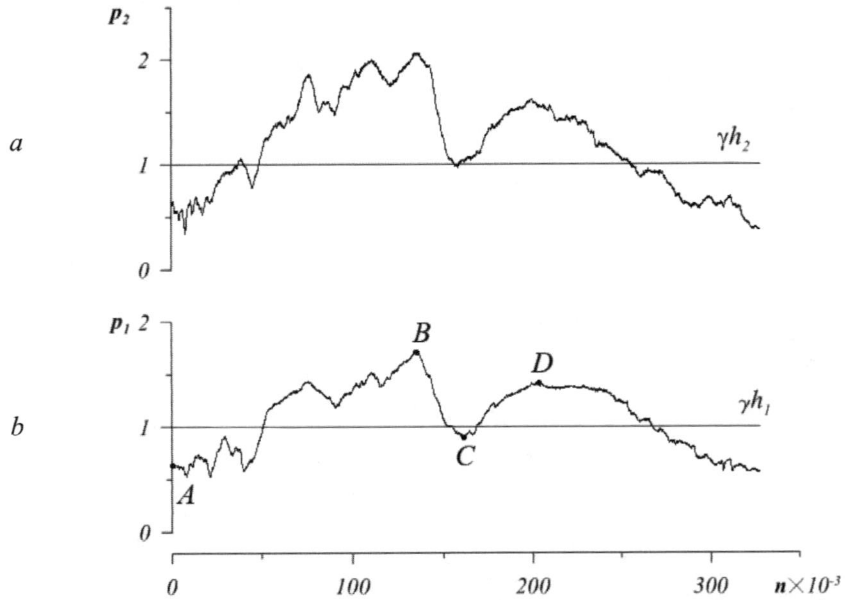

Figure. 2. Graphs of normalized pressure on the sensors: *a* – the second sensor; b – the first one

As we can see, the pressure on the sensors is not constant but varies with the number of hits in a non-monotone way. The pressure increases dropping and surging locally up to 120 thousand of hits (AB segment). It falls sharply on BC segment, then, it grows more smoothly up to the local maximum at D. Then it decreases slightly dropping locally.

For general description of the obtained data series we used the following values: $\bar{p} = \frac{1}{K}\sum_{j=1}^{K} p_j$ - mean pressure on the sensor, *j* – number of reading, p_j - reading *j* on the sensor, K - total readings (series length); $\sigma^2 = \frac{1}{K}\sum_{j=1}^{K}\left(p_j - \bar{p}\right)^2$ - dispersion of the series; $R = p_{\max} - p_{\min}$ - amplitude of the series, $p_{\min}$ и $p_{\max}$ - minimum and maximum pressure recorded by the sensor over the period of experiment. Numerical values of the mentioned parameters are shown in Table 1.

IOP Conf. Series: Materials Science and Engineering **91** (2015) 012089 doi:10.1088/1757-899X/91/1/012089

The graphs of pressure demonstrate slow changing elements (trends) and quicker changes against their background. The trends indicate that the series are unsteady at least over the time of the experiment. Let us expand the series into harmonic components by Fourier transform to remove trends. Then, according to Parseval's theorem [5], the mean-square pressure $\frac{1}{K}\sum_{j=1}^{K}(p_j)^2$ or, in terms of electric engineering, mean power dispersed by the signal p can be expand into the components generated by each harmonic. The mean-square p relative to the mean $\bar{p}$ is the most suitable measure. It simply equals to dispersion or mean power.

$$\sigma^2 = \frac{1}{K}\sum_{j=1}^{K}(p_j - \bar{p})^2 = 2\sum_{j=1}^{K/2-1} A_j^2 + A_{K/2}^2 \quad (1)$$

where A_j - amplitude of *j* harmonic.

Table 1. Major parameters of $P_i = P_i(t)$ series

Series (sensor)	Pressure of the head of material, kPa, γh_i	Mean pressure, $\bar{p}$	Dispersion, σ^2	Minimum pressure, p_{min}	Maximum pressure, p_{max}	Amplitude, R
1	2.26	1.11	0.10	0.58	0.62	1.20
2	1.68	1.21	0.21	0.86	0.86	1.72

Fig. 3 shows calculated according to Fourier transform contributions $2A_j^2$ of low frequencies to the power of signals p and graphs of its accumulation.

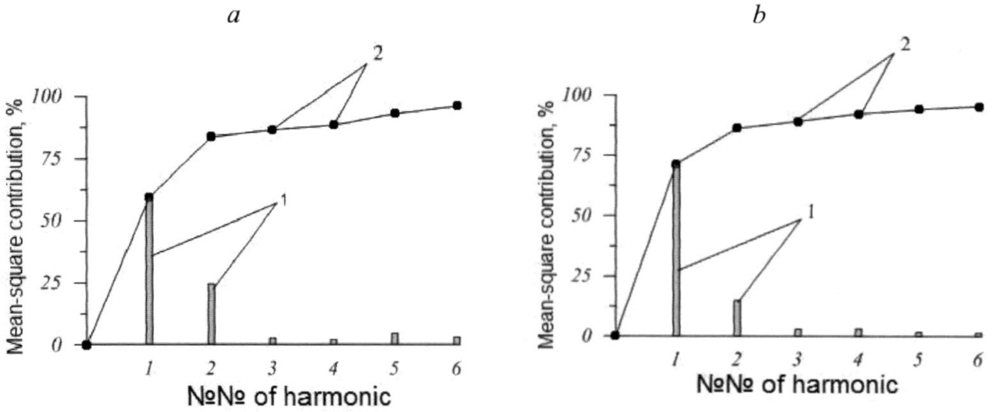

Figure 3. Periodograms (1) and cumulative graphs of mean-square pressure (2):
a – lower sensor; *b* – upper sensor

As we can see the main contribution to the power, which equals to 83.9 and 86.3% for lower and upper sensor, respectively, is made by two first harmonics with periods N_j (j=1, 2 – number of harmonic) and 327.1 and 163.6 thousand hits. After deducting mean $\bar{p}_i$ and two most important harmonics from the series $p_i = p_i(n)$ we obtain stationary (or approximately stationary) series $p_i^* = p_i^*(n)$, depicted in Fig. 4.

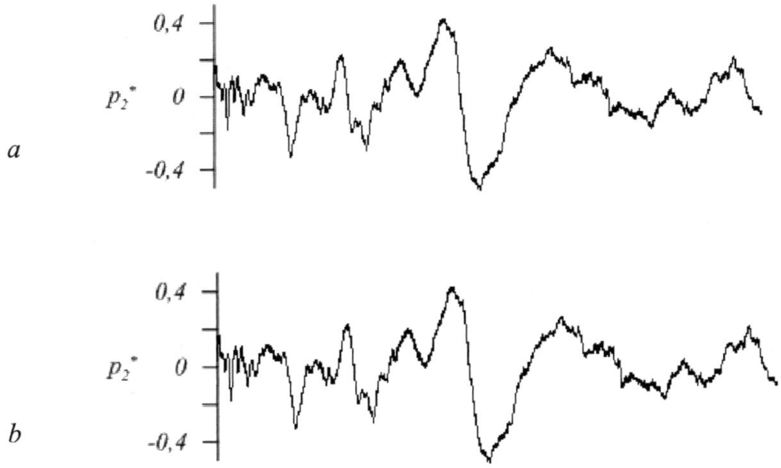

Figure. 4. Graphs of pressure obtained after trend deduction: *a* – the second sensor; b – the first one

Fig. 4 shows that in obtained series p_i^* there are oscillations relatively to zero line. Dispersion and maximum amplitude of deviations p_i^* are provided in Table 2.

Table 2. Characteristics of modified pressure series $p_i^* = p_i^*(n)$

Series (sensor)	Dispersion, σ^2	Minimum pressure, p^*_{min}	Maximum pressure, p^*_{max}	Amplitude, R
1	0.02	-0.33	0.39	0.72
2	0.03	-0.52	0.43	0.95

If we compare Table 1 with Table 2, we can see that dispersion has decreased 5-7 times, and the amplitude – 1.7-2.4 times.

There is the question, whether these series are absolutely random or somehow determined, have inner "memory"? To answer it we make sample autocorrelation functions $r_i = r_i(n)$ of series p_i^* according to equation [6, 7]:

$$r = \frac{1}{K}\sum_{j=1}^{K-u}\frac{(p_j^* - \bar{p})(p_{j+u}^* - \bar{p})}{\sigma^2}, \tag{2}$$

where r - autocorrelation function, u – shift.

The graphs of autocorrelation functions are depicted in Fig. 5, where shifts u measured in thousand of hits are shown on abscissa axis.

Fig. 5 demonstrates that the process under consideration is distinguished by the long-term "memory". Reducing the correlation function up to zero, i.d. it takes the sample 15.5 thousand hits for

IOP Conf. Series: Materials Science and Engineering **91** (2015) 012089 doi:10.1088/1757-899X/91/1/012089

"forgetting" of its current state. Furthermore, a periodic component appears when pressure changing. It means that the series are not absolutely random.

Spectral analysis of temporal series is widely applied in research experience [8]. The distribution of spectral density *S* depending on the frequency is rather informative. According to these distributions one can estimate regular characteristics of the processes. In our experiment this analysis was carried out with mathematical system Math Lab. The obtained results are depicted in Fig. 6. A decimal logarithm of spectral density is shown on ordinate axis, and on abscissa axis there is a frequency logarithm *1/N*, where *N* – a period of oscillation measured in number of hits.

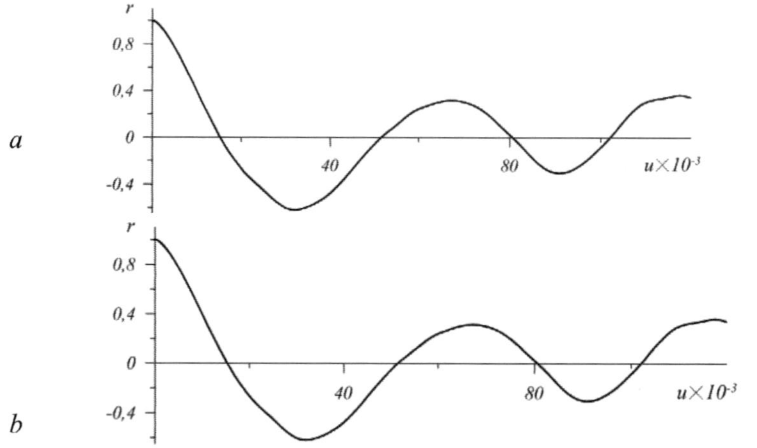

Figure. 5. Autocorrelation functions: *a* – the second sensor; b – the first one

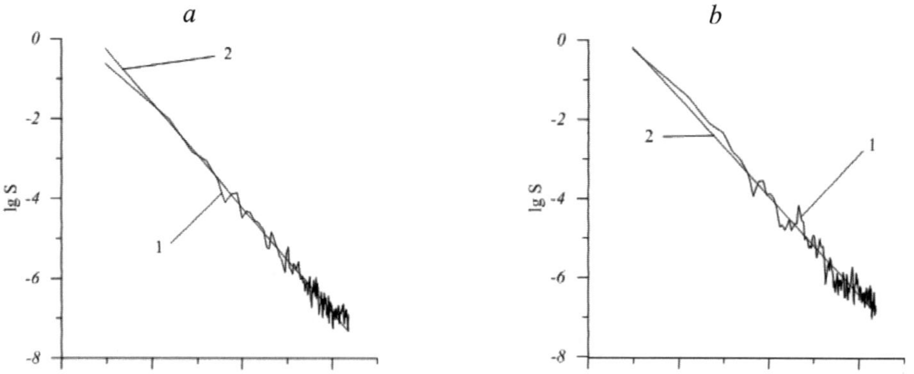

Figure 6. Spectral density of pressure on the lower sensor (*a*) and on the upper one (*b*): 1 – experiment; 2 – approximation

Fig. 6 shows that the logarithmic functions of spectral density in frequency logarithm corresponding with periods $N = 500..320 \cdot 10^3$ hits are well placed on straight lines. That is to say that $S \sim (1/N)^{\alpha}$, i.d. the dependencies are power ones. As far as it is known [9], proportion of spectral density S and $(1/f)^{\alpha}$, where f - frequency, α - exponent, is typical for colored noise. The noises are classified according to α . So, if $\alpha = 2$, the noise is called a brown one and corresponds with the process of Brownian movement. If $\alpha > 2$ the noise is black. In nature we know a lot of processes of black noise. For instance in [10] it is mentioned that «Black noise occurs in long-term cyclical data of observations like river-levels, number of sunspots, thickness of growth rings, and changing prices on stock market. Provided that the system increased over the previous period it will probably grow in the next one». These processes are distinguished by long-term memory; therefore, "long-term correlations of current events and future ones are possible".

In the process under consideration the exponent α is 2.62 for the lower sensor and 2.48 for the upper sensor. Thus, the change in pressure of granular materials caused by long-term weak dynamic effects is the process of black noise.

The sample of granular material consists of a great number of particles, in our case about 10^7. It is quite natural that the state of the system obtained when sample forming is a metastable one. For example, for sand, which had been used in experiments, we obtained initial pressure on the sensor within the range of $(0.4 \div 1.8)\gamma h$ by various formation procedures of packages of particles. Therefore, an initial material isn't in a thermodynamic equilibrium. A somehow energized sample, for example, by light shocks can attain stability in different ways. While a sample is approaching the state of equilibrium dissipative structures can occur [11], including formation and destruction of clusters (zones of compression and decompression). Clusters are unavoidably attend by rearrangement of power chains [12] and repacking of material, the latter is the cause of changing pressure with power range of black noise.

Thus, direct experiments demonstrate alternating reaction of granular material to long-term weak effects. Its comparison with alternating response of rocks to explosion effects [1, 2] brings us to a conclusion that the nature of mentioned phenomena is similar in both cases – unlimited forms of equilibrium are possible in geomaterials.

Conclusions

- The phenomenon of alternating reaction of rocks to explosion effects is similar to the response of granular material to long-term weak effects.
- In both cases the phenomenon is caused by unlimited forms of equilibrium in geomaterials.
- Long-term weak effects result in stress evolution from one state to another. Here, in granular materials stresses can change up to 30 – 150% of initial values.

References

[1] Kurlenya M.V., Oparin V.N., Revuzhenko A.F., Shemyakin E.N. On some peculiarities of rock reaction to near explosions // Proceedings of Academy of Sciences of USSR. – 1987. – V. 293 - T 1. – pp. 67 – 70.

[2] Kurlenya M.V., Adushkin V.V., Garnov V.V., Oparin V.N., Revuzhenko A.F., Spivak A.A. Alternating reaction of rocks to dynamic effect // Proceedings of Academy of Sciences of USSR. – 1993. – V. 323 - № 2. – pp. 263 – 265.

[3] Kosykh V.P. Experimental research into the changing local density of granular material when cyclic deforming // Journal of Mining Sciences. - 2012. –№6. – pp. 56 – 62.

[4] Lavrikov S.V., Revuzhenko A.F. On one experimental model of rocks // Journal of Mining Sciences, 1991, N 4, pp. 24-30

[5] Jenkins G. Watts D. Spectral analysis and its applications. – M.: World, 1971. – Issue 1. – 316 p; 1972. – Issue 2. – 287 p.

[6] Boks J., Jenkins G. Analysis of temporal series, forecasting and control: – M.: World, 1974, Vol. 1. – 406 p.

[7] Loskutov A.Yu., Mikhailov A.S. Fundamentals of the theory of complex systems. – M.-Izhevsk: Institute of computer research. – 2007. – 620 p.

[8] Kanasevich E.R. Analysis of temporal sequences in geo-physics. – M.: Depths. – 1985. – 300 p.

[9] Shreder M. Fractals, chaos, power laws. Miniatures from endless paradise. – Izhevsk: Research

IOP Conf. Series: Materials Science and Engineering **91** (2015) 012089 doi:10.1088/1757-899X/91/1/012089

center «Regular and chaos dynamics». – 2001. – 528 p.

[10] Peters E. Fractal analysis of financial markets: Chaos theory application for investments in economy. M.: Internet-trading. – 2004. – 304p.

[11] Prigozhin I. End of definiteness. Time, chaos and new laws of nature. — Izhevsk: Research center «Regular and chaos dynamics». — 2000. – 208 p.

[12] Drescher A., de Josselin de Jong G. Photoelastic verification of a mechanical model for the flow of a granular material // J. Mech. Phys. Solids. 1972. V. 20, No.5. P. 337-351

IOP Conf. Series: Materials Science and Engineering **91** (2015) 012090 doi:10.1088/1757-899X/91/1/012090

On the theory of plasticity, associated with a new integral characteristic of shearing stresses

A A Kazantsev[1], O A Mikenina[2], A F Revuzhenko[2]

[1] Yugra Institute of Technology, TPU affiliate, Yurga, RU
[2] Institute of Mining of Siberian branch of RAS, Novosibirsk, RU

E-mail: kazantsev@tpu.ru, olgarev@yandex.ru, revuzhenko@yandex.ru

Abstract. A new invariant of stress tensor is introduced – the mean shearing stress, resulted from the integration with respect to Mohr's circle. The invariant is used to lay down the terms of plasticity. Determining equations are written on the basis of associated flow law. Rigid variants of the model and elastoplastic ones are obtained. Characteristic surfaces with normals, coinciding with main stresses direction are demonstrated for the rigid-plastic variant.

Introduction

Plastic deformation of solids has a shearing character and is associated with shearing stresses, existing in the solid. This fact is confirmed by the data of numerous experiments. It is reflected in corresponding theoretical constructs. For instance, the equation of plane deformation of perfectly plastic bodies is of a hyperbolic type. Characteristics of stress field are similar to those of field of velocities – glide lines. For the latter ones Geiringer's ratios are characteristic [1]. These properties of the model confirm the shearing nature of plasticity and are arguments in support of adequacy of classical theory [2-4].

The situation is more complicated in three-dimensional space. In general, the conditions of plasticity are to consider both maximum shearing stress (Tresca criterion), and intermediate stresses, effecting in other places of a voluentary unit. In this respect numerous experimental [5-9] and theoretical [10] investigations have been conducted for various construction materials and for rocks. Nevertheless, the problem has not been solved completely yet. Efforts are being made both experimentally and theoretically [11, 12]. This paper provides the consideration of plasticity theory variant, which is based on a new integral characteristic of shearing stresses, effecting in various places of a voluentary unit of a body to be deformed. A classical integral characteristic of shearing stresses is associated with the second invariant of stress tensor. It appears rather naturally when investigating proper values of stress tensor. The well-known paper [13] has demonstrated that this measure of shearing stresses can be made as the root mean square calculated according to the surface of infinitely small elementary sphere. It is the sphere, which has all symmetrical and equivalent directions of places. One more method of averaging can be noted, and involves all quite equivalent values of shearing stresses τ_n. It is averaging in the plane of Mohr's circle (σ_n, τ_n). Here σ_n, τ_n are normal and shearing stresses on the place with the normal $\overline{n}$. We know, that every possible $\overline{n}$ is in line with the area restricted by three Mohr's semicircumferences. Averaging in regard with mentioned semicircumferences has the following result [14, 15]:

IOP Conf. Series: Materials Science and Engineering **91** (2015) 012090 doi:10.1088/1757-899X/91/1/012090

$$\langle \tau_n \rangle = \frac{1}{2\pi} \frac{(\sigma_1 - \sigma_3)^2 + (\sigma_1 - \sigma_2)^2 + (\sigma_2 - \sigma_3)^2}{\sigma_1 - \sigma_3}. \quad (1)$$

Here $\sigma_3 < \sigma_2 < \sigma_1$ are ranged main stresses. Let us consider the model of plastic deformation, associated with the loading surface or yield surface of the form:

$$f(\sigma_1, \sigma_2, \sigma_3) = \langle \tau_n \rangle - k = 0, \quad (2)$$

where k is either a set constant, or a pre-set hardening function. As it is easy to see, if a point $(\sigma_1^0, \sigma_2^0, \sigma_3^0)$ is an element of the surface (2), so the point $(\sigma_1^0 + C, \sigma_2^0 + C, \sigma_3^0 + C)$ is an element of this surface too, where $C = const$. Therefore, the surface (2) is a cylinder with a straight parallel line $\sigma_1 = \sigma_2 = \sigma_3$. The section of the cylinder by the orthogonal plane is depicted in Fig. 1. In comparison with

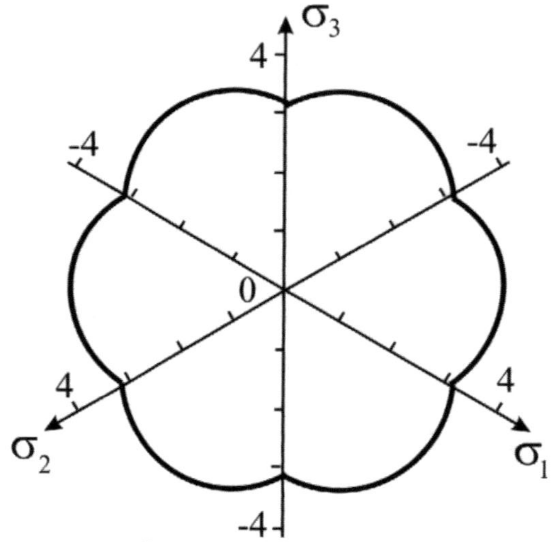

Figure. 1. Loading surface section by deviatoric plane

Tresca prism and Mises cylinder the surface (2) is nonconvex one. Therefore, Mises principle of the maximum is not relevant for it. [11]. However, it can't interfere with development of the theory associated with loading surface (2). The principle of maximum isn't as obligatory as conservation law. Furthermore, it isn't observed by materials with internal friction meeting the unassociated flow law. Moreover, this principle can be reformulated to make solely associated flow law follow from it but not convexity of a loading surface. Let us admit that the rate of energy dissipation in the course of plastic deformation is maximal for actual stress state among all stress states, which are close to the actual one and are in line with the given condition of plasticity. The main point of it is, that we don't consider all states meeting the plasticity condition, but those ones, which are close to the actual stress state. In this context we can say that the introduced principle is a local one as if compared to Mises principle.

Plastic model

Let us consider the equation of deformation. Let a particular deflected mode be obtained at a definite point of time t. This mode can be thought as the known one. Let us set up an equation of stress, deformation and shearing increment. Let us suppose that the body is a work-hardening one. It means that parameter k can change in the condition (2) in the course of deformation. The conditions of dynamic loading and un-loading are as follows:

$$\Delta f(\sigma_1, \sigma_2, \sigma_3) > 0 \text{ – loading, } \Delta f(\sigma_1, \sigma_2, \sigma_3) < 0 \text{ – un-loading.} \quad (3)$$

IOP Conf. Series: Materials Science and Engineering **91** (2015) 012090 doi:10.1088/1757-899X/91/1/012090

As usual $\Delta f = 0$ coincides with neutral loading. When un-loading, there is no change in plastic deformation. When dynamic loading we have an associated flow law according to (3) and the introduced above reformulated principle of maximum. In terms of this law for work-hardening body we have

$$\Delta\varepsilon_1^p = \frac{\partial f}{\partial\sigma_1}\Delta\lambda = \left[\frac{1}{\pi}\frac{2\sigma_1-\sigma_2-\sigma_3}{\sigma_1-\sigma_3} - \frac{k}{\sigma_1-\sigma_3}\right]\Delta\lambda,$$
$$\Delta\varepsilon_2^p = \frac{\partial f}{\partial\sigma_2}\Delta\lambda = \frac{2\sigma_2-\sigma_1-\sigma_3}{\sigma_1-\sigma_3}\Delta\lambda, \quad (4)$$
$$\Delta\varepsilon_3^p = \frac{\partial f}{\partial\sigma_3}\Delta\lambda = \left[\frac{1}{\pi}\frac{2\sigma_3-\sigma_1-\sigma_2}{\sigma_1-\sigma_3} + \frac{k}{\sigma_1-\sigma_3}\right]\Delta\lambda,$$

where $\Delta\lambda$ is a parameter, its equations are given consideration to later, $\Delta\varepsilon_i^p$ – increments of main plastic deformations, $i = 1,2,3$.

As it can be seen from (4), the material is practically incompressible:

$$\Delta\varepsilon_1^p + \Delta\varepsilon_2^p + \Delta\varepsilon_3^p = 0.$$

In terms of associated flow law plastic incompressibility is caused by the absence of internal friction in material, i.d. plasticity condition doesn't depend on the first invariant of stress tensor. A special case, when deformation comes practically to the plane one, is of high relevance for applications and the theory.

Let us admit for the case of plane deformation $\Delta\varepsilon_2^p = 0$. Then, according to (4) we have $\sigma_2 = (\sigma_1 + \sigma_3)/2$. Plasticity condition (2) transforms into Tresca condition:

$$\langle\tau_n\rangle = \frac{3}{2\pi}\frac{\sigma_1-\sigma_3}{2} = k.$$

Therefore, the theory under consideration in a particular case transforms into classical one [1-4]. Thus, the consistency requirement, which is usually specified for a new theory is met in this case. Let us address the issue of similarity of stress tensor to the plastic deformation rate. The definition states that tensor similarity is determined by comparison of Lode-Nadai coefficients for various tensors. It can be seen from (4):

$$\mu_{\dot\varepsilon} = \frac{2\Delta\varepsilon_2^p - \left(\Delta\varepsilon_1^p + \Delta\varepsilon_2^p\right)}{\Delta\varepsilon_1^p - \Delta\varepsilon_3^p} = \frac{2\sigma_2-\sigma_1-\sigma_3}{\sigma_1-\sigma_3-\frac{2}{3}\pi k}.$$

Therefore, in conditions $k \neq 0$, $\mu_\varepsilon \neq \mu_\sigma$, i.d. there is no similarity of tensors. The degree of deviation from similarity is determined by the obtained limit of plasticity k. The combination of tensor components relevant for Lode-Nadai coefficient is important for the theory of plasticity. It is widely applied when processing experimental data and developing theories; and, to be more precise, as or instead of the third invariant of stress tensor. However, no papers (except [14, 15]) are at authors' disposal, where this combination grounds on reasonably general theoretical pre-conditions. In [14, 15] and equalities (4) this combination is the result of admitted method to average shearing stresses. Moreover, we have above accepted $\langle\tau_n\rangle$ as an integral characteristic of shearing stresses. Hence, the condition of hardening can be naturally written in terms $\langle\tau_n\rangle - \Delta\lambda$:

$$\Delta\langle\tau_n\rangle = G\Delta\lambda.$$

Here G is a local module of plastic shear. Explicitly, the condition takes on form:

IOP Conf. Series: Materials Science and Engineering **91** (2015) 012090 doi:10.1088/1757-899X/91/1/012090

$$\left[\frac{2\sigma_1-\sigma_2-\sigma_3}{\pi(\sigma_1-\sigma_3)}-\frac{k}{\sigma_1-\sigma_3}\right]\Delta\sigma_1+\frac{2\sigma_2-\sigma_1-\sigma_3}{\pi(\sigma_1-\sigma_3)}\Delta\sigma_2+$$
$$+\left[\frac{2\sigma_3-\sigma_1-\sigma_2}{\pi(\sigma_1-\sigma_3)}+\frac{k}{\sigma_1-\sigma_3}\right]\Delta\sigma_3=G\Delta\lambda. \quad (5)$$

In terms of associated flow law we suppose that tensor of plastic deformation increment is coaxial to stress tensor. Let us make use of [11] and formulate a condition of two tensors σ_{ij} and $\Delta\varepsilon_{ij}^p$ coaxiality as follows:

$$\begin{aligned}&(\sigma_{11}\Delta\varepsilon_{12}^p-\sigma_{12}\Delta\varepsilon_{11}^p)+(\sigma_{12}\Delta\varepsilon_{22}^p-\sigma_{22}\Delta\varepsilon_{12}^p)+(\sigma_{13}\Delta\varepsilon_{23}^p-\sigma_{23}\Delta\varepsilon_{13}^p)=0,\\&(\sigma_{12}\Delta\varepsilon_{13}^p-\sigma_{13}\Delta\varepsilon_{12}^p)+(\sigma_{22}\Delta\varepsilon_{23}^p-\sigma_{23}\Delta\varepsilon_{22}^p)+(\sigma_{23}\Delta\varepsilon_{33}^p-\sigma_{33}\Delta\varepsilon_{23}^p)=0.\\&(\sigma_{13}\Delta\varepsilon_{11}^p-\sigma_{11}\Delta\varepsilon_{13}^p)+(\sigma_{23}\Delta\varepsilon_{12}^p-\sigma_{12}\Delta\varepsilon_{23}^p)+(\sigma_{33}\Delta\varepsilon_{13}^p-\sigma_{13}\Delta\varepsilon_{33}^p)=0.\end{aligned} \quad (6)$$

Now let us address to the formula of mean shearing stress (1). The structure of this formula predetermines the associated flow law and all further constructs. In the numerator (1) there is the square of stress intensity. It is rater easy to be calculated through stress components in any axes. Unfortunately, the denominator can't be written as visually graspable formulae through stresses in random coordinates. That is why, the increment of main stresses are to be considered as unknown functions. We have to apply the following equation in order to state the relation of main stresses increment to the increment in random axes:

$$\sigma^3-J_1\sigma^2+J_2\sigma-J_3=0, \quad (7)$$

where

$$\begin{aligned}&J_1=\sigma_{11}+\sigma_{22}+\sigma_{33},\\&J_2=\sigma_{11}\sigma_{33}+\sigma_{22}\sigma_{33}+\sigma_{11}\sigma_{22}-\sigma_{13}^2-\sigma_{12}^2-\sigma_{23}^2,\\&J_3=\sigma_{11}\sigma_{22}\sigma_{33}+2\sigma_{23}\sigma_{13}\sigma_{12}-\sigma_{11}\sigma_{23}^2-\sigma_{22}\sigma_{13}^2-\sigma_{33}\sigma_{12}^2.\end{aligned} \quad (8)$$

first, second and third stress invariants.

As far as we know, this equation has three real roots, which are the main stresses. Root values can be expressed principally trough stresses (8) according to Cardano formula. Then we deal with increments and use the result in the restricted model of plasticity. The needed computations are quite cumbersome. It is better to address to increments in equation (7) and then use equations (8). So, we have:

$$3\sigma^2\Delta\sigma-2J_1\sigma\Delta\sigma-\sigma^2\Delta J_1+J_2\Delta\sigma+\sigma\Delta J_2-\Delta J_3=0.$$

Hence,

$$\Delta\sigma=A(\sigma)\Delta J_1+B(\sigma)\Delta J_2+C(\sigma)\Delta J_3,$$

where

$$A(\sigma)=\frac{\sigma_1}{3\sigma^2-2\sigma J_1+J_2},$$
$$B(\sigma)=-\frac{\sigma}{3\sigma^2-2\sigma J_1+J_2},$$
$$C(\sigma)=\frac{1}{3\sigma^2-2\sigma J_1+J_2}.$$

Addressing to increments in (8) we obtain

$$\Delta\sigma_i=D_{11}^i\Delta\sigma_{11}+D_{22}^l\Delta\sigma_{22}+D_{33}^i\Delta\sigma_{33}+D_{12}^i\Delta\sigma_{12}+D_{13}^i\Delta\sigma_{13}+D_{23}^i\Delta\sigma_{23}, \quad (9)$$

where

$$D_{11}^{i} = A^{i} + B^{i}(\sigma_{22} + \sigma_{33}) + C^{i}(\sigma_{22}\sigma_{33} - \sigma_{23}^{2}),$$
$$D_{22}^{i} = A^{i} + B^{i}(\sigma_{11} + \sigma_{33}) + C^{i}(\sigma_{11}\sigma_{33} - \sigma_{13}^{2}),$$
$$D_{33}^{i} = A^{i} + B^{i}(\sigma_{11} + \sigma_{22}) + C^{i}(\sigma_{11}\sigma_{22} - \sigma_{12}^{2}),$$
$$D_{12}^{i} = -2B^{i}\sigma_{12} + 2C^{i}(\sigma_{23}\sigma_{13} - \sigma_{33}\sigma_{12}),$$
$$D_{13}^{i} = -2B^{i}\sigma_{13} + 2C^{i}(\sigma_{23}\sigma_{12} - \sigma_{22}\sigma_{13}),$$
$$D_{23}^{i} = -2B^{i}\sigma_{23} + 2C^{i}(\sigma_{13}\sigma_{12} - \sigma_{11}\sigma_{23}),$$
$$A^{i} = A(\sigma_{i}),\ B^{i} = B(\sigma_{i}),\ C^{i} = C(\sigma_{i}),\ i = 1,2,3.$$

When stating the relations (9) we took into account that stresses σ_{ij} formed a symmetric tensor. Therefore, the same relations (9) are relevant for plastic deformations, where σ is to be replaced by ε^{p} with corresponding indexes:

$$\Delta\varepsilon_{i}^{p} = D_{11}^{i}(\varepsilon_{i}^{p})\Delta\varepsilon_{11}^{p} + D_{22}^{i}\Delta\varepsilon_{22}^{p} + D_{33}^{i}\Delta\varepsilon_{33}^{p} + D_{12}^{i}\Delta\varepsilon_{12}^{p} + D_{13}^{i}\Delta\varepsilon_{13}^{p} + D_{23}^{i}\Delta\varepsilon_{23}^{p} \tag{10}$$

Three equations of equilibrium are to be added to the obtained equations

$$\frac{\partial\Delta\sigma_{ij}}{\partial x_{i}} + \Delta X_{i} = 0, \tag{11}$$

where ΔX_{i} – components of mass force increment. If we exclude elastic deformations, the equations can be closed by relations of deformation increment to displacement vector components:

$$\Delta\varepsilon_{ij}^{p} = \frac{1}{2}\left(\frac{\partial\Delta u_{i}}{\partial x_{j}} + \frac{\partial\Delta u_{j}}{\partial x_{i}}\right). \tag{12}$$

Thus, a restricted model is formulated and includes 22 equations in the following 22 unknowns: $\Delta\sigma_{ij}$, $\Delta\varepsilon_{ij}^{p}$, $\Delta\sigma_{i}$, $\Delta\varepsilon_{i}^{p}$, Δu_{i}, $\Delta\lambda$. This model is a rigidly plastic one because elastic deformations aren't taken into consideration.

Type of equations

The fact that the system of equations is kinematically determined in a rigidly plastic variant is of some interest. In fact, we substitute relations (12) into coaxiality equation (6). As the result we obtain three equations in variables $\Delta u_1, \Delta u_2, \Delta u_3$. We know all coefficients in this system from the previous stage. The system is invariant with respect to coordinate rotation. Therefore, we can assume that characteristic surface touches the plane $x_3 = 0$ in the origin of coordinates. Let us know Cauchy data on the surface. Consequently, we know in-situ derivatives with respect to x_1 and x_2. Therefore [3], the system of differential equations can be treated as the algebraic system with respect to derivatives $\Delta u_1, \Delta u_2, \Delta u_3$ of normal $0x_3$. So, we need to equate to zero the determinant of three derivatives $\partial\Delta u_i / \partial x_3$ to calculate a characteristic surface. As the result we obtain

$$\sigma_{13}\sigma_{23}(\sigma_{22} - \sigma_{11}) + \sigma_{12}(\sigma_{13}^{2} - \sigma_{23}^{2}) = 0.$$

The equation has one solution of the kind $\sigma_{13} = 0,\ \sigma_{23} = 0$. Thus, the normal to the surface coincides with the direction of main stress.

Consideration of elastic deformations

In some cases, when plastic flows are in the first stages of their development, elastic deformations can be easily compared with plastic deformations. Therefore, they are necessary to pay attention to. Here, one more important fact necessary for theoretical description of the process is to be mentioned. If we consider elastic deformations, conditions of consistency (Saint Vincent identities) are relevant for full deformations only. In particular, the system of equations isn't kinematically determined any more.

Moreover, elastic deformations, (with no regard to their values) are principally important to determine the type of the system and to set boundary problems. We confine ourselves with the simplest case, when the body is isotropic and homogenous to elastic deformations. In terms of Hook's law,

$$\Delta\sigma_{ij} = \lambda \cdot \left(\Delta\varepsilon_{11}^{e} + \Delta\varepsilon_{22}^{e} + \Delta\varepsilon_{33}^{e}\right)\delta_{ij} + 2\mu\Delta\varepsilon_{ij}^{e}. \qquad (13)$$

Here and later $\Delta\varepsilon_{ij}^{e}$ is increment of elastic deformations, λ, μ are elastic Lame parameters. Increments of full deformations are made of two components:

$$\Delta\varepsilon = \Delta\varepsilon_{ij}^{e} + \Delta\varepsilon_{ij}^{p}$$

Here

$$\frac{1}{2}\left(\frac{\partial\Delta u_i}{\partial x_j} + \frac{\partial\Delta u_j}{\partial x_i}\right) = \Delta\varepsilon_{ij}^{e} + \Delta\varepsilon_{ij}^{p}. \qquad (14)$$

Therefore, the closed model of rigid plastic body is enriched by 6 new unknowns $\Delta\varepsilon_{ij}^{e}$ and 6 new equations (13). 6 equations (12) were replaced by 6 equations (14). Therefore, equations and unknowns are balanced – the system of equations for elastoplastic body is closed.

Conclusions

1. Implementation of the new integral characteristic of shearing stresses results in the condition of plasticity, which is proportional to the relation of the octahedral shearing stress square to the maximal shearing stress.
2. There are characteristic surfaces with normals in a rigidly plastic variant of the model, which are in line with the main stresses direction.
3. In general, the accepted condition of plasticity is the reason for a closed consistent model, which can be used to solve various boundary problems.

The work is done under financial support of Russian Foundation for Basic Research, project № 13-05-00432-a.

References

[1] A. Freudental, N. Geiringer. Mathematical theory of inelastic continuum. M, pic., 1962, 432 p.
[2] D.D. Ivlev. Theory of ideal plasticity. M, Nauka, 1966, 232 p.
[3] L.M. Kachanov. Fundamentals of plasticity theory. M, Nauka, 1969, 420p.
[4] D.D. Ivlev, G.I. Bykovetsev. The theory of work-hardening plastic body. M., Nauka, 1972, 232p.
[5] A.N. Stavrogin, A.G. Protosenya. Strength of rocks and deep mine working stability. M., Nedra, 1985, 271p.
[6] V.V. Vinogradov. Geo-mechanics of rock mass near mine workings. Edit. by Zorin A.N., AS USSR. Institute of geo-technical mechanics. Kiev, Nauk. dumka, 1982, 192p.
[7] A.N. Stavrogin, A.G. Protosenya. Mechanics of deformation and disintegration of rocks. M. Nedra, 1992, 224 p.
[8] B.D. Aninin, V.M. Zhigalkin. Behaviour of materials in condition of complex loading, Published at SB RAS, 1999, 342 p.
[9] Yu.P. Chshuplensov. Strength and deformability of rocks. Yekaterinburg. UB RAS, 2003. 195 p.
[10] G.I. Bykovsev, D.D. Ivlev. Theory of plasticity. Vladivostok. Dalnauka. 1998. 528 p.
[11] A. Yu. Ishlinky, D.D. Ivlev. Mathematical theory of plasticity. M., Fismatlit, 2001, 704 p.
[12] Problems of mechanics of deformed solids and rocks. Collective papers for the 75-th anniversary of E.I. Shemyakin / Edit. By D.D. Ivlev and N.F. Morozov. M. Fismatlit, 2006. 864 p.
[13] V.V. Novozhilov. On physical nature of stress invariants. PMM, 1951, V. 15, issue 2.

[14] O.A. Mikenina, A.F. Revuzhenko. Criteria of limit state and decomposition of ideally bounded granular materials. Physical and technical problems of natural resources exploration, №4, 2014, pp. 55-60.
[15] A.F. Revuzhenko, O.A. Mikenina. Criteria of limit state and failure of geomaterials based on the new system of invariants of stress tensor. Institute of Mining SB RAS, Novosibirsk, International Conference of Physical Mesomechanics.
[16] A. Nadai. Plasticity and decomposition of solids. V.2. Published. Mir, M, 1969, 863 p.

IOP Conf. Series: Materials Science and Engineering **91** (2015) 012091 doi:10.1088/1757-899X/91/1/012091

Improving of the operation efficiency of the vehicle due to using of the neodymium magnets inside the vibration isolation devices

E G Gurova

Novosibirsk State Technical University,
Av. K. Marx, 20, Novosibirsk, 630073, Russia

e-mail: lena319@mail.ru

Abstract. In this paper the isolation suspension with stiffness compensator based on neodymium magnets is suggested. It was found that the passive vibration isolators not completely sufficient of modern requirement of the vibration isolation. It was determined that the neodymium magnets with the same initial parameters are most effective in comparison with DC current electromagnets. The mathematical model of the vibration isolation suspension has been developed. In this research the traction characteristics for given magnets are presented. Also the design of the vibration isolation suspension with compensator of the stiffness based on neodymium magnets has been developed. This research has been performed under support of the President scholarship for young scientists under the order of Russian Federation Ministry of the education and science № 184 from 10th of March 2015.

Introduction

Nowadays the one of the problem in many areas of the industry, vehicle and agriculture is large level of the vibration which does not meet the sanitary standards. Vibration renders negative influence on the human body, environment, operation of the mechanisms and vehicles. Long influence of the vibration and noise leads to disturbance of the human body normal condition, causing the disturbances of the physiological condition, and so functional changing of the inner organs. The man reacts on vibration depends from common duration of its influence. The greatest impact of the overall vibration affects the vision system processes duy to fluctuations of the eyeballs and head. Long term influence of the big enough intensity can negatively impact on the spine. In addition the vibration affect on normal course of the physiological processes, that causes varicose veins, coronary heart disease and hypertension. Excessive exposure of the local vibration can lead to the deceases of the blood vessels, nerves, muscles, bones and joints of the upper extremities.

Mechanical oscillations can cause malfunctions of the equipment operation, accumulations of the damages in materials, damages of the normal functioning of the different objects, defects of couplings, improper alignment of the motor with the driven machine, unbalance motor and machinery.

All these facts lead to necessity to found the way of the solutions directed on decreasing of the vibration levels, the transport industry is not an exception. The sources of the vibration in vehicles are electroengines, combustion engines, bad conditions of the roads, direct movement of the transport. At the present time the main methods of the protection can be separated in two groups. First group is directed on decreasing of the vibration inside of the vibration source. Prevention of the negative

IOP Conf. Series: Materials Science and Engineering **91** (2015) 012091 doi:10.1088/1757-899X/91/1/012091

influence of the vibration on the human organism, and also equipment operation, in that case it reduces to minimization of the mechanical vibration intensities and duration of its activity. At first we need to give an attention on the technique process, precisely on operation of the equipment, which cause of the vibration. Reduction of the level of the vibration from the rotating elements and mechanisms of transport, can be achieved whereby a precise fitting of the details and its operation: balancing, deskew, reducing to a minimum tolerances between the parts to be joined, timely lubrication components and assemblies. Second group, is directed on decreasing of the vibration parameters by the way of its elimination from the source to protected object. In that case for protection from vibration are applied the next methods: decreasing of vibration activity of the machines, the detuning from the resonance frequencies, damping, vibration isolation, vibration dampening.

Nowadays the most promising way to reduce the vibration levels to be considered the application of the devices with floating area of the zero stiffness [1,2,3].

In paper [4] the authors presented the vibration isolation suspension with electromagnet compensator of the stiffness. Electromagnet compensator of the stiffness was made as two counter-connected DC current electromagnets. Compensator which was suggested has a several advantages in comparison with mechanical stiffness compensator, as the absence of the friction force, and as a consequence and the wear, the compensator has a fast response adjustment system which redistributes the voltage at the electromagnets under the load changes, provides zero stiffness.

However suggested system because of presence of the electromagnets supposes the using of the voltage regulator, position sensors and also supposes the tracking system, which finally leads to increasing of the system cost and also to necessity of periodic service. In practice the application of the suggested vibration isolation device is impeded with big enough mass-size parameters of the electromagnets stiffness compensator, despite their high ability to reduce the vibration.

In research [5] the design of the compensator based on supermagnets was suggested. Vibration isolators with supermagnets stiffness compensator are most practically applicable, efficiency of reducing of the vibrations is high enough, with the dimensions and weight is significantly less than when using electromagnetic stiffness compensators.

Simulations

Based on the mathematical description, taken in [5], we can assemble the set of the equation for the suspension:

$$\begin{cases} m\dfrac{d^2x}{dt^2} + c \cdot x - F_k = F(t); \\ F_k = (ax^3 + cx) - F_p; \\ F_p = f(t) \cdot (ax^3 + cx) \cdot x. \end{cases} \tag{1}$$

On the basis of the set of the equations (1) is composed block diagram of the vibration isolation suspension with regulator (see figure 1).

IOP Conf. Series: Materials Science and Engineering **91** (2015) 012091 doi:10.1088/1757-899X/91/1/012091

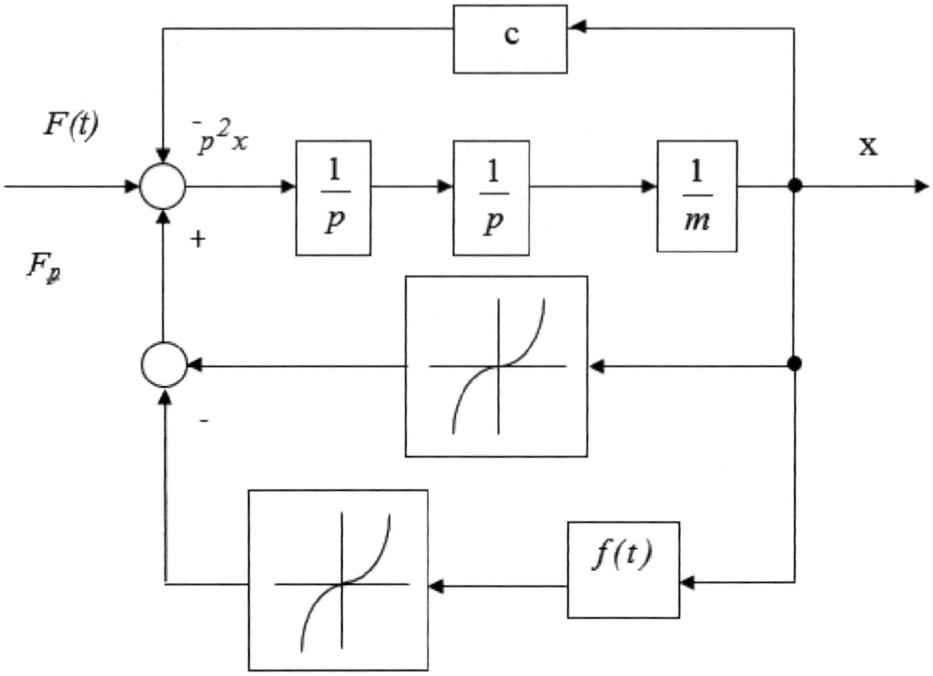

Figure 1. Schematic diagram of the vibration isolation suspension

The model of the vibration isolation suspension has been composed on the basis of this block diagram (see figure 2).

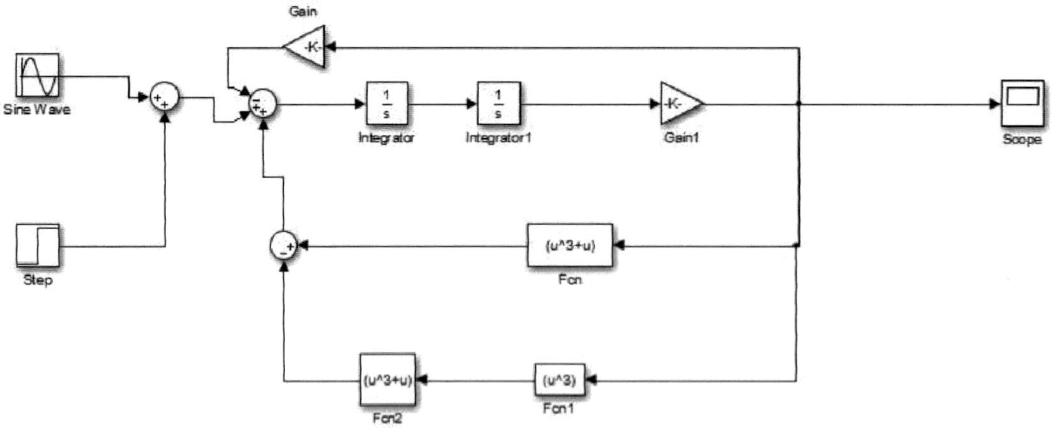

Figure 2. Model of the vibration isolation suspension with tuning system

Simulation of the vibration isolation suspension with nonlinear regulator was carried out in Simulink. For presented model the input signal is sine (Sine Ware) with parameters: amplitude is 1; frequency, another parameters is 0. The step signal (Step) is used as the load. On the oscilloscope ("Scope") the reaction of the system on sinusoidal impact is shown. Results of the simulation of the vibration isolation system without correction link are shown on figure 3.

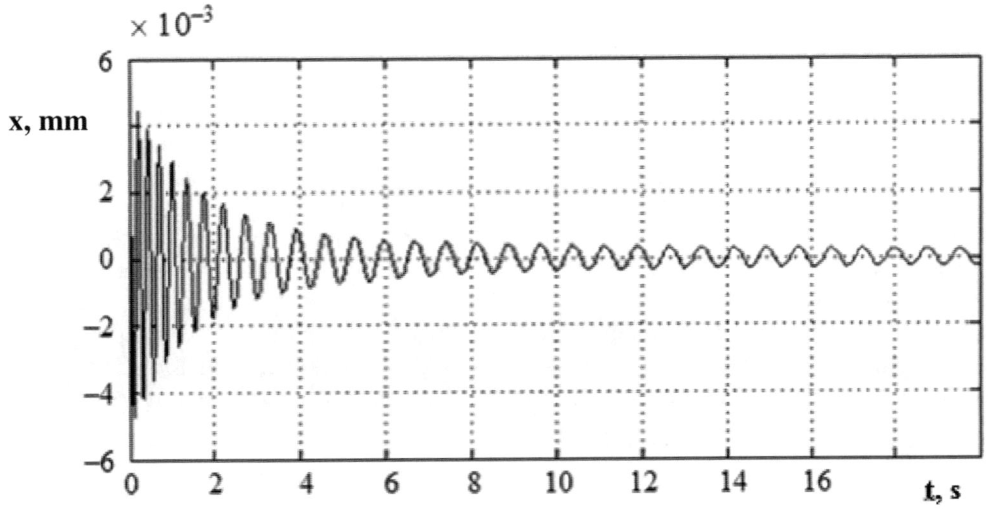

Figure 3. Oscillogram of the vibration isolator operation with tuning system without correction link

The model of the designed stiffness vibration isolator, which has been done in AutoCAD is shown on figure 4.

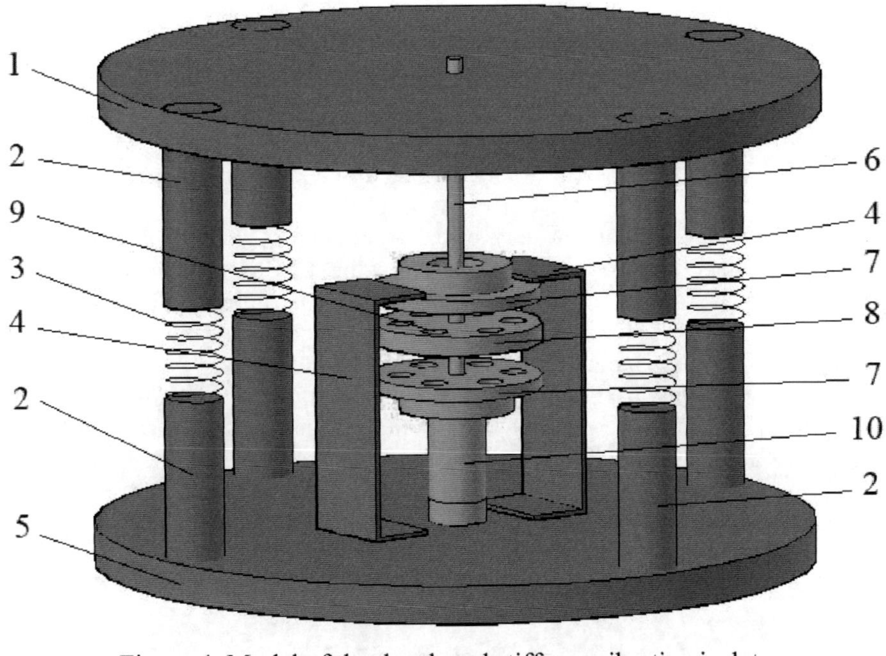

Figure 4. Model of the developed stiffness vibration isolator
1- protected base; 2- basis for resilient element; 3- resilient elements (springs); 4- duralumin rods; 5-vibroisolation base; 6- stock; 7- duralumin disk; 8-anchor; 9-mounting place for neodymium magnets; 10- duralumin basis
2-

Magnetic compensator of the stiffness is mounted in parallel to resilient element, which has been made as four springs 3, which are mounted with help of duralumin bases 2 for resilient element by one side to vibrating base 5 and by another side to protected base 1. Magnetic compensator is the two disks 7, one of them is rigidly mounted on the vibration base 5 through duralumin base 10, and another disk 7 is mounted on the vibration base 5 with help of duralumin rods 4. Stock 6 is rigidly coupled with protected element 1. Neodymium magnets are mounted on the bottom and top disks 7 inside the deepenings 9, and so on the both sided of the anchor 8. Magnets are placed in plane of the anchor and sides of the disk have different polarity. In that case, the total stiffness of the vibration isolator will be sum of all the stiffness's of each resilient element (spring) and magnetic stiffness compensator. Thus the stiffness can be reduced to zero, and this provides ideal vibration isolation.

Principle of operation is based on movement of the anchor with changing of the magnitude of the external force, so the power characteristic became falling, that provide the zero-stiffness of all vibration isolator.

Supermagnets, which are used in suggested compensator, are produced from alloy of rare-earth material neodymium (Nd). Main advantage of such magnets is big traction forces under the small sizes. The traction characteristic of the used neodymium magnet is shown on the figure 5.

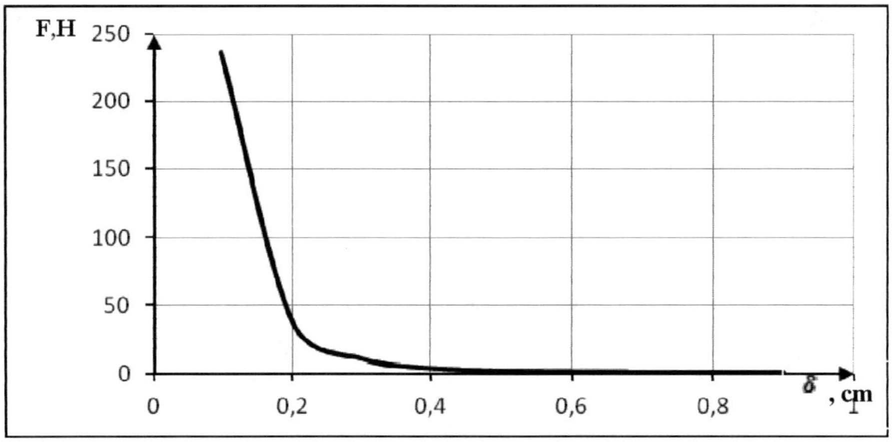

Figure 5. Traction characteristic of the neodymium magnet

Magnet characteristics are laid on the stage of the production of the magnet and can not be changed further. Main parameters of the magnets are residual magnetic induction and stability to demagnetization (coercivity force). Neodymium magnets can be produced in shape of the disk, cube, rail, cylinder, rod, ring, sector or ball. The view of the used neodymium magnets is presented on figure 6. Basic characteristics of the magnets have been summarized in table 1.

It is necessary to know the characteristics of the given neodymium magnets for calculation of the vibration isolation suspension with compensator of the stiffness based on the neodymium magnets. These characteristics are shown in table 1.

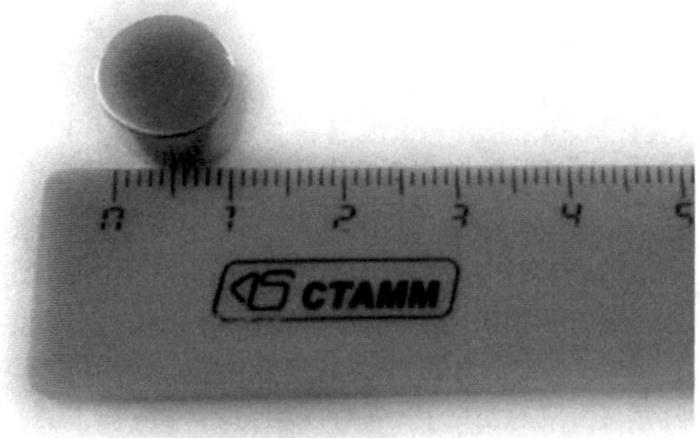

Figure 6. Size of the neodymium magnets

Table 1 – Characteristics of the magnet

Type	N42
Shape	disk
Height (h)	8 mm
Diameter (D)	12 mm
Tolerance	+/-0,1 mm
Properties	B_r>1170 mT; H_{cb}>868 kA/m; H_{cj}>955 kA/m; $(BH)_{max}$>263 kJ/m^3; T_w<80ºC
Coating	Nickel (Ni)
Max temperature	80 ºC
Direction of the magnetization	axial

Neodymium magnets are placed thus that there are two magnets on each axis (see figure 7). This is due to the fact that the all vibration isolation supports must have not less 6 degrees of freedom. The duralumin disk with supermagnets inside the mounting places is shown on the figure 8.

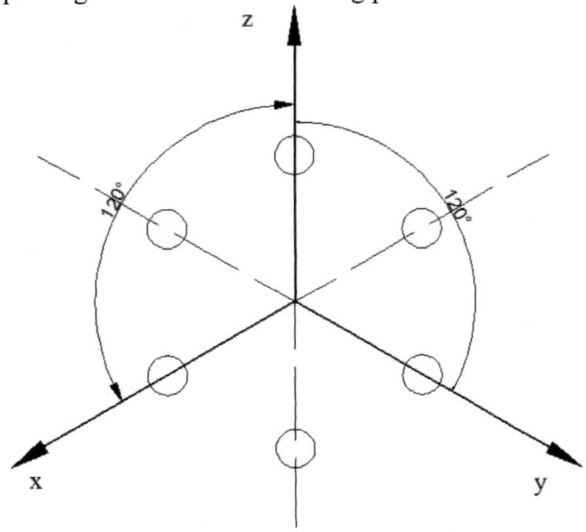

Figure 7. Layout of the neodymium magnets

a) b)
Figure 8. Model of the stiffness compensator based on neodymium magnets
a) duralumin disk with installed supermagnets inside the mounting places
b) supermagnet stiffness compensator

Conclusion

The represented vibration isolator is based on compensator of the stiffness with supermagnets can be used in engineering and vehicle area. The suggested solution can be efficient in prevention of negative influence of the vibration on the human body and in improving of the efficiency of the exploitation of the vehicle.

This research has been performed under support of the President scholarship for young scientists under the order of Russian Federation Ministry of the education and science № 184 from 10th of March 2015.

References

[1] Gurova E G Power characteristics of the vibration isolators E G Gurova M G Gurov V V Ivanov Applied Mechanics and Materials 2015 Vol 698 : *Electrical Engineering Energy Mechanical Engineering* EEM 2014 pp 575-579

[2] Gurova E G Determination of traction characteristic of stiffness compensator for vibroisolation device E G Gurova M G Gurov S V Makarov *Life Science Journal* 2014 Vol 11 №12s pp 624-625

[3] Features of the power characteristics of the vibration isolations E G Gurova S V Makarov M G Gurov A A Sergeev Advanced Materials Research 2014 Vol 1040 : *High technology: research and applications* pp 678-681

[4] Development of Spatial Vibration Protection Devices E G Gurova V Y Gross V S Kurbatov S V Makarov A A Sergeev N I Shchurov *World Applied Sciences Journal* 2013 Vol 22 (Special Issue on Techniques and Technologies) pp 44-48

[5] Gurova E G Vibroisolator with neodymium magnets compensator of the stiffness E G Gurova M G Gurov Applied Mechanics and Materials 2014 Vol 682 : *Innovative Technologies and Economics in Engineering* pp 118-121

IOP Conf. Series: Materials Science and Engineering **91** (2015) 012092 doi:10.1088/1757-899X/91/1/012092

Research of the pre-launch powered lubrication device of major parts of the engine D-240

M Korchuganova, A Syrbakov, A Tkachev, T Zorina

Yurga Technological Institut branch of Tomsk Polytechnic University, Russia

E-mail: kma77@list.ru , kormar@tpu.ru

Abstract. In the publication, the issues have been considered concerning combustion engine start wear of mobile machines in case of outside storage in the conditions of low environmental temperature. Based on the analysis of existing methods and constructions of powered lubrication devices for contact surfaces of engines, a design of a combined device has been suggested which unites the functions of hydraulic and heat accumulators. On the basis of the elaborated design, preparatory tests have been conducted in order to evaluate the effectiveness of pre-start oil circulation in the engine D-240, as well as the effectiveness rate of thermoinsulation and the heating device of the hydraulic accumulator. The findings of the survey have shown that the pre-start powered lubrication device for major parts of the engine is effective.

1. Introduction

Winter conditions are a great challenge for the reliability of motive power, the quality of engineering maintenance materials and the production machinery for pre-launch processing [3].

In Siberia, the temperature in winter often drops to the mark -20 °C and lower, which, in the conditions of outside storage of mobile machinery, makes a big problem for the cold launch of diesel engines [2].

In this context, the improvement of diesel engine start characteristics and the creation of effective pre-launch processing methods make an important and multidisciplinary task.

Another important issue of low-temperature exploitation is the decreasing of start engine wear [5].

The analysis of influencing factors of low temperatures on the engine lifetime has shown that the increasing of start wear and the general launch reliability are massively affected by the lubrication retard of major parts of the engine [2].

2. Overview of research methods

In order to solve the problem of start wear in lubrication systems of modern engines, various constructive and exploitation steps are made which provide the powered lubrication of contact surfaces within the first after-launch operation period [1].

These devices and methods can be divided in 3 groups [1]:

1) Devices providing pre-launch lubrication of contact parts of the engine via oil pumps of various types and constructions.

2) Devices, in which hydraulic accumulators are used for the lubricant supply. The oil is accumulated therein while engine is running and then given under pressure right before the launch.

3) Different constructive features of the lubrications system which help decrease the oil supply retard during the launch of the engine (using of gate valves in the lubrication system, increasing of the relieve valve opening pressure etc.).

The considered devices for pre-launch motor oil circulation are only limitedly used in highly boosted diesel engines of specialized machinery.

3. The measurement procedure

According to the analysis of the considered devices for pre-launch motor oil circulation, the most effective and the easiest device is based on hydraulic accumulators [4].

This is why we suggest to design a device for powered pre-launch lubrication of contact elements of the engine D-240 based on the hydraulic accumulator, for the tractor MTZ-80 which is widely used during the winter period.

In order to decrease calorific loss of the motor oil in hydraulic accumulators and to maintain the temperature of service fluid within the period of its interim storage, we suggest to put the device body into a heating jacket and to install an additional heating unit.

The main purpose of the modernization of engine lubrication system (figure 1, 2) is to decrease the time of oil flow to the parts of the engine D-240 during the process of cold launch by means of maintenance of positive oil temperature in the hydraulic accumulator within the period of interim storage as well as by means of pre-launch oil supply from the hydraulic accumulator to the contact parts of the engine.

Thus, the operation time of crankshaft bearings and turbocharges in unfavourable terms (in unlubricated and boundary friction modes) is shortened.

The given device has following features (figure 1, 2):

1. The usage of standard bores in the engine lubrication system allows to install the pre-launch lubrication device without any special adjustments in the engine construction. Thus, the exploited tractors can be upgraded too.
2. The appliance of thermoinsulation for the elaborated device helps enhance the effectiveness of heat conservation of the motor oil within the period of interim storage of machinery; this facilitates the launch in the conditions of low temperatures.
3. The usage of a self-regulated heating unit in the elaborated device provides automatic motor oil heating in the hydraulic accumulator.

The device functions as follows (figure 1). Upon completion of the tractor shift, machine operators open tap 8 and, while the motor is still running, by means of standard oil pump, the heated motor oil is partially pumped from the crankcase into the heat accumulator (about 60-70% of the total liquid volume in the oil crankcase), where, in the period of interim machinery storage, the motor oil warmth is accumulated.

Launching the engine, we open tap 8 and, under the effect of the additional air pressure (0.3-0.5 MPa) within the hydraulic accumulator (by means of a pneumatic pump, a receiver), a part of the motor oil is directed into the oil passage and completes the pre-launch oil circulation during 20-40 s; the rest of the motor oil flows through additional hosepipes and a drain hole directly into the engine crankcase.

The process of charging and discharging of the accumulator is controlled by reference to the standard oil pressure sensor which is situated on the dashboard of the tractor MTZ-80 (figure 3). The results of the conducted tests showed that if the oil temperature in the hydraulic accumulator equals to 40°°C, and the air pressure equals to 0.3 MPa, the oil pressure in the oil passage amounts to 0.22 MPa at the moment of power fluid feed (within the limits of the safe working pressure) and does not change for 40-60 s.

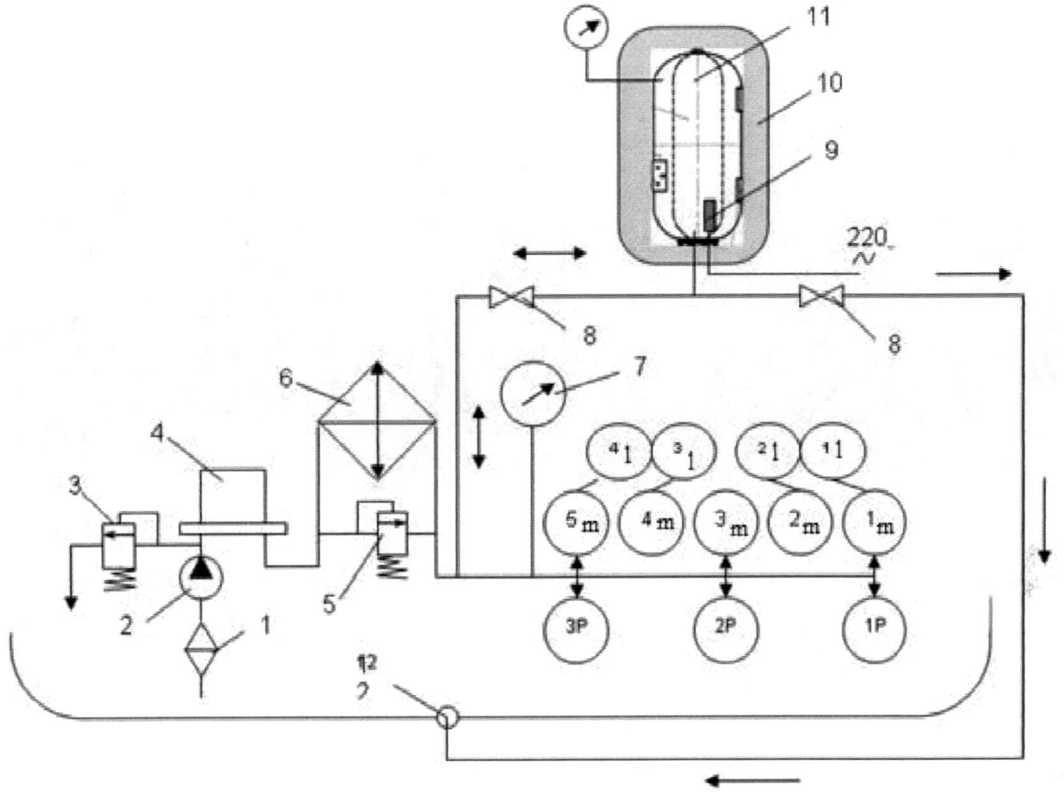

Figure 1. The plan of the pre-launch oil supply device into the lubrication system of the engine D-240:
1 - oil intake; 2 - oil pump; 3 - protection valve; 4 - rotary accelerator; 5 - thermostatic valve of the heat radiator; 6 - heat radiator; 7 - pressure gage indicator; 8 - tap; 9 - oil heater; 10 - thermoinsulation; 11 - hydraulic accumulator; 12 - drain hole of the oil crankcase; m and l - main and large end bearings; P - bearings of the countershaft.

Pre-launch oil circulation helps decrease the feed time of the oil into the contact parts and increase engine lifetime.

In the course of further improvement of the suggested design we have suggested to install a self-regulated heating unit 9 (figure 1) and (figure 4) within the hydraulic accumulator. This heating unit is energized from an external source, in order to maintain the power fluid temperature and, if necessary, to heat up motor oil before launching the engine.

As a heating device, we chose an electric bar heater with a heat controller, with a capacity of 30 W and the maximum fluid heating temperature 55 °C.

In the process of experiment, the effectiveness of thermoinsulation has been registered in reference to the oil cooling rate in the hydraulic accumulator, as well as the dynamics of the fluid heating by means of a heating unit 9 (figure 1) in the suggested device. As thermoinsulation, we used basalt fiber, penofol, and extruded polystyrene foam (figure 5).

IOP Conf. Series: Materials Science and Engineering **91** (2015) 012092 doi:10.1088/1757-899X/91/1/012092

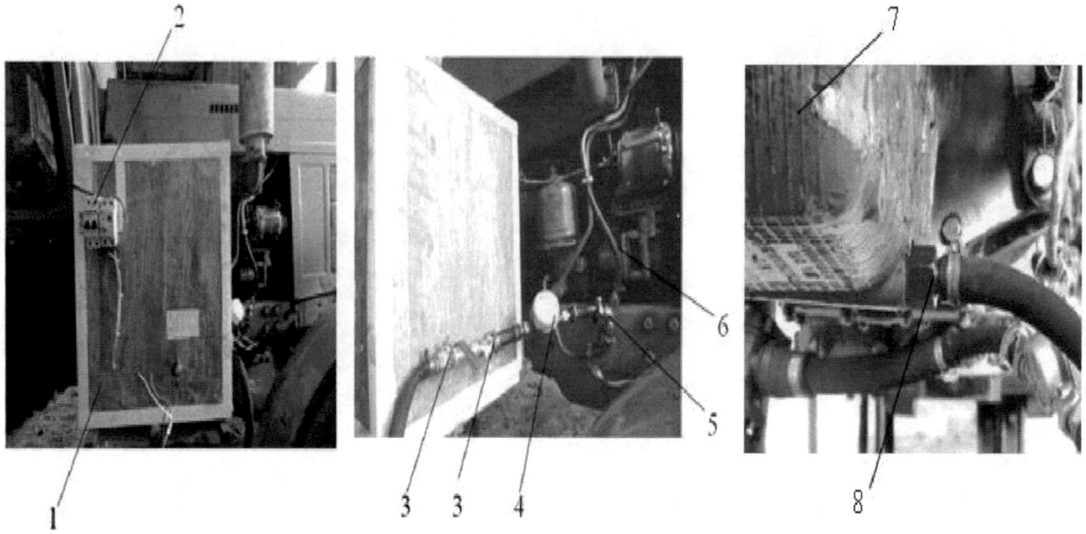

Figure 2. Pre-launch oil supply device into the lubrication system of the engine D-240:
1 - thermoinsulated hydraulic accumulator; 2 - module for the connection of the oil heater with the external network; 3 - tap; 4 - fluid meter; 5 - tee-joint; 6 - main oil passage; 7 - oil crankcase of the engine D-240; 8 - connector of the drain hole.

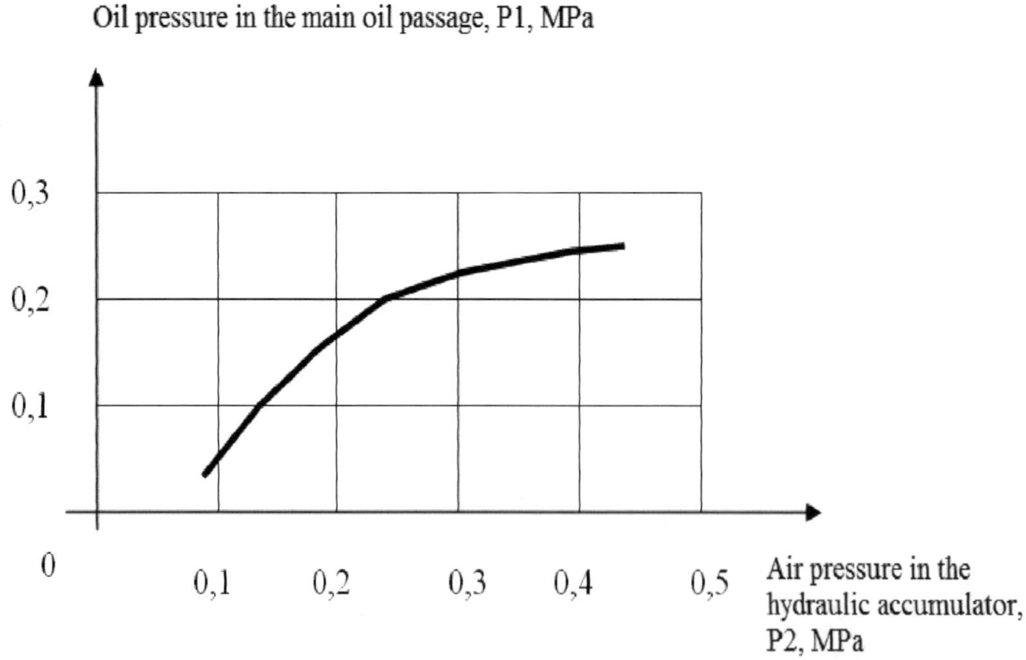

Figure 3. Dependence of the pressure in the main oil passage from the air pressure in the hydraulic accumulator (motor oil temperature in the hydraulic accumulator equals to 38-42 °C, the environmental temperature equals to -18 ... - 24 °C, motor oil grade is M-10G_2).

IOP Conf. Series: Materials Science and Engineering **91** (2015) 012092 doi:10.1088/1757-899X/91/1/012092

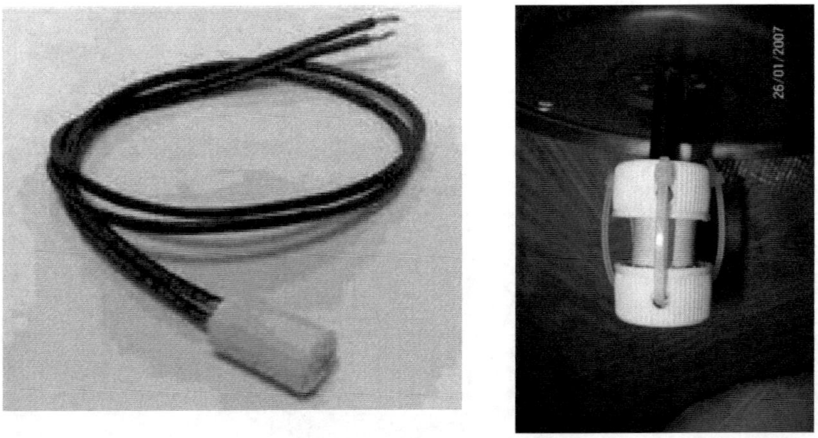

Figure 4. Self-regulated heating unit with a capacity of 30 W.

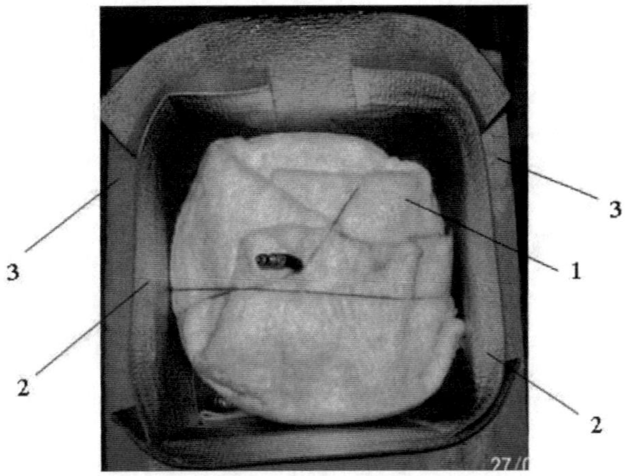

Figure 5. Thermoinsulated hydraulic accumulator:
1 - basalt fiber; 2 - penofol; 3 - extruded polystyrene foam.

4. Results and Discussion

An average motor oil cooling rate in the heat accumulator amounted to 1.5-2 °C per hour (figure 6) by the environmental temperature of -18 ... -23 °C; the motor oil heating rate in the heat accumulator by means of an electric heating device amounted to 3-5 °C per hour (figure 7), which proves the effectiveness of the elaborated device.

5. Conclusion

Based on the results of the tests, following preliminary conclusions have been made concerning the effectiveness of the pre-launch lubrication device:

1) The device provides a sufficient lubricant material supply to the crankshaft bearings before the launch of the engine, help decrease the oil feed time to the parts during a cold launch, as well as prevent their start wear.

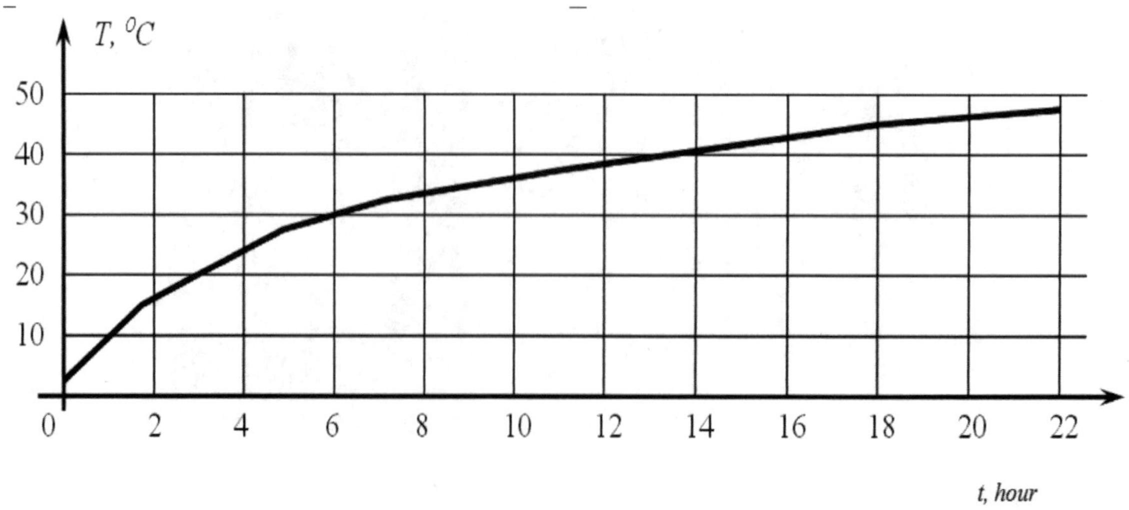

Figure 6. The diagram of the motor oil heating in the hydraulic accumulator (the capacity of the heating unit N=30 W, the environmental temperature T=-20 ... - 25 °C)

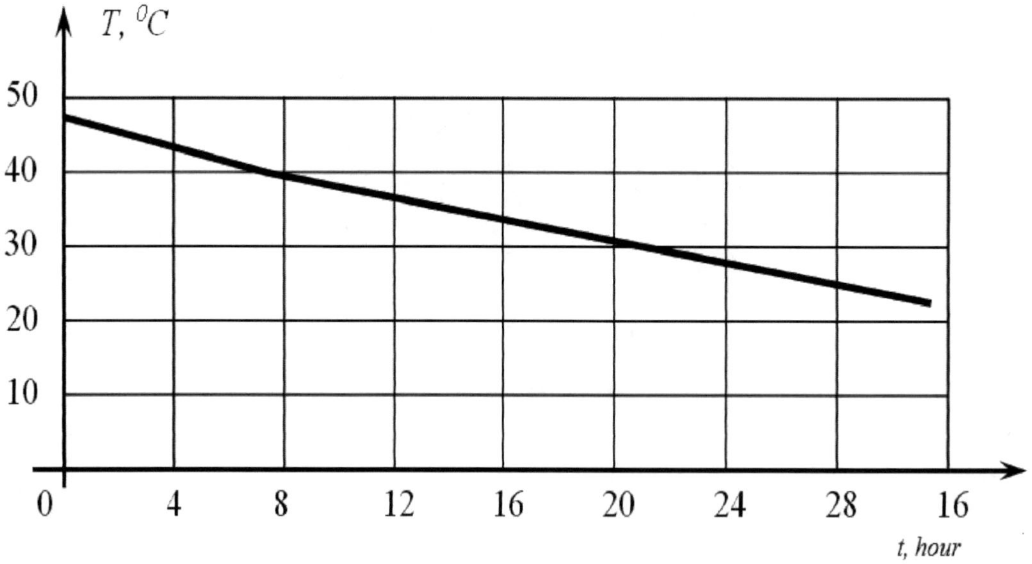

Figure 7. The diagram of motor oil cooling in the hydraulic accumulator (environmental temperature T= -18 ... -23 °C)

2) The device helps maintain the pressure in the lubrication system before the launch within the limits of 0.1-0.2 MPa depending on the initial motor oil temperature in the device and in the engine crankcase, as well as maintain the pressure in the hydraulic accumulator.

3) The usage of thermoinsulation and a heating device increases the effectiveness of the hydraulic accumulator in the pre-launch oil circulation in the conditions of subzero temperatures.

Reference

[1] Syrbakov AP, Korchuganova MA Job diesel fuel injection equipment in subzero temperatures. - Saarbrucken: LAP LAMBERT, 155 (2011).

[2] Xian Ting Ma A Review of the Present Status of Hydraulic Accumulator, Applied Mechanics and Materials 246-247 (2013), 629-634.

[3] Yi Bin Guo, Xi Qun Lu, De Quan Zou, Tao He, Wan You Li, Zi Jun Li, Analysis of Piston Ring Lubrication with Different Lubricant Supply, Advanced Materials Research, 199–200 (2011); 700-706.
[4] Wen Jie Zhai, Chang Xiong Liu, Pei Lian Feng Hydrodynamic Analysis of Circular Translational Polishing under Mixed Lubrication Key Engineering Materials, 264 (2007); 359-360.
[5] Yu Yu Zuo, Gas Turbine Engines Lubrication System Design, Advanced Materials Research 900 (2014); 773-776.
[6] Guo Ping Yang, Jun Hao Gao, Bo Chen Computer Simulation of Controlled Hydraulic Impact or System, Advanced Materials Research, 179-180 (2011), 122-127.

Formation of the composition and properties of dumps on the open-pit mines of Kuzbass

Y V Lesin, S Y Luk'yanova, M A Tyuleneva

Yurga Technological Institute (branch) of Tomsk Polytechnical University
Kuzbass State Technical University

e-mail: [a] tma.geolog@kuzstu.ru

Abstract. In 2012 the total volume of coal mining in the Kuzbass was about 200 million tons, including 116 million tons (60.3%) produced open way, Figure 1. In 2014, it was produced more than 203 million tons of coal. In addition, there is no reason to believe that in the near future volume of coal will decline. Accordingly, the volume of wastewater discharged by enterprises will increase. So, from 2006 to 2009 there was an increase of polluted water discharge coal mines from 217 to 245 million m3. Therefore, the problem of water pollution mines governmental waters and career is very important.

Despite its advantages, open coal mining does more damage to the environment than underground. The most significant impact has opened a way to develop on the hydrosphere and land resources, where the share of the total environmental damage to all elements of the biosphere is 66.4% versus 33.6% for underground mining.

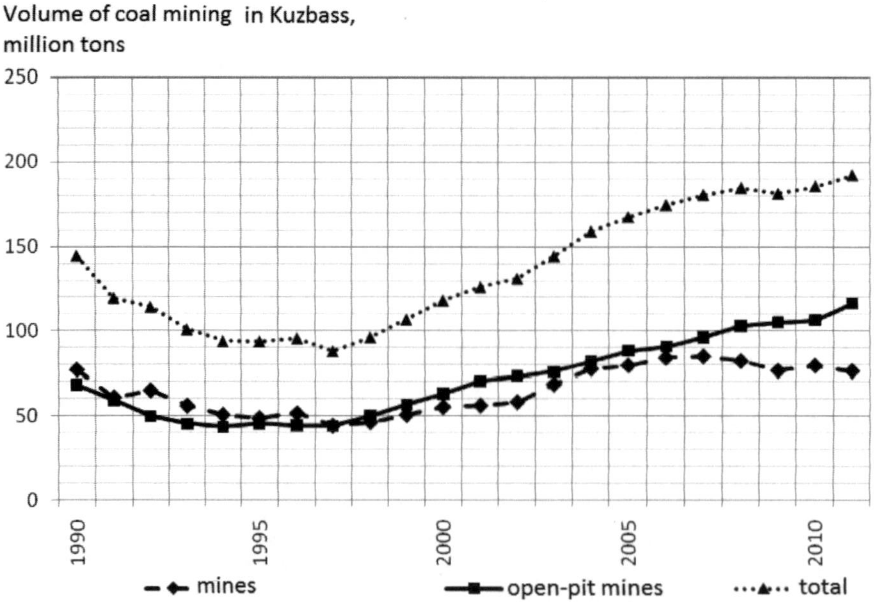

Figure 1. Dynamics of mining of coal in Kuzbass

IOP Conf. Series: Materials Science and Engineering **91** (2015) 012093 doi:10.1088/1757-899X/91/1/012093

Annually during open-pit mining on Kuzbass coal deposits hundreds of millions m^3 of overburden lay down in the dumps. To ensure the safety of the dumping process, use [1, 2] and subsequent effective recovery the rock mass it needs to know their material composition, structure and properties.

The material composition, in this case, is a content of different lithotypes and their particle size in the dump. Lithological composition of technogenic massives of overburden is determined, first of all, by the composition of rocks developed coal deposits and order removal of the blasted rock faces. A composition and properties of open-pit wastes around the world has been studied by many scientists [3-9]. We are considering this problem for mining and geological conditions of Kuzbass. In waste dumps of open-pit mines developed seams Kolchugino series of Kuzbass coal-bearing deposits, rocks prevails by siltstone (up to 51.2%), and Balakhonka series – sandstones (up to 53.5%), Table 1.

Table 1. Consist of overburden on Kuzbass coal-bearing deposits [%]

Series, subseries	Conglomerate, gritstone	Sandstone	Siltstone	Mudstone	Carbonaceous rocks
Kolchugino series	3.5	26.0	49.6	20.8	0.1
Erunakovo subseries	5.0	26.2	47.8	20.9	0.1
Ilyin subseries	1.8	25.8	51.2	21.1	0.1
Balahonka series	1.4	45.9	39.3	11.3	2.1
Upper subseries	1.3	53.5	37.2	5.5	2.5
Bottom subseries	1.6	39.4	39.1	17.0	2.9

The particle size distribution in dumps of overburden characterized by high heterogeneity: heterogeneity coefficient ($C_h = d_{60} / d_{10}$) is greater than 5. It is also established that during the construction of the overburden dump occurs segregation of particle size that allows to select its different in particle size layers: the bottom – large-sized, middle – medium-sized and upper layer – small-sized. Researches on recently paved waste dump on a open-mine "Kedrovsky" found that the average diameter (d_{50}) of clasts in the bottom layer is 1.3 m, in the middle – 0.8 m and in the upper – 0.2 m. As a result of mechanical stress, and exposure to water and weathering size distribution is constantly changing. Thus, the content of fractions plus-25 mm top layer of waste dump in less than a year decreased 1.8 times, and minus-1 mm fraction increased 3.3 times, Table 2. Two years later granulometric composition of different layers of overburden dump is significantly changed (Fig. 2).

Table 2. Particle size distribution in the upper layers of dumps

Dump age	Particle size [mm]						
	+25	25-10	10-7	7-5	5-3	3-1	-1
	Percentage fractions [%]						
3 months	33.4	15.9	10.4	7.9	9.7	15.2	7.5
12 months	18.5	19.3	7.6	5.9	8.6	15.4	24.7

Fineness different lithotypes overburden by blasting, loading, transporting and dumping depends on their mechanical properties. Most durable and resistant to water are sandstones. Their compression strength can be 100 MPa or more, and usually greater than the tensile strength of siltstones and mudstones more than one and a half to two times.

Table 3 shows the results of research strength and water resistance of rocks with a few open-pit mines of Kuzbass. The sandstones are more resistant to water. It should also be borne in mind that soaking of argillaceous rocks is much higher than sandstone.

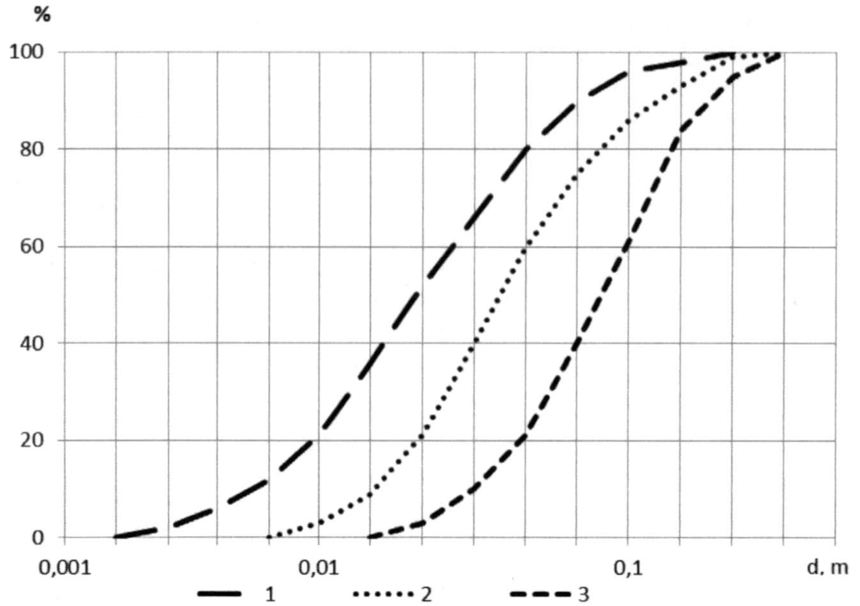

Figure 2. Curves of granulometric composition of different layers of dump paved two years ago: 1 – upper layer, 2 – middle layer, 3 – bottom layer.

Table 3. Strength and water-resisting of bedrock overburden

Place of sampling	Lithotype of rock	Strength of dry rocks [MPa]	Strength of water-saturated rocks [MPa]	Strength reduction ratio S_r
Open-pit "Kedrovsky"	sandstone	67.1	43.1	0.64
	siltstone and mudstone	38.7	19.2	0.50
Open-pit "Mohovsky" (Mohovsky section)	sandstone	60.2	30.9	0.51
	siltstone and mudstone	27.6	17.7	0.64
Open-pit "Mohovsky" (Sartakinsky section)	sandstone	36.7	36.7	1.0
	siltstone and mudstone	32.3	15.0	0.45
Open-pit "Krasnobrodsky"	sandstone	100.5	75.8	0.75

Assessment of the degree of grinding various lithotypes overburden at the same mechanical actions performed on a specially designed drop-hammer, Fig. 3.

Figure 3. Drop-hammer for assessment of the degree of grinding various lithotypes overburden

These researches were carried out on a samples of overburden from the open-pit mine "Krasnobrodsky". 200 samples of rocks (100 - sandstone, 100 - siltstone) were used. Fractional composition of broken rocks was determined by sieve method. Granulometric composition of sandstone is given in Table 4, of siltstone – in Table 5.

Indicators of granulometric composition of the rock after fracture are given in Table 6.

Table 4. Granulometric composition of sandstone

Fraction size [mm]	Percentage fractions [%]	Total (integrated) content [%]
> 40	-	-
40 – 30	2.34	100
30 – 20	43.49	97.66
20 – 10	33.39	54.17
10 – 7	5.42	20.78
7 – 5	3.18	15.36
5 – 3	2.58	12.18
3 – 2	1.49	9.6
2 – 1	1.95	8.11
1 – 0.5	1.00	6.16
0.5 – 0.25	0.27	5.16
< 0.25	4.89	4.89
Total	100	-

Table 5. Granulometric composition of siltstone

Fraction size [mm]	Percentage fractions [%]	Total (integrated) content [%]
> 40	-	-
40 – 30	-	-
30 – 20	10.62	100
20 – 10	44.31	89.38
10 – 7	21.02	45.07
7 – 5	8.59	24.05
5 – 3	6.84	15.46
3 – 2	2.73	8.62
2 – 1	2.75	5.89
1 – 0.5	1.01	3.14
0.5 – 0.25	0.22	2.13
< 0.25	1.91	1.91
Total	100	-

Table 6. Indicators of granulometric composition of the rock

Lithotype of rock	Controlling particle diameter d_{60} [mm]	Median particle diameter d_{50} [mm]	Acting (effective) particle diameter d_{10} [mm]	Heterogeneity coefficient C_h
sandstone	21	19	3,5	6
siltstone	12	11	3,5	3,4

As can be seen from Table 6, the sandstone is less prone to crushing under mechanical stress than siltstone. Controlling diameter d_{60} and average diameter d_{50} of sandstone clasts in the 1.7-1.75 times more than siltstone. It should also be noted that an increase in the sandstone fracture minus-0.25 mm fraction is more than double. This is due to the presence in the sandstone cement component. Thus, it follows from this conclusion that the content of sandstone in the lower layers of waste dumps is significantly higher than the average content in backfilled overburden. In the upper layers of dumps, on the contrary, clay lithotypes prevails: siltstones and mudstones.

Grading and lithological heterogeneity dump layer of indigenous overburden is the cause and filtration heterogeneity. Permeability lower blade layers composed of larger rock clasts is significantly higher than the upper fine-grained [10].

The researches found that the filtration coefficients of the upper, middle and bottom layers of the recently paved waste dump are respectively 4, 7 и 84 · 10^{-3} m/s [11]. The movement of water in the upper layers of the dump corresponds to a linear Darcy's law, i.e. $V = K_f \cdot J$, and in the bottom layers of the dump – quadratic: $V = K_f \cdot \sqrt{J}$, where V – filtration velocity, J – hydraulic gradient; K_f – coefficient of filtration.

Overburden dumps consisting primarily of sand types less susceptible to softening and soaking and more resistant to water, and hence water filtering them does not significantly change theirs permeability. Dumps of mudstones and siltstones impacted by water change their properties and can form aquitard. Permeability of bottom layers that largely represented by sandstones, eventually reduced to 0.012-0.017 m/s, while the filter coefficient of the upper layers is reduced hundreds of times.

IOP Conf. Series: Materials Science and Engineering **91** (2015) 012093 doi:10.1088/1757-899X/91/1/012093

Conclusion

Thus, heterogeneity of the material composition at dump height is revealed. This heterogeneity is the cause of spatial and temporal variations of filtration and mechanical properties of the waste dumps that must be considered when designing the filter arrays of overburden dump sewage waters, as well as the development of measures to prevent landslides in the dumps.

References

[1] Certificate 1223958 USSR MKI B01 D23/10 A method of manufacturing a water filter / Ryzhkov YA Lesin YV Kretov BK etc - Publ 04/15/86 - Bull Num 14 Opening Invention - 1986 - № 14 - S 24

[2] Russian Federation Patent number 2225743 MPK7 B01 D24/20 A method of manufacturing a water filter / YuV Lesin MA Tyulenev etc - Publ 20032004 - Bull Num 8

[3] Arya LT Paris JF 1981 A physicoempirical model to predict the soil moisture characteristic from particle-size distribution and bulk density data Soil Sci Soc Am J 45 1023–1030

[4] Lefebvre R Ge´linas PJ 1995 Numerical modelling of AMD production in waste rock dumps Sudbury '95 —Mining and the Environment Conference May 29–June 1st 1995 Sudbury Canada Paper NT2 vol 3 pp 869–878

[5] Plumlee GS 1999 The environmental geology of mineral deposits; *in* Plumlee GS and Logsdon MJ (eds) The Environmental Geochemistry of Mineral Deposits Part A Processes Techniques and Health Issues: Society of Economic Geologists Reviews in Economic Geology v 6A pp 71–116

[6] du Bray EA (ed) 1995 Preliminary compilation of descriptive geoenvironmental mineral deposit models: US Geological Survey Open-File Report 95–831 272 pp (also available online at http://mineralscrusgsgov)

[7] Alpers CN and Nordstrom DK 1999 Geochemical modeling of water-rock interactions in mining environments; in Plumlee GS and Logsdon MJ (eds) The Environmental Geochemistry of Mineral Deposits Part A Processes Techniques and Health Issues: Society of Economic Geologists Reviews in Economic Geology v 6A pp 289–323

[8] Smith KS Plumlee GS and Ficklin WH 1994 Predicting water contamination from metal mines and mining waste: Notes Workshop #2 International Land Reclamation and Mine Drainage Conference and Third International Conference on the Abatement of Acidic Drainage: US Geological Survey Open-File Report 94–264 112 pp

[9] Millington RJ Quirk JP 1961 Permeability of porous solids Faraday Soc Trans 57 1200–1206

[10] Lesin YuV Luk'yanova SYu and Tyulenev MA Mass transfer of dispersed particles in water filtration in macro-grained media / Journal of Mining Science January 2010 Vol 46 Issue 1 pp 78-81

[11] Tyulenev MA Lesin YV Waste water technology on the open-pit mines of Kuzbass / Mining Information-Analytical-newsletter Separate issue number 6 Mining engineer 3 Industrial safety and labor protection / Moscow - 2012 - № S6 - P 104-109

Institute of Physics
Dirac House, Temple Back
Bristol BS1 6BE UK

ISSN: 1757-8981
ISBN 978-1-5108-1312-0